AF794863

Heterocyclic Compounds as Pharmaceutical Ingredients

Scrivener Publishing
100 Cummings Center, Suite 541J
Beverly, MA 01915-6106

Publishers at Scrivener
Martin Scrivener (martin@scrivenerpublishing.com)
Phillip Carmical (pcarmical@scrivenerpublishing.com)

Heterocyclic Compounds as Pharmaceutical Ingredients

Edited by

Divya Bajpai Tripathy

Priyanka Chhabra

and

Anjali Gupta

Scrivener Publishing

WILEY

This edition first published 2026 by John Wiley & Sons, Inc., 111 River Street, Hoboken, NJ 07030, USA and Scrivener Publishing LLC, 100 Cummings Center, Suite 541J, Beverly, MA 01915, USA
© 2026 Scrivener Publishing LLC
For more information about Scrivener publications please visit www.scrivenerpublishing.com.

All rights reserved. No part of this publication may be reproduced, stored in a retrieval system, or transmitted, in any form or by any means, electronic, mechanical, photocopying, recording, or otherwise, except as permitted by law. Advice on how to obtain permission to reuse material from this title is available at http://www.wiley.com/go/permissions.

Wiley Global Headquarters
111 River Street, Hoboken, NJ 07030, USA

For details of our global editorial offices, customer services, and more information about Wiley products visit us at www.wiley.com.

The manufacturer's authorized representative according to the EU General Product Safety Regulation is Wiley-VCH GmbH, Boschstr. 12, 69469 Weinheim, Germany, e-mail: Product_Safety@wiley.com.

Limit of Liability/Disclaimer of Warranty
While the publisher and authors have used their best efforts in preparing this work, they make no representations or warranties with respect to the accuracy or completeness of the contents of this work and specifically disclaim all warranties, including without limitation any implied warranties of merchantability or fitness for a particular purpose. No warranty may be created or extended by sales representatives, written sales materials, or promotional statements for this work. The fact that an organization, website, or product is referred to in this work as a citation and/or potential source of further information does not mean that the publisher and authors endorse the information or services the organization, website, or product may provide or recommendations it may make. This work is sold with the understanding that the publisher is not engaged in rendering professional services. The advice and strategies contained herein may not be suitable for your situation. You should consult with a specialist where appropriate. Neither the publisher nor authors shall be liable for any loss of profit or any other commercial damages, including but not limited to special, incidental, consequential, or other damages. Further, readers should be aware that websites listed in this work may have changed or disappeared between when this work was written and when it is read.

Library of Congress Cataloging-in-Publication Data

ISBN 9781394284801

Cover image: Generated with AI using Adobe Firefly
Cover design by Russell Richardson

Set in size of 11pt and Minion Pro by Manila Typesetting Company, Makati, Philippines

Printed in the USA

10 9 8 7 6 5 4 3 2 1

Contents

Preface **xvii**

1 Heterocyclic Molecules as an Active Pharmaceutical Ingredient: An Introduction **1**

Anushka Sharma, Awaneet Kaur, Pranav Gupta, Debashis Paramanick and Abdulsalam Alhalmi

1.1 Introduction 2
1.2 Synthetic Approach for Heterocyclic Compound 4
1.2.1 Microwave-Assisted Synthesis 4
1.2.1.1 Pyrazole Derivatives 5
1.2.1.2 Benzo[b]furans Synthesis 5
1.2.1.3 Tetrazole Derivatives 5
1.2.1.4 Benzimidazole Derivatives 6
1.2.2 Solvent-Free Synthesis 7
1.2.3 Synthesis of Heterocyclic Compounds Using Nanocatalysts 7
1.2.3.1 Indole Synthesis 7
1.2.3.2 Imidazole Synthesis 8
1.2.4 Ultrasound-Mediated Synthesis 9
1.3 Medicinal Impact of Heterocyclic Moieties 9
1.4 Pharmaceutical Applications of Heterocyclic Compounds 11
1.4.1 Heterocycles as Drugs 11
1.4.2 Heterocycles as Polymers 12
1.4.3 Heterocycles as Dyes 13
1.5 Advantages of Using Heterocycles in Drug Design 15
1.5.1 Versatility and Reactivity 15
1.5.2 Enhanced Selectivity and Potency 16
1.6 Challenges in Developing Heterocyclic Active Pharmaceutical Ingredients 17
1.7 Recent Advances in Heterocyclic Chemistry 18

1.8 Conclusion 19
References 20

2 Biologically Active Nitrogen Heterocyclic Compounds 27
Diwakar Chauhan, Monika Chauhan, Priya Tyagi and Amrita Kaushik
2.1 Introduction 28
2.2 History of Nitrogen-Containing Heterocycles 29
2.3 Classification of Nitrogen Heterocycles 30
2.3.1 Nitrogen Heterocycles with Four Members 30
2.3.2 Nitrogen Heterocycles with Five Members 31
2.3.3 Nitrogen Heterocycles with Six Members 31
2.3.4 Fused or Condensed Nitrogen Heterocycles 31
2.4 N-Heterocycles Containing Medicinally Important Drugs 32
2.5 Biological Activity 34
2.5.1 Anti-Alzheimer's Disease Activity 34
2.5.2 Anticancer Activity 37
2.5.3 Antibacterial Activity 40
2.5.4 Anti-Parkinson's Activity 42
2.5.5 Anti-Inflammatory Activity 44
2.5.6 Antidiabetic Agent 46
2.5.7 Antiviral Activity 49
2.5.8 Antioxidant Activity 52
2.6 Conclusion 54
Abbreviations 54
References 55

3 Heterocyclic Compounds as Antimicrobial (Antibacterial, Antifungal, Antiviral, and Anthelmintics) Agent 61
Gaurank Pal, Devaang Mathur, Lakshit Srivastav, Vishal Mehare, Ajay Gupta, Shruti Khanna Ahuja and Puja Prasad
3.1 Introduction 62
3.2 Heterocyclic Compounds as Antimicrobial Agent 63
3.2.1 Application of Heterocyclic Compounds as Antibacterial Agent 63
3.2.1.1 Pyrrole 64
3.2.1.2 Furan 65
3.2.1.3 Thiophene 67
3.2.1.4 Indole 69
3.2.1.5 Quinoline 70

3.2.1.6 Imidazole 71
3.2.1.7 Pyrazole 73
3.2.1.8 Oxazole 74
3.2.1.9 Thiazole 76
3.2.1.10 Pyrimidine 77
3.2.1.11 Pyridine 79
3.2.2 Antifungal Activity and Structure–Activity Relationship of Heterocyclic Compounds 79
3.2.2.1 Triazole 81
3.2.2.2 Imidazoles 84
3.2.2.3 Oxazoles 87
3.2.2.4 Thiazoles 88
3.2.2.5 Pyrimidines 90
3.2.2.6 Benzimidazoles 91
3.2.2.7 Benzothiazoles 92
3.2.3 The Antiviral Activity and SAR of N-Heterocyclic Compounds 93
3.2.3.1 Triazoles 93
3.2.3.2 Thiadiazoles 96
3.2.3.3 Pyrazoles 97
3.2.3.4 Pyrimidine 98
3.2.3.5 Imidazole 99
3.2.3.6 Indole 100
3.2.3.7 Thiazoles 103
3.2.4 Application of Heterocyclic Compound as Anthelmintics Agent 104
3.2.4.1 Pyrrolidine 104
3.2.4.2 Imidazole 105
3.2.4.3 Triazole 106
3.2.4.4 Thiazole 107
3.3 Conclusion 107
References 108

4 Heterocyclic Compounds as Anticancer Agents 123
Aadya Singh, Priyanshi Kamboj, Pragya Swami, Puja Prasad and Shruti Khanna Ahuja
4.1 Introduction 124
4.2 Classification of Heterocyclic Compounds 126
4.2.1 Monocyclic versus Polycyclic Compounds 126
4.2.2 Different Types of Heterocyclic Compounds Based on Ring Size 130

4.2.2.1 Five-Membered Heterocycles 130
4.2.2.2 Six-Membered Heterocycles 131
4.2.2.3 Polycyclic Heterocycles 132
4.2.3 Heterocyclic Compounds Based on the Heteroatom 132
4.2.3.1 Nitrogen-Based Heterocycles 134
4.2.3.2 Oxygen-Based Heterocycles 134
4.2.3.3 Sulfur-Based Heterocycles 134
4.3 Mechanism of Action of Heterocyclic Compounds 135
4.3.1 Interaction with Nucleic Acids: Intercalation and Alkylation of DNA 135
4.3.2 Inhibition of Key Enzymes and Enzymatic Pathways (Targeting Specific Cancer Pathways) 141
4.3.3 Modulation of Cell Signaling Pathways and Cell Cycle 149
4.3.4 Interaction with Immune System 154
4.3.5 Inhibition of Tumor Growth and Metastasis 156
4.3.6 Role of Heterocycles in Modulating the Tumor Microenvironment 159
4.4 Key Heterocyclic Anticancer Agents 160
4.4.1 Alkaloids: Vincristine and Vinblastine, CPT Derivatives 160
4.4.2 Camptothecin Derivatives 162
4.4.3 Flavonoids 164
4.4.4 Flavonols 167
4.4.5 Nitrogen-Containing Heterocycles: Purines and Pyrimidines, Imidazoles, and Benzimidazoles 167
4.4.6 Taxanes 171
4.4.7 Anthracycline 172
4.5 Challenges in Heterocyclic Anticancer Drug Development 173
References 175

5 Heterocyclic Compounds as Anti-Inflammatory Agents 179
Hemlata Nimesh, Sagar Kumar and Nidhi Srivastava
5.1 Introduction 180
5.2 Classification 184
5.3 Various Heterocyclic Derivatives as Anti-Inflammatory Drugs 186
5.3.1 Pyrazole and Pyrazolines Derivatives as Anti-Inflammatory Drugs 186
5.3.2 Pyrimidine Derivatives as Anti-Inflammatory Drugs 189
5.3.3 Indole Derivatives as Anti-Inflammatory Drugs 190

5.3.4 Imidazole and Benzimidazole Derivatives as Anti-Inflammatory Drugs 192
5.3.5 Pyrrole as Anti-Inflammatory Drugs 193
5.3.6 Furanone as Anti-Inflammatory Drugs 194
5.3.7 Oxazole and Isoxazole as Anti-Inflammatory Drugs 196
5.3.8 Thiazole as Anti-Inflammatory Drugs 196
5.3.9 Pyridine as Anti-Inflammatory Drugs 198
5.4 Conclusion and Future Prospects 199
References 201

6 Heterocyclic Compounds as Antioxidant Agents 207
Niranjan Kaushik, Shristi Singh, Ajita Paliwal and Sachin Chaudhary
6.1 Introduction: Antioxidants and Oxidative Stress 208
6.2 Mechanism of Antioxidant Agent 209
6.3 Structures and Classifications of Heterocyclic Compound 212
6.3.1 Role of Heterocyclic Structure Enhancing Antioxidant Activity 213
6.3.2 Existing Heterocyclic Antioxidant Compounds 215
6.4 Types of Heterocyclic Compounds with Antioxidant Properties 218
6.4.1 Nitrogen-Containing Heterocyclic 218
6.4.1.1 Pyridine 219
6.4.1.2 Pyrimidine 221
6.4.2 Oxygen-Containing Heterocycles 222
6.4.2.1 Furan 223
6.4.2.2 Benzofuran 223
6.4.2.3 Coumarin 225
6.4.3 Sulfur-Containing Heterocycles 226
6.4.3.1 Thiophene 226
6.5 Antioxidant Effect and Mechanism of Action of Polyphenolic Compound 228
6.5.1 Flavonoids 229
6.5.1.1 Catechins 230
6.5.1.2 Flavonols 231
6.5.2 Phenolic Acids 232
6.5.3 Stilbenes 233
6.6 Factors Influencing the Antioxidant Capacity of Polyphenols 234
6.6.1 Structure 234
6.6.2 Stability 235

6.6.3 Bioavailability 235
6.6.4 Environmental Factor 236
6.7 Conclusion 236
References 237

7 Heterocyclic Compounds as Anticonvulsants, Antipyretics, Antiallergic Agents, and Antileprosy Agents 249
Anushka Sharma, Awaneet Kaur, Pranav Gupta, Debashis Paramanick and Abdulsalam Alhalmi
7.1 Introduction 250
7.2 Heterocyclic Compounds as Anticonvulsant Agents 251
7.3 Heterocyclic Compounds as Antipyretic Agents 254
7.4 Heterocyclic Compounds as Antiallergic Agents 255
7.5 Heterocyclic Compounds as Antileprosy Agents 256
7.6 Recent Advances and Future Prospects 259
7.6.1 Novel Synthetic Approaches 259
7.6.2 Drug Design and Molecular Modeling 259
7.7 Conclusion 260
References 261

8 Heterocyclic Compounds as Antihistamine Agents 269
Avinash Jha, Saurabh Sharma, Deepika Paliwal, Aman Thakur and Saurabh Srivasatva
8.1 Introduction 270
8.1.1 Overview of Histamine 270
8.1.2 Histamine Role in Disease 272
8.1.3 Antihistamine Agents 272
8.1.3.1 H1 Antihistamine 272
8.1.4 Role of Heterocyclic Compound in Drug Development 276
8.2 Histamine Receptor 277
8.2.1 H1 Receptor 279
8.2.2 H2 Receptor 281
8.2.3 H3 Receptor 284
8.2.4 H4 Receptor 286
8.3 Imidazole as Antihistamine Agents 287
8.4 Piperazine as Antihistamine Agents 289
8.5 Quinazoline as Antihistamine Agents 290
8.6 Piperadine as Antihistamine Agents 291
8.7 Ethanolamines as Antihistamine Agents 293
8.8 Conclusion 294
References 294

9 Heterocyclic Compound Having Herbicides Activity **307**
Niranjan Kaushik, Mridul Singh Sengar, Deepika Paliwal and Nadeem Ahmad Siddique
9.1 Introduction 308
9.1.1 Historical Background 309
9.1.2 Herbicides 310
9.2 Mechanisms of Herbicide Action and Their Role in Plant Growth Inhibition 314
9.2.1 Inhibitors for Lipid Synthesis 314
9.2.2 Inhibitors of Amino Acid Synthesis 315
9.2.3 Photosynthesis Inhibitors 316
9.2.4 Growth Regulators 317
9.2.5 Nitrogen Metabolism Inhibitors 317
9.2.6 Pigment Inhibitors 318
9.2.7 Cell Membrane Disrupter 318
9.2.8 Seedling Shoot Growth Inhibitors 319
9.2.9 Seedling Root Growth Inhibitors 319
9.2.10 Several Innovative Ways That Herbicides Work 320
9.3 Heterocyclic Having Herbicides Activity 320
9.3.1 Triazines 321
9.3.2 Triazinone 323
9.3.3 Pyridine 324
9.3.4 Pyrazole Benzophenone Derivatives 325
9.3.5 Imidazolines 326
9.4 Environmental Impact and Resistance Management 327
9.4.1 Environmental Considerations 327
9.4.2 Fundamentals of How Weeds React to a Climate with Higher CO_2 and Temperature 330
9.4.3 Resistance Develop in Transgenic Plants 330
9.5 Field Applications 331
9.6 Future Prospects of Heterocyclic Herbicides 335
9.7 Conclusion 336
References 338

10 Heterocyclic Compounds as Antihypertensive Agents **349**
Manjot Singh, Deepika Paliwal, Aman Thakur, Manvi Karayat and Saurabh Srivasatva
10.1 Introduction 350
10.1.1 Marketed Drugs for Hypertension 353
10.1.2 Significance of Heterocyclic Compounds 357

10.1.3 Structural Versatility of Heterocyclic Compounds 357
10.1.4 Pharmacophoric Potential of Heterocyclic Rings 357
10.1.5 Enhancement of Bioavailability and Stability 358
10.1.6 Specific Mechanistic Action 358
10.2 Heterocyclic Moieties against Hypertension 359
10.2.1 Indole and Indazole Moiety 359
10.2.2 Imidazole 368
10.2.3 Furan 370
10.2.4 Tetrazole 371
10.2.5 Thiadiazole 373
10.2.6 Quinazoline 375
10.2.7 Thiazolidine 376
10.2.8 Triazole 378
10.2.9 Pyridine 379
10.3 Future Prospective 380
10.4 Conclusion 381
References 382

11 *In Silico* Studies on Heterocyclics as an Active Pharmaceutical Agent 389
Anurag Paul, Debanjali Adhikary, Abdul Kareem Parchur, Gayatri Sharma and Apoorva Mishra
11.1 Introduction 390
11.2 *In Silico* Approaches for Heterocyclic Drug Discovery 390
11.2.1 Molecular Docking 390
11.2.2 Molecular Dynamic Simulations 392
11.2.3 Quantitative Structure–Activity Relationship 393
11.2.4 Pharmacophore Modeling 394
11.2.5 ADMET and Toxicity Prediction 396
11.3 Heterocyclic Compounds as Active Pharmaceutical Agents 397
11.3.1 Antifungal Activity 397
11.3.2 Antibacterial Activity 399
11.3.3 Anticancer Activity 399
11.3.4 Herbicidal Activity 399
11.3.5 Anticonvulsant 400
11.4 Challenges and Future Trends 400
11.5 Conclusion 401
References 401

12 Piperazine Derivatives: Potential Antimicrobial Agents 407
R. Prathap, Rashmi Dubey, Anjali Gupta and Divya Bajpai Tripathy
12.1 Introduction 408
12.2 Synthetic Strategies for Piperazine Derivatives 409
12.2.1 Synthesis of 1-Benzyl-4-Phenylpiperazine 411
12.2.2 Synthesis of 1-Phenylpiperazine 412
12.2.3 Synthesis of 1-[(4-Chlorophenyl)sulfonyl]piperazine 414
12.2.4 Synthesis of Piperazine-Phthalimide Derivatives 415
12.2.5 Synthesis of Sulfonyl Piperazine 416
12.2.6 Synthesis of 1-(2,5-Dimethoxyphenyl)piperazine 418
12.3 Structural Relationship of Piperazine Derivatives 419
12.4 Antimicrobial Activities of Piperazine Derivatives 421
12.4.1 Antibacterial Activity 421
12.4.2 Antifungal Activity 422
12.4.3 Antiparasitic Activity 422
12.4.4 Anticancerous Activities of Piperazine Derivatives 423
12.5 Synergistic Activity of Piperazine Derivatives with Other Antimicrobial Agents, Enhancing Their Effectiveness against Certain Pathogens 426
12.6 Mechanism of Action 427
12.6.1 Inhibition of Cell Wall Synthesis 427
12.6.2 Inhibition of Protein Synthesis 427
12.6.3 Disruption of Membrane Integrity 427
12.6.4 Inhibition of Nucleic Acid Synthesis 428
12.7 Applications of Piperazine Derivatives as Microbial Agents 428
12.7.1 In human medicine 428
12.7.2 Treatment of Bacterial Infections 428
12.7.3 Treatment of Fungal Infections 428
12.7.4 Treatment of Parasitic Infections 428
12.7.5 Disinfection and Sanitation 429
12.7.6 As Anticancerous Agents 429
12.8 In Veterinary Medicine 430
12.8.1 Treatment of Bacterial Infections in Livestock 430
12.8.2 Treatment of Parasitic Infections in Pets 431
12.8.3 Use as a Disinfectant 431
12.9 In Agriculture 431
12.9.1 Control of Plant Pathogens 431
12.9.2 Preservation of Crops 432
12.9.3 Control of Nematodes 432

12.10 Challenges and Future Prospects 432
12.11 Conclusion 433
References 434

13 Synthetic Drug Discovery Strategy of Benzimidazole Derivatives and Their Applications 443
Prasad Vijay Korake, Diwakar Chauhan and Subhalaxmi Pradhan
Abbreviations 444
13.1 Introduction 444
13.2 Benzimidazole as an Antifungal Agent 447
13.3 Benzimidazole as an Antimicrobial Agent 451
13.4 Benzimidazole as an Anticancer Agent 457
13.5 Conclusion 471
References 472

14 Synthesis and Biological Evaluation of Thiophene- and Piperazine-Based Active Compounds 475
Ayyagari Subramanya Sharma, Subhalaxmi Pradhan and Prashant Kumar
14.1 Introduction 476
14.2 Chemistry of Thiophene and Synthesis of Its Derivative 478
14.2.1 Metal-Catalyzed Synthetic Approaches 478
14.2.2 Metal-Free Synthetic Approaches 478
14.2.3 Multicomponent Synthetic Approaches 478
14.3 Biological Application of Thiophene and Its Derivatives 480
14.3.1 Mono-Substituted Thiophene Derivatives as Anticancer Agents 482
14.3.2 Di-Substituted Thiophene Derivatives as Anticancer Agents 483
14.3.3 Tri-Substituted Thiophene Derivatives as Anticancer Agents 484
14.3.4 Biological Application of Thiophene Derivatives as Antimalarial Agent 485
14.3.5 Biological Application of Thiophene-Based Compounds as Anti-Inflammatory Agent 487
14.3.6 Biological Applications Thiophene-Based Compounds as Antidepressant 487
14.3.7 Biological Applications of Thiophene Analogs as Anti-Inflammatory and Analgesic 489
14.3.8 Biological Applications of Thiophene Analogs as Anticonvulsant 490

14.3.9 Biological Applications of Thiophene-Based Compound as Antidiabetic Agent 491
14.4 Biological Application of Piperazine-Based Compounds 491
14.4.1 Piperazine as an Antimicrobial Agent 493
14.4.2 Anticancer Activity of Piperazine 493
14.4.3 Antioxidant Activity of Piperazine 495
14.4.4 Antitumor Activity of Piperazine 495
14.4.5 Piperazine: Insecticidal Activity 497
14.4.6 Piperazine: Antidiabetic Activity 497
14.4.7 Piperazine Derivatives as Anti-Inflammatory and Analgesic Agents 498
14.5 Conclusion 499
References 500

15 Heterocyclic Compounds as Antileprotic Agents 509
Sanjita Das, Anjali Singh, Shalini Gautam and Gyas Khan
15.1 Introduction 510
15.2 Historical Perspectives 510
15.3 Pathophysiology 511
15.4 Classification of Heterocyclic Antileprotic Drugs 512
15.4.1 Characteristics of Structure 512
15.4.2 Heterocyclic Ring's Role in Mechanism of Action 513
15.5 Important Heterocyclic Antileprotic Drugs 514
15.5.1 Clofazimine 514
15.5.2 Rifampin 515
15.5.3 Ethionamide 516
15.5.4 Dapsone or Di-Amino-Diphenyl-Sulfone 517
15.5.5 Fluoroquinolones 518
15.5.6 Minocycline 520
15.5.7 Clarithromycin 521
15.6 Advances in Antileprotic Agents 524
15.7 Obstacles and Prospects for the Development of Heterocyclic Antileprotic Drugs 525
References 526

16 Bromodomains: Biomedical Applications of Active Ingredients 531
Sweta Raj, Anjan Kumar Nayak, Harsh Vardhan and Divya Bajpai Tripathy
16.1 Introduction 532
16.1.1 Significance in Medical Researches 532

16.2 History and Discovery of BRDs 533
16.2.1 Evolution of BRD Research 533
16.2.2 Structure of BRDs 533
16 2.3 Interaction with Acetylated Lysines 534
16.3 BRDs as Drug Targets 535
16.3.1 Drug Design Strategies Targeting BRDs 535
16.4 BRD Inhibitors 536
16.5 Next-Generation BRDs Inhibitors 537
16.5.1 Two Classes of BET Inhibitors 537
16.6 BET Inhibitors in Clinical Trials 537
16.6.1 Key Insights on Selectivity between BRD4 (BRD1) and BRD3 (BRD2) 538
16.6.2 BD2 Selectivity (Why Target Selective Compounds?) 538
16.7 Resistance toward BET Inhibitors 540
16.8 Toxicity of BET Inhibitors Used Clinically 541
16.9 BRD Inhibition and Human Diseases 541
16.9.1 Cancer 541
16.9.2 Cardiovascular Disease 542
16.10 Conclusion and Future Perspectives 542
References 543

Index 547

Preface

Heterocyclic compounds, defined by the inclusion of heteroatoms such as nitrogen, oxygen, or sulfur atom in the form of the ring, are central in the field of pharmaceutical chemistry. Their precise structural flexibility and capability to engage biological targets with accuracy have rendered them prospective candidate in the drug discovery, supporting therapies for an extensive range of disorders, ranging from the infectious diseases to complicated health ailments such as cancer and cardiovascular diseases.

This volume, *Heterocyclic Compounds as Pharmaceutical Ingredients*, presents an in-depth tour of these molecules, combining fundamental principles and the latest breakthroughs to be a primary resource for pharmaceutical sciences students, researchers, and industry professionals as well as for organic chemistry and medicinal chemistry professionals.

Introduction to the book is initiated with the presentation of heterocyclic molecules as active pharmaceutical ingredients and setting the ground as Chapter 1 to realize their chemical and biological significance. This is then succeeded by a comprehensive overview on biologically active nitrogen heterocycles, Chapter 2 is highlighting their central role in therapeutic interventions. The study continues with focus on the antimicrobial potential of these compounds and their potency against bacteria, fungi, viruses, and helminths, which are still ongoing challenges in global medicine. The book goes on with the uses of heterocycles as anticancer agents, detailing their mechanisms of action and efficacy against malignancies.

The book further discusses other heterocyclic derivatives such as benzimidazole in Chapter 3 with respect to their synthetic methods and applications as antifungal, antimicrobial, and anticancer agents. Synthesis and biological activity of thiophene and piperazine derivatives are also presented in the subsequent chapters with a stress on their applications in healing cancer, malaria, inflammation, and diabetes, among others. In addition, a vibrant discussion of heterocyclic compounds as antileprotic agents discusses their chronological history, classification, and recent developments; addressing some of the existing challenges in leprosy treatment is offered. Subsequent chapters illustrate the anti-inflammatory and antioxidant properties of heterocyclic compounds, identifying their application in the cure of chronic inflammatory diseases and oxidative stress–related disorders. Their employment as anticonvulsant, antipyretic, antiallergic, and antileprosy drugs is discussed at length to create an image of their therapeutic significance.

The book also provides insights on the antihistamine activity of heterocycles, highlighting the role of certain moieties such as imidazole, piperazine, and quinazoline, and their action on histamine receptors to treat allergy.

The scope extends to embrace herbicidal activity of heterocyclic compounds, on the basis of their agroindustrial significance and as antihypertensive drugs, where structural diversity is helpful in bioavailability and mechanism specificity. *In silico* heterocyclic drug discovery processes are comprehensively discussed, such as molecular docking, dynamics simulations, and pharmacophore modeling, along with their application in antifungal, antibacterial, anticancer, and herbicidal activities. The antimicrobial action of piperazine derivatives is reviewed, considering their synthesis, structure correlations, and synergism with other medicinal drugs, covering a range of uses in human and veterinary medicine and agriculture.

Ultimately, the book examines bromodomains as new drug targets, describing their structure, inhibitors, and biomedical uses in cancer and cardiovascular disease with clues for clinical trials and future directions. By this logical sequence, the book covers theoretical understanding with real-world applications, presenting an integrated view of challenges and possibilities of using heterocyclic compounds for therapeutic development. We acknowledge with gratitude the contributions of experts whose work has enriched this book and the readers whose interest fuels progress in this exciting discipline. This publication will stimulate further investigation and development, propelling the discovery of new medicines to meet global health demands.

Editors
Divya Bajpai Tripathy
Priyanka Chhabra
Anjali Gupta

1

Heterocyclic Molecules as an Active Pharmaceutical Ingredient: An Introduction

Anushka Sharma[1], Awaneet Kaur[2*], Pranav Gupta[2], Debashis Paramanick[3] and Abdulsalam Alhalmi[4]

[1]Faculty of Pharmacy, Vidya University, Meerut, Uttar Pradesh, India
[1]Department of Pharmacy, School of Medical and Allied Sciences, Galgotias University, Greater Noida, Uttar Pradesh, India
[2]School of Medical and Allied Sciences, K.R. Mangalam University, Gurugram, India
[3]Department of Pharmaceutics, College of Pharmacy, University of Aden, Aden, Yemen

Abstract

The extensive range of biological activities and structural diversity of heterocyclic compounds underscores their significance in contemporary pharmaceutical chemistry. This chapter elucidates the significance of heterocycles as active pharmaceutical ingredients (APIs), focusing particularly on their presence in both natural sources and synthetic medicinal compounds. Several synthetic processes are addressed, including environmentally friendly or ecologically beneficial organic procedures such as solvent-free reactions, microwave-assisted synthesis, and ultrasound-assisted approaches. These methods offer advantages such as reduced reaction times, decreased solvent consumption, and improved yields. Heterocyclic moieties play a significant role in medicine, serving as essential components in a range of pharmaceuticals designed to combat infections, cancers, viruses, and disorders of the central nervous system. Heterocycles are utilized in drug design because of their straightforward chemical manipulation and reactivity, allowing for targeted interactions at the molecular level, while ensuring enhanced selectivity and effectiveness. In addition to drug molecules, they play a crucial role in the production of functional polymers and synthetic dyes, highlighting their

**Corresponding author*: awaneet.kaur@galgotiasuniversity.edu.in

Divya Bajpai Tripathy, Priyanka Chhabra and Anjali Gupta (eds.) Heterocyclic Compounds as Pharmaceutical Ingredients, (1–26) © 2026 Scrivener Publishing LLC

importance across various fields. In the realm of drug development, the incorporation of heterocycles encounters ongoing challenges, including issues related to synthesis, bioavailability, and regulatory frameworks. Recent advancements, such as C–H activation methods, integrating green chemistry principles with conventional approaches, and computer-assisted drug design, present new avenues for the development of innovative heterocyclic pharmaceuticals.

Keywords: Heterocyclic compounds, active pharmaceutical ingredients (APIs), green synthesis, medicinal chemistry, drug design

1.1 Introduction

Heterocyclic chemistry is a major area within organic chemistry, accounting for roughly 50% of known organic compounds. A notable class of organic chemical compounds is characterized by the existence of ring structures within their molecules, incorporating at least one atom of an element other than carbon [1]. The term "cyclic" in heterocyclic indicates the presence of at least one ring structure within the compound, whereas "hetero" refers to the inclusion of noncarbon atoms, or heteroatoms, in the ring [2]. Heterocyclic compounds generally exhibit a structure similar to cyclic organic compounds that consist exclusively of carbon atoms in their rings. The presence of heteroatoms introduces physical and chemical properties that often differ markedly from those of all-carbon-ring analogs [3–5]. Common heterocycles generally consist of five- or six-member rings containing heteroatoms, including nitrogen (N), oxygen (O), or sulfur (S). Simple heterocyclic compounds that are widely recognized include pyridine, pyrrole, furan, and thiophene (Figure 1.1). Pyridine is a molecule defined by a six-member ring consisting of five carbon atoms and one nitrogen atom. Pyrrole, furan, and thiophene are five-member ring compounds, each comprising four carbon atoms and one heteroatom: nitrogen, oxygen, or sulfur, respectively [6–9].

The development of heterocyclic chemistry traces back to the 1800s, emerging concurrently with the field of organic chemistry. Significant advancements encompass the isolation of alloxan from uric acid in 1818,

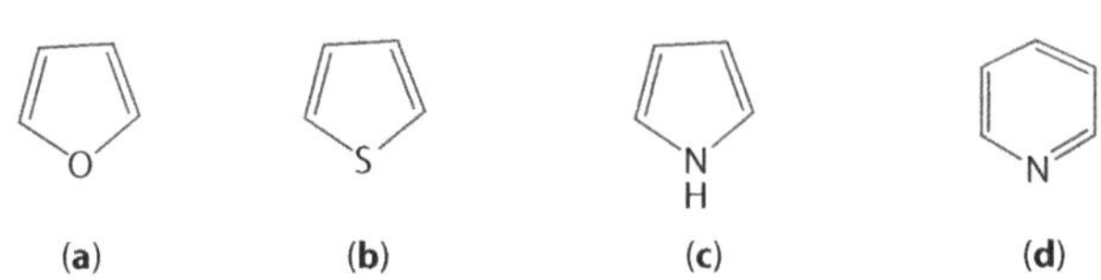

Figure 1.1 Heterocyclic compounds: (a) furan, (b) thiophene, (c) pyrrole, (d) pyridine.

the production of furfural in 1832, the extraction of pyrrole in 1834, the synthesis of indigo dye in 1906, the discovery of chlorophyll derivatives in 1936, and Chargaff's rules in 1951, which underscored the importance of heterocyclic compounds in the genetic code [10]. The progress made in this area played a significant role in the evolution of synthetic chemistry. A significant aspect of the architecture of numerous heterocyclic compounds is the ability to integrate functional groups either as substituents or as integral components of the ring structure. For instance, fundamental nitrogen atoms can be integrated as amino substituents as well as components of a ring [11–13]. This indicates that the structures possess significant versatility in offering or simulating a functional group.

Heterocyclic compounds exhibit a broad presence throughout nature. The significance of these compounds to living systems cannot be overstated; it is remarkable how frequently a heterocyclic compound serves as a crucial element in biological processes. The nucleic acid bases, specifically the purine derivatives adenine and guanine, along with the pyrimidine derivatives thymine and cytosine, play a vital role in the replication mechanism [14–16]. Certain purines and pyrimidines possess antibiotic properties by disrupting DNA synthesis, exemplified by puromycin. Histamine, a monosubstituted imidazole, is produced *in vivo* from the amino acid histidine through decarboxylation facilitated by the enzyme histidine decarboxylase. This compound is a biogenic amine that plays a role in local immune responses and functions as a neurotransmitter [17, 18]. This substance plays a role in allergic responses and is released from skin cells upon injury, while also contributing to the regulation of gastric acid secretion. Beginning in the 1940s, a number of synthetic drugs emerged that function as histamine antagonists. Two distinct types of histamine receptors were identified in the body, designated as H_1 and H_2 receptors [19]. Cetirizine, chlorpheniramine, and similar compounds function as antagonists of H_1 receptors. A significant number of pharmaceuticals, along with various heterocyclic compounds that have practical applications, are predominantly synthesized rather than extracted from natural sources. The beginnings of organic chemistry are rooted in the examination of natural products [20, 21]. The foundation laid by these has been pivotal in the design of numerous beneficial compounds that followed. Notable instances include the initial creation of vat dyes inspired by the indigo structure and the ongoing innovation of novel antibacterial agents derived from the β-lactam framework of penicillin. Examples of drugs that contain a β-lactam moiety include cephalosporin C, amoxicillin, clavulanic acid, and penicillin, among others. One of the methods that has been thoroughly examined is the exploration of analogs of the nucleosides. Nucleic acid fragments are composed of a heterocyclic nitrogenous base, which can be either a purine or a pyrimidine, connected to a sugar, for instance, 2′-deoxythymidine [22, 23]. Analogs can be synthesized with modifications to either the structure of the heterocyclic nitrogenous base, the sugar, or both. A molecule of this nature may disrupt the viral replicative cycle by substituting for the native nucleoside [24]. The primary issue is selectivity, as most of these chemicals are likely to be hazardous to normal cells as well. Nonetheless, some significant pharmaceuticals of this category have been developed. Zidovudine/azidothymidine, commonly known as ZDV, is a nucleoside analog reverse transcriptase inhibitor utilized in the treatment of AIDS and is an analog of 2′-deoxythymidine. Acyclovir is a derivative of 2′-deoxyguanosine, wherein the acyclic side chain resembles the sugar component of the original nucleoside [24–26].

1.2 Synthetic Approach for Heterocyclic Compound

Heterocyclic compounds are used extensively in medicines, agrochemicals, and materials research. There has been a recent uptick in the pursuit of novel, environmentally benign synthetic processes with the goals of reducing waste, boosting energy efficiency, and improving environmental effect. Improved yields and selectivity have been achieved using methods such solvent-free conditions, ultrasonic irradiation, and microwave-assisted synthesis. A growing number of nitrogen- and oxygen-containing heterocycles are being synthesized using ecologically friendly methods that do not involve catalysts and are carried out in an aqueous phase. C–H activation approaches, multicomponent reactions (MCRs), and transition-metal catalysis are all part of modern synthetic procedures. A more precise and efficient area of study, heterocycle synthesis has recently undergone a revolution thanks to advancements in organocatalysis, computational design, and flow chemistry. Efficient and sustainable synthesis of heterocyclic compounds is being propelled forward by the continual application of cutting-edge methodology in conjunction with ecologically friendly principles.

1.2.1 Microwave-Assisted Synthesis

Microwave radiation encompasses a frequency range of 30 GHz to 300 MHz. From 1990 onward, the distinctive combination of microwave-assisted organic reaction and organic reaction in water/neat conditions was demonstrated, and it was subsequently elaborated upon by others [27].

1.2.1.1 *Pyrazole Derivatives*

Pyrazole derivatives demonstrate significant efficacy in drug design due to their therapeutic importance. This unique structural feature has led to various bioactivities, such as antihypertensive, antiviral, and neuroprotective effects. Gomha and colleagues [28] synthesized a variety of novel pyrazole-based azoles *via* an MCR that incorporated three substrates, using microwave heating. This study aimed to evaluate the anticarcinogenic properties of the compounds against hepatocellular carcinoma (HepG2) and A-549 (human lung cancer) cell lines, using cisplatin as the reference drug in an MTT assay [28].

1.2.1.2 *Benzo[b]furans Synthesis*

A solvent-free method for the synthesis of benzo[b]furans has been proposed. A phase transfer catalyst enabled the reaction of salicylaldehyde derivatives with various esters of chloroacetic acid to generate benzo[b]furans 11 under neat conditions, resulting in favorable outcomes (Figure 1.2). The vessel's dimensions and configuration affected the product yield. When the reaction was carried out in a conical flask with a capacity greater than the volume of the reaction mixture, favorable yields were recorded. The process used a catalytic amount of tetrabutylammonium bromide and adsorbed the reaction mixture onto potassium carbonate [29].

1.2.1.3 *Tetrazole Derivatives*

A wide range of medical and therapeutic possibilities can be derived from tetrazole derivatives. The structural basis of many medications that have been approved by the Food and Drug Administration (FDA) and are now on the market is a tetrazole scaffold. A direct route to pharmaceutically active tetrazole was disclosed *via* the aryl nitrile-sodium azide cycloaddition process with a glycerol solvent, and a heterogeneous catalyst ZnBr2-SiO_2 was performed under microwave irradiation (Figure 1.3). Radical scavenging activity is a useful indicator of the produced compounds' antioxidant properties. When 1,1-diphenyl-2-picrylhydrazyl

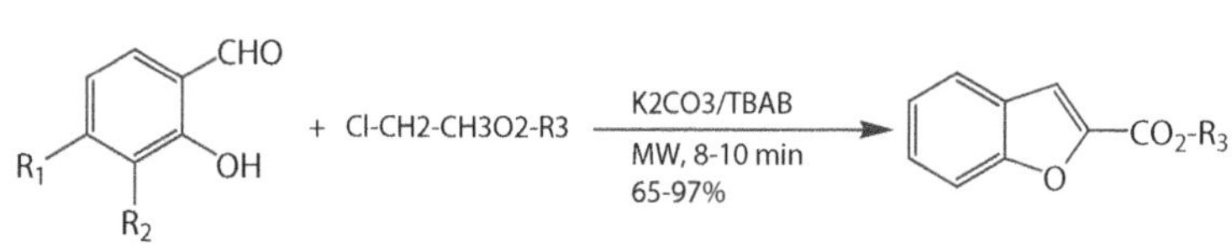

Figure 1.2 Preparation of benzo[b]furans under neat microwave (MW) conditions.

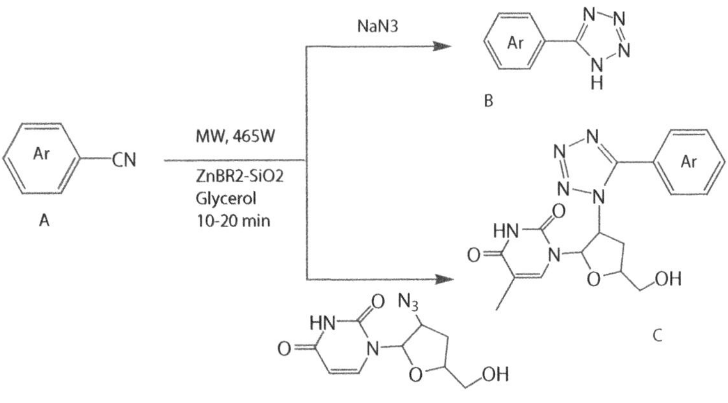

Figure 1.3 Synthesis of pyrazole derivatives under microwave heating.

(DPPH) showed a strong absorption peak at 517 nm, it was determined that radicals were present. The standard antioxidant used in the research was butylated hydroxytoluene (BHT) [30].

1.2.1.4 Benzimidazole Derivatives

Benzimidazole derivatives, akin to pyrazole, are recognized as highly effective bioactive pharmacophores, with extensive applications in the pharmaceutical field. Consequently, extensive efforts were dedicated to broadening the collection of diverse substituted benzimidazole derivatives. Researchers have devised a microwave-assisted two-step protocol for the synthesis of 9-aryl-6-chloropurines. This method entails the reaction of aniline with 4,6-dichloro-5-aminopyrimidines, succeeded by cyclization, resulting in an excellent yield. The initial phase occurred in isobutanol at 150°C for 1 hour. This was followed by a second phase in which the synthesized compound reacted with trimethylformate in acetic anhydride at 120°C for 1 hour, leading to the formation of target molecules (Figure 1.4) [31].

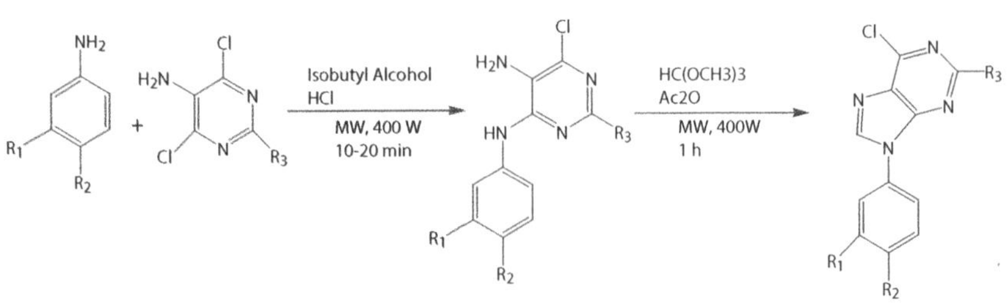

Figure 1.4 Microwave-assisted benzimidazole derivatives.

1.2.2 Solvent-Free Synthesis

Tetrahydropyrimidines stand out among the diverse range of heterocycles as valuable components in synthetic organic chemistry. Their significant applications in medicinal and biological domains stem from their wide array of biopharmaceutical activities. Mariki and colleagues [32] developed a method for synthesizing novel tetrahydropyrimidine-4-yl pyrimidine derivatives without the use of solvents, aimed at inhibiting cholinesterase enzymes. A catalytic amount of the ionic liquid $[Et_3NH]$ $[HSO_4]$ was introduced to facilitate the reaction. The synthesized tetrahydropyrimidines exhibit notable inhibitory activities against BChE, demonstrating greater potency than donepezil, which serves as the standard for comparison. Their AChE inhibition activities are noteworthy, with IC_{50} values ranging from 0.08 to 0.1 μM. It was observed that the 4-methyl–substituted derivatives exhibit the highest potency, with an IC_{50} value of 0.082 μM, surpassing all other compounds [32].

1.2.3 Synthesis of Heterocyclic Compounds Using Nanocatalysts

Nanocatalysis is emerging as a pivotal focus in synthetic chemistry, involving the integration of nanoparticles into chemical methodologies for the synthesis of heterocyclic and polycyclic compounds. Metal nanoparticles exhibit significant potential due to their swift chemical transformation and exceptional catalytic activity. This approach incorporates metals such as silver, gold, palladium, copper, ruthenium, and nickel, which are combined with different supports and subsequently transformed into nanoparticles to enhance their catalytic properties [33].

1.2.3.1 Indole Synthesis

The pharmacological activities of the indole ring are diverse. Tryptophan, which is present in proteins, is another one of them. Polycyclic molecules are produced when it reacts with other heterocyclic compounds due to the presence of the NH group. Yamane *et al.* [34] used Au nanoparticles impregnated in iron (III) oxide to study the production of indoles triggered by (2-nitrophenyl) alkynes. Through optimization experiments, a 63% yield of phenyl indole was produced by hydrogenating 1-nitro-2-(2-2-phenylethynyl) benzene at 120°C with Au/TiO_2 acting as a catalyst and 1.2 MPa H_2. Likewise, $AuCO_3O_4$ yielded lower amounts of the required indoles. However, the yield increased to 67% when Fe_2O_3 nanoparticles

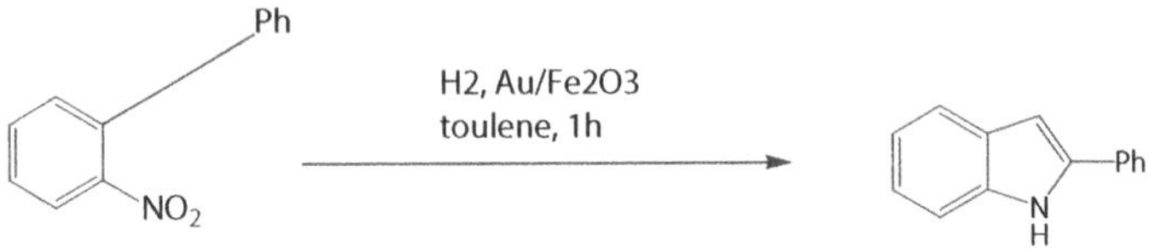

Figure 1.5 Au/Fe_2O_3 catalyzed synthesis of indoles.

were used as a catalyst at a slightly higher pressure (2 MPa). However, the yield dropped by 51% when the temperature was raised to 160°C. Additionally, optimization experiments at different temperatures showed that 110°C was the ideal temperature for maximum yield. 1-Nitro2-(2-phenylethynyl)benzene was used in two stages of the improved procedure (Figure 1.5) [34].

1.2.3.2 *Imidazole Synthesis*

Ganji *et al.* [35] synthesized imidazoles from α-diketones using ruthenium nanoparticles supported by phosphine. Although the method was designed to synthesize pyrazines from α-diketones, it was only fortunate that imidazoles were formed from alkyl α-diketones, even if it was effective in producing pyrazines from aryl α-diketones. In order to create the ruthenium nanoparticles required for imidazole synthesis, xantphos ligand (phosphine support) and degassed tetrahydrofuran were used in an argon atmosphere. A temperature of 110°C was then used to introduce [Ru(COD)(COT)]. At 3 bar of pressure, the reaction mixture was stirred for 16 hours. When the reaction mixture became black, ruthenium nanoparticles were precipitated using degassed pentane. A rotary evaporator was used to evaporate the solvent, and it was then vacuum-dried for a whole night to accomplish the separation. Trisubstituted imidazoles were produced in excellent yields by the reaction of α-diketone substrates with ruthenium nanoparticles coated with xantphos and ammonium formate, a nitrogen source (Figure 1.6) [35].

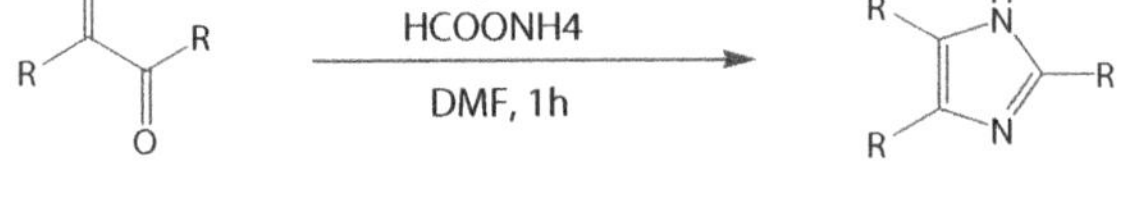

Figure 1.6 Synthesis of imidazoles from xantphos impregnated Ru nanoparticles.

1.2.4 Ultrasound-Mediated Synthesis

The importance of ultrasonography in promoting ecofriendly practices is demonstrated by its application in synthetic techniques for the direct synthesis of N-heterocyclic compounds. This intriguing green synthesis heavily relies on N-substituted pyrrole derivatives produced by ultrasound-promoted reactions. Pyrroles are the basic structural element of several types of compounds of pharmacological significance. One conventional method for creating pyrroles is the Pall–Knorr reaction. Because of their growing chemical and biological significance in a variety of fields, pyrimidine and its derivatives are recognized as important heterocyclic scaffolds. In a one-pot, three-component reaction involving 5-(4-chlorophenyl)-1,3,4-thiadiazol-2-amine, aromatic aldehydes, and malononitrile, Nikalje *et al.* reported the ultrasound-mediated synthesis of novel pyrimidine derivatives, obtaining good to excellent yields with a catalytic amount of NaOH [36]. Using 5-fluorouracil as the reference medication, the produced compounds were then evaluated for their anticancer efficacy against human tumor cell lines. The synthesized compounds were subsequently assessed for their anticancer activity against human tumor cell lines, utilizing 5-fluorouracil as the reference drug [37].

1.3 Medicinal Impact of Heterocyclic Moieties

Medicinal chemistry is a discipline that integrates traditional sciences, including organic chemistry, biology, and physics. The absence of specificity generally restricts the utilization in human and veterinary medicine. A substantial array of natural and synthesized heterocyclic compounds have notable therapeutic effects. Heterocyclic compounds have enhanced chemical versatility and can effectively respond to the varied requirements of biological systems [38]. Heterocyclic compounds demonstrate a variety of pharmacological actions [39]. Hormones, vitamins, antibiotics, and other biological substances in living beings consist of heterocyclic molecules. Although heterocyclic compounds containing sulfur constitute the majority of FDA-approved pharmaceuticals, those containing nitrogen are considered the predominant category of chemical substances used in medicinal chemistry [40]. Numerous physiologically active natural materials utilized in conventional therapies or sanctioned prescription drugs comprise nitrogen-containing heterocyclic molecules. A multitude of synthetic counterparts are accessible on the therapeutic market for the treatment of diverse diseases (Table 1.1).

Table 1.1 Medicinal impact of heterocyclic compounds.

Heterocyclic scaffold	Drug/ compound	Biological activities	Mechanism of action	References
Pyridine	Isoniazid, nicotinamide	Antitubercular	Identified in vitamin B_3, involved in NAD+/ NADP+ synthesis; acts as an inhibitor of mycolic acid	[41]
Pyrimidine	5-Fluorouracil, zidovudine	Anticancer	Blocks the synthesis of DNA and RNA; utilized in the treatment of HIV and in cancer chemotherapy	[42]
Quinoline	Chloroquine, quinine	Antimalarial, antibacterial	Disrupts heme polymerization in *Plasmodium*	[43, 44]
Imidazole	Metronidazole, ketoconazole	Antiprotozoal, antifungal	Interferes with DNA synthesis; obstructs ergosterol synthesis	[45, 46]
Thiazole	Ritonavir, thiamine	Antiviral	HIV protease inhibitor; crucial for carbohydrate metabolism	[47]
Indole	Indomethacin, serotonin	Anti-inflammatory, neurotransmitter	Influences central nervous system functions; cyclooxygenase inhibition in inflammatory processes	[48]
Qxazole	Oxaprozin, linezolid	Anti-inflammatory, antibacterial	Blocks the process of protein synthesis; utilized in cases of resistant bacterial infections	[49]
Triazole	Fluconazole, itraconazole	Antifungal	Blocks fungal cytochrome P450; addresses systemic fungal infections	[50, 51]
Benzodiazepines	Diazepam, lorazepam	Anxiolytic, anticonvulsant	Interacts with GABA-A receptors; functions as a central nervous system depressant	[52]

1.4 Pharmaceutical Applications of Heterocyclic Compounds

The advancement of society has been significantly influenced by heterocycles, both in medical and industrial contexts. A significant portion of pharmaceuticals, dyes, cosmetics, polymers, agrochemicals, and similar compounds are heterocyclic in nature [53, 54]. The heterocyclic ring system holds significant value in industry, as many heterocycles act as fundamental templates for the synthesis of a range of compounds that are useful in various industrial applications. Heterocycles are found in a wide range of substances, including pharmaceuticals, natural products, vitamins, biomolecules, dyes, and agrochemicals. Certain heterocycles exhibit biochemiluminescence and photochromic characteristics. Heterocycles demonstrate a wide range of applications in material sciences, including their use as analytical reagents, dyes, information storage solutions, sensors, and more. Furthermore, their application extends to polymer chemistry, particularly in the realm of conjugated polymers [1, 55, 56]. Additionally, they can function as semiconductors, organic conductors, light-emitting diodes, photovoltaic cells, light-harvesting systems, and liquid crystalline compounds. Heterocycles have demonstrated significant interest due to their synthetic utility in various applications such as catalysts, protecting groups, and metal ligands in inorganic synthesis. Consequently, heterocycles warrant significant focus, and effective strategies for the development of new heterocycles represent one of the most notable fields within organic chemistry.

1.4.1 Heterocycles as Drugs

Nitrogen-containing heterocycles are some of the most important structural components in medicines. The FDA did a comprehensive review of N-heterocycles and discovered that 59% of small-molecule medications contain nitrogen heterocycles, making them the most prominent and regarded structures in the heterocycle arena. Piperazine, morphinan, cepham, penam, or phenothiazine moiety are the most prominent chemicals used in medicines [57, 58]. A series of pyrido-benzothiazine compounds has strong antibacterial activity. Rufluoxacin, a fluoro-substituted chemical, has comparable *in vivo* potency to ofloxacin and ciprofloxacin when administered orally. The methyl and pyrido-substituted benzothiazines have increased absorbance and maintain stable levels in plasma and urine. Nonetheless, the inclusion of bigger groupings resulted in a dramatic drop in activity. Structural modifications of thiazines, such as N-methyl and N-acetyl piperazine derivatives, have substantial antifungal action against *Penicillium marneffei* and *Aspergillus niger*. The study investigated the incorporation of a benzothiazine moiety into the ketoconazole molecule to act as a biosteric substitute for the 2,4-dichlorophenyl group. Notably, the *trans*-analog demonstrated greater activity [59–62].

1.4.2 Heterocycles as Polymers

Heterocycles are a structural feature of both naturally occurring and manufactured polymers. Protein production in living organisms is regulated by natural polymers having heterocycles, such as polysaccharides and nucleic acids. Photolithography, high-temperature applications, and synthetic polymers having heterocycles on the backbone find widespread use. These polymers have unique photophysical, electrical, and mechanical characteristics. Aromatic heterocyclic polymers, multicomponent polymerization, and direct polymerization of heterocycles are some of the synthetic ways to make these polymers. A wide range of industries find application for these polymers due to their exceptional mechanical, insulating, and thermal qualities. Beginning with a vinyl-heterocycle as the monomer or nucleophilic displacement on polymer derivatives with the heterocycle, synthetic polymers with heterocycles not on the backbone are created. Acid scavengers, ion-exchange resins, catalyst supports, antibacterials, surface cleansers, inhibitors of color transfer, and pollutants control are some of their uses [63].

The versatility of nitrogen heterocycles as a polymer stems from their ability to form hydrogen bonds with a wide variety of molecules, including metals, ligands, proteins, and medicines. There is a wide range of uses for these polymers. For instance, hydrogels made from methylimidazolium-based ionic liquids (Figure 1.7) have several useful bioactive uses

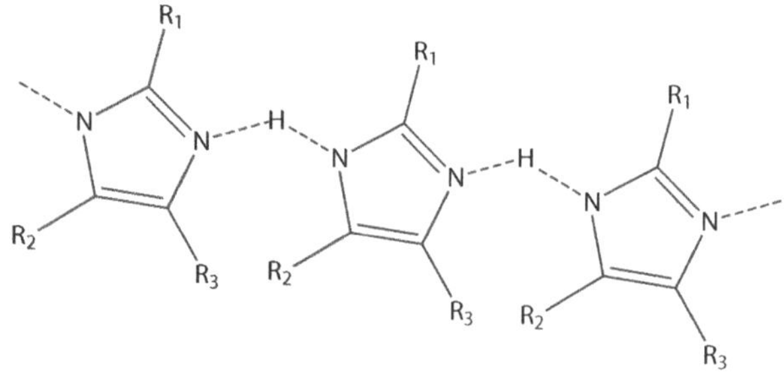

Figure 1.7 Polymerized methylimidazole derivative.

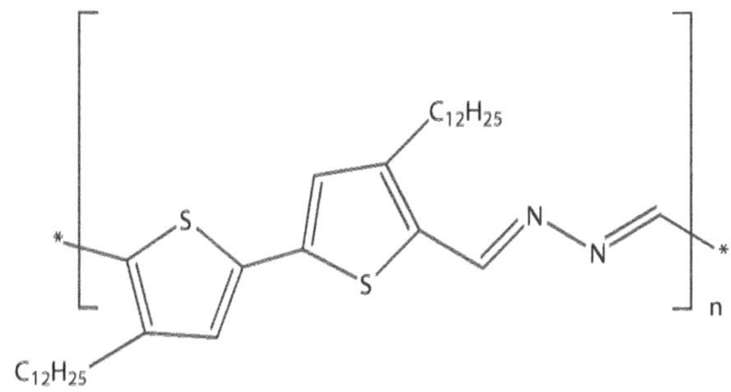

Figure 1.8 Poly(4,40-didodecyl-2,20-bithiophene-azine) (PDDBTA).

because the polymers in them self-assemble into lamellae with unique swelling characteristics [64].

Among the most important types of electroactive polymers are nitrogen heterocycles. These materials display electrical and ionic properties when subjected to a particular electrolyte solution. A film-forming cationic polymer based on low-molecular-weight piperazine is utilized as a hair conditioner for cosmetic purposes in concentrations ranging from approximately 0.15 to 5 weight percent in solvents such as alcohol or water. Miniature electrochemical immunosensors made of nitrogen heterocycles, such as polypyrroles, have shown great promise [65]. Polyazoles play a key role in the manufacture of polymer membranes and films, which have improved mechanical characteristics and preserve polymer membranes beyond the working temperature of 100°C. Acid-doped polymer membranes based on polyazoles are used in a variety of fuel cells, battery systems, capacitors, and other devices [66]. Recently, azine-based π-conjugated polymers have been suggested for use in research as a viable semiconductor for electronics. The findings of the synthesis of PDDBTA, a new azine-based polymer that functions as a p-type semiconductor channel in organic thin-film transistors and a donor in organic photovoltaics, are displayed in Figure 1.8 [67, 68].

1.4.3 Heterocycles as Dyes

Heterocyclic compounds serve as both naturally occurring and synthetic food dyes, making up a substantial part of the colorants utilized in inkjet printing, photography, and xerography. Additionally, they are prominently featured among the commercially important cellulosic and textile dyes [69]. Azo dyes represent a crucial category of chromophores, exhibiting a wide range of applications across scientific, industrial, and pharmaceutical fields. Azo chromophores represent a category of organic colorants distinguished by the incorporation of azo groups within their primary structural framework. It is possible to encounter two azo groups (dis-azo); for example, 6-hydroxy-1,4-dimethyl-2-oxo-5-((4-(phenyldiazenyl)phenyl)diazenyl)-1,2-dihydropyridine-3-carbonitrile [2] contains two fundamental azo structures. Additionally, there may be three groups (tris-azo), four groups (tetrakis-azo), or even more (poly-azo) in uncommon instances (Figure 1.9). Besides serving as colorants in more than half of all commercial dyes, they have found applications in various fields, including inkjet printing and thermal transfer printing [70, 71].

Heterocycle dyes containing nitrogen have several uses in a variety of industries, such as analytical chemistry, medicine, leather, textiles, and cosmetics. With more than 50% of the entire yearly output, this class of dyes now makes up the largest component. The luminous characteristics of benzo analogs of indolenes are known to be influenced by dye structural components. A variety of visible and near-infrared dyes that are now on the market are produced by benzoindolium derivatives. For a long time, symmetrical indocyanines have been used as well-known dyes. Because of their increased sensitivity to external stimuli, the dyes with asymmetrical structures that include two different heterocycles are of special interest for use as chemo-sensors. Moreover, these dyes' structural alterations allow for accurate manipulation of several chemical and physical properties, including solubility and fluorescence parameters. A suitable polyene precursor and an indolium salt undergo a sequential addition reaction in the widely used synthesis technique. In TiO_2 dye-sensitized nanocrystalline solar

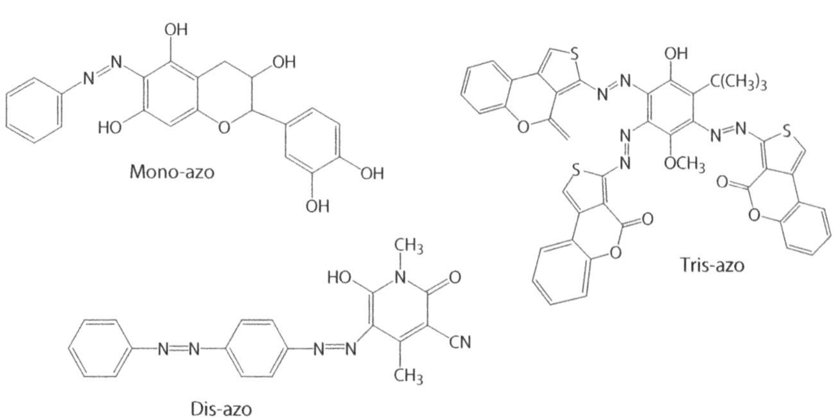

Figure 1.9 Azo dye chemical structures with varying azo group counts.

cells, this chromophore showed a power conversion efficiency of around 3.0% when combined with several cyanine dyes [8, 72].

1.5 Advantages of Using Heterocycles in Drug Design

1.5.1 Versatility and Reactivity

Fundamental to contemporary drug discovery, heterocycles offer unparalleled structural diversity, chemical flexibility, and the capacity to influence a great range of biological targets. Their striking frequency in medicines: at least one heterocyclic moiety can be found in more than 85% of all physiologically active chemicals and more than 60% of unique small-molecule medications sanctioned by regulatory authorities [59, 73, 74]. Heteroatoms such as nitrogen, oxygen, and sulfur give distinct electrical and physicochemical qualities that help particular interactions with enzymes, receptors, and nucleic acids, hence improving both medication potency and selectivity [59, 75, 76].

Heterocyclic compounds' flexibility lets medical chemists customize medication candidates for best pharmacokinetic and pharmacodynamic properties. Antibacterial drugs are designed around five-member heterocycles such thiazole, triazole, and thiazolidinone; their inclusion can enhance both the pharmacokinetic profile of the medication and the spectrum of action [76]. Particularly, nitrogen-based heterocycles are essential in the creation of anti-cancer medications because they appear in almost three-quarters of all heterocyclic anticancer medicines authorized by the FDA between 2010 and 2015 [75]. Especially in oncology, where indole and pyridine derivatives have shown the ability to cause cell death in several cancer cell lines, their therapeutic effectiveness is mostly determined by their capability to create hydrogen bonds with DNA and protein targets [59, 75].

By allowing the fast and efficient synthesis of complicated heterocyclic scaffolds with great atom economy and stereoselectivity, recent developments in synthetic techniques—including MCRs—have transformed the field [77, 78]. These developments not only increase the available chemical area but also help to find new medication candidates with better activity and safety profiles. The use of green chemistry methods—including solvent-free and microwave-assisted reactions—highlights even more the sustainable development of heterocycle synthesis [77]. A constant problem in both infectious diseases and cancer is overcome by heterocycles, which also help to combat drug resistance. Their structural adaptability lets different substituents be added, hence allowing the creation of multi-targeted molecules capable of bypassing resistance mechanisms, as shown in the evolution of hybrid heterocyclic drugs and kinase inhibitors [75, 76]. Moreover, the inclusion of heterocycles into sophisticated drug delivery systems, including nanoparticles and hydrogels, has enhanced the bioavailability and targeted distribution of therapies for diseases including diabetes and cancer [77].

Because of their extraordinary structural adaptability and chemical reactivity, heterocycles are fundamental to drug design. Their presence lets medical chemists directly affect the absorption, distribution, metabolism, excretion, and toxicity (ADME/Tox) profiles of drug candidates by fine-tuning vital drug characteristics including solubility, lipophilicity, polarity, and hydrogen bonding capability. The production of very varied chemical libraries is made feasible by the broad spectrum of possible ring widths, heteroatom kinds (nitrogen, oxygen, sulfur), and substitution patterns, hence enabling the fast identification and optimization of lead compounds [59, 76].

In the creation of multitargeted medicines, this flexibility is particularly useful because heterocyclic scaffolds can be methodically changed to interact several biological pathways at once, hence providing novel approaches for treating complicated disorders and avoiding medication resistance [6]. Moreover, contemporary synthetic methods such as MCRs allow for the quick and effective construction of heterocyclic frameworks, hence speeding the drug development process and enabling the precise adjustment of molecular characteristics to reach the best biological activity [79].

1.5.2 Enhanced Selectivity and Potency

One of the main benefits of heterocycles is their capacity to improve the strength and selectivity of pharmaceutical candidates. Heteroatoms' distinctive electrical and steric characteristics let heterocyclic compounds create particular interactions with biological targets such as enzymes, receptors, and nucleic acids: hydrogen bonding and π-π stacking. Minimizing off-target effects and enhancing therapeutic efficacy depend on this selectivity. For instance, nitrogen-containing heterocycles such as indoles and pyridines are common in drugs because they efficiently interact with protein binding sites, hence increasing target affinity and selectivity [75, 80].

Structure–activity relationship (SAR) studies show that modest changes to the heterocyclic core or its substituents can significantly affect a compound's potency and selectivity, hence guiding the rational design of medications with better safety and efficacy [75, 78]. Reflecting their capacity to engage with various enzyme binding sites and disturb important cancer pathways, approximately two-thirds of new molecular agents authorized by the FDA between 2010 and 2015 included heterocyclic rings. Particularly, five-member heterocycles have shown to be crucial as pharmacophores in antibacterial and antiviral medications, where their natural qualities can be maximized for enhanced action, pharmacokinetics, and safety [78].

Advanced computational and crystallographic research shows that, by maximizing electrostatic interactions and using small variations in target protein structures, the selectivity of heterocyclic medicines may be fine-tuned. Depending on therapeutic demands, such accuracy in molecular design is essential for creating medications with narrow or broad selectivity profiles. Heterocycles' adaptability and reactivity allow for the fast creation and optimization of therapeutic candidates; their unique structural characteristics improve selectivity and potency. These benefits confirm heterocycles as necessary scaffolding in the continuous creation of safer, more potent treatments over a broad spectrum of disease categories [38, 75, 80].

1.6 Challenges in Developing Heterocyclic Active Pharmaceutical Ingredients

Modern drug discovery is fundamentally based on the creation of heterocyclic active pharmaceutical ingredients (APIs), which is not without great scientific, technical, and legal hurdles. Many heterocyclic APIs' natural structural complexity, which usually includes several ring systems and stereogenic centers, is one of the main challenges because it makes synthesis, purification, and scale-up very difficult. Often, high regioselectivity and stereoselectivity call for sophisticated synthetic techniques such as enantioselective catalysis or MCRs, which might not be generally relevant and could call for particular reagents or settings. Furthermore, functional group sensitivity and the volatility of particular intermediates call for exact control over reaction conditions, which complicates the synthetic process and may even call for the modification of synthetic pathways [38]. Although good for drug-like qualities, the movement toward more three-dimensional heterocyclic scaffolds increases complexity for SAR research and late-stage functionalization because these changes are less clear than with conventional planar heteroaromatics [81]. Scaling up laboratory syntheses to industrial production raises further difficulties of safety, cost, reproducibility, and the requirement for specialized equipment, particularly when dangerous reagents or harsh environments are involved [38]. Environmental and regulatory concerns are also mounting as more sustainable manufacturing procedures that reduce waste and toxic byproducts—requirements not always consistent with the chemistry of particular heterocycles [37, 38], become more in demand. The existence of closely similar isomers and contaminants makes analytical characterization and purification of heterocyclic APIs and their intermediates especially difficult, usually requiring expensive and labor-intensive purification techniques. These difficulties taken together highlight the necessity of ongoing invention in synthetic chemistry, process development, and analytical methods to guarantee the effective and sustainable manufacture of heterocyclic APIs.

1.7 Recent Advances in Heterocyclic Chemistry

Recent developments in heterocyclic chemistry have been characterized by the appearance of creative skeletal editing and single-atom molecular editing techniques, which are changing the synthesis and diversity of heterocyclic frameworks. Single-atom editing lets scientists customize heterocyclic scaffolds for improved pharmacological characteristics and material functions by permitting exact atomic level changes including atom insertion, deletion, and substitution. Significantly expanding the variety of available heterocyclic compounds and allowing the creation of new architectures that were formerly difficult to manufacture, this strategy has resulted in the creation of effective techniques for ring contractions, expansions, and atom insertions supported by developments in transition-metal catalysis, organocatalysis, and photoredox catalysis [82, 83].

Recent advances include photochemical techniques for ring expansion and contraction as well as metal carbenoids for asymmetric Buchner reactions and carbene/alkyne metathesis. A 2024 study, for example, proposed a photochemical technique using O-(mesitylsulfonyl) hydroxylamine to enlarge pyridines into 1,2-diazepines; serendipitous finds during these transformations have allowed the contraction of rings to produce pyrazoles. Practical boron insertion techniques also let fused BN-heterocycles be built without the use of powerful reducing agents, hence significantly broadening the range of these processes for uses in materials science and medical chemistry.

Notwithstanding these developments, improving the selectivity, diversity, and generality of skeleton editing techniques still presents difficulties. Current methods tend to emphasize carbon and nitrogen; more strong tactics are needed to include a larger spectrum of heteroatoms. The pragmatic use of these methods in complicated molecule production also depends on enhanced chemoselectivity, regioselectivity, and stereoselectivity [84].

Alongside these synthetic developments, the inclusion of heterocyclic structures into covalent organic frameworks is breaking fresh ground in materials science by allowing the creation of porous, crystalline materials with adjustable electronic and catalytic characteristics for sophisticated uses [85]. Novel heterocyclic scaffolds such as benzimidazole derivatives have shown encouraging therapeutic potential in medicinal chemistry, especially in the creation of new antitubercular agents with high efficacy and low cytotoxicity, thus highlighting the continuing relevance of heterocycle-based drug discovery [83].

With future directions aimed at increasing molecular diversity, enhancing selectivity, and investigating multidisciplinary applications in medicine and materials science, the field is generally moving quickly toward more accurate, efficient, and sustainable ways for building and modifying heterocyclic frameworks [83, 84, 86].

1.8 Conclusion

Heterocyclic compounds are extensively utilized in the design and development of APIs due to their notable versatility in bioactivity, a trend anticipated to continue. The introduction of contemporary synthetic techniques, such as microwave-assisted synthesis, solvent-free methods, and ultrasound-mediated reactions, has significantly enhanced efficiency, yield in heterocycle production, and environmentally friendly practices. These materials fulfill roles beyond that of a drug, as they are also functionalized into polymers and dyes, underscoring their significance across multiple disciplines. The adjustable biological activity of heterocycles renders them highly effective for selective and tailored rational drug design. Despite the challenges presented by complex synthesis pathways, intricate molecular structures, and pharmacokinetic limitations, research in these areas, coupled with technological advancements, is increasingly addressing these issues. The advancement of green chemistry and heterocyclic computational modeling, in conjunction with late-stage functionalization, is expanding the range of heterocyclic APIs. Future trends indicate that multidisciplinary research fields will increasingly converge within pharmaceuticals, with heterocyclic compounds remaining central to pharmaceutical innovation and advancements.

AI disclosure statement: The author acknowledges the use of AI-based tools for minor language editing or grammatical correction. However, all intellectual contributions, analyses, and conclusions presented in this chapter are entirely the author's own.

References

1. Arora, P., Arora, V., Lamba, H.S., Wadhwa, D., Importance of heterocyclic chemistry: A review. *Int. J. Pharm. Sci. Res.*, 3, 9, 2947–2954, 2012.
2. Dze, K.C. and Samad, F., Heterocycles, their synthesis and industrial applications: A review. *Int. J. Appl. Sci. Eng. Technol.*, 2020.
3. Kumar, A., Singh, A.K., Singh, H., Vijayan, V., Kumar, D., Naik, J., Thareja, S., Yadav, J.P., Pathak, P., Nitrogen-containing heterocycles as anticancer agents: A medicinal chemistry perspective. *Pharmaceuticals*, 16, 2, Article 299, 2023.
4. Stępień, M., Gońka, E., Żyła, M., Sprutta, N., Heterocyclic nanographenes and other polycyclic heteroaromatic compounds: Synthetic routes, properties, and applications. *Chem. Rev.*, 117, 4, 3479–3716, 2017.
5. Sahu, R., Verma, D.K., Verma, C., Heterocyclic compounds: Fundamental and corrosion inhibition, in: *Heterocyclic organic corrosion inhibitors*, D.K. Verma and C. Verma (Eds.), pp. 1–15, CRC Press, Boca Raton, FL, USA, 2023.
6. Aljamali, N.M., Review in cyclic compounds with heteroatom. *Asian J. Res. Chem.*, 7, 11, 975–984, 2014.
7. Patnayak, S., Heterocyclic compounds and potential drug and its biological activity: A review. *J. Nonlinear Anal. Optim.*, 15, 1, 346–354, 2024.
8. D'Souza, A., D'Souza, A.S., Mendonca, A.P., Nayak, P., Poddar, R.K., Sadananda, V., Gowrish, S., A review on green multicomponent synthesis of heterocyclic compounds. *Int. J. Health Sci.*, 6, S6, 10865–10880, 2022.
9. Quin, L.D. and Tyrell, J.A., *Fundamentals of heterocyclic chemistry: Importance in nature and in the synthesis of pharmaceuticals*, John Wiley & Sons, 2010 Jul 8.
10. dos Santos, G.C., Martins, L.M., Bregadiolli, B.A., Moreno, V.F., da Silva-Filho, L.C., da Silva, B.H.S.T., Heterocyclic compounds as antiviral drugs: Synthesis, structure–activity relationship and traditional applications. *J. Heterocycl. Chem.*, 58, 12, 2226–2260, 2021.
11. Mąkosza, M. and Wojciechowski, K., Nucleophilic substitution of hydrogen in heterocyclic chemistry. *Chem. Rev.*, 104, 5, 2631–2666, 2004.
12. Mallappa, Chahar, M., Choudhary, N., Yadav, K.K., Qasim, M.T., Zairov, R., Patel, A., Yadav, V.K., Jangir, M., Recent advances in the synthesis of nitrogen-containing heterocyclic compounds via multicomponent reaction and their emerging biological applications: a review. *J. Iran. Chem. Soc.*, 22, 1, 1–33, 2025 Jan.

13. Jacobi, P.A., *Introduction to heterocyclic chemistry*, John Wiley & Sons, 2018 Oct 22.
14. Rosemeyer, H., Nucleolipids: Natural occurrence, synthesis, molecular recognition, and supramolecular assemblies as potential precursors of life and bioorganic materials. *Chem. Biodivers.*, 2, 8, 977–1063, 2005.
15. Lythgoe, B., Some aspects of pyrimidine and purine chemistry. *Q. Rev. Chem. Soc.*, 3, 2, 181–203, 1949.
16. Décout, J.L. and Maurel, M.C., Purine Chemistry in the Early RNA World at the Origins of Life: From RNA and Nucleobases Lesions to Current Key Metabolic Routes. *ChemBioChem.*, 26, 11, e202500035, 2025 Jun 3.
17. Shahid, M., Tripathi, T., Sobia, F., Moin, S., Siddiqui, M., Khan, R.A., Histamine, histamine receptors, and their role in immunomodulation: An updated systematic review. *Open Immunol. J.*, 2, 9–41, 2009.
18. Jain, K.P., Utekar, M.M., Shinde, Y., Sinwal, G., A review on heterocyclic compounds. *Int. J. Sci. Eng. Sci.*, 7, 10, 23–30, 2023.
19. Parsons, M.E. and Ganellin, C.R., Histamine and its receptors. *Br. J. Pharmacol.*, 147, S1, S127–S135, 2006.
20. Khan, M.F., Antihistamines, Proton Pump Inhibitors, and Related Drugs, in: *Medicinal Chemistry for Pharmacy Students*, pp. 160–219, Bentham Science Publishers, 2024 Feb 19.
21. Heravi, M.M. and Zadsirjan, V., Prescribed drugs containing nitrogen heterocycles: An overview. *RSC Adv.*, 10, 72, 44247–44311, 2020.
22. Lima, L.M., da Silva, B.N.M., Barbosa, G., Barreiro, E.J., β-lactam antibiotics: An overview from a medicinal chemistry perspective. *Eur. J. Med. Chem.*, 208, Article 112829, 2020.
23. Beatriz, A., Mondino, M.G., de Lima, D.P., Lactams, azetidines, penicillins, and cephalosporins: An overview on the synthesis and their antibacterial activity, in: *N-Heterocycles: Synthesis and Biological Evaluation*, pp. 97–142, Springer, 2022 Apr 24.
24. Romeo, G., Chiacchio, U., Corsaro, A., Merino, P., Chemical synthesis of heterocyclic–sugar nucleoside analogues. *Chem. Rev.*, 110, 6, 3337–3370, 2010.
25. Ugbaja, S.C., Omerigwe, S.A., Ndlovu, S.M., Ngcobo, M., Gqaleni, N., Evaluating the efficacy of repurposed antiretrovirals in hepatitis b virus treatment: a narrative review of the pros and cons. *Int. J. Mol. Sci.*, 26, 3, 925, 2025 Jan 23.
26. Ghodke, Y., Anderson, P.L., Sangkuhl, K., Lamba, J., Altman, R.B., Klein, T.E., PharmGKB summary: Zidovudine pathway. *Pharmacogenet. Genomics*, 22, 12, 891–894, 2012.
27. Varma, R.S., Solvent-free organic syntheses using supported reagents and microwave irradiation. *Green Chem.*, 1, 1, 43–55, 1999.
28. Gomha, S.M., Edrees, M.M., Faty, R.A.M., Muhammad, Z.A., Mabkhot, Y.N., Microwave-assisted one pot three-component synthesis of some novel pyrazole scaffolds as potent anticancer agents. *Chem. Cent. J.*, 11, Article 37, 2017.

29. Gupta, S., Singh, S., Kathuria, A., Kumar, M., Sharma, S., Kumar, R., Sharma, S.K., Ammonium derivatives of chromenones and quinolinones as lead antimicrobial agents. *J. Chem. Sci.*, 124, 437–449, 2012.
30. Krishna, C.R., Reddivari, S.R., Devineni, J.K., Venkateshwarulu, M., Baki, V.B., Chippada, A.R., Wudayagiri, R., ZnBr2-SiO2 catalyzed green synthesis of tetrazoles: Molecular docking and antioxidant activity studies. *Eur. J. Chem.*, 8, 1, 66–75, 2017.
31. Aguado, L., Canela, M.D., Thibaut, H.J., Priego, E.M., Camarasa, M.J., Leyssen, P., Neyts, J., Efficient synthesis and anti-enteroviral activity of 9-arylpurines. *Eur. J. Med. Chem.*, 49, 279–288, 2012.
32. Mariki, A., Anaeigoudari, A., Zahedifar, M., Pouramiri, B., Ayati, A., Lotfi, S., Design, green synthesis, and biological evaluation of new substituted tetrahydropyrimidine derivatives as acetylcholinesterase inhibitors. *Polycycl. Aromat. Compd.*, 42, 8, 5231–5241, 2022.
33. Cong, H. and Porco, J.A., Jr., Chemical synthesis of complex molecules using nanoparticle catalysis. *ACS Catal.*, 2, 1, 65–70, 2012.
34. Yamane, Y., Liu, X., Hamasaki, A., Ishida, T., Haruta, M., Yokoyama, T., Tokunaga, M., One-pot synthesis of indoles and aniline derivatives from nitroarenes under hydrogenation condition with supported gold nanoparticles. *Org. Lett.*, 11, 22, 5162–5165, 2009.
35. Ganji, P. and van Leeuwen, P.W.N.M., Phosphine supported ruthenium nanoparticle catalyzed synthesis of substituted pyrazines and imidazoles from α-diketones. *J. Org. Chem.*, 82, 3, 1768–1774, 2017.
36. Nikalje, A.P., Tiwari, S.V., Sangshetti, J.N., Damale, M.D., Ultrasound-mediated synthesis, biological evaluation, docking and *in vivo* acute oral toxicity study of novel indolin-2-one coupled pyrimidine derivatives. *Res. Chem. Intermed.*, 44, 5, 3031–59, 2018 May.
37. Majee, S., Shilpa, Sarav, M., Banik, B.K., Ray, D., Recent advances in the green synthesis of active N-heterocycles and their biological activities. *Pharmaceuticals*, 16, 6, Article 873, 2023.
38. Kabir, E. and Uzzaman, M., A review on biological and medicinal impact of heterocyclic compounds. *Results Chem.*, 4, Article 100506, 2022.
39. Brown, D.J., Ellman, J.A., Taylor, E.C., *The Pyridazines*, vol. 57, Supplement 1, John Wiley & Sons, 2000 Jan 3.
40. Qadir, T., Amin, A., Sharma, P.K., Jeelani, I., Abe, H., A review on medicinally important heterocyclic compounds. *Open Med. Chem. J*, 16, Article e187410452202280, 2022.
41. Chauhan, P.M.S., Sunduru, N., Sharma, M., Recent advances in the design and synthesis of heterocycles as anti-tubercular agents. *Future Med. Chem.*, 2, 9, 1469–1500, 2010.
42. Kwatra, A. and Negi, B., A review of recent progress on the anticancer activity of heterocyclic compounds. *SynOpen*, 8, 3, 185–210, 2024.
43. Senerovic, L., Opsenica, D., Moric, I., Aleksic, I., Spasić, M., Vasiljevic, B., Quinolines and quinolones as antibacterial, antifungal, anti-virulence,

antiviral and anti-parasitic agents. *Adv. Microbiol. Infect. Dis. Public Health*, 14, 37–69, 2019 Sep 13.
44. Kucharski, D.J., Jaszczak, M.K., Boratyński, P.J., A review of modifications of quinoline antimalarials: Mefloquine and (hydroxy)chloroquine. *Molecules*, 27, 3, Article 1003, 2022.
45. Adeyemi, O.S., Eseola, A.O., Plass, W., Atolani, O., Sugi, T., Han, Y., Batiha, G.E., Kato, K., Awakan, O.J., Imidazole derivatives as antiparasitic agents and use of molecular modeling to investigate the structure–activity relationship. *Parasitol. Res.*, 119, 6, 1925–1941, 2020.
46. Rani, N., Sharma, A., Gupta, G.K., Singh, R., Imidazoles as potential antifungal agents: A review. *Mini-Rev. Med. Chem.*, 13, 11, 1626–1655, 2013.
47. Farghaly, T.A., Alsaedi, A.M.R., Alenazi, N.A., Harras, M.F., Anti-viral activity of thiazole derivatives: An updated patent review. *Expert Opin. Ther. Pat.*, 32, 7, 791–815, 2022.
48. Dawood, S. and Bano, S., Inflammation and serotonin deficiency in major depressive disorder: Molecular docking of antidepressant and anti-inflammatory drugs to tryptophan and indoleamine. *Biosci. Rep.*, 42, 5, Article BSR20220426, 2022.
49. Hussain, E.A., Ghani, A., Sadiq, Z., *Oxazole, Isoxazole, Benzoxazole-Based Drug Discovery*, Elsevier, 2024 Sep 17.
50. Lass-Flörl, C., Triazole antifungal agents in invasive fungal infections: A comparative review. *Drugs*, 71, 18, 2405–2419, 2011.
51. Peyton, L.R., Gallagher, S., Hashemzadeh, M., Triazole antifungals: a review. *Drugs Today*, 51, 12, 705–18, 2015 Dec 1.
52. Brigo, F. and Lattanzi, S., Anticonvulsant agents: Benzodiazepines (clobazam, clonazepam, diazepam, lorazepam, midazolam), in: *NeuroPsychopharmacotherapy*, pp. 1–8, Springer International Publishing, Cham, 2020 Dec 17.
53. Gupta, R.R. (Ed.), *Phenothiazines and 1,4-benzothiazines: Chemical and biomedical aspects*, 1988.
54. Marcos, C.F., Polo, C., Rakitin, O.A., Rees, C.W., Torroba, T., Thiazines in one pot: The fast route to highly sulfurated heterocycles. *Angew. Chem. Int. Ed.*, 36, 23, 2697–2699, 1997.
55. Asif, M., A mini review: Biological significances of nitrogen hetero atom containing heterocyclic compounds. *Int. J. Bioorg. Chem.*, 2, 3, 146–152, 2017.
56. Al-Mulla, A., A review: Biological importance of heterocyclic compounds. *Der Pharma Chem.*, 9, 13, 141–147, 2017.
57. Barraza, S.J. and Denmark, S.E., Synthesis, reactivity, functionalization, and ADMET properties of silicon-containing nitrogen heterocycles. *J. Am. Chem. Soc.*, 140, 21, 6668–6684, 2018.
58. Liu, J., Jiang, J., Zheng, L., Liu, Z.Q., Recent advances in the synthesis of nitrogen heterocycles using arenediazonium salts as nitrogen sources. *Adv. Synth. Catal.*, 362, 22, 4876–4895, 2020.

59. Amin, A., Qadir, T., Sharma, P.K., Jeelani, I., Abe, H., A review on the medicinal and industrial applications of N-containing heterocycles. *Open Med. Chem. J.*, 16, Article e187410452209010, 2022.
60. Khera, M., Cliffe, I., Mathur, T., Synthesis and *in vitro* activity of novel 1,2,4-triazolo pyrimidine oxazolidinone antibacterial agents. *Bioorg. Med. Chem. Lett.*, 21, 10, 2885–2889, 2011.
61. Charris, J., Barazarte, A., Domínguez, J.E., Lobo, G., Camacho, J., Ferrer, R., Gamboa, N., Synthesis and antimalarial activity of ethyl 3-amino-4-oxo-9-(phenylsubstituted)thienoquinoline-2-carboxylate derivatives. *J. Heterocycl. Chem.*, 44, 3, 639–643, 2007.
62. Swarnkar, P.K., Kriplani, P., Gupta, G.N., Ojha, K.G., Synthesis and antibacterial activity of some new phenothiazine derivatives. *J. Chem.*, 4, 1, 14–20, 2007.
63. Murugan, R., Heterocycles in polymers, in: *Progress in Heterocyclic Chemistry*, vol. 35, pp. 93–118, Elsevier, 2023 Jan 1.
64. Reddinger, J.L., Sotzing, G.A., Katritzky, A.R., Reynolds, J.R., Electroactive and luminescent polymers: New fluorene-heterocycle-based hybrids. *J. Mater. Chem.*, 9, 9, 2169–2176, 1999.
65. Anderson, E.B. and Long, T.E., Imidazole- and imidazolium-containing polymers for biology and material science applications. *Polymer*, 51, 12, 2447–2454, 2010.
66. Tsai, C.H., Fortney, A., Qiu, Y., Gil, R.R., Yaron, D., Kowalewski, T., Noonan, K.J.T., Conjugated polymers with repeated sequences of group 16 heterocycles synthesized through catalyst-transfer polycondensation. *J. Am. Chem. Soc.*, 138, 21, 6798–6804, 2016.
67. Shum, J., Leung, P.K.K., Lo, K.K.W., Luminescent ruthenium(II) polypyridine complexes for a wide variety of biomolecular and cellular applications. *Inorg. Chem.*, 58, 4, 2231–2247, 2019.
68. Wang, G., Kumar, P., Zhang, Z., Hendsbee, A.D., Liu, H., Li, X., Wang, J., Li, Y., Facile synthesis of a semiconducting bithiophene-azine polymer and its application for organic thin film transistors and organic photovoltaics. *RSC Adv.*, 10, 17, 10068–10075, 2020.
69. Murphree, S.S., Heterocyclic dyes: Preparation, properties, and applications, in: *Progress in Heterocyclic Chemistry*, vol. 22, pp. 21–58, Elsevier, 2011 Jan 1.
70. Mezgebe, K. and Mulugeta, E., Synthesis and pharmacological activities of azo dye derivatives incorporating heterocyclic scaffolds: A review. *RSC Adv.*, 12, 41, 26577–26597, 2022.
71. Benkhaya, S., M'rabet, S., El Harfi, A., Classifications, properties, recent synthesis and applications of azo dyes. *Heliyon*, 6, 1, Article e03271, 2020.
72. Cai, S., Hu, X., Han, J., Zhang, Z., Li, X., Wang, C., Su, J., Efficient organic dyes containing dibenzo heterocycles as conjugated linker part for dye-sensitized solar cells. *Tetrahedron*, 69, 11, 2390–2398, 2013.
73. Kaur, N., *Metal and Nonmetal Assisted Synthesis of Six-Membered Heterocycles*, Elsevier, 2020 Apr 29.

74. Huggins, D.J., Sherman, W., Tidor, B., Rational approaches to improving selectivity in drug design. *J. Med. Chem.*, 55, 4, 1424–1444, 2012.
75. Rusu, A., Moga, I.M., Uncu, L., Hancu, G., The role of five-membered heterocycles in the molecular structure of antibacterial drugs used in therapy. *Pharmaceutics*, 15, 11, Article 2554, 2023.
76. Kumar, N. and Goel, N., Heterocyclic compounds: Importance in anticancer drug discovery. *Anti-Cancer Agents Med. Chem.*, 22, 19, 3196–3207, 2022.
77. Tripathy, D.B. and Mishra, A., Microwave synthesis and characterization of waste soybean oil-based gemini imidazolinium surfactants with carbonate linkage. *Surf. Rev. Lett.*, 24, 5, Article 1750062, 2017.
78. Pearce, S., The importance of heterocyclic compounds in anti-cancer drug design. *Drug Discovery World*, 18, 3, 39–44, 2017.
79. Nadal Rodríguez, P., Ghashghaei, O., Bagán, A., Escolano, C., Lavilla, R., Heterocycle-based multicomponent reactions in drug discovery: From hit finding to rational design. *Biomedicines*, 10, 7, Article 1488, 2022.
80. Zhang, T.Y., The evolving landscape of heterocycles in drugs and drug candidates, in: *Advances in Heterocyclic Chemistry*, vol. 121, pp. 1–12, Academic Press, 2017 Jan 1.
81. Hill, S.A., Steinfort, R., Hartmann, L., Progress, challenges and future directions of heterocycles as building blocks in iterative methodologies towards sequence-defined oligomers and polymers. *Polym. Chem.*, 12, 31, 4439–4456, 2021.
82. Barbuceanu, S.F. and Olaru, O.T., Synthesis and evaluation of biologically active compounds from heterocycles class. *Molecules*, 30, 2, Article 394, 2025.
83. Brishty, S.R., Hossain, M.J., Khandaker, M.U., Faruque, M.R., II, Osman, H., Rahman, S.M.A., A comprehensive account on recent progress in pharmacological activities of benzimidazole derivatives. *Front. Pharmacol.*, 12, Article 762807, 2021.
84. Moghimi, P., Sabet-Sarvestani, H., Moghimi, V., Norozi-Shad, N., Szostak, M., Single-atom molecular editing: Transformative advances in carbocyclic and heterocyclic frameworks. *Org. Biomol. Chem.*, 23, 10, 2036–2056, 2025.
85. Torabi, M., Yarie, M., Tavassoli, A.M., Zarei, N., Vatannavaz, L., Zolfigol, M.A., Azizian, S., Heterocyclic-linked covalent organic frameworks: Design, synthesis and applications. *Coord. Chem. Rev.*, 520, Article 216159, 2025.
86. Sharma, R., Arisawa, M., Takizawa, S., Salem, M.S.H., Remodelling molecular frameworks *via* atom-level surgery: Recent advances in skeletal editing of (hetero)cycles. *Org. Chem. Front.*, 12, 5, 1633–1657, 2025.

2

Biologically Active Nitrogen Heterocyclic Compounds

Diwakar Chauhan[1]*, Monika Chauhan[1], Priya Tyagi[1] and Amrita Kaushik[2]

[1]*School of Basic Sciences, Galgotias University, Greater Noida, India*
[2]*School of Environmental and Civil Engineering, Seneca College, Ontario, Canada*

Abstract
Nitrogen heterocycles are the most imperative and common classes in medicinal chemistry. According to statistics from drugs that have received Food and Drug Administration approval, almost 75% contain nitrogen. The numerous types of alkaloids that have these heterocycles in them are found throughout nature. They also have a significant role in biomolecular structures as well. They also aid in treating multiple illnesses as they are helpful in current biomedical approaches, such as anticancer, antibacterial, antidiabetic, anti-inflammatory, and antiproliferative, which are attributed to nitrogen heterocycles. The present chapter potentially highlights the medicinal merits and effectiveness of 4-, 5-, and 6-membered and fused nitrogen heterocycles. In this chapter, we examined the biological impacts of these nitrogen-containing heterocycles, including their anti-Alzheimer's disease, anticancer, anti-Parkinson's disease, antiviral, anti-inflammatory, antidiabetic, and antioxidant qualities.

Keywords: Anti-Alzheimer, antibacterial, anticancer, antidiabetic, carbazole, nitrogen heterocyclic compounds

**Corresponding author*: chauhan.diwakar@gmail.com

Divya Bajpai Tripathy, Priyanka Chhabra and Anjali Gupta (eds.) Heterocyclic Compounds as Pharmaceutical Ingredients, (27–60) © 2026 Scrivener Publishing LLC

2.1 Introduction

Heterocycles are the representative compounds in organic chemistry highlighting a major class and having an important role in designing various pharmaceuticals. These are cyclic compounds that contain one or more carbon atoms together with one nitrogen, oxygen, or sulfur atom. Heterocycles account for more than 50% of all known compounds [1]. Among them, heterocycles containing nitrogen have shown their significance in the biomedical and pharmacological fields. They draw the interest of researchers due to their various biological applications. Nitrogen-based heterocycles belong to the class of heterocycles, which have a ring structure with at least one nitrogen atom. They can be aromatic and aliphatic. These comprise azetidine, carbazole, pyrrole, pyrimidine, piperidine, imidazole, pyrazole, indole, pyrazine, quinoline, and isoquinoline.

According to the data on Food and Drug Administration (FDA)-approved drugs, it was found that 75% of drug molecules contain nitrogen heterocycles [2]. These are abundantly found in nature and frequently isolated from various alkaloids; e.g., piperidine was isolated from piperine, and indole was isolated from the indigo [3, 4]. Apart from this, these heterocycles are a building block of many biomolecules, i.e., DNA, RNA, proteins, vitamins, nucleic acids, nitrogenous bases, amino acids, enzymes, coenzymes, and hormones [5, 6]. Because these heterocycles demonstrate comprehensive biological actions, having analgesic, antitumor, antidepressant, antiproliferative, antibacterial, antiviral, antidepressant, and antidiabetic properties, they are important pharmaceuticals. Due to this, scientists are eager to explore these heterocycles. Many medicinally important drugs synthesized from these heterocycles are easily available in the market, e.g., axitinib, chloroquine, captopril, and tivozanib [7, 8].

In medicinal chemistry, N-based heterocyclic molecules are recognized for their diverse pharmacological properties. This chapter is mainly focused on studying the therapeutic role of N-based heterocyclic organic compounds (especially 4-, 5-, and 6-membered nitrogen heterocycles and condensed nitrogen heterocycles) having various medicinal activities that help in the treatments of various medical disorders such as anti-Alzheimer's agent, antibacterial agent, anticancer agent, antifungal agent, antidiabetic agent, anti-inflammatory agents, anti-Parkinson's, and antiviral agent (Figure 2.1).

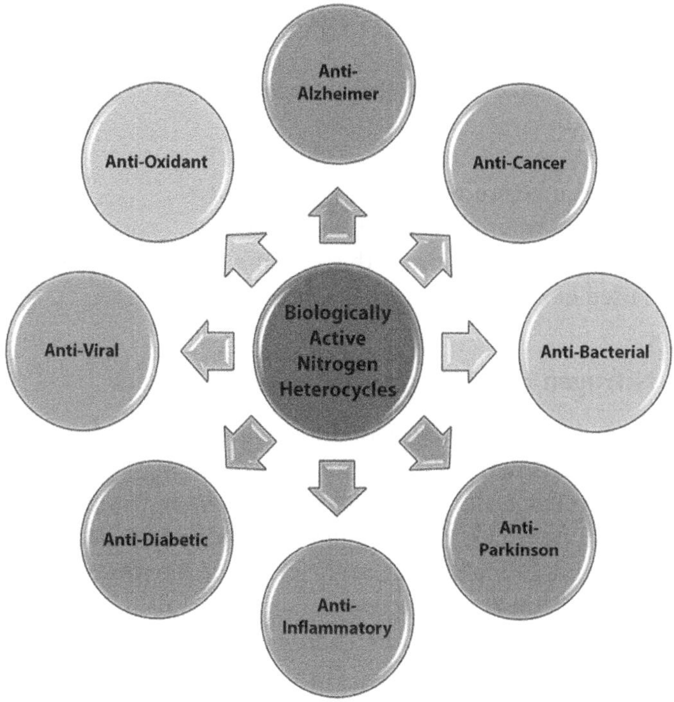

Figure 2.1 Biological activities of nitrogen-containing heterocycles.

2.2 History of Nitrogen-Containing Heterocycles

The most diverse class of organic compounds comprises heterocyclic compounds. The name heterocycle comes from the Greek word "heteros," which means "other or different," and "kyklos," which means "circle." With the growth of organic chemistry domains, the history of heterocyclic chemistry commenced in the 1800s. By oxidizing uric acid, Brugnatelli in 1818 discovered alloxan [9]. In 1834, pyrrole was isolated by Runge by dry distillation of bones. It was also known as fiery oil [10].

Wohler and Liebig subsequently reported purines and pyrimidines, which are uric acid derivatives, in 1838 [11]. Pyridine, another nitrogen-containing organic compound, was also isolated from bone oil. It was isolated by Anderson in 1849 [12]. In 1866, indole was discovered by Baeyer and Knoop as one of the products obtained when they studied the structure of indigo dye and reduced it to isatin [13]. In 1951, Chargaff's rules were described, which highlighted the role of purine and pyrimidines in the genetic code [14].

2.3 Classification of Nitrogen Heterocycles

Based on the structure, nitrogen heterocycles are classified into the following types:

- Nitrogen heterocycles with four members
- Nitrogen heterocycles with five members
- Nitrogen heterocycles with six members
- Fused or condensed nitrogen heterocycles

2.3.1 Nitrogen Heterocycles with Four Members

The heterocyclic analogs of cyclobutane with four members are known as four-membered heterocycles, and they are thought to be formed by substituting a nitrogen atom for a $-CH_2$ (methylene group), for example, azetidines, 2-azetidines. Four-membered cyclic lactam (beta-lactam) skeleton azetidin-2-one has been recognized as the most effective thing that serves as a fundamental component for crafting a diverse array of chemical entities by using the energy of strain associated with it, in addition to its implementation in the synthesis of various beta-lactam antibiotics [15] (Figure 2.2).

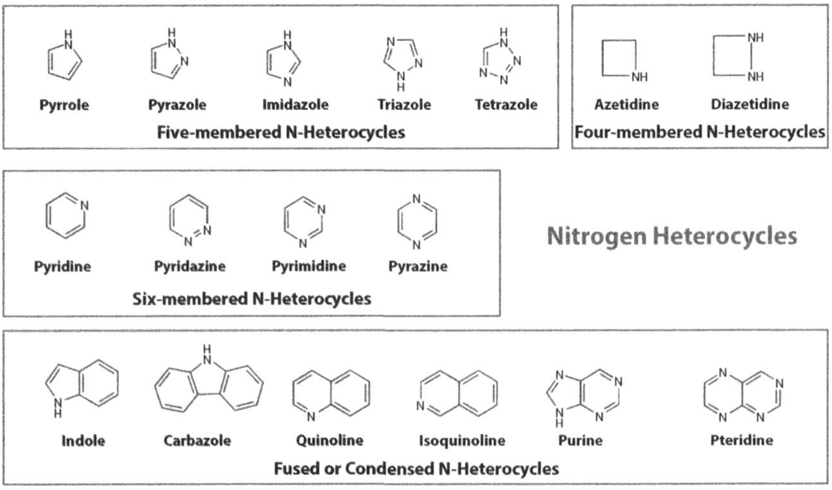

Figure 2.2 Different nitrogen-based heterocycles.

2.3.2 Nitrogen Heterocycles with Five Members

These heterocyclic compounds can be described as coming from benzene by swapping a nitrogen atom and a lone pair of electrons into one C=C link, for example, pyrrole, triazole, tetrazole, pyrazole, and imidazole. Pyrazole is a ring with five members and two nitrogen atoms that is an aromatic nitrogen heterocycle. Numerous biological activities, including antibacterial and antiviral, antidepressant, and anti-inflammatory, have been associated with pyrazole nucleus [16]. Lonazolac is an anti-inflammatory that has pyrazole in its central core [17]. In 1834, coal tar and bone oil were found to contain pyrrole, a five-membered heterocyclic aromatic ring. Drugs belonging to the imidazole class, such as miconazole, econazole, and ketoconazole, have been utilized for the treatment of fungal-related ailments [18] (Figure 2.2).

2.3.3 Nitrogen Heterocycles with Six Members

This class of compounds can be understood as the outcome of replacing a carbon atom in benzene with an isoelectronic atom, for example, pyridine, pyridazine, pyrimidine, and pyrazine. Pyrimidine nitrogen heterocycle is a pyranose ring constituting two nitrogen atoms at "1" and "3" positions. Brugnatelli successfully synthesized the first pyrimidine derivative in 1818, paving the way for further research and development in the field. As subsequent pyrimidine's nitrogen bases possess connection to DNA and RNA, they are constantly being developed for use in biological systems [19]. For the synthesis of desired medicines and agrochemicals, pyridine is utilized as a precursor [20] (Figure 2.2).

2.3.4 Fused or Condensed Nitrogen Heterocycles

The nitrogen heterocycle in this class may have two or more fused rings, some of which may be carbocyclic, and some of them could be heterocyclic. Common examples of the carbocyclic category of heterocyclic compounds are indole, quinoline, isoquinoline, carbazole, etc., and common examples of the completely heterocyclic category of N-heterocyclic compounds are purine and pteridine. An aromatic ring called an indole contains a lone pair of nitrogen and 10 π electrons. Indole is famous for its various biological properties, i.e., antiviral, antidiabetic, antimicrobial, and so on [21]. Benzene with a fused pyridine ring makes up quinoline, a nitrogen heterocycle present in coal, petroleum, and shale oil. Essential tools in the fight against malaria are the antimalarial drugs quinine, mefloquine, and chloroquine, which all contain quinolines [22] (Figure 2.2).

2.4 N-Heterocycles Containing Medicinally Important Drugs

Approximately 75% of medications incorporate nitrogen heterocycles in their structures, according to statistics provided by the FDA. It makes them one of the most important and privileged heterocycles as compared to other classes of heterocycles. Because of the many biological actions that nitrogen heterocycles have demonstrated, including antibacterial, antimalarial, antioxidant, anticancer, and so on, they are becoming increasingly important components in drug design. They have shown their usefulness in treating common medical problems as well as severe medical problems.

A variety of nitrogen heterocycles with distinctive structural changes exhibiting a variety of biological activities are shown in (Table 2.1).

Table 2.1 Various medically important N-heterocycle drugs.

S. no.	Structure	Name of drug	Medicinal uses
1.	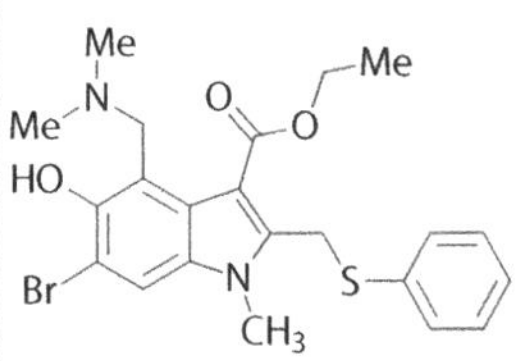	Arbidol	Antiviral [23]
2.	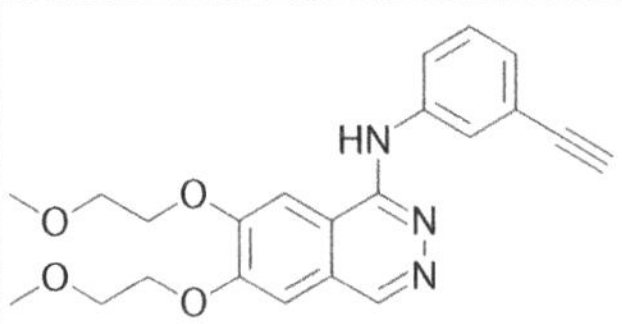	Erlotinib	Anticancer [24]
3.	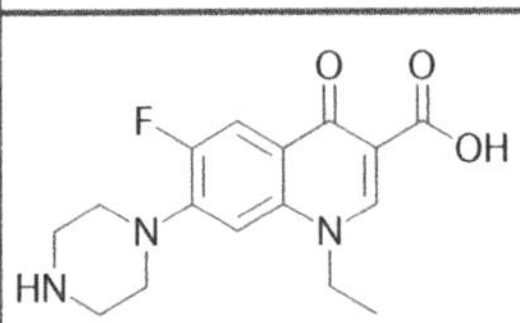	Norfloxacin	Antibacterial [25]
4.	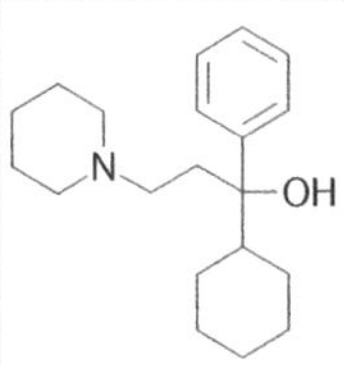	Trihexyphenidyl	Anti-Parkinson's [26]

(Continued)

Table 2.1 Various medically important N-heterocycle drugs. (*Continued*)

S. no.	Structure	Name of drug	Medicinal uses
5.	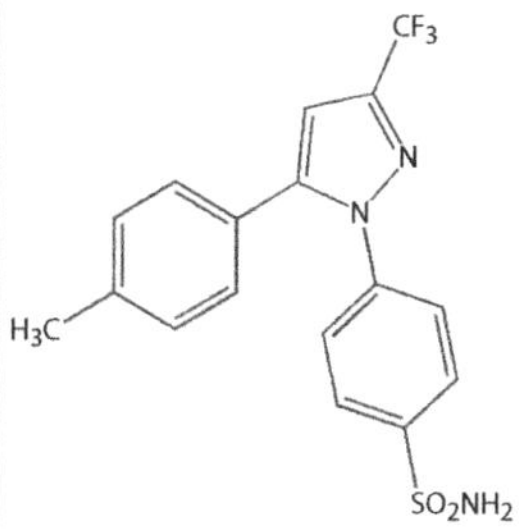	Celecoxib	Anti-inflammatory [27]
6.	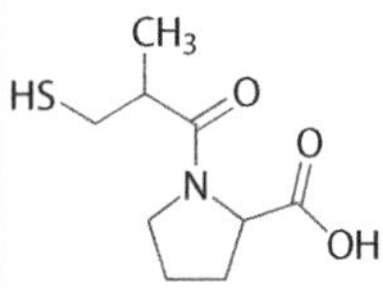	Captopril	Hypertension [28]
7.	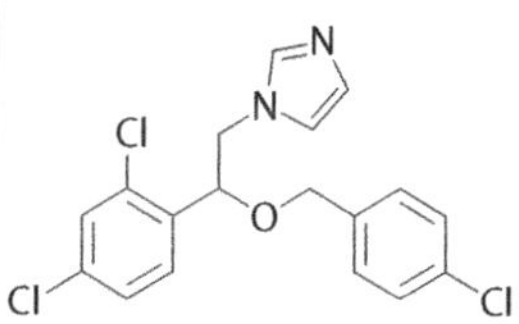	Econazole	Antifungal [29]
8.	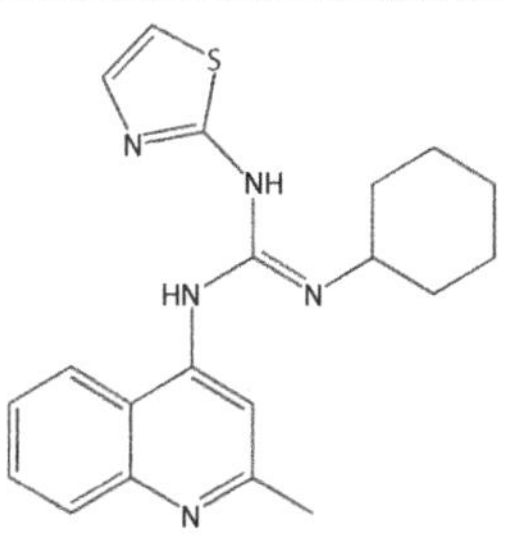	Timegadine	Leukotriene agonist [30]
9.	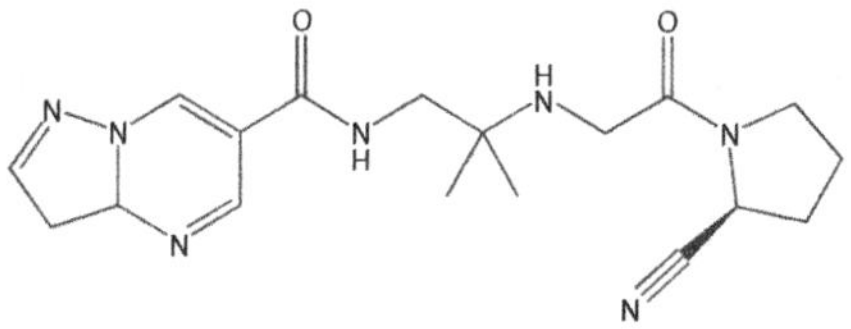	Anagliptin	Antidiabetic [31]
10.	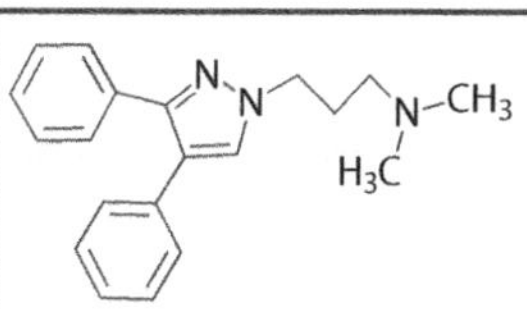	Fezolamine	Antidepressant [32]

(*Continued*)

Table 2.1 Various medically important N-heterocycle drugs. (*Continued*)

S. no.	Structure	Name of drug	Medicinal uses
11.	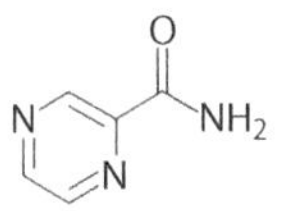	Pyrazinamide	Antituberculosis [33]
12.	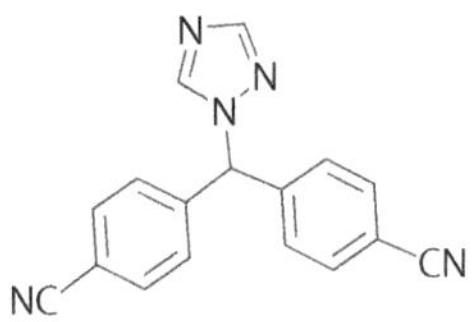	Letrozole	Aromatase inhibitor [34]
13.	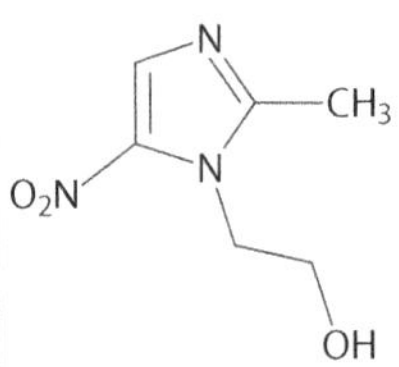	Metronidazole	Antiprotozoal [35]
14.	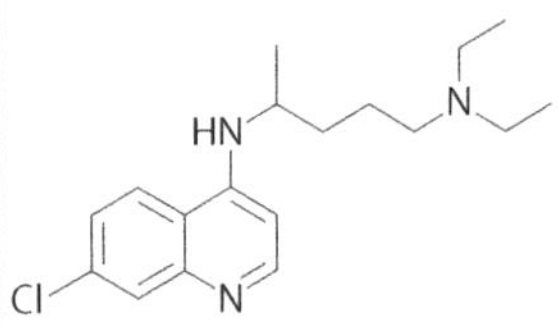	Chloroquine	Antimalarial [36]

2.5 Biological Activity

2.5.1 Anti-Alzheimer's Disease Activity

It is regarded as a complex, progressive neurological condition and can be recognized by the amyloid β (Aβ) protein build-ups and slow degeneration of cholinergic neuronal cells in brain regions such as the cerebral cortex and hippocampus. Loss of short-term memory is the initial indication of development, which then gradually leads to total loss of cognitive function, poor performance, the inability to reason, and ultimately death [37]. Patients with Alzheimer's disease have few clinical alternatives, which improve their symptoms only temporarily.

New 3-aminomethyl indole derivatives with neuroprotective potential were synthesized and pharmacologically investigated by Cuesta *et al.* These derivatives have a multitarget activity that is directed toward phosphatase 2A (PP2A) inhibition prevention and reduction of neuronal Ca^{2+} overload, two significant initial pathological and physiological events observed in neurodegenerative conditions. Compound 1, a derivative of 1-(1-benzyl-5-chloro-1H-indol-3-yl)-N, N-dimethylmethanamine, stood out for its highest level of PP2A activity recovery and voltage-gated Ca^{2+} channel blockage, as well as its positive neuroprotective profile. These results qualify this compound for additional drug screening to place novel, cutting-edge medications for Alzheimer's disease treatment [38] (Figure 2.3).

By joining the carbazole ring with the stilbene scaffold, Patel *et al.* synthesized a new set of multitarget-directed ligands in 2020. These ligands were investigated for their anti-Alzheimer's properties, such as the inhibition of cholinesterase and Aβ aggregation. Compound 2 with pyrrolidine and thiourea moiety in it, which had good inhibitory effect against the acetylcholinesterase (AChE) (IC_{50} value = 2.64 M) and the butyrylcholinesterase (BuChE) (IC_{50} value = 1.29 M), proved to be the greatest candidate to produce anti-Alzheimer's medications among all of them. Moreover, it demonstrated positive in silico ADMET characteristics [39] (Figure 2.4).

A distinct family of carbazole-based benzylpiperidine compounds was synthesized and assessed as multitarget antidepressants in 2020 by Sadeghian *et al.* The compounds with the strongest inhibitory effect in contrast to AChE and BuChE were Compound 3a (IC_{50} = 16.5 M for AChE and IC_{50} = 0.59 M for BuChE) and Compound 3b (IC_{50} = 26.5 M for AChE and 0.18 M for BuChE), respectively. Compound 3b also reported moderate inhibition action toward the BACE1 (24.5% at 50 M) [40] (Figure 2.5).

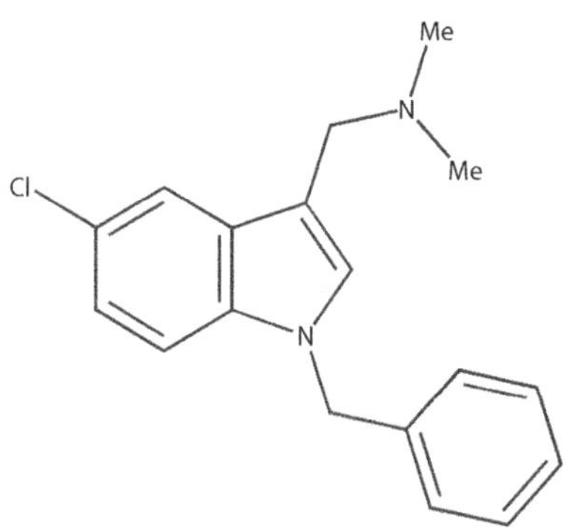

Compound 1

Figure 2.3 3-aminomethyl indole derivatives.

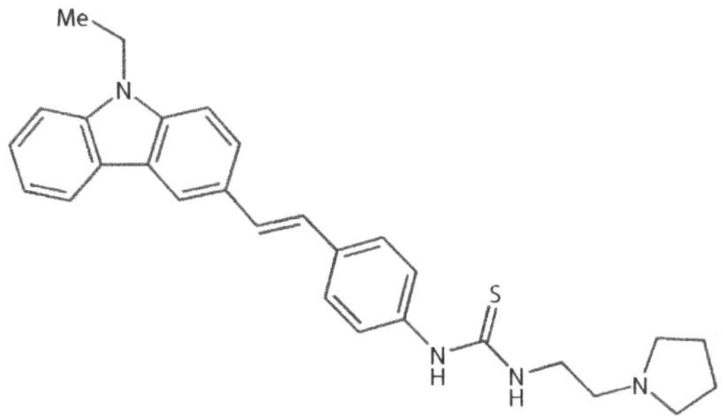

Compound 2

Figure 2.4 Carbazole-based stilbene derivatives.

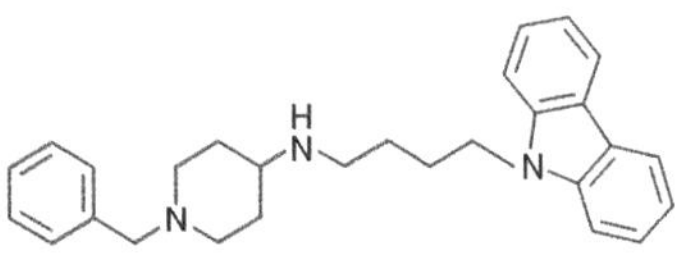

Compound 3a

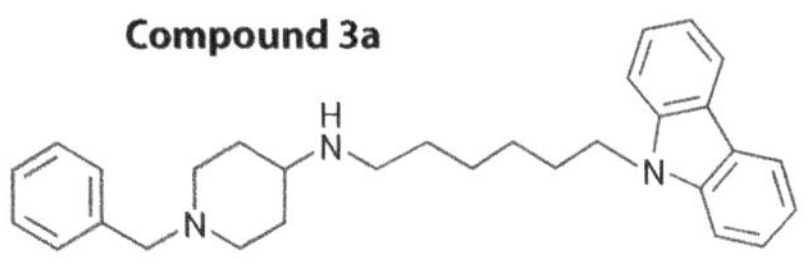

Compound 3b

Figure 2.5 Carbazoles-based benzylpiperidine derivatives.

Manzoor *et al.* reported on a novel class of phenyl sulfonyl-pyrimidine carboxylate compounds in 2021 and looked into them as possible AD medications. Compound 4 demonstrated considerable BuChE inhibition (IC_{50} = 159.43 ± 0.72 nM) and better AChE inhibition (IC_{50} = 47.33 ± 0.02 nM) in molecular docking tests [41] (Figure 2.6).

The 1,2,3-triazole-chromenone compounds were studied as possible AD medications by Askarani *et al.* in 2020. Every target material was developed and studied as a potential inhibitor of AChE or BuChE. Subsequently, the most potent compounds were evaluated based on their extensive biological activities, such as their ability to prevent Aβ1–42 aggregation, neuroprotective properties, and metal-chelating properties against PC12 cell damage caused by A25–35. The findings showed the compounds to be extremely potent and selective BuChE inhibitors, with Compound 5 having the highest IC_{50} value of 21.71 μM among them. Using a molecular modeling analysis, it has also been found that it significantly binds the

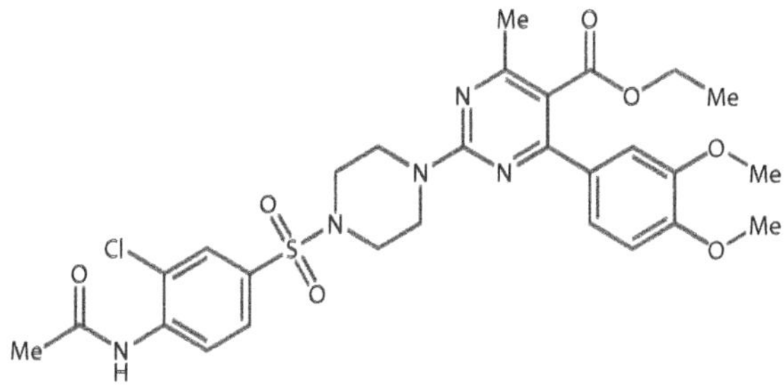

Compound 4

Figure 2.6 Phenyl sulfonyl-pyrimidine carboxylate derivatives.

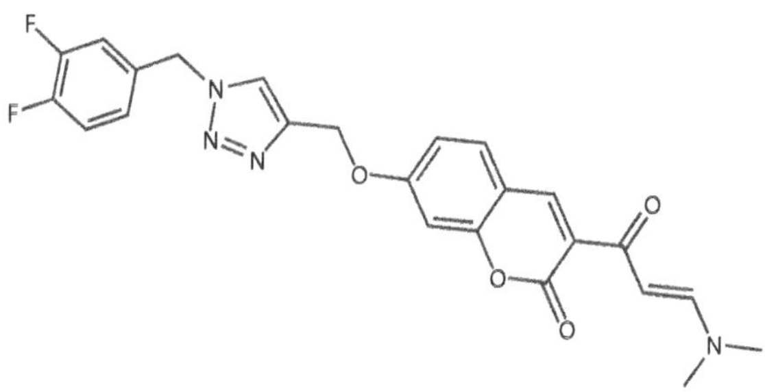

Compound 5

Figure 2.7 1,2,3-triazole-chromenone derivative.

peripheral anionic site and catalytic active site of BuChE. It also worked as a metal chelator and showed a potent anti-Aβ aggregation activity. These findings suggested it is a potential candidate for promising multifunctional AD candidates [42] (Figure 2.7).

2.5.2 Anticancer Activity

"Cancer" describes a condition marked by inhibition of contact inhibition in normal cells leading to abnormal proliferations in cells. This leads to the capacity of cells to infiltrate or malignantly spread to different organs in the body. Around 19.3 million case reports of cancer have been reported in 2020, and 28.4 million cases are predicted by 2040, significantly around the globe [43]. Cancer can be of many types, i.e., lung cancer, breast cancer, blood cancer, etc. Nitrogen-based heterocycles such as pyridine, pyrimidine, indole, triazoles, and carbazoles are used to treat different kinds of cancer. Being a strong BCL-2 protein-targeting drug that has efficient anticancer action against A549 cells, a carbazole–piperazine hybrid compound (EUCPU-0001) was produced by Mongre *et al.* [44]. This molecule was created in a three-step process by the authors, who first produced a chloroethyl urea derivative by reacting carbazole amine with chloroethyl isocyanate in anhydrous THF. Then, it was combined with 2-furoyl piperazine and refluxed in the tetrahydrofuran with the catalytic quantity of NaI and $NaHCO_3$ to create the final compound, ECPU-0001 (Compound 6). Molecular mechanisms of its action had also been examined, and it was discovered that ECPU-0001 effectively caused cellular death in A549 cells by focusing on BCL-2 to initiate the inherent apoptosis pathway. When tested on xenograft mice, it showed remarkable *in vivo* anticancer efficacy and dramatically inhibited the capacity of A549 cells to form colonies. Molecular docking and MD simulation analysis revealed a favorable interaction between ECPU-0001 and the BCL-2 active site, having a −8.35 kcal/mol binding energy (ΔG) and a Ki value of 5.72 M.

The significant interaction of it with BCL-2 was further confirmed by a thermal shift experiment (Figure 2.8).

Huo *et al.* in 2021 reported the discovery of derivatives of the [1,2,4] triazolo[1,5-a] pyrimidines as possible cancer-fighting compounds against MGC-803. Of all the synthesized compounds, Compound 7 showed 31-fold selectivity toward GES-1 and was capable of inhibiting MGC-803 (IC_{50} = 0.96 mM). Further research on the underlying mechanisms revealed that it prevented MGC-803 cells from migrating and forming colonies, and it also had an antiproliferative response by causing G0/G1 phase halt. Its activation of the intrinsic mitochondrial route and the extrinsic death receptor pathway led to the induction of cell apoptosis. By raising the level of ROS, it brought about cell death. It showed excellent

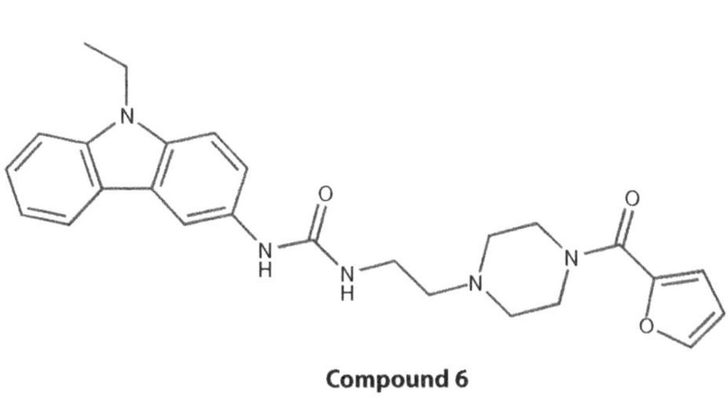

Compound 6

Figure 2.8 Carbazole–piperazine hybrid compound.

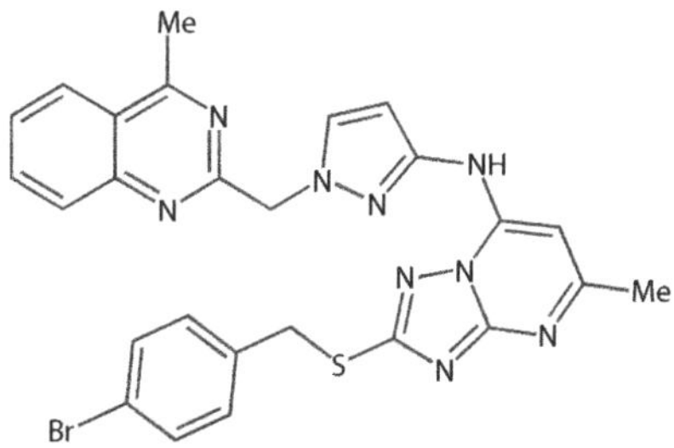

Compound 7

Figure 2.9 [1,2,4] triazolo[1,5-a]pyrimidine derivative.

in vivo safety, according to an experiment on acute toxicity [45] (Figure 2.9). According to Wang *et al.*, the compound was synthesized using the earlier reaction scheme [46].

In 2019, Cheng *et al.* synthesized novel indole-based HDAC/BRD4 dual inhibitors to treat leukemia. GI50 values ranged from 7.82 to 24.21 mM, indicating that the human acute monocytic leukemia cell line THP-1 was significantly suppressed by all synthesized compounds. Compound 8 was the most potent in inhibiting BRD4 and HDAC3, with 88% inhibition at 10 μM and 5 nM, respectively. Western blot study revealed that it was found to be a promising dual HDAC/BRD4 inhibitor that warranted further development because of its capacity to decrease c-Myc expression and increase acetylated histone H3 (Ac-H3) expression [47] (Figure 2.10).

Recent research has shown that targeting Pim-1 kinase is a key part of breast cancer therapy. In 2020, Nafie *et al.* created novel anti-breast cancer drugs that precisely target Pim-1 kinase and are pyrazolo[3,4-b]

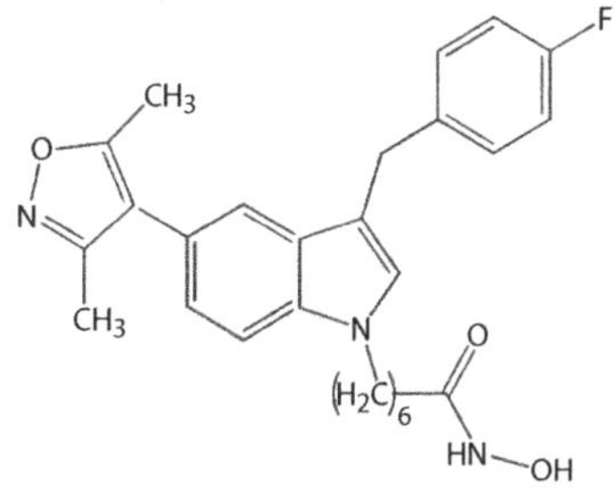

Compound 8

Figure 2.10 Indole containing scaffold.

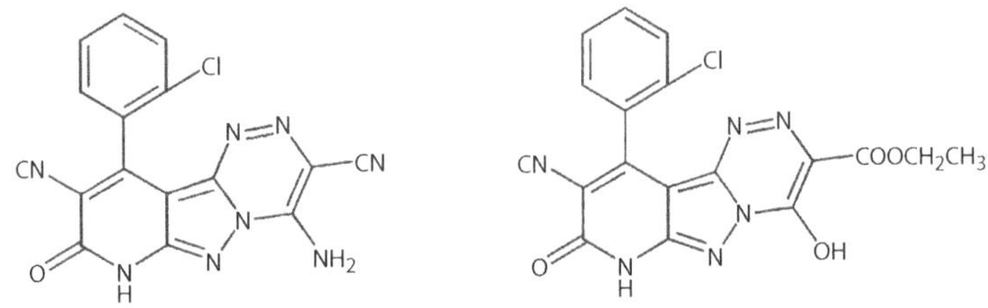

Compound 9 **Compound 10**

Figure 2.11 Indole-based scaffolds.

pyridine scaffolds based. The noncancerous MCF-10A cells and the two breast cancer cell lines, MCF-7 and MDA-MB-231, have been used to test the cytotoxic effects of each synthesized compounds *in vitro*. Compounds 9 and 10 exhibited high potency and specificity against MCF-7 cells by effectively triggering apoptosis in these cells. Both compounds exhibited significant cytotoxicity against MCF-7 cells, with IC_{50} values of 5.98 and 5.61 M, respectively, although they were ineffective against MDA-MB-231. In enzymatic, reverse transcriptase–polymerase chain reaction, and in silico experiments, both were effective inhibitors of Pim-1 kinase [48] (Figure 2.11).

2.5.3 Antibacterial Activity

Bacterial diseases are caused by bacteria. A group of quinoline-3-carbonitrile compounds was reported and assessed as possible bacterial agents in 2019 by Khan *et al.* The compounds were synthesized using a single-reactor multicomponent reaction strategy for streamlined synthesis, which entailed the reactions of ammonium acetate, ethyl cyanoacetate, 6-methoxy-1,2,3,4-tetrahydro-naphthalin-1-one, and certain aldehydes. Compound 11 showed

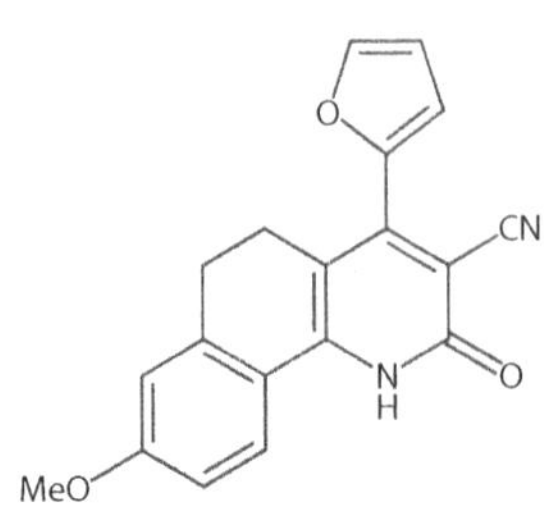

Compound 11

Figure 2.12 Quinoline-3-carbonitrile derivative.

good antibacterial activity as it represents interaction with the targeted gyrase without any severe cytotoxicity [49] (Figure 2.12).

Mandal *et al.* produced a novel class of substituted phenyl azetidine-2-one sulfonyl compounds in 2020. The produced compounds' *in vitro* antifungal and antibacterial properties were assessed. Compounds 12d, 12e, 12f, 12h, 12i, and 12j showed superior antibacterial activity in comparison to ampicillin out of all the tested derivatives [50] (Figure 2.13).

In 2021, Adamovich *et al.* produced a new series of silatrane pyrrole-2-carboxamide hybrids evaluated for their antibacterial activity. They had been obtained by reacting 1-(3-aminopropyl)silatrane and 2-(trichloro acetyl)pyrrole. These compounds of penta-coordinated silicon were equipped with the 1H-R-pyrrole-2-carboxamide group. The antibacterial activity constituted with substances has been assessed *in vitro* against

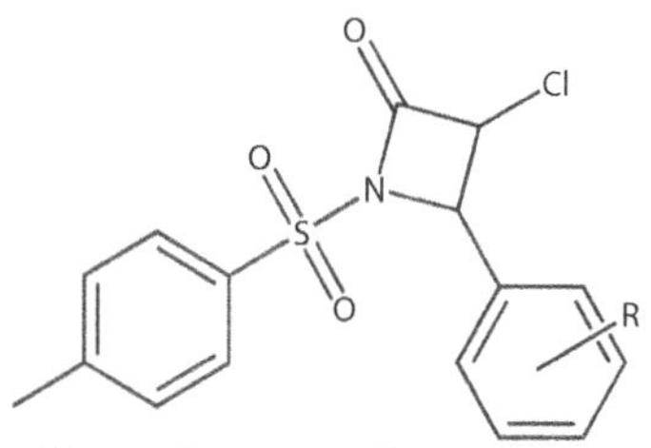

Compound	R
12 d	4-Br
12 e	2-Cl
12 f	3-Cl
12 h	2-F
12 i	3-F
12 j	4-F

Figure 2.13 Substituted phenyl azetidine-2-one sulfonyl derivatives.

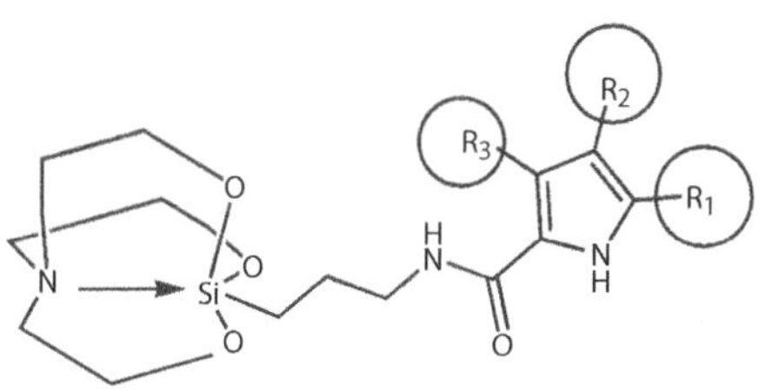

Compound 13-16

13- $R_1 = R_2 = R_3 = H$
14- $R_1 = R_2 = R_3 = Me$
15- $R_1 = Ph, R_2 = R_3 = H$
16- $R_1 + R_2 = (CH_2)_4, R_3 = H$

Figure 2.14 Silatrane pyrrole-2-carboxamide derivatives.

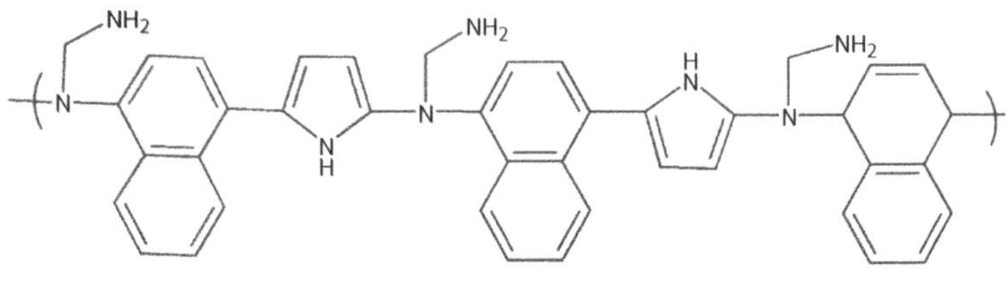

Compound 17

Figure 2.15 Polypyrrole grafted with chitosan.

gram-positive *Bacillus subtilis* B-407, gram-negative *Escherichia coli* B-1238, and gram-positive *Enterococcus durans* B-603. Compounds 13, 14, 15, and 16 showed moderate to excellent activity against gram-positive bacteria, whereas Compound 16 was also effective against gram-negative bacteria [51] (Figure 2.14).

As the antibacterial characteristic of the polymer chitosan is well known, Maruthapandi *et al.* reported the formation of the polypyrrole grafted with the chitosan. They did so with the aid of carbon dots and a microwave-assisted process. With varying degrees of intensity, the polymer exhibited antibacterial action against *E. coli* and *Staphylococcus aureus*. The bacterial cell membrane's negative charge triggered a lethal electrostatic bond with the positively charged polymer, disrupting the cell's structural integrity and leading to its demise. Its synthesis was cost-friendly, and also its potential antimicrobial activities will guarantee its future as an antibacterial agent (Compound 17) [52] (Figure 2.15).

2.5.4 Anti-Parkinson's Activity

Parkinson's disease is abruptly an increasingly debilitating neurodegenerative condition marked by persistence death of dopaminergic neurons specifically present in the substantia nigra during postmortem tissue examination. There is now no treatment available for this disorder [53]. Physical manifestations of the condition include stiffness in the muscles, rigidity, tremors, instability of posture, and poor sensory processing.

It is a well-known fact that selective monoamine oxidase (MAO) B inhibitors can significantly improve motor dysfunction and have long-term benefits in people with Parkinson's disease. Wu *et al.* in 2020 synthesized several procaine–imidazoles and subsequently used an enzymatic assay to determine their capacity to inhibit MAO-A and MAO-B. Among all the synthesized compounds, Compound 18 was able to selectively inhibit

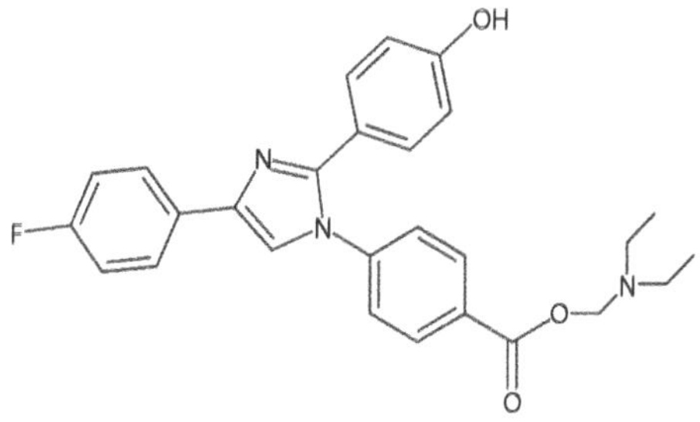

Compound 18

Figure 2.16 Procaine–imidazole derivatives.

MAO-B with IC_{50} of 0.032. Additionally, it was assessed in an animal-based model of Parkinson's disease caused by MPTP. The footprint and horizontal wire test have shown that it considerably improves the motor behavior of mice, which is a method of motor behavioral assessment. These findings made them a potential treatment for Parkinson's disease [54] (Figure 2.16).

Also in 2021, Anastassova *et al.* produced a limited number of benzimidazole derivatives with hydroxy and methoxy arylhydrazone components to create multitarget Parkinson's disease medications, which were capable of MAO-B inhibition while having antioxidant as well as neuroprotective protective properties. An excellent safety profile was found when the novel benzimidazole derivatives were assessed for their safety *in vitro* using rat brain synaptosomes and SH-SY5Y cells. Compound 19 demonstrated the highest *in vitro* IC_{50} value on SH-SY5Y cells, measuring 311.43 μM. The rat brain synaptosomes were also found to have the lowest toxicity of this compound when synaptosomal viability and glutathione levels were measured. It demonstrated the most inhibitory impact on hMAO-B of all investigated compounds, comparable to that of the reference Parkinson's disease medications rasagiline and selegiline [55] (Figure 2.17).

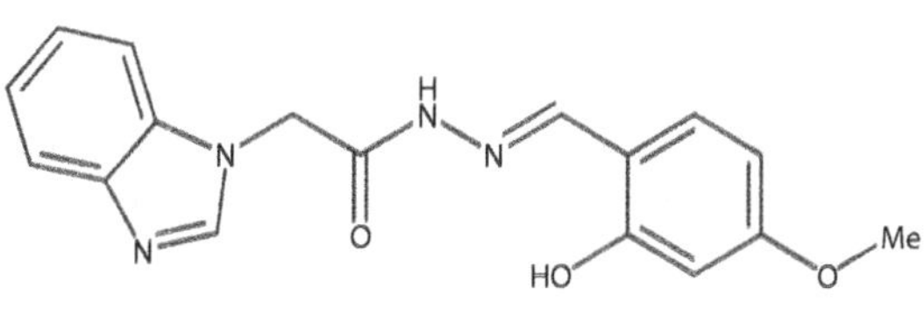

Compound 19

Figure 2.17 Benzimidazole derivative.

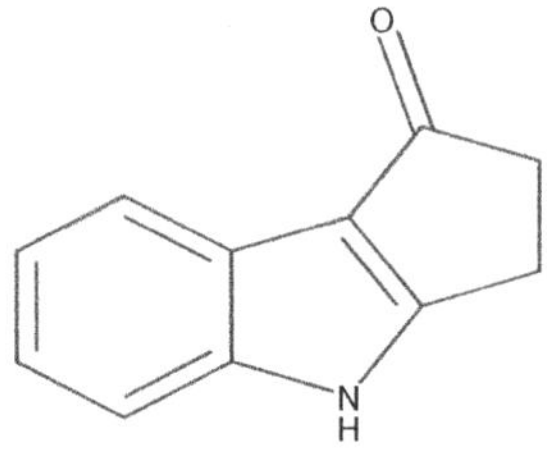

Compound 20

Figure 2.18 Indole derivative.

A recent study suggests that antioxidants and oxidative stress may aid in the pathophysiology and treatment of Parkinson's disease. Wei *et al.* in 2019 examined the antioxidative properties of NC001-8, an indole derivative (Compound 20), in DAergic neurones obtained from SH-SY5Y cells and Parkinson's disease-specific pluripotent stem cells (Parkinson's disease–iPSCs) generated with a PARKIN ex5del mutation. The DAergic neurones derived from Parkinson's disease–iPSCs showed decreased NRF2 and NQO1 expression levels, along with an increased susceptibility to environmental risks such as H_2O_2. These characteristics made them excellent models for evaluating the impacts of NC001-8. As a result of H_2O_2 toxicity, NC001-8 treatment decreases the enhanced counts of reactive oxygen species as well as cleaved caspase 3 enzyme expression. It also increased NRF2 and NQO1 in Parkinson's disease–iPSC–derived DAergic neurones, indicating the drug's efficacy for neurological protection and antioxidation in DAergic nerve cells of Parkinson's disease patients [56] (Figure 2.18).

2.5.5 Anti-Inflammatory Activity

Inflammation is a necessary immune reaction that supports tissue homeostasis and permits survival in a variety of dangerous conditions, including infection and tissue injury [57].

To produce new compounds that inhibit the cyclooxygenase 2 (COX-2) enzyme and prevent the unfavorable side effects of anti-inflammatory drugs that are not specific, Mohamed *et al.* synthesized new pyrazoles and pyrazolo[3,4-b] pyridines in 2019 and then investigated their anti-inflammatory properties *in vivo*. Comparison with reference drugs indomethacin and celecoxib revealed that the pyrazole ring analogs were the most effective analogs, whereas the pyrazolo[3,4-b] pyridines were shown to have very moderate efficacy. The assay demonstrated

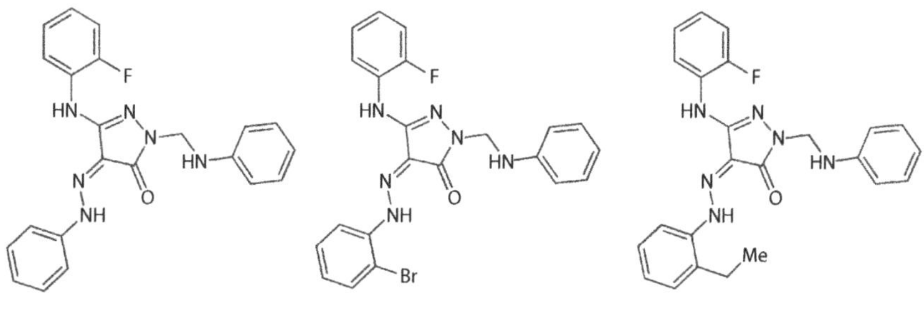

Compound 21 **Compound 21a** **Compound 21b**

Figure 2.19 Pyrazole–pyridine derivatives.

that Compounds 21, 21a, and 21b were shown to be the most active and selective COX-2 inhibitors, as well as the most useful in preventing edema; they also had the lowest ulcerogenic effect of all derivatives in an *in vivo* assay to measure their influence on paw edema rates and ulcerogenicity [58] (Figure 2.19).

The benzimidazole ring, in particular, is one of the preferred scaffolds used in drug design as it satisfies the minimal structural specifications that are necessary for anti-inflammatory drugs. Maghraby *et al.* synthesized and created novel molecular hybrids of 2-methylthiobenzimidazole linked to several anti-inflammatory pharmacophores through a 2-aminothiazole linker in 2020 to investigate the effects of such molecular variation on inhibition of COX and 15-lipoxygenase (15-LOX) enzymes *in vivo.* Because compound 22 offered an ensemble of steric, electronic, hydrogen bonding, and features specific to the possible guaranteed molecular interactions with specific biological targets and are also used to prevent their immediate biological responses, compound 22 produced the most active and most selective COX-2 and also the 15-LOX inhibitors in this study. At this point, it is declared the significance of the benzimidazole-thiazole hybrid

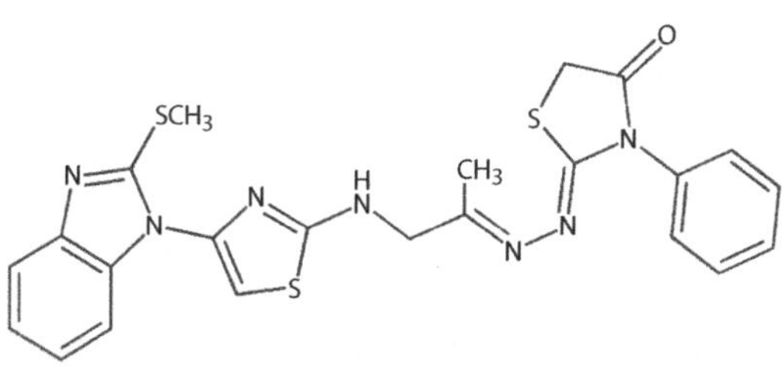

Compound 22

Figure 2.20 2-methylthiobenzimidazole hybrid.

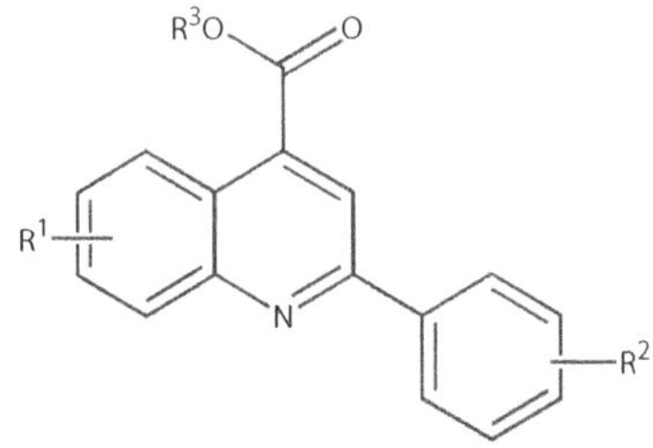

Compound 23

Figure 2.21 Quinoline derivative.

is associated with various aromatic as well as heterocyclic moieties as a potential framework for systematically developing new anti-inflammatory agents with dual COX-2 or 15-LOX enzymes inhibitory activity. In addition, hybrid 16 demonstrated the greatest edema inhibition at both the 3- and 4-hour periods (119% and 102%, respectively, compared to indomethacin) [59] (Figure 2.20).

In 2022, Huang *et al.* reported a brand-new, metal-free, three-component technique by reacting the anilines, alkyl lactates, and α-keto acids to synthesize 2,4-disubstituted quinolines. Without creating any organic mass-based side products, the reactions move forward with the help of tert-butyl peroxy benzoate and p-toluene sulfonic acid to produce a variety of quinoline products by creating new C-C single, C=C, and C=N double bonds. The quinolines' anti-inflammatory action was studied by evaluating their ability to block NO release by lipopolysaccharide (LPS)-induced RAW264.7 cells, resulting in the identification of 23a, 23b, and 23c as powerful anti-inflammatory compounds *in vitro* [60] (Figure 2.21).

2.5.6 Antidiabetic Agent

One of the most prevalent metabolic disorders, diabetes mellitus, has an alarmingly rising prevalence around the globe. This persistent health condition arises from either inadequate insulin secretion by the pancreas or the body's ineffective utilization of the insulin that is produced. It falls into three categories: gestational diabetes, type 1, and type 2. Around 90% of the cases of diabetes are of type 2, which is the most widespread type. Diabetes is an incurable disease that can be controlled by implementing self-care in daily life, educating people about diabetes, and achieving substantial progress in administration, understanding, attitudes, and abilities [61].

Monitoring blood sugar levels is one of the most important strategies for preventing diabetic complications. One of the most attractive therapeutic targets for the production of T2D drugs has been identified as α-glucosidase inhibitors because they decrease the absorption of carbohydrates, which inhibits the synthesis of glucose and lowers blood sugar levels [62]. In 2019, Pogaku *et al.* designed novel hybrids of pyrazole–triazolopyrimidine by the multicomponent reaction. All of the synthesized derivatives had been investigated against acarbose, a common medicine, for their ability to block the enzyme α-glucosidase. In comparison to the benchmark medication acarbose (IC_{50}:12.6 μM), Compound 24a has shown excellent glucosidase enzyme inhibitory efficiency with an IC_{50} value of 12.45 μM. Strong activity has also been shown by Compounds 24b and 24c, whose IC_{50} values were reported to be 14.47 and 17.27 μM, respectively. All of the title compounds' structure–activity relationship studies were established. How the glucosidase enzyme and the substances interacted during binding contacts were investigated. For the target compounds, similar to drug characteristics, Lipinski parameters and in silico ADME features have been anticipated. These compounds were reported to be prospective hits for antidiabetic activators due to their glucosidase inhibition, molecular docking, and drug-likeness features [63] (Figure 2.22).

In 2021, Azimi *et al.* created, synthesized, and assessed new pyrazole–phenyl semicarbazone compounds to determine their potential to inhibit *in vitro* α-glucosidase. Pharmacophore modification was established to promote the α-glucosidase inhibitory activity. All newly synthesized substances demonstrated good *in vitro* yeast glucosidase inhibition,

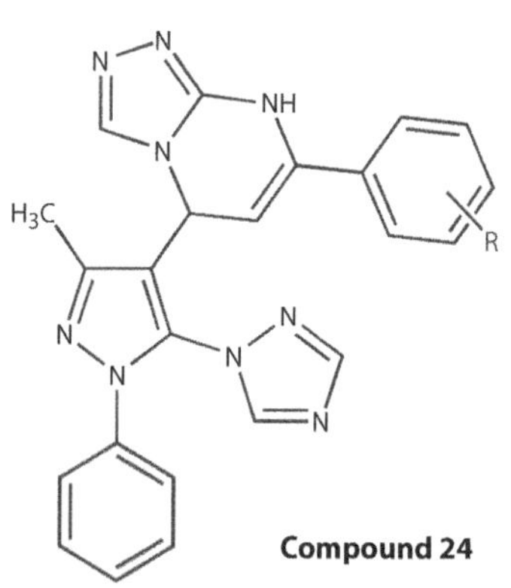

R
24a= 4-Cl
24b= 4-F
24c= 3-NO_2

Compound 24

Figure 2.22 Pyrazole–triazolopyrimidine derivatives.

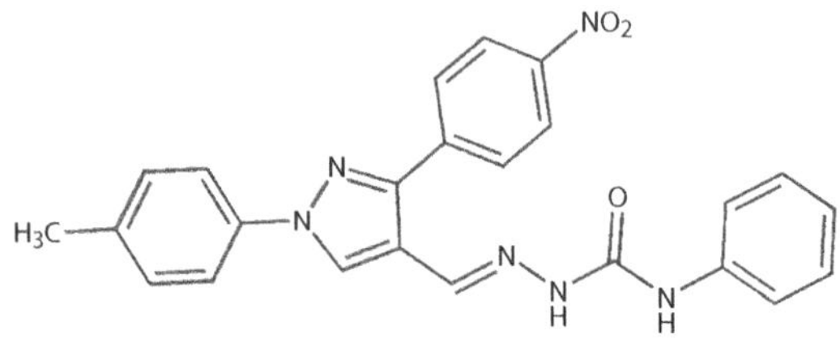

Compound 25

Figure 2.23 Pyrazole–phenyl semicarbazone derivative.

which was even more potent than the commonly used drug acarbose. The strongest α-glucosidase inhibitory action was shown by Compound 25 (IC_{50} = 65.1 ± 0.3 μM) [64] (Figure 2.23).

In 2022, Ali *et al.* produced a group of 13 new derivatives of benzimidazoles and investigated them as a novel α-glucosidase inhibitory capability. The enzyme was significantly inhibited by all of the compounds (IC_{50} range = 0.64 ± 0.05 μM to 343.10 ± 1.62 μM) in comparison to acarbose (IC_{50} = 873.34 ± 2.21), it was found that Compound 26 exhibited the greatest activity (IC_{50} = 0.64 ± 0.05 μM). According to the docking score, binding energy, and affinity, molecular docking provided additional evidence that these drugs effectively inhibited enzymes. The discovery of drugs to treat type 2 diabetes mellitus may be aided by these encouraging findings [65] (Figure 2.24).

Due to its ability to lower postprandial hyperglycemia, α-amylase has been viewed as a key therapeutic option for treating the type 2 diabetes mellitus. Using simple, adaptable, and straightforward synthesis techniques, Shabaan *et al.* created a class of novel imidazo[4,5-c]pyrazoles Compound 27, after which their *in vitro* amylase inhibitory activity was evaluated. The findings indicated that the derivatives of diphenylaminomethyl and phenylcarbamodithioate both induced strong-amylase inhibitory

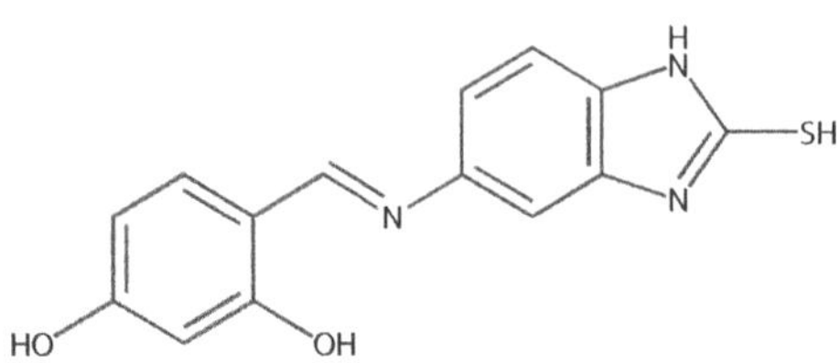

Compound 26

Figure 2.24 Benzimidazole derivative.

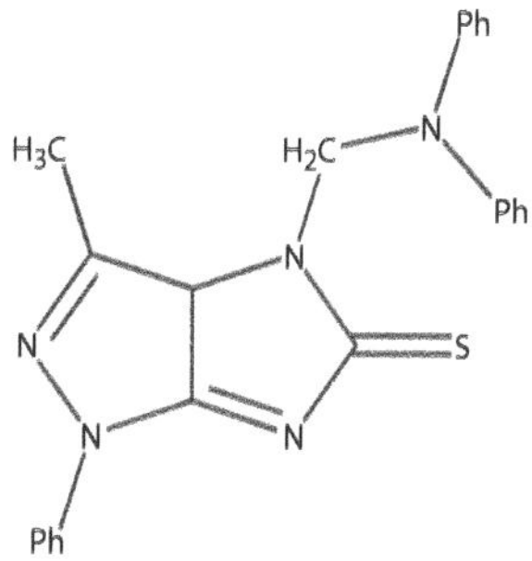

Compound 27

Figure 2.25 Imidazo[4,5-c]pyrazole derivative.

effects, with IC_{50} values of 0.071 and 0.083 μM, respectively, which were comparable to those of acarbose (IC_{50} = 0.060 μM). Molecular docking along with in silico studies showed that they had good binding affinities to α-amylase with binding free energies similar to acarbose. These compounds demonstrated significant biological activity, physicochemical properties, drug-likeness, and good pharmacokinetics, providing an opportunity for further development and optimization of the imidazopyrazole scaffold [66] (Figure 2.25).

2.5.7 Antiviral Activity

The single biggest factor contributing to infectious disease morbidity and death globally is viruses, which inflict a terrible toll on the human population. When harmful viruses infiltrate a living thing's body and infectious virus particles (virions) adhere to and penetrate vulnerable cells, this condition is known as a viral infection. In addition to replicating inside host cells, viruses can also create poisons that can lead to illness. Examples of such infections are COVID, influenza, HIV/AIDS, herpes, and flu [67].

Several novel thiosemicarbazone-based benzimidazole derivatives were synthesized in 2020 and reported by Francesconi *et al.* as potential antiviral drugs against human respiratory viruses, including respiratory syncytial virus, coronavirus 229E, influenza A/H1N1, and influenza A/H3N2. Compounds 28, 29, and 30, which are all benzimidazole derivatives, might have antiviral properties [68] (Figure 2.26).

Abu-Zaied *et al.* reported a new innovative way to synthesize the novel pyrimidine glucoside derivatives in 2021 and evaluated them for their

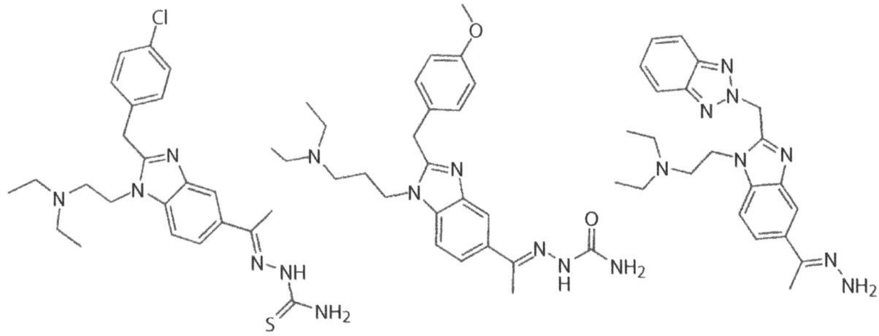

Compound 28 Compound 29 Compound 30

Figure 2.26 Benzimidazole derivatives.

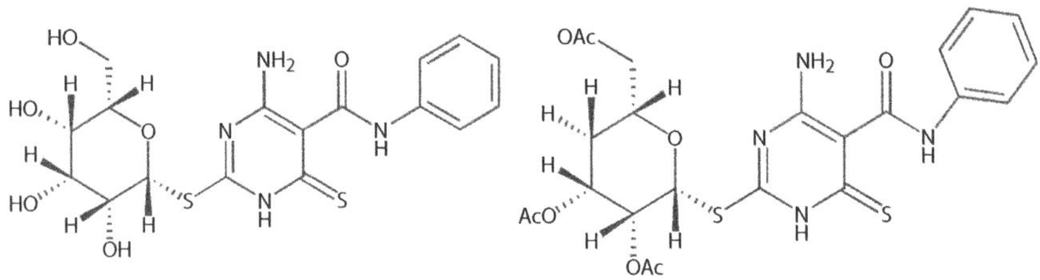

Compound 31 Compound 32

Figure 2.27 Pyrimidine glucoside derivatives.

activities against SARS-CoV-2 and avian influenza H5N1 virus. *In vitro*, pharmacological testing of these compounds against SARS-CoV-2 and avian influenza H5N1 virus strain showed that Compounds 31 and 32 have promising activity [69] (Figure 2.27).

In 2021, Hattori *et al.* showed that GRL-1720 (Compound 33) and Compound 34, each containing an indoline and an indole moiety, potently blocked the infectivity of SARS-CoV-2 by targeting Mpro. Both drugs suppressed SARS-CoV-2 infectivity with EC_{50} values of 15 ± 4 and 4.2 ± 0.7 μM, according to VeroE6 cell–based RNA–quantitative polymerase chain reaction, immunocytochemistry tests, and cytopathic assays. An X-ray structural investigation revealed that Compound 34 forms a covalent connection with Mpro and interacts polarly with a number of the active site amino acid residues, which results in a synergistic effect of it and remdesivir against SARS-CoV-2. In the creation of SARS-CoV-2 infection treatments, Compound 34 might serve as a key Mpro inhibitor [70] (Figure 2.28).

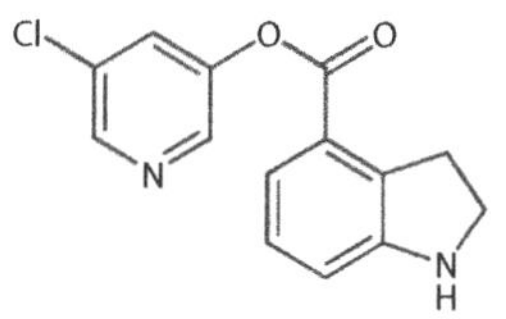

Compound 33

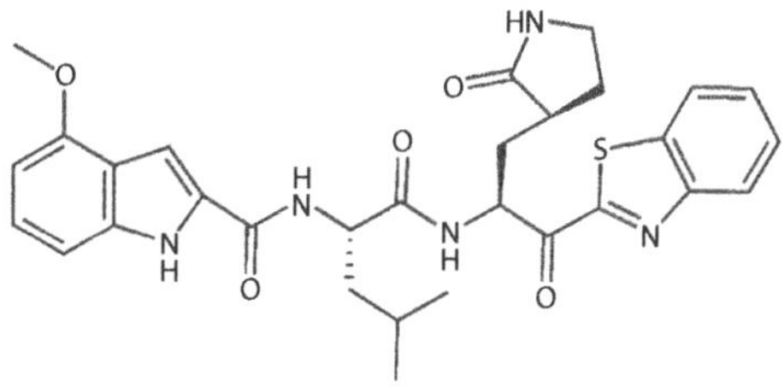

Compound 34

Figure 2.28 Indoline and indole derivatives.

HIV/AIDS continues to be a significant worldwide health problem. Targeting various phases of viral DNA replication over the long term is essential for HIV/AIDS medicines to be effective. Malancona *et al.* synthesized the 5,6-dihydroxypyrimidine scaffolds for precisely hitting major HIV nucleocapsid (NC), which is a protein with multiple functions specifically required for amino acid mediated interactions, enabling the sequence-specific attachment during reverse transcription. According to their previous work, nordihydroguaiaretic acid exhibits poor drug-like properties and functions as an NC inhibitor because it contains catechol moieties. It also exhibits mild antiviral activity, but these catechol moieties have lower physicochemical properties. It was evident to look forward a catechol-like scaffold with better physicochemical qualities that could bind NC protein to solve this issue. As a result, Compound 35, a 5,6-dihydroxypyrimidine-4-carboxamide, was recognized as a novel class of HIV-1 NC inhibitors. Hit validation efforts resulted in the discovery of an optimized analog, Compound 36, which exhibits enhanced NC inhibition and antiviral activity in addition to having favorable ADME and PK characteristics [71] (Figure 2.29).

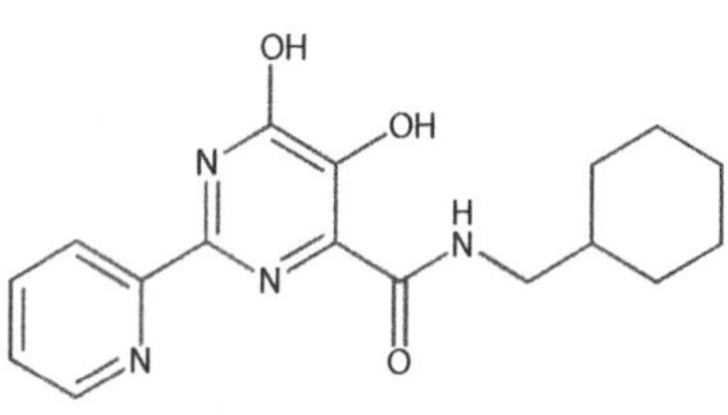

Compound 35

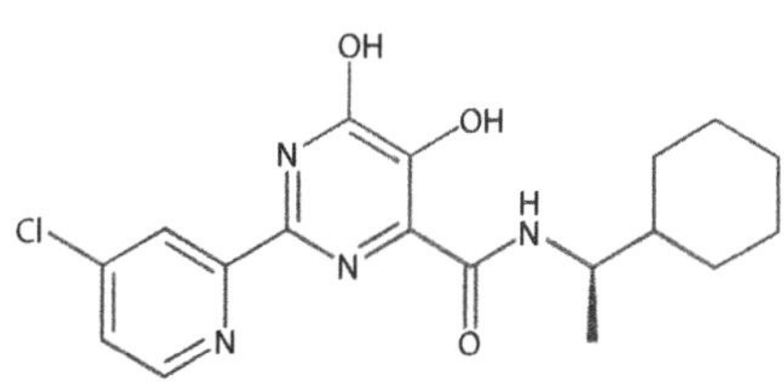

Compound 36

Figure 2.29 Pyrimidine derivatives.

2.5.8 Antioxidant Activity

Compounds known as antioxidants have the potential to neutralize free radicals by taking one or more electrons away from the radical's unpaired state. In addition to directly neutralizing reactive radicals, antioxidant molecules can change into free radicals that are new and possess less potential, more resilient, and less hazardous in comparison to the original radicals they have neutralized. Other antioxidants or other processes may be able to neutralize them and end their radical condition [72].

Ali *et al.* reported a novel series for pyrazolo[4′,3′:5,6]pyrano [2,3-d] pyrimidines with a bioactive chromone ring in 2020, along with functionalized pyrano[2,3-c]pyrazoles. The procedure comprised reacting different electrophile reagents with the compound 6-amino-4-(4-oxo-4H-chromen-3-yl)-3-phenyl-1,4-dihydropyrano [2,3-c]pyrazole-5-carbonitrile. All the synthesized compounds have been extensively characterized with the help of spectroscopic methods by using spectroscopic data and therefore tested for *in vitro* antioxidant activities using DPPH radical scavenging techniques. An initial examination of the structure–activity connection revealed that Compound 37's total antioxidant capacity was enhanced by adding the pyrano-pyrazole framework and both NH and OH groups to the pyrimidine moiety [73] (Figure 2.30).

In 2022, Gomaa *et al.* synthesized 2,3-dihydropyrazino[1,2-a]indole-1,4-dione derivatives and investigated them as one of the antiproliferative and antioxidant agent. Compound 38 showed good antioxidant properties

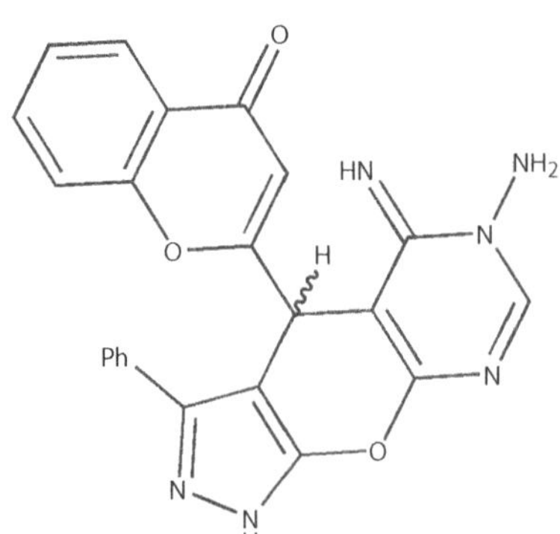

Compound 37

Figure 2.30 Pyrano–pyrazole derivative.

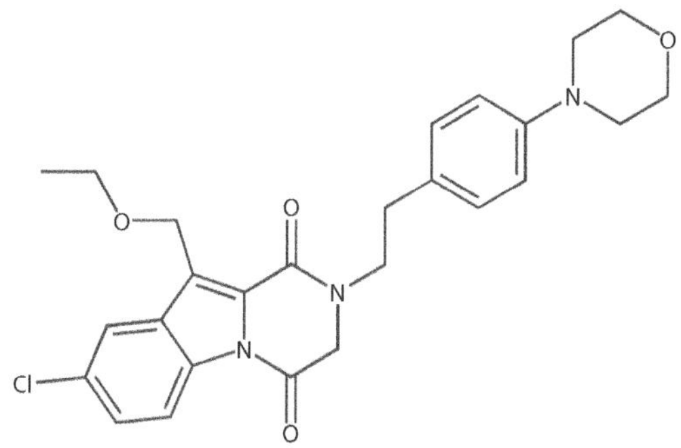

Compound 38

Figure 2.31 Pyrazino–indole derivative.

when compared to the reference Trolox. Its radical scavenging activity was found to be 73% [74] (Figure 2.31).

In 2023, Selvakumaran *et al.* synthesized new benzimidazole semicarbazide derivatives by multistep synthesis and evaluated them for their antibacterial and antioxidant activities. Using BHT as a reference molecule, the DPPH test was used to assess the *in vitro* antioxidant activity of the synthesized compounds. By monitoring the drop in the absorbance of the DPPH radical at 517-nm wavelength, the antioxidants' antiradical efficacy is assessed. Of all the compounds, Compound 39 showed the best antioxidant activity, DPPH radical scavenging; however, it showed less effect than BHT [75] (Figure 2.32).

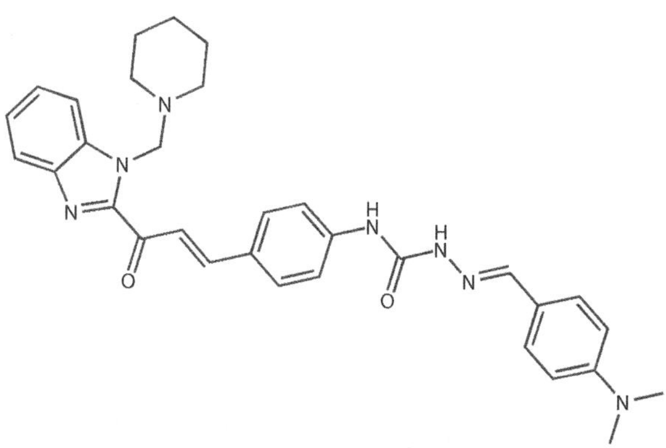

Compound 39

Figure 2.32 Benzimidazole–semicarbazide derivative.

2.6 Conclusion

Nitrogen heterocycles are abundantly found in nature and contribute to one of the largest classes of organic compounds, which are suitable for synthesizing various types of pharmaceuticals. They showed a great number of medicinal bioactivities, which are helpful in treating common infections to severe kinds of infections. In comparison to other chemicals, nitrogen heterocycles have the advantages of being widely available, easy to synthesize, and highly biocompatible. This book chapter intends to offer useful information about the different medicinal activities possessed by various nitrogen-based heterocycles (five, six, and fused N-heterocycles) that help researchers in designing new drugs based on them.

Abbreviations

AD	Alzheimer's disease
PP2A	phosphatase 2A
AChE	acetylcholinesterase
BuChE	butyrylcholinesterase
BACE1	beta-site amyloid precursor protein cleaving enzyme
BCL-2	B-cell leukemia
ROS	reactive oxygen species
HDAC3	histone deacetylases 3
BRD4	bromodomain containing protein 4
c-Myc	cellular myelocytomatosis oncogene
PIM-1 kinase	proviral Integration site for Moloney murine leukemia virus kinase
PD	Parkinson's disease
DAergic neurons	dopaminergic neurons
GSH	glutathione
hMAO-B	human monoamine oxidase B
NRF2	nuclear factor erythroid 2-related factor 2
NQO1	NAD(P)H quinone dehydrogenase1
COX-2	cyclooxygenase
T2D	type 2 diabetes
ADME	absorption, distribution, metabolism, and excretion
PK	pharmacokinetics
DPPH	2,2-diphenyl-1-picrylhydrazyl
BHT	butylated hydroxytoluene

AI disclosure statement: The author confirms that no artificial intelligence (AI)-generated content or AI-assisted writing tools were used in the preparation of this chapter (including summaries and the abstract). All text, analysis, and conclusions are the result of the author's original work.

References

1. Rees, C.W., Polysulfur-nitrogen heterocyclic chemistry. *J. Heterocycl. Chem.*, 29, 639–651, 1992.
2. Vitaku, E., Smith, D.T., Njardarson, J.T., Analysis of the structural diversity, substitution patterns, and frequency of nitrogen heterocycles among U.S. FDA-approved pharmaceuticals. *J. Med. Chem.*, 57, 10257, 2014.
3. Park, K.J., Kim, D.H., Kim, C.S., *et al.*, Isolation of indole alkaloid and anthranilic acid derivatives from Pulverata Levis. *Tetrahedron Lett.*, 59, 50, 4380, 2018.
4. Mitra, S., Anand, U., Jha, N.K., *et al.*, Pharmacological Properties of Piperidine and Piperine: A Comprehensive Review on Molecular Mechanisms and Therapeutic Perspectives. *Front. Pharmacol.*, 12, 772418, 2022.
5. Kaur, R., Chaudhary, S., Kumar, K., Gupta, M.K., Rawal, R.K., Recent synthetic and medicinal perspectives of dihydropyrimidinones. A review. *Eur. J. Med. Chem.*, 132, 108, 2017.
6. Saini, M.S., Kumar, A., Dwivedi, J., Singh, R., A review: biological significances of heterocyclic compounds. *Int. J. Pharma. Sci. Res.*, 4, 3, 66, 2013.
7. Jurgenson, C.T., Begley, T.P., Ealick, S.E., The Structural and Biochemical Foundations of Thiamin Biosynthesis. *Annu. Rev. Biochem.*, 78, 569, 2009.
8. Heravi, M.M. and Zadsirjan, V., Prescribed drugs containing nitrogen heterocycles: an overview. *RSC Adv.*, 10, 44247, 2020.
9. Brugnatelli, G., Prepn by direct oxidation of uric acid. *Ann. Chim.*, 8, 2, 201, 1818.
10. Aljamali, N.M., Review paper in heterocyclic compounds. *J. Plast. Polym. Tech.*, 1, 1, 49, 2015.
11. Wohler, F. and Liebig, J., Untersuchungen über die Natur der Harnsäure. *Ann. Pharm.*, 26, 241, 1838.
12. Udoh, G. E., Ikenna, K., Effiong, B., Duke, O. E. E., Synthesis and characterization of pyridine transition metal complexes of Ni(II), Cu(I) and Ag(I). *JSM Chem.*, 10, 1, 1063, 2023.
13. Van Order, R.B. and Lindwall, H.G., Indole. *Chem. Rev.*, 30, 1, 69, 1942.
14. Elson, D. and Chargaff, E., On the deoxyribonucleic acid content of sea urchin gametes. *Experientia*, 8, 4, 143, 1952.
15. Anaya, J. and Sánchez, R.M., 4 – Four-membered ring systems. *Prog. Heterocycl. Chem.*, 34, 79, 2023.

16. Yerragunta, V., Suman, D., Swamy, K., Anusha, V., Patil, P., Naresh, M., Pyrazole and Its Biological Activity. *Pharma. Tutor.*, 2, 1, 40, 2014.
17. Gudimani, P., Hebbar, N.U., Shastri, S.L., *et al.*, A new approach for the synthesis of tri-substituted pyrazole propionic acids derivatives: Anti-inflammatory antimicrobial and molecular docking studies. *J. Mol. Struct.*, 1285, 135405, 2023.
18. Meghana, K. and Sivaprasad, Comparative Analysis of Anti-Fungal Properties of Carbohydrates Derivatives Over Commercial Antifungal Drugs. *Cardiometry*, 25, 1570, 2022.
19. Sykorova, V., Tichy, M., Hocek, M., Polymerase Synthesis of DNA Containing Iodinated Pyrimidine or 7-Deazapurine Nucleobases and Their Post-synthetic Modifications through the Suzuki-Miyaura Cross-Coupling Reactions. *ChemBioChem*, 23, 3, e202100608, 2022.
20. Zakharychev, V.V., Kuzenkov, A.V., Martsynkevich, A.M., Good pyridine hunting: a biomimic compound a modifier and a unique pharmacophore in agrochemicals. *Chem. Heterocycl. Comp.*, 56, 1491, 2020.
21. Tripathy, D.B. and Mishra, A., Microwave synthesis and characterization of waste soybean oil-based gemini imidazolinium surfactants with carbonate linkage. *Surf. Rev. Lett.*, 24, 05, 1750062, 2017.
22. Foley, M. and Tilley, L., Quinoline antimalarials: mechanisms of action and resistance. *Int. J. Parasitol.*, 27, 2, 231, 1997.
23. Boonma, T., Soikudrua, N., Nutho, B., Rungrotmongkol, T., Nunthaboot, N., Insights into binding molecular mechanism of hemagglutinin H3N2 of influenza virus complexed with arbidol and its derivative: A molecular dynamics simulation perspective. *Comp. Biol. Chem.*, 101, 107764, 2022.
24. Shagufta, A., I., An insight into the therapeutic potential of quinazoline derivatives as anticancer agents. *Med. Chem. Commun.*, 8, 871, 2017.
25. Zhang, J., Wang, S., Ba, Y., Xu, Z., 1,2,4-Triazole-quinoline/quinolone hybrids as potential anti-bacterial agents. *Eur. J. Med. Chem.*, 174, 1, 2019.
26. Lazdon, E., Stolero, N., Frenkel, D., Microglia and Parkinson's disease: footprints to pathology. *J. Neural Transm.*, 127, 2, 149, 2020.
27. Baumann, M., Baxendale, I.R., Ley, S.V., Nikbin, N., An overview of the key routes to the best-selling 5-membered ring heterocyclic pharmaceuticals. *Beilstein J. Org. Chem.*, 7, 442, 2011.
28. Man, S., Yang, L., Xiang, H., *et al.*, Antihypertensive and renal protective effect of Shunaoxin pill combined with captopril on spontaneous hypertension rats. *Biomed. Pharmacother.*, 125, 1, 2020.
29. Montazeri, M., Razzaghi-Abyaneh, M., Nasrollahi, S.A., Maibach, H., Nafisi, S., Enhanced topical econazole antifungal efficacy by amine-functionalized silica nanoparticles. *Bull. Mater. Sci.*, 43, 13, 1, 2020.
30. George, S., McBurney, A., Ward, J.W., The pharmacokinetics of timegadine and two of its metabolites after multiple oral dosing and the effects of concomitant administration of ibuprofen. *Eur. J. Clin. Pharmacol.*, 29, 581, 1986.

31. Nishio, S., Abe, M., Ito, H., Anagliptin in the treatment of type 2 diabetes: safety efficacy and patient acceptability. *Diabetes Metab. Syndr. Obes.*, 8, 163, 2015.
32. Gouda, M.A. and Qurban, J., Recent progress in the chemical reactivity of 3-Amino-1H-pyrazol-5(4H)-one derivatives (part II). *Synth. Commun.*, 51, 16, 2445, 2021.
33. Su, R., Li, G., Xiao, X., Ag/Poly(N-isopropylacrylamide)-laponite Hydrogel Surface-Enhanced Raman Membrane Substrate for Rapid Separation Concentration and Detection of Hydrophilic Compounds in Complex Sample All-in-One. *Anal. Chem.*, 95, 15, 6399, 2023.
34. Sakai, N., Kamimura, K., Miyamoto, H., *et al.*, Letrozole ameliorates liver fibrosis through the inhibition of the CTGF pathway and 17β-hydroxysteroid dehydrogenase 13 expression. *J. Gastroenterol.*, 58, 1, 53, 2023.
35. Rafique, B., Khalid, A.M., Akhtar, K., Iqbal, M., Investigation of Metronidazole–DNA Interactions by Using Electrochemical and Spectroscopic Techniques. *Russ. J. Electrochem.*, 58, 8, 704, 2022.
36. McCubrey, J.A., Abrams, S.L., Follo, M.Y., *et al.*, Effects of chloroquine and hydroxychloroquine on the sensitivity of pancreatic cancer cells to targeted therapies. *Adv. Biol. Regul.*, 87, 100917, 2023.
37. Gallaway, P.J., Miyake, H., Buchowski, M.S., *et al.*, Physical Activity: A Viable Way to Reduce the Risks of Mild Cognitive Impairment Alzheimer's Disease and Vascular Dementia in Older Adults. *Brain Sci.*, 7, 22, 2017.
38. Lajarín-Cuesta, R., Arribas, R.L., Nanclares, C., García-Frutos, E.M., Gandía, L., de Los Ríos, C., Design and synthesis of multipotent 3-aminomethylindoles and 7-azaindoles with enhanced protein phosphatase 2A-activating profile and neuroprotection. *Eur. J. Med. Chem.*, 157, 294, 2018.
39. Patel, D.V., Patel, N.R., Kanhed, A.M., *et al.*, Novel carbazole-stilbene hybrids as multifunctional anti-Alzheimer agents. *Bioorg. Chem.*, 101, 1, 2020.
40. Sadeghian, B., Sakhteman, A., Faghih, Z., *et al.*, Design, synthesis, and biological activity evaluation of novel carbazole-benzylpiperidine hybrids as potential anti-Alzheimer agents. *J. Mol. Struct.*, 1221, 128793, 2020.
41. Manzoor, S., Prajapati, S.K., Majumdar, S., *et al.*, Discovery of new phenyl sulfonyl-pyrimidine carboxylate derivatives as potential multi-target drugs with effective anti-Alzheimer's action: Design, synthesis, crystal structure, and *in-vitro* biological evaluation. *Eur. J. Med. Chem.*, 215, 1, 2021.
42. Askarani, H.K., Iraji, A., Rastegari, A., *et al.*, Design and synthesis of multi-target directed 1,2,3-triazole-dimethylaminoacryloyl-chromenone derivatives with potential use in Alzheimer's disease. *BMC Chem.*, 14, 64, 2020.
43. Sung, H., Ferlay, J., Siegel, R.L., *et al.*, Global cancer statistics 2020: GLOBOCAN estimates of incidence and mortality worldwide for 36 cancers in 185 countries. *CA Cancer J. Clin.*, 71, 3, 209, 2021.
44. Mongre, R.K., Mishra, C.B., Prakash, A., *et al.*, Novel carbazole-piperazine hybrid small molecule induces apoptosis by targeting BCL-2 and inhibits

tumor progression in lung adenocarcinoma *in vitro* and xenograft mice model. *Cancers*, 11, 9, 1245, 2019.
45. Huo, J.L., Wang, S., Yuan, X.H., Yu, B., Zhao, W., Liu, H.M., Discovery of [1,2,4]triazolo[1,5-a]pyrimidines derivatives as potential anticancer agents. *Eur. J. Med. Chem.*, 211, 1, 2021.
46. Wang, S., Zhao, L., Shi, X.J., *et al.*, Development of highly potent, selective, and cellular active triazolo[1,5-a]pyrimidine-based inhibitors targeting the DCN1–UBC12 protein–protein interaction. *J. Med. Chem.*, 62, 5, 2772, 2019.
47. Cheng, G., Wang, Z., Yang, J., *et al.*, Design, synthesis, and biological evaluation of novel indole derivatives as potential HDAC/BRD4 dual inhibitors and anti-leukemia agents. *Bioorg. Chem.*, 84, 410, 2019.
48. Nafie, M.S., Amer, A.M., Mohamed, A.K., Tantawy, E.S., Discovery of novel pyrazolo [3,4-b]pyridine scaffold-based derivatives as potential PIM-1 kinase inhibitors in breast cancer MCF-7 cells. *Bioorg. Med. Chem.*, 28, 24, 1, 2020.
49. Khan, S.A., Asiri, A.M., Basisi, H.M., *et al.*, Synthesis and evaluation of quinoline-3-carbonitrile derivatives as potential antibacterial agents. *Bioorg. Chem.*, 88, 102968, 2019.
50. Mandal, M.K., Ghosh, S., Bhat, H.R., Naesens, L., Singh, U.P., Synthesis and biological evaluation of substituted phenyl azetidine-2-one sulfonyl derivatives as potential antimicrobial and antiviral agents. *Bioorg. Chem.*, 104, 104320, 2020.
51. Adamovich, S.N., Sadykov, E.K., Ushakov, I.A., Oborina, E.N., Belovezhets, L.A., Antibacterial activity of new silatrane pyrrole-2-carboxamide hybrids. *Mendeleev Commun.*, 31, 2, 204, 2021.
52. Maruthapandi, M., Sharma, K., Luong, J.H.T., Gedanken, A., Antibacterial activities of microwave-assisted synthesized polypyrrole/chitosan and poly(pyrrole-N-(1-naphthyl) ethylenediamine) stimulated by C-dots. *Carbohyd. Polym.*, 243, 116474, 2020.
53. Charvin, D., Medori, R., Hauser, R.A., Rascol, O., Therapeutic strategies for Parkinson's disease: beyond dopaminergic drugs. *Nat. Rev. Drug Discov.*, 17, 11, 804, 2018.
54. Wu, J., Liu, Q., Hu, Y., Wang, W., Gao, X., Discovery of novel procaine-imidazole derivative as inhibitor of monoamine oxidase-B for potential benefit in Parkinson's disease. *ChemistrySelect.*, 5, 35, 10928, 2020.
55. Anastassova, N., Aluani, D., Kostadinov, A., *et al.*, Evaluation of the combined activity of benzimidazole arylhydrazones as new anti-Parkinsonian agents: monoamine oxidase-B inhibition, neuroprotection, and oxidative stress modulation. *Neural. Regen. Res.*, 16, 11, 2299, 2021.
56. Wei, P.C., Lee-Chen, G.J., Chen, C.M., *et al.*, Neuroprotection of Indole-Derivative Compound NC001-8 by the Regulation of the NRF2 Pathway in Parkinson's Disease Cell Models. *Oxid. Med. Cell. Longev.*, 2019, 5074367, 2019.

57. Chen, L., Deng, H., Cui, H., *et al.*, Inflammatory responses and inflammation-associated diseases in organs. *Oncotarget.*, 9, 6, 7204, 2017.
58. Mohamed, L.W., Shaaban, M.A., Zaher, A.F., Alhamaky, S.M., Elsahar, A.M., Synthesis of new pyrazoles and pyrozolo[3,4-b]pyridines as anti-inflammatory agents by inhibition of COX-2 enzyme. *Bioorg. Chem.*, 83, 47, 2019.
59. Maghraby, M.T.E., Abou-Ghadir, O.M.F., Abdel-Moty, S.G., Ali, A.Y., Salem, O.I.A., Novel class of benzimidazole-thiazole hybrids: The privileged scaffolds of potent anti-inflammatory activity with dual inhibition of cyclooxygenase and 15-lipoxygenase enzymes. *Bioorg. Med. Chem.*, 28, 7, 115403, 2020.
60. Huang, L., Yang, L., Wan, J.P., Zhou, L., Liu, Y., Hao, G., Metal-free three-component assemblies of anilines, α-keto acids, and alkyl lactates for quinoline synthesis and their anti-inflammatory activity. *Org. Biomol. Chem.*, 20, 4385, 2022.
61. Alam, S., Hasan, M.K., Neaz, S., Hussain, N., Hossain, F., Rahman, T., Diabetes Mellitus: Insights from Epidemiology, Biochemistry, Risk Factors, Diagnosis, Complications, and Comprehensive Management. *Diabetology*, 2, 2, 36, 2021.
62. Hossain, U., Das, A.K., Ghosh, S., Sil, P.C., An overview on the role of bioactive α-glucosidase inhibitors in ameliorating diabetic complications. *Food Chem. Toxicol.*, 145, 111738, 2020.
63. Pogaku, V., Gangarapu, K., Basavoju, S., Tatapudi, K.K., Katragadda, S.B., Design, synthesis, molecular modeling, ADME prediction, and antihyperglycemic evaluation of new pyrazole-triazolopyrimidine hybrids as potent α-glucosidase inhibitors. *Bioorg. Chem.*, 93, 103307, 2019.
64. Azimi, F., Ghasemi, J.B., Azizian, H., *et al.*, Design and synthesis of novel pyrazole-phenyl semicarbazone derivatives as potential α-glucosidase inhibitor: Kinetics and molecular dynamics simulation study. *Int. J. Biol. Macromol.*, 166, 1082, 2021.
65. Ali, S., Ali, M., Khan, A., *et al.*, Novel 5-(Arylideneamino)-1H-Benzo[d]imidazole-2-thiols as Potent Anti-Diabetic Agents: Synthesis Vitro α-Glucosidase Inhibition, and Molecular Docking Studies. *ACS Omega*, 7, 48, 43468, 2022.
66. Shabaan, S.N., Alshehri, F.S., AL-Rasheed, J.F., El-Sebaey, S.A., Husseiny, E.M., Synthesis and Exploration of New Imidazo[4,5-c]Pyrazoles as Potent α-Amylase Inhibitors. *ChemistrySelect*, 8, 14, e202204757, 2023.
67. Graham, B.S. and Sullivan, N.J., Emerging viral diseases from a vaccinology perspective: preparing for the next pandemic. *Nat. Immunol.*, 19, 1, 20, 2018.
68. Francesconi, V., Cichero, E., Schenone, S., Naesens, L., Tonelli, M., Synthesis and Biological Evaluation of Novel (thio)semicarbazone-Based Benzimidazoles as Antiviral Agents against Human Respiratory Viruses. *Molecules*, 25, 7, 1487, 2020.

69. Abu-Zaied, M.A., Elgemeie, G.H., Mahmoud, N.M., Anti-Covid-19 Drug Analogues: Synthesis of Novel Pyrimidine Thioglycosides as Antiviral Agents Against SARS-CoV-2 and Avian Influenza H5N1 Viruses. *ACS Omega*, 6, 26, 16890, 2021.
70. Hattori, S., Higashi-Kuwata, N., Hayashi, H., *et al.*, A small molecule compound with an indole moiety inhibits the main protease of SARS-CoV-2 and blocks virus replication. *Nat. Commun.*, 12, 1, 668, 2021.
71. Malancona, S., Mori, M., Fezzardi, P., *et al.*, 5,6-dihydroypyrimidine scaffold to target HIV-1 nucleocapsid protein. *ACS Med. Chem. Lett.*, 11, 5, 766, 2020.
72. Lu, J.M., Lin, P.H., Qizhi, Y., Chen, C., Chemical and molecular mechanisms of antioxidants: experimental approaches and model systems. *J. Cell. Mol. Med.*, 14, 4, 840, 2010.
73. Ali, T.E., Bakhotmah, D.A., Assiri, M.A., Synthesis of some new functionalized pyrano[2,3-c]pyrazoles and pyrazolo[4′,3′:5,6] pyrano[2,3-d]pyrimidines bearing a chromone ring as antioxidant agents. *Synth. Commun.*, 50, 21, 3314, 2020.
74. Gomaa, H.A.M., Shaker, M.E., Alzera, S.I., *et al.*, Optimization and SAR investigation of novel 2,3-dihydropyrazino[1,2-a]indole-1,4-dione derivatives as EGFR and BRAFV600E dual inhibitors with potent antiproliferative and antioxidant activities. *Bioorg. Chem.*, 120, 105616, 2022.
75. Selvakumaran, M., Predhanekar, M.I., Kubaib, A., Visagaperumal, D., Novel Benzimidazole linked piperidine derivatives screened for antibacterial and antioxidant properties with Density Functional and Molecular Mechanic tools. *Results Chem.*, 5, 100765, 2023.

3

Heterocyclic Compounds as Antimicrobial (Antibacterial, Antifungal, Antiviral, and Anthelmintics) Agent

Gaurank Pal[1‡], Devaang Mathur[2‡], Lakshit Srivastav[2‡], Vishal Mehare[1], Ajay Gupta[3], Shruti Khanna Ahuja[2*] and Puja Prasad[1†]

[1]*Amity Institute of Click Chemistry Research and Studies, Amity University, Noida, Uttar Pradesh, India*
[2]*Amity Institute of Biotechnology, Amity University, Noida, Uttar Pradesh, India*
[3]*Institute for Drug Research, School of Pharmacy, Faculty of Medicine, The Hebrew University of Jerusalem, Jerusalem, Israel*

Abstract
Heterocyclic compound is one of the widely studied organic compounds due to its diverse synthetic and biological application. Advanced synthetic chemistry and structure–activity relationship studies enabled the development of novel heterocyclic scaffolds providing a wide range of drug-like synthetic compounds in drug discovery. In addition, the solubility and lipophilicity of heterocyclic compounds can be modulated to improve the potency, selectivity, and pharmacokinetic profiles of drug-like molecules. Various heterocyclic scaffolds have been studied as antimicrobial agents to overcome microbial resistance. In this chapter, we would discuss recent discoveries of heterocyclic compounds as antibacterial, antifungal, antiviral, and anthelmintics agents.

***Keywords*:** Antibacterial, antifungal, antiviral, anthelmintics, heterocyclic compounds, antimicrobial resistance

**Corresponding author*: skahuja@amity.edu
†*Corresponding author*: pprasad@amity.edu
‡*Equal contributors*

Divya Bajpai Tripathy, Priyanka Chhabra and Anjali Gupta (eds.) Heterocyclic Compounds as Pharmaceutical Ingredients, (61–122) © 2026 Scrivener Publishing LLC

3.1 Introduction

Infectious diseases remain one of the major global threats in the modern era, largely due to the rise of antimicrobial resistance (AMR). The antibiotic era began in the year 1910 with the discovery of Salvarsan, used to treat syphilis caused by *Treponema pallidum* [1]. The subsequent discovery of penicillin by Alexander Fleming in 1928 marked a revolutionary breakthrough in medicine and saved countless lives, particularly during World War II. The period from 1950 to 1970 is often referred to as the "golden era of antibiotics," during which approximately half of the antibiotics in use today were discovered [2].

However, the misuse and empirical overuse of antibiotics have contributed significantly to the emergence of AMR, leading to a growing number of bacteria that are resistant to currently available treatments [3, 4]. Alarmingly, no new class of antibiotics has been discovered in the past 35 years, whereas resistant bacteria continue to rise [5]. This has ushered in what is now being referred to as the "post–antibiotic era," where even minor infections or injuries could become life-threatening [6].

Current statistics indicate that presently more than 700,000 deaths are caused globally each year due to rise of drug-resistant infections. If no action is taken by 2050, this number could rise to 10 million annually [3, 4]. There is an urgent and unmet need for the discovery and development of new antimicrobial agents to combat this escalating crisis.

In this regard, heterocyclic compounds have emerged as lifesaving compounds. Heterocyclic compounds, containing atoms such as nitrogen, oxygen, or sulfur in their ring structures, are widely explored as antibacterial agents due to their diverse biological activities [7]. Due to their structural diversity and ability to interact with a wide range of biological targets, heterocycles are central to many pharmaceuticals and agrochemicals [8].

The rising threat of AMR has intensified the search for novel and effective antimicrobial agents. The versatility of heterocycles allows for extensive chemical modifications, enabling the design of new derivatives to overcome resistance mechanisms [8, 9]. Many widely used antibiotics, such as penicillins, cephalosporins, tetracyclines, and fluoroquinolones, contain heterocyclic cores that are crucial for their biological activity [10]. Heterocyclic scaffolds, such as imidazoles, triazoles, quinolines, pyridines, and β-lactams, have demonstrated potent activity against a broad spectrum of bacteria, fungi, and viruses [11]. Heterocyclic compound has the ability to inhibit cell wall synthesis, disruption of membrane function, inhibition of protein, and nucleic acid synthesis or inhibit essential enzymes, making them ideal candidates for antimicrobial drug development [12].

Natural products and synthetic derivatives containing heterocycles have shown promising antimicrobial profiles, and many have been developed into clinically approved drugs, such as ciprofloxacin (a quinolone) and metronidazole (an imidazole) [13]. The continued modification of heterocyclic cores and the development of hybrid molecules offer a promising strategy to overcome current limitations in antimicrobial therapy.

3.2 Heterocyclic Compounds as Antimicrobial Agent

There are several heterocyclic compounds that have been approved by the Food and Drug Administration (FDA) for clinical use. The following section highlights the recent development of heterocyclic compounds as antimicrobial agents.

3.2.1 Application of Heterocyclic Compounds as Antibacterial Agent

Heterocyclic compounds, containing atoms such as nitrogen, oxygen, or sulfur in their ring structures, are widely explored as antibacterial agents due to their diverse biological activities [7, 11]. Compounds such as quinolones, β-lactams, and azoles target bacterial enzymes, cell wall synthesis, or DNA replication, disrupting essential processes. For instance, fluoroquinolones inhibit DNA gyrase, whereas penicillins block peptidoglycan cross-linking [14, 15]. Their structural versatility allows for modifications to enhance potency, specificity, and resistance to bacterial degradation. Additionally, heterocyclic scaffolds such as pyridine and pyrimidine derivatives show promise in combating multidrug-resistant strains, offering new avenues for developing effective antibiotics in response to growing AMR challenges [16].

Heterocyclic compounds encompass a broad range of molecular frameworks, from simple five-membered rings such as pyrrole, furan, and thiophene to six-membered rings such as pyridine, pyrimidine, and piperazine, as well as fused systems such as quinoline and indole [17]. The presence of heteroatoms imparts unique electronic and chemical properties, enabling these compounds to form hydrogen bonds, coordinate with metal ions, and engage in π-π interactions with biological targets [17]. These properties enhance their ability to bind selectively to bacterial enzymes, receptors, or nucleic acids, disrupting critical processes essential for bacterial survival [18].

Heterocyclic compounds exert antibacterial effects through various mechanisms, targeting essential bacterial processes such as cell wall synthesis, protein synthesis, DNA replication, and metabolic pathways [12, 19]. Below are some prominent classes of heterocyclic antibacterial agents and their modes of action.

3.2.1.1 Pyrrole

Pyrrole is a key heterocyclic aromatic compound with the chemical formula C_4H_4NH, consisting of four carbon atoms and one nitrogen atom within its ring. It serves as a biologically significant scaffold due to its diverse properties and broad applicability in pharmaceuticals, agriculture, and related fields [20]. Nitrogen-containing heterocycles have been extensively studied for their biological activities, particularly as core structures in antibacterial agents used to combat bacterial infections [21]. Medicinal chemists worldwide have synthesized a variety of heterocyclic compounds, leading to the development of potent antibacterial drugs [7].

Over recent years, significant advances have been made in pyrrole chemistry. Numerous pyrrole-based derivatives have been discovered with strong biological activity, particularly against both Gram-positive and Gram-negative bacteria (Figure 3.1) [22, 23].

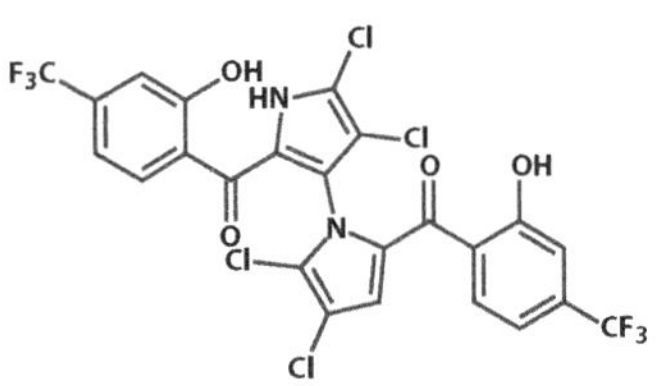

Marinopyrrole (**1**)

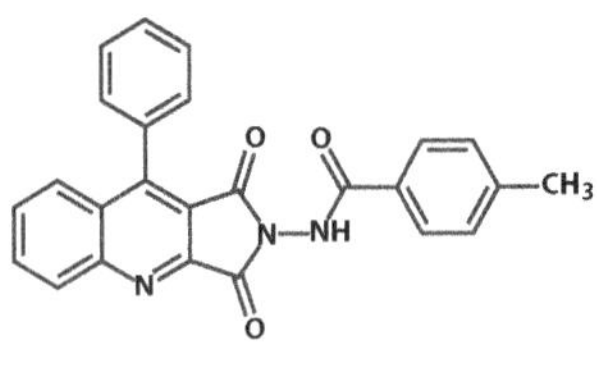

2

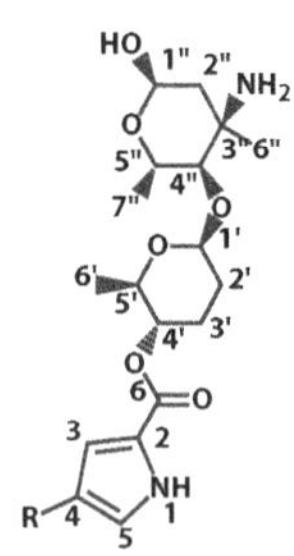

3: R = Cl
4: R = Br

3, 4

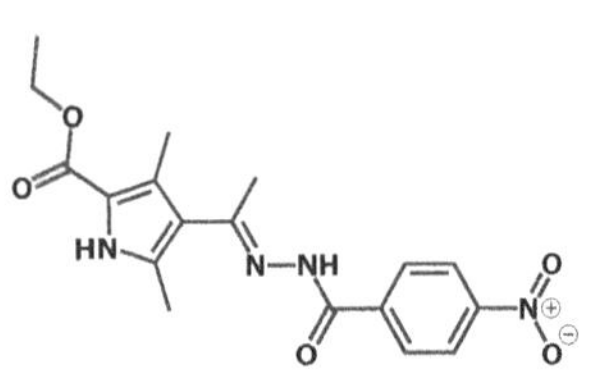

5

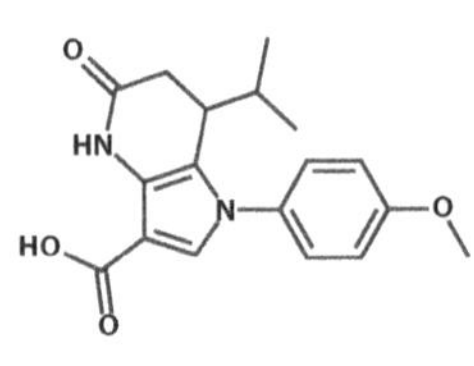

6

Figure 3.1 Pyrrole-based heterocyclic compounds (**1–6**) as antibacterial agent.

In 2013, Cheng *et al.* conducted a structural analysis revealing that marinopyrrole (Compound **1**) and its derivatives demonstrated potent antibacterial activity (Figure 3.1) [24]. The compound's efficacy was attributed to the presence of an ortho-hydroxy group and the electron-withdrawing CF_3 group. It showed strong inhibition against *Staphylococcus aureus* with a minimum inhibitory concentration (MIC) value of 8 μg/mL. In a different investigation, Largani *et al.* used benzohydrazide and 9-phenylfuro [3,4-b]quinoline-1,3-diones to create pyrrolo[3,4-b]quinolin-2(3H)-yl benzamides. The resulting Compound **2** was tested against *S. aureus*, *Escherichia coli*, and *Enterococcus faecalis*, demonstrating strong antibacterial activity with MIC values of 0.25, 0.25, and 0.5 mg/mL, respectively [25]. In 2019, Zhang *et al.* isolated five new pyrrole-derived alkaloids, named phallusialides, from the marine bacterium *Micromonospora* sp. Among them, Compounds **3** and **4** exhibited notable activity against methicillin-resistant *S. aureus* (MRSA) and *E. coli* with MIC values of 32 and 64 μg/mL, respectively [26].

In 2022, Rawat *et al.* synthesized pyrrole-2-carboxylate derivatives and assessed their efficacy against *Mycobacterium tuberculosis*. A standout compound, ethyl-4-{[-(1-(2-(4-nitrobenzoyl)hydrazono)ethyl]}-3,5-dimethyl-1H-pyrrole-2-carboxylate (Compound **5**), showed strong activity with an MIC of 0.7 μg/mL [27]. In 2023, Veselov *et al.* reported new derivatives of 5-oxo-4H-pyrrolo[3,2-b]pyrimidine with significant antibacterial potential. Among them, the most active Compound **6**, demonstrated potent activity against *E. coli* with an MIC of 3.35 μg/mL [28].

3.2.1.2 Furan

Furan is an important heterocyclic aromatic organic compound that contains one oxygen atom and four carbon atoms in the ring having chemical formula C_4H_4O [29]. The word furan came from a Latin word furfural, which means bran. Compounds containing furan derivatives exhibit a variety of pharmacological characteristics and hence recognized as an important compound in pharmaceutical and in other industries [29].

The antibacterial efficiency of 3-aryl-3[furan-2-yl] propanoic acid derivatives were studied. The studies concluded that Compound **7** showed high inhibition against *E. coli* growth at an MIC value 64 μg/mL (Figure 3.2) [30]. In recent years, Benzofuran and its derivatives have been evaluated for their antibacterial efficiency against Gram-positive bacteria such as *S. aureus*, *Bacillus subtilis*, etc., and Gram-negative bacteria such as *E. coli* [31]. The result concluded that various derivatives showed antibacterial potency against these bacterial strains. In another study done by

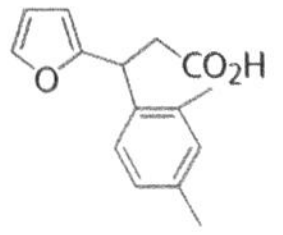

Compound **7**

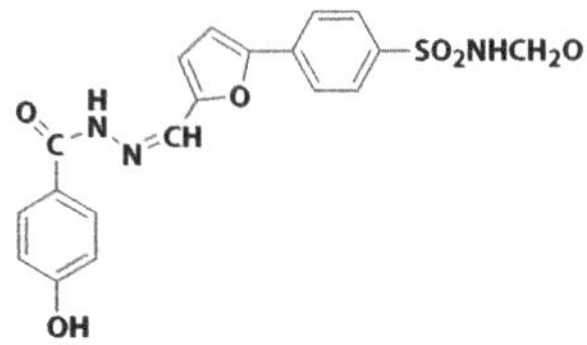

Compound **8**

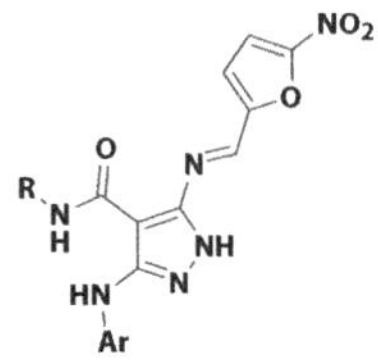

a; R = H
b; R = Ph
c; R = 4-CH_3-C_6H_4
d; R = 4-Cl-C_6H_4

Compound **9a-d**

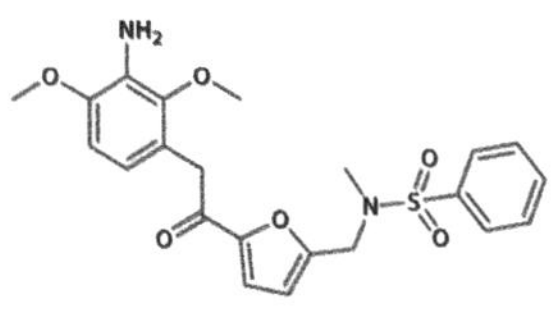

N-(3-amino-2,4-dimethoxy phenyl)-5-((N-methyl phenyl sulfonamido)methyl) furan-2-carboxamide

Compound **10**

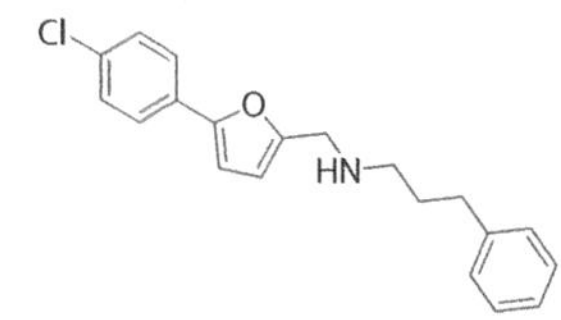

N-((5-(4-Chlorophenyl)furan-2-yl)methyl)-3phenylpropan-1-amine

Compound **11**

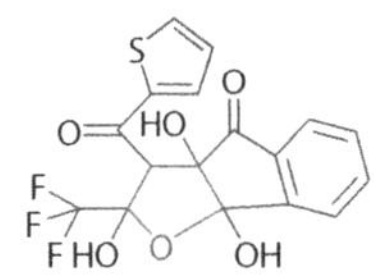

Compound **12**

Figure 3.2 Furan-based heterocyclic Compounds **1–12** as antibacterial agent.

researcher, some furan-substituted compounds were synthesized and evaluated for their antibacterial efficiency against strains. The study showed that Compound **8** was found to be more potent against *Bacillus cereus, S. aureus*, etc., in comparison to ampicillin as standard. In year 2020, Hassan and colleagues have synthesized nitrofurantoin analogs containing furan and pyrazole scaffolds as N-aryl-3-(arylamino)-5-(((5-substituted furan-2-yl)methylene)amino)-1H-pyrazole-4-carboxamide and studied its antibacterial property against Gram-negative bacteria such as *E. coli*, etc., and Gram-positive bacteria such as *S. aureus*, etc [32]. The result showed that Compounds **9a**, and **9b**, etc., exhibit good antibacterial properties against *E. coli*, whereas Compounds **9c** and **9d** exhibit moderate activity in comparison to nitrofurantoin. It was reported that compound **(a-d)** showed remarkable activity against *Salmonella typhimurium*.

Certain derivatives of compound furan 2,4-disubstituted compound **10** exhibited high potential activity against bacteria such as *E. coli* and *Proteus vulgaris* [33]. Andrade *et al.* reported arylfuran, Compound **11**, which demonstrated high antibacterial activity against *E. coli* (MIC = 49 μM) and *S. aureus* (98 μM) and modest activity against *Pseudomonas aeruginosa*

(MIC = 770 μM) [34]. Obafemi *et al.* have synthesized Compound **12** and studied their antibacterial activity. This compound exhibited broad spectrum antibacterial activity against 13 different strains type of bacteria and showed better activity against *Pseudomonas fluorescence* having MIC of 0.065 mg/mL compared to streptomycin (0.25 mg/mL) [35].

3.2.1.3 Thiophene

Thiophene is an important heterocyclic aromatic organic compound having chemical formula C_4H_4S. Its structure includes four carbon atoms and one sulfur atom. Thiophene and its derivatives were reported to possess diverse pharmacological properties, which makes it garner attention of various industries as well as scientists across the globe [36]. Coskun *et al.* synthesized derivatives of thiophene containing diketoxime. By the use of standard dilution method, the MIC values of derivatives were determined against *S. aureus*, *E. coli*. The test concluded that derivatives were able to inhibit the growth of bacterial strains with MIC values between 32 and 128 μg/mL. The compound 1,1-(2,5-thiophenedyl)bis[1-(2-benzofuranyl) methanone],1,1-bis[O-(4-ethenylphenyl) methyl oxime] (compound **13**) showed good activity against *E. coli* and *S. aureus* [37].

Mabkhot *et al.* synthesized thiophene derivatives from ethyl (E)-5-(3-dimethylamino) acryloyl)-4-methyl-2-(phenylamino) thiophene-3-carboxylate, Compound **14** and evaluated their antibacterial property against *E. coli*, *P. aeruginosa*, and *Streptococcus pneumoniae* (Figure 3.3) [38]. The activities of the synthesized compounds revealed that presence of substituent at position 2 is essential for activity of thiophene. Popiolek *et al.* synthesized new compounds by cyclization reaction of N-substituted thiophene-2 carboxamide derivatives with thiosalicyclic acid and tested their antibacterial activity *S. aureus*, *Staphylococcus epidermidis*, *Micrococcus luteus*, *B. subtilis*, *B. cereus*, *Bordetella bronchiseptica*, *E. coli*, *Candida albicans*, *Candida parapsilosis*, etc [39]. Among all the synthesized compounds, only N-[(E)-(3-bromo-4-methoxyphenyl)methylidene] thiophene-2-carbohydrazide (Compound **15**) showed remarkable activity against Gram-positive strains with MIC values 1.25–7.81 μg/mL.

A novel 4-(1-(3-chlorobenzo[b]thiophene-2-carbonyl)-1H-indol-3-yl)-7, 7-dimethyl-3,4,7,8-tetrahydroquinazoline 2,5(1H,6H) dione thiophene derivatives were created by Mehta *et al.* [40]. These compounds were tested against a variety of bacterial species, including S. aureus, *P. aeruginosa*, and *E. coli*. Ampicillin was used as a standard drug giving MIC value of 50 to 100 μg/mL against these bacteria. Whereas Compound **16** showed MIC values ranging from 100 to 500 μg/mL. Adiwish *et al.* synthesized

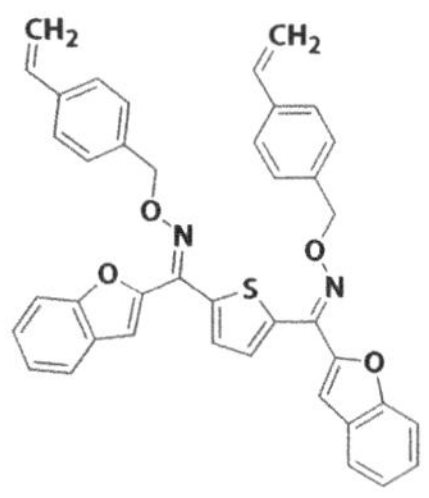

13

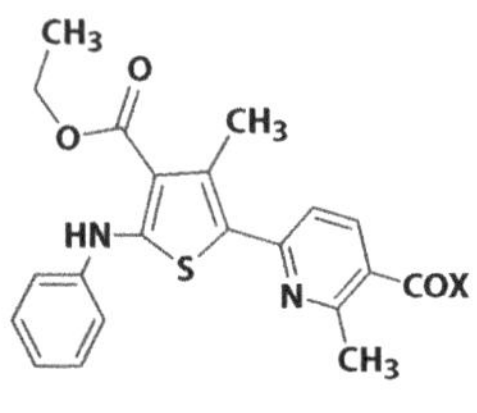

X = CH_3, OEt

14

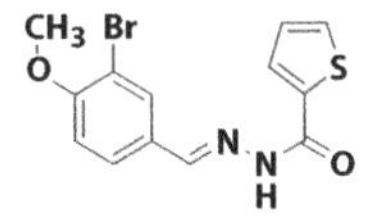

15

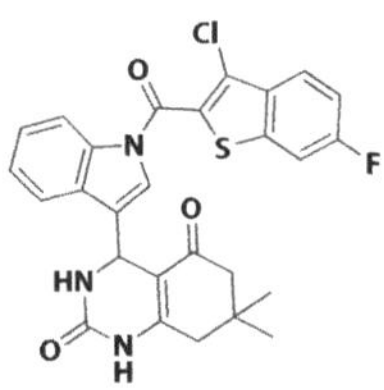

16

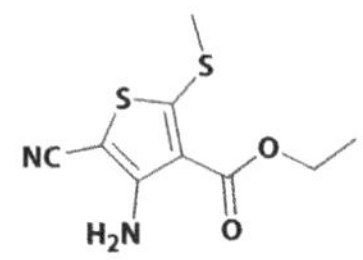

Ethyl 4-amino-5-cyano-2-(methylthio)thiophene-3-carboxylate

17

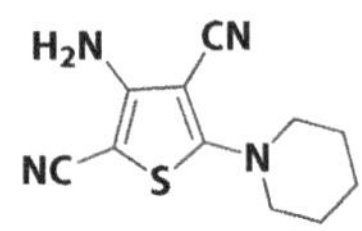

3-Amino-5-(1-piperidinyl)-2,4-thiophenedicarbonitrile

18

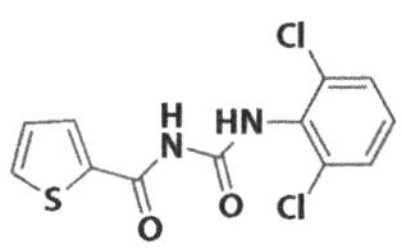

19

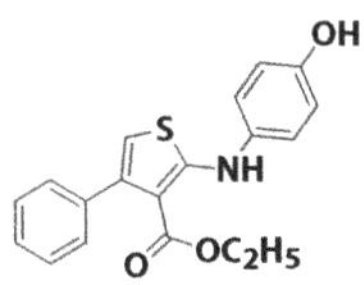

Ethyl 2-(4-hydroxyphenylamino)-4-phenylthiophene-3-carboxylate

20

Figure 3.3 Thiophene-based heterocyclic Compounds **13–20** as antibacterial agent.

tetrasubstituted thiophene from ketene dithioacetals [41]. The antibacterial activity of Compounds **17** and **18** was evaluated against Gram-negative bacteria such as *E. coli* and *Klebsiella pneumoniae* and Gram-positive bacteria such as *S. aureus*, etc., *via* disc diffusion method. Compound **18** showed larger inhibition zone than Compound **17** for *E. coli* and *S. aureus*. For *K. pneumonia*, both the compounds were ineffective.

In another example, Badiceanu *et al.* synthesized Compound **19** by reacting 3-thenoyl-isothiocyanate with primary aromatic amine, 3-thenoyl-isothiocyanate amines, and its antibacterial efficiency was evaluated against bacterial strains and fungi [42]. Evaluation of these derivatives showed significant activity against these strains with MIC values ranging from 15.6 to 500 μg/mL, because of the presence of electron withdrawing group, -Cl (present in Compound **19**). However, Compound **19** showed better antifungal property than antibacterial. Prasad and their colleagues synthesized new compounds of ethyl 2-amino-4-phenylthiophene-3-carboxylate [43]. Using ampicillin and streptomycin as reference drugs. The antibacterial efficacy of these compounds was examined against various strains. Only Compound **20** had a strong inhibitory impact against bacteria such as *E. coli* and *B. subtilis*.

3.2.1.4 *Indole*

Indole is an important heterocyclic aromatic organic compound having chemical formula C_8H_7N. Its structure involves a benzene ring fused with a five-membered pyrrole ring [44]. Indole is recognized as an important structural component in various pharmaceutical drugs. Their ability to bind with receptors with high affinity makes them unique [45]. Leena *et al.* used synthetic techniques to generate a variety of spirofused tryptanthrin-thiopyrnao [2,3-b] indole hybrids and assessed their antibacterial effectiveness against vancomycin-resistant *S. aureus* and MRSA. The MIC value of Compound **21** containing a nitrogen group was 0.25 μg/mL, indicating strong antibacterial activity (Figure 3.4) [46].

Using a variety of marine-indole derivatives, Li *et al.* assessed the antibacterial efficacy of these compounds against *S. aureus*, *P. aeruginosa*, *E. coli*, and other microorganisms [47]. The findings showed that Compounds **22** and **23** were very effective against *S. aureus*, with MIC values of 0.021 and 0.031 mg/mL, respectively. 3D-QSAR analysis revealed that hydrogen bonding, hydrophobic interaction, and steric effect were the important parameters for their antibacterial activity. This was further verified through molecular docking.

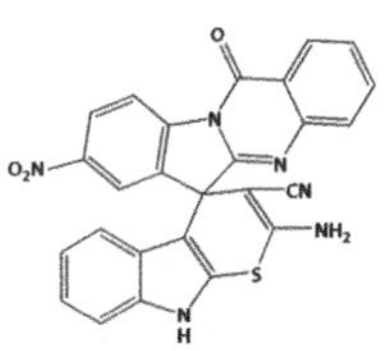

Compound 21

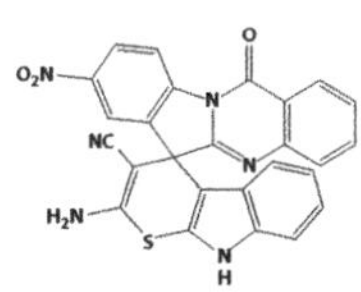

Compound 22

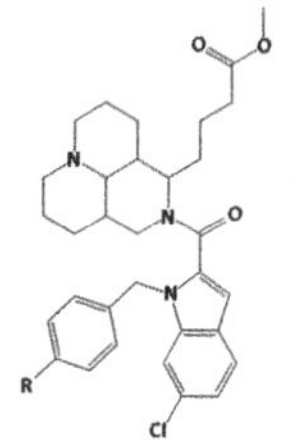

Compound 23; $R=CF_3$
Compound 24; $R=NO_2$

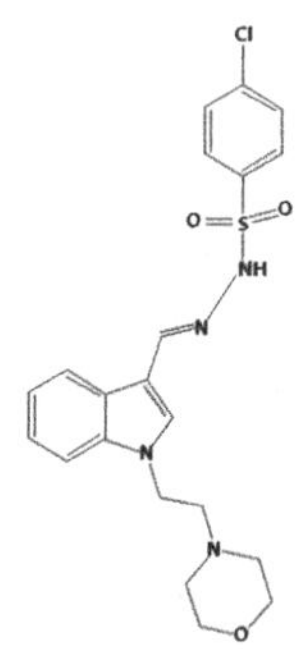

Compound 25

Figure 3.4 Indole-based heterocyclic Compounds **21–25** as antibacterial agent.

In 2018, Shirinzadeh *et al.* synthesized novel indole derivatives containing 1,2,4-triazole, 1,3,4-thiadiazole and carbothioamide [48a]. These derivatives were evaluated to determine their antibacterial potential against strains such as *E. coli*, *C. albicans*, etc. In 2020, Shaker *et al.* synthesized derivatives of 2-(4-methylsulfonylphenyl) indole, to evaluate their antibacterial potential [48b]. Among all the compounds, Compound **25** showed high potential against bacterial strains such as *K. pneumonia*, *E. coli*, *P. aeruginosa*, etc.

3.2.1.5 Quinoline

Quinoline is a heterocyclic aromatic organic compound having chemical formula C_9H_7N. It has a conjugated structure, which includes both benzene and pyridine. Quinoline and its derivatives were reported to possess diverse biological and pharmaceutical properties, which makes them important for drug discoveries [49]. Zeleke *et al.* synthesized 10 quinoline-based derivatives and evaluated them against various bacterial strains (Figure 3.5) [50]. These derivatives were screened against Gram-positive bacteria such as *B. subtilis*, etc., and Gram-negative bacteria such as *E. coli*, etc.; ciprofloxacin was used as a standard drug. The experimental result concluded that the derivatives showed high antibacterial potency against all strains. The derivatives **26** and **27** exhibited excellent antibacterial activity against *P. aeruginosa* (zone of inhibition of derivative **26** and **27** is 9.67 and 10.00 mm, respectively), whereas derivative **28** was potent against *E. coli* with a mean inhibition value of 9.00 mm as compared to other derivatives.

Various novel quinoline derivatives **29** and **30** were evaluated for their antibacterial activity against *E. coli*, *P. aeruginosa*, *S. aureus*, etc., where chloramphenicol and ampicillin were used as a reference drug [51]. Compound **29** exhibited high antibacterial activity against *S. aureus* than Compound **30** with MIC value of 6.25 μg/mL. However, compound **30** showed better activity than chloramphenicol. Kharb and Kaur synthesized nine quinoline compounds having carbothioamide-based quinoline motif [52a]. Compound **31** showed the highest potential against *P. aeruginosa* showing 20 mm of zone of inhibition. Sun *et al.* synthesized a series of derivatives of quinolinium iodide salts (**32–43**) and studied their antibacterial actions against *E. coli* and *S. aureus* [52b]. The result concluded that all the synthesized compounds showed remarkable activity against bacterial strains. The derivatives **32–43** showed high activity against *E. coli* with MIC value of 3.125–6.25 nmol/mL, which is four to eight times better than amoxicillin and ciprofloxacin [53]. Novel quinoline-3-carbonitrile derivatives (**44a–e**) were synthesized to determine their antibacterial potential against bacterial strains.

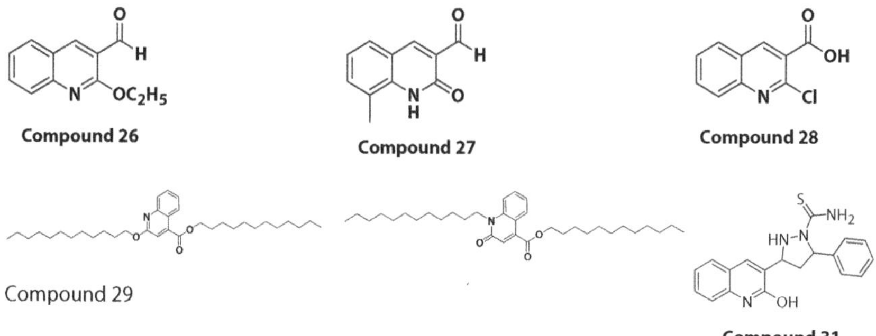

Compound 26 Compound 27 Compound 28

Compound 29 Compound 30 Compound 31

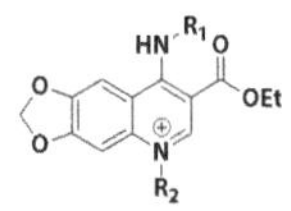

	R1	R2
32	$-CH_2CH_2Ph$	-
33	$-CH_2CH_2Ph(OCH_3)_2$	-
34	$-CH_2Ph$	-
35	$PhOCH_2Ph$	-
36	$-CH_2CH_2Ph$	$-CH_3$
37	$-CH_2CH_2Ph(OCH_3)_2$	$-CH_3$
38	$-CH_2Ph$	$-CH_3$
39	$-PhOCH_2Ph$	$-CH_3$
40	$-CH_2CH_2Ph$	$-CH_2CH_2CH_3$
41	$-CH_2CH_2Ph(OCH_3)_2$	$-CH_2CH_2CH_3$
42	$-CH_2Ph$	$-CH_2CH_2CH_3$
43	$-PhOCH_2Ph$	$-CH_2CH_2CH_3$

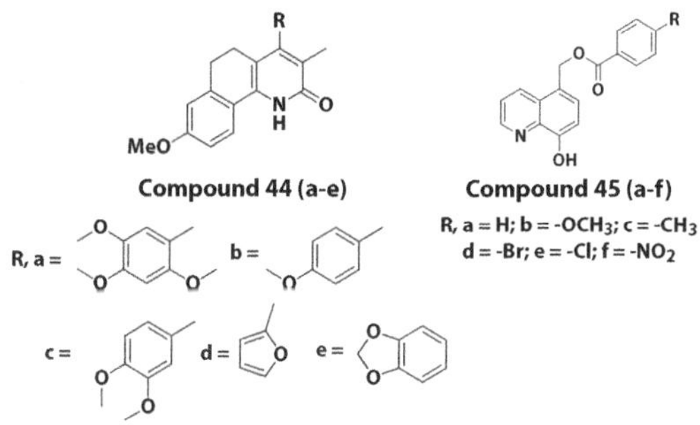

Compound 44 (a-e)

Compound 45 (a-f)

R, a = H; b = $-OCH_3$; c = $-CH_3$ d = -Br; e = -Cl; f = $-NO_2$

Figure 3.5 Quinoline-based heterocyclic Compounds **26–45** as antibacterial agent.

All the synthesized derivatives showed brilliant antibacterial activity with MIC value ranging from 3.13 to 100 μM [54]. Rbaa *et al.* have synthesized 8-hydroxyquinoline derivatives (**45a–f**) and studied antibacterial activity against various bacteria using disk diffusion assay [55]. Compound **45a** showed no antibacterial activity against tested bacteria species. However, presence of electron-donating group increased the antibacterial effect.

3.2.1.6 Imidazole

Imidazole is a five-membered heterocyclic organic compound with two nitrogen atoms in its ring. The chemical formula of imidazole is $C_3N_2H_4$. It is a highly polar compound and soluble in water. This compound can also show amphoteric nature.

In the recent years with the advancement in medicinal chemistry, imidazole and its derivatives have garnered attention due to its pharmacological properties and applications against various diseases [56]. Compounds such as [2-(substituted phenyl)-imidazol-1-yl]-methanone and 2-(substituted phenyl)-1H-imidazole analogs were studied for their antibacterial activity and antifungal activity taking norfloxacin as reference drug (Figure 3.6). Patil *et al.* synthesized and evaluated the antibacterial properties of several imidazole derivatives, testing their efficacy against *E. coli*, *S. aureus*, *B. subtilis*, and *Micrococcus* using ciprofloxacin and sulfamethoxazole as the reference medications [57]. Compounds **46** and **47** with electron-donating groups had excellent activity against *Micrococcus* and *E. coli* but low activity against *B. subtilis*, whereas compound **49** had the best activity against all strains. Compound **48** remained inert against *B. subtilis* while demonstrating comparable effectiveness against *Micrococcus* and *E. coli*.

Choudhari *et al.* in 2020 reported the synthesis of imidazole compounds on the basis of 1,4-naphthpoquinones (**50-53**) and (**54-57**) (Figure 3.6). Compound 50 showed a high MIC value against all strains, ranging from 128 to 512 μg/mL [58]. MIC values for Compounds **51** and **52** against *P. aeruginosa* NCIM 5029 were 32 and 64 g/mL, respectively. With an MIC of 8 μg/mL, Compound **57** has shown the most efficacy against *S. aureus* NCIM 5021.

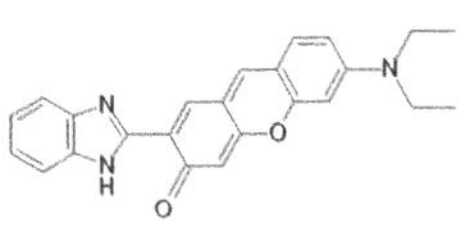

Compound 46

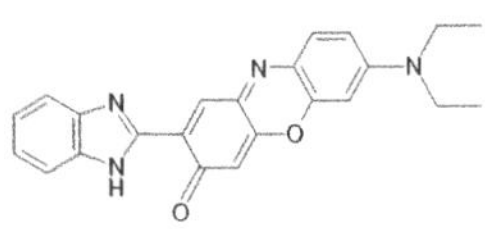

Compound 47

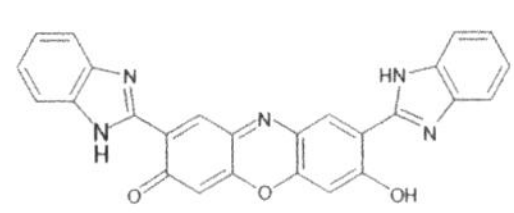

Compound 48

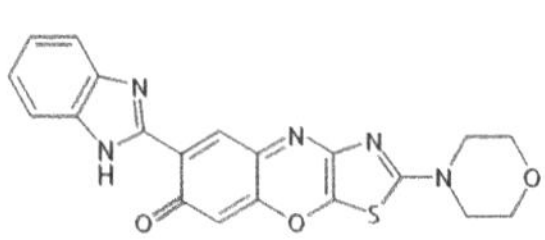

Compound 49

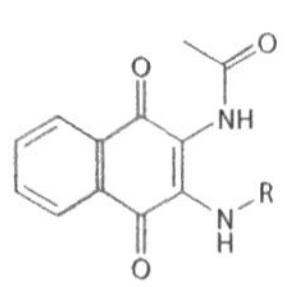

Compound 50-53

50: R=Me,**51:** R=C_2H_5

52: R=C_3H_7, **53: R**=C_4H_9

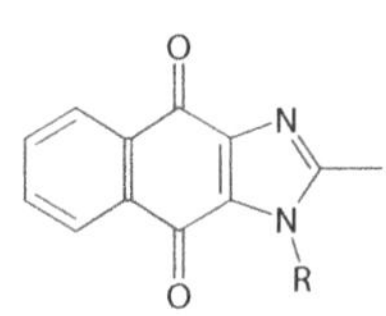

Compound 54-57

54: R=Me, **55:** R=C_2H_5

56: R=C_3H_7, **57: R**=C_4H_9

Figure 3.6 Imidazole-based heterocyclic Compounds **46–56** as antibacterial agent.

3.2.1.7 *Pyrazole*

Pyrazole is a five-membered heterocyclic aromatic organic compound with two nitrogen atoms, which comprises of a basic and a neutral atom. Its chemical formula is $(CH)_3N_2H$. The compound is basic in nature.

Pyrazole motifs have application in various fields such as pharmaceuticals, food, petrochemicals, etc. [59a] These compounds are notable members of azole family, which grabbed the attention of scientists due to structure and biological activities. Pyrazole has a tendency to exist in reduced forms such as 1-pyrazoline, 2-pyrazoline, 3-pyrazoline, etc.

Two pyrazole containing compounds such as cefoselis and ceftlozane were FDA-approved antibiotics to treat bacterial infection (Figure 3.7). The mode of mechanism for antibacterial activity is *via* penetrating through the outer membrane of Gram-negative bacteria. Cefoselis is a fourth-generation, whereas ceftlozane is a fifth-generation of cephalosporin antibiotics. These compounds have been widely used to treat bacterial infection including MRSA and *P. aeruginosa*. [59b] Researchers have

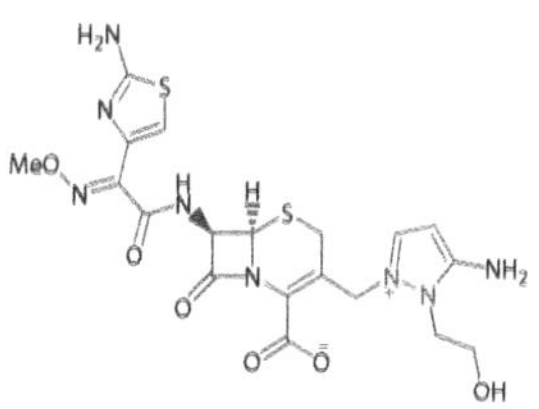

Cefoselis

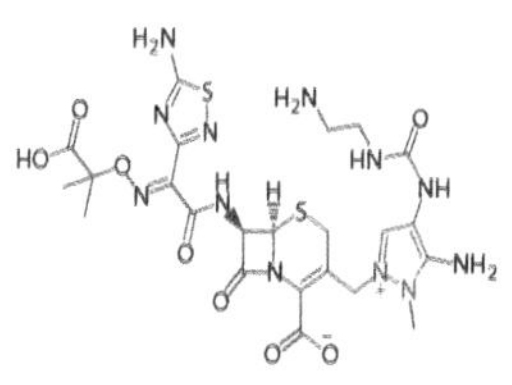

Ceftlozane

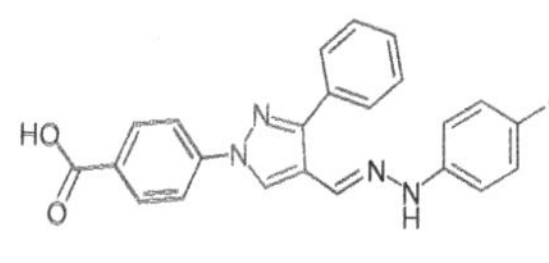

Compound **58**

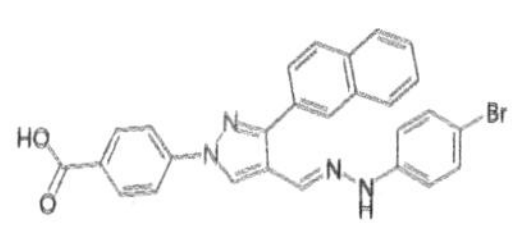

Compound **59**

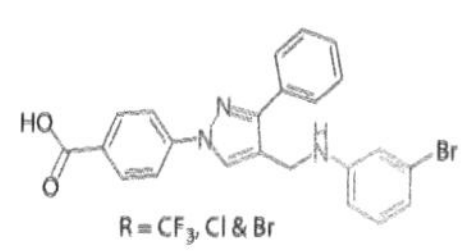

Compound **60**

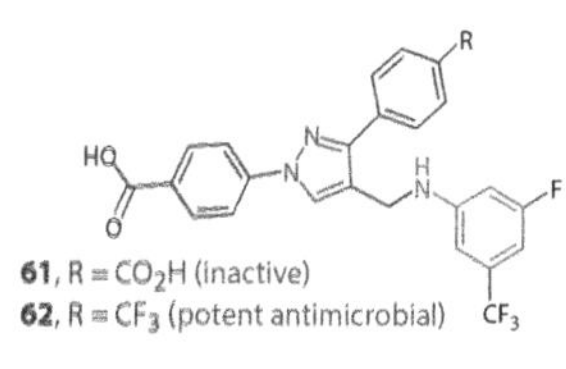

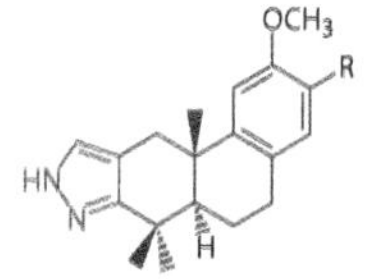

R = *n*Pr & *n*Bu

Compound **63**

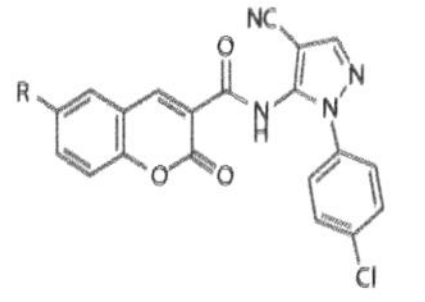

R=Br, NO_2

Compound **64**

Figure 3.7 Pyrazole-based heterocyclic Compounds **58–64** as antibacterial agent.

developed Compound **58**, *N*-benzoic acid–derived pyrazole hydrazone, as a potent growth inhibitor *Acinetobacter baumannii* with MIC value as low as 4 μg/mL [60]. Further, naphthyl substituted pyrazole; Compound **59**–derived hydrazone was synthesized and found to be potent against Gram-positive strain and *A. baumannii* with MIC value in the range of 0.78 to 1.56 μg/mL [61]. These naphthyl derivatives were capable of disrupting biofilms. The mechanism for antibacterial activity was found to be cell wall disruption. Compound **59** was found to be bacteriostatic for *A. baumannii* and bactericidal for *S. aureus*. Aniline-derived pyrazoles were found to be active against Gram-positive bacteria such as *S. aureus*. Compound **60**, 3-phenyl pyrazoles had moderate antibacterial activity against *S. aureus* with MIC value of 16 μg/mL [62]. The functional moiety on pyrazole has great influence on antimicrobial activity; benzoic acid–derived hydrazones (Compound **58**) showed specific activity toward *A. baumannii* and Compound **60** showed moderate activity toward *S. aureus*. Replacing benzoic acid (Compound **61**) with trifluorophenyl-substituted Compound **62**, showed potent antibacterial activity against *S. aureus* including resistant strain MRSA with low MIC value of 0.39 μg/mL.

In general pyrazole-fused diterpenoids have been reported as potent growth inhibitors of *S. aureus* strains with MIC values as low as 0.71 μg/mL. The fused terpenoid is a rare example of pyrazole-fused natural product. Compound **63** shows activity against *S. aureus* Newman strain with MIC value of 1 μg/mL and now showed toxicity in human fibroblast (HAF) cells at this concentration [63]. In 2018, Liu *et al.* have synthesized coumarin–pyrazole carboxamide; Compound **64** showed excellent activity against *S. aureus*, *Salmonella*, and *Escherichia* excellent antibacterial activity [64].

3.2.1.8 Oxazole

Oxazole is a heterocyclic aromatic organic compound, which consists of one nitrogen atom and an oxygen atom separated by a carbon atom. Its chemical formula is C_3H_3NO. All the atoms are sp2 hybridized and planar. It is a conjugate acid with a pka value of 0.8. Oxazole and its derivative such as isoxazole play a valuable role in medicinal chemistry.

Arshad had synthesized 1,2-oxazole derivatives (**65–70**) and studied computationally to determine their physicochemical properties and drug potential (Figure 3.8). All the compounds showed similar activity as reference drug, ciprofloxacin, in Gram-positive (*S. aureus* and *S. epidermidis*) and Gram-negative bacteria (*E. coli and P. mirabilis*) with MIC value 3.125 to 12.5 μg/mL) [65].

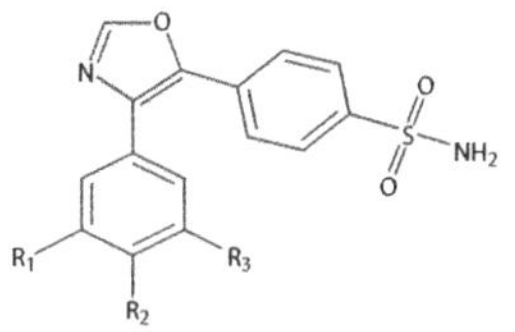

65: R1 = H, R2 = OMe, R3 = H

66: R1 = OMe, R2 = OMe, R3 = H

67: R1 = OMe, R2 = OMe, R3 = OMe

68: R1 = H, R2 = OH, R3 = H,

69: R1 = OH, R2 = OH, R3 = H

70: R1 = OH, R2 = OH, R3 = OH

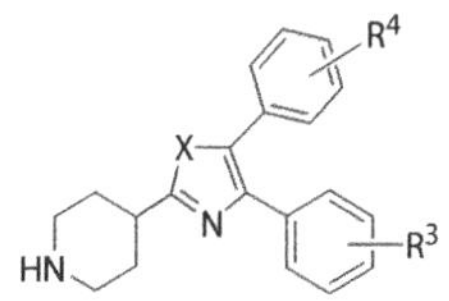

71: X = O, R3 = 3-Cl, 3-$PhCH_2O$

72: X = N, R3 = 3-Cl, 3-$PhCH_2O$

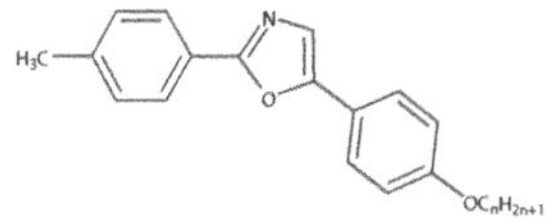

73: n=4

74: n=6

75: n=8

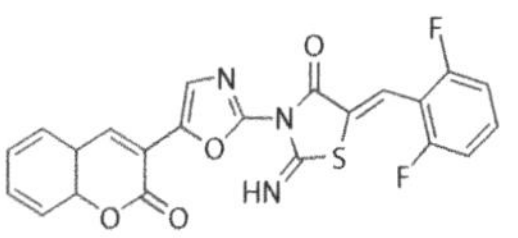

Compound 76

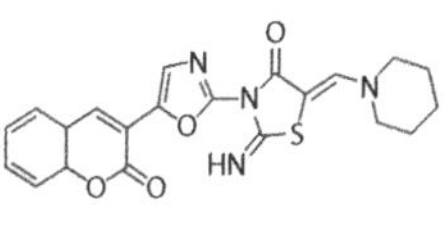

Compound 77

Figure 3.8 Oxazole-based heterocyclic Compounds **65–77** as antibacterial agent.

Tanitame *et al.* prepared an oxazole Compound **71**, and Compound **72** (imidazole derivative of 71) and studied their antibacterial activity against drug susceptible strain, *S. aureus* and resistant strain MRSA and Gram-negative, *E. coli* NIHJ JC-2 and multidrug efflux pump mutant, *E. coli* W3110 DacrA. Compound 71 show activity against resistant and susceptible *S. aureus* with MIC value 2 μg/mL. However, oxazole derivative show more antibacterial effect on resistant and less potent for susceptible *E. coli* strain. However, Compound **72** shows same antibacterial activity against resistant and susceptible *S. aureus* and *E. coli* with MIC value of 2 - 4 μg/mL. Compounds **71** and **72** shows more potent antibacterial activity than sparfloxacin against clinically isolated quinolone-resistant Gram-positive bacteria. Compounds **71** and **72** showed the same inhibitory effect against DNA gyrase and topoisomerase IV [66].

Tomi *et al.* in 2015 synthesized oxazole and benzothiazole derivatives. The studies showed that oxazole Compound **75** showed greater activity against *P. aeruginosa*. However, Compounds **73** to **75** showed similar activity against *E. coli* and *S. aureus*. In addition, the studies also showed that benzothiazole derivative showed better result than oxazole derivatives [67a].

A scientist Reddy *et al.* synthesized a series of novel oxazole derivatives and evaluated their antibacterial activity against Gram-positive (*B. subtilis*, *S. aureus*, and *M. luteus*) and Gram-negative bacteria (*P. vulgaris*, *S. typhimurium*, and *E. coli*) using penicillin and streptomycin as reference drugs. The result concluded that Compounds **76** and **77** were detected to possess high antibacterial potency in all bacteria [67b, 67c].

3.2.1.9 Thiazole

Thiazole is also known as 1,3-thiazole, a heterocyclic aromatic organic compound containing sulfur and nitrogen atoms in the ring. Its chemical formula is C_3H_3NS. The compound belongs to a member of the azole family. It most commonly occurs in natural compounds such as vitamin B (thiamine).

Thiazole and its based derivatives are a class of heterocyclic compound, which plays a significant role in pharmaceutical industry and medicinal chemistry. Recent studies on thiazole have concluded that their scaffolds are of paramount importance because of their active biological properties. The rise of antibacterial resistant pathogens in the world has led to the necessity for discovering more antimicrobial compounds.

Karegoudar *et al.* synthesized trichlorophenyl thiazole derivatives (**78–80**) showed excellent inhibitory activity against Gram-positive and Gram-negative bacteria including *E. coli.* having MIC values = 6.25 μg/mL, same as standard drug ciprofloxacin (Figure 3.9) [68].

Various strains of Gram-positive and Gram-negative bacteria, including *K. pneumoniae*, *P. aeruginosa*, *S. aureus*, and *E. coli*, were tested against Compounds **81–86**. The testing result concluded that Compound **83**, **81**, **85**, and **84** showed remarkable antibacterial activity against *K. pneumoniae.* Compounds **81**, **82**, **85**, and **86** showed excellent activity against *E. coli.* Compounds **81**, **82**, and **86** exhibited remarkable activity against *P. aeruginosa.* Certain compounds had more action than ciprofloxacin [68]. About 30 thiazole derivatives have been synthesized by Zha *et al.*, who have also investigated antibacterial properties against *E. coli*, *S. aureus*, *K. pneumoniae*, and *B. subtilis* [69]. According to the results, Compounds **87** and **88** had significant action against every bacterial strain, whereas Compounds **89** and **90** demonstrated efficacy against *S. aureus* and *E. coli* [69].

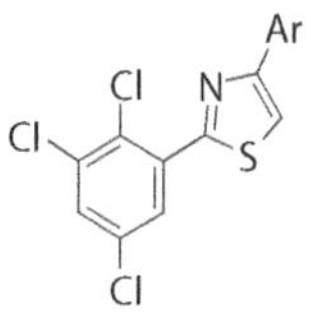

Compound **78**

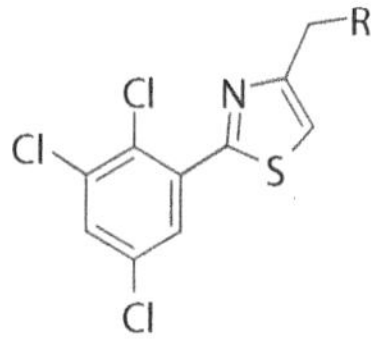

Compound **79**

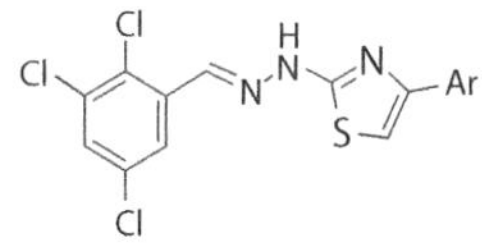

Compound **80**

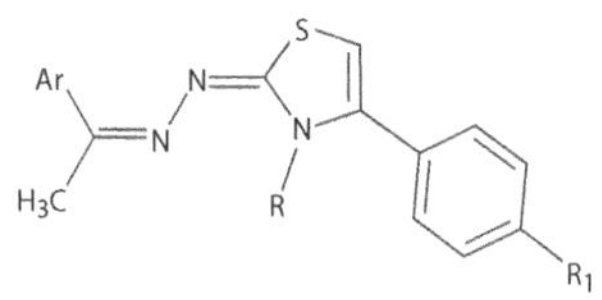

81: Ar = Ph; R = H; R1 = H

82: Ar = Pyridin-3-yl; R = H; R1 = H

83: Ar = Ph; R=H; R1 = Br

84: Ar = Furan-2-yl; R = H; R1 = Br

85: Ar = Pyridin-3-yl, R = H; R1 = Ph

86: Ar = Ph, R=PhCH_2; R1 = Br

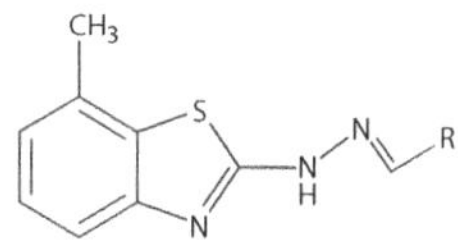

87: R =1-benzyl-4-methyl benzene

88: R =1-(benzyloxy)-4-methyl benzene

89: R =3-Br, 4= -OH-Ph

90: R =3-F, 4= -OH-Ph

Figure 3.9 Thiazole-based heterocyclic Compounds **78–90** as antibacterial agent.

3.2.1.10 *Pyrimidine*

Pyrimidine is an important heterocyclic aromatic organic compound with chemical formula $C_4H_4N_2$, containing two nitrogen atoms at the first and third positions.

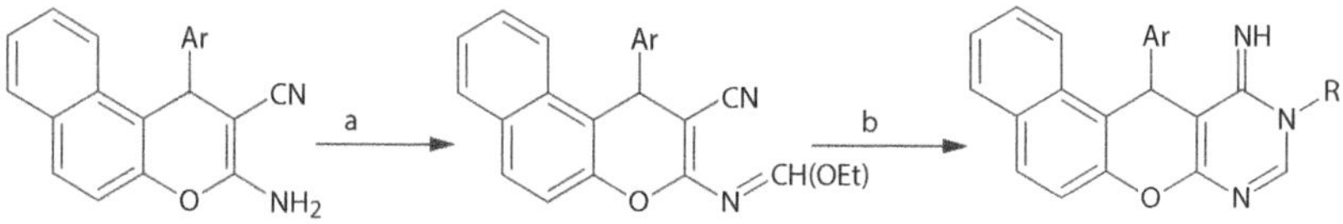

Compound **91** Compound **92-94** Compound **95-101**

(a)=HC(OET)$_3$/Ac$_2$O/Reflux, (b)=MeOH/R-NH$_2$

91: Ar = 3-NO2Ph; **92**: Ar = 4-OHPh; **93**: Ar = 4-ClPh; **94**: Ar = 2-ClPh; **95**: Ar = 3-NO$_2$Ph, R = Me, **96**: Ar-4-OHPh, R = Me; **97**: Ar = 4-ClPh, R = Me; **98**: Ar = 3-NO$_2$Ph, R = Benzyl; **99**: Ar = 3-NO$_2$Ph, R = Ph, **100**: Ar = 4-OHPh, R = Ph; **101**: Ar = 2-Cl-C_6H_4, R = Ph

Figure 3.10 Scheme showing synthesis of benzochromeno[2,3-*d*]pyrimidine (**91–101**).

In 2017, Ameli *et al.* have synthesized benzochromeno[2,3-*d*]pyrimidine (**91–101**) as shown in Figure 3.10 and studied the antibacterial effect against *S. aureus*, *B. subtilis*, *E. coli*, and *S. typhimurium* [70]. The result demonstrated that Compounds **91–94** (MIC value >250) were less effective compared to Compounds **95–101**. This clearly signifies the role of pyrimidine moiety in antibacterial activity.

Compounds (**96, 97, 100**, and **101**) containing -Cl and -OH showed greater inhibitory effect against *E. coli*, *S. aureus*, *S. typhimurium*, and *B. subtilis* strains. In 2021, Bassyouni *et al.* synthesized a series of fused pyrimidine compounds and evaluated for their antibacterial efficiency [71]. The results concluded that Compounds **102** and **103** showed good zone of inhibition against bacterial strains such as *B. cereus*, *E. coli*, *P. aeruginosa*, and *S. typhimurium* compared to standard drug cephradine (Figure 3.11). It was reported that enhanced activity of the compound may be due to the presence of methyl, amine, and sulfur substituted pyrimidine moiety. Molecular docking studies showed that protein could be the potential target for their antibacterial activity. Mahdavi *et al.* studied various pyrimidine compounds and analyzed their antibacterial efficiency against *E. coli* and *S. aureus* [72]. The result concluded that Compounds **104** and **105** showed high zone of inhibition against *S. aureus*, whereas Compounds **106** and **107** showed high zone of inhibition against *E. coli*.

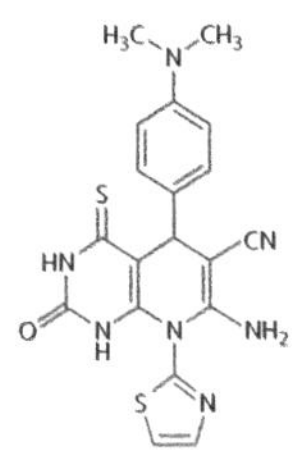

Compound **102**

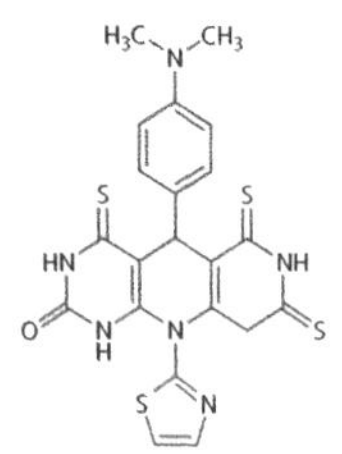

Compound **103**

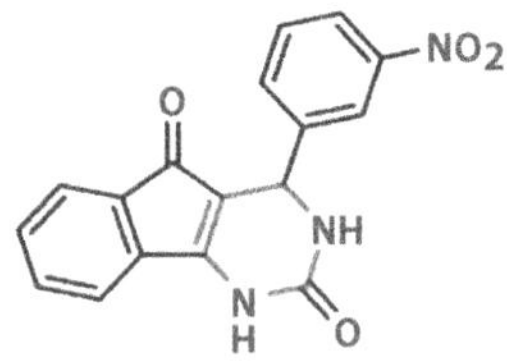

Compound **104**

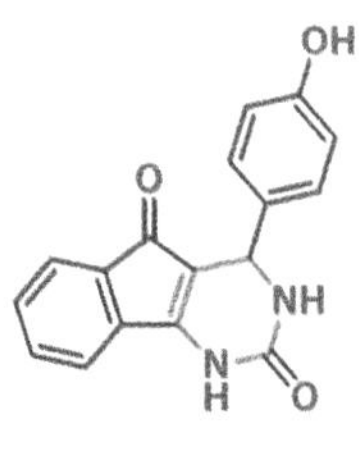

Compound **105**

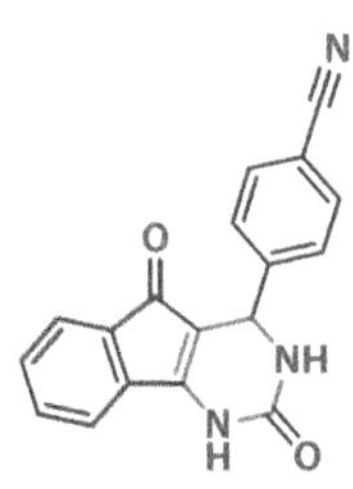

Compound **106**

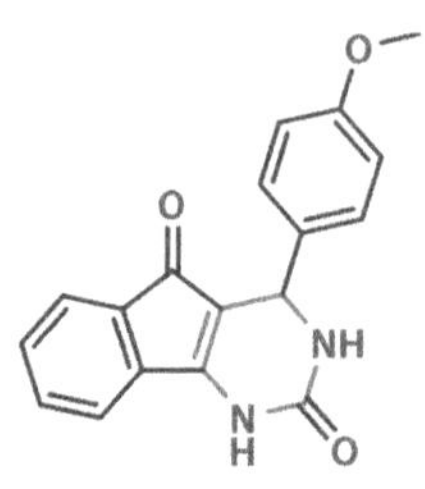

Compound **107**

Figure 3.11 Pyrimidine-based heterocyclic compound as antibacterial agent.

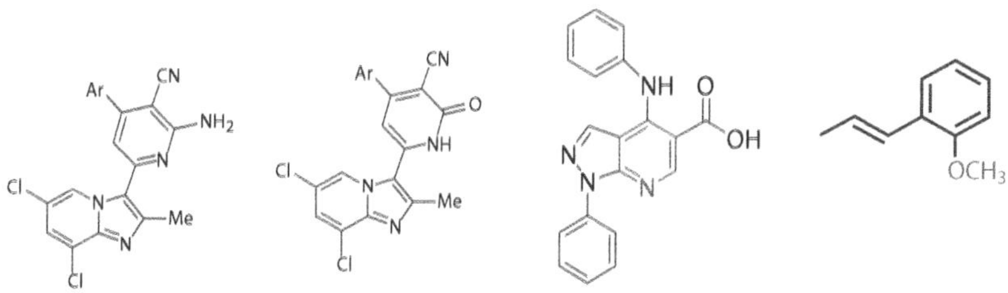

Compound **108-110**

108: Ar=MePh

109: Ar=OMePh

110: Ar=NO_2Ph

Compound **111-113**

111: Ar=MePh

112: Ar=OMePh

113: Ar=NO_2Ph

Compound **114**

Compound **115**

Figure 3.12 Pyrimidine-based heterocyclic Compounds **108–115** as antibacterial agent.

3.2.1.11 Pyridine

Pyridine is an important heterocyclic aromatic organic compound having chemical formula C_5H_5N. It is considered as a privileged medical scaffold with numerous pharmacological properties, making it immensely important for various industries.

In 2021, Althagafi and Latif synthesized derivatives of 6,8-dichloro-imidiazo[1,2-a] pyridine (Compounds **108–113** (Figure 3.12)). The synthesized compounds were evaluated against strains such as *K. pneumoniae, E. coli, B. subtilis*, etc [73]. The screening concluded that the synthesized compound showed significant activity against antibacterial strains. Leal *et al.* have synthesized 4-(arylamino)-1-phenyl-1H-pyrazolo [3,4-*b*]pyridine-5-carboxylic acids (Compound **114**) and exhibited potent antibacterial efficacy against a drug-resistant *S. epidermidis* [74a, 74b]. The MIC of the prepared compound showed similar activity in comparison with oxacillin and twofold better activity compared to chloramphenicol. Ali *et al.* synthesized a series of N-alkylated pyridine salts and tested them for their antibacterial properties [74c]. The result concluded that Compound **115** showed high antibacterial activity with MIC values of 56 μg/mL against *S. aureus* and 55 μg/mL against *E. coli.*

3.2.2 Antifungal Activity and Structure–Activity Relationship of Heterocyclic Compounds

Various fungal species invade humans and cause infections with increased mortality rate in human patients with a weak immune system, such as HIV and cancer patients [75–77]. Thus, we develop and do research on drugs to prevent such invasions.

Fungi are known to be eukaryotes with similar cell structure to that of mammalian cells. This makes the synthesis of drugs a challenging task [78, 79]. To date, the antifungal agents are divided into four categories: azoles (e.g., itraconazole and fluconazole), polyenes (e.g., amphotericin B and nystatin), antimetabolites (e.g., 5-fluorocytosine), and echinocandins (e.g., micafungin and caspofungin) (Figure 3.13) [80].

Fungi [81] affect not only humans but also the plants. For example, fungi such as *Oryza sativa*, *Sclerotinia sclerotiorum*, and *Fusarium graminearum* are known to cause disease such as rice sheath blight, rape sclerotinia rot, and *Fusarium* head blight. These fungi reduce the crop yield and increase agricultural economic losses [82, 83]. With the development and use of fungicides, carbendazim, valinomycin, and pyraoxystrobin helps in the growth of plant pathogens [84–86]. These antifungal medications still have a number of adverse consequences, such as pharmacokinetics, safety,

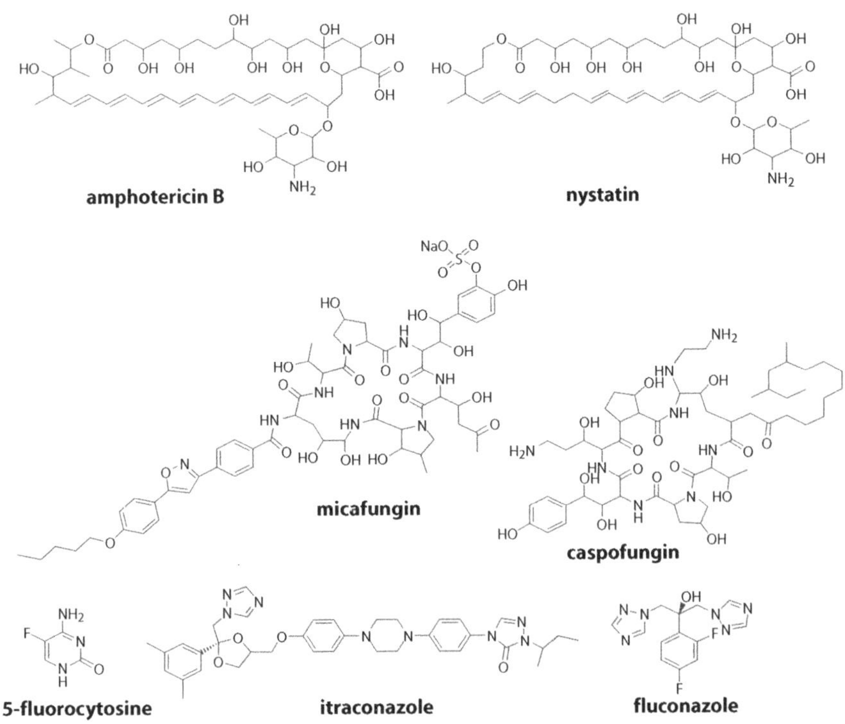

Figure 3.13 Chemical structure of clinically used polyenes, echinocandins, antimetabolite, and azole.

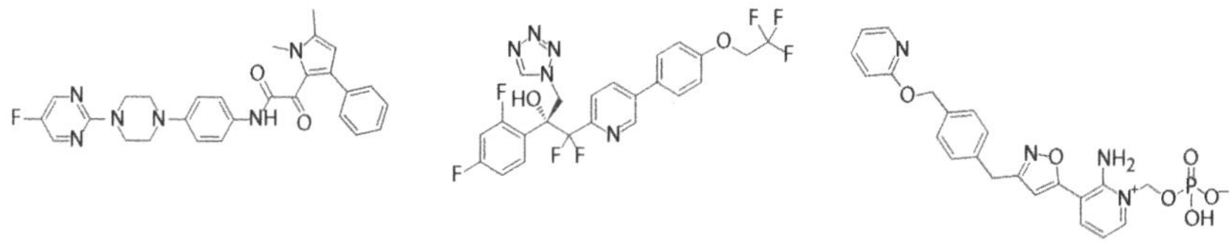

Olorofim (F-901318) **Oteseconazole** **Fosmanogepix (APX001)**

Figure 3.14 Current under development of antifungal compounds.

and toxicity [87–89]. Resistance to fungicides has developed as a result of the abuse and overuse of antifungal medications. This has resulted in widespread fungus-like *Candida* infections [90]. Therefore, either new antifungal drugs or optimization of existing antifungal drugs must be discovered, on an urgent basis [91]. For the following reasons, various antifungal drugs, such as olorofim (F-901318), oteseconazole, and fosmanogepix (APX001), have been used for clinical research in recent years (Figure 3.14).

Olorofim exhibits antifungal efficacy against a variety of mold species, such as *Aspergillus* sp., *Penicillium* spp., *Coccidioides* spp., and *Scopulariopsis brevicaulis*. Oral and injectable olorofim versions are presently undergoing Phase 2 studies [92]. Oteseconazole, an oral tetrazole antifungal medication, is undergoing Phase 2 and Phase 3 studies to treat onychomycosis and recurrent vulvovaginal candidiasis, respectively [93]. A broad-spectrum antifungal drug called fosmanogepix works well against *Candida* species, *Cryptococcus neoformans*, and *Coccidioides immitis* [94].

3.2.2.1 Triazole

Triazole-containing compounds have been discovered to exhibit a wide range of pharmacological characteristics, such as antiviral, hypoglycemic, anticancer, antitubercular, anti-inflammatory, antifibrosis, antihypertensive, and antitubercular [95–119]. There are two forms of this five-membered ring compound: 1,2,3-triazoles and 1,2,4-triazoles. This azole antifungal drug inhibits the production of ergosterol by binding to the active site of CYP51 (lanosterol 14a-demethylase).

(a) 2,4,5-Trisubstituted 1,2,3-triazole compounds were synthesized by Venugopala *et al.*, [120] and their antifungal efficacy against five fungal strains—*C. parapsilosis*, *C. albicans*, *Candida tropicalis*, *Aspergillus niger*, and *Trichophyton rubrum*—was assessed. Every synthesized compound demonstrated antifungal activity against the strains that were tested (Figure 3.15).

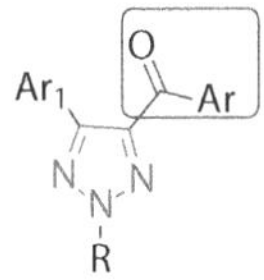

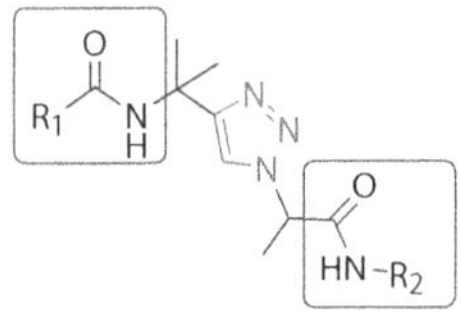

116: R=Methyl; Ar=thiophene; Ar_1=4-ethylpenyl

117: R=methyl; Ar=p-methoxyphenyl; Ar1=4-bromophenyl

118: R=Methyl; Ar=thipophene; Ar1=3,4,5-trimethoxyphenyl

119: R1=4-$CH_3OC_6H_4$; R2=4-$NO_2C_6H_4$

120: R1=4-$CH_3OC_6H_4$; R2=4-FC_6H_4

121: R1=4-$CH_3OC_6H_4$; R2=a-$C_{10}H_7$

122: R1=4-$NO_2C_6H_4$; R2=C_6H_5

123: R1=4-$NO_2C_6H_4$; R2=4-$CH_3OC_6H_5$

124: R1=4-$NO_2C_6H_4$; R2=4-$NO_2C_6H_4$

125: R1=4-$NO_2C_6H_4$; R2=a-$C_{10}H_7$

126: R1=C_4H_3S; R2=4-$NO_2C_6H_4$;

Figure 3.15 1,2,3-Triazoles as antifungal agent.

Interestingly, Compounds **116**, **117**, and **118** displayed MIC values that were similar to the common medication fluconazole, ranging from 0.49 to 1.95 μg/mL. The inclusion of a methyl group at the 2-position reduced steric hindrance, making it a viable lead structure, even if the impact of substituents at the 2-, 4-, and 5-positions on antifungal activity was not very noticeable. Kaushik and colleagues [121] developed a series of 1,4-disubstituted 1,2,3-triazoles featuring two amide linkages (Compounds **119–126**) using a rapid one-pot synthesis method. These compounds were tested for antifungal activity against *C. albicans* (MTCC 227) and *A. niger* (MTCC 282). Several compounds demonstrated potent antifungal effects. In particular, Compounds **119**, **121**, **123**, **124**, **125**, and **126** displayed MIC values of 0.0302–0.0552 μmol/mL against *C. albicans*. Against *A. niger*, Compounds **119**, **123**, **124**, **125**, and **126** were effective, with MIC values ranging from 0.0151 to 0.0288 μmol/mL. The study concluded that the presence of electron-withdrawing groups, along with naphthyl and thienyl substituents, contributed significantly to the observed antifungal potency.

(b) **1,2,4-Triazoles**: Zoidis along with his team synthesized various 1,2,4-triazole derivatives against different strains of *C. albicans*, *Candida krusei*, and *Aspergillus fumigatus* was investigated (Figure 3.16) [122]. According to the results obtained *in vitro*, it was observed that all compounds played a significant role in antifungal activity. The amide analogs (**127**b1–b10) had stronger antifungal activity as compared to the less bulky triazoles (**127**a–a8 and **127**a10–a13). Because of the presence of less bulky

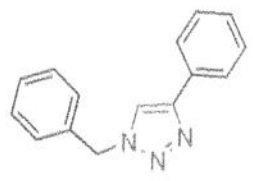

127a

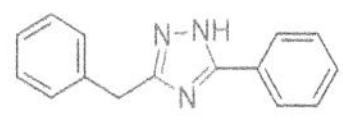

127a1

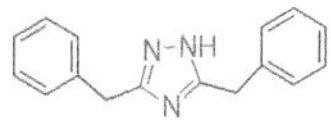

127a13

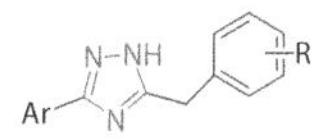

127a2-127a12

127a2-a5: Ar=phenyl

127a6-a10: Ar=4-chlorophenyl

127a2: R=H
127a3: R=4-OCH3
127a4: R=4-Br
127a5: R=3,4-diCl
127a6: R=2-F-6-Cl
127a7: R=4-OCH_3

127a11: Ar=thien-2yl
127a12: Ar=pyrid-3-yl

127a8: R=4-NH2
127a9: R=4-Ph
127a10: R=3,4-(OCH2O)
127a11: R=H
127a12: R=H

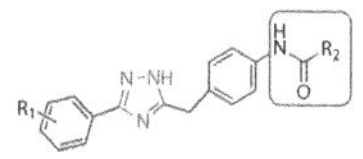

127b1-127b10

127b1-b5: R1=H
127b1: R2=Ph
127b2: R2=2,4, diClPh
127b3: R2=3-ClPh
127b4: R2=CH2Ph
127b5: R2=CH2Naphyh-2yl

127b6: R1=4-Cl; R2=Ph
127b7: R1=H; R2=(CH2)2Ph
127b8: R1=H; R2=CyHexyl
127b9: R1=4-Cl; R2=Ph
127R1=3Cl; R2=Ph

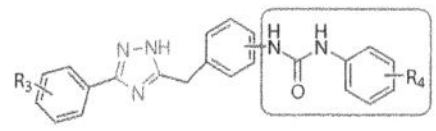

127c4: R3=4-Cl; R4=3-Cl

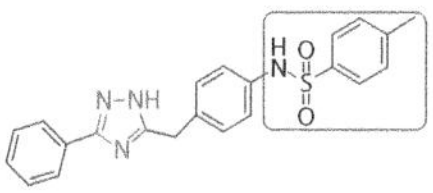

127c5

Figure 3.16 1,2,4-Triazoles as antifungal agent.

triazoles (**127**a–a8), Compound **127**a8 displayed strong antifungal activity. For *A. fumigatus* ATCC9197, compounds examined displayed similar or higher antifungal activity as keto-conazolea part from Compounds **127b8** and **127c5**. But after comparing these compounds with itraconazole, it was observed that they were less effective, i.e., showed lower antifungal activity. Compounds **127**a3–5 and **127**c4 has the highest amount of antifungal activity against *A. fumigatus*. Upon replacing aromatic ring with a heterocycle (thiophene or pyridine) in these compounds, the activity decreased by 3.2- and 1.5-fold. This led to the observation that the 1,2,4-triazole ring can further be developed and optimized for future fungicides.

This led to the observation that 1,2,4-triazole ring can further be developed and optimized for antifungal properties through appropriate substitution. Bai *et al.* [123] developed and synthesized 1,2,4-triazole derivatives (**128**a1–18 and **128**b1–9) with the presence of an oxime ether and a phenoxy pyridine moiety (Figure 3.17). The antifungal activity of these derivatives against eight phytopathogens was then assessed using these chemicals. Methyl oxime ether series **128**a compounds (22.9%–90.7%) were shown to have more antifungal activity than benzyl oxime ether series **128**b compounds (0%–90.0%). The presence of electron-withdrawing groups at the benzene ring's second position in these compounds was crucial in boosting their antifungal activity. Another finding was that, regardless of where it is located in the benzene ring, the chlorine atom contributes significantly

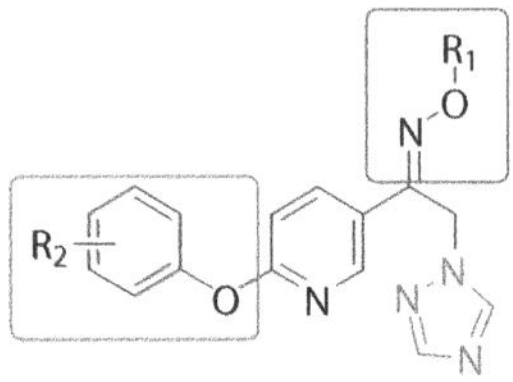

128 a: R_1 = Me
128 b: R_1 = Bn

128 a1: R_2 = H	**128 a10:** R_2 = 4-Me	**128 b1:** R_2 = H
128 a2: R_2 = 2-Cl	**128 a11:** R_2 = 4-*t*-BuI	**128 b2:** R_2 = 2-Cl
128 a3: R_2 = 2-Br	**128 a12:** R_2 = 4-Cyclohexyl	**128 b3:** R_2 = 2-Br
128 a4: R_2 = 2-Cl-4-Br	**128 a13:** R_2 = 4-OMe	**128 b4:** R_2 = 2, 4-$(t\text{-Bu})_2$
128 a5: R_2 = 2-*t*-Bu-4-Me	**128 a14:** R_2 = 4-Br	**128 b5:** R_2 = 2-*t*-Bu-4-Me
128 a6: R_2 = 2-Me	**128 a15:** R_2 = 4-Cl	**128 b6:** R_2 = 2-*t*-Bu
128 a7: R_2 = 3-Me	**128 a16:** R_2 = 4-F	**128 b7:** R_2 = 3-Me
128 a8: R_2 = 3-OMe	**128 a17:** R_2 = 4-CF_3	**128 b8:** R_2 = 4-Me
128 a9: R_2 = 3-Cl	**128 a18:** R_2 = 4-NO_2	**128 b9:** R_2 = 4-CN

Figure 3.17 1-Substitute 1,2,4-triazole derivatives (128a1–a18 and 128b1–b9) containing an oxime ether and a phenoxy pyridine moiety.

to the compounds' increased antifungal activity [80]. Compounds **128**a2–5, **128**a7, **128**a14, **128**a15, **128**a17, **128**a18, and **128**b2 shown antifungal activity against *Rhizoctonia solani*, *S. sclerotiorum*, *Botrytis cinerea*, and *Phytophthora infestans*, according to their EC_{50} values. These substances were shown to exhibit potent inhibitory effects against *S. sclerotiorum*. Furthermore, **128**a4 and **128**b2 were shown to be able to fit into the CYP51 substrate binding cavity and exhibit a comparable binding orientation inside the cavity, according to modeling and molecular docking results [80].

3.2.2.2 Imidazoles

One of the most important pharmacophores is imidazole with a variety of activities, including anticancer, [124] anti-infective, [125] and antidengue virus activity [126]. Two series of imidazole derivatives [127] containing dithiocarbamate and (benz)azolethiol side chains were synthesized, and their antifungal activity was examined against *C. parapsilosis*, *C. krusei*, *Candida glabrata*, and *C. albicans* (Figure 3.18). Compounds having dithiocarbamate side-chain series showed stronger antifungal activity (MIC 50 = 0.78–12.5 μg/mL) as compared to compound having (benz)azole thiol side chains (MIC 50 = 25– 200 μg/mL). Among all the compounds, **129** showed the strongest antifungal activity against *C. krusei* having MIC value of 0.78 μg/mL.

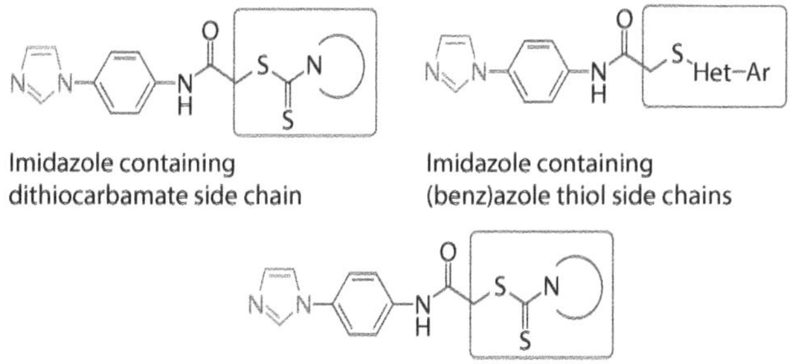

129=4-methoxy-benzylpiperazine

Figure 3.18 Imidazole derivatives containing dithiocarbamate and (benz) azole thiol side chains.

Compound **129** demonstrated twofold more activity than positive control compounds (ketoconazole and fluconazole having MIC 50 = 1.56 μg/ mL) [80]. It was observed that the antifungal effect of Compound **129** was due to its ability to block the production of ergosterol in *C. krusei.*

Al-Wabli *et al.* have studied various benzodioxole imidazole hybrids (**130a–r**) containing ester groups and examined their antifungal activity against *C. albicans, C. tropicalis, C. parapsilosis,* and *A. niger* (Figure 3.19) [128]. Compounds **130l** and **130m** with the presence of 3- or 4-trifluoromethylphenyl analogs exhibited MIC value of 0.297 μmol/mL and 0.148 and against *A. niger* and *C. albicans,* whereas Compounds **130b** and **130o,** which have 4-bromrophenyl or 3,4,5-trimethoxyphenyl derivatives, show antifungal activity against *C. parapsilosis* and *C. tropicalis* with MIC value of 0.565 and 0.145 to 0.579 μmol/mL. The antifungal activity of these benzodioxole imidazole hybrids was found to be lower than marketed drugs, ketoconazole (MIC = 0.02 μmol/mL) and fluconazole (MIC = 0.045–0.051 μmol/mL) [129]. Elejalde *et al.* synthesized a series of N-substituted 4-aryl-2-methylimidazoles (**131a–e** and **132a–d**) and evaluated their antifungal activity against *Candida albicans* and *Cryptococcus neoformans.* The findings revealed that these compounds exhibited strong antifungal activity against *C. neoformans* (MIC_{50} = 15.6–250 μg/mL), but only weak activity against *C. albicans* (MIC_{50} >250 μg/mL). Among them, the ketone-containing compounds (**131** series) demonstrated greater potency compared to the alcohol-containing analogs (**132** series), suggesting that reducing the carbonyl group to a hydroxyl diminished antifungal efficacy.

The majority of the produced substances exhibited broad-spectrum antifungal properties. A different set of imidazole derivatives, **133**a–b, **133**f–g, **133**q, and **135**, demonstrated MIC values ranging from 0.03125

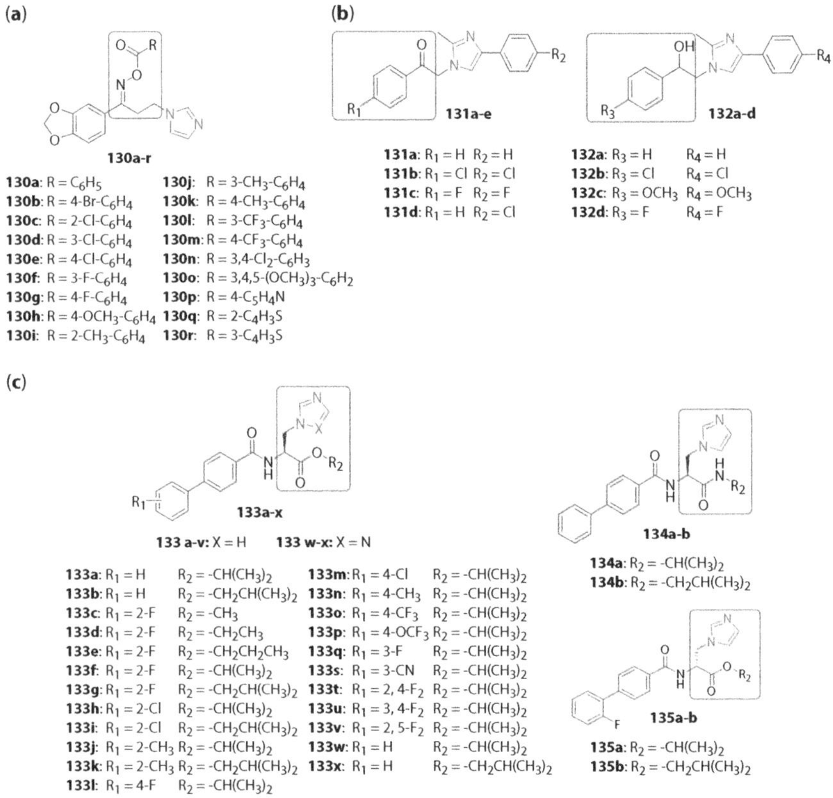

Figure 3.19 (a) Benzodioxole–imidazole hybrids (**130a–r**) having ester moiety, (b) *N*-substituted 4-aryl-2-methylimidazoles (**131a–e** and **132a–d**), and (c) biphenyl imidazole derivatives (**133a–x**, **134a–b**, and **135a–b**).

to 2 μg/mL against *C. tropicalis*, *C. albicans*, and *Candida neoformans*. The action was similar to that of antifungal medications such as itraconazole and fluconazole that are utilized in clinical settings. However, there was less antifungal activity when an amide group was substituted for the alkyl ester side chain (as in **135**a–b). In contrast, there was no discernible change in activity when a triazole (**133**w–x) was substituted for the imidazole moiety.

To assess the impact of chirality, the R-enantiomers (**135a–b**) were synthesized. However, both R- and S-enantiomers (**134a–b and 135a–b**) showed comparable antifungal activity, indicating that the presence of a chiral center did not significantly influence potency [80]. Compounds **133f–g** and **135a–b** also displayed moderate efficacy (MIC = 2–9 μg/mL) against fluconazole-resistant strains of *C. albicans*. Additionally, Compounds **133g** and **134b** demonstrated selectivity for fungal targets over human cytochrome P450 enzymes (CYP1A2, CYP2C9, CYP2C19, and CYP2D6), with IC_{50} values greater than 1 μM. Structure–activity relationship (SAR) analysis of these compounds ultimately led to the identification of Compound **135a** as a promising lead.

3.2.2.3 Oxazoles

Zhang *et al.* synthesized a series of cuminaldehyde derivatives (**136**a1–12 and **136**b1–12), which included isoxazoline and pyrazoline analogs (Figure 3.20) [130]. These compounds were evaluated for their antifungal activity against six different plant–pathogenic fungi. The results demonstrated that many of these derivatives exhibited strong antifungal effects, with greater than 50% inhibition observed across the tested fungal strains. Based on SAR analysis, several key observations were made:

1. The isoxazoline-based compounds (**136a1–12**) exhibited superior antifungal activity compared to the pyrazoline-based compounds (**136b1–12**).
2. Compounds containing electron-withdrawing substituents, such as halogens, showed higher antifungal activity than those with methyl groups. Among the halogens, the order of activity was found to be F > Cl > Br > Me.
3. Substituents positioned at the ortho-position of the phenyl ring conferred better antifungal activity than those at the meta- or para-positions.

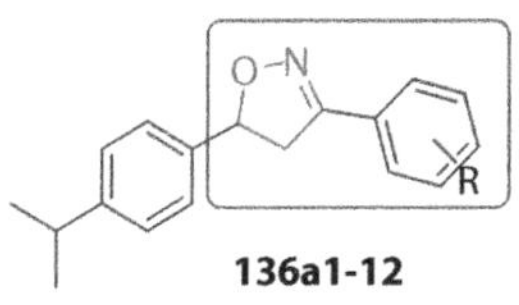

136a1-12

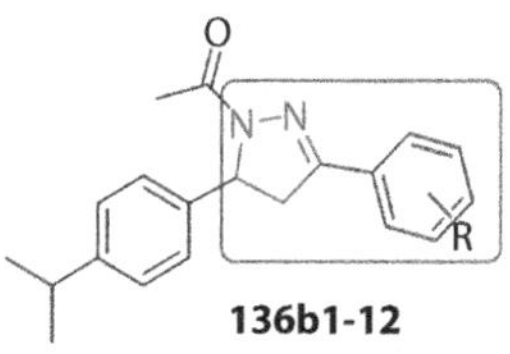

136b1-12

136a1: R = *o*-CH_3	**136a7**: R = *m*-Cl	**136b1**: R = *o*-CH_3	**136b7**: R = *m*-Cl
136a2: R = *o*-Br	**136a8**: R = *m*-F	**136b2**: R = *o*-Br	**136b8**: R = *m*-F
136a3: R = *o*-Cl	**136a9**: R = *p*-CH_3	**136b3**: R = *o*-Cl	**136b9**: R = *p*-CH_3
136a4: R = *o*-F	**136a10**: R = *p*-Br	**136b4**: R = *o*-F	**136b10**: R = *p*-Br
136a5: R = *m*-CH_3	**136a11**: R = *p*-Cl	**136b5**: R = *m*-CH_3	**136b11**: R = *p*-Cl
136a6: R = *m*-Br	**136a12**: R = *p*-F	**136b6**: R = *m*-Br	**136b12**: R = *p*-F

Figure 3.20 Cumialdehyde derivatives (**136a1–12 and 136b1–12**) having isoxazoline and pyrazoline moieties.

3.2.2.4 Thiazoles

Thiazole-based compounds are known for their broad spectrum of biological activities, including anticancer, antitubercular, and antiviral properties. Hagrasa *et al.* [131] synthesized a range of oxadiazolyl-thiazole derivatives and evaluated their antifungal efficacy [80]. Among them, Compound **137** (MIC = 1 μg/mL), which adopts a *cis* conformation, exhibited potent antifungal activity (Figure 3.21). It was effective against various strains of *C. albicans* and non-*albicans* species, including fluconazole-resistant isolates.

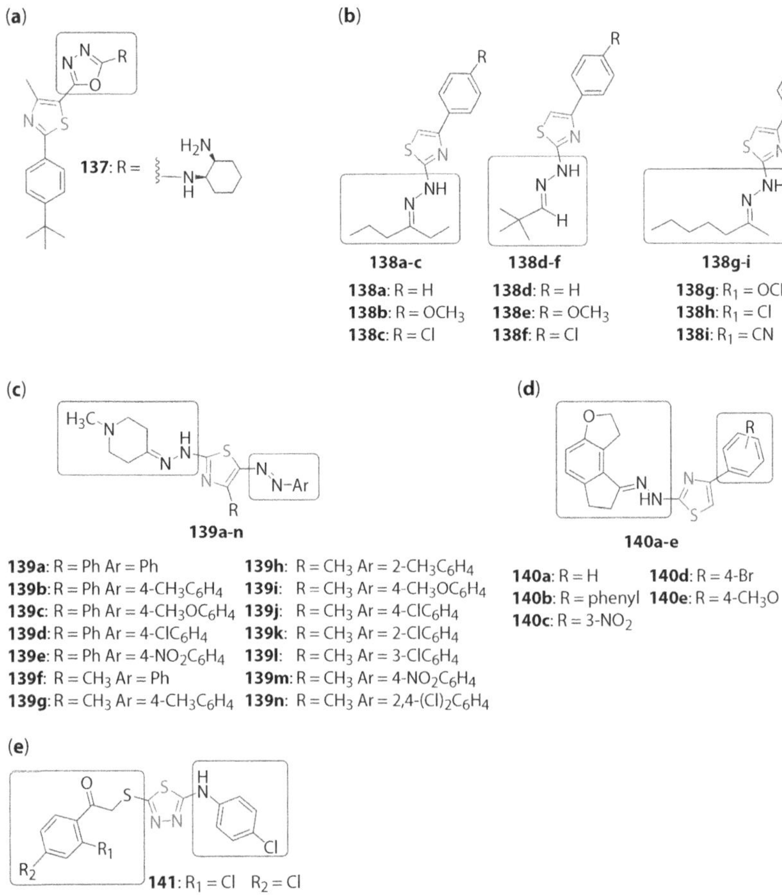

Figure 3.21 (a) Oxadiazolythiazoles **137**, (b) hydrazine-thiazole derivatives **(138a-i)**, (c) arylazothiazole derivatives (**139a–n**), (d) 2-(2-hydrazinyl)thiazole derivatives (**140a–e**), (e) 2,5-disubstituted 3,4-thiadiazole derivatives, **141**.

Additionally, Compound **137** demonstrated excellent tolerability in mammalian cells. Notably, Compound **137** showed better antifungal activity than fluconazole in inhibiting *C. albicans* biofilm formation and did not induce rapid drug resistance, even after 10 successive passages.

In a separate study, Lino *et al.* [132] developed a series of hydrazine-thiazole derivatives (**138a–i**) and tested their antifungal activity against multiple fungal species, including *C. neoformans*, *C. parapsilosis*, *C. albicans*, *C. tropicalis*, *C. krusei*, *C. gattii*, and *P. brasiliensis*. Eight compounds (**138a–f**, **138h**, and **138i**) demonstrated notable antifungal activity, with MIC values ranging from 0.45 to 31.2 mM [80]. Some of these compounds exhibited activity comparable to, or even exceeding, that of fluconazole and amphotericin B. The presence of a hydrophobic aliphatic chain in the hydrazine moiety was found to enhance antifungal activity, whereas aryl group substitutions had minimal impact. Furthermore, active compounds showed reduced cytotoxicity toward human embryonic kidney cells [80, 133].

Ouf *et al.* [133] synthesized a series of novel arylazothiazole derivatives (**139a–n**) and investigated their antifungal potential against *C. albicans*, *M. gypseum*, and *T. mentagrophytes*. Most of these derivatives exhibited strong antifungal activity, surpassing the efficacy of fluconazole. Among them, the 4-phenyl-substituted thiazoles (**139a–e**) demonstrated greater *in vitro* activity compared to their 4-methyl-substituted counterparts (**139f–i**). Introducing electron-donating groups such as $-OCH_3$ or $-CH_3$ on the phenyl ring at the 4-position significantly enhanced activity, whereas electron-withdrawing substituents such as chlorine or nitro groups had the opposite effect, reducing antifungal potency.

Adole and colleagues [134] developed a series of 2-(2-hydrazinyl)thiazole derivatives (**140a–e**) and tested their activity against a panel of fungi including *R. oryzae*, *C. albicans*, *M. mucido*, and *A. niger*. These compounds showed broad-spectrum antifungal activity at concentrations as low as 1 μM. Compounds such as **140a** and **140d** were active against all tested fungal strains. Several compounds, including **140c**, **140d**, and **140e**, exhibited MIC values ranging from 1.95 to 7.81 μg/mL. Substituents such as 3-nitro, 4-methoxy, and 4-bromo were associated with enhanced antifungal activity. Cytotoxicity assays against HL-60 cells revealed that these compounds had low toxicity.

A number of 2,5-disubstituted 1,3,4-thiadiazole derivatives were synthesized and their antifungal properties evaluated by Karaburun *et al.* [135] in a different investigation. With an MIC_{50} of 5 μg/mL, Compound **141**, which has a 2,4-dichlorophenyl substituent, showed the most effectiveness against *C. albicans* [80]. Compound **141** also exhibited superior activity

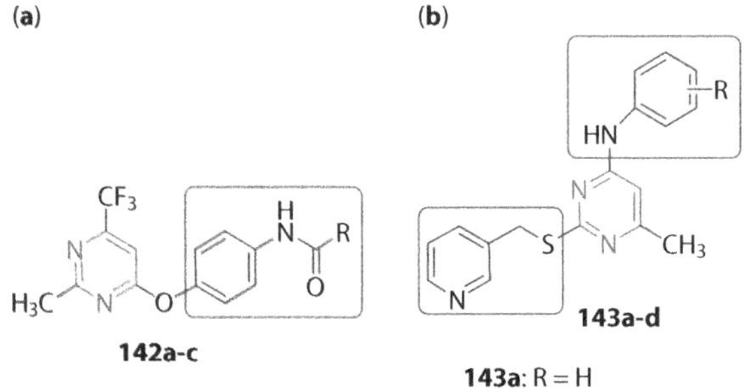

142a: R = 4-Cl-pyridine
142b: R = 2-F-5-Br-C_6H_3
142c: R = 2-I-4-Br-C_6H_3

143a: R = H
143b: R = 4-$CH(CH_3)_2$
143c: R = 4-CH_3O
143d: R = 2-OH-5-Cl

Figure 3.22 (a) Pyrimidine derivatives (**142a-c**) containing an amide moiety and (b) pyrimidine derivatives **(143a–d)** with (pyridin-3-ylmethyl)thio and phenylamino moiety.

compared to fluconazole across all tested *Candida* strains. The addition of halogens, such as chlorine or fluorine, was found to significantly enhance antifungal potency. Furthermore, the antifungal mechanism of Compound **141** was linked to the inhibition of ergosterol biosynthesis in *C. albicans*.

3.2.2.5 Pyrimidines

The pyrimidine ring is a key structural motif in many biologically active compounds, known for its antibacterial, insecticidal, and herbicidal properties. A series of pyrimidine derivatives (**142a–c**) containing amide moieties were synthesized and evaluated for their antifungal potential (Figure 3.22). Previous studies have demonstrated that compounds in this series (**142**) exhibit notable antifungal activity against plant–pathogenic fungi, with EC_{50} values ranging from 1.4 to 10.9 μg/mL, significantly better than standard antifungal agents. With corresponding EC_{50} values of 1.8, 1.4, and 1.7 μg/mL, compounds **142a**, **142b**, and **142c** demonstrated exceptional efficacy against *Phomosis* species, outperforming the reference fungicide pyrimethanil (EC_{50} = 32.1 μg/mL).

In another study, Wang *et al.* [136] developed a different series of pyrimidine derivatives (**143a–d**), incorporating (pyridin-3-ylmethyl)thio and phenylamino groups. These compounds were assessed for antifungal activity against *B. cinerea* and *S. sclerotiorum*. Among them, Compound **143a** exhibited strong antifungal activity against *B. cinerea*, achieving an inhibition rate of 87.5% at a concentration of 100 μg/mL. Additionally, Compounds **143a** through **143d** demonstrated significant activity against *S. sclerotiorum*, with inhibition rates ranging from 86.6% to 93.7% at the same concentration.

3.2.2.6 *Benzimidazoles*

Benzimidazole is a highly significant heterocyclic scaffold, with its derivatives exhibiting a wide range of biological activities including anticancer, [137] antibacterial, [138] and anti-VEGFR properties [139].

A number of benzimidazole–oxadiazole derivatives of compound **144** were synthesized by Karaburun *et al.*, [141], who then assessed the antifungal efficacy of these compounds against *C. albicans*, *C. krusei*, and *C. parapsilosis*. The majority of the synthesized compounds had no discernible antifungal action, according to the data. However, Compounds **144a** (MIC_{50} = 1.95–62.5 μg/mL) and **144b** (MIC_{50} = 1.95–31.25 μg/mL) exhibited antifungal activity comparable to standard reference drugs (MIC_{50} = 1.95–7.8 μg/mL) (Figure 3.23). These active compounds contained two ortho-hydroxyl groups forming a catechol moiety, which acts as a hydrogen bond donor and may contribute to their antifungal potency. Additionally, both compounds demonstrated low toxicity against the tested fungal strains. It was proposed that their mechanism of action might involve the inhibition of ergosterol biosynthesis in fungal cell membranes.

Huang *et al.* [142] synthesized 2-aryl benzimidazole derivatives (**145**a–c and **146**b–c) and tested them against strains of *Botrytis cinerea* that were

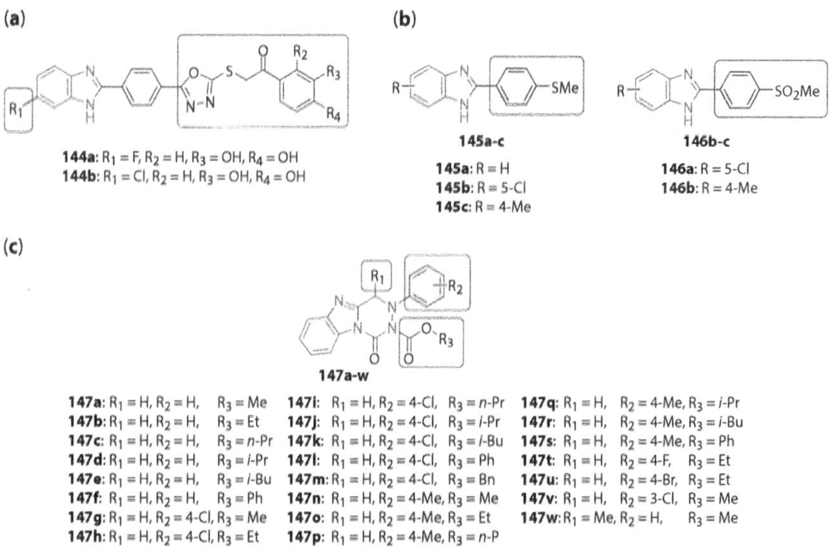

Figure 3.23 (a) Benzimidazole-oxadiazole compounds, **144**; (b) 2-aryl benzimidazoles derivatives (**145a–c** and **146b–c**); and (c) benzo[4,5]imidazo[1,2-d][1,2,4]triazine derivatives (**147a–w**).

susceptible to benzimidazole and strains that were highly resistant to it. Compounds **146**b and **146**c, which contained sulfoxide groups, displayed moderate antifungal activity with inhibition rates between 41.33% and 53.33%. In contrast, Compounds **145a–c**, featuring methylthio substituents, exhibited no activity—highlighting the crucial role of the sulfoxide group in antifungal efficacy.

Using phytopathogenic fungi such *B. cinerea*, *C. capsici*, and *R. solani*, Wang *et al.* [140] synthesized a series of benzo[4,5]imidazo[1,2-d][1,2,4] triazine derivatives (**147a–w**) and evaluated their antifungal ability. Most compounds showed inhibition rates ranging from 11.4% to 76.7% at a concentration of 50 μg/mL. SAR analysis revealed the following insights:

1. Antifungal activity increased with the introduction of electron-withdrawing groups at the R2 position.
2. Small alkyl groups (e.g., methyl, ethyl, n-propyl, isopropyl) at the R3 position enhanced activity.
3. The addition of a methyl group at the R1 position negatively impacted antifungal activity.

3.2.2.7 Benzothiazoles

Benzothiazole derivatives possess a wide range of biological activities, including anticancer, [143] antidiabetic, [144] and anticonvulsant [145] properties. Sirgamalla *et al.* [146] synthesized a series of 2-substituted benzothiazoles (**148a–f**) and evaluated their antifungal activity against *A. niger* and *C. albicans* (Figure 3.24).

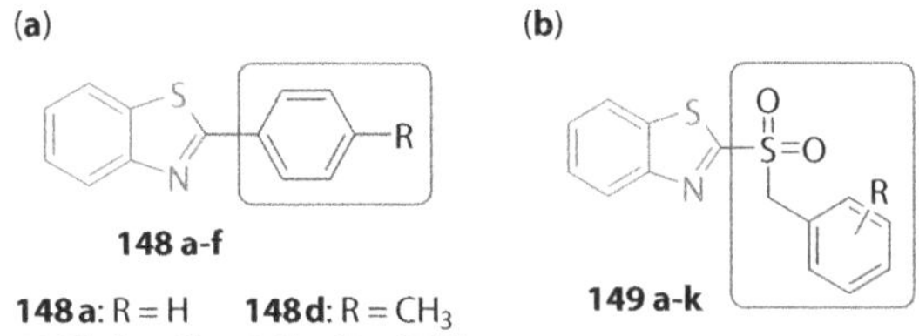

148 a-f

148a: R = H **148d**: R = CH_3
148b: R = Cl **148e**: R = OCH_3
148c: R = F **148f**: R = NO_2

149 a-k

149a: R = H **149g**: R = 4-Br
149b: R = 4-F **149h**: R = 2-Br
149c: R = 4-CF_3 **149i**: R = 4-I
149d: R = 4-CN **149j**: R = 2-I
149e: R = 4-NO_2 **149l**: R = 4-CH_3
149f: R = 2-Cl

Figure 3.24 (a) 2-Substituted benzothiazoles (**148a–f**) and (b) 2-(benzylsulfonyl) benzothiazole derivatives (**149a–k**).

When tested against *A. niger*, the antifungal effectiveness of **148c** (100 μg/mL: 25.0 mm; 200 μg/mL: 29.3 mm) and **148e** (100 μg/mL: 27 mm; 200 μg/mL: 28.6 mm) was similar to that of the reference medication voriconazole (100 μg/mL: 27 mm; 200 μg/mL: 28.6 mm) [80]. Similar inhibitory effects were also observed against *C. albicans*. Furthermore, benzoxazole compounds substituted at the 2-position were also tested for antifungal potential. However, they exhibited lesser activity compared to their 2-substituted benzothiazole counterparts.

A series of 2-(benzylsulfonyl)benzothiazole derivatives (**149a–k**) were also synthesized by Ballari *et al.*, [147] and their antifungal effectiveness against six phytopathogenic fungi—*B. cinerea*, *F. oxysporum*, *A. fumigatus*, *A. niger*, *Aspergillus terreus*, and *Aspergillus ustus*—was examined. Compared to their unoxidized counterparts, these sulfonyl derivatives demonstrated more potent antifungal properties. Particularly against *A. fumigatus* and *A. niger*, Compounds **149f** (IC_{50} = 3–155 μM) and **149k** (IC_{50} = 2.3–39 μM) showed the most action.

3.2.3 The Antiviral Activity and SAR of N-Heterocyclic Compounds

There have been various important advances in therapy of acquired immunodeficiency syndrome patients with HIV-1 virus (AIDS) with the help of highly active antiretroviral therapy. An important effect is displayed by nonnucleoside reverse transcriptase inhibitors (NNRTIs) in highly active antiretroviral therapy because of their better antiviral activity, low cytotoxicity, and very high specificity. The longtime effect of NNRTIs is reduced because of drug resistance showcased by mutated strains of the virus. Thus, there is an urgent need for the development of new heterocyclic compounds to overcome the drug resistance displayed by these strains.

3.2.3.1 Triazoles

A group of researchers has synthesized 5′-silylated nucleoside derived from 3′-azidothymidine derivatives (Compounds **150–161**) and studied antiviral effect against West Nile virus (WNV) and dengue virus (DENV) (Figure 3.25). The effect of attaching different functionalities on triazole group on antiviral properties was investigated. The modeling studies and competitive S-adenosyl-L-methionine binding assay indicated that these derivatives showed antiviral activity due to binding to methyltransferase (MTase). For reference, lycorine (an alkaloid, i.e., naturally occurring

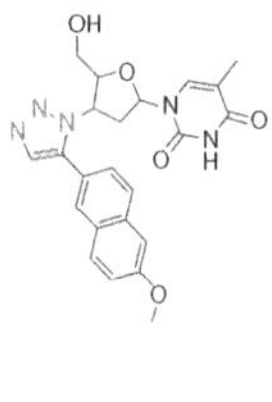

Compound **150**

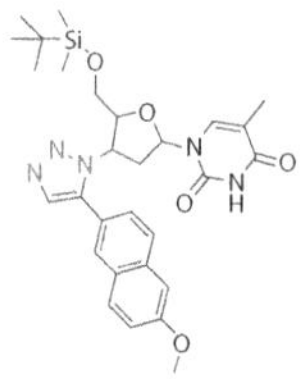

Compound **151**

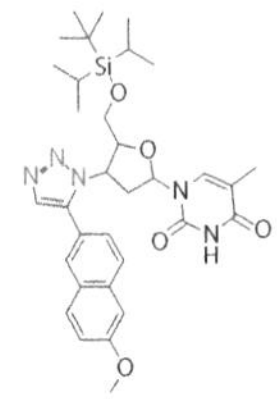

Compound **152**

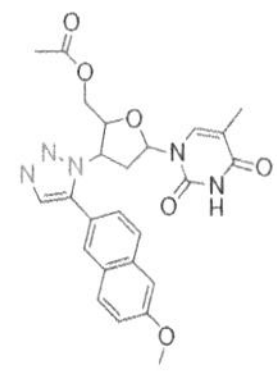

Compound **153**

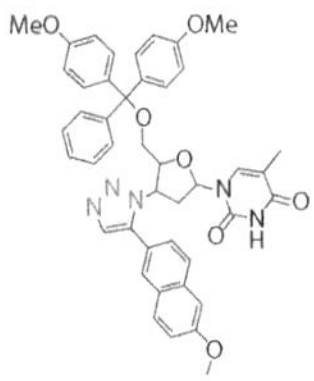

Compound **154**

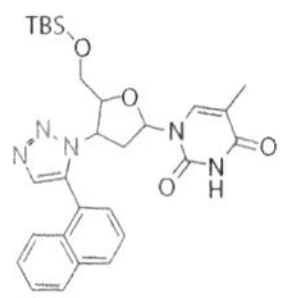

Compound **155**

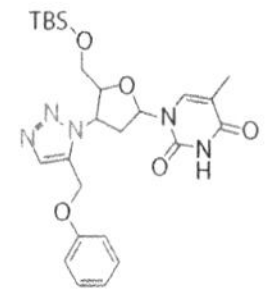

Compound **156**

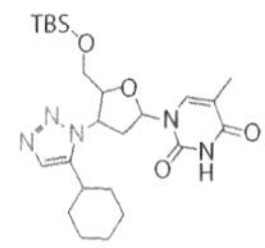

Compound **157**

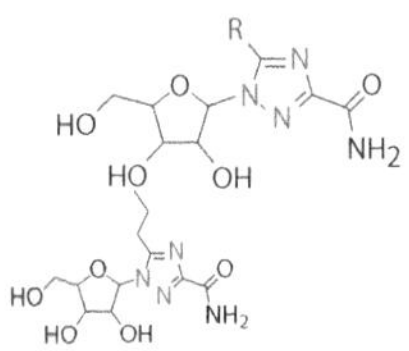

R=alkyl and aryl groups
Compound **158**

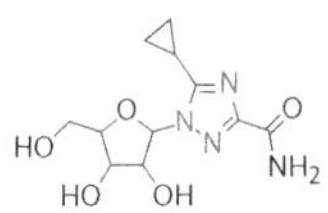

Compound **159**

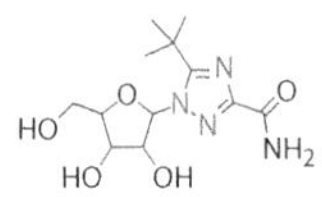

Compound **160**

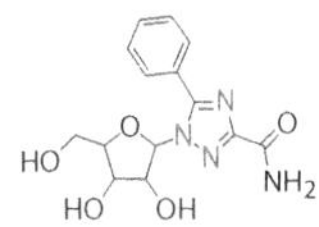

Compound **161**

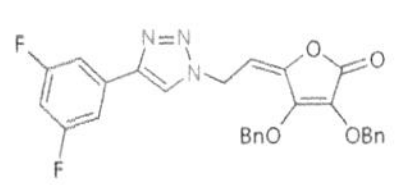

Compound **162**

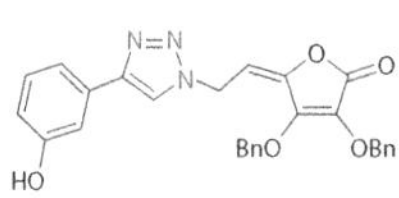

Compound **163**

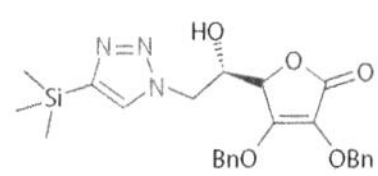

Compound **164**

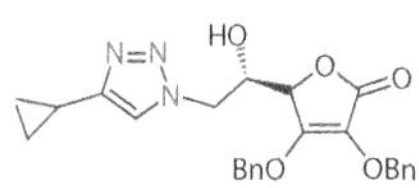

Compound **165**

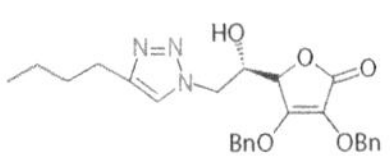

Compound **166**

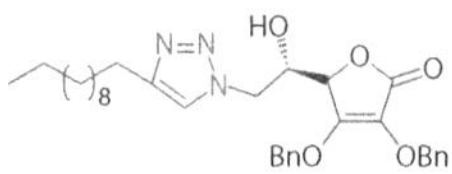

Compound **167**

Figure 3.25 Triazole-based compounds (**150–167**) showing antiviral properties.

organic compound that mostly contains basic nitrogen atoms) for WNV inhibitor (substance that slows down or prevents a particular chemical reaction) and MPA (mycophenolic acid, used to slow down or prevent the replication of DENV) for DENV inhibitor were used.

Compound **150** showed no antiviral activity, indicating a lack of inhibitory effect against both the viruses (WNV and DENV) tested. In contrast, **151** demonstrated effective antiviral properties, successfully inhibiting viral activity. Compound **152**, which contains a bulky silyl group, exhibited strong antiviral efficacy, inhibiting both viruses by 85% to 100%. However, **153**, which lacks the silyl group, displayed no antiviral activity. Compound **154**, featuring a large non-silyl group, demonstrated moderate antiviral activity—23% against WNV and 46% against DENV. Among other triazole derivatives, most compounds with unprotected 5′-hydroxyl groups (R = H), with the exception of Compound **155**, showed neither cytotoxicity nor antiviral activity against WNV and DENV.

Substitution at the 5-position of the triazole ring with a cyclohexyl group significantly enhanced antiviral efficacy against both viruses. Replacing this group with a phenoxymethyl moiety led to reduced activity; for example, compound **156**, which contains phenoxymethyl, showed only 28% inhibition of DENV, whereas compound **157**, bearing a cyclohexyl group, achieved approximately 75% inhibition.

Furthermore, substitution at the triazole ring with 5-alkyl or 5-aryl groups generally reduced antiviral potency. Compound **161**, which features an aryl-substituted derivative, exhibited the strongest activity against the hepatitis C virus. In comparison, the replacement of the alkyl chain with a tert-butyl group in **158** and **160** led to diminished activity. Compound **159**, with a cyclopropyl group attached to the triazole ring, showed no inhibitory effect.

Ethylidene linkers (Compounds **162** and **163**) and 4-substituted-1,2,3-triazole-2,3-dibenzyl-L-ascorbic acid derivatives were evaluated for their antiviral activity against a range of viruses, including cytomegalovirus (CMV), herpes simplex virus types 1 and 2 (HSV-1 and HSV-2), influenza A/H1N1, influenza B, vaccinia virus, varicella-zoster virus (VZV), thymidine kinase–deficient (TK^-), thymidine kinase–positive (TK^+), and adenovirus-2 [148]. Compounds **162** and **163**, which had electron-withdrawing groups at the 4-position of the triazole ring, exhibited enhanced antiviral activity against CMV (Davis and AD-169 strains) and VZV compared to other derivatives.

Compound **164**, containing trimethylsilyl and cyclopropyl groups at the C-4 position of the 1,2,3-triazole ring, showed potent activity against CMV, with an EC_{50} of approximately 12.20 μM against the Davis strain. In comparison, **165** had an EC_{50} of 44.72 μM against the AD-169 strain.

Extension of the alkyl chain at the triazole C-4 position influenced TK^+ and TK^- activity. Compounds **166** and **167**, with C4 and C9 alkyl chains, respectively, showed strong TK-inhibiting effects, with EC_{50} values of 7.31 and 7.76 μM.

3.2.3.2 Thiadiazoles

Thiadiazoles are five-membered heterocyclic compounds containing two nitrogen atoms and one sulfur atom. They exist in several isomeric forms, including 1,2,3-thiadiazole, 1,2,4-thiadiazole, 1,2,5-thiadiazole, and 1,3,4-thiadiazole. Due to their diverse biological and chemical properties, thiadiazole derivatives are widely utilized in research. Among these isomers, 1,2,4- and 1,3,4-thiadiazoles are reported to exhibit greater biological activity compared to the others.

A series of five-membered N-heterocyclic compounds featuring halogen-substituted aryl groups at the C-5 position of both thiazolidinone and thiadiazole scaffolds were synthesized to develop inhibitors against the **DENV** [149]. Compounds **169**, **170**, and **172** emerged as the most potent, primarily due to the 3-fluorophenyl group attached to the 4-thiazolidinone core (Figure 3.26). Compound **168**, bearing a 2-fluorophenyl group at the C-5 position of the thiadiazole ring and a 3-fluorophenyl group on the thiazolidinone ring, demonstrated antiviral activity with an EC_{50} of 7.2 ± 0.3 μM against DENV.

Substituting fluorine with chlorine on the phenyl ring linked to the thiadiazole increased antiviral activity nearly threefold. Likewise, replacing the 2-fluoro group with a 2-chloro group on the 1,3,4-thiadiazole ring

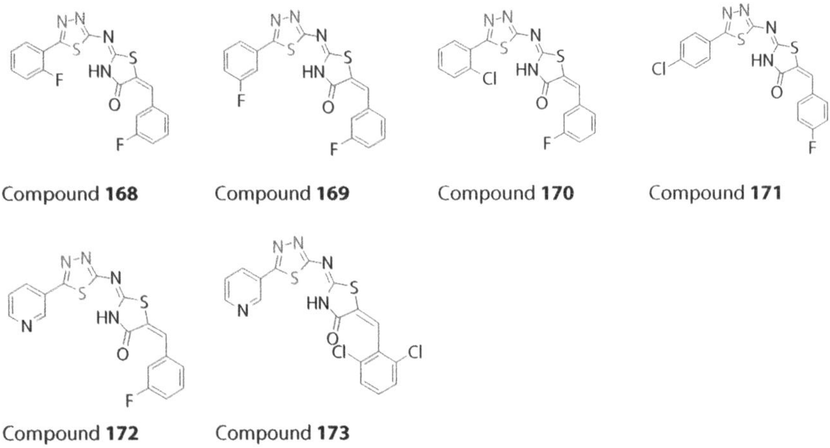

Compound **168** Compound **169** Compound **170** Compound **171**

Compound **172** Compound **173**

Figure 3.26 Thiadiazoles derivatives (**168–173**) showing antiviral properties.

enhanced activity by approximately 3.5 times compared to Compounds **168** and **170**. However, introducing a 2,6-dichlorophenyl group (**173**) in place of the 3-fluorophenyl on the thiazolidinone core significantly reduced activity, by nearly fivefold.

Among the tested compounds, **172** (IC_{50} = 5.5 ± 0.4 μM) and **173** (IC_{50} = 29.3 ± 0.8 μM) showed notable efficacy. According to SAR data, 3-fluorobenzylidene at the C-5 position of thiazolidinone provided comparable antiviral activity. Additionally, 2-chlorophenyl and 3-fluorophenyl substituents at the C-5 position of the 1,3,4-thiadiazole core demonstrated nearly equal effectiveness against the virus, with IC_{50} values of 2.3 ± 0.5 μM (**169**) and 2.1 ± 0.4 μM (**170**), respectively.

3.2.3.3 *Pyrazoles*

In 2016, Saudi *et al.* synthesized two series of compounds: pyrazine-2,3-dicarboxamides (**174–176**) and phthalimide-linked pyrazoles (**177–180**), which were evaluated for their antiviral activity against DENV and yellow fever virus (YFV) (Figure 3.27). Among them, compound A demonstrated promising antiviral activity, with EC_{50} values of 2.5 μM for DENV and 3.5

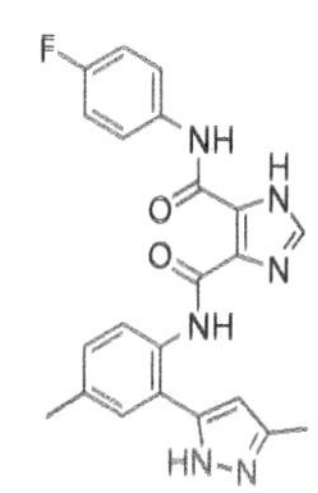

Compound **A**

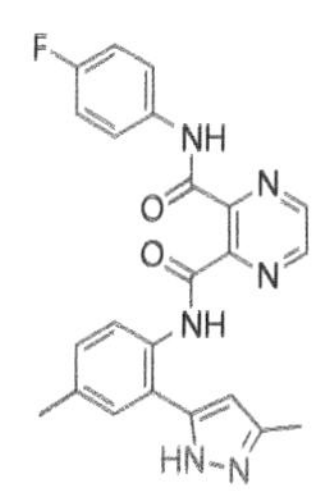

Compound **174**

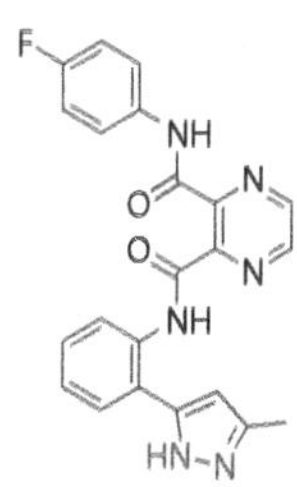

Compound **175**

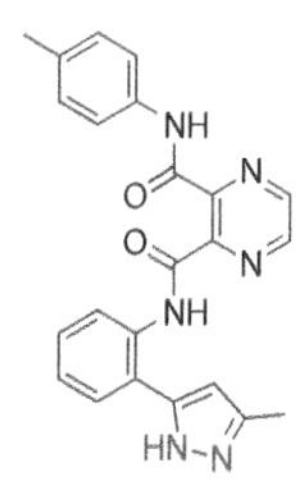

Compound **176**

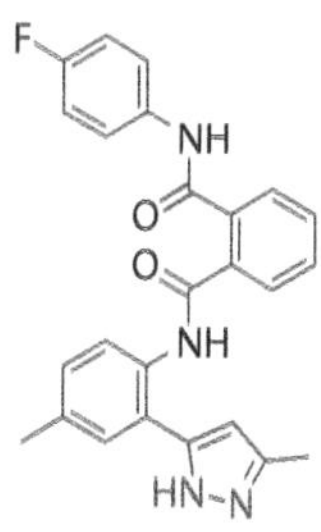

Compound **177**

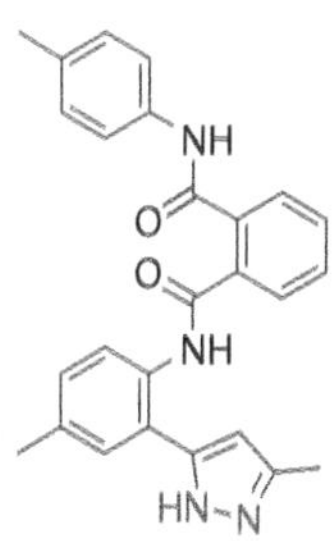

Compound **178**

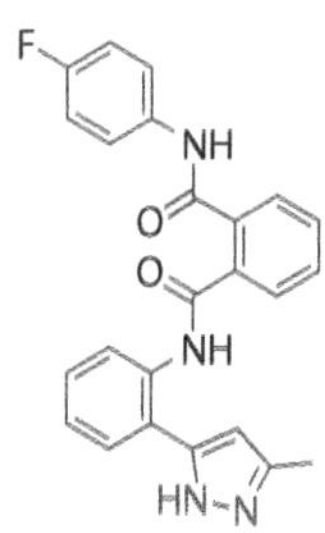

Compound **179**

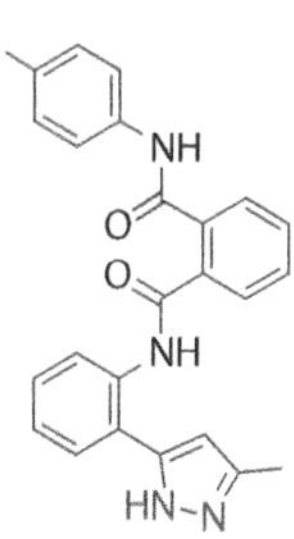

Compound **180**

Figure 3.27 Pyrazole derivatives (**174–180**) showing antiviral properties.

μM for YFV. However, Compound **177**, which contains a phthalimide moiety, showed the most potent effect with an EC_{50} of 0.5 μM, making it five times more effective than the reference drug. In contrast, Compound **174**, bearing a pyrazine-2,3-dicarboxamide group, was significantly less active, with an EC_{50} of 26.5 μM, nearly 10 times less effective than the reference compound against DENV.

SAR analysis revealed that replacing a para-fluoro group with a para-methyl group notably affected antiviral activity against DENV, but had little impact on YFV. Specifically, replacing the fluorine in **177** with a methyl group to produce Compound **178** led to a 58-fold decrease in DENV activity (EC_{50} = 29 μM).

Additionally, Compound **174** (pyrazine-substituted, EC_{50} = 7.8 μM) showed lower activity compared to Compound **179** (phthalimide-substituted, EC_{50} = 12 μM). Altering the linker groups in Compounds **176** and **180** also resulted in reduced activity against both viruses. Antiviral activity improved when a non-substituted phenyl ring was used instead of a methyl-substituted phenyl ring; for example, Compound **175** was 13 times more active than Compound **174**.

Moreover, substituting the electron-withdrawing fluorine group in **179** with an electron-donating methyl group in **180** led to enhanced antiviral activity against YFV, although the effect against DENV was less pronounced. Specifically, **179** had EC_{50} values of >120 μM (DENV) and 12 μM (YFV), whereas Compound **180** showed EC_{50} values of 14.1 μM (DENV) and 5.8 μM (YFV).

3.2.3.4 Pyrimidine

In 2018, Jin *et al.* synthesized a series of substituted piperazine-1-yl-pyrimidine derivatives as potential HIV-1 inhibitors (Figure 3.28) [150]. These compounds featured various N-substituted aromatic groups attached to the piperazine ring and were evaluated for their antiviral efficacy against resistant strains of HIV-1. Among the synthesized compounds, **181** and **182** exhibited significant inhibitory activity, with EC_{50} values of 3.36 and 31.50 μM, respectively.

Molecular docking studies of Compound **181** provided valuable insights into the development of new HIV-1 antiviral agents. It was found that altering the position of nitrophenyl and amidobenzene substituents from meta to para reduced antiviral potency. Additionally, the introduction of trifluoromethyl or methoxy groups into the phenyl ring resulted in a marked decrease in activity. Compounds containing methyl or halogen substituents also displayed reduced efficacy compared to other analogs.

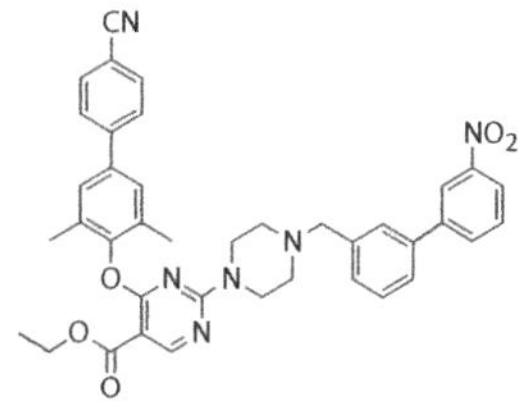

Compound **181**

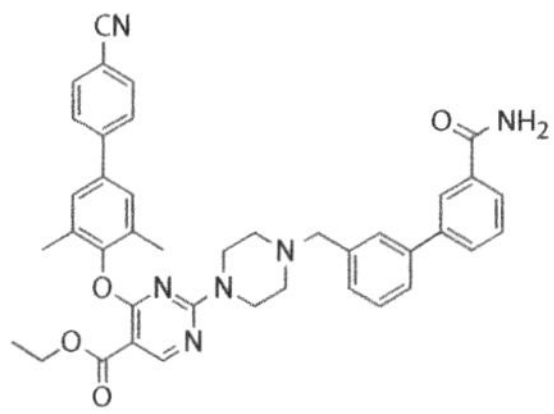

Compound **182**

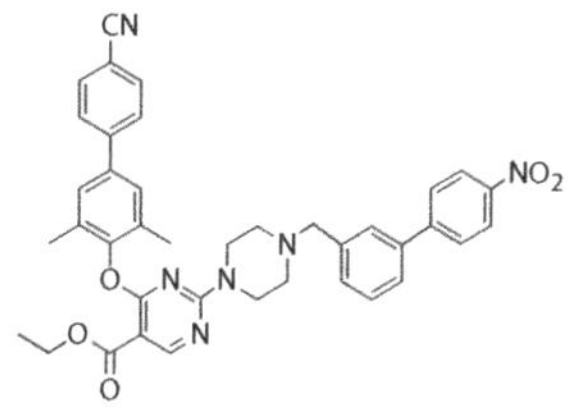

Compound **183**

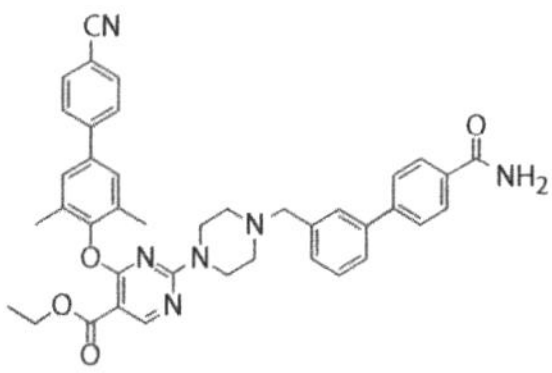

Compound **184**

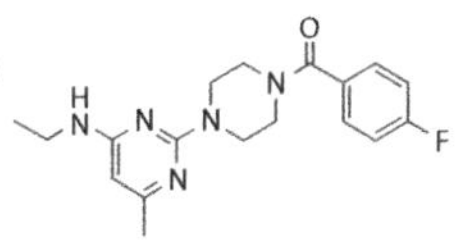

Compound **185**

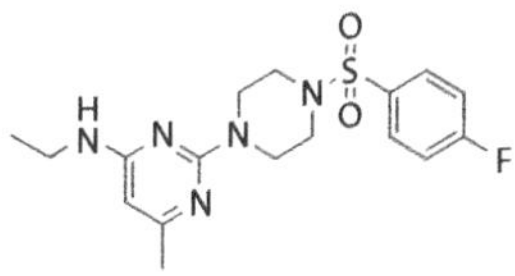

Compound **186**

Figure 3.28 Pyrimidine derivatives (**181–186**) showing antiviral properties.

In 2020, Moesslacher *et al.* synthesized a novel class of Chikungunya virus (CHIKV) inhibitors, including a promising lead, Compound **185** [151]. This molecule featured a carbonyl linker between the piperazine and phenyl rings, along with a fluorine atom on the phenyl ring, which contributed to its strong activity (EC_{50} = 3.95 μM). However, replacing the carbonyl with a sulfoxide group reduced its antiviral effect by half. For example, **189**, which contains the sulfoxide linkage, showed a weaker EC_{50} of 8.68 μM against CHIKV.

As shown in Figure 3.28, pyrimidine derivatives **181–186** exhibit varying levels of antiviral activity. Substituting the fluorine atom in **189** with a nitrile group at the para-position led to a significant loss in efficacy. A range of para-substituted phenyl groups, including bromine, fluorine, chlorine, and iodine, were also evaluated. Although these halogenated compounds retained some antiviral activity, all showed weaker potency than **185**.

3.2.3.5 *Imidazole*

Researchers have synthesized a series of compounds derived from two nucleoside-based imidazole derivatives that demonstrated antiviral activity against the DENV virus, specifically Compounds **187** and **188** (Figure 3.29). All synthesized compounds exhibited some degree of antiviral activity, with Compounds **187** and **181** showing the strongest effects among them. Compound **189**, which included a sulfur atom and a nitrile

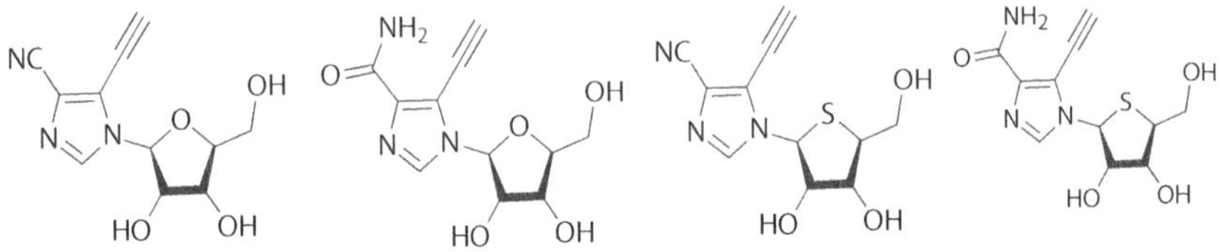

Compound **187** Compound **188** Compound **189** Compound **190**

Figure 3.29 Imidazole derivatives (**187**–**190**) showing antiviral properties.

substituent on the imidazole ring, showed enhanced antiviral activity. However, when **189** was compared to the standard reference drug, ribavirin, with an IC_{50} of 6.57 μM, did not showed better activity.

Compound **190**, which was generated by replacing the nitrile group with an amide, exhibited strong antiviral activity, ranking as the second most potent in the series. Its IC_{50} value was 14.63 μM, making it approximately twofold as weak as ribavirin.

Another set of compounds was developed by substituting the hydroxy group with PPh_3, but these compounds did not exhibit any antiviral activity.

3.2.3.6 *Indole*

Indole derivatives have been widely studied for antiviral activity (Figure 3.30). In 2016, Brigg *et al.* synthesized a series of indole-based derivatives, which were identified as potent HIV-1 NNRTIs (Figure 3.30) [152]. Two main indole derivatives were explored: indole sulfides and indole methyl ethers. The synthesized compounds also varied in the functional groups attached to the phenyl ring linked to the indole core. Among them, **191** showed the highest antiviral activity, primarily due to the presence of a methyl ether between the indole and phenyl rings, along with an additional substituent on the phenyl ring (IC_{50} = 0.016 μM).

Replacing the methyl group with hydrogen in **192** led to a twofold decrease in activity (IC_{50} = 0.034 μM). Introducing halogen or methyl groups on the phenyl ring did not significantly affect activity; for example, **193**, with a meta-bromine substituent, maintained strong activity (IC_{50} = 0.02 μM). However, replacing the methyl ether with a methyl sulfide group resulted in reduced potency **194**, which contains a methyl sulfide and a meta-chlorine on the phenyl ring, showed lower activity than **195** with IC_{50} of 0.042 μM.

The impact of an ester group at the 2-position of the indole ring was also assessed. Its removal drastically reduced antiviral activity, as seen in **196**,

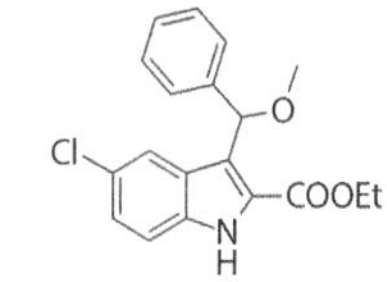

Compound **191**

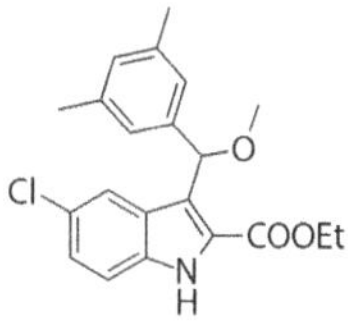

Compound **192**

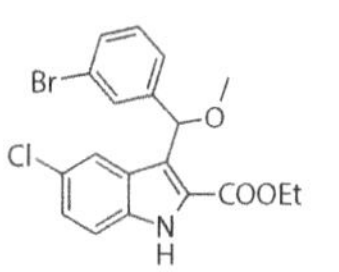

Compound **193**

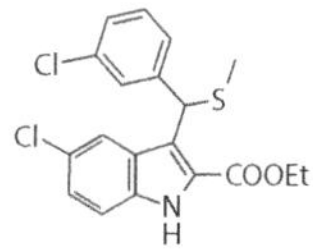

Compound **194**

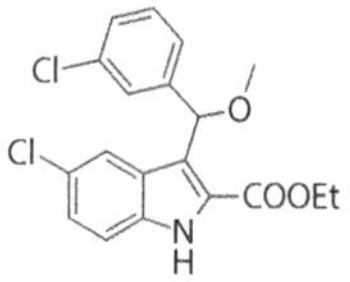

Compound **195**

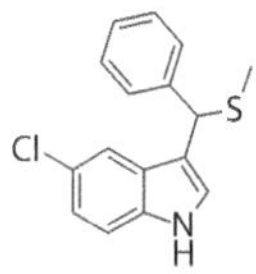

Compound **196**

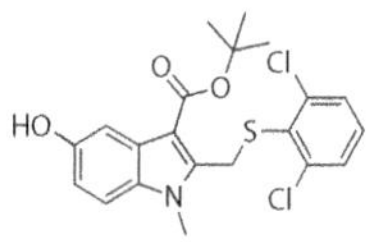

Compound **197**

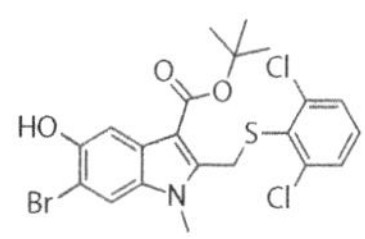

Compound **198**

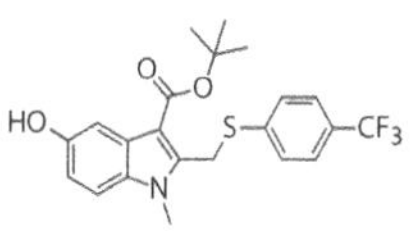

Compound **198**

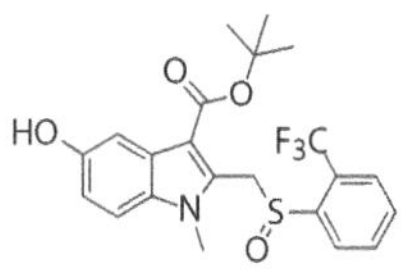

Compound **199**

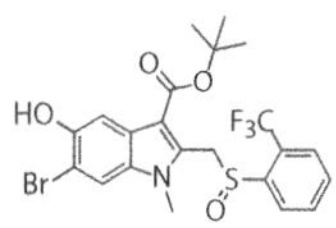

Compound **200**

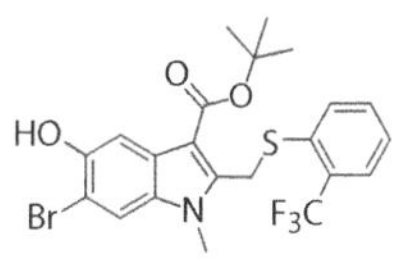

Compound **201**

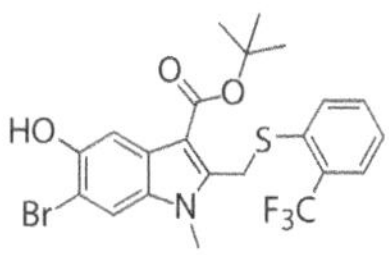

Compound **202**

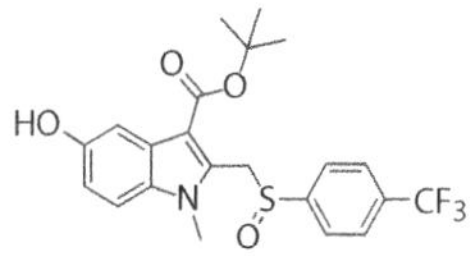

Compound **203**

Figure 3.30 Indole derivatives (**191–203**) showing antiviral properties.

which exhibited an IC_{50} of 9.95 μM, making it 255-fold less potent than its ester-containing counterpart.

In another example, Scuotto *et al.* synthesized two distinct groups of compounds based on the Arbidol scaffold, which contains an indole ring, and evaluated their antiviral activity against the CHIKV [153]. The key difference between these two groups was the presence of either a sulfur or sulfoxide substituent at the 2-position of the indole ring. Substituents at the C-6 position of the indole ring and on the phenyl ring were found to play a crucial role in antiviral efficacy.

A total of 12 compounds exhibited strong inhibitory activity against CHIKV, with EC_{50} values ranging from 2.9 to 20.0 μM, making them 2- to 17-fold more effective than Arbidol. Among the sulfur-containing derivatives, **197**, which contains a 2,6-dichlorophenyl group, a sulfur atom, and a hydrogen atom at the C-6 position, showed the highest potency with an

EC_{50} of 2.9 μM. When the hydrogen at C-6 was replaced by bromine (compound **198**), the antiviral activity decreased by half.

Compound **199**, featuring a sulfur atom and a para-CF_3-substituted phenyl ring, showed an EC_{50} of 3.9 μM, which slightly improved to 3.4 μM when the CF_3 group was moved to the ortho position. However, in a similar molecule with a bromine atom at the C-6 position of the indole ring, the ortho-CF_3 substitution led to reduced activity (EC_{50} = 9.6 μM).

It was also observed that phenyl rings bearing chlorine or fluorine at the para-position were more effective than those substituted with bromine on the indole ring.

Compound **200** demonstrated exceptional anti-CHIKV activity, being approximately 10 times more potent with an EC_{50} of 6.5 ± 1 μM. Compounds **201** and **202**, each bearing a bromine at C-6 of the indole and an ortho-CF_3 group on a sulfoxide-linked phenyl ring, displayed similar antiviral activity (EC_{50} = 9.6 μM). Compound **203**, with an EC_{50} of 13.0 μM, was less effective, and it was concluded that the combination of bromine substitution on the indole ring and oxidation of sulfur (from sulfide to sulfoxide) contributed to the reduced antiviral potency.

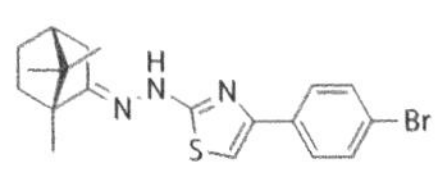

Compound **204**

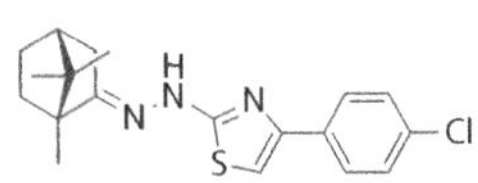

Compound **205**

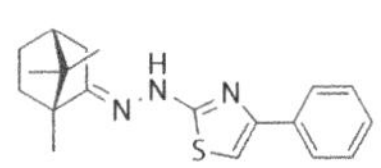

Compound **206**

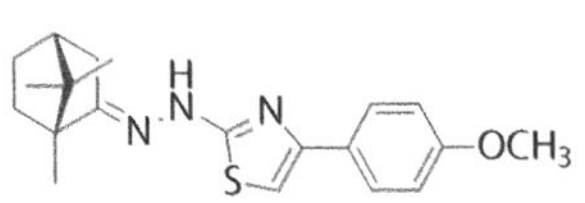

Compound **207**

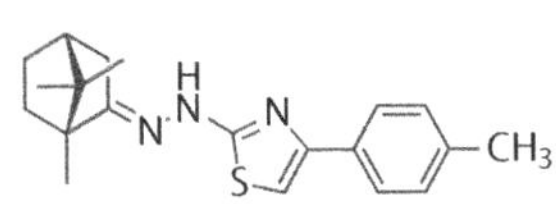

Compound **208**

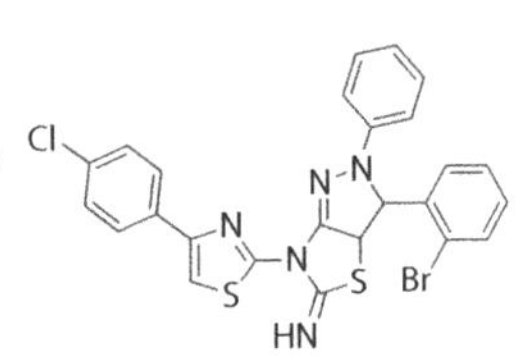

Compound **209**

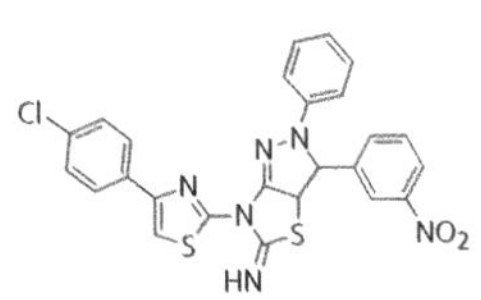

Compound **210**

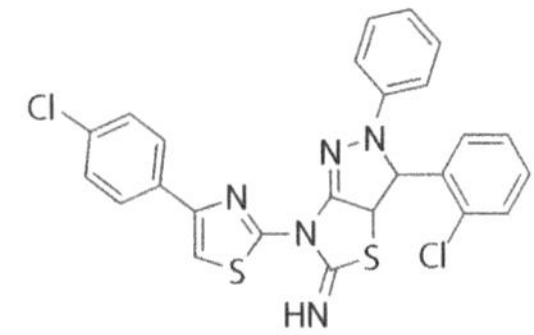

Compound **211**

Figure 3.31 Thiazole derivatives (**204–2011**) showing antiviral properties.

3.2.3.7 Thiazoles

In 2018, Sokolova and colleagues synthesized a series of camphor-based thiazole derivatives (compounds **204–208**) to investigate their potential as inhibitors of *Orthopoxvirus* replication (Figure 3.31) [154]. These five thiazole compounds varied in their para-substituents on the phenyl ring at the C-4 position of the thiazole ring, incorporating both electron-donating and electron-withdrawing groups such as nitro, methyl, fluoro, bromo, and chloro [7].

Among these, compounds bearing 4-fluorophenyl and 4-nitrophenyl substituents at the C-4 position of the thiazole ring exhibited higher antiviral activity compared to cidofovir, a standard antiviral. Compound **204** (with a bromo group) and Compound **205** (with a chloro group) showed potent activity, recording IC_{50} values of 2.4 ± 1.2 and 2.7 ± 0.5 μM, respectively.

In contrast, **206**, which featured an unsubstituted phenyl ring at the same position, showed moderate activity, but still outperformed the 4-fluorophenyl derivative, with an IC_{50} of 3.2 ± 1.1 μM.

Compounds **207** and **208**, containing electron-donating methoxy and methyl groups at the para-position of the phenyl ring, also exhibited antiviral activity with IC_{50} values of 3.7 ± 1.4 and 4.5 ± 1.1 μM, respectively. All active compounds were found to be 8 to 17 times more effective than the reference drug.

Separately, Bhusare *et al.* synthesized pyrazolo[3,4-d]thiazole hybrids to evaluate their anti–HIV-1 NNRTI potential [7]. Following the synthesis of 2-amino-4-aryl thiazoles, the team also attempted to create 5-arylidine and thiazol-2-yl-substituted 2-imino-thiazolidin-4-ones, although these lacked antiviral activity [7].

Among the successful derivatives, Compound **209** stood out with a 2-bromophenyl group at the C-2 position of the pyrazolo[3,4-d] ring and a para-chlorophenyl group at the C-4 position of the thiazole ring. It exhibited strong activity against HIV-1 IIIB (EC_{50} = 0.74 ± 0.02 μM) and also showed inhibition against HIV-1 ADA5 (EC_{50} = 1.08 ± 0.04 μM). Other active compounds included **210 and 211**, in which the 2-bromophenyl group was replaced with 3-nitrophenyl and 2-chlorophenyl, respectively. These compounds demonstrated anti–HIV-1 activity with EC_{50} values of 1.04 ± 0.12 and 1.12 ± 0.14 μM against HIV-1 IIIB, and 0.86 ± 0.09 and 0.95 ± 0.01 μM against HIV-1 ADA5, respectively [7].

3.2.4 Application of Heterocyclic Compound as Anthelmintics Agent

Anthelmintic agents are those drugs, which can be used to treat parasitic infections caused by helminths, which are parasitic worms. There are various types of worms such as tapeworm (cestodes), roundworms (nematodes) which can cause infection. These drugs are proven to be indispensable because of rising cases of parasitic diseases, which stands as a major global health concern.

Late 19th century marked the inception of development of anthelmintic agents, although these agents were not very potent and often had toxic effects. In the 20th century, the synthesis of drugs began with the selection and isolation of compounds, which were potent against parasites. The discovery of thiabendazole was considered a major breakthrough.

3.2.4.1 Pyrrolidine

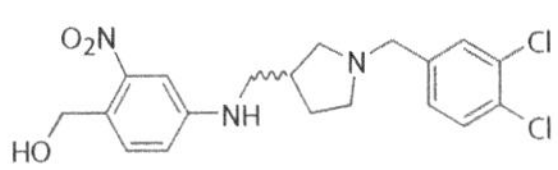

Compound **212**

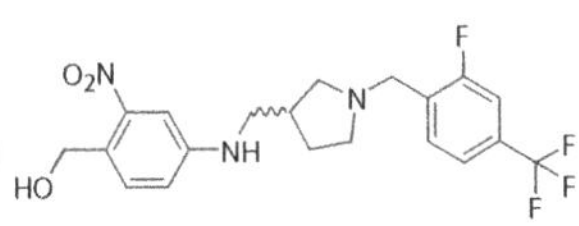

Compound **213**

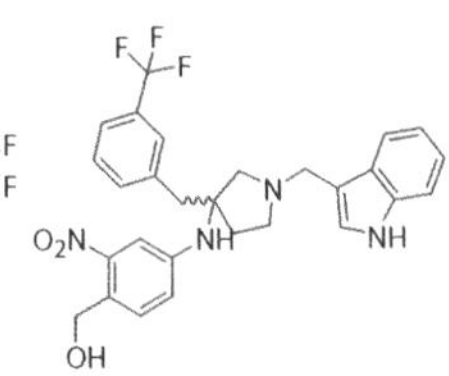

Compound **214**

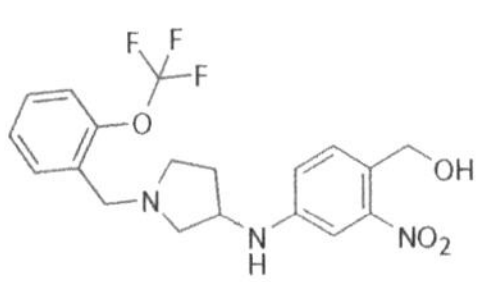

Compound **215**

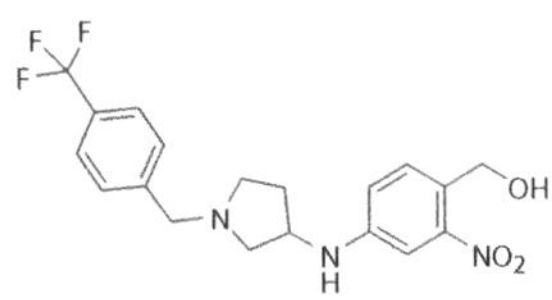

Compound **216**

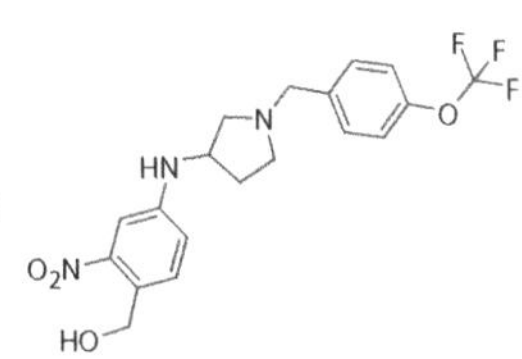

Compound **217**

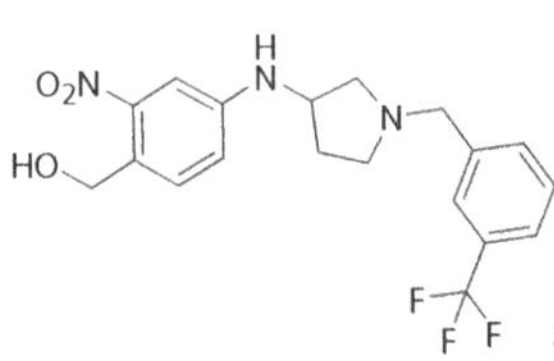

Compound **218**

Figure 3.32 Pyrrolidine derivatives (**212–218**) as anthelmintics activity.

In 2020, Guzman and colleagues synthesized various new pyrrolidine analogs (compounds **212–214**) and were assessed *in vitro* and results concluded that the compounds showed remarkable anthelmintics activity [155] (Figure 3.32). Compound **214** showed high potential against *Schistosoma haematobium* and *Schistosoma japonicum*, whereas Compounds **212** and **213** were undetectable.

Fifteen pyrrolidine-based derivatives were synthesized by Rugel *et al.* synthesized and their anthelmintics effects were studied [156]. All the compounds were evaluated against *S. haematobium* and *S. japonicum*. Among them, Compounds **215–218** had moderate effects against *S. haematobium* whereas the effects of other compounds were undetectable.

3.2.4.2 Imidazole

In 2014, Wang *et al.* synthesized an imidazolidine-based compound and conducted *in vivo* studies against *Schistosoma mansoni* (SM) [157] (Figure 3.33). The most potent compound, **219**, demonstrated significant anthelmintic activity, reducing juvenile worm populations by 95% and adult worms by 64% at a dosage of 100 mg/kg. Mechanistic studies using a ligand competition assay revealed that the compound's activity was not associated with androgen receptor binding [39].

In 2019, Osei and colleagues evaluated an imidazole-based alkaloid conjugate, Compound **220**, for its activity against SM cercariae [158]. The compound exhibited time-dependent lethality, with minimum lethal concentration (MLC) values recorded as 33.7 μM (1 h), 25.0 μM (1.5 h), 20.2 μM (2 h), 16.44 μM (2.5 h), and 15.08 μM (3 h).

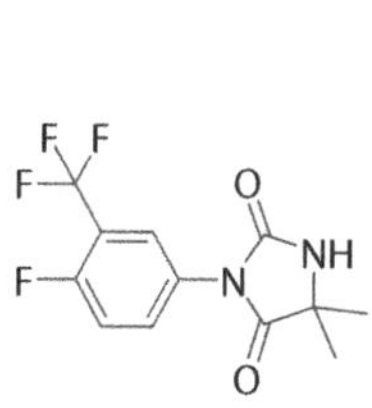

Compound **219**

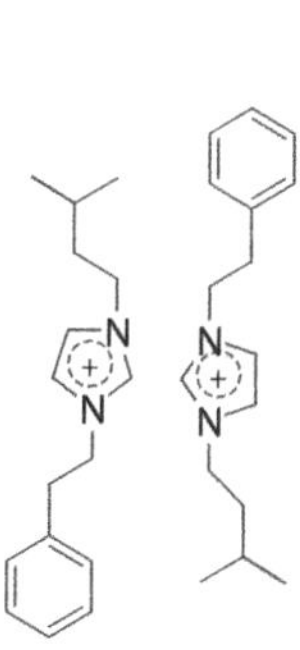

Compound **220**

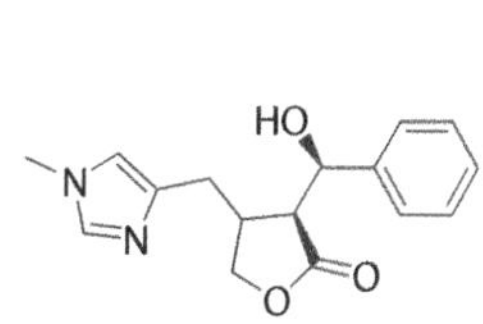

Compound **221**

Figure 3.33 Imidazole derivatives (**219–221**) as anthelmintics agent.

In 2015, Guimarães *et al.* investigated the anthelmintic potential of the alkaloid epiisopiloturine using murine models [159]. Compound **221** was tested against both juvenile and adult stages of SM worms. *In vivo* results showed that the compound reduced juvenile worm burden by 50.2% at 40 mg/kg and 46.3% at 300 mg/mL. For adult worms, mortality rates were 70% at 40 mg/kg, 30% at 100 mg/kg, and 46.7% at 300 mg/kg.

3.2.4.3 Triazole

In 2020, Kalinin and colleagues synthesized a series of triazole-based Compounds **222–224** and evaluated them against *Schistosoma mansoni* [160] (SM) (larvae, schistosomula) (Figure 3.34). At 10 μm, derivatives of compounds **222–224** eliminated 64% to 90% SM worms. At 20 μm, Compound **224** could eliminate 49% of SM worms, whereas the activity of Compounds **222** and **223** were undetectable.

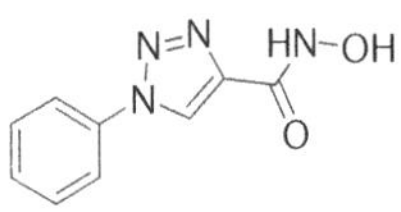
Compound **222**

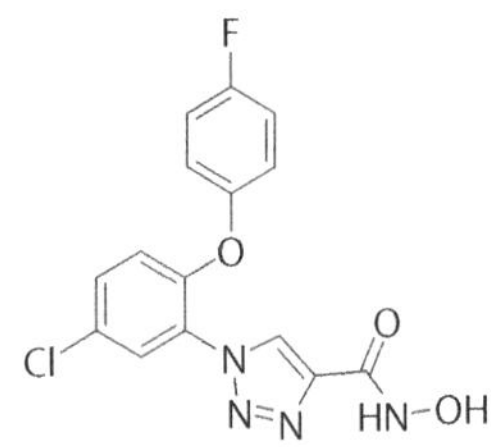
Compound **223**

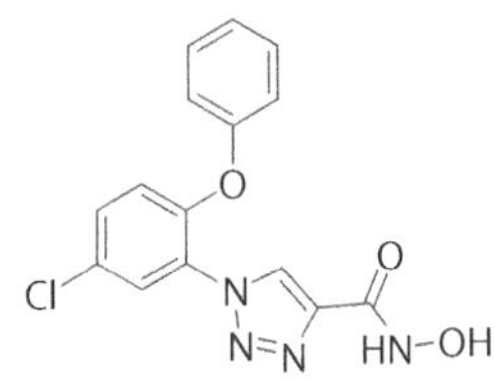
Compound **224**

Figure 3.34 Triazole derivatives (**222–224**) as anthelmintics agent.

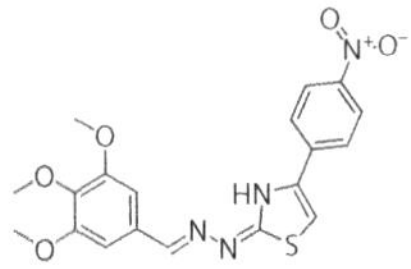
Compound **225**

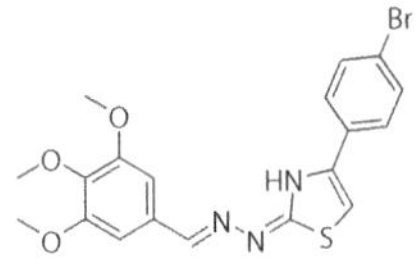
Compound **226**

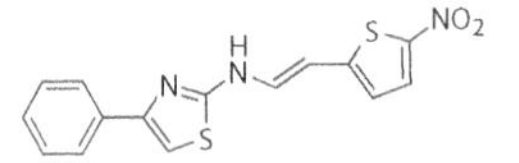
Compound **227**

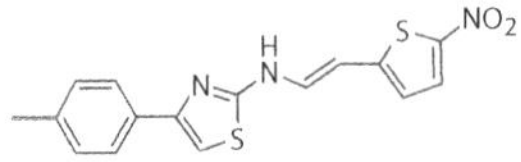
Compound **228**

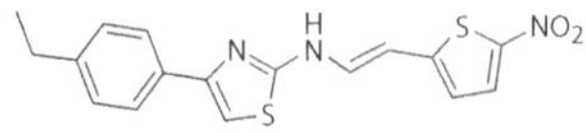
Compound **229**

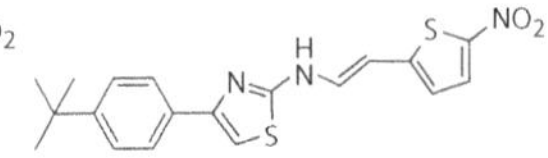
Compound **230**

Figure 3.35 Thiazole derivatives (**225–230**) as anthelmintics agent.

3.2.4.4 *Thiazole*

In 2019, Pereira and colleagues investigated the anthelmintic activity of aryl-based thiazole compounds (**225** and **226**) against SM [161] (Figure 3.35). These compounds were capable of eliminating between 10% to 40% of SM worms. At a concentration of 6.25 μM, both compounds individually demonstrated approximately 30% lower worm-killing efficiency compared to their activity at 50.0 μM. However, when used in combination at the same lower concentration (**6.25 μM**), they exhibited a 70% reduction in efficacy. Despite this, Compounds **225** and **226** still showed promising potential as antiparasitic agents.

Santiago *et al.* reported the synthesis and evaluation of a series of 4-phenyl-2-aminothiazole derivatives (**227–230**) targeting adult SM worms [162]. Among these, the compounds displayed notable IC_{50} values: 29.44 μM (**227**), 60.17 μM (**228**), 33.96 μM (**229**), and 38.02 μM (**230**), demonstrating varied but significant levels of activity.

3.3 Conclusion

The emergence of drug-resistant bacteria has made the management of pathogenic infections increasingly difficult, particularly in cases involving minor surgeries, chemotherapy, or organ transplantation. Heterocyclic compounds, known for their structural diversity, straightforward synthesis, and wide range of biological activities, have been widely studied for tackling pathogens causing infection. Numerous studies have demonstrated that both naturally occurring and synthetically derived heterocycles exhibit significant activity against various microorganisms. Nevertheless, the continual evolution of resistant bacterial strains underscores the need for the rational design and development of new heterocyclic compounds capable of addressing the growing challenge of AMR.

AI disclosure statement: The author acknowledges the use of AI-based tools for minor language editing or grammatical correction. However, all intellectual contributions, analyses, and conclusions presented in this chapter are entirely the author's own.

References

1. Hutching, M.I., Truman, A.W., Wilkinson, B., Antibiotics: past, present and future. *Curr. Opin. Microbiol.*, 51, 72–80, 2019.
2. Gaynes, R., The Discovery of Penicillin-New Insights After More Than 75 Years of Clinical Use. *Emerg. Infect. Dis.*, 23, 849–853, 2017.
3. Prasad, P. and Gupta, S., Nanobioconjugates: weapons against antibacterial resistance. *ACS Appl. Bio Mater.*, 3, 8271–8285, 2020.
4. Prasad, P., Gupta, A., Sasmal, P.K., Aggregation-induced emission active metal complexes: A promising strategy to tackle bacterial infections. *Chem. Commun.*, 57, 174–186, 2021.
5. Akram, F., Imtiaz, M., Haq, I., Emergent crisis of antibiotic resistance: A silent pandemic threat to 21st century. *Microb. Pathog.*, 174, 105923, 2023.
6. Gupta, A., Prasad, P., Gupta, S., Sasmal, P.K., Simultaneous ultrasensitive detection and elimination of drug-resistant bacteria by cyclometalated irid ium (III) complexes. *ACS Appl. Mater. Interfaces*, 12, 35967–35976, 2020.
7. Mermer, A., Keles, T., Sirin, Y., Recent studies of nitrogen containing heterocyclic compounds as novel antiviral agents: A review. *Bioorg. Chem.*, 114, 105076, 2021.
8. Kabir, E. and Uzzaman, M., A review on biological and medicinal impact of heterocyclic compounds. *Results Chem.*, 4, 100606, 2022.
9. Pibiri, I., Recent Advances: Heterocycles in Drugs and Drug Discovery. *Int. J. Mol. Sci.*, 25, 9503, 2024.
10. Zinner, S.H., Antibiotic use: present and future. *New Microbiol.*, 30, 321–325, 2007.
11. Mallappa, M., Choudhary, N., Qasim, M.T., Zairov, R., Patel, A., Yadav, V.K., Jangir, M., Recent advances in the synthesis of nitrogen-containing heterocy clic compounds *via* multicomponent reaction and their emerging biological applications: a review. *J. Iran. Chem. Soc.*, 22, 1–33, 2025.
12. Borah, P., Hazarika, S., Chettri, A., Sharma, D., Deka, S., Venugopala, K.N., Shinu, P., Al-Shar'i, N.A., Bardaweel, S.K., Deb, P.K., Bagchi, D., Heterocyclic compounds as antimicrobial agents, in: *Viral, Parasitic, Bacterial, and Fungal Infections*, 2023, https://doi.org/10.1016/B978-0-323-85730-7.00068-0.
13. Heravi, M.M. and Zadsirjan, V., Prescribed drugs containing nitrogen het erocycles: an overview. *RSC Adv.*, 72, 44247–44311, 2020.
14. Spencer, A.C. and Panda, S.S., DNA Gyrase as a target for quinolones. *Biomedicines*, 11, 371, 2023.
15. Sauvage, E., Kerff, F., Terrak, M., Ayala, J.A., Charlier, P., The penicillin-bind ing proteins: structure and role in peptidoglycan biosynthesis. *FEMS Microbiol. Rev.*, 32, 234–258, 2008.
16. Kumar, S. and Narasimhan, B., Therapeutic potential of heterocyclic pyrimi dine scaffolds. *Chem. Cent. J.*, 12, 38, 2018.
17. Gou, K., Heterocyclic compounds: The Diverse World of Ringed Molecules. *J. Med. Org. Chem.*, 6, 108–110, 2023.

18. Rusu, A., Moga, I.-M., Uncu, L., Hancu, G., The Role of Five-Membered Heterocycles in the Molecular Structure of Antibacterial Drugs Used in T herapy. *Pharmaceutics*, 15, 2554, 2023.
19. Aleam, R.H.A., George, R.F., Georgey, H.H., Abdel-Rahman, H.M., Bacterial virulence factors: a target for heterocyclic compounds to combat bacterial resistance. *RSC Adv.*, 11, 36459–36482, 2021.
20. Gholap, S.S., Pyrrole: An emerging scaffold for construction of valuable ther apeutic agents. *Eur. J. Med. Chem.*, 110, 13–31, 2016.
21. Aatif, M., Raza, M.A., Javed, K., Nashre-ul-Islam, S.M., Farhan, M., Alam, M.W., Potential Nitrogen-Based Heterocyclic Compounds for Treating Infectious Diseases: A Literature Review. *Antibiotics*, 11, 1750, 2022.
22. Ivan, B.-C., Barbuceanu, S.-F., Hotnog, C.M., Anghel, A.I., Ancuceanu, R.V., Mihaila, M.A., Brasoveanu, L.I., Shova, S., Draghici, C., Olaru, O.T., Nitulescu, G.M., Dinu, M., Dumitrascu, F., New Pyrrole Derivatives as Promising Biological Agents: Design, Synthesis, Characterization, *In Silico,* and Cytotoxicity Evaluation. *Int. J. Mol. Sci.*, 23, 8854, 2022.
23. Rusu, A., Oancea, O.-L., Tanase, C., Uncu, L., Unlocking the Potential of Pyrrole: Recent Advances in New Pyrrole-Containing Compounds with Antibacterial Potential. *Int. J. Mol. Sci.*, 25, 12873, 2024.
24. Cheng, C., Liu, Y., Song, H., Pan, L., Li, J., Qin, Y., Li, R., Marinopyrrole Derivatives as Potential Antibiotic Agents against Methicillin-Resistant Staphylococcus aureus (II). *Mar. Drugs*, 11, 2927–2948, 2013.
25. Largani, T.H., Imanzadeh, G., Zahri, S., Noroozi Pesyan, N., Sahin, E., Afacile synthesis, E., and antibacterial activity of novel pyrrolo[3,4-b]quino lin-2(3H)-yl) benzamides. *Green Chem. Lett. Rev.*, 10, 387–392, 2017.
26. Zhang, F., Braun, D.R., Chanana, S., Rajski, S.R., Bugni, T.S., Phallusialides, A.E., Pyrrole-Derived Alkaloids Discovered from a Marine-Derived Micromonospora sp. Bacterium Using MS-Based Metabolomics Approaches. *J. Nat. Prod.*, 82, 3432–3439, 2019.
27. Rawat, P., Gautam, A., Singh, R.N., Synthesis, spectral, structural and anti microbial activities of Ethyl-4-{[-(1-(2-(4-nitrobenzoyl)hydrazono)eth yl]}-3,5-dimethyl-1H-pyrrole-2-carboxylate. *J. Mol. Struct.*, 1255, 132405, 2022.
28. Veselov, M.S., Ivanenkov, Y.A., Yamidanov, R.S., Osterman, I.A., Sergiev, P.V., Aladinskiy, A.V., Aladinskaya, V.A., Ayginin, A.A., Skvortsov, D.A., *et al.*, Identification of Pyrrolo-Pyridine Derivatives as Novel Class of Antibacterials. *Mol. Diversity*, 24, 233–239, 2020.
29. Pal, P., Recent advances in syntheses and antibacterial activity of novel furan derivatives, in: *Heterocyclic Chemistry - New Perspectives*, 2024, https://doi.org/10.5772/intechopen.1004117.
30. Kalyaev, M.V., Ryabukhin, D.S., Borisova, M.A., Ivanov, A.Y., Boyarskaya, I.A., Borovkova, K.E., Nikiforova, L.R., Salmova, J.V., Ul'yanovskii, N.V., Kosyakov, D.S., Vasilyev, A.V., Synthesis of 3-Aryl-3-(Furan-2-yl)Propanoic

Acid Derivatives, and Study of Their Antimicrobial Activity. *Molecules*, 27, 4612, 2022.
31. Xu, Z., Zhao, S., Lv, Z., Feng, L., Wang, Y., Zhang, F., Bai, L., Deng, J., Benzofuran derivatives and their anti-tubercular, anti-bacterial activities. *Eur. J. Med. Chem.*, 162, 266–276, 2019.
32. Hassana, A.S., Moustafab, G.O., Morsya, N.M., Abdoud, A.M., Hafez, T.S., Design, synthesis and antibacterial activity of N-Aryl-3- (arylami no)-5-(((5-substituted furan-2-yl)methylene)amino)-1H-pyrazole-4- car boxamide as nitrofurantoin® analogues. *Egypt. J. Chem.*, 63, 4469–4481, 2020.
33. Malladi, S., Nadh, R.V., Babu, K.S., Babu, P.S., Synthesis and antibacterial activity studies of 2,4-di substituted furan derivatives. *Beni-Seuf Univ. J. Appl. Sci.*, 6, 345–353, 2017.
34. Andrade, M.M.S., Protti, I.F., Maltarollo, V.G., *et al.*, Synthesis of arylfuran derivatives as potential antibacterial agents. *Med. Chem. Res.*, 30, 1074–1086, 2021.
35. Obafemi, C.A., Adelani, P.O., Fadare, O.A., Akinpelu, D.A., Famuyiwa, S.O., Synthesis, crystal structure and *in-vitro* antibacterial activity of 2,3a,8b trihy- droxy-3-(thiophen-2-ylcarbonyl)-2-(trifluoromethyl)-2,3,3a,8b-tetra hydro-4Hindeno[1,2-b]furan-4-one. *J. Mol. Struct.*, 1049, 429–435, 2013.
36. Mishra, R., Sachan, N., Kumar, N., Mishra, I., Chanda, P., Thiophene Scaffold as Prospective Antimicrobial Agent: A Review. *J. Heterocycl. Chem.*, 55, 2019–2034, 2018.
37. Coskun, D., Gur, S., Coskun, M.F., Synthesis, characterization and antimi crobial activity of novel benzofuran- and thiophene-containing diketoxime derivatives. *J. Serb. Chem. Soc.*, 82, 367–377, 2017.
38. Mabkhot, Y.N., Kaal, N.A., Alterary, S., Al-Showiman, S.S., Farghaly, T.A., Mubarak, M.S., Antimicrobial activity of thiophene derivatives derived from ethyl Q19 (E)-5-(3-(dimethylamino)acryloyl)-4-methyl-2-(phenylamino) thio phene-3-carboxylate. *Chem. Cent. J.*, 11, 75, 2017.
39. Popiołek, L., Biernasiuk, A., Malm, A., Design, synthesis, and *in vitro* Antimicrobial activity of new Furan/Thiophene-1,3-Benzothiazin-4-One hybrids. *J. Heterocycl. Chem.*, 53, 479, 2016.
40. Mehta, A., Bhatt, R., Sharma, S., Patidar, A.K., Rathore, K.K., Senwar, R.C., Synthesis, characterization and antimicrobial evaluation of some tetrahydro quinazoline derivatives of benzo[b]thiophene. *Int. J. Pharm. Sci. Drug Res.*, 7, 417–420, 2015.
41. Adiwish, W.M., Yaacob, W.A., Adan, D., Nazlina, I., Synthesis and antibacte rial activity of thiophenes. *Int. J. Adv. Sci. Eng. Inf. Technol.*, 2, 27–30, 2012.
42. Badiceanu, C.D., Missir, A.V., Chifiriuc, M.C., Dracea, O., Raut, J., Larion, C., Dity, L.M., Mihaescu, Synthesis, Characterization and Antimicrobial Evaluation of Some New Thioureas derived from 3-thiophenecarboxylic Acid, G. *Rom. Biotechnol. Lett.*, 15, 5545, 2010.

43. Prasad, K.C., Angothu, B.N., Latha, T.M., Nagulu, M., Synthesis of some novel 2-aminothiophene derivatives and evaluation for their antimicrobial activity. *Int. J. Pharm. Bio. Sci.*, 7, 1–7, 2017.
44. Naim, M.J., Alam, O., Alam, M.J., Bano, F., Alam, P., Shrivastava, N., Recent review on indole: A privileged structure scaffold. *Int. J. Pharm. Sci. Res.*, 7, 51–62, 2016.
45. Mo, X., Rao, D.P., Kaur, K., Hassan, R., Abdel-Samea, A.S., Farhan, A.M., Brase, S., Hashem, H., Indole Derivatives: A Versatile Scaffold in Modern Drug Discovery—An Updated Review on Their Multifaceted Therapeutic Applications (2020–2024). *Molecules*, 29, 4770, 2024.
46. Leena, S.S., Kaul, G., Akhir, A., Saxena, D., Chopra, S., Deepthi, A., Green synthesis and antibacterial evaluation of spiro fused tryptanthrin-thiopyra no[2,3-b]indole hybrids targeting drug-resistant S. aureus. *Bioorg. Chem.*, 128, 106046, 2022.
47. Li, Y., Kowah, J.A.H., Jiang, M., Wu, Y., Wang, L., Yang, F., Synthesis, antibac terial activity, and 3D-QASR studies of matrine-indole derivatives as poten tial antibiotics. *Bioorg. Med. Chem. Lett.*, 102, 129671, 2024.
48. (a) Shirinzadeh, H., Süzen, S., Altanlar, N., Westwell, A.D., Antimicrobial Activities of New Indole Derivatives Containing 1,2,4-Triazole, 1,3,4-Thiadiazole and Carbothioamide. *Turk. J. Pharm. Sci.*, 15, 291–297, 2018. (b) Shaker, A.M.M., Abdelall, E.K.A., Abdellatif, K.R.A., Abdel-Rahman, H.M., Synthesis and biological evaluation of 2-(4-methylsulfonyl phenyl) indole derivatives: multi-target compounds with dual antimicrobial and anti-inflammatory activities. *BMC Chem.*, *14*, 1, 23, 2020.
49. Kumar, N., Khanna, A., Kaur, K., Kaur, H., Sharma, A., Bedi, P.M.S., Quinoline derivatives volunteering against antimicrobial resistance: rational approaches, design strategies, structure activity relationship and mechanistic insights. *Mol. Diversity*, 1–30, 2022.
50. Zeleke, D., Eswaramoorthy, R., Belay, Z., Melaku, Y., Synthesis and Antibacterial, Antioxidant, and Molecular Docking Analysis of Some Novel Quinoline Derivatives. *J. Chem.*, 1–16, 2020.
51. Bouzian, Y., Karrouchi, K., Sert, Y., Lai, C.-H., Ahabchane, N.H., Talbaoui, A., Mague, J.T., Essassi, E.M., Synthesis, spectroscopic characterization, crys tal structure, DFT, molecular docking and *in vitro* antibacterial potential of novel quinoline derivatives. *J. Mol. Struct.*, 1209, 127940, 2020.
52. (a) Kharb, R. and Kaur, H., Therapeutic significance of quinoline derivatives as antimicrobial agents. *Int. Res. J. Pharm.*, 4, 63–69, 2013. (b) Jin, G., Xiao, F., Li, Z., Qi, X., Zhao, L., Sun, X., Design, Synthesis, and Dual Evaluation of Quinoline and Quinolinium Iodide Salt Derivatives as Potential Anticancer and Antibacterial Agents. *ChemMedChem*, 15, 7, 600–609, 2020, doi: 10.1002/cmdc.202000002.
53. Behera, S., Mohanty, P., Behura, R., Nath, B., Barick, A.K., Jali, B.R., Antibacterial Properties of Quinoline Derivatives: A MiniReview. *Biointerface Res. Appl. Chem.*, 12, 6078–6092, 2022.

54. Khana, S.A., Asiri, A.M., Basisi, H.M., Asad, M., Zayed, M.E.M., Sharma, K., Wani, M.Y., Synthesis and evaluation of Quinoline-3-carbonitrile derivatives as potential antibacterial agents. *Bioorg. Chem.*, 88, 102968, 2019.
55. Rbaa, M., Jabli, S., Lakhrissi, Y., Ouhssine, M., Almalki, F., Hadda, T.B., Messgo-Moumene, S., Zarrouk, A., Lakhrissi, B., Synthesis, antibacterial properties and bioinformatics computational analyses of novel 8- hydroxy quinoline derivatives. *Heliyon*, 5, e02689, 2019.
56. Alghamdi, S.S., Suliman, R.S., Almutairi, K., Kahtani, K., Aljatli, D., Imidazole as a Promising Medicinal Scaffold: Current Status and Future Direction. *Drug Des. Devel. Ther.*, 15, 3289–3312, 2021.
57. Patil, V.S., Padalkar, V.S., Phatangare, K.R., Umape, P.G., Borase, B.N., Sekar, N., Synthesis, Characterization, and Antibacterial Activity of Novel (1H-Benzo [d] imidazole-2-yl)-6-(diethylamino)-3H-one-xanthene, Phenoxazine, and Oxazine. *J. Heterocycl. Chem.*, 52, 124–129, 2015.
58. Choudhari, D., Salunke-Gawali, S., Chakravarty, D., Shaikh, S.R., Lande, D.N., Gejji, S.P., Rao, P.K., Satpute, S., Puranik, V.G., Gonnade, R., Synthesis and biological activity of imidazole based 1, 4-naphthoquinones. *New J. Chem.*, 44, 6889–6901, 2020.
59. (a). Li, G., Cheng, Y., Han, C., Song, C., Huang, N., Du, Y., Pyrazole containing pharmaceuticals: target, pharmacological activity, and their SAR studies. *RSC Med. Chem.*, 13, 1300–1321, 2022. (b). Yusuf, E., Bax, H.I., Verkaik, N.J., Van Westreenen, M., An update on eight 'new' antibiotics against multidrug-resistant Gram negative bacteria. *J. Clin. Med.*, 10, 1068, 2021.
60. Allison D, D., Delancey, E., Ramey, H., *et al.*, Synthesis and antimicrobial studies of novel derivatives of 4-(4-formyl-3-phenyl-1H-pyrazol-1-yl)ben zoic acid as potent anti-Acinetobacter baumannii agents. *Bioorg. Med. Chem. Lett.*, 27, 387–392, 2017.
61. Alnufaie, R., Alsup, N., Kc, H.R., *et al.*, Design and synthesis of 4-[4-formyl-3-(2-naphthyl)pyrazol-1-yl]benzoic acid derivatives as potent growth inhibitors of drug-resistant Staphylococcus aureus. *J. Antibiot.*, 73, 818–827, 2020.
62. Brider, J., Rowe, T., Gibler, D.J., *et al.*, Synthesis and antimicrobial studies of azomethine and N-arylamine derivatives of 4-(4-formyl-3-phenyl-1H pyrazol-1-yl)benzoic acid as potent anti-methicillin-resistant Staphylococcus aureus agents. *Med. Chem. Res.*, 25, 2691–2697, 2016.
63. Yu, L.-G., Ni, T.-F., Gao, W., *et al.*, The synthesis and antibacterial activity of pyrazole-fused tricyclic diterpene derivatives. *Eur. J. Med. Chem.*, 90, 10–20, 2015.
64. Liu, H., Z., -L., Wang, W., Gong, J.-X., Chu, M.-J., Ma, Q.-W., Wang, J.-C., Lv, X.-H., Novel coumarin-pyrazole carboxamide derivatives as poten tial topoisomerase II inhibitors: Design, synthesis and antibacterial activity. *Eur. J. Med. Chem.*, 157, 81–87, 2018.
65. Arshad, M., Synthesis, characterization, antimicrobial and computational studies of some sulfonamide derivatives posesing thiadiazole and indole nucleus. *Eur. J. Pharm. Med. Res.*, 4, 511–517, 2017.

66. Tanitame, A., *et al.*, Synthesis and antibacterial activity of novel and potent DNA gyrase inhibitors with azole ring. *Bioorg. Med. Chem.*, 12, 5515–5524, 2004.
67. (a) Tomi, I.H.R., Tomma, J.H., Al-Daraji, A.H.R., Al-Dujaili, A.H., Synthesis, characterization and comparative study the microbial activity of some het erocyclic compounds containing oxazole and benzothiazole moieties. *J. Saudi Chem. Soc.*, 19, 392–398, 2015. (b) Kakkar, S. and Narasimhan, B., A comprehensive review on biological activities of oxazole derivatives. *BMC Chem.*, 13, 16, 2019, https://doi.org/10.1186/s13065-019-0531-9. (c) Reddy, C.S., Rao, L.S., Devi, M.V., Kumar, G.R., Nagaraj, A., Synthesis of some new 3-[5-(2-oxo-2H-3-chromenyl)-1, 3-oxazol-2-yl]-1, 3-thiazolan-4-ones as antimicrobials. *Chin. Chem. Lett.*, *21*, 9, 1045–1048, 2010.
68. Karegoudar, P., *et al.*, Synthesis of some novel 2,4-disubstituted thiazoles as possible antimicrobial agents. *Eur. J. Med. Chem.*, 43, 261–267, 2008.
69. Zha, G.F., *et al.*, Synthesis, SAR and molecular docking studies of benzo[d] thiazole-hydrazones as potential antibacterial and antifungal agents. *Bioorg. Med. Chem. Lett.*, 27, 3148–3155, 2017.
70. Ameli, S., Pordel, M., Davoodnia, A., Jajarmi, M., Synthesis and antibacte rial activity of some new benzo[5,6]chromeno[2,3-d]pyrimidines. *Russ. J. Bioorg. Chem.*, 43, 429–434, 2017.
71. Bassyouni, F., Tarek, M., Salama, A., Ibrahim, B., Salah El Dine, S., Yassin, N., Hassanein, A., Moharam, M., Abdel-Rehim, M., Promising antidiabetic and antimicrobial agents based on fused pyrimidine derivatives: molecular modeling and biological evaluation with histopathological effect. *Molecules*, *26*, 8, 2370, 2021.
72. Mahdavi, M., Ghasemzadeh, M.A., Javadi, A., Synthesis of ZIF-8/ZnFe2O4/ GO-OSO3H nanocomposite as a superior and reusable heterogeneous catalyst for the preparation of pyrimidine derivatives and investigation of their antimicrobial activities. *Heliyon*, *10*, 4, e26339, 2024, https://doi.org/10.1016/j.heliyon.2024.e26339.
73. Althagafi, I. and Abdel-Latif, E., Synthesis and Antibacterial Activity of New Imidazo[1,2-*a*]pyridines Festooned with Pyridine, Thiazole or Pyrazole Moiety. *Polycyclic Aromat. Compd.*, *42*, 7, 4487–4500, 2022, https://doi.org/1 0.1080/10406638.2021.1894185.
74. (a) De, S., K, A. K. S., Shah, S.K., Kazi, S., Sarkar, N., Banerjee, S., Dey, S., Pyridine: the scaffolds with significant clinical diversity. *RSC Adv.*, *12*, 24, 15385–15406, 2022, https://doi.org/10.1039/d2ra01571d. (b) Leal, B., Afonso, I.F., Rodrigues, C.R., Abreu, P.A., Garrett, R., Pinheiro, L.C.S., Azevedo, A.R., Borges, J.C., Vegi, P.F., Santos, C.C., Da Silveira, F.C., Cabral, L.M., Frugulhetti, I.C., Bernardino, A.M., Santos, D.O., Castro, H.C., Antibacterial profile against drug-resistant Staphylococcus epidermidis clin- ical strain and structure–activity relationship studies of 1H-pyrazolo[3,4-b] pyridine and thieno[2,3-b]pyridine derivatives. *Bioorg. Med. Chem.*, *16*, 17, 8196–8204, 2008, https://doi.org/10.1016/j.bmc.2008.07.035. (c) Ali, I.,

Burki, S., El-Haj, B.M., Shafiullah, N., Parveen, S., Nadeem, H.S., Nadeem, S., Shah, M.R., Synthesis and characterization of pyridine-based organic salts: Their antibacterial, antibiofilm and wound healing activities. *Bioorg. Chem.*, *100*, 103937, 2020b, https://doi.org/10.1016/j.bioorg.2020.103937.

75. Xu, H., Su, X., Guo, M.B., An, R., Mou, Y.H., Hou, Z., Guo, C., Eur., J., Design, synthesis, and biological evaluation of novel miconazole analogues containing selenium as potent antifungal agents. *Med. Chem.*, 198, 112360, 2020, https://doi.org/10.1016/j.ejmech.2020.112360.
76. Dominique, S., Emerging threats in Antifungal-Resistant fungal pathogens. *Front. Med.*, 3, 11, 2016, 10.3389/fmed.2016.00011.
77. Brown, G.D., Denning, D.W., Levitz, S.M., Tackling Human Fungal Infections. *Science*, 336, 647, 2012, https:// doi.org/10.1126/science.1222236.
78. Liu, N., Tu, J., Dong, G., Wang, Y., Sheng, C., Emerging New Targets for the Treatment of Resistant Fungal Infections. *J. Med. Chem.*, 61, 5484–5511, 2018, https://doi.org/10.1021/acs.jmedchem.7b01413.
79. Campoy, S. and Adrio, J.L., Antifungal. *Biochem. Pharmacol.*, 133, 86–96, 2017.
80. Yang, L., Xu, W.B., Sun, L., Zhang, C., Jin, C.H., SAR Analysis of Heterocyclic Compounds with Monocyclic and Bicyclic Structures as Antifungal. *ChemMedChem*, 17, e20220022, 2022, https://doi.org/10.1002/cmdc.202200221.
81. Kruijt, M., De Kock, M.J.D., De Wit, P.J.G.M., Receptor-like proteins involved in plant disease resistance. *Mol. Plant Pathol*, 6, 1, 85–97, 2005, https://doi.org/10.1111/j.1364-3703.2004.00264.
82. Emami, S., Ghobadi, E., Saednia, S., Hashemi, S.M., Current advances of triazole alcohols derived from fluconazole: Design, *in vitro* and *in silico* studies. *Eur. J. Med. Chem.*, 170, 173–194, 2019, https://doi.org/10.1016/j.ejmech.2019.03.020.
83. Monk, B.C., Sagatova, A.A., Hosseini, P., Ruma, Y.N., Wilson, R.K., Keniya, M.V., Fungal Lanosterol 14α-demethylase: A target for next-generation antifungal design. *BBA- Proteins Proteom.*, 1868, 140206, 2020, https://doi.org/10.1016/j. bbapap.2019.02.008.
84. Sathyanarayana, R., Poojary, B., Chin, J., Exploring recent developments on 1,2,4-triazole: Synthesis and biological applications. *Chem. Soc.*, 67, 459–477, 2019, https://doi.org/10.1002/jccs.201900304.
85. El-Sherief, H.A., Youssif, B.G., Abdelazeem, A.H., Abdel-Aziz, M., Abdel Rahman, H.M., Design, Synthesis and Antiproliferative Evaluation of Novel 1,2,4-Triazole/Schiff Base Hybrids with EGFR and B-RAF Inhibitory Activities. *Med. Chem.*, 19, 697–706, 2019, ://doi.org/10.2174/1871520619666181224115346.
86. Wang, Z., Ni, X., Peng, Q., Hou, Y., Fang, Y., Mu, W., Liu, C., Liu, P., Liu, X., Pestic. The novel fungicide SYP-14288 acts as an uncoupler against Phytophthora capsica. *Biochem. Physiol.*, 147, 83–89, 2018, https://doi.org/10.1016/j. pestbp.2018.01.014.

87. Qian, H., Duan, M., Sun, X., Chi, M., Zhao, Y., Liang, W., Du, J., Huang, J., Li, B., The binding mechanism between azoles and FgCYP51B, sterol 14α-demethylase of Fusarium graminearum. *Pest. Manage. Sci.*, 74, 126–134, 2018, https://doi.org/10.1002/ps.4667.
88. Baugh, S.D.P., Chaly, A., Weaver, D.G., Pelletier, J.C., Thanna, S., Freeman, K.B., Reitz, A.B., Scott, R.W., Highly potent, broadly active antifungal agents for the treatment of invasive fungal infections. *Bioorg. Med. Chem. Lett.*, 33, 127727, 2021, https://doi.org/10.1016/j.bmcl.2020.127727.
89. Zhao, L., Tian, L., Sun, N., Sun, Y., Chen, Y., Wang, X., Zhao, S., Su, X., Zhao, D., Cheng, M., Design, synthesis, and structure-activity relationship studies of l-amino alcohol derivatives as broad-spectrum antifungal agents. *Eur. J. Med. Chem.*, 177, 374–385, 2019, https://doi. org/10.1016/j.ejmech.2019.05.047.
90. Wang, T., Shao, J., Da, W., Li, Q., Shi, G., Wu, D., Wang, C., Design, synthesis, and structure-activity relationship studies of l-amino alcohol derivatives as broad-spectrum antifungal agents. *Front. Microbiol.*, 9, 2892, 2018, https://doi.org/10.3389/fmicb.2018.02892.
91. Chandrika, K.V.S.M. and Sharma, S., Promising antifungal agents: A mini-review. *Bioorg. Med. Chem.*, 28, 115398, 2020, https://doi.org/10.1016/j.bmc.2020.115398.
92. Gintjee, T.J., Donnelley, M.A., Thompson, G.R., Aspiring Antifungals: Review of Current Antifungal Pipeline Developments. *J. Fungi*, 6, 1–14, 2020, https://doi.org/10.3390/jof6010028.
93. Nishimoto, A.T., Whaley, S.G., Wiederhold, N.P., Zhang, Q., Yates, C.M., Hoekstra, W.J., Schotzinger, R.J., Garvey, E.P., Rogers, P.D., Impact of the Major Candida glabrata Triazole Resistance Determinants on the Activity of the Novel Investigational Tetrazoles VT-1598 and VT-1161. *Antimicrob. Agents Chemother.*, 63, e01304-19, 2019, https://doi.org/10.1128/aac.01304-19.
94. Rauseo, A.M., Coler-Reilly, A., Larson, L., Spec, A., Hope on the Horizon: Novel Fungal Treatments in Development. *Open Forum Infect. Dis.*, 7, ofaa016, 2020, 10.1093/ofid/ofaa016.
95. El-Sayed, W.A., Khalaf, H.S., Mohamed, S.F., Hussien, H.A., Kutkat, O.M., Amr, A.E., Synthesis and antiviral activity of 1,2,3-triazole glycosides based substituted pyridine *via* click cycloaddition. *Russ. J. Gen. Chem.*, 87, 2444–2453, 2017.
96. Cao, X., Wang, W., Wang, S., Bao, L., Asymmetric synthesis of novel triazole derivatives and their *in vitro* antiviral activity and mechanism of action. *Eur. J. Med. Chem.*, 139, 718–725, 2017.
97. Wu, J., Ma, S., Zhang, T.Y., Wei, Z.Y., Wang, H.M., Guo, F.Y., Zheng, C.J., Piao, H.R., Synthesis and biological evaluation of ursolic acid derivatives containing an aminoguanidine moiety. *Med. Chem. Res.*, 28, 959–973, 2019.
98. Zhang, T.Y., Li, C., Li, Y.R., Li, X.Z., Sun, L.P., Zheng, C.J., Piao, H.R., Synthesis and Antimicrobial Evaluation of Aminoguanidine and

3-amino- 1,2,4-triazole Derivatives as Potential Antibacterial Agents. *Lett. Drug Des. Discovery*, 13, 1063–1075, 2016.
99. Wei, Z.Y., Liu, J.C., Zhang, W., Li, Y.R., Li, C., Zheng, C.J., Piao, H.R., Synthesis and Antimicrobial Evaluation of Aminoguanidine and 3-amino-1,2,4-triazole Derivatives as Potential Antibacterial Agents. *Med. Chem.*, 12, 751–759, 2016.
100. Song, M.X., Deng, X.Q., Wei, Z.Y., Zheng, C.J., Wu, Y., An, C.S., Piao, H.R., Synthesis and Antibacterial Evaluation of (S,Z)-4-methyl-2-(4-oxo-5-((5-substituted phenylfuran-2-yl) methylene)-2-thioxothiazolidin-3-yl) Pentanoic Acids. *Iran. J. Pharm. Res.*, 14, 89–96, 2015.
101. Bozorov, K., Zhao, J., Aisa, H.A., 1,2,3-Triazole-containing hybrids as leads in medicinal chemistry: A recent overview. *Bioorg. Med. Chem.*, 27, 3511-3531, 2019.
102. Xu, Z., Zhao, S.J., Liu, Y., 1,2,3-Triazole-containing hybrids as potential anticancer agents: Current developments, action mechanisms and structure-activity relationships. *Eur. J. Med. Chem.*, 183, 111700, 2019.
103. Akhtar, M.J., Khan, A.A., Ali, Z., Haider, R., Yar, M.S., Structure-activity relationship (SAR) study and design strategies of nitrogen-containing heterocyclic moieties for their anticancer activities. *Eur. J. Med. Chem.*, 125, 143–189, 2017.
104. Guo, Z., Song, X., Zhao, L.M., Piao, M.G., Quan, J., Piao, H.R., Jin, C.H., Synthesis and biological evaluation of novel benzo[c][1,2,5]thiadiazol-5-yl and thieno[3,2-c]- pyridin-2-yl imidazole derivatives as ALK5 inhibitors. *Bioorg. Med. Chem. Lett.*, 29, 2070–2075, 2019.
105. Liu, Y.Y., Guo, Z., Wang, J.Y., Wang, H.M., Da Qi, J., Ma, J., Piao, H.R., Jin, C.H., Jin, X., Synthesis and evaluation of the epithelial-to-mesenchymal inhibitory activity of indazole-derived imidazoles as dual ALK5/p38α MAP inhibitors. *Eur. J. Med. Chem., 216*, 113311, 2021.
106. Wang, Z., Li, M.Y., Zhang, Z.H., Zuo, H.X., Wang, J.Y., Xing, Y., Ri, M.H., Jin, H.L., Jin, C.H., Xu, G.H., Piao, L.X., Jiang, C.G., Ma, J., Jin, X., Panaxadiol inhibits programmed cell death-ligand 1 expression and tumour proliferation *via* hypoxia-inducible factor (HIF)-1α and STAT3 in human colon cancer cells. *Pharmacol. Res.*, 155, 104727, 2020.
107. Zhang, Z.H., Mi, C., Wang, K.S., Wang, Z., Li, M.Y., Zuo, H.X., Xu, G.H., Li, X., Piao, L.X., Ma, J., Jin, X., GABA-modulating phytomedicines for anxiety: A systematic review of preclinical and clinical evidence. *Phytother. Res.*, 32, 65–75, 2018.
108. Xing, Y., Mi, C., Wang, Z., Zhang, Z.H., Li, M.Y., Zuo, H.X., Wang, J.Y., Jin, X., Ma, J., Fraxinellone has anticancer activity *in vivo* by inhibiting programmed cell death ligand 1 expression by reducing hypoxia inducible factor 1α and STAT3. *Pharmacol. Res.*, 135, 166–180, 2018.
109. Mi, C., Ma, J., Wang, K.S., Zuo, H.X., Wang, Z., Li, M.Y., Piao, L.X., Xu, G.H., Li, X., Quan, Z.S., Jin, X., Imperatorin suppresses proliferation and

angiogenesis of human colon cancer cell by targeting HIF-1α *via* the mTOR/p70S6K/4E-BP1 and MAPK pathways. *J. Ethnopharmacol.*, 203, 27–38, 2017.

110. Venugopala, K.N., Dharma Rao, G.B., Bhandary, S., Pillay, M., Chopra, D., Aldhubiab, B.E., Attimarad, M., Alwassil, O.I., Harsha, S., Mlisana, K., Design, synthesis, and characterization of (1-(4-aryl)-1H-1,2,3-triazol-4-yl)methyl, substituted phenyl-6-methyl-2-oxo-1,2,3,4-tetrahydropyrimidine-5-carboxylates against Mycobacterium tuberculosis. *Drug Des. Dev. Ther.*, 10, 2681–2690, 2016.
111. Mohassab, A.M., Hassan, H.A., Abdelhamid, D., Mohamed Abdel-Aziz, M., Dalby, K.N., Kaoud, T.S., Novel quinoline incorporating 1,2,4□triazole/oxime hybrids: Synthesis, molecular docking, anti inflammatory, COX inhibition, ulceroginicity and histopathological investigations. *Bioorg. Chem.*, 75, 242–259, 2017.
112. Shang, Y., Li, X.F., Jin, M.J., Li, Y., Wu, Y.L., Jin, Q., Zhang, Y., Li, X., Jiang, M., Cui, B.W., Lian, L.H., Nan, J.X., Leucodin attenuates inflammatory response in macrophages and lipid accumulation in steatotic hepatocytes *via* P2X7 receptor pathway: A potential role in alcoholic liver disease. *Biomed. Pharmacother.*, 107, 374–381, 2018.
113. Jiang, S., Zhang, Y., Zheng, J.H., Li, X., Yao, Y.L., Wu, Y.L., Song, S.Z., Sun, P., Nan, J.X., Lian, L.H., Potentiation of hepatic stellate cell activation by extracellular ATP is dependent on P2X7R-mediated NLRP3 inflammasome activation. *Pharmacol. Res.*, 117, 82–93, 2017.
114. Lu, Y., Xu, X., Jiang, T., Jin, L., Zhao, X.D., Cheng, J.H., Jin, X.J., Ma, J., Piao, H.N., Piao, L.X., Sertraline ameliorates inflammation in CUMS mice and inhibits TNF-α-induced inflammation in microglia cells. *Int. Immunopharmacol.*, 67, 119–128, 2019.
115. Zhu, W.J., Cui, B.W., Wang, H.M., Nan, J.X., Piao, H.R., Lian, L.H., Jin, C.H., Design, synthesis, and antifibrosis evaluation of 4-(benzo-[c][1,2,5]thiadiazol-5-yl)-3(5)-(6-methyl- pyridin-2-yl)pyrazole and 3(5)-(6-methylpyridin- 2-yl)-4-(thieno-[3,2,-c]pyridin-2-yl)pyrazole derivatives. *Eur. J. Med. Chem.*, 180, 15–27, 2019, https://doi.org/10.1016/j. ejmech.2019.07.013.
116. Zhang, Y., Jiang, M., Cui, B.W., Jin, C.H., Wu, Y.L., Shang, Y., Yang, H.X., Wu, M., Liu, J., Qiao, C.Y., Zhan, Z.Y., Ye, H., Zheng, G.H., Jin, Q., Lian, L.H., Nan, J.X., Inhibition of erythropoietin-producing hepatoma receptor B4 (EphB4) signalling suppresses the vascularisation and growth of endometriotic lesions. *J. Pharmacol.*, 177, 2793–2811, 2020, https://doi.org/10.1111/bph.15044.
117. Miao, Y., Wu, Y., Jin, Y., Lei, M., Nan, J., Wu, X., Benzoquinone derivatives with antioxidant activity inhibit activated hepatic stellate cells and attenuate liver fibrosis in TAA-induced mice. *Chem. Biol. Interact.*, 317, 108945, 2020.
118. Li, X., Zhang, Y., Jin, Q., Xia, K.L., Jiang, M., Cui, B.W., Wu, Y.L., Song, S.Z., Lian, L.H., Nan, J.X., Facilitation of ischaemia-induced ventricular fibrillation by catecholamines is mediated by β1 and β2 agonism in the rat heart *in vitro*. *Br. J. Pharmacol.*, 175, 1451–1470, 2018.

119. Dixit, D., Verma, P.K., Marwaha, R.K., A review on 'triazoles': their chemistry, synthesis and pharmacological potentials. *J. Iran. Chem. Soc.*, 18, 2535–2565, 2021.
120. Venugopala, K.N., Khedr, M.A., Girish, Y.R., Bhandary, S., Chopra, D., Morsy, M.A., Aldhubiab, B.E., Deb, P.K., Attimarad, M., Nair, A.B., Sreeharsha, N., Kandeel, R.V.M., Akrawi, S.H., Reddy, M., Shashi kanth, M.B.S., Alwassil, O.I., Mohanlall, V., Crystallography, *in silico* studies, and *in vitro* antifungal studies of 2,4,5-trisubstituted 1,2,3-triazole analogues. *Antibiotics*, 9, 350–362, 2020.
121. Kaushik, C.P. and Luxmi, R., Synthesis and antimicrobial evaluation of ester-linked 1,4-disubstituted 1,2,3-triazoles with a furyl/thienyl moiety. *Synth. Commun.*, 47, 2225–2231, 2017.
122. Zoidis, G., Kritsi, E., Lecinska, P., Ivanov, M., Zoumpoulakis, P., Sokovic, M., Catto, M., The Triazole Ring as a Privileged Scaffold for Putative Antifungals: Synthesis and Evaluation of a Series of New Analogues. *ChemMedChem*, 16, 134–144, 2021.
123. Bai, H., Liu, X., Chenzhang, P., Xiao, Y.M., Fu, B., Qin, Z., Design, Synthesis and Fungicidal Activity of New 1,2,4-Triazole Derivatives Containing Oxime Ether and Phenoxyl Pyridinyl Moiety. *Molecules*, 25, 5852–5864, 2020.
124. Ali, I., Lone, M.N., Aboul-Enein, H.Y., Imidazoles as potential anticancer agents. *MedChemComm*, 8, 1742, 2017.
125. Tabassum, K., Ekta, P., Kavitkumar, P., Imidazole and Pyrazole: Privileged Scaffolds for Anti-Infective Activity. *Mini-Rev. Org. Chem.*, 15, 459–475, 2018.
126. Okano, Y., Saito-Tarashima, N., Kurosawa, M., Iwabu, A., Ota, M., Watanabe, T., Kato, F., Hishiki, T., Fujimuro, M., Minakawa, N., Synthesis and biological evaluation of novel imidazole nucleosides as potential anti-dengue virus agents. *Bioorg. Med. Chem.*, 27, 2181–2186, 2019.
127. Işık, A., Acar Çevik, U., Sağlık, B.N., Özkay, Y., Synthesis, Characterization, and Molecular Docking Study of Some Novel Imidazole Derivatives as Potential Antifungal Agents. *J. Heterocycl. Chem.*, 56, 142–152, 2019.
128. Al-Wabli, R.L., Al-Ghamdi, A.R., Ghabbour, H.A., Al-Agamy, M.H., Attia, M.L., Synthesis, structure elucidation, and antifungal potential of certain new benzodioxole–imidazole molecular hybrids bearing ester functionalities. *Drug Des. Dev. Ther.*, 13, 775–789, 2019.
129. Elejalde, N.R., Macıás, M., Castillo, J.C., Sortino, M., Svetaz, L., Zacchino, S., Portilla, J., Synthesis and *in vitro* Antifungal Evaluation of Novel N-Substituted 4-Aryl-2-methylimidazoles. *ChemistrySelect*, 3, 5220–5227, 2018.
130. Zhang, T., Dong, M., Zhao, J., Zhang, X., Mei, X., Synthesis and antifungal activity of novel pyrazolines and isoxazolines derived from cuminaldehyde. *J. Pestic. Sci.*, 44, 181–185, 2019.

131. Hagrasa, M., Salama, E.A., Sayed, A.M., Abutaleb, N.S., Kotb, A., Seleem, M.N., Mayhoub, A.S., Oxadiazolylthiazoles as novel and selective antifungal agents. *Eur. J. Med. Chem.*, 189, 112046, 2020.
132. Lino, C.I., Goncalves de Souza, I., Borelli, B.M., Silvério Matos, T.T., Santos Teixeira, I.N., Ramos, J.P., Maria de Souza Fagundes, E., de Oliveir-a Fernandesd, P., Maltarolloa, V.G., Johann, S., de Oliveira, R.B., Synthesis, molecular modeling studies and evaluation of antifungal activity of a novel series of thiazole derivatives. *Eur. J. Med. Chem.*, 151, 248–260, 2018, https://doi.org/10.1016/j.ejmech.2018.03.059.
133. Ouf, S.A., Gomha, S.M., Ewies, M.M., Sharawy, I.A.A., Synthesis, Characterization, and Antifungal Activity Evaluation of Some Novel Arylazothiazoles. *J. Heterocycl. Chem.*, 55, 258–264, 2018.
134. Adole, V.A.A., More, R.A., Jagdale, B.S., Pawar, T.B., Chobe, S.S., Efficient Synthesis, Antibacterial, Antifungal, Antioxidant and Cytotoxicity Study of 2-(2-Hydrazineyl)thiazole Derivatives. *ChemistrySelect*, 5, 2778–2786, 2020, 10.1002/slct.201904609.
135. Karaburun, A.C., Çevik, U.A., Osmaniye, D., Sağlık, B.N., Çavusoğlu, B.K., Levent, S., Özkay, Y., Koparal, A.S., Behçet, M., Kaplancıklı, Z.A., Synthesis and Evaluation of New 1,3,4-Thiadiazole Derivatives as Potent Antifungal Agents. *Molecules*, 23, 3129–3146, 2018, 10.1016/j.mencom.2018.07.017, https://doi.org/10.3390/molecules23123129.
136. Wang, S.C., Gao, J.G., Zhang, S., Liu, S., Jiang, L., Synthesis of novel pyrimidine derivatives with (pyridin-3-ylmethyl)thio and phenylamino moQ18ieties and evaluation of their antifungal activity. *Phosphorus Sulfur*, 193, 245–248, 2018.
137. Mahender, T., Update and Literature Review: Benzimidazoles as Anti-Cancer Agents. *World J. Pharm. Res.*, 9, 632–647, 2020, 10.20959/ wjpr20205-17329.
138. Devkate, C.G., Warad, K.D., Bhalerao, M.B., Gaikwad, D.D., Siddique, M.I., Synthesis and antibacterial activity of benzimidazole derivatives. *Pharm. Chem. J.*, 9, 115–118, 2017.
139. Pathak, A., Zulfiqar, R., Alam, S., Benzimidazoles as Anti-VEGFRs: A Review. *Int. J. Pharm. Res. Scholars*, 7, 82, 2018.
140. Li, L.X., Jiao, J., Wang, X.B., Chen, M., Fu, X.C., Si, W.J., Yang, C.L., Synthesis, Characterization, and Antifungal Activity of Novel Benzo[4,5]imidazo[1,2-d][1,2,4]triazine Derivatives. *Molecules*, 23, 746, 2018.
141. Karaburun, A.C., Kaya Çavuşoğlu, B., Acar Çevik, U., Osmaniye, D., Sağlık, B.N., Levent, S., Özkay, Y., Atlı, O., Koparal, A.S., Kaplancıklı, Z.A., Synthesis and Antifungal Potential of Some Novel Benzimidazole-1,3,4-Oxadiazole Compounds. *Molecules*, 24, 191–204, 2019.
142. Huang, D., Qiu, F., Zhang, Z., Shi, L., Cao, C., Ke, S., Synthesis and antifungal activity of substituted 2□aryl benzimidazole derivatives. *J. Heterocycl. Chem.*, 56, 2494–2498, 2019.
143. Irfan, A., Batool, F., Naqvi, S.A.Z., Islam, A., Osman, S.M., Nocentini, A., Alissa, S.A., Supuran, C.T., Benzothiazole derivatives as anticancer agents.

J. Enzyme Inhib. Med. Chem., 35, 1, 265–279, 2019, https://doi.org/10.1080/14756366.2019.1698036.

144. Kumar, R. and Abdullah, M.M., Synthesis, Characterization and Anticonvulsant Potential of 2,5-Disubstituted 1,3,4-Oxadiazole Analogues. *Asian J. Chem.*, 31, 1389–1397, 2019.
145. Dik, B., Coskun, D., Bahcivan, E., Uney, K., Potential antidiabetic activity of benzimidazole derivative albendazole and lansoprazoledrugs in different doses in experimental type 2 diabetic rats. *Turk. J. Med. Sci.*, 51, 1578–1585, 2021.
146. Sirgamalla, R., Kommakula, A., Konduru, S., Ponakanti, R., Devaram, J., Boda, S., Copper-catalyzed an efficient synthesis, characterization of 2-substituted benzoxazoles, 2-substituted benzothiazoles derivatives and their anti-fungal activity. *Chem. Data Collect.*, 27, 100362, 2020.
147. Ballari, M.S., Cano, N.H., Wunderlin, D.A., Feresin, G.E., Santiago, A.N., One-pot sequential synthesis and antifungal activity of 2-(benzylsulfonyl) benzothiazole derivatives. *RSC Adv.*, 9, 29405–29413, 2019.
148. Macan, A.M., Harej, A., Cazin, I., Klobučar, M., Stepanić, V., Pavelić, K., Pavelić, S.K., Schols, D., Snoeck, R., Andrei, G., Raić-Malić, S., Antitumor and antiviral activities of 4-substituted 1,2,3-triazolyl-2,3-dibenzyl-L-ascor bic acid derivatives. *Eur. J. Med. Chem.*, 184, 2019.
149. Manvar, D., Küçükgüzel, I., Erensoy, G., Tatar, E., Deryabaşoğullari, G., Reddy, H., Talele, T.T., Cevik, O., Kaushik-Basu, N., Discovery of conjugated thiazolidinonethiadiazole scaffold as anti-dengue virus polymerase inhibitors. *Biochem. Biophys. Res. Commun.*, 469, 743–747, 2016.
150. Jin, K.J., Sang, Y.L., De Clercq, E., Pannecouque, C., Meng, G., Design and synthesis of a novel series of non-nucleoside HIV-1 inhibitors bearing pyrimidine and Nsubstituted aromatic piperazine. *Bioorg. Med. Chem. Lett.*, 28, 3491–3495, 2018.
151. Moesslacher, J., Battisti, V., Delang, L., Neyts, J., Abdelnabi, R., Pürstinger, G., Urban, E., Langer, T., Identification of 2-(4-(Phenylsulfonyl)piperazine-1-yl) pyrimidine Analogues as Novel Inhibitors of Chikungunya Virus. *ACS Med. Chem. Lett.*, 11, 906–912, 2020.
152. Brigg, S., Pribut, N., Basson, A.E., Avgenikos, M., Venter, R., Blackie, M.A., van Otterlo, W.A.L., Pelly, S.C., Novel indole sulfides as potent HIV-1 NNRTIs. *Bioorg. Med. Chem. Lett.*, 26, 1580, 2016.
153. Scuotto, M., Abdelnabi, R., Collarile, S., Schiraldi, C., Delang, L., Massa, A., Ferla, S., Brancale, A., Leyssen, P., Neyts, J., Filosa, R., Discovery of novel multi-target indolebased derivatives as potent and selective inhibitors of chi kungunya virus replication. *Bioorg. Med. Chem.*, 25, 327–337, 2017.
154. Sokolova, A.S., Yarovaya, O.I., Bormotov, N.I., Shishkina, L.N., Salakhutdinov, N.F., Synthesis and antiviral activity of camphor-based 1,3-thiazolidin-4-one and thiazole derivatives as Orthopoxvirus-reproduction inhibitors. *MedChemComm*, 9, 1746–1753, 2018.

155. Guzman, M.A., Rugel, A.R., Tarpley, R.S., Alwan, S.N., Chevalier, F.D., Kovalskyy, D.P., Cao, X., Holloway, S.P., Anderson, T.J., Taylor, A.B., McHardy, S.F., An iterative process produces oxamniquine derivatives that kill the major species of schistosomes infecting humans. *PLoS Negl.Trop. Dis.*, *14*, 8, e0008517, 2020.
156. Rugel, A., Tarpley, R.S., Lopez, A., Menard, T., Guzman, M.A., Taylor, A.B., Cao, X., Kovalskyy, D., Chevalier, F.D., Anderson, T.J., Hart, P.J., Design, synthesis, and characterization of novel small molecules as broad range antischistosomal agents. *ACS Med. Chem. Lett.*, *9*, 10, 967–973, 2018.
157. Wang, C., Zhao, Q., Min, J., Muniyan, S., Vargas, M., Wang, X., Dong, Y., Guy, R.K., Lin, M.F., Keiser, J., Vennerstrom, J.L., Antischistosomal versus antiandrogenic properties of aryl hydantoin Ro 13-3978. *Am. J. Trop. Med. Hyg.*, *90*, 6, 1156, 2014.
158. Osei, E., Kwain, S., Mawuli, G.T., Anang, A.K., Owusu, K.B.A., Camas, M., Camas, A.S., Ohashi, M., Alexandru-Crivac, C.N., Deng, H., Jaspars, M., Paenidigyamycin, A., potent antiparasitic imidazole alkaloid from the Ghanaian Paenibacillus sp. DE2SH. *Mar. Drugs*, *17*, 1, 9, 2019.
159. Guimaraes, M.A., de Oliveira, R.N., Véras, L.M., Lima, D.F., Campelo, Y.D., Campos, S.A., Kuckelhaus, S.A., Pinto, P.L., Eaton, P., Mafud, A.C., Mascarenhas, Y.P., Anthelmintic activity *in vivo* of epiisopiloturine against juvenile and adult worms of Schistosoma mansoni. *PLoS Negl.Trop. Dis.*, *9*, 3, e0003656, 2015.
160. Kalinin, D.V., Jana, S.K., Pfafenrot, M., Chakrabarti, A., Melesina, J., Shaik, T.B., Lancelot, J., Pierce, R.J., Sippl, W., Romier, C., Jung, M., Structure-based design, synthesis, and biological evaluation of triazole-based smHDAC8 inhibitors. *ChemMedChem*, *15*, 7, 571–584, 2020.
161. Pereira, A.S., Silveira, G.O., Amaral, M.S., Almeida, S.M., Oliveira, J.F., Lima, M.C., Verjovski-Almeida, S., *In vitro* activity of aryl-thiazole derivatives against Schistosoma mansoni schistosomula and adult worms. *PLoS One*, *14*, 11, e0225425, 2019.
162. Santiago, E.D.F., de Oliveira, S.A., de Oliveira Filho, G.B., Moreira, D.R.M., Gomes, P.A.T., da Silva, A.L., de Barros, A.F., da Silva, A.C., Dos Santos, T.A.R., Pereira, V.R.A., Gonçalves, G.G.A., Evaluation of the anti-Schistosoma mansoni activity of thiosemicarbazones and thiazoles. *Antimicrob. Agents Chemother.*, *58*, 1, 352–363, 2014.

4

Heterocyclic Compounds as Anticancer Agents

Aadya Singh[1], Priyanshi Kamboj[1], Pragya Swami[2,3], Puja Prasad[4*] and Shruti Khanna Ahuja[1†]

[1]Centre for Medical Biotechnology, Amity Institute of Biotechnology, Amity University Uttar Pradesh, Noida, India
[2]Institute for Innovation in Imaging, Department of Radiology, Massachusetts General Hospital, Charlestown, Massachusetts, MA, USA
[3]Center for Systems Biology, Massachusetts General Hospital and Harvard Medical School, Boston, MA, USA
[4]Amity Institute of Click Chemistry Research and Studies, Amity University Uttar Pradesh, Noida, India

Abstract
The medical field relies heavily on heterocyclic compounds: rings that have at least one non-carbon atom such as nitrogen or sulfur; extend significant structural possibilities with extensive biological functions. The complex mechanisms of heterocyclic compounds' action on cancer cells, which include DNA intercalation and topoisomerase blocking, along with the modulation and interference of vital signaling routes and metabolism, are elucidated herein. Details about structural characteristics that enhance therapeutic properties are discussed in this chapter; these characteristics involve ring dimensions, saturation, and electronic characteristics that determine pharmacokinetic profiles, targeted selectivity, and bioavailability rates. This analysis reviews the heterocyclic elements that support anticancer efficacy in certified drugs including anthracyclines and camptothecin derivatives and kinase inhibitors. Also, recent therapeutic agents that include quinoxalines and pyrazoles are examined. It is also demonstrated how heterocyclic agents affect tumor microenvironments and reduce metastasis through vimentin binding and activate immunogenic cell death to serve either chemotherapy or immunotherapy purposes. This chapter also provides details on how understanding heterocyclic

**Corresponding author*: pprasad@amity.edu
†*Corresponding author*: skahuja@amity.edu

Divya Bajpai Tripathy, Priyanka Chhabra and Anjali Gupta (eds.) Heterocyclic Compounds as Pharmaceutical Ingredients, (123–178) © 2026 Scrivener Publishing LLC

scaffolds' functions for anticancer drug development allows scientists to create advanced cancer treatment compounds.

***Keywords*:** Heterocyclic, anticancer, mechanism of action, tumor, therapeutics

4.1 Introduction

One of the most important branches of organic chemistry is heterocyclic compounds having deep connections in organic synthesis and medicinal chemistry. According to the International Union of Pure and Applied Chemistry, these are cyclic compounds in which at least one atom in the ring structure is an element except carbon. Nitrogen, oxygen, and sulfur are the most common heteroatoms present in these rings, and their presence has been the cause of a wide spectrum of chemical and biological properties of these molecules. Heterocycles can be classified based on the type of heteroatom(s) incorporated as nitrogen, oxygen or sulfur based. Furthermore, their classification also applies to the ring size itself, because this size has a great influence on their physicochemical behavior, stability, and biological activity.

This structural diversity of heterocycles forms the basis of a large body of biologically active molecules, and hence, heterocycles are fundamental in medicinal chemistry (Figure 4.1). They are widely used in drug discovery and its therapeutic development due to their capacity to interact with biological targets. Broad pharmacological activities and the presence of the heterocyclic compounds in a number of clinically important drugs such as antibacterial, antiviral, antifungal, anti-inflammatory, and anticancer agents have been observed. As such, they are an important structural component of most marketed drugs. As examples, all the top 5 highest-selling

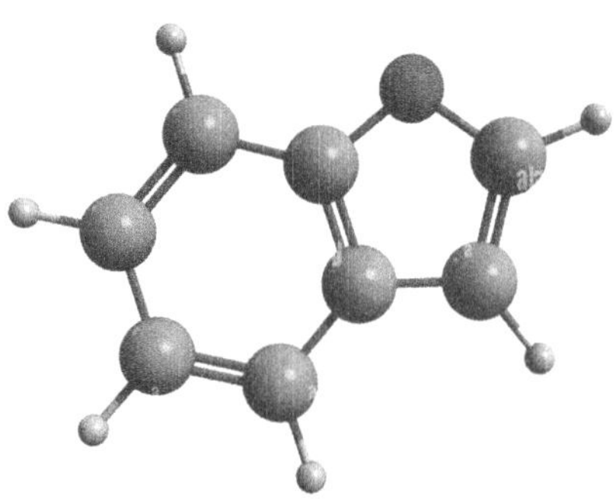

Figure 4.1 General structure of a heterocyclic compound.

small molecule drugs in the United States of 2014 had heterocyclic structure, which topped $27.4 million (almost 80%) of the total income from these top drugs [1].

Unique electronic and steric properties give rise to these molecules as competent components in the design of innovative molecules bearing tailor made functionalities. Heterocycles have a huge influence on drug discovery and development; their study continues to be a subject of increasing interest to medicinal chemistry and represents a focal area, where great efforts have been and are still being made in designing novel heterocyclic derivatives with improved therapeutic activity and reduced side effects.

Strategic incorporation of heterocyclic fragments can well serve the principles of drug design going hand in hand with the deep intertwining between the underlying physicochemical properties of the heterocycles and the need of small numbers of these moieties to establish the desired potency and selectivity to ensure overall drug performance [2]. Researchers replace one functional group with a second one that possesses similar biological properties to maintain drug efficacy while reducing undesired side effects. In addition, these biological properties, such as lipophilicity, polarity, and aqueous solubility, determine the drug's interaction with its target, their simple efficient crossing of biological membranes, their adsorption distribution, and extraction from the body through biological mechanisms such as: absorption, distribution, metabolism, excretion, and transportation (ADMET). Such characteristics enable the modulation of their nature, thereby permitting the biochemical control of the drug's mechanism of action, thus the development of molecularly targeted agents, whose molecules are specifically interacting with disease associated biomolecules.

Although heterocyclic compounds are versatile and hold a great therapeutic potential, they have not been widely used in the clinic due to several difficulties. For instance, in oncology, chemotherapy currently still represents the main therapeutic strategy; however, it is usually accompanied by very strong limitations. Traditional chemotherapeutic agents tend to be very nonselective and do not discriminate between healthy and diseased tissues, thus causing severe side effects from the toxicity in healthy tissues. In addition, cancer cells are able to develop resistance to treatment, adding an unnecessarily complicated step in therapeutic outcomes. This has made one of the most promising strategies to overcome these limitations, i.e., selective targeting of drugs to tumor cells, be either passive or active. The enhanced permeability and retention effect is exploited by passive targeting, which is based on the fact that tumor vasculature enhances the accumulation of nanoparticles or macromolecules in cancerous tissues. In contrast, active targeting makes drug carriers functionalized with ligands that recognize and bind to specific receptors overexpressed on cancer cells to improve drug delivery precision and reduce the toxicity to normal tissues. However, only a few examples of the success stories in molecularly targeted cancer therapy exist, for instance, imatinib (inhibitor of BCR-ABL protein in leukemia) [3].

4.2 Classification of Heterocyclic Compounds

Heterocyclic compounds are organic molecules with at least one hetero atom (nitrogen, oxygen, or sulfur) in the ring, and therefore, their characteristic is one reactive center (heterolytic) rather than two (diolytic). These compounds are grouped with respect to several structural and electronic factors that affect their chemical behavior and applications [2]. The first and the most striking difference is between monocyclic and polycyclic heterocycles, which contain one or more nonring atoms, and polycyclic heterocycles comprising multiple fused or fused rings that confer stability and biological activity. There is also another important classification of saturated (without π-electron delocalization) versus unsaturated (with aromaticity) heterocycles, which makes saturated heterocycles more reactive, and unsaturated heterocycles are responsible for their unique electronic properties. In addition, heterocycles are divided by ring size: five-membered (i.e., pyrrole, furan, thiophene), six-membered (pyridine, pyran), and polycyclic heterocycles (purine, quinoline) with increasing complexity modulating reactivity and function. Second, heterocycles have different electronic and steric effects, depending on the heteroatom present, i.e., nitrogen, sulfur, or oxygen based, and the heterocycles include imidazole, pyrimidine, thiophene, benzothiophene, furan, chromone, and so on. These classifications are important to understand those roles of heterocycles in medicinal chemistry, material science, and synthetic applications [3, 4].

4.2.1 Monocyclic versus Polycyclic Compounds

Electronic effects and aromaticity

Heterocyclic compounds are those having one or more heteroatoms (O, N, S, Se, P, As) in one or more rings, and thus the name (heterocycle). Monocyclic heterocycles either have one ring (e.g., pyrrole, furan, thiophene) or more than one single ring system (Figure 4.2). Basicity, nucleophilicity, and aromaticity are influenced by the delocalization of π electrons, and as a result, their electronic properties depend on this. For example,

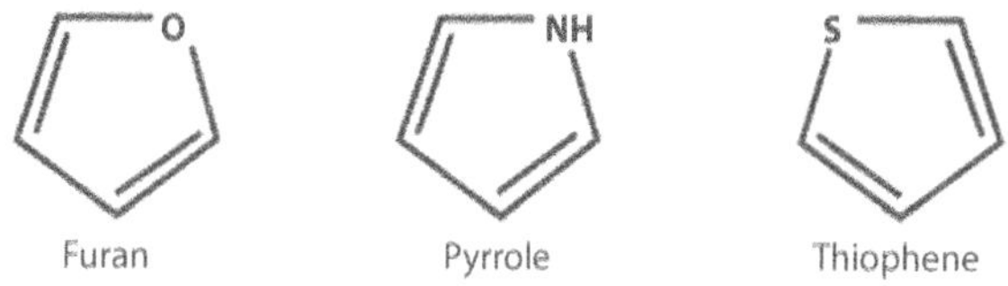

Figure 4.2 Examples of monocyclic compounds with heteroatoms.

pyrrole is more electron-rich than pyridine because its lone pair participates in aromaticity [5]. Unlike the monocyclic compounds, polycyclic heterocyclic compounds possess more than one fused aromatic and nonaromatic regions (e.g., quinoline, purine, benzothiophene) [6]. Conjugation is extended to fusion of rings, which affects electron density distribution, increases rigidity, and alters interactions in the π stacking. Often, this extended delocalization increases stability and changes the redox properties leading to more stable systems in the polycyclic versus the monocyclic systems.

Stability and reactivity

Structurally, monocyclic heterocycles are inferior to that due to fewer steric constraints and localized electron density. Furan is very easily oxidized, whereas pyridine's lone pair renders it less reactive in electrophilic aromatic substitution reactions but more susceptible to electrophilic attack. But in polycyclic systems, the increase in stability enhances in the case of heteroaromatic systems. Multiple rings can affect resonance structures, and one position in the ring system can be more reactive than another. For example, electrophilic substitution of quinoline takes place at C5 and C8 primarily, because of the electron distribution pattern [7].

Basicity and protonation

Basicity in polycyclic heterocycles can be enhanced or be reduced by the introduction of additional rings (Figure 4.3). For example, in purines, two nitrogen atoms with different electronic environments result in selective protonation or coordination in biological systems. The basicity of

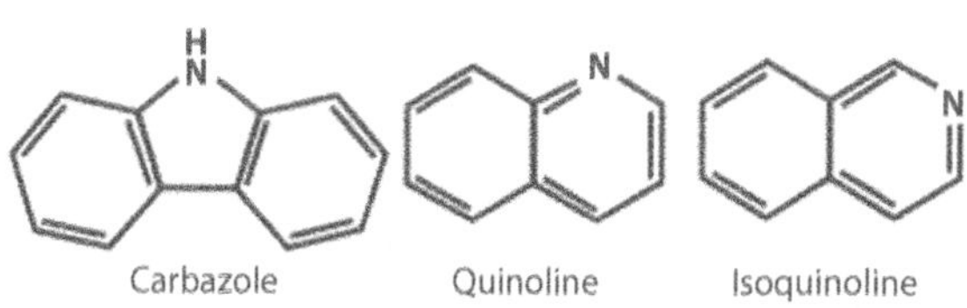

Figure 4.3 Examples of polycyclic compounds with heteroatoms.

monocyclic heterocycles depends on their position and nature of the heteroatom. Unlike pyrrole, the nitrogen lone pair of pyridine cannot contribute to the aromatic system of the ring, and thus, pyridine is more basic, because its nitrogen lone pair is protonable [4].

Solubility and lipophilicity

Particularly, solubility in polar solvents can be increased for monocyclic heterocycles because of hydrogen bonding capabilities of heteroatoms. The solubility of such molecules is, however, smaller with increased hydrophobicity of substituents. Larger hydrophobic surfaces of polycyclic heterocycles result in increased lipophilicity (lower water solubility, greater membrane permeability of drugs is significant in drug design). For use in medicinal chemistry, many polycyclic heterocycles are used for their propensity to intercalate with DNA or interact with hydrophobic protein pockets [5].

Biological activity and pharmacokinetics

Much stronger and more specific biological interactions often occur for polycyclic heterocycles because rigid structures are better able to find their way into enzyme active sites or into the DNA grooves. Symmetrical polycyclic heterocycles give many alkaloids (quinine, caffeine), and anticancer agents (anthracyclines) enhanced stability and specificity. Heterocycles with two rings (polycyclics) are known to be often utilized as bioactive cores in drug discovery (e.g., pyridine in nicotinamide, furan in furosemide) (Figure 4.4). As they are simpler structures, they are ideal for synthetic modification, yet may need additional functionalization to improve selectivity and potency [7].

Chemical difference between saturated and unsaturated heterocycles

Saturated heterocyclic compounds bear many chemical properties that differ greatly from their unsaturated counterparts, thus differentiating their reactivity, stability, and multidrug applications in cancer, particularly in anticancer drug design. Tetrahydrofuran and piperidine do not possess delocalized π electrons; hence, they are not aromatic. The reactivity toward electrophiles and nucleophiles is determined by their electron density; that is, it is localized. Although the conjugated π systems are present in unsaturated heterocycles such as pyridine, thiophene and furan, they are usually aromatic. Hückel's rule ($4n + 2$ π electrons) holds for aromatic heterocycles because of the extra stability they boasted by delocalization of electrons [4, 5].

These compounds are highly reactive. Resonance stabilization is the reason saturated heterocycles are more prone to oxidation and hydrolysis, behaving more as the acyclic counterparts in reactions such as nucleophilic substitution and hydrogenation. However, on the other side, unsaturated heterocycles are usually more stable because of aromaticity, but are electrophilic and nucleophilic substitution rather than stable. For example, pyrrole is more susceptible to electrophilic substitution than pyridine because it is electron-rich. One other very important point is fundamental to their solubility and lipophilicity. Saturated heterocycles tend to be more soluble in the aqueous solvents because they can participate in hydrogen bonding, especially nitrogen-containing structures such as morpholine and piperazine. On the contrary, lipophilicity of unsaturated heterocycles is higher, thus contributing to better membrane permeability, and it is a very important feature for drug bioavailability and targeting intracellular cancer pathways.

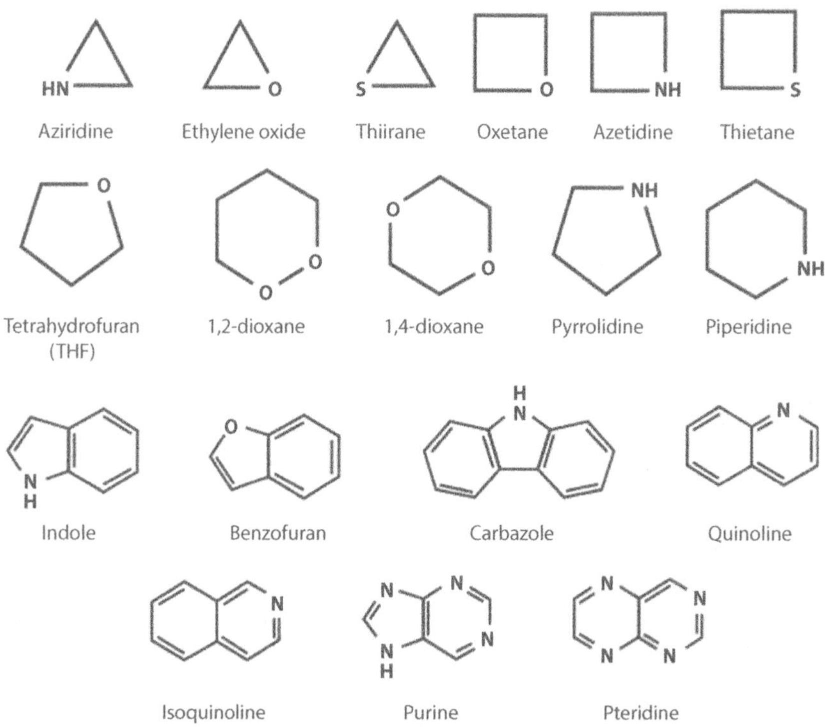

Figure 4.4 Structures of different types of heterocyclic compounds.

Both these properties have been shown to be critical for anticancer drug design because they directly impact the bioactivity and therapeutic potential of drug candidates. The structural scaffold for many of the drugs under development is saturated heterocycles, which are used to increase solubility and eliminate metabolic instability. For example, piperazine and morpholine rings are used commonly in anticancer drugs to tune drug-likeness and shield from unwanted interactions [7].

Optimizing pharmacokinetics (i.e., improved aqueous solubility and stability) of these saturated rings provides for optimal pharmacokinetics: efficient drug delivery, as well as prolonged circulation time in the body. On the other hand, unsaturated heterocycles, in particular, the aromatic ones such as purines, pyrimidines, and quinolines, are important for hitting cancer at the molecule level. Thus, many anticancer agents functioning as pyrimidine analogs, such as 5-fluorouracil (5-FU), are directly involved in halting cancer cell proliferation by interfering with DNA synthesis. Just as endothelial growth factor receptor (EGFR) inhibitors based on quinazoline disrupt pathways essential for tumor growth, the cyclen emerging covalent quinolines, which we are now developing, do the same. It has been shown that other unsaturated heterocycles, such as indoles and isoquinolines, possess potent cytotoxic activity with the capacity to initiate apoptosis and antiangiogenic activity. Conjugated π system in these heterocycles makes them interact more effectively with biomolecules (i.e., enzymes and receptors) that are part of a molecularly targeted cancer therapy.

4.2.2 Different Types of Heterocyclic Compounds Based on Ring Size

4.2.2.1 Five-Membered Heterocycles

Heterocyclic compounds with at least one heteroatom and a ring with five atoms together form a five-membered ring (Figure 4.5). Common members are pyrrole (nitrogen based), furan (oxygen based), and thiophene (sulfur based). The heterocycles often show aromaticity, which lead to stabilization and bear influence on the electronic properties. For instance, pyrrole has a nitrogen lone pair participating in resonance, thus being a weakly basic molecule uninterruptedly reactive in electrophilic substitution. Electron-rich furan and thiophene can also be used in drug design.

Pyrrole scaffold–based compounds are used in kinase inhibitors of cell signaling used in anticancer drug development, and thiophene is used in vascular endothelial growth factor receptor (VEGFR) inhibitors preventing tumor angiogenesis. Often, the furan derivatives are utilized in redox

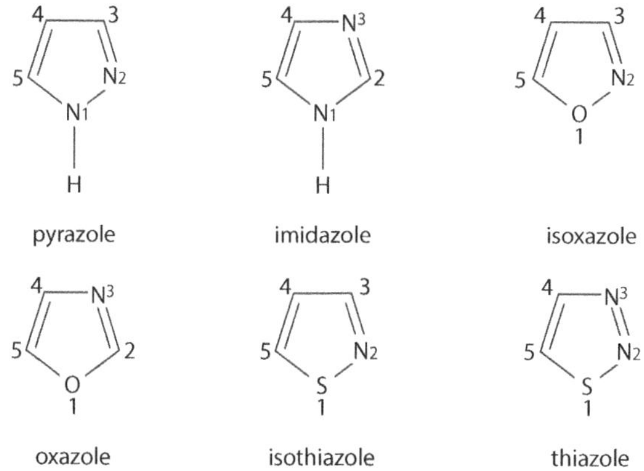

Figure 4.5 Structures of five-membered heterocyclic compounds.

active anticancer agents to provide reactive oxygen species (ROS) to trigger cancer cell apoptosis. Heterocycles present in these two compounds are crucial as fundamental building blocks for the development of targeted therapies seeking to arrest cancer progression [5].

4.2.2.2 Six-Membered Heterocycles

Six-membered heterocycles such as pyridine, pyran, and thiopyran allow greater stability and diversity of functionalization (Figure 4.6). For example, pyridine is an electron-deficient aromatic base with moderate basicity

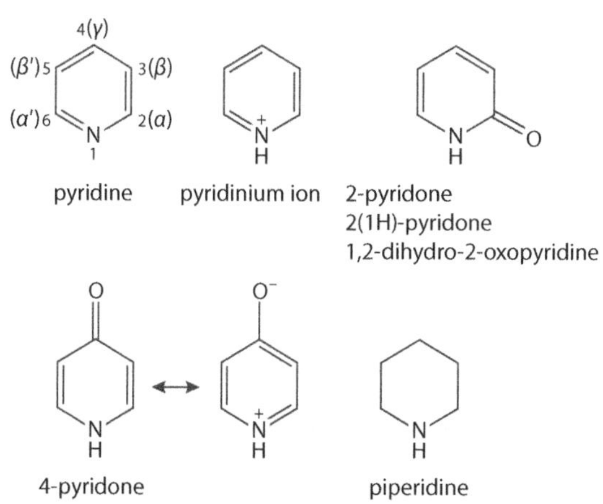

Figure 4.6 Structures of six-membered heterocyclic compounds.

and therefore participates in hydrogen bonding and improves solubility of the drug. One useful feature of a reduced reactivity compared to five-membered heterocycles is its stability and long-acting ability for stable drugs.

The cancer therapy needs three significant heterocycles with six-membered rings due to these properties. 5-FU and other pyrimidine derivatives are nucleobase analogs that interfere with DNA synthesis in rapidly proliferating cancer cells. In particular, imatinib, an effective chronic myeloid leukemia (CML) treatment, is among the class of pyridine-based drugs used extensively in kinase inhibitors. Thiopyran-containing structures also carry some role in the modulation of ROS, as well as oxidative stress pathways in cancer therapy. Such heterocycles are further enhanced in their pharmaceutical importance by their capability to be fine-tuned for metabolic stability and targeted action [1].

4.2.2.3 Polycyclic Heterocycles

Fused rings provide polycyclic heterocycles the ability to extend conjugation and electronic interactions of the molecule. Examples of such are purines, quinolines, and indoles, all of which have strong aromaticity, increase lipophilicity, and have greater DNA-binding potential. Polycyclic heterocycles are of an extraordinary importance in antitumor drugs as they are capable of intercalating into DNA, of inhibiting enzymes, and of interference with cellular processes.

Polycyclic heterocycles form the basis for a number of highly potent anticancer drugs. Mercaptopurine and other purine-based drugs inhibit the synthesis of nucleotides in leukemias and are purine-based drugs acting as antimetabolites. Topoisomerases such as quinoline and isoquinoline derivatives (e.g., the camptothecin [CPT]) cause inhibition of cancer cell replication. Being indole-containing, the structures in question target microtubules interfering specifically with mitotic events, thus leading to cell cycle arrest in tumors. The extended conjugation in these polycyclic systems enhances accessibility to the drug to the target, which makes these polycyclic systems crucial drug-targeting components of chemotherapy [1, 2].

4.2.3 Heterocyclic Compounds Based on the Heteroatom

Broad categories could be made into heterocyclic compounds based on the type of heteroatom in their ring system; there are three main types of heterocycles including: nitrogen based, sulfur-based, and oxygen-based heterocycles (Table 4.1). Structures based on nitrogen atoms (heterocycles)

Table 4.1 Illustration of classification of heterocyclic compounds.

	Saturated			Unsaturated		
Heteroatom	Nitrogen	Oxygen	Sulphur	Nitrogen	Oxygen	Sulphur
Three-membered Heterocyclic Compounds	 Aziridine	 Oxirane	 Thiirane	 Azirine	 Azirine	 Thiirene
Four-membered Heterocyclic Compounds	 Azetidine	 Oxetane	 Thietane	 Azete	 Oxete	 Thiete
Five-membered Heterocyclic Compounds	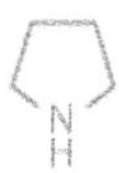 Pyrrolidine	 Oxolane	 Thiolane	 Pyrrole	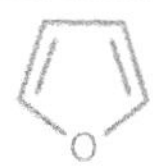 Furan	 Thiophene
Six-membered Heterocyclic Compounds	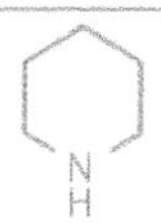 Piperidine	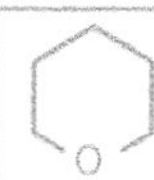 Oxane	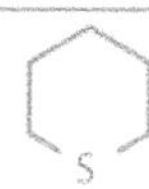 Thiane	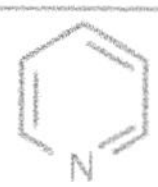 Pyridine	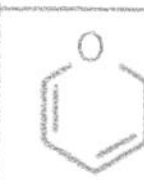 Pyran	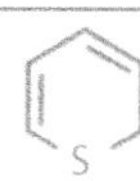 Thiopyran

such as pyridine, pyrimidine, and purine are used in a large number of pharmaceuticals, because heterocyclics can form hydrogen bonds and may take part in π–π interactions, and thus are useful in inhibition of enzyme and binding to receptors. In contrast, sulfur-based heterocycles such as thiophene and benzothiophene are electron donor and lipophilic; these properties represent good stability and ability interacting with hydrophobic part of a protein, which are useful features in designing anticancer agents that will target specific intracellular pathways. Of particular note are oxygen-based heterocycles such as furan and chromone, which are reagent with diverse reactivity profiles, as well is the ability to participate in such things as oxidation–reduction reactions, often responsible for their antioxidant properties as well as influencing their pharmacokinetic properties. The heterocyclic chemistry constitutes a powerful means of fine tuning unique electronic and steric attributes, which have broad practical utility for the design of new therapeutic agents, presented in each class being characterized by drug efficacy, selectivity and bioavailability optimized through optimization of different classes [3].

4.2.3.1 Nitrogen-Based Heterocycles

Medicinal chemistry has numerous iterates of nitrogen-based heterocycles, which are notorious for versatile reactivity and biological importance. To mention one, pyridine, a six-membered aromatic ring with one nitrogen, is a key precursor in a large number of agrochemical and pharmaceutical synthesis due to its electron-deficient character, which in turn makes it a good reactant in different chemical reactions. Like pyrimidine, nucleic acids, DNA and RNA in particular, are built from a six ring molecule with two nitrogen atoms, and thus a foundation molecule for these molecules as a target for anticancer and antiviral drugs. As purine is a fusion of a six- and five-membered ring system with plenty of nitrogen, it is vital to genetic material, as well as energy carriers such as ATP. These nitrogenous systems are important molecular building-blocks of many natural products and form a highly reliable chemo scaffold for the rational design of drugs, specific to the enzymes and receptors involved in cancer and other diseases [5].

4.2.3.2 Oxygen-Based Heterocycles

Heterocycles are oxygen based; oxygen atoms are incorporated within the ring systems and result in different reactivity and physical properties that are utilized in both natural products and synthetic drugs. Amongst them, furan is a representative example of a five-membered aromatic ring containing one oxygen atom, because of the electron-rich nature and the ability to undergo electrophilic substitution reactions, which cause it to be used for synthesis of pharmaceuticals and agrochemicals. Especially important is pyran, six-membered ring with an oxygen atom, which is a backbone for many carbohydrates and natural products and therefore involves biological recognition and metabolism. Additionally, the six-membered ring containing two oxygen atoms, dioxane, is widely used as a solvent in chemical industries due to its favorable solvation properties. These heterocycles are also made more polar and have the potential to increase water solubility and bioavailability due to the presence of oxygen [1].

4.2.3.3 Sulfur-Based Heterocycles

Heterocycles exhibiting sulfur-based heterocycles are composed of sulfur atom(s) in its ring structure that imparts significant electronic and lipophilic properties necessary to several industrial as well as medicinal applications.

Thiophene: Due to its five-membered aromatic ring containing one sulfur atom, it is used in organic electronics; it also exists in natural pigments and has a relatively high electron-donating capacity to offer robust π–π interactions in existing anticancer drugs targeting specific protein domains. Also, thiazole has been described, a five-membered ring with both a nitrogen and a sulfur atom, which is a crucial component of several drugs, including vitamin B1 (thiamine) and also an important feature of many therapeutic agents' pharmacophore design. In addition, the six-membered ring with two sulfur atoms, dithiane, is often utilized in organic synthesis as a protecting group for carbonyl compounds because dithiane is stable and reactive under numerous reaction conditions. Enhanced hydrophobic character from the incorporation of sulfur into these heterocyclic systems is often utilized for the purpose of improving cell membrane permeability as well as targeting intracellular pathways in anticancer therapy [3, 4].

4.3 Mechanism of Action of Heterocyclic Compounds

4.3.1 Interaction with Nucleic Acids: Intercalation and Alkylation of DNA

Anthracycline derivatives, doxorubicin, daunorubicin, epirubicin, and mitoxantrone, are a group of heterocyclic compounds of great importance used in cancer chemotherapy particularly in transition between induction and maintenance including in the double conjunction of AC (doxorubicin plus cyclophosphamide) for repetitive breast cancers (Figure 4.7) [8, 9]. They are characterized by an oxygenated tetracyclic anthraquinone core with sugar moiety, including daunosamine and nitrogen (such as

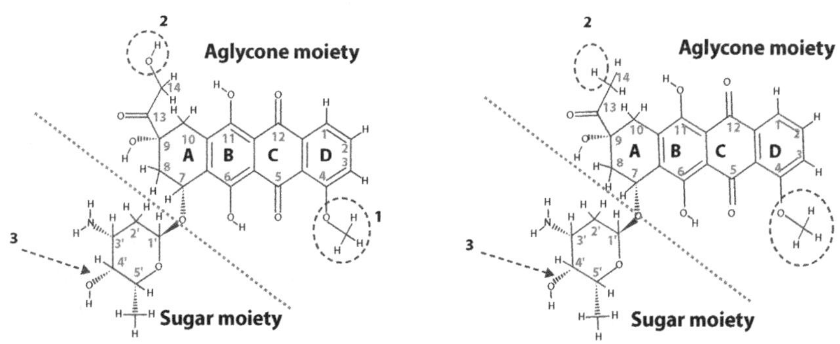

Figure 4.7 Chemical structure of doxorubicin and daunorubicin.

pyrrolidine C nucleosides). These heteroatoms are important for the anticancer activity of these halogenated congeners, as they are able to intercalate into DNA, inhibit topoisomerase II, and generate oxidative stress through redox cycling. The alkylating agent cyclophosphamide removes alkylated bases, thereby inducing AP sites in DNA, and recent studies have also found that anthracyclines, such as mitoxantrone, can bind to these AP sites *via* Schiff base forming a stable drug - DNA adduct that interferes with transcription and replication.

In these molecules, electron transfer takes place by means of the two quinone and hydroquinone groups present in adjacent rings, enabling the conversion of quinone to a semiquinone radical. This semiquinone radical reverts under aerobic conditions to quinone, yielding ROS on superoxide anion and hydrogen peroxide. These free radicals are also known to produce excess amount, so as to increase lipid peroxidation in cell membrane and DNA damage and retort to cell death. Because of this mechanism, these compounds show remarkable nonselective anticancer activity [10, 11] and are used in treatment of small cell lung cancer, breast cancer, and both lymphoblastic and myeloblastic leukemia.

A standard chemotherapy, known as AC chemotherapy, contains two key drugs: the cyclophosphamide and doxorubicin. The cyclophosphamide acts directly on the DNA of cancer cells by adding alkyl groups to it, leading to mispairing of DNA bases, which destabilizes the DNA and more often than not leads to cell death. Whereas doxorubicin acts as a stabilizer of the topoisomerase II and DNA complex, these two drugs work in conjugation such that the doxorubicin prevents the resolution of damage created by cyclophosphamide. Cyclophosphamide creates AP sites through deglycosylation, which are highly reactive, and anthracyclines such as doxorubicin contain amino groups, which form Schiff base adducts, which prevent DNA replication by creating bulky lesions.

This interaction of the two drugs was tested on various cell lines including, a 12-mer DNA duplex, which is a short, synthetic DNA sequence, a calf thymus DNA, and MDA-MB-231 human breast cancer cells. All the cell lines showed the interaction such that the doxorubicin and cyclophosphamide together prevented DNA replication, hence causing eventual cell death (Figure 4.8) [12].

The anticancer effects of anthracyclines are exerted through multiple mechanisms including topoisomerase II poisoning through increased topoisomerase II cleavable complex, ROS, and formation of DNA adducts (Figure 4.9). These drugs attach to DNA, disrupt its structure, and cause important cellular functions to stop working. The major mode of action is topoisomerase II poisoning, in which anthracyclines stabilize its

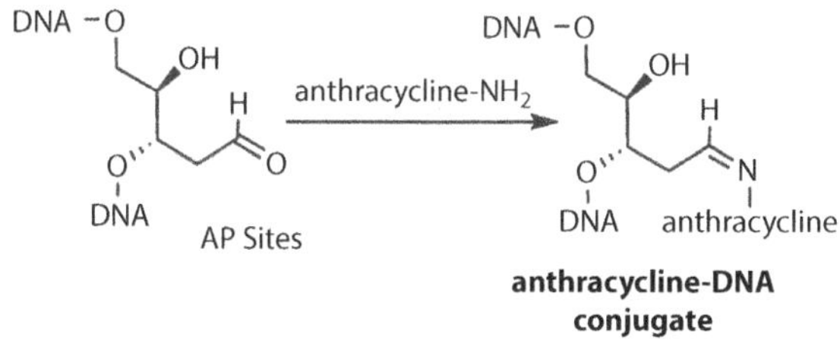

Figure 4.8 Interaction of anthracycline with DNA.

topoisomerase II–DNA complex and thereby prevent enzyme resealing of DNA breaks. Thus, this results in persistent double-stranded breaks that ultimately triggers apoptosis. Structural modification of anthracyclines in this pathway affects the effectiveness of these ligands in interacting with DNA and thus their effectiveness in this pathway [13–15].

The relation of oxidative stress to another major cytotoxic mechanism is also discussed. Anthracyclines are known to generate ROS [7] *via* redox cycling at the anthracycline quinone's moiety, which induces DNA damage and lipid peroxidation. In this process, semiquinone radical is formed, which donates electrons to oxygen and produces superoxide anions, hydroxyl radicals, and hydrogen peroxide, which are all culprits responsible for cell damage. In this process, mitochondria are involved and can finally cause apoptosis through oxidative stress. This fails further to counteract anthracycline cytotoxicity because the action of antioxidant defenses such as superoxide dismutase and catalase is overcome by these actions.

Other than these mechanisms, anthracyclines bind covalently to DNA with formaldehyde present. These bulky lesions generated by these adducts interfere with DNA replication and transcription. It has been shown that anthracycline-induced DNA adducts contribute *via* DNA strand breaks independent of topoisomerase II inhibition [14]. The process by which these adducts are formed is similar to that of cisplatin–DNA adducts and can be catalyzed by the cellular formaldehyde to which it can be raised by prodrug administration of anthracyclines to improve anthracycline efficacy.

Overall, the anthracyclines are highly effective because of the combined effects of topoisomerase II poisoning, oxidative stress, and DNA adduct formation, but this also contributes to their toxicity. These mechanisms then provide insight into why anthracyclines are so useful in chemotherapy; however, they also help explain why methods to reduce adverse effects of anthracyclines are needed.

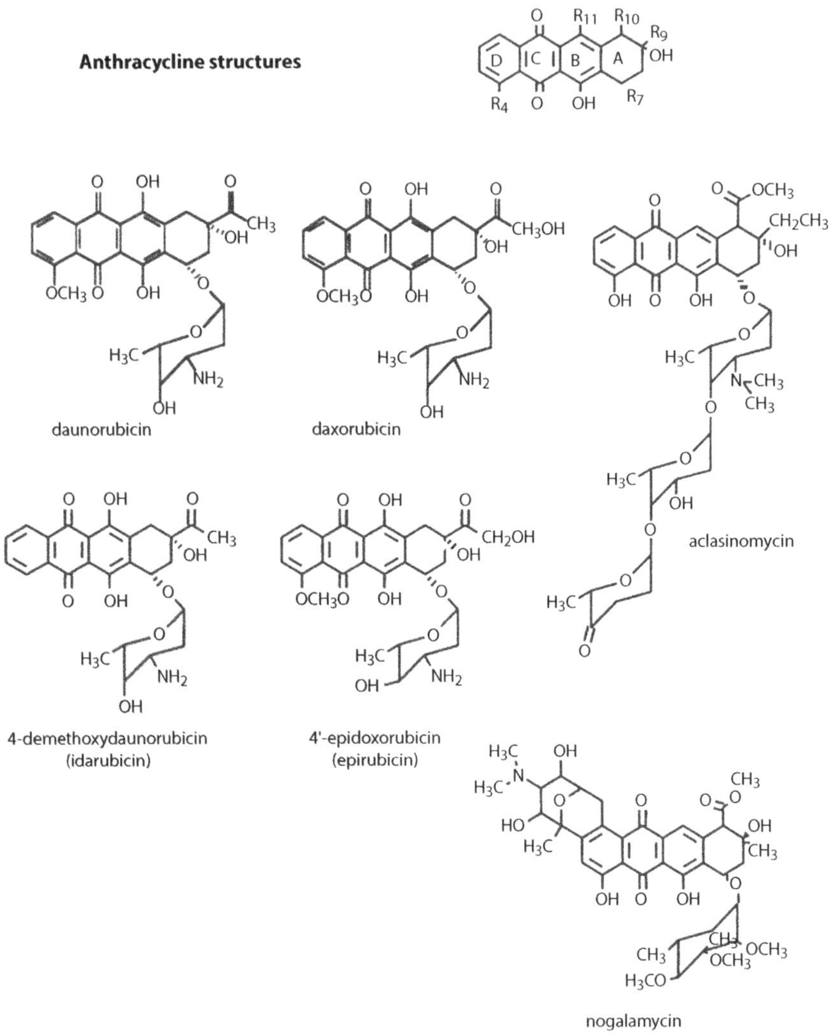

Figure 4.9 Anticancer compounds with anthracycline structures.

Chromophore is a planar aromatic system, which is essential for incorporation into DNA, which is a key component in intercalation of DNA by heterocyclic moieties. DNA intercalators are small molecules that incorporate into the base pairs stacked on each other, hence disrupting its structure and stability and furthermore preventing transcription and translation. The basic center of the intercalators contains a protonated nitrogen, which interacts with the DNA phosphates, and, finally, the groove binding moiety of the intercalator, which stabilizes the complex by binding to the major or minor grooves of the DNA [19].

The drug, doxorubicin, as discussed previously is an antibiotic containing all three components necessary for DNA intercalation, and hence, it is used to intercalate in various combinations with other drugs. Amsacrine, another nitrogen containing heterocyclic moiety, contains tricyclic acridine as its chromophore and a sulfonamide group for groove binding. Similar to doxorubicin, amsacrine also inhibits topoisomerase II along with DNA intercalation. Also, neutral ellipticine, another nitrogen-containing compound, performs intercalation combined with various drugs (Figure 4.10).

The rational design of quinoxaline-based derivatives (6a–h and 7a–g) as molecular hybrids of first-generation DNA intercalators, doxorubicin, amsacrine, and ellipticine, is depicted in the diagram. These newly synthesized compounds are designed to increase anticancer activity by improving their ability to intercalate DNA and inhibit topoisomerase II. Primary intercalating chromophore is a [1,2,4]triazolo[4,3-a]quinoxaline core that provides strong π–π interactions with DNA base pairs. Moreover, a

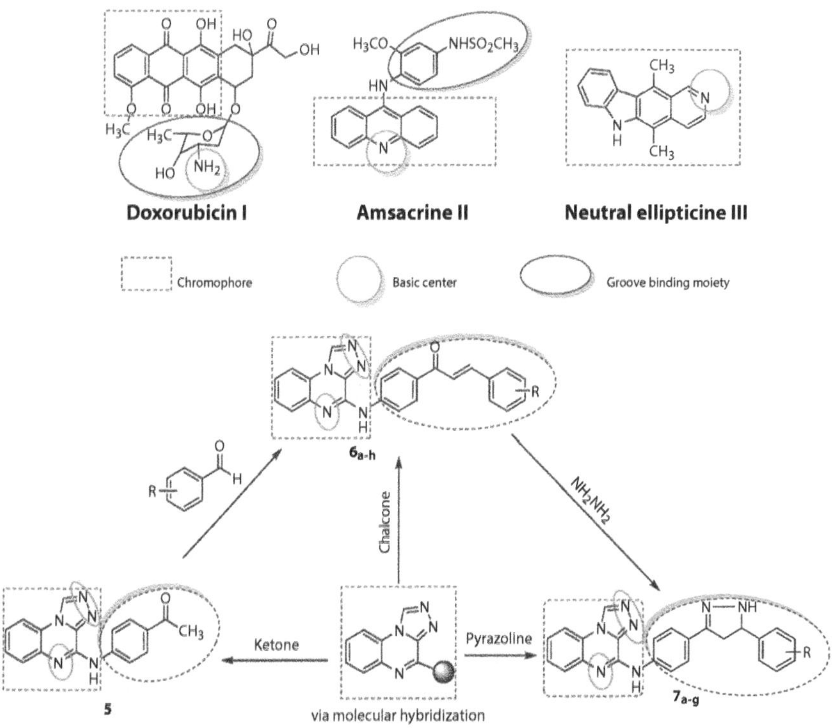

Figure 4.10 Approved DNA intercalators derivatives main pharmacophoric groups [19].

prop-2-en-1-one (chalcone) or pyrazoline linker is also present that controls the electronic properties of the molecule and thus affects its binding affinity and biological activity. Additionally, these linkers allow the compounds to disrupt DNA processes and hence are effective cytotoxic agents. Additionally, the substitution of a phenyl moiety enhances lipophilicity and electronic nature of the molecules and thereby affects their selectivity and potency against cancer cells. Integration of these structural modifications in quinoxaline derivatives allows their optimal DNA binding and topoisomerase II inhibition, thus making them promising candidates for heterocyclic anticancer agent development.

The anticancer potential of quinoxaline-based heterocyclic derivatives, namely, substituted [1,2,4]triazolo[4,3-a]quinoxalines, was investigated; two structural groups of quinoxaline derivatives were synthesized and their cytotoxicity effects to several cancer cell lines, such as MCF-7 (breast cancer), HCT-116 (colon cancer), and HepG2 (liver cancer). The first group 6a–h had a prop 2-en-1 one linker, and the second group 7a–g had a pyrazoline linker. Excessive modifications, especially substitution nature and position around the quinoxaline core, were significantly important to anticancer activity. Among the tested derivatives, 6e, 7b, 7c, 7e, and 7g displayed the greatest potency as anticancer agents with cytotoxicity stronger than the reference drug doxorubicin. The IC_{50} of these compounds was low, which showed that the tumor cell proliferation could be inhibited by them effectively. Further mechanistic studies further demonstrated that these derivatives acted on cancer cells by DNA intercalation and topoisomerase II inhibition, important mechanisms for killing cancer cells and causing cancer cell apoptosis.

Quinoxaline derivatives, specifically substituted [1,2,4]triazolo[4,3-a]quinoxalines, of two types based on their linker types, prop-2-en-1-one and pyrazoline linkers, were tested for various anticancer activities; one of which was the ability to inhibit cell cancer growth by preventing cell division. MTT assay, which is a colorimetric test used to test cell cytotoxicity and cell viability by measuring the amount of purple formazan product formed when interacting with cell lines [16–18]. Stronger intensity of purple color refers to a higher number of viable cells. MCF-7 human breast adenocarcinoma cell line, HCT-116 human colorectal carcinoma cell line, and HepG2 human hepatocellular carcinoma (HCC) cell line, as well as VERO, a noncancerous cell line from African green monkey kidney cells, were all tested against quinoxaline derivatives. 7e, 7c, and 7b showed greater potency than common anticancer drugs, and when tested on VERO cells, it showed lower toxicity, hence indicating that the derivatives are selective for cancer cells.

It is believed that the electronic properties, hydrophobic interactions, and capability of targeting rapidly dividing cancer cells with high DNA replication activity aid in the high selectivity of the best-performing derivatives toward cancer cells.

Their capacity to serve as DNA intercalators and topoisomerase II inhibitors, as well as a disruption of cancer cell proliferation, is the basis for these cancer cell interactions that the synthesized heterocyclic derivatives exhibit with them. When the DNA double helix is stacked, a planar aromatic or heterocyclic molecule intercalates and inserts itself between the stacked base pairs. The result of this process is to distort the DNA structure, much to what essential biological functions such as replication and transcription rely upon. The [1,2,4]triazolo[4,3-a]quinoxaline derivatives possess conjugated systems and hydrophobic substituents that allow strong intercalative interactions with DNA that lead to potent cytotoxicity toward cancer cell. Quantitative measurement of the binding affinity for these compounds was performed by obtaining their IC_{50} values, from which compounds such as 7e showed the highest potency as they intercalate DNA at lower concentrations than doxorubicin, a well-known chemotherapeutic agent.

These derivatives beyond intercalation target topoisomerase II [20], an essential enzyme that untangles the strands and relieves intrastrain supercoiling during replication. Topoisomerase II has the ability to introduce transient double strand breaks to relax supercoiled DNA before resealing the strands. However, the synthesized compounds inhibit topoisomerase, and this leads to accumulation of DNA breaks because it does not bind to a re-ligated form. Once triggered, DNA damage responses (DDRs) trigger an end for the fate of cancer cells, sending them toward apoptosis. IC_{50} values were also evaluated from inhibition potency, and compounds such as 7e and 7c exhibited such potency equal to that of doxorubicin. The molecular docking studies validate that these compounds fit reasonably well inside the topoisomerase II active site, supporting their mechanism of action, and the findings are consistent with what is observed [21].

4.3.2 Inhibition of Key Enzymes and Enzymatic Pathways (Targeting Specific Cancer Pathways)

Our understanding of cancer cells' unique metabolic adaptation referred to as the Warburg effect is the redirection of ATP production using aerobic glycolysis instead of oxygen even in its presence. However, when mitochondrial respiration and ATP production are inhibited, cancer cells make up for the loss by stimulating increased glycolysis to support their energy supply [22–24]. However, such a metabolic shift can be prevented by blocking hexokinase, the glycolytic enzyme at the moment of the first step. Oxidative stress then results and ROS production is upregulated that disrupts mitochondrial function and activates AMPK. Any effect of AMPK phosphorylation includes activating p53 and Bax, cell arrest, mammalian target of rapamycin (mTOR) inhibition, and reduced protein synthesis. The net result of steps like these is that it damages cancer cells, thereby making them more vulnerable to oxidative damage and impaired energy status. In some instances, cells go into a nonproliferative state slowing down the tumor progression, whereas some cells undergo apoptosis if they are not able to compensate for the loss of ATP.

In the design of these novel anticancer compounds, they have targeted both mitochondrial complex I and hexokinase, more specifically Schiff base compounds containing certain heterocyclic structures such as quinazolinone and indole (Figure 4.11). Molecular docking studies were also carried out to evaluate the compounds binding with key enzymes in cancer metabolism. In addition, pharmacokinetic, distribution, metabolism, excretion, and toxicity *in silico* studies were performed to predict ADMET. Further analysis of the structure–activity relationship (SAR) was carried to determine the most promising compounds in affecting complex

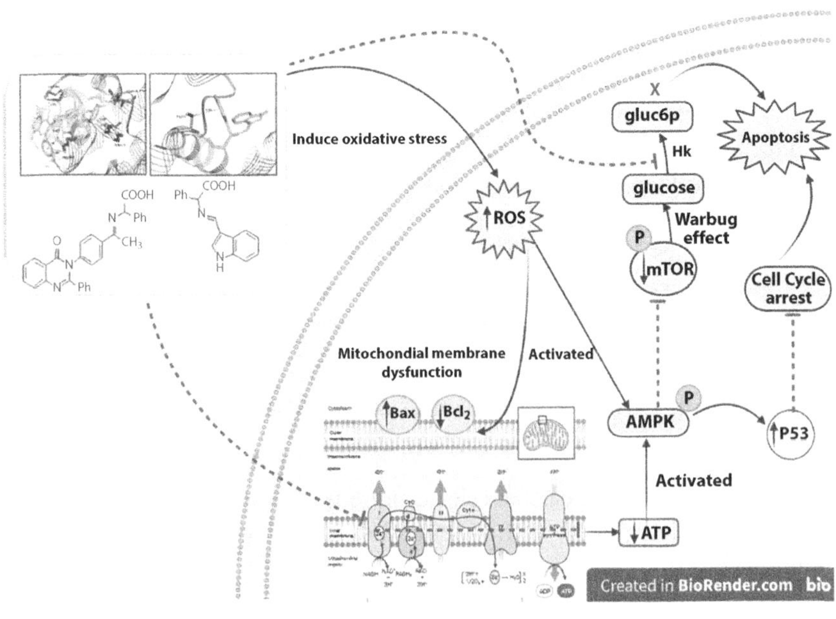

Figure 4.11 Schematic diagram of the signaling pathways induced mitochondria-associated hexokinase mediated apoptosis [25].

I mitochondrial inhibition or hexokinase inhibition. *In vitro* experiments were finally performed to confirm their ability to disrupt cell metabolism of cancer cells interfering with the AMPK/mTOR signaling pathway and inhibiting the Warburg effect [25].

Oxidative stress is assessed by measuring malondialdehyde (MDA) levels and reduced glutathione (GSH) levels in the cells. The compounds were shown to increase levels of MDA in cancer treated cells and decrease GSH levels. To assess hexokinase activity, MCF-7 and MDA-MB-231 breast cancer cells were cultured in six-well plates for 24 hours to adhere under optimal conditions. After 48 hours of treatment of the compounds to the cells, the cells were harvested, and their lysates were extracted. Human enzyme-linked immunosorbent assay kits (Sigma-Aldrich, MAK091) were used to measure the hexokinase activity levels of cells as instructed by the manufacturer. Hexokinase is a key enzyme in glycolysis, and its activity is often abnormally high in tumor stem cells for supporting rapid energy production. The same cell lines were cultured, treated, washed with phosphate-buffered saline, scraped, and cell collected. The collected cells were lysed in a lysis buffer containing Tris HCl, NaCl, Na_2EDTA, Triton, and sodium pyrophosphate to break the cells and release intracellular components. After this, cellular debris were separated from the supernatant by centrifugation of lysates at 15,000 rpm and 4°C for 15 minutes. MDA, a marker for lipid peroxidation, and reduced GSH, the key antioxidant, were measured. Furthermore, the protein concentration in the samples was determined by the Bradford assay, displaying the total protein content *via* dye binding [25].

Pyrazole-thiazole-coumarin 10 (PTC10) was found among tested compounds with the most significant anticancer activity as dual drug-targeting multiple cancerogenic pathways, including RAS/MAP kinase and PI3K-AKT pathways, essential for survival, proliferation, and apoptosis in HCC cells (Figure 4.12) [26]. This indicates that the pathways involving PTC10

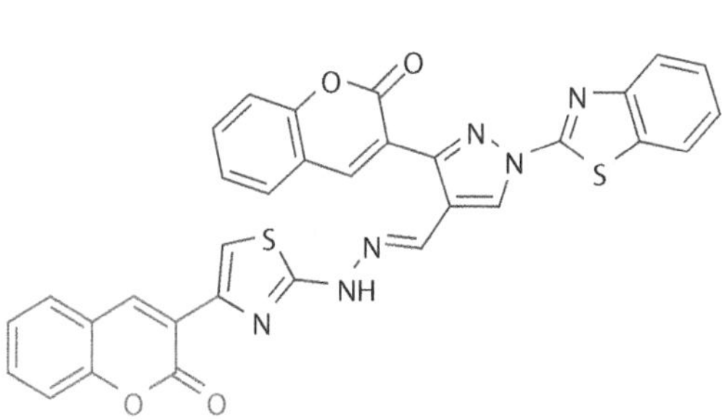

Figure 4.12 Structure of PTC10 [26].

are well suppressed and are effective in disrupting the major signaling networks controlling tumor growth and progression. Three key biological effects were established on the mechanism of action of PTC10. It first arrested the cell cycle at the G2 phase, a crucial checkpoint between the cell cycle and mitosis (M phase).

PTC10 was able to halt tumor expansion by inducing uncontrolled division of the cell by stopping the progression of the cell cycle. Second, the compound induced apoptosis, a process that cancer cells generally do away with to keep growing out of control. The results showed that treatment of cells with PTC10 produced apoptotic markers and apoptosis characteristic morphological changes. Third, PTC10 effectively inhibits the heat shock protein network, including its most significant member, as a molecular chaperone, HSP90, which protects cancer cells from stress-induced apoptosis. Interaction of HSP90 with Cdk4/Cdc37, a complex for cell cycle progression, has been known for stabilizing oncogenic proteins. PTC10 inhibited cancer cell survival mechanisms by interfering with cell cycle arrest as well as apoptosis by inhibiting HSP90. To validate that PTC10 could indeed inhibit HSP activity, ATP-hydrolyzing and aggregation assays were performed to show its ability to disrupt ATP-dependent HSP function, proving that PTC10 indeed drives cancer cell death. PTC10 holds a high promise as HCC therapy candidate due to its ability to target multiple oncogenic pathways, induce apoptosis, and impede HSP-mediated survival pathways. PTC10's multitargeting nature is notably different from conventional treatments that restrict their treatment's scope and often lead to resistance, and it is much less likely for cancer cells to develop resistance and yet maximize therapeutic efficacy.

In addition, HSP inhibition does not only promote apoptosis but also prevents the stabilization of additional oncogenic proteins and thus further enhances its antitumor activity. The results of this study highlight the possibility to use pyrazole–thiazol–coumarin derivatives as a therapeutic strategy to MHC progression and PTC10 as a highly potent multitarget inhibitor of the disease. PTC10 represses RAS/MAP kinase and PI3K–AKT pathways, arrests cancer cells in G2 phase, blocks HSPs functions, and offers a novel and putting drug template for HCC future treatments.

Induction of HSP90 aggregation is the mechanism of PTC10. If HSP90 cannot hydrolyze the ATP, then, when the protein is HSP90, conformation becomes unstable and results in the protein misfolding and aggregation. It may induce endoplasmic reticulum (ER) stress leading to unfolded protein response (UPR) and finally promoting cancer cell apoptosis. HSP90 acts as a multicomponent complex with co-chaperones, such as Cdc37 and p23, associated with stabilizing client kinases. Disrupting the hydrogen bonding and hydrophobic areas between HSP90 and its co-chaperone prevents the stabilization of cancer-promoting proteins such as Cdk4, Akt, and Raf1. PTC10 is likely to damage the hydrogen bonding and hydrophobic interactions between HSP90 and its co-chaperone.

PTC10 is believed to bind the ATPase domain and antagonize this reaction, thus preventing ATP hydrolysis, a requisite HSP10's chaperone activity. HSP90 cannot undergo conformational changes that are necessary for stabilizing oncogenic client proteins without ATP hydrolysis, and they are degraded through the ubiquitin–proteasome pathway (Figure 4.13).

Hospital pharmacology views nitrogenous heterocycles including isoxazoles and their counterparts as essential in medicinal chemistry due to their capability to serve as ATP mimics to bind key biomolecules, thus becoming strong inhibitors for EGFR-TK and VEGFR-2 TK [27, 28]. The heterocyclic structures freeze EGFR and VEGFR-2 kinase action by three processes including competitive competition once they connect to the ATP receptor through hydrogen bonds and π–π stacking and van der Waals contacts to stop ATP binding, which switches on cancer-causing signaling

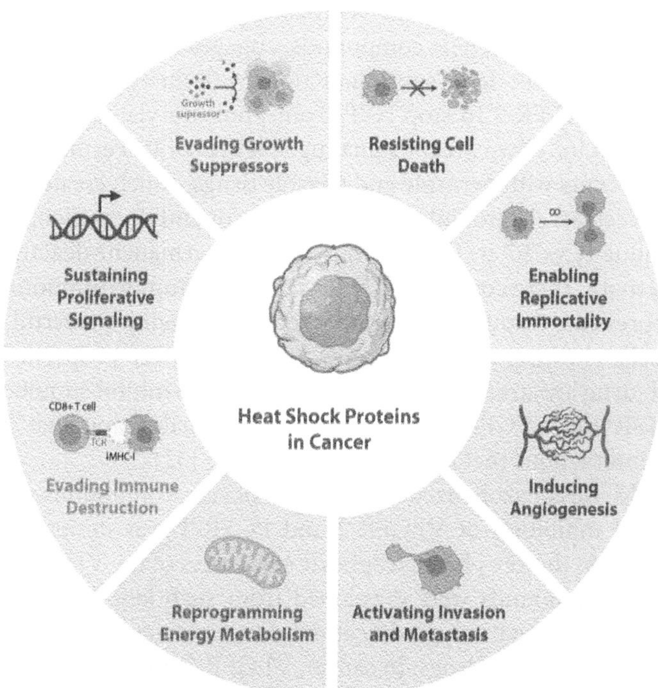

Figure 4.13 Schematic illustration of effect of heat shock proteins in cancer.

routes [29]. Letrozole and anastrozole in the triazole drug group show specific affinity for Thr790 Lys745 Met793 ATP-binding site residues, which stops EGFR phosphorylation. Heterocyclic compounds with acrylamide functional groups permanently inactivate protein kinase activity after forming covalent bonds with cysteine residues such as Cys797 in EGFR [30]. The ATP-binding site of VEGFR-2 becomes inhibited through the binding mechanism of triazine- and azobenzene-based inhibitors, which results in suppressed angiogenesis and tumor growth prevention [31]. Dual inhibitors based on the pyrazole compound family (including sorafenib) use common features from EGFR and VEGFR-2 to establish simultaneous blocking capabilities for enhanced anticancer properties. Photo-induced *cis–trans* transformations of azobenzene derivatives trigger selective kinase inhibition control in photo-controlled PROTAC systems that target protein degradation. These special molecular interactions combined with structural changes establish nitrogenous heterocycles as primary anticancer medications that provide targeted therapy with minimal harm to promote the synergistic potential needed to block cancer cell pathways.

Heterocycles exhibit two types of inhibitory mechanisms, which include competitive and covalent inhibition. This group of inhibitors including triazole-based drugs (letrozole, anastrozole, vorozole) and triazine derivatives, as well as azobenzene compounds, engages in hydrogen bonds and π–π stacking and van der Waals interactions with the ATP-binding pocket of EGFR or VEGFR-2 to stop ATP from phosphorylating tyrosine residues essential for downstream signaling (Figure 4.14). Permanent inhibitor action occurs with pyrazole and triazole drugs, which create a covalent bond between an electrophilic acrylamide agent and the nucleophilic cysteine residue (Cys797 in EGFR), thus causing permanent deactivation of kinase activity. These compounds generate lasting tumor-promoting pathway suppression through inhibition therefore proving powerful against cancer cells showing drug resistance. The anticancer properties of the sorafenib drug are strengthened due to its ability to inhibit simultaneously both EGFR and VEGFR-2 pathways through dual action that prevents pathway interactions that lead to drug resistance [32].

A novel series of azobenzene-based heterocyclic compounds with affinity for dual inhibition of VEGFR-2 and EGFR-T was designed (Figure 4.15). Based on the key pharmacophoric structures of the VEGFR-2 inhibitors, these compounds were developed from such key pharmacophoric structures with structural modifications from sorafenib and pazopanib, two well-known cancer drugs. In cancer, VEGFR-2 inhibition can suppress blood vessel formation (angiogenesis) and therefore limit tumor growth; four pharmacophoric features are shared by many VEGFR-2 inhibitors including sorafenib. There is the terminal hydrophobic part that occupies the allosteric hydrophobic fissure in the drug–receptor complex by hydrophobic interaction, as well as pharmacophore linker amino/urea groups, which binds and interacts with the critical amino acids Glu883 and Asp1044 by forming two hydrogen bonds necessary for receptor inhibition. Central hydrophobic spacer will help with the compound fitting nicely into the receptor's binding pocket. A flat heteroaromatic ring system greatly enhances binding affinity and selectivity. To retain drug-likeness, the researchers applied bioisosteric modifications, structural additions, or substitutions made on a natural molecule to improve selectivity, potency, or pharmacokinetics of a given molecule. Inspired by sorafenib and pazopanib, these modifications were based on replacing functional groups to increase VEGFR 2 binding. Interactions are improved with the introduction of heterocyclic moieties, because heterocycles are highly biologically compatible. Focusing on modification of the hydrophobic and hydrogen-bonding regions for pharmacologically better effects can thereforelead

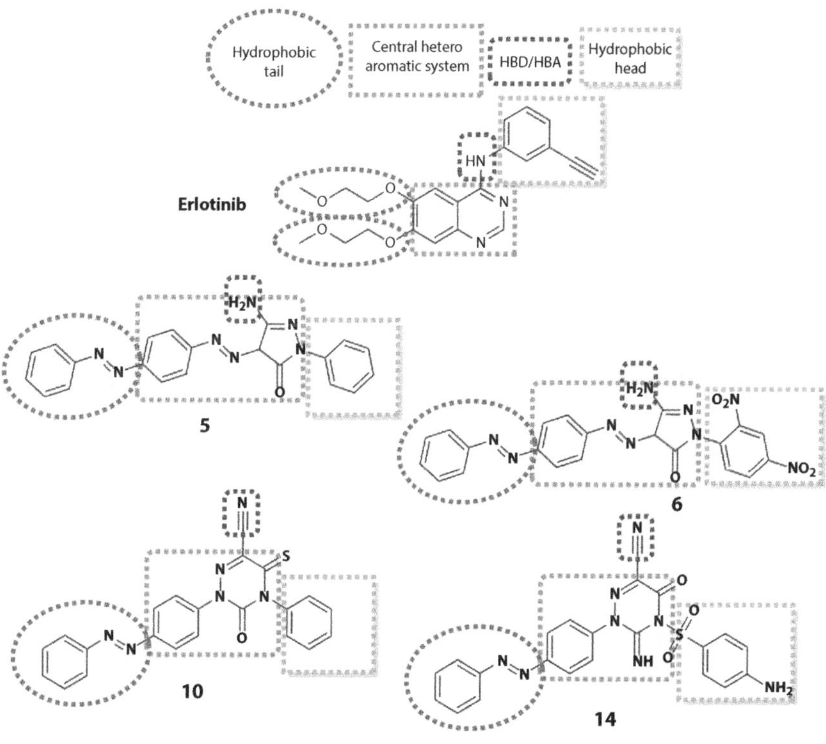

Figure 4.14 Design rationale based on structural characteristics of EGFR inhibitors [32].

to enhanced binding affinity and overall therapeutic efficacy. In addition to VEGFR-2, the newly synthesized compounds had EGFR-TKI (epidermal growth factor receptor tyrosine kinase inhibitor)–type phosphorylating domains, and these compounds were dual antagonists. Regarding these novel compounds, these four key features of EGFR inhibitors are all present: lipophilic head occupies hydrophobic pockets as first-generation EGFR TKIs (4.14). EGFR has an adenine adherence pocket interaction, such that the heteroaromatic ring of adenosine fits within the adenine binding region of ATP.

Strong hydrogen bond formed by drug–receptor interaction: The linker domain with the hydrogen bond donor/acceptor (e.g., NH fragment).

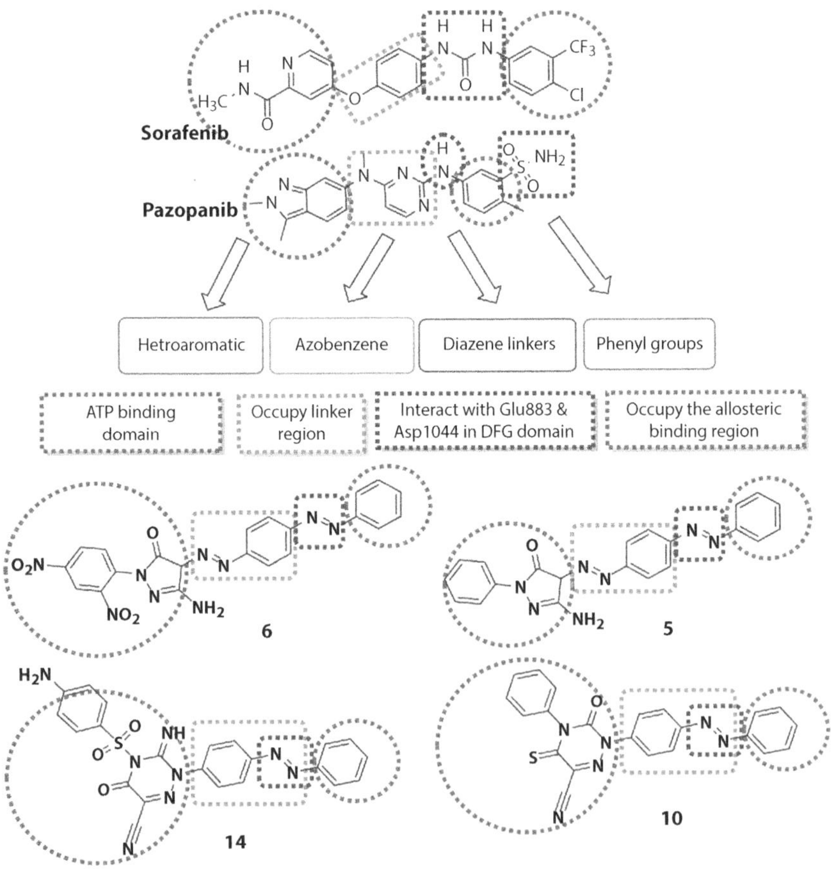

Figure 4.15 Pharmacophoric and structural similarities of some designed compounds and VEGFR-2 inhibitors [32].

Hydrophobic tail: Connects to hydrophobic region II, crucial for EGFR inhibition [32].

4.3.3 Modulation of Cell Signaling Pathways and Cell Cycle

The aggressive nature of prostate cancer along with castration-resistant prostate cancer (CRPC) shows resistance to conventional chemotherapy treatments. Research on natural compounds with anticancer capabilities has grown extensively because (+)-brevipolide H (Figure 4.16) shows long-lasting anticancer activity against CRPC PC-3 cells. A combination of pathways links together to express the effects of (+)-brevipolide H through cell cycle arrest and apoptosis development and oxidative damage and calcium regulation disturbances and metabolic mitochondrial breakdown.

The intracellular signaling cascade known as Akt/mTOR/p70S6K pathway strongly controls cell cycle procedures as well as protein synthesis and cellular metabolic processes. The pathway becomes unusually active in different types of cancer such as prostate cancer, thus promoting constant cellular growth as well as prolonged survival capabilities.

Akt (protein kinase B) receives activation signals from upstream signals until it phosphorylates numerous substances, which promote cell survival and growth. Phosphorylated Atk activates mTOR. mTOR regulates all the protein synthesis and cell growth by integrating signals related to nutrient

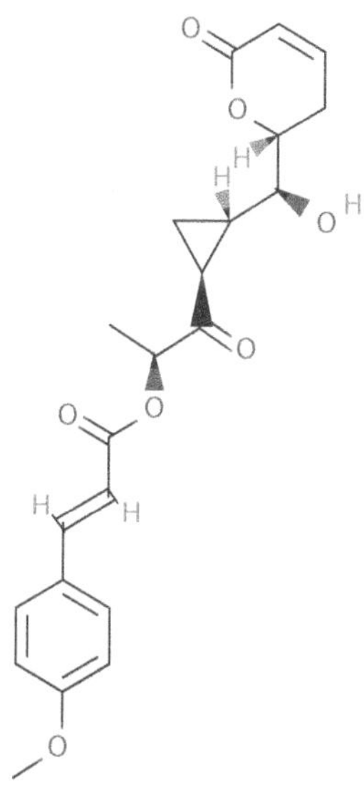

Figure 4.16 Structure of brevipolide H ($C_{22}H_{24}O_7$).

availability. The downstream effector p70S6K functions as a biochemical element of mTOR by phosphorylating S6 ribosomal protein for required cell cycle progression through enhanced protein synthesis.

(+)-Brevipolide H decreases phosphorylation at both Thr308 and Ser473 residues on Akt, which are essential sites for its full activation (Figure 4.17). Akt's ability to forward survival and growth signals becomes diminished because of this reducing effect. The substance decreases mTOR Ser2448 phosphorylation, which demonstrates an activated state of mTOR. The compound interferes with mTOR activation, making it unable to regulate cell growth and protein production activities. The activation process of p70S6K at its Thr389 regulatory site becomes highly suppressed after receiving (+)-brevipolide H medication. The compound limits S6 ribosomal protein phosphorylation, which subsequently suppresses the necessary protein synthesis for cellular growth.

The severity of mitochondrial stress determines whether cells will die during this condition. ROS are reactive oxygen molecules that include three distinct forms of oxygen-based species, which are superoxide radicals and hydrogen peroxide and hydroxyl radicals. The compound (+)-brevipolide

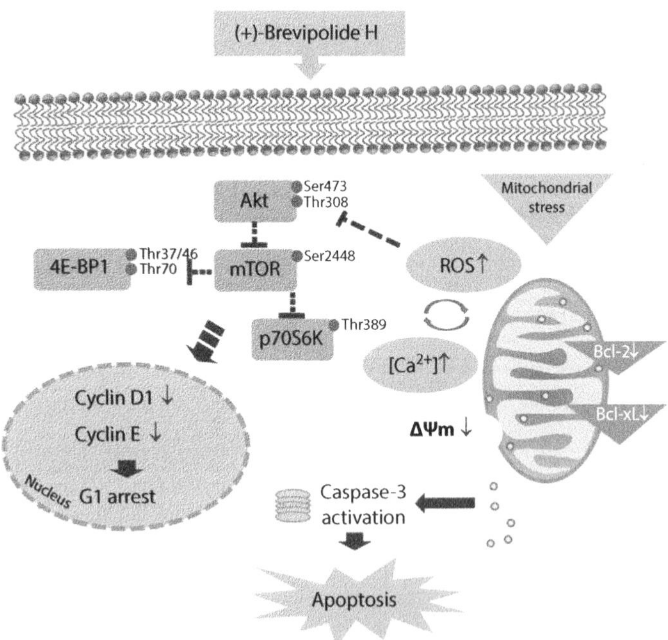

Figure 4.17 Mechanism of action of (+)-brevipolide H leading to apoptosis and G1 arrest in cell [33].

H increases ROS production, thereby causing oxidative stress that leads to mitochondrial damage. The cellular process of increasing cytosolic Ca^{2+} by (+)-brevipolide H disrupts mitochondrial functions as well as leads to muscle contraction failure alongside neurotransmission breakdown and apoptosis development. High levels of Ca^{+2} within the cell produce mitochondrial permeability transition pore opening, which results in lowering mitochondrial membrane potential and cell death. The damage on mitochondrial membranes generated by ROS results in unrestricted Ca^{2+} entry. The continuous increase of Ca^{2+} within the cell stimulates ROS production in an escalating cycle, which intensifies stress to both cellular and apoptotic processes.

Bcl-2 (B-cell lymphoma 2) and Bcl-xL (Bcl-extra long) work as antiapoptotic factors to protect mitochondrial structure by stopping cytochrome c from escaping, which would begin apoptosis. The action of (+)-brevipolide H leads to the suppression of Bcl-2 and Bcl-xL proteins, thus enabling proapoptotic protein Bax and Bak to perform their function of forming pores in the mitochondrial membrane. When Bcl-2 and Bcl-xL become suppressed, the following events take place: the mitochondrial membrane permeability rises, and cytosolic release of cytochrome c activates caspases (apoptosis-driving proteins) along with a loss of mitochondrial membrane potential that disrupts ATP production, which results in cell death. The signaling component Akt functions as a protein kinase B, which preserves cells through growth and survival functions in addition to regulating metabolic activities. During its activation, phase Akt phosphorylates multiple proteins to enable them both cell survival mechanisms and proliferation triggers.

mTOR: A central regulator of cell growth, protein synthesis, and metabolism. The protein kinase Akt activates mTOR, which allows the enzyme to control cell cycle progression along with elevating the translation of growth-promoting proteins. The activated mTOR targets p70S6K, which functions as a ribosomal protein S6 kinase that promotes cell proliferation through ribosomal protein phosphorylation [33].

The cell cycle requires strict control to conduct its sequential phases during cell growth together with cell proliferation [16]. The cell cycle contains three stages including the G0/G1 phases and S phases and G2/M phases based on research by Scudiero *et al.* [17]. DNA replication happens in the G0/G1 phase because the phase needs mRNA and protein synthesis for this process [18]. Cell cycle disruption serves as a major sign to detect cancer. Recent investigations demonstrate that proteins known as cyclins and CDK complex and other proteins act as vital elements in cell cycle progression [19]. Numerous clinical trials provided laboratory data about how many anticancer medications block tumor development through cell cycle arrest processes [20]. The formation of cell cycle arrest in tumors leads to suppressed cell proliferation and triggered programmed cell death [21]. Programmed cell death through cell apoptosis functions as a mechanism linked with tumor genesis and progression. Different mechanisms including gene regulation and signal transduction appeared in this process. The procedure of inducing cell death functions as an effective therapeutic strategy for cancer management.

The protein caspase-3 serves as a primary driver of programmed cell death through apoptosis to maintain proper cellular balance. When activated caspase-3 performs an operation that cleaves the vital cellular substrate poly(ADP-ribose) polymerase (PARP) until apoptosis occurs. Many drugs used in anticancer therapies activate caspase systems to achieve their apoptotic effects because of this apoptotic mechanism. During cancer development, the essential role belongs to the nuclear factor κB (NF-κB) signaling pathway. The transcription factor family NF-κB facilitates inflammatory responses and plays a role in the formation of lung cancers along with breast and liver and pancreatic cancers as well as lymphomas. The NF-κB signaling factor exists in an inactive state in cellular cytoplasm bound to its IκB inhibitor protein. The IκB kinase (IKK) complex triggers IκB phosphorylation and then allows ubiquitination, which results in proteasome destruction. NF-κB becomes available when the cell releases this factor, which enables the protein to move into the nucleus where it activates proinflammatory and survival-related gene expression of interleukin-6 (IL-6), tumor necrosis factor α (TNF-α), and intercellular adhesion molecule 1 (ICAM-1). Research indicates that NF-κB plays a dominant role as a therapeutic target because it regulates immune response signaling and controls cellular survival as well as prevent cell death pathways. Research has evidenced that NF-κB pathway suppression in lung cancer makes *Polygonatum sibiricum* polysaccharides anticancer agents, yet low selenium doses can stop prostatitis-caused prostate cancer through blocking NF-κB nuclear translocation. NF-κB control of tumorigenesis makes its inhibition into a potential therapeutic approach for treating cancer diseases. The exploration of natural products as anticancer pharmaceuticals has grown ring the last several decades because these substances attack various pathways effectively but overcome limitations present in single-drug therapies including secondary adverse effects and drug resistance. Fungi fibrinolytic compound 1 (FGFC1) represents a unique molecule discovered in marine fungus *Stachybotrys longispora* FG216 that consists of 869-Da molecular weight (Figure 4.18).

The fungi-produced fibrinolytic compound 1 (FGFC1) effectively interacts with cells from non–small cell lung cancer (NSCLC) (Figure 4.19). The anti-inflammatory agent FGFC1 works through inhibition of the NF-κB

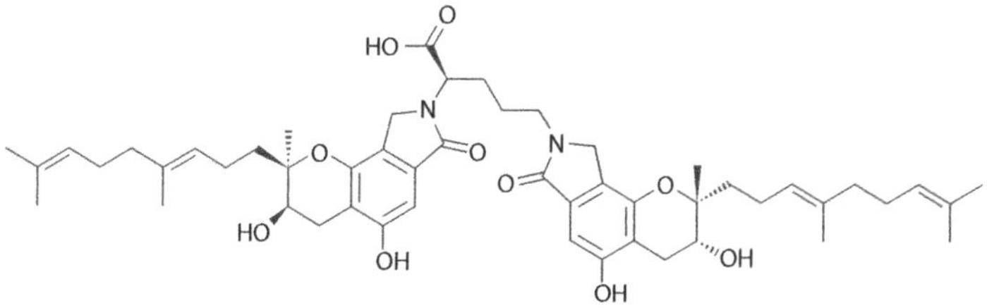

FGFC1

Figure 4.18 Structure of heterocyclic compound "fibrinolytic compound 1" or FGFC1.

signaling pathway, which controls both inflammatory reactions and cell survival and cancer development. IKK elements IKKα and IKKβ function as the typical initiators of the pathway through IKK complex phosphorylation. Activation makes these kinases phosphorylate IκBα after which IκBα becomes ubiquitinated and then gets degraded. The p65 subunit of NF-κB protein is able to enter the nucleus where it activates the transcription of inflammation-related genes including IL-6, IL-1β, and TNF-α as well as promoting cell cycle regulation genes such as cyclin D1 and CDK4 that both support cancer cell proliferation.

Through its action of suppressing the phosphorylation of IKKα, IKKβ, and IκBα, FGFC1 stops NF-κB from activating. The inhibition by FGFC1 blocks inflammatory cytokine (IL-6, IL-1β, and TNF-α) expression,

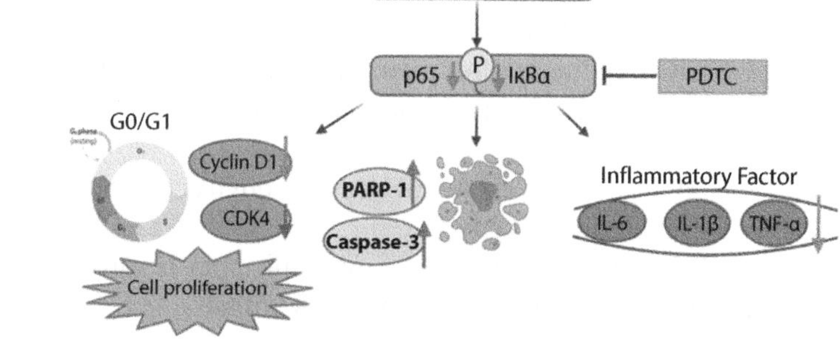

Figure 4.19 Mechanism of effect of FGFC1 on NSCLC cells [34].

which decreases the cancer-promoting inflammatory microenvironment. Through its mechanism, FGFC1 reduces the cellular expression of cyclin D1 and CDK4, which normally drive G1/S transition in the cell cycle. FGFC1 creates a cell cycle detention at the G0/G1 phase, which halts cancer cell division.

The inhibition of NF-κB signaling through FGFC1 treatment creates an increase in apoptotic activity based on the elevated expression of caspase-3 and PARP-1. Programmed cell death execution is primarily carried out by caspase-3, but PARP-1 cleavage serves as a diagnostic marker of that death process. Apoptosis in NSCLC cells results from the cumulative action of these cellular effects that lead to their elimination. Research shows that the inhibitor PDTC (pyrrolidine dithiocarbamate) stops IκBα phosphorylation that the diagram demonstrates as depicted in the diagram. FGFC1 demonstrates potential as an NSCLC therapeutic candidate because it combats cancer through mechanisms that involve inhibiting NF-κB signaling and both cell cycle arrest and inflammation suppression [34].

4.3.4 Interaction with Immune System

Medical experts believed chemotherapy inhibited immune responses in the past until modern research revealed selected anticancer drugs including those built with heterocyclic scaffolds trigger tumor-specific immune responses by means of immunogenic cell death (ICD). The process reestablishes immunosurveillance because it enables the body to detect and eliminate cancerous cells. ICD induction demonstrates exceptional rarity because only less than 1% of anticancer agents can trigger this effect. Heterocyclic ICD inducers show great potential in cancer immunotherapy because their rational design provides effective therapeutic solutions. Mice achieving tumor regression became the foundation that established the concept of ICD during 2005 through studies using doxorubicin, which is a heterocyclic anthracycline. The development of heterocyclic ligand-containing platinum and ruthenium and copper complexes represents promising new ICD inducers after platinum was discovered in 2005. The compounds belong to two groups recognized as type I and type II.

- Type I ICD inducers create ER stress through actions that do not target the ER directly.
- Type II ICD inducers have improved efficiency because they directly generate ROS that activate ER stress.

Researchers have proven the promising qualities of heterocyclic gold-based compounds in this field during recent developments. Research on the cytotoxic effects of gold(I) complexes with heterocyclic ligands exists, yet their ability to induce ICD emerged as a recent investigation topic (Figure 4.20). Researchers have revealed that a gold(I) complex containing two N-heterocyclic carbene (NHC) ligands induces tumor apoptosis and immune activation to provide novel heterocycle-based ICD inducer methods. Two parallel processes enable this mechanism to function.

1. The gold complex interacts with thioredoxin reductase 2 (TrxR2) in mitochondria to maintain cancer cell redox balance as part of its antioxidant system targeting mechanism. When TrxR2 becomes inhibited, ROS concentrations grow high enough to create oxidative stress that culminates in cell death through apoptosis.
2. The heterocyclic gold complex triggers excessive redox cycling that produces excessive ROS, which destroys the proper protein folding mechanism within the ER. The UPR starts when the protein misfolding takes place, which eventually results in cancer cell death while releasing damage-associated molecular patterns that activate the immune system.

The process of cell death by dying tumor cells creates a sort of "anticancer vaccine," which stimulate dendritic cells and T cells to destroy

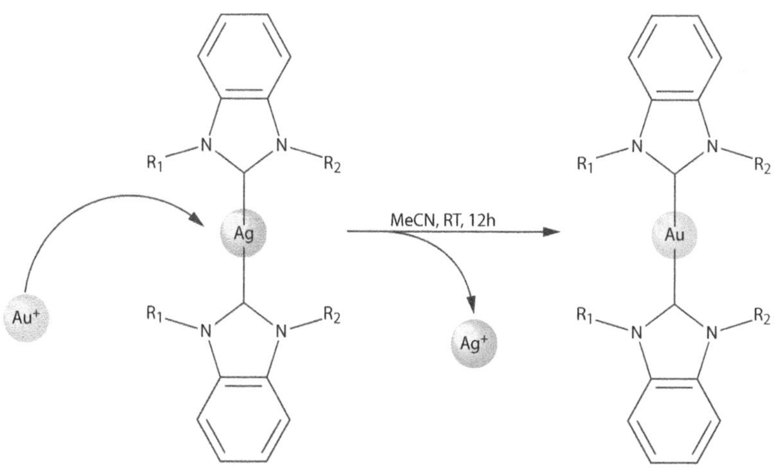

Figure 4.20 Transmethylation to create gold complexes with heterocyclic ligands [35].

the tumor. The heterocyclic core within gold complexes serves essential roles in redox functions and stability and selectivity that promotes its potential as a framework to develop future ICD inducers. The study demonstrates heterocycles have great promise as anticancer agents because they simultaneously kill tumor cells while activating immune responses, which creates possibilities for combined chemotherapy and immunotherapy treatments [35].

4.3.5 Inhibition of Tumor Growth and Metastasis

The main causes behind cancer relapse and unacceptably high mortality exist through invasion and metastasis progression. The initiation point for metastasis begins when epithelial cells undergo a biological change called epithelial–mesenchymal transition (EMT) that causes them to acquire mesenchymal cell characteristics for enhanced cellular motility and invasive capabilities and stem cell–like properties. During transition, cancer cells can travel through blood or lymphatic pathways to create secondary growths of tumors in different body locations.

Vimentin (VIM) functions as a critical EMT biomarker because it serves as a protein that keeps cytoskeleton structures organized and ensures correct organelle placement and facilitates cell movements and signal-conduction processes within the cell. The expression of VIM protein at elevated levels acts as a biomarker to identify aggressive epithelial cancers frequently appearing in lung cancer and breast cancer and prostate cancer together with malignant melanoma cells. The elevated amounts of VIM protein create more mobile cells with higher invasive capabilities during the metastatic cascade that transports cancer cells between tumors and distant organs. High VIM expression levels act as an independent indicator of poor prognosis because these levels predict unfavorable outcomes for both NSCLC patients and triple-negative breast cancer (TNBC) patients.

Hitting VIM presents great potential as a therapeutic strategy because advanced-stage cancers show marked resistance to chemotherapy and antihormonal therapy. Very few inhibitors and disruptors exist to target VIM, and researchers need to develop these compounds further by studying their biomolecular interactions. The spreading of cancer throughout the body remains the main cause behind cancer relapses as well as high mortality levels. EMT represents a biological process through which cancer cells from epithelial origins become mesenchymal-like cells that enhance their mobility and invasiveness in addition to acquiring stem cell attributes. Through this process, cancer cells break away from their original tumor then travel through bloodstream or lymphatic system while creating secondary tumors in different organs.

The intermediate filament protein VIM serves as an essential biomarker of EMT because it supports cell cytoskeletal structure and signal transduction pathways as well as organelle positioning and cell movement functions. Scientific studies demonstrate that aggressive cancer tumors showing high VIM levels exist in lung cancer, breast cancer, prostate cancer, and malignant melanoma. Higher expression levels of VIM create conditions that enable cancer cells to move toward distant sites and enter the multistep metastatic cascade that involves tumor progression to other organs. Higher VIM levels function as indicators for unfavorable disease outcomes when these levels appear in patients managing NSCLC and TNBC.

Advanced-stage cancers become susceptible to treatment when VIM becomes the pharmaceutical target due to their resistance to typical chemotherapy protocols and antihormonal therapies. Few VIM inhibitors and disruptors exist in the pharmaceutical field, although research exploring biomolecular drug interactions with VIM remains poorly developed. Scientists studied several organic compounds, which demonstrated VIM-targeting properties and *in vitro* anticancer capabilities, yet their ability to stop metastasis in live patients remains unknown. This research evaluates the VIM binding potential and anticancer properties of the platinum(II) complex known as [PtII(C^N^N)(NHC2Bu)]PF6 (Pt1a) (Figure 4.21).

The Pt1a complex features a pincer scaffold referred to as [PtII(C^N^N)] together with an NHC ligand (Figure 4.22). The binding orientation of the NHC ligand at a right angle to the pincer plane in this complex type allows VIM interaction through noncovalent bonding, whereas VIM-targeting compounds normally form covalent bonds. The unique structure of Pt1a

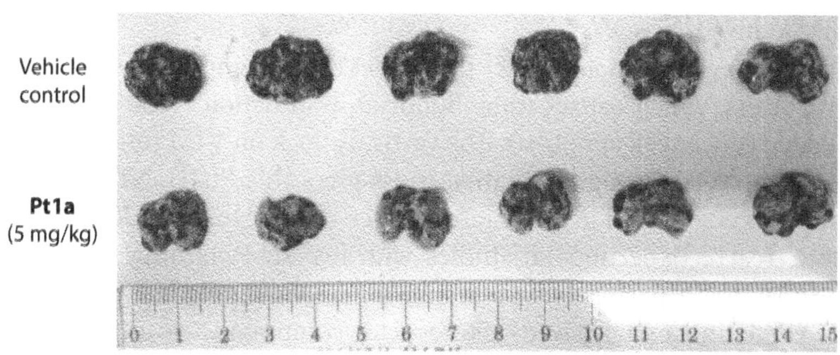

Figure 4.21 Photograph of excised lung tissues from mice bearing B16-F10 lung metastases (black nodules) from vehicle and Pt1a groups (n = 6) throughout a 14-d treatment. (I) The numbers of black nodules in excised lungs were counted [36].

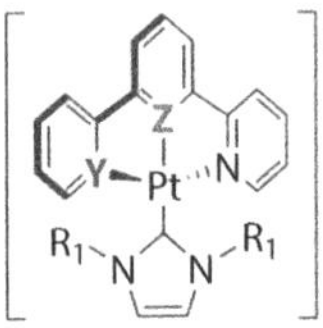

	Y	Z	R_1	n	X
Pt1a			^nBu		PF_6
Pt1b	C	N	Me	1	
Pt1c			^nBu		Cl
Pt2	N	C	^nBu		PF_6
Pt3	N	N		2	$2PF_6$

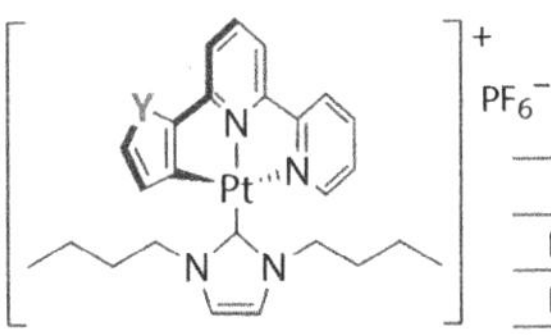

	Y
Pt4	S
Pt5	Se

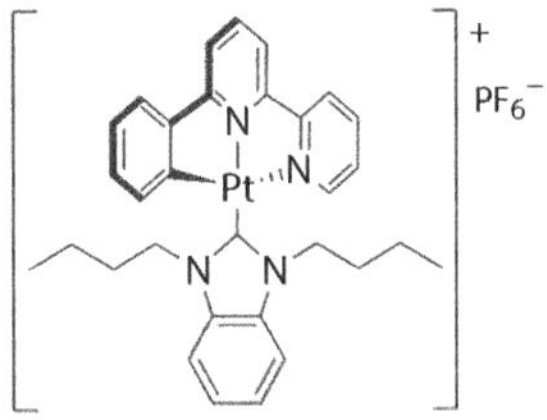

Pt6

Figure 4.22 Chemical structures of different pincer cyclometalated platinum (II) N-heterocyclic carbene complexes [36].

lets it specifically attach to VIM, which subsequently blocks its metabolic operations, thus cutting off cancer cell metastasis and migration capability.

Cancer cells interact mainly with platinum(II) NHC complex through its specificity toward VIM, which controls cancer cell migration and metastasis formation. The noncovalent binding of the Pt(II)–NHC complex to VIM occurs because of its distinctive PtII(C^N^N) pincer framework and orthogonal NHC ligand arrangement. When bound noncovalently to VIM, the complex disrupts structural arrangements and functional capability of the protein, thereby disabling cytoskeletal organization alongside intracellular transport and cell motility that significantly influence metastatic potential. The Pt(II)–NHC complex blocks VIM activity, which results in suppression of EMT tumors cells acquire while reducing their ability to invade tissue.

The redox-active metal center properties of the compound enhance oxidative stress, which causes mitochondrial dysfunction resulting in ROS-mediated cancer cell apoptosis. The vertical attack on VIM functions while generating ROS through oxidative stress gives this Pt(II)–NHC complex strong efficacy to tackle metastatic cancer cells beyond traditional therapeutic reach. Experimental studies with live subjects indicate that this molecular bond performs well at slowing down cancer cell progression, thus validating metal-based heterocyclic compounds as new potential therapeutic agents against challenging cancers that resist standard treatments [36].

4.3.6 Role of Heterocycles in Modulating the Tumor Microenvironment

A tumor microenvironment (TME) consists of cancer cells alongside immune cells and fibroblasts, as well as endothelial cells and adipocytes, while including extracellular matrix (ECM) along with cytokines and chemokines and growth factors. The continuous interaction between these TME components drives tumor development at different stages and uselessly guides therapeutic treatment responses. Macrophages and lymphocytes among immune cells within the TME assist in cancer cell elimination at first but undergo reprogramming, which enables them to facilitate tumor progression, immune escape mechanisms, and metastasis development. The functioning of fibroblasts modifies the ECM to help cancer cells migrate, but endothelial cells advance tumor angiogenesis to deliver oxygen and nutrients and facilitate metastasis through vessel entry points. Anticancer therapy success highly depends on the TME because it creates penetration-blocking physical barriers and induces drug resistance through soluble factors and immune responses while facilitating cell–cell interactions.

Metabolic reprogramming stands as a prominent characteristic of TME because it allows cancer cells to preserve high energy requirements and biosynthetic needs by regulating their oxidative stress and antioxidant systems. The metabolic alteration functions through signaling pathway modifications and protein expression variations and metabolic socioeconomic exchanges between cancer and stromal cells. Glycolytic cancer cells along with fibroblasts release lactate to provide fuel for tumor cells with low glycolytic activity, as well as enable metastasis and facilitate angiogenesis and immune evasion. Research has developed tumor metabolism within the TME as a promising therapeutic approach because of its primary role in cancer development.

The heterocyclic derivative Epigallocatechin gallate (EGCG) found in green tea expresses strong anticancer functionality through its broad action against several cell mechanisms. The polyphenolic heterocyclic compound EGCG receives widespread scientific evaluation because it displays antiproliferative together with antioxidant as well as anti-inflammatory and anti-angiogenic properties in numerous cancer types. The mechanism of action of this heterocyclic compound involves hindering tumor cell metabolism through reduced activity of glycolysis-related regulators including GLUT1 and HIF-1α that enable cancer cell survival and proliferation. The antiproliferative effects of EGCG occur through enzyme blockage, which reduces cellular energy generation, thus causing cellular death. Metabolic profiling analysis shows how EGCG affects lipid pathways and GSH pathways as well as mitochondrial functions to enhance its anticancer properties.

The antiangiogenic properties of endothelial cells arise from EGCG's action on nucleotide and amino acid synthesis and energy-producing enzymes and membrane signaling pathways needed for neovascularization. The monoclonal antibody drug bevacizumab targets vascular endothelial growth factor (VEGF) signaling in a narrow capacity, whereas EGCG demonstrates a wider range of metabolic impact that blocks cancer growth along with enabling protective pathways to sustain cellular integrity. The multicultural targets engaged by heterocyclic compounds prove their usefulness as adaptable anticancer molecules. Such heterocyclic compounds including EGCG control important metabolic pathways, thus enabling dual suppression of cancer progression and improved benefit from existing cancer therapies along with promising potential for novel anticancer treatment development [37].

4.4 Key Heterocyclic Anticancer Agents

4.4.1 Alkaloids: Vincristine and Vinblastine, CPT Derivatives

Phenolic metabolism products known as alkaloids demonstrate extraordinary properties through toxic effects that destroy diverse types of cancer cells and other diseases when tested both in Petri dishes and living organisms. These organic nitrogen bases occur naturally as bases that display the capacity to stop or restrict carcinogenesis development. Herbal anticancer drugs processed from *Catharanthus roseus* possess well-defined drug moieties, so physicians widely prescribe them as anticancer treatments. The anticancer properties of *Vinca* alkaloids consist of naturally occurring indole compounds. Natural scientists obtain these compounds by extracting them from pink periwinkle plant of *C. roseus* G. Don. *Vinca roseus* G. Don originates from the Mediterranean as a permanent flowering plant that grows throughout the northern part of the globe. The regions of southern Asia and tropics constitute the natural habitats of these species. *C. roseus* L. from Madagascar obtains multiple applications. Dimeric alkaloids are found in plant leaves and stems, whereas antihypertensive serpentine alkaloids together with ajmalicine can be isolated from roots of *C. roseus*. These compounds demonstrate hypotensive sedative effects and anticancerous properties. The *Vinca* alkaloids are also important for being cancer fighters. Specifically *C. roseus* is a decorative and curing plant of enormous pharmaceutical interest as it is nothing less than a chemical factory, producing more than about 130 different terpenoid indole alkaloids, some of which are exhibiting strong pharmacological activities. Even in varieties selected for the highest alkaloid content, the concentrations of these antitumor agents are very low, as much as 1 kg of dried leaves are being needed to obtain 3 mg of vincristine [38].

Mechanism of antitumor action

Vinca alkaloid treatment creates cytotoxicity by binding to tubulin structures, which disrupts cellular microtubule functions. The drug interferes with microtubule especially the metaphase-spindle apparatus microtubules, which leads to permanent metaphase arrest during cell division. Tubules form a microtubule basic unit using α- and β-tubulin subunits that become mitotic spindle structures, which attach to mitotic chromosomes. The cell poles receive chromosomal movement through either pushing or pulling actions from microtubules based on their current state of expansion speed and shrinkage speed. The attachment points of *Vinca* alkaloids exist on microtubules because tubulin demonstrates strong binding to the microtubule surface at plus ends yet weak binding to the sides. The binding process starts immediately, but it has the ability to return to previous states. Both *Vinca* alkaloid–binding sites exist within each mole of tubulin dimer based on existing evidence (Figure 4.23) [39].

Each microtubule possesses around 16 to 17 strong binding positions that exist at the ends of each microtubule structure. The bond between *Vinca* alkaloids and these sites leads to disruption of microtubule congregation.

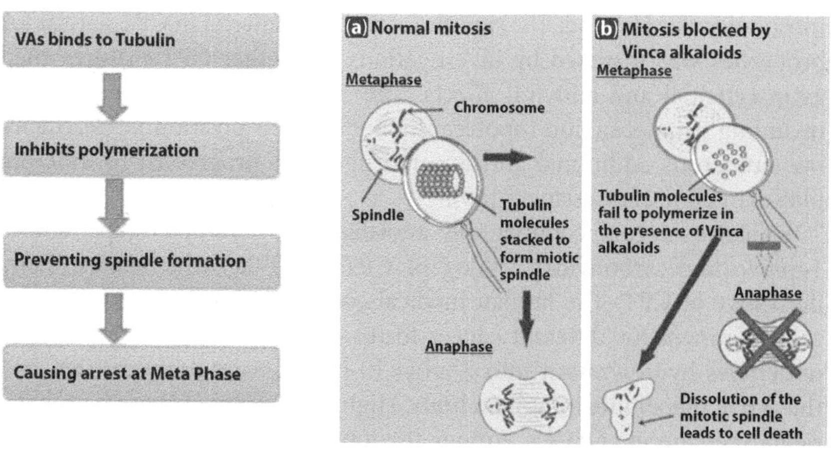

Figure 4.23 Mechanism of *Vinca* alkaloids blocking polymerization and preventing cell division leading to cell death [39].

The growth and shortening rates at the assembly end of microtubules slow down substantially when drug levels remain low resulting in a suppression of functional capabilities [38]. The microtubule regulative process of *Vinca* alkaloids leads to metaphase arrest by producing disturbances in spindle end function when drug concentration remains beneath the mass reduction threshold. The destabilizing compound category of *Vinca* alkaloids disrupts microtubule development by both splitting microtubules apart while stopping cell progression through mitosis, which leads to cell death and cycle arrest.

4.4.2 Camptothecin Derivatives

The natural alkaloid CPT maintains its structure as a ring system containing five subunits (pyrrolo[3, 4–b]quinoline) under its indole alkaloid classification. The plant compound comes mostly from *Camptotheca acuminata*, which belongs to the Nyssaceae group. This alkaloid is also the plant species *Ervatamia heyeano* as well as *Nothapodytes feotida*, and several species of *Ophiorrhiza* produce CPT. several species of the genus *Ophiorrhiza*. Scientists determined CPT chemical structure in 1966 through an analysis that included a combination of x-ray and nuclear magnetic resonance testing approach. CPT has two remarkable chemical properties. The chemical substance has a neutral character because it does not contain important alkaline elements. The molecule remains neutral because of its chemical structure, whereas the C-20 tertiary alcohol bestows an uncommon attribute to the lactone carbonyl group through an intramolecular H-bond, electrophilicity to the lactone carbonyl group, possibly through a strong intramolecular H-bond. The significant research interest in CPT antitumor properties was matched by investigations into other CPT features such as insecticidal and antiviral effects. The semisynthetic versions of CPT including irinotecan and topotecan maintain their position in treatment, but numerous additional anticancer drugs have progressed to different phases of clinical assessment [40].

Evidence suggests commercially available 7-ethyl-10-[4–(1–piperidino)–1–piperidino]carbonyloxy analog of CPT (CPT-11), the semisynthetic derivative of CPT. The human medical community uses this compound as a treatment for different cancer forms. Compound CPT-11 generally undergoes hydrolyzation into 7-ethyl-10-hydroxycamptothecin (SN-38), the active metabolite in human body. High top-1 activity levels have been established among OSCCs. Among the top-1 inhibitors, irinotecan (CPT-11) has been extensively investigated regarding its antitumor activity, though this agent has limited effect in cancer cells.

Activity in cancer of the head and neck: To date, US FDA has authorized only two forms of CPT analogs, which include irinotecan and topotecan, for cancer treatment. The treatment method is better because of its effectiveness, but the side effects remain an important issue. Evidence pharmacophoric moieties should be incorporated. Combining imino and azetidinone rings with different organic compounds improves their antitumor effects [41].

1. **Target: DNA topoisomerase I**: DNA topoisomerase I functions as an essential enzyme that facilitates the DNA coiling relief needed during the replication and transcription processes. It does this by as follows:
   - The enzyme creates short-lived breaks in the DNA single strands
   - Allowing rotation to relieve supercoiling
   - The enzyme re-ligates DNA breaks to maintain its original structure
   - The activity of CPT blocks DNA strand re-ligation through its ability to maintain Top-1–DNA cleavage complex stability.
2. **Mechanism of action**
   a. **Formation of the CPT–Top-1–DNA ternary complex**: CPT reversibly connects with Top-1–DNA cleavage complexes to keep the enzyme within an intermediate state. Further, DNA resealing is inhibited through this process, which leaves only one strand of DNA broken.
   b. **Collision with replication forks**: The replication machinery turns SSBs into DSBs when it finds a cleavage complex stabilized by CPT. Double-strand breaks created by CPT treatment prove devastating to cells because they produce genomic instability that causes either cell death through apoptosis or cell division failure called mitotic catastrophe.
   c. **Induction of DDR**: The DSBs activate ATM/ATR kinase, which leads to cell cycle arrest; p53 activation triggers apoptosis in cells with irreparable DNA damage. The activation of Chk1/Chk2 functions to stop cell cycle progression, thus preventing more replication stress.
   d. **Selective cytotoxicity in cancer cells**: The rapid cell replication of cancer cells depends heavily on Top-1, which makes them vulnerable to the DNA damaging effect of

CPT treatment. The efficient repair machinery and slow division rate of normal cells protects them from the most damaging effects.

3. **Resistance mechanisms**
   Multiple mechanisms allow cells to develop drug resistance against CPT, even though the treatment remains effective. Mutations of Top-1 enable the enzyme to avoid the binding of CPT by creating abnormal shapes. The expression of P-glycoprotein and ABC transporters leads to higher drug removal from cells, which minimizes CPT availability inside the cells. The activation of both homologous recombination (HR) and nonhomologous end joining (NHEJ) aids proper DSB DNA repair. The regulation of the cell cycle becomes altered when cancer cells shut down p53 functions to avoid apoptosis.
4. **Clinical derivatives and applications**
   The development of irinotecan (CPT-11) and topotecan occurred because CPT showed limited water solubility and high toxicity levels. Irinotecan acts as a prodrug that generates the effective metabolite SN-38 within the body. The treatment of colorectal, ovarian, lung, and cervical cancers uses this medication.

Future directions and research

Future research should focus on developing novel CPT analogs with better stability and reduced resistance. Efficacy enhancement becomes possible through DDR inhibitor treatment with PARP inhibitors. Nanoparticle formulations for targeted delivery for achieving precise and targeted delivery of CPT derivatives.

4.4.3 Flavonoids

Research shows that citrus fruit, together with flavonoid compounds present in citrus peel, demonstrates effectiveness for cancer treatment. Flavonoids function as low-molecular-weight compounds, which generate fruit peels and leaves together with pulp their vibrant colors [48]. The natural distribution of these compounds extends throughout citrus fruits along with seeds and olive oil acquisitions and appears in tea and red wine products. Scientists have identified more than 9000 flavonoids that exist to this day. Flavonoids feature a basic C6–C3–C6, 15-carbon skeleton. Two benzene rings (A and B) form the basis of this chemical structure, which connects to a heterocyclic pyran ring. The polyphenolic compound group known as flavonoids contains six different subdivisions called isoflavonoids [42].

The six types of flavonoids include flavanones, flavanols, flavonols, and flavones and anthocyanidins that are distributed within numerous plant species [43].

Anticancer effects of flavonoids

Numerous studies have proven that flavonoids show their ability to eliminate free radicals as well as control cell metabolism while protecting against diseases from oxidative stress. The anticancer properties of flavonoids demonstrate their ability to modify ROS-scrubbing enzyme capacities and their participation in cell cycle arrest and apoptosis promotion along with autophagy stimulation and their role in stopping cancer cell growth and invasion [42, 43].

Mechanism of flavonoids is linked with cell cycle inhibition

The development of cancer directly links to problems within the cell cycle regulation. The cell cycle operates through four distinct phases known as G1, S, and G2, along with M (122). Cell motion between different phases takes place through the active complex formation of cyclin-dependent kinases (CDKs) with their cyclin subunits. The development of an active complex depends on CDK inhibitors for regulation purposes. The cell cycle in normal cells puts itself into wait mode during periods of damaged DNA repair needs or nonessential cell mitosis [44].

The requirement for DNA repair or the discontinuation of cellular replication defines this specific cell condition. For cancer applications, the cell cycle arrest of malignant cells achieved through CPEs enables physicians to reduce or treat tumor or metastatic cancer masses effectively.

The cells of malignant tumors do not manifest these features. The main compound auraptene within CPE-treated cells demonstrated G1 phase arrest according to cell cycle distribution analysis [42].

Mechanism of action of CPE flavonoids

Different cancer lines exhibit diverse efficacious levels of CPEs, which depend on their specific composition together with the sensitivity of the cancer cells.

Suppression of proliferation: Normal cells lack three characteristics found in cancer cells: uncontrolled growth, apoptotic resistance, and new blood vessel formation and the ability to spread to other body areas.

Cellular proliferation is suppressed through multiple protein-modulating effects of the flavonoids within CPEs, which ultimately result in proliferation inhibition of oncoproteins (Figure 4.24).

Laboratory tests demonstrated that Korean *Citrus aurantium* fruit peel's methanol extract stopped human A549 lung carcinoma cell proliferation at multiple concentrations while creating apoptotic conditions for the cells. Flavonoids obtained from Korean *C. aurantium* peel inhibited cell growth in the same manner when administered to A549 cancer cells. The basic compound within polyhydroxylated flavonoids known as aglycone exists as quercetin.

Polyhydroxylated compound quercetin (flavonols), which exists in onions and berries together with grapes and green vegetables and apples, stands as the leading flavonoid type in clinical studies of cell proliferation effects. Research has shown that this substance functions as a growth inhibitor against various cancer cell types including immortal human HeLa cells and human epidermoid carcinoma (A431) cells and NK/LY ascites tumor cells, as well as stomach cancer cells NUGC-2, HGC-27, MKN-28 and MKN-7, colon (COLO 320 DM) cells, and breast, squamous, gliosarcoma

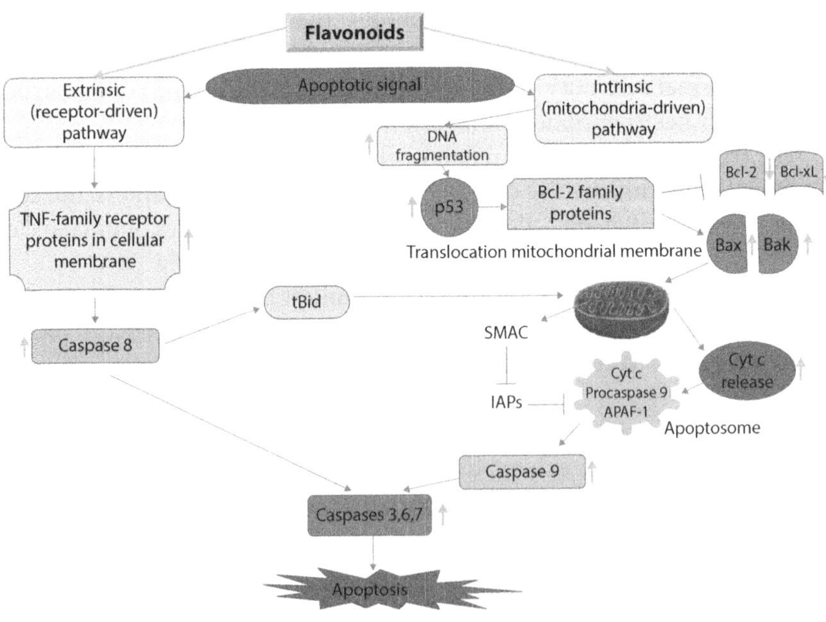

Figure 4.24 Two pathways (intrinsic and extrinsic) of flavonoids, which induce apoptosis through increasing levels of caspases 3, 6, and 7 in cell [43].

cells and ovarian cells and pancreatic and liver (HepG2) cancer cells. Strong antiproliferative action of quercetin seems to work by blocking the protein kinase C pathway [42].

4.4.4 Flavonols

EGCG, an ingredient found in tea, shows antiangiogenic properties because it stops VEGF-induced signaling through VEGFR-2 in human umbilical vein endothelial cells. This inhibitory action occurs because the VEGF peptide directly binds with EGCG in solution. The internal cellular signaling process triggered by VEGF becomes inhibited because EGCG suppresses VEGF expression and hinders growth factor binding to VEGFR-2 and blocks the receptor's phosphorylation step. The combination of extra effects strengthens endostatin production along with VEGF-blocking activities by EGCG. EGCG stops VEGF production and secretion triggered by IL-6 and simultaneously blocks gastric carcinoma cells from forming new blood vessels through STAT3 activity suppression. STAT3/VEGF signaling pathway becomes a target when these reactions occur, thus leading to reduced VEGF levels.

Level induced by IL-6: The process of STAT3 protein movement between cell compartments and its access to the nucleus and binding to VEGF promoter in gastric tumors becomes reduced because of the inhibition mechanism, which also decreases VEGF protein levels and secretion and mRNA expression from inhibited STAT3 activity in gastric carcinoma cells [42].

4.4.5 Nitrogen-Containing Heterocycles: Purines and Pyrimidines, Imidazoles, and Benzimidazoles

Pyrimidine derivatives as anticancer agents

Organic chemistry literature recognized pyrimidines through their historical name of "m-diazine" after uric acid breakdown. During a study of uric acid treatment with nitric acid, Brugnatelli initially isolated the pyrimidine derivative alloxan in 1818 [45]. The aromatic six-membered heterocyclic ring of pyrimidine positions its first and third nitrogen atoms. Mutual transformation of pyrimidine occurs at temperatures between 22.5°C for melting and 124°C for boiling. Several classes of bicyclic heterocyclic compounds with nitrogen functions include purines as well as quinazolines and pteridines and pyridopyrimidines.

Bicyclic compounds function as well-known pharmacological components in the field of medicinal chemistry. Examples of commercial medications with a bicyclic include quinazoline group in gefitinib drug function as an EGFR inhibitor for cancer treatment, whereas acyclovir derived from purines serves in antiviral drug applications, and pteridine-based methotrexate serves as a chemotherapeutic agent by inhibiting dihydrofolate reductase. Drug designers commonly use bicyclic heterocycles because of their dual advantages to bind biological targets while altering enzymatic activities and presenting many pharmacological properties.

The tyrosine kinase inhibitors gefitinib and erlotinib serve as quinazoline derivatives, which form the main body of these pharmaceutical products. Several physicians prescribe these medicine compositions to treat NSCLC cases [45, 46].

Tyrosine kinase

The signaling cascade requires tyrosine kinases, which serve as crucial structural elements. Cell division together with metabolism and migration and death falls under this signaling cascade's control mechanism, the death of cells. The enzyme requires ATP for its role in tyrosine residue phosphorylation. Tyrosine kinase enzymes function fundamentally during various stages of cancer development and progression stages of the emergence and spread of cancer. The signal transduction pathway controls both excessive cell multiplication and death sensitivity through its normal mechanisms and increases apoptotic sensitivity. Malfunctioning of signaling pathways, caused by genetic or epigenetic changes, often turns normal cells into cancerous cells. Specially tailored benefits are provided to cancer cells through this mechanism. As a result, it is not surprising that excessive tyrosine signaling would be observed during cancer cell experimentations. These enzymes receive oncogenic status due to their increased activity of phosphorylation through the tyrosine kinase pathway, resulting in signaling network failure [46].

Mechanism of tyrosine kinase inhibitors

A subset of amino acids located in substrate enzymes gets phosphorylated through different changes resulting from substrate enzyme phosphorylation by tyrosine kinases (downstream cellular biology and signaling). These enzymes are subject to modification by tyrosine kinases and they further control cell growth, differentiation and death and also affect apoptosis and migration. Triggering downstream signal transduction. Tyrosine kinase inhibitors trigger continuous activation or blocking of pathways because of either genetic mutations or alternative causes. The improper use of pathways leads to disorder in signal cascades, resulting in various diseases including cancer. The dysregulation of tyrosine kinase functions causes cancer development together with multiple other diseases. These medicines work by blocking the abnormal signal transmission pathway.

Transcription in both genetic amplification and mutation: The disease-causing signals develop because of Compounds 47–51, according to research. Excellent results were recorded by these compounds during assay testing of multiple kinases.

TKs, PI3K, and CDK4/6: In molecule 47, the presence of the addition of dichloro and tertiary amine improved the level of activity. In molecule 48, piperazine with amine linkage and acetyl group at pyridine, the combination of acetyl group at pyridine ring with thiophene and phenyl with fluorine produced greater activity (Figure 4.25) [46].

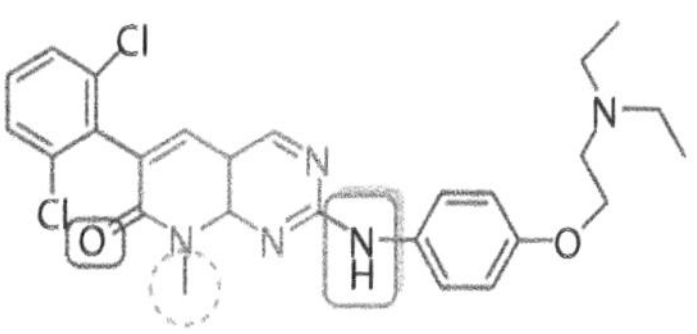

47

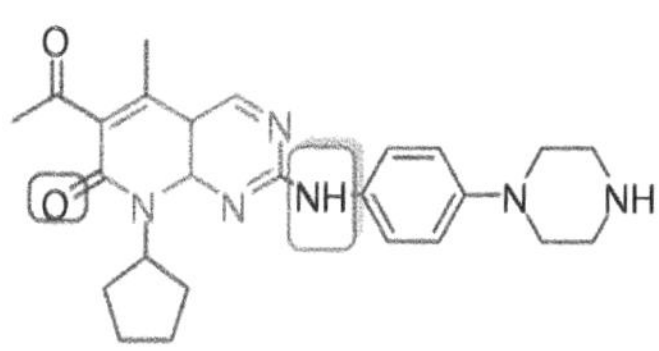

48

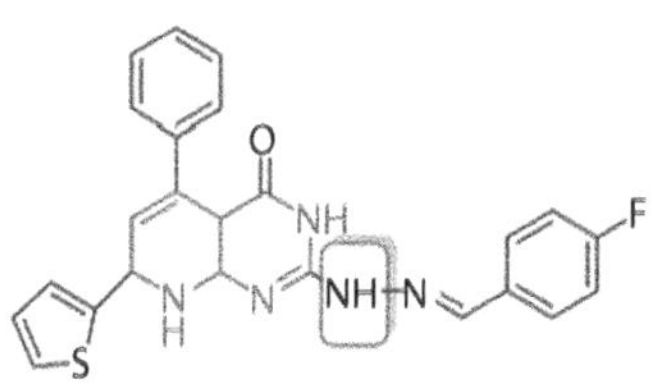

49

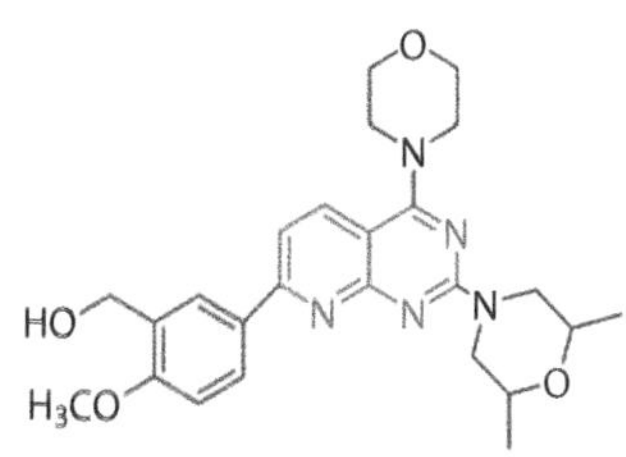

50

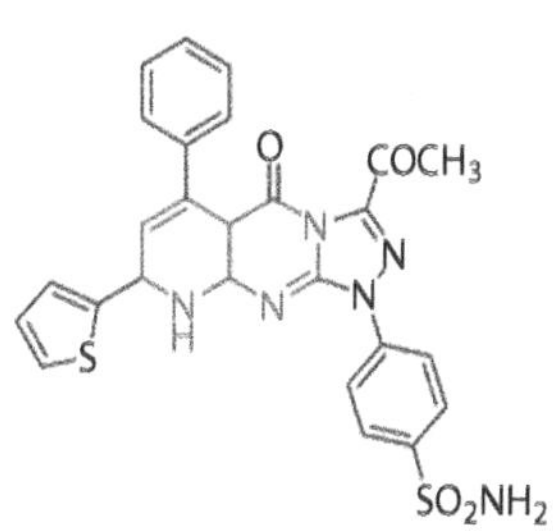

51

Figure 4.25 Reported derivatives of pyrido[2, 3–d]pyrimidine, which function as kinase inhibitors [46].

Benzimidazoles

The marketed compounds containing imidazole and benzimidazole rings serve as a design template to enhance water solubility. The molecule creates hydrogen bonds because of its two nitrogen atoms, which serves as the essential structural factor. The chemical ring structure known as imidazole, together with benzimidazole ring, serves as a vital replacement for multiple structural components such as oxazole, tetrazole, pyrazole, amide, thiazole, and triazole in medicinal molecule creation.

Patients benefit from chronic lymphoma treatment with the alkylating agent bendamustine (I), which contains nitrogen mustard combinations with benzimidazole structures. The oral medicine dovitinib (II) shows antineoplastic properties through its ability to block multiple receptor tyrosine kinases (RTKs) as a benzimidazole–quinolinone hybrid compound. The compound displays potent binding to receptor 1 of fibroblast growth factor (FGFR-1) at IC_{50} ¼ 8 nM and equally targets VEGFR-2 at IC_{50} ¼ 13 nM and additional RTKs that drive tumor development along with angiogenesis including FGFR-3, VEGFR-1, VEGFR-3, and platelet-derived growth factor receptor (Figure 4.26) [47].

Research indicates that the VEGFR-inhibiting properties of the indole-2-one–based 1,2,3-triazole scaffold demonstrate high effectiveness against cancer cells and low toxicity effects on regular human cells. The improved inhibition capability of this scaffold stands out when compared to current treatments based on molecular docking results, showing its effective binding to the protein pocket and VEGFR-2 inhibition. A different triazole derivative showed superior topoisomerase IIB inhibitory activity compared to conventional chemotherapeutic drugs, thus presenting significant

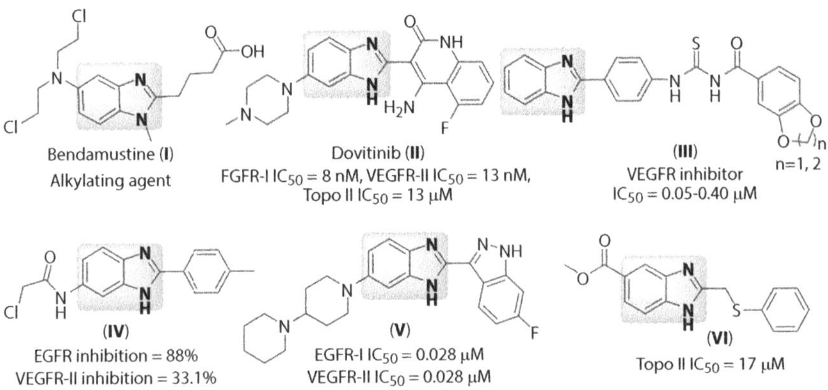

Figure 4.26 Multienzyme targeting derivatives of benzimidazoles [47].

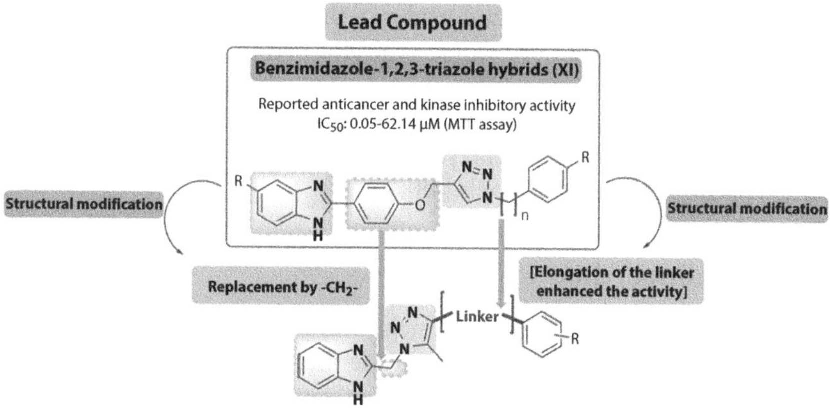

Figure 4.27 Mechanism of function of benzimidazole-1,2,3-triazole hybrids through replacement by CH_2 and elongation of linker [47].

possibilities for anticancer therapy. The research has evolved to combine 2-substituted benzimidazole with a 1,2,3-triazole backbone into hybrid pharmacophore structures for creating advanced anticancer compounds. The hybrid compounds offer multiple mechanisms of action, which enables their application to drug-sensitive and drug-resistant cancers (Figure 4.27). Scientific assessment reveals that these compounds display substantial toxicity properties for eliminating different types of cancer cells starting from lung cells through cervical cells as well as pancreatic cells and metastatic colorectal cancer cells. Accurate distance measurements between the triazole and benzimidazole parts through SAR studies show that condensed alkyl chains give optimal effectiveness for the compounds. Anticancer therapy development benefits substantially from heterocyclic hybrid compounds because of their established significance according to these research findings.

4.4.6 Taxanes

Taxanes represent a vital medication collection, which medical professionals use both in primary cancer treatment and after recurrence of several solid tumors including metastatic cancer.

Recurrent cancer for several major solid tumors, especially metastatic cancer: The most recent taxane mechanism (paclitaxel) operates through micronucleus production that results in immediate nuclear membrane breakdown. Studies confirm that cancer cells in culture show nuclear envelope rupture. Studies found evidence that the broken nuclear membrane exists in a condition, which typically permits repair. The defects present in micronuclei lead to frequent detrimental effects on these structures. Our findings indicate that taxane-induced disruption and permanent tearing of micronuclei represent an essential oxidative mechanism by which paclitaxel kills cancer cells [48].

4.4.7 Anthracycline

These drugs maintain their treatment importance because they have proved effective for cancer therapies since the 1960s by working through multiple cellular pathways despite causing cardiotoxicity and developing resistance-related to efflux transporters. Scientists continue to develop and utilize novel anthracyclines because they gained better knowledge of these medications while making essential structural modifications, which enhanced their pharmacokinetic and pharmacodynamic properties [47].

The clinical practice often utilizes doxorubicin and epirubicin as anthracyclines. The drug action of anthracycline works through topoisomerase II enzymes by inhibiting tumor DNA, which blocks downstream protein synthesis and results in cell membrane dysfunction in cancer cells (Figure 4.28). The improved patient outcomes from anthracycline treatments create

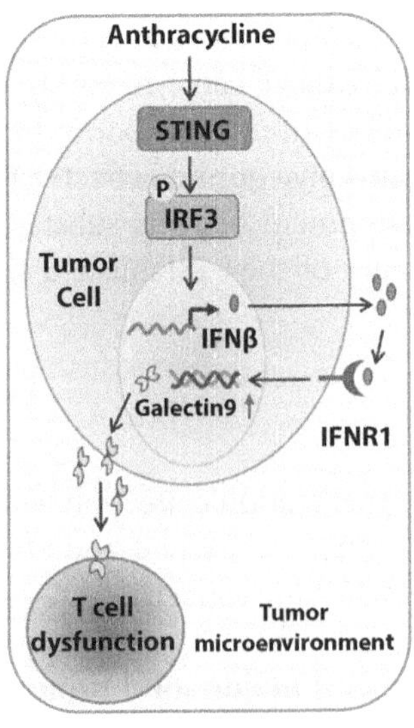

Figure 4.28 Action of anthracycline on cells by increasing the levels of Gal9 and STING proteins [49].

a clinical challenge because it leads to higher rates of treatment resistance in breast cancer cases. Research shows that uniting doxorubicin with anti–Gal-9 therapy effectively blocked tumor development while boosting mice survival rates in immune-active syngeneic mouse model experiments. The human and murine tumor cells of diverse origins increased Gal-9 expression after exposure to doxorubicin treatment. The STING/interferon β pathway actuated by doxorubicin caused the production of tumoral Gal-9. Anthracycline treatment of breast cancer patients resulted in elevated Gal-9 and p-STING protein levels within their tumor tissues. The research shows that anthracyclines drive Gal-9 expression to enable immune escape through new mechanisms, thus opening new opportunities for combining Gal-9 blockade with anthracyclines as effective cancer therapies [49].

4.5 Challenges in Heterocyclic Anticancer Drug Development

Scientific literature classifies therapeutic resistance in cancer into two distinct groups according to how long resistance develops. The categories are intrinsic and acquired resistance. The intrinsic resistance, also known as primary resistance, is mediated by the endogenous factors that are present in tumor cells or tissues before therapeutic applications, which provide cancer cells with survival advantages and adaptability to primary therapeutic stress. However, acquired drug resistance is developed after receiving cancer treatment, which is generally mediated by the adaptive alterations against the given therapy in initially sensitive tumors, resulting in compromised treatment effectiveness [50].

The main reason chemotherapy and targeted therapy treatments fail in cancer treatment arises from drug resistance. Drug resistance exists in two forms as primary or acquired based on its fundamental mechanisms. Drug resistance arises from several specific mechanisms, which involve both drug efflux enhancement and drug metabolism variations along with drug target mutations and DNA damage repair pathways and signal transduction modifications along with anti–cell death mechanisms. ROS levels together with metabolic changes act as the primary nonnegligible causes that underpin drug resistance. New therapy targets emerge from this development, which eases the strains of cancer therapy [51].

The most essential tumor treatment methods consist of radiotherapy together with chemotherapy and targeted therapy and immunological agents. The development of drug resistance occurs in every tumor type,

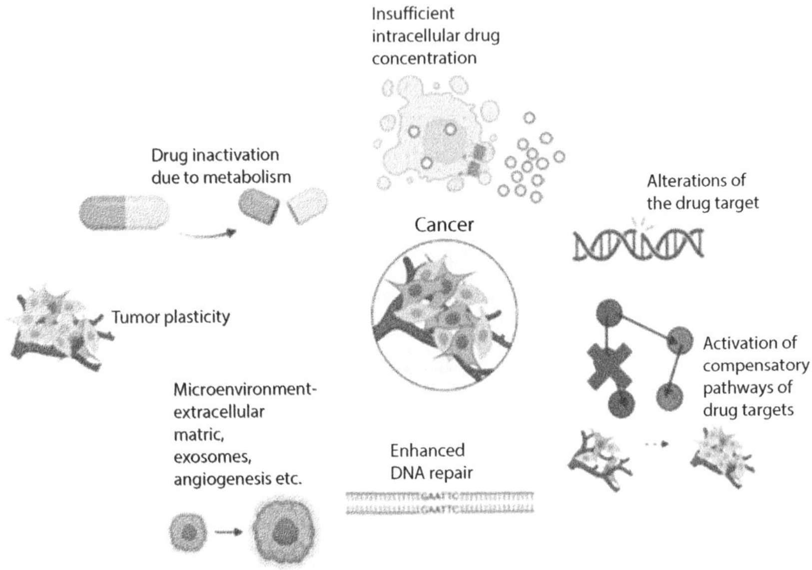

Figure 4.29 Common challenges faced in developing anticancer drugs [50].

thus reducing drug potency until final therapeutic benefit becomes zero. Metabolic reprogramming allows tumors to obtain needed nutrients throughout their progression even when they face difficult environmental challenges including drug treatment. Redox metabolism demonstrates tight connections with the tumor occurrence and its progression together with resistance to medications. The FDA has approved antitumor drugs whose mechanism of tumor cell destruction depends on ROS production enhancement. The effective treatment of cancer requires identifying how tumor metabolism impacts anticancer drug outcomes while creating metabolic-based resistance reduction methods. This section provides a summary of drug resistance processes in tumors alongside an explanation of how metabolic-redox circuits affect the drug resistance of tumors (Figure 4.29).

AI disclosure statement: The author confirms that no artificial intelligence (AI)-generated content or AI-assisted writing tools were used in the preparation of this chapter (including summaries and the abstract). All text, analysis, and conclusions are the result of the author's original work.

References

1. Thakur, D., A Review on the Medicinal Significance of Heterocyclic Compounds. *Int. J. Res. Appl. Sci. Eng. Technol.*, 2023. https://doi.org/10.22214/ijraset.2023.54569.
2. Slman, D., Satar, H., Ketan, Z., Jawad, A., Heterocyclic Compounds: A Study of its Biological Activity. *Al-Nahrain J. Sci.*, 2024. doi.org/10.22401/anjs.27.5.03.
3. Joule, J., Mills, K., Smith, G., Heterocyclic Chemistry, 2020. doi.org/10.1007/978-1-4899-3222-8.
4. Shvaika, O. and Korotkikh, M., The Principles of Chemistry of Heterocyclic Compounds, 2020. doi.org/10.15407/akademperiodyka.415.194.
5. Kabir, E. and Uzzaman, M., A review on biological and medicinal impact of heterocyclic compounds. *Results Chem.*, 2022. doi.org/10.1016/j.rechem.2022.100606.
6. Charushin, V.N., *et al.*, The chemistry of heterocycles in the 21st century. *Russ. Chem. Rev.*, 2024. doi.org/10.59761/rcr5125.
7. Alvarez-Builla, J. and Barluenga, J., Heterocyclic Compounds: An Introduction. *ChemInform*, 43, 2011. https://doi.org/10.1002/CHIN.201201253.
8. Bachur, N.R., Anthracycline antibiotic pharmacology and metabolism. *Cancer Treat. Rep.*, 63, 8, 7–820, 1979.
9. Beretta, G.L. and Zunino, F., Molecular mechanisms of anthracycline activity, in: *Anthracycline Chemistry and Biology II*, pp. 1–19, Springer, Berlin, Heidelberg, 2007.
10. Fujiwara, A., Hoshino, T., Westley, J.W., Anthracycline antibiotics. *Crit. Rev. Biotechnol.*, 3, 2, 133–57, 1985.
11. Krohn, K., *Anthracycline Chemistry and Biology I: Biological Occurrence and Biosynthesis, Synthesis and Chemistry*, Springer, Berlin, Heidelberg, 2009.
12. Anthracyclines React with Apurinic/Apyrimidinic Sites in DNA, American Chemical Society, pp. 1554–8929. doi: 10.1021/acschembio.3c00033.
13. Gewirtz, D., A critical evaluation of the mechanisms of action proposed for the antitumor effects of the anthracycline antibiotics Adriamycin and daunorubicin. *Biochem. Pharmacol.*, 57, 7, 727–41, 1999. 11 Advances in Precision Medicine Oncology.
14. Pommier, Y., Leo, E., Zhang, H., Marchand, C., DNA topoisomerases and their poisoning by anticancer and antibacterial drugs. *Chem. Biol.*, 17, 5, 421–33, 2010.
15. Visone, V., Szabó, I., Perugino, G., Hudecz, F., Bánóczi, Z., Valenti, A., Topoisomerases inhibition and DNA binding mode of daunomycin oligoarginine conjugate. *J. Enzyme Inhib. Med. Chem.*, 35, 1, 1363–71, 2020.
16. Mosmann, T., Rapid colorimetric assay for cellular growth and survival: application to proliferation and cytotoxicity assays. *J. Immunol. Methods*, 65, 55–63, 1983.

17. Scudiero, D.A., Shoemaker, R.H., Paull, K.D., *et al.*, Evaluation of a soluble tetrazolium/formazan assay for cell growth and drug sensitivity in culture using human and other tumor cell lines. *Cancer Res.*, 48, 4827–33, 1988.
18. Freimoser, F.M., Jakob, C.A., ebi, M., Tuor, U., The MTT [3-(4,5-Dimethylthiazol-2-yl)-2,5-diphenyltetrazolium bromide] assay is a fast and reliable method for colorimetric determination of fungal cell densities. *Appl. Environ. Microbiol.*, 65, 3727–9, 1999.
19. Elwan, A., *et al.*, Triazoloquinoxalines-based DNA intercalators-Topo II inhibitors: design, synthesis, docking, ADMET and anti-proliferative evaluations. *J. Enzyme Inhib. Med. Chem.*, 37, 1, 1556–1567, 2022. doi: 10.1080/14756366.2022.2080205.
20. El-Adl, K., El-Helby, A.A., Sakr, H., Elwan, A., Design, synthesis, molecular docking and anti-proliferative evaluations of [1,2,4]triazolo[4,3-a]quinoxaline derivatives as DNA intercalators and Topoisomerase II inhibitors. *Bioorg. Chem.*, 105, 104399, 2020.
21. El-Adl, K., El-Helby, A.A., Sakr, H., Elwan, A., [1,2,4]Triazolo[4,3-a]quinoxaline and [1,2,4]triazolo[4,3-a]quinoxaline-1-thiol derived DNA intercalators: design, synthesis, molecular docking, *in silico* ADMET profile and anti-proliferative evaluations. *New J. Chem.*, 45, 881–97, 2021.
22. Griss, T., Vincent, E.E., Egnatchik, R., Chen, J., Ma, E.H., Faubert, B., Viollet, B., DeBerardinis, R.J., Jones, R.G., Metformin Antagonizes Cancer Cell Proliferation by Suppressing Mitochondrial-Dependent Biosynthesis. *PLoS Biol.*, 13, e1002309, 2015.
23. Zhou, G., Myers, R., Li, Y., Chen, Y., Shen, X., Fenyk-Melody, J., Wu, M., Ventre, J., Doebber, T., Fujii, N., *et al.*, Role of AMP-activated protein kinase in mechanism of metformin action. *J. Clin. Investig.*, 108, 1167–1174, 2001.
24. Bogachus, L.D. and Turcotte, L.P., Genetic downregulation of AMPK-isoforms uncovers the mechanism by which metformin decreases FA uptake and oxidation in skeletal muscle cells. *Am. J. Physiol.—Cell Physiol.*, 299, 1549–1561, 2010.
25. Noser, A.A., Abdelmonsef, A.H., El-Naggar, M., Salem, M.M., NewAminoAcidSchiff Bases as Anticancer Agents *via* Potential Mitochondrial Complex I-Associated Hexokinase Inhibition and Targeting AMP-Protein Kinases/mTOR Signaling Pathway. *Molecules*, 26, 5332, 2021. https://doi.org/10.3390/molecules26175332.
26. Gümüş, M., *et al.*, Triad pyrazole–thiazole–coumarin heterocyclic core effectively inhibit HSP and drive cancer cells to apoptosis. *J. Biomol. Struct. Dyn.*, 41, 23, 14382–14397, 2023. doi: 10.1080/07391102.2023.2181643.
27. Tripathy, D.B. and Mishra, A., Microwave synthesis and characterization of waste soybean oil-based gemini imidazolinium surfactants with carbonate linkage. *Surf. Rev. Lett.*, 24.05, 1750062, 2017.
28. Warda, E.T., Shehata, I.A., El-Ashmawy, M.B., ElGohary, N.S., New series of isoxazole derivatives targeting EGFRTK: synthesis, molecular modeling

and antitumor evaluation. *Bioorg. Med. Chem.*, 28, 21, 115674, 2020. DOI: 10.1016/j.bmc.2020.115674.
29. Saleh, N.M., El-Gazzar, M.G., Aly, H.M., Othman, R.A., Novel Anticancer Fused Pyrazole Derivatives as EGFR and VEGFR-2 Dual TK Inhibitors. *Front. Chem.*, 7, 917, 2020. DOI: 10.3389/fchem.2019.00917.
30. Dahmani, R., Manachou, M., Belaidi, S., Chtita, S., Boughdiri, S., Structural characterization and QSAR modeling of 1,2,4-triazole derivatives as a-glucosidase inhibitors. *New J. Chem.*, 45, 1253–1261, 2021. DOI: 10.1039/D0NJ05298A.
31. El-Faham, A., Farooq, M., Almarhoon, Z., Abd Alhameed, R., Wadaan, M.A.M., de la Torre, B.G., Albericio, F., Diand trisubstituted s-triazine derivatives: Synthesis, characterization, anticancer activity in human breastcancer cell lines, and developmental toxicity in zebrash embryos. *Bioorg. Chem.*, 94, 103397, 2020. DOI: 10.1016/j.bioorg.2019.103397.
32. Anwer, Kurls, E., *et al.*, Five and six membered heterocyclic rings endowed with azobenzene as dual EGFRT790M and VEGFR-2 inhibitors: Design, synthesis, *in silico* ADMET profile, molecular docking, dynamic simulation and anticancer evaluations. *RSC Adv.*, 13, 35321 - 35338, 2023.
33. Sheng, Y., Leu, W., Chen, C., Hsu, J., Liu, Y., Hsu, L., Hou, D., Guh, J., The (+)-Brevipolide H Displays Anticancer Activity against Human Castration-Resistant Prostate Cancer: The Role of Oxidative Stress and Akt/mTOR/p70S6K-Dependent Pathways in G1 Checkpoint Arrest and Apoptosis. *Molecules*, *25*, 2020.
34. Feng, J., Li, S., Zhang, B., Duan, N., Zhou, R., Yan, S., Elango, J., Liu, N., Wu, W., FGFC1 Exhibits Anti-Cancer Activity *via* Inhibiting NF-κB Signaling Pathway in EGFR-Mutant NSCLC Cells. *Mar. Drugs*, 20, 76, 2022. https://doi.org/10.3390/md20010076.
35. Sen, S., Hufnagel, S., Maier, E.Y., Aguilar, I., Selvakumar, J., DeVore, J.E., Arambula, J.F., Rationally Designed Redox-Active Au(I) N-Heterocyclic Carbene: An Immunogenic Cell Death Inducer. *J. Am. Chem. Soc.*, 2020. doi:10.1021/jacs.0c09753.
36. Wan, P., Tong, K., Lok, C., Zhang, C., Chang, X., Sze, K., Wong, A.S.T., Che, C., Platinum(II) N-heterocyclic carbene complexes arrest metastatic tumor growth. *Proc. Natl. Acad. Sci. U.S.A.*, 118, 17, e2025806118, 2021. https://doi.org/10.1073/pnas.2025806118.
37. Dias, A.S., Helguero, L., Almeida, C.R., Duarte, I.F., Natural Compounds as Metabolic Modulators of the Tumor Microenvironment. *Molecules*, 26, 3494, 2021. https://doi.org/10.3390/molecules26123494.
38. Tembhe, H. and Tasgaonkar, R., Vinca Alkaloids – Anti cancer drugs. *Int. J. Res. Appl. Sci. Eng. Technol.*, *11*, 1, 408–416, 2023. https://doi.org/10.22214/ijraset.2023.48559.
39. Verma, N.V., Sharma, N.S., Gaur, N.K., Kumar, N.N., Role of vinca alkaloids and their derivatives in cancer therapy. *World J. Adv. Res. Rev.*, *16*, 3, 794–800, 2022. https://doi.org/10.30574/wjarr.2022.16.3.1378.

40. Ghanbari-Movahed, M., Kaceli, T., Mondal, A., Farzaei, M.H., Bishayee, A., Recent Advances in Improved Anticancer Efficacies of Camptothecin Nano-Formulations: A Systematic Review. *Biomedicines*, *9*, 5, 480, 2021. https://doi.org/10.3390/biomedicines9050480.
41. Ugrappa, S., Jagadeesan, D., Lalitha, P., Ravichandran, M., Solyappan, M., Khor, G., Wu, Y., Balakrishnan, V., Thangeswaran, D., Fuloria, S., Synthesis, Characterization, Docking Studies and *in vitro* Response of New Camptothecin Derivatives towards Oral Squamous Cell Carcinoma. *Asian J. Chem.*, *36*, 3, 731–739, 2024. https://doi.org/10.14233/ajchem.2024.31038.
42. Koolaji, N., Shammugasamy, B., Schindeler, A., Dong, Q., Dehghani, F., Valtchev, P., Citrus peel flavonoids as potential cancer prevention agents. *Curr. Dev. Nutr.*, *4*, 5, nzaa025, 2020. https://doi.org/10.1093/cdn/nzaa025.
43. Kopustinskiene, D.M., Jakstas, V., Savickas, A., Bernatoniene, J., Flavonoids as anticancer agents. *Nutrients*, *12*, 2, 457, 2020. https://doi.org/10.3390/nu12020457.
44. Wei, Q. and Zhang, Y., Flavonoids with Anti-Angiogenesis Function in Cancer. *Molecules*, *29*, 7, 1570, 2024. https://doi.org/10.3390/molecules29071570.
45. Kumar, A., Singh, A.K., Singh, H., Vijayan, V., Kumar, P., Nitrogen containing heterocycles as anticancer agents: A Medicinal Chemistry perspective. *Pharmaceuticals*, *16*, 2, 299, 2023. https://doi.org/10.3390/ph16020299.
46. Kumar, A., Bhagat, K.K., Singh, Kumar, P., Medicinal chemistry perspective of pyrido[2,3-d]pyrimidines as anticancer agents. *RSC Adv.*, *13*, 10, 6872–6908, 2023. https://doi.org/10.1039/d3ra00056g.
47. Othman, D., II, Hamdi, A., Tawfik, S.S., Elgazar, A.A., Mostafa, A.S., Identification of new benzimidazole-triazole hybrids as anticancer agents: Multi-target recognition, *in vitro* and *in silico* studies. *J. Enzyme Inhib. Med. Chem.*, *38*, 1, 2023. https://doi.org/10.1080/14756366.2023.2166037.
48. Xu, A.P., Xu, L.B., Smith, E.R., Fleishman, J.S., Chen, Z., Xu, X., Cell death in cancer chemotherapy using taxanes. *Front. Pharmacol.*, *14*, 2024. https://doi.org/10.3389/fphar.2023.1338633.
49. Sun, X., Wang, Hung, M., Inhibition of Galectin-9 sensitizes tumors to anthracycline treatment *via* inducing antitumor immunity, 2023. https://doi.org/10.7150/ijbs.84108.
50. Lei, Z., Tian, Q., Teng, Q., Wurpel, J.N.D., Zeng, L., Pan, Y., Chen, Z., Understanding and targeting resistance mechanisms in cancer. *MedComm.*, 4, 3, 2023. https://doi.org/10.1002/mco2.265.
51. Wang, Y., He, J., Lian, S., Zeng, Y., He, S., Xu, J., Luo, L., Yang, W., Jiang, J., Targeting Metabolic–Redox Nexus to Regulate Drug Resistance: From Mechanism to tumor therapy. *Antioxidants*, 13, 7, 828. https://doi.org/10.3390/antiox13070828.

5

Heterocyclic Compounds as Anti-Inflammatory Agents

Hemlata Nimesh[1], Sagar Kumar[1] and Nidhi Srivastava[2]*

[1]*Amity Institute of Pharmacy, Amity University, Noida, Uttar Pradesh, India*
[2]*Amity Institute of Biotechnology, Amity University, Noida, Uttar Pradesh, India*

Abstract
Heterocyclic compounds have emerged as a major area of interest in the discovery of new anti-inflammatory drugs due to their potential to target important molecular targets involved in inflammation. Inflammation is an extremely complex and multifaceted process that is a major cause of many chronic diseases, such as rheumatoid arthritis, osteoarthritis, cardiovascular disease, inflammatory bowel disease, and neurodegenerative diseases. Although conventional anti-inflammatory therapy in the form of corticosteroids and nonsteroidal anti-inflammatory drugs is available, they are associated with severe side effects, including cardiovascular disease, gastrointestinal disturbance, and addiction. There is therefore an immediate need for safer and more selective agents to effectively suppress inflammation. Heterocyclic compounds with their varied and flexible structure offer an attractive potential. They have the ability to act on a range of molecular targets involved in the inflammatory process including cyclooxygenase (COX), lipoxygenase (LOX), nuclear factor κB (NF-κB), cytokines, and other inflammatory mediators. By acting on these targets, heterocyclic compounds are able to inhibit the synthesis of inflammatory molecules such as prostaglandins, leukotrienes, and cytokines, all of which are crucial to the inflammatory process. This chapter presents a comprehensive overview of heterocyclic compounds as anti-inflammatory agents, their mechanisms of action, and recent developments in their design and synthesis as anti-inflammatory agents. Pyrazoles, thiazoles, oxazoles, indoles, quinolines, and benzothiazoles have been highly promising in preclinical and clinical trials. These compounds have substantial anti-inflammatory activity *via* inhibition of key enzymes COX and LOX and *via* regulation of essential transcription factors NF-κB, which is involved in regulating the inflammation-related genes.

**Corresponding author*: nsrivastava@amity.edu

Divya Bajpai Tripathy, Priyanka Chhabra and Anjali Gupta (eds.) Heterocyclic Compounds as Pharmaceutical Ingredients, (179–206) © 2026 Scrivener Publishing LLC

The chapter also discusses challenges and opportunities in heterocyclic anti-inflammatory drug design. These include enhancing their pharmacokinetic and pharmacodynamic properties, optimizing their bioavailability, and maximizing their specificity toward the target molecular pathways to reduce off-target activities. Lastly, the chapter discusses future research directions, such as the use of sophisticated computational tools for drug design and virtual screening. The chapter also discusses some of the potentialities of the use of heterocyclic compounds as combination therapy agents for the targeting of more than one inflammatory pathway. As technology continues to advance and get more optimized, heterocyclic compounds have an enormous potential to produce more potent and safer anti-inflammatory drugs and a new ray of hope for patients with numerous inflammation-related diseases.

Keywords: Inflammation, heterocyclic compounds, nonsteroidal anti-inflammatory drugs (NSAIDs), cyclooxygenase inhibitors (COX inhibitors), prostaglandin synthesis inhibitors

5.1 Introduction

Inflammation, from the Latin "inflammare," is the body's first reaction to injury or harm to its tissues or cells, caused by such factors as injury, infection, or other noxious stimuli. It is the body's first line of defense, which protects against injury or infection. It is essentially a protective response that is necessary for healing, consisting of the release of proinflammatory signals, activation of immune cells, and sometimes damage to tissues [1].

The four main signs of inflammation—heat, redness, swelling, and pain—are usually followed by a loss of tissue function. Inflammation also induces alterations in microcirculation, which include heightened blood vessel permeability, white blood cell recruitment, and release of inflammatory mediators. Inflammation is either acute or chronic, with acute inflammation being the body's attempt to reestablish tissue balance. But if the acute inflammation worsens or prolongs, then it may give rise to chronic inflammation [2]. Interleukin (IL) is a significant soluble mediator in inflammation with IL-1 as a highly potent proinflammatory cytokine, which has an important role to play in the immune response of injury or infection [3].

Blood monocytes secrete IL, which induces a broad array of responses during the acute-phase reaction. IL, even in small amounts, can induce signs such as hypotension, fever, and the release of adrenocorticotropic hormone. It also induces the synthesis of other cytokines, including IL-6, which further stimulate the liver to produce acute-phase proteins such as serum amyloid A and C-reactive protein [4]. This results in a rise in white blood cells and platelets and the synthesis of adhesion molecules in endothelial cells. Apart from IL, other significant mediators of inflammation include vasoactive amines such as histamine, bradykinin, serotonin, and eicosanoids. Basophils contribute to inflammation through releasing histamine, which sustains the acute-phase response [5].

Arachidonic acid (AA), a central structural element of cell membrane phospholipids, is an essential precursor for the synthesis of eicosanoids, which are critical inflammatory mediators. Enzymes such as 5-lipoxygenase (5-LOX), 12-lipoxygenase (12-LOX), and cyclooxygenase (COX) are included among these. As a result of the high side effects caused by steroidal anti-inflammatory drugs, their application has declined steadily over the years. In response to these challenges, nonsteroidal anti-inflammatory drugs (NSAIDs) were created as they presented a safer option with lower side effects, although they also pose risks such as renal damage and gastric ulcers. The finding of COX and its coenzyme was a significant milestone in the understanding of inflammation and its central role in propelling the inflammatory reaction (Figure 5.1) [6].

Inhibition of the enzymes would prove to be an effective therapeutic application for the cure of inflammatory conditions. COX comes in two forms in the human body, namely, COX-1 and COX-2. Both have the same catalytic reaction and use the same substrate AA. COX enzymes are heme

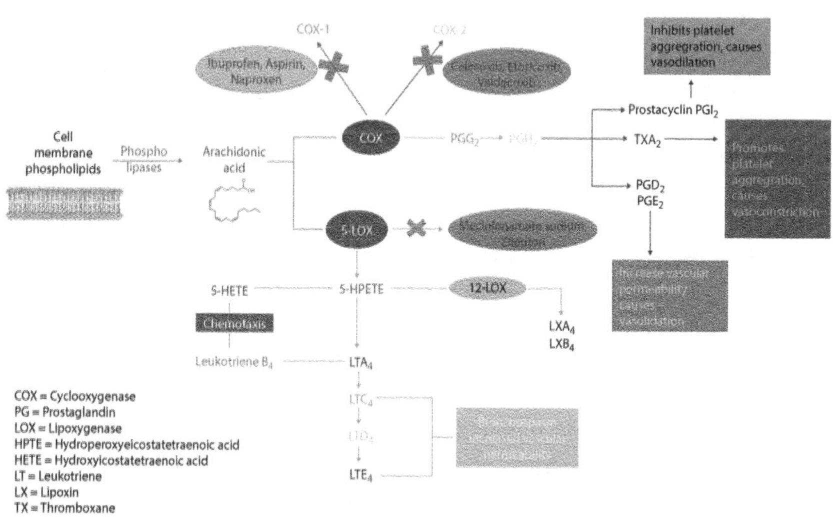

Figure 5.1 Inflammatory pathway and several important anti-inflammatory drugs [1].

proteins that possess different patterns and are involved in different physiological events. COX-1 is the constitutive form, being present in most parts of the gastrointestinal (GI) tract, renal collecting tubules, and platelets. It is thought to be required to maintain physiological processes such as renal integrity and GI protection, primarily through the production of protective prostaglandins (PGs). COX-2, on the other hand, is an inducible isoform, produced in response to tissue injury and stimulated by several inflammatory mediators. It is thought to be involved in producing proinflammatory PGs.

When both COX isoforms are inhibited by typical NSAIDs, the production of gastroprotective PGs through the pathway of COX-1 is reduced as well. Whereas such inhibition, despite being against the harmful COX action, has a tendency to produce GI side effects such as irritation, bleeding, and ulceration, which are commonly observed in the case of NSAID usage, COX-2, which is an inducible isoenzyme, is the one that produces the PGs of inflammation. Therefore, COX-2 selective inhibitors were developed to provide a reduced risk of gastric ulcers when compared to the conventional NSAIDs. Although selective COX-2 inhibitors are safer for the GI system, excessive usage of these agents has been associated with severe side effects, including cardiovascular disturbances. These disturbances are mainly caused by the decrease in antithrombotic PGs [7].

Heterocyclic chemistry has gained significant importance because of its useful properties, which render it applicable in a variety of applications, including the synthesis of therapeutic agents, drugs, copolymers, antioxidants, and dyes. More than 75% of the drugs that are being used in clinical practice are heterocyclic compounds. These nitrogen-containing heterocyclic compounds are especially important, as they are vital building blocks in most pharmacologically active compounds. These drugs have been investigated as therapeutic agents for centuries, with such compounds as morphine and vincristine being familiar [8]. Of the nitrogen-containing heterocycles investigated in anti-inflammatory drug research, pyrazole/pyrazoline, benzimidazole, indole, and pyrimidine are significant. These drugs are of interest because they closely resemble critical components used by the human body. Inflammation is a critical mechanism of defense in the immune system, facilitating elimination of noxious stimuli such as pathogens or damage and healing. During an acute inflammatory response, a number of cellular and molecular processes work in concert to mitigate possible damage or infection. Important among these are changes in vascular permeability, recruitment and migration of leukocytes, and inflammatory mediator release [9, 10].

During inflammation, some cell surface receptors are stimulated, which subsequently activate phospholipase A2, which releases AA into the cell membrane. Upon release, AA is quickly metabolized by COXs into active metabolites, which in turn lead to PGs, prostacyclin, and thromboxane formation, or by LOXs into leukotriene (LT) formation. The metabolites have varied physiological actions. Prostacyclin and PGs are powerful vasodilators, whereas thromboxanes are vasoconstrictors and platelet proaggregatory agents. LTs, on the other hand, cause bronchoconstriction. Additionally, PGs and LTs are involved in other inflammatory conditions, allergic reactions, and neoangiogenesis [11].

COX-1 is constitutively expressed in all tissues and produces PGs participating in normal body functions, including the preservation of the lining of the stomach, platelet activation, and the regulation of renal blood flow. COX-2 is, on the other hand, induced transiently by proinflammatory stimuli and plays a predominant role in pain, fever, and inflammation [12]. The traditional NSAIDs, including aspirin and the potent indomethacin, inhibit COX-1 and COX-2 and thus quench inflammation. But as they inhibit COX-1 as well, they can cause side effects such as gastric bleeding and renal problems. On the other hand, selective COX-2 inhibitors, or coxibs (e.g., celecoxib [CXB], rofecoxib, and valdecoxib), are powerful anti-inflammatory drugs that are less likely to produce GI side effects, as they selectively inhibit the induced COX-2 enzyme.

However, the use of COX-2–specific inhibitors has been linked with an increased risk of myocardial thrombotic events, which led to the removal of rofecoxib and valdecoxib from the market. Inhibition of COX enzymes channels the metabolism of AA toward the LOX pathway, leading to overproduction of LTs [13]. LTs are bioactive compounds exerting promigration actions on neutrophils and eosinophils, aggregation of neutrophils and monocytes, leukocyte adherence to the endothelium, capillary hyperpermeability, and contraction of smooth muscles. All of these result in inflammation, edema, bronchial constriction in asthmatic patients, and cardiovascular diseases [14]. Thus, novel anti-inflammatory agents that could inhibit both COX-2 and 5-LOX may be beneficial. These agents would inhibit the formation of both LTs and PGs with synergistic anti-inflammatory effects and a broader mechanism of action for treating conditions such as allergies and asthma. This two-way blockade would also inhibit mucosal damage to the GI mucosa and provide a more cardiology-friendly profile [15].

5.2 Classification

A. Classification of anti-inflammatory drugs based on the heterocyclic ring

S. no.	Heterocyclic ring name	Heterocyclic ring structure	Heterocyclic ring IUPAC name	Drugs example
1	Pyrrole	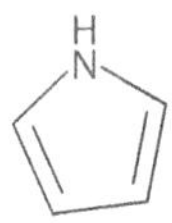	^{1}H-pyrrole	Ketorolac Etodolac
2	Pyrazole	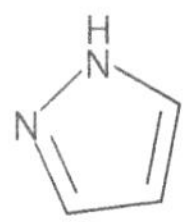	1,2-Diazacyclopenta2,4-diene	Celecoxib Ramifenazone Rimonabant Lonazolac
3	Imidazole	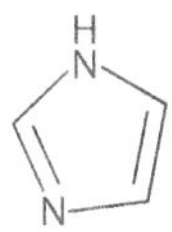	1,3-Diazole	Ketoconazole
4	Pyridine	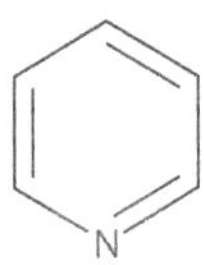	Azinine	Piroxicam Niflumic acid
5	Pyrimidine	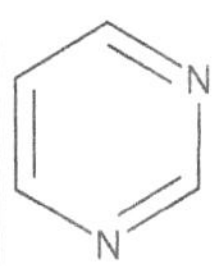	1,3-Diazabenzene	Efiriazole Fursultiamine S-adenosyl methionine Afloqualone Proquazone Tofacitinib
6	Indole	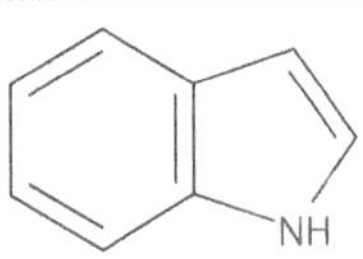	^{1}H-indole	Indomethacin
7	Furanone	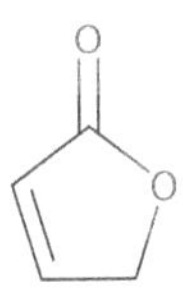	2,5-Dihydrofuran-2-one	Rofecoxib

(*Continued*)

(*Continued*)

S. no.	Heterocyclic ring name	Heterocyclic ring structure	Heterocyclic ring IUPAC name	Drugs example
8	Thiophene		Thiophene	Tinoridine Tiaprofenic acid Tenidap zileuton
9	Triazine	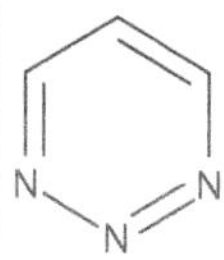	1,2,3-Triazine	Altretamine Enasidenib
10	Quinazoline	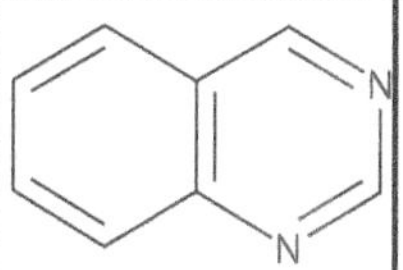	1,3-Diazanaphthalene	Proquazone Fluproquazone
11	Thiazole	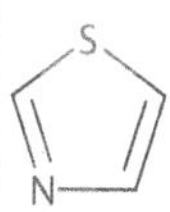	1,3-Thiazole	Meloxicam Fentiazac Fanetiazole Sudoxicam
12	Oxazole	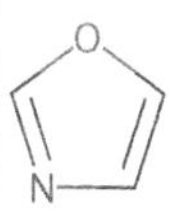	1,3-Oxazole	Oxaprozin
13	Isoxazole	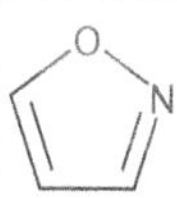	1,2-Oxazole	Valdecoxib Mofezolac
14	Pyrazolidine	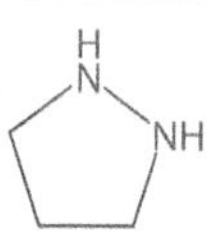	1,2-Biazacyclopentane	Phenylbutazone Oxyphenbutazone
15	Quinoline	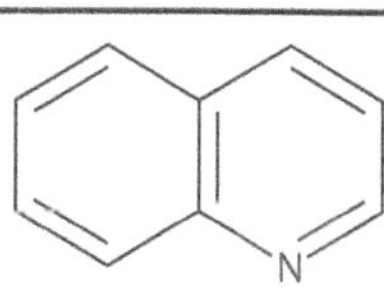	1-Benzopyridine	Chloroquine

B. Classification of heterocyclic rings anti-inflammatory drugs based on the pharmacological action

S. no.	Drugs class		Drug examples
1	**Nonselective COX inhibitors**		
	i.	Acetic acid derivatives	Ketorolac Indomethacin
	ii.	Pyrazolone derivatives	Phenylbutazone Oxyphenbutazone
	iii.	Enolic acid derivatives	Piroxicam Meloxicam
2	**Preferential COX-2 inhibitors**		Meloxicam Etodolac
3	**Selective COX-2 inhibitors**		Celecoxib Etoricoxib Parecoxib
4	**Analgesic antipyretics with poor anti-inflammatory action**		
	i.	Pyrazolone derivatives	Metamizole Propyhenazone
	ii.	Benzoxazocine derivative	Nefopam

5.3 Various Heterocyclic Derivatives as Anti-Inflammatory Drugs

5.3.1 Pyrazole and Pyrazolines Derivatives as Anti-Inflammatory Drugs

Pyrazoles are, in fact, a wonderful family of heterocyclic compounds, and this description captures their versatile uses and significance in medicinal chemistry as well as the pharmaceutical industry. That pyrazoles, and their derivatives such as pyrazolin-3-ones and pyrazolidine-3,5-diones, exhibit such a wide range of biological activities—from anti-inflammatory to anti-cancer and neuroprotection—makes them an extremely valuable target for drug discovery [16].

The pyrazole synthesis history, starting with Knorr's in 1883, is a good foundation to understand how this class of compounds has evolved and how its therapeutic potential has been exploited over the years. Pyrazole derivatives have been used for chronic gout, inflammation, analgesia, and even tuberculosis and other infection treatment [17].

The specific examples of pyrazole drugs, antipyrine, phenazone, metamizole (dipyrone), and phenylbutazone illustrate their continuing use in the clinic. Use as NSAIDs testifies to the activity and versatility of pyrazole derivatives in therapy [18].

NSAIDs' therapeutic use is primarily due to two major actions: (a) suppression of PG formation from AA by blocking the COXs and thromboxane synthase, in a differential manner, and (b) altering AA metabolism by 5-LOX to yield inflammatory mediators such as LTs and PGs [19].

The anti-inflammatory efficacy of the pyrazoles and pyrazolines that were synthesized was determined *in vitro* by the bovine serum albumin assay. This assay is premised on the fact that external molecules or stressors can lead to a loss of the protein's tertiary or secondary structure, and one of the major contributing factors to inflammation is protein denaturation. The capacity to prevent protein denaturation underlies this assay. CXB, a typical anti-inflammatory drug, was used as a comparison and inhibited maximally at 81.67% at the highest concentration (0.5 mg/mL) tested [20].

Some NSAIDs such as CXB, ramifenazone, lonazolac, and rimonabant, possessing a pyrazole moiety, are designed particularly as very potent COX-2 inhibitors (Figure 5.2). Although the initial results of the COX-2 selective medications (the coxibs) were promising, prolonged therapy set off alarms, and alternatives safer than they were have been searched for. Chemists and pharmacists have therefore focused on designing newer NSAIDs with fewer side effects, although the search for an ideal, side-effect–free anti-inflammatory continues [21].

Phenylbutazone, or bute, is a widely used NSAID in horses for the management of musculoskeletal pain. It is especially common in mares on assisted reproductive technology programs, where it is given for extended periods to manage chronic lameness. Previous work has shown that treatment with NSAIDs such as bute or flunixin meglumine during ovulation in mares increases the risk of not ovulating, further adding to the development of anovulatory hemorrhagic follicles. Phenylbutazone, as a nonselective COX inhibitor, exerts its effect through the reduction of PG synthesis, causing not only pain relief but also anti-inflammatory effect [22]. Classic NSAIDs that are more COX-1 preferentially bound are more associated with GI bleeding, ulcers, and nephrotoxicity than the COX-2 selective blocker drugs. Research has therefore resulted in the discovery of selective

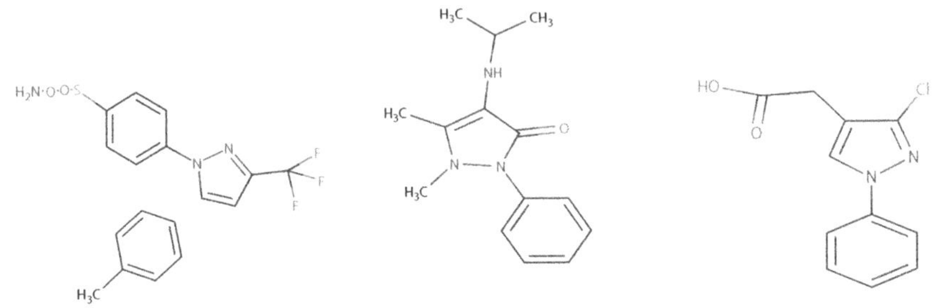

Celecoxib **Ramifenazone** **Lonazolac**

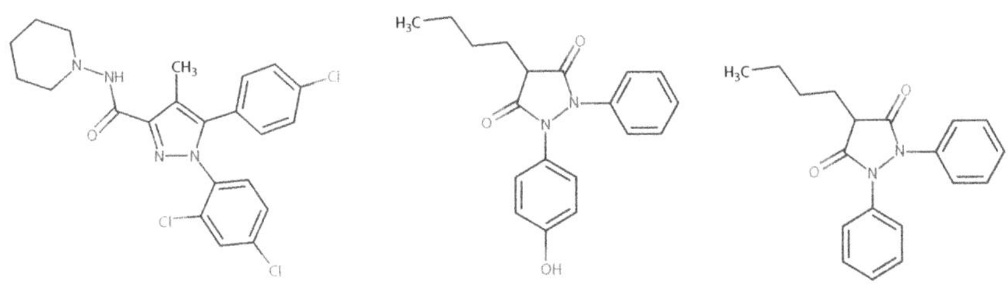

Rimonabant **Oxyphenbutazone** **Phenylbutazone**

Figure 5.2 Pyrazole and pyrazolines derivatives as anti-inflammatory drugs [25].

COX-2 inhibitors (coxibs) with the hope of enhancing GI safety over nonselective NSAIDs [23].

CXB and rofecoxib were the first two selective COX-2 inhibitors to receive US Food and Drug Administration (FDA) approval for treating osteoarthritis, rheumatoid arthritis, and pain. Clinical experience showed that the coxibs were equally effective as nonselective NSAIDs in efficacy and renal toxicity with the added advantage of enhanced GI safety [24]. But cardiovascular side effects such as elevation of blood pressure and risk of myocardial infarction due to the highly selective COX-2 inhibitors rofecoxib and valdecoxib resulted in withdrawal of these medications from the market. Conversely, CXB, which is a moderately selective COX-2 drug, did not elevate the risk of cardiovascular events over placebo or nonselective NSAIDs. Hence, there remains a need for the creation of new NSAIDs with moderate COX-2 selectivity and improved GI safety [25].

Oxyphenbutazone exerts its effects mainly as an anti-inflammatory agent by inhibiting COX enzymes, COX-1, and COX-2. This action lowers the synthesis of PGs, which are central in inflammation, pain, and fever. PGs play a role in mechanisms such as vasodilation, blood vessel permeability increase, and attraction of immune cells to sites of inflammation. Reducing PG concentration decreases pain, swelling, and stiffness in conditions of inflammation such as arthritis. Its action is the same as other NSAIDs, targeting to relieve pain and inflammation at the site of disease or injury [26].

Based on this information and our earlier studies on the synthesis of anti-inflammatory compounds, we here describe the synthesis and *in vitro* activity of a number of new pyrazole and pyrazoline derivatives. The compounds were examined for antioxidant activity, LOX inhibition, and activity as anti-inflammatory and analgesic drugs in adjuvant-induced arthritis [27].

5.3.2 Pyrimidine Derivatives as Anti-Inflammatory Drugs

Pyrimidine is an aromatic heterocyclic molecule that is structurally related to pyridine. It is a member of the diazine family, which consists of six-membered rings with two nitrogen atoms. In pyrimidine, the nitrogen atoms are at positions 1 and 3. Heterocyclic molecules that incorporate pyrimidine rings are of vital importance because they constitute an important family of natural and synthetic molecules. Most of these molecules possess important clinical uses and bioactivities. Substituted purines and pyrimidines occur copiously in living organisms and are some of the most investigated compounds in chemistry [28]. Pyrimidines constitute the most prevalent members of the diazine class and are represented by thymine, uracil, and cytosine, which are fundamental constituents of DNA and RNA. Moreover, the pyrimidine structure occurs in some natural compounds, e.g., vitamin B_1 (thiamine), and in certain synthetic compounds, such as barbituric acid and veronal, which are used as sedative medicines [29].

The action of pyrimidine-derived anti-inflammatory drugs is generally attributed to the inhibition of the synthesis of PGE_2 by COX enzymes. Like all other NSAIDs, these drugs inhibit the functioning of both the COX-1 and COX-2 enzymes, causing a decrease in the production of PGE_2 [30].

Epirizole inhibits inflammation and pain primarily by blocking COX enzymes, especially COX-2 to a certain degree. Besides, it fortifies lysosomal membranes and inhibits the mobility of inflammatory cells, which aids in its therapeutic action. The knowledge of its action informs us of its therapeutic applications and potential side effects, making it critical to use under supervision and careful control in treating inflammatory diseases. Proquazone, a typical NSAID, eases pain and inflammation by inhibiting COX enzymes, which play an important role in the synthesis of PGs (Figure 5.3). PGs are lipid substances that are responsible for inflammation and pain in

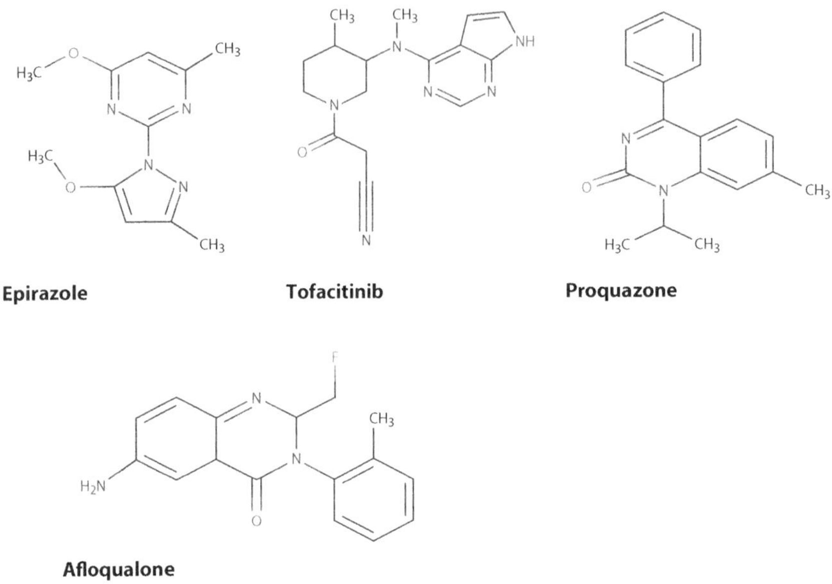

Figure 5.3 Pyrimidine derivatives as anti-inflammatory drugs [28].

the body. To learn how proquazone functions, we must examine its action on COX enzymes, how the body metabolizes it (pharmacokinetics), its action on the body (pharmacodynamics), and how it differs from other NSAIDs. By inhibiting these enzymes, proquazone decreases PG levels, thereby decreasing inflammation, pain, and fever [31].

5.3.3 Indole Derivatives as Anti-Inflammatory Drugs

Indoles also exist naturally in a few plants and fungi and are one of the most useful and ubiquitous nitrogen-containing heterocyclic frameworks. Indoles possess four reactive sites: a nitrogen atom at position 1, a carbon atom at position 3, a sigma bond between C2 and N, and the π-bond between C2 and C3. They also accept protons from strong acids such as HCl, and the carbon atom at position 3 is more protonatable than the nitrogen atom. Indole derivatives are of significant pharmacological interest [32].

Oxypertine, for instance, is an indole derivative, which is an antipsychotic and antidepressant and is used to treat schizophrenia. Further, indole3-carbinol, a bioactive molecule present in cruciferous vegetables, has also been investigated for its ability to prevent several types of cancers such as breast, prostate, colorectal, lymphoma, and transplacental cancer in offspring. Being of medicinal importance, the present study seeks to explore the gastric protective effects of new synthetic derivatives of indoles against indomethacin-induced gastric damage. These compounds, by virtue of their antipyretic, analgesic, and anti-inflammatory activities, are speculated to be gastroprotective against gastric ulcers caused by indomethacin [33].

A number of derivatives, including indomethacin (Indocin), acemetacin (Emflex), and etodolac (Etodine)—indole derivatives—and ketorolac (Ketolac) (Figure 5.4), a pyrrole derivative, have been synthesized. These agents act by inhibiting PG synthesis. Some, including indomethacin, acemetacin, tolmetin, and ketorolac, do this through nonselective inhibition of COX-1 and COX-2. Etodolac, on the other hand, selectively inhibits COX-2 [34].

Indomethacin, like all other NSAIDs, produces its effect by inhibiting PG synthesis. These drugs, synthesized predominantly by the COX enzymes, are prime movers in inflammation, fever, and pain. They also participate in the conservation of renal function, GI mucosa, and platelet function.

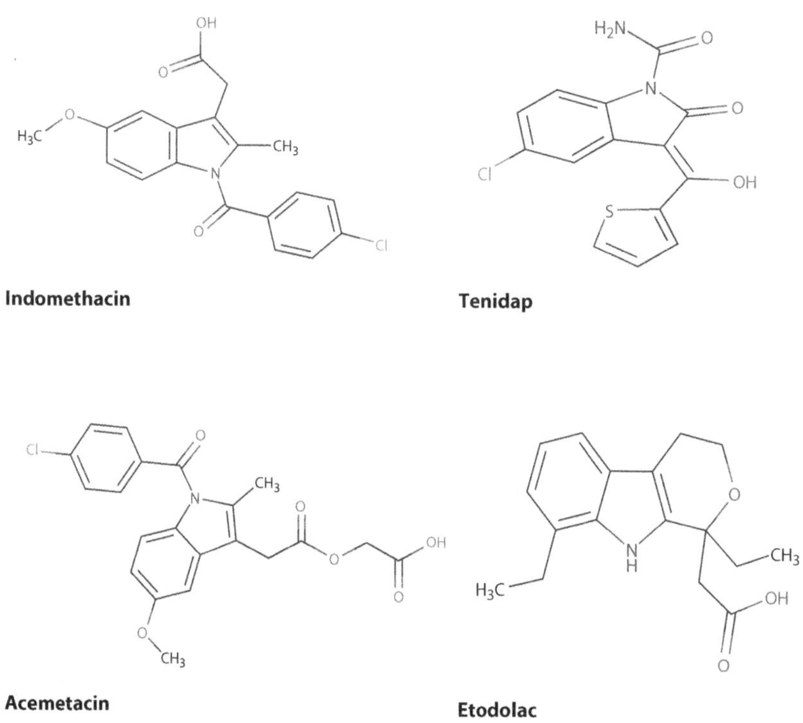

Figure 5.4 Indole derivatives as anti-inflammatory drugs [34].

Inhibition of COX enzymes by NSAIDs, nevertheless, results in numerous side effects [35].

Etodolac is an NSAID that is used to treat conditions such as rheumatoid arthritis and osteoarthritis. Its therapeutic effect is primarily due to the blocking of COX-1 and COX-2 enzymes, which reduces peripheral PGs that lead to inflammation. This blocking, however, results in undesirable side effects, primarily GI issues. To reverse these side effects and regulate the drug release rate, alginate microbeads and chitosan-coated alginate microbeads have been developed. Controlled-release etodolac microbeads were developed by extrusion/external Gelation technique with Box–Behnken experimental design [36].

In addition to its anti-inflammatory activity, indomethacin is also antiviral in nature as it blocks viral replication. It has been found to act against a number of viruses such as the rhabdovirus vesicular stomatitis virus, hepatitis B virus, and coronavirus. Although preliminary evidence suggests indomethacin can be useful against COVID-19, additional studies are needed to confirm its effectiveness [37].

Indomethacin is also used in ductus arteriosus closure in preterm neonates with respiratory distress syndrome. In preterm neonates, the ductus arteriosus fails to close due to excessive levels of PGE_2, which are responsible for its maintenance in a patent state. Indomethacin, by its action on inhibiting the synthesis of PGE_2, will close an open ductus arteriosus and prevent differential cyanosis as a complication [38].

5.3.4 Imidazole and Benzimidazole Derivatives as Anti-Inflammatory Drugs

Imidazoles and benzimidazoles are the key groups of N-heterocycles of contemporary studies. Imidazole is a five-membered heterocycle, containing three carbon atoms and two nitrogen atoms with nitrogen at the first and third ring positions. It has a planar structure. The imidazole ring is a significant unit in a number of significant natural compounds such as histamine, purine, histidine, and nucleic acids [39]. Conversely, benzimidazole is a six-membered benzene ring fused with a five-membered imidazole ring. The benzimidazole core continues to be a main area of research in the field of heterocyclic compounds because of its diverse pharmacological and biological activities. Benzimidazoles are especially known for their stability, good bioavailability, and wide range of biological activity. This article discusses the biological activities of imidazole and benzimidazole scaffolds, which are used extensively as anti-inflammatory agents such as flumizole and emedastine are shown in Figure 5.5 [40].

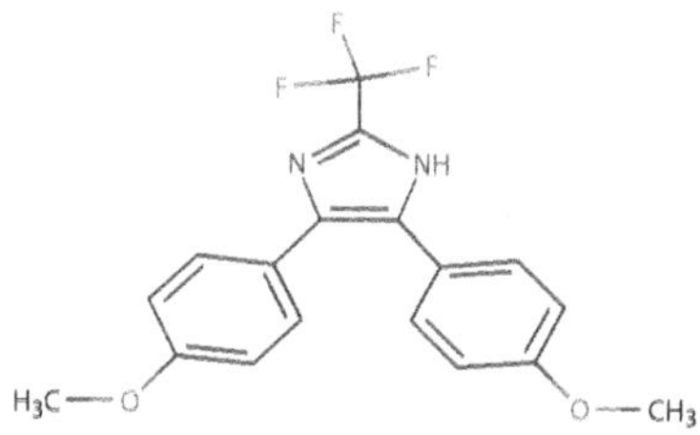

Flumizole

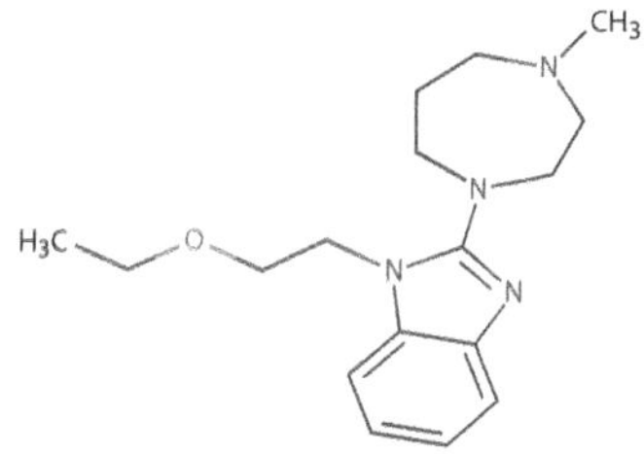

Emedastine

Figure 5.5 Imidazole and benzimidazole derivatives as anti-inflammatory drugs [40].

5.3.5 Pyrrole as Anti-Inflammatory Drugs

Pyrrole, which is one of the heterocycles, occurs not naturally but naturally occurring analogs are found in all cofactors, as well as natural products such as vitamin B_{12}, bile pigments including bilirubin and biliverdin, as well as those porphyrins that occur in heme, chlorophyll, chlorins, bacteriochlorins, and porphyrinogens [41]. Pyrrole-containing secondary metabolites such as makaluvamine M, ryanodine, rhazinilam, lamellarin, prodigiosin, myrmicarin, and sceptrinare also exhibit considerable biological activity. Pyrrole and pyrrolidine analogs also possess numerous therapeutic applications such as anti-inflammatory drugs, cholesterol-lowering agents, antitubercular drugs, and antitumor agents [42].

Nitrogen heterocycles are very well known for their medicinal value. Pyrrole and pyrrolidine moieties are present in many medicinal drugs. Some of the drugs with pyrrole and pyrrolidine structures are already on the market, whereas others are in the process of clinical trials [43].

Ketorolac is another NSAID belonging to this group (Figure 5.6). It comes in different forms, such as oral tablets, injections, nasal sprays, and

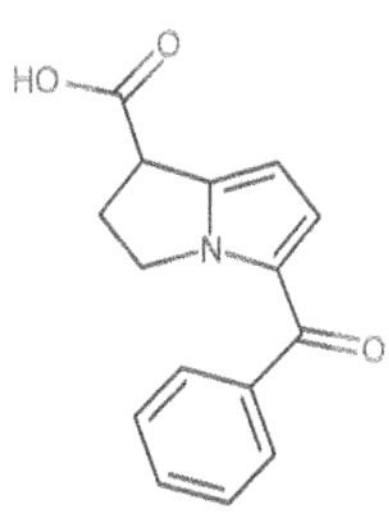

Ketorolac

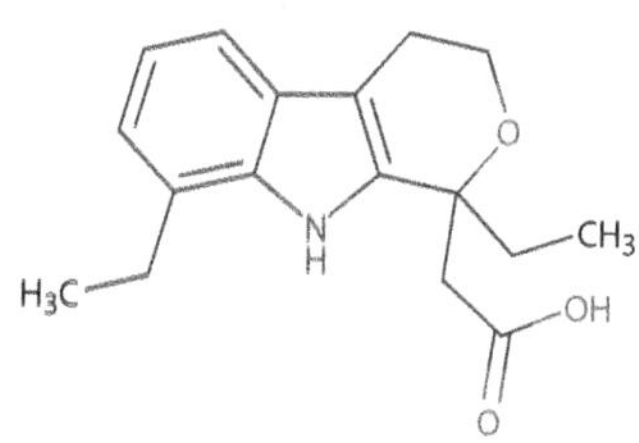

Etodolac

Figure 5.6 Pyrrole as anti-inflammatory drugs [45].

eye drops. Due to its analgesic properties, ketorolac is used to manage conditions such as rheumatoid arthritis, postoperative pain, osteoarthritis, menstrual disorders, and spondylitis [44].

Ketorolac is an NSAID of chemical structure comparable to indomethacin and tolmetin. Ketorolac tromethamine occurs as a racemic mixture with the [−]S and [+]R-enantiomers, and it is the S-enantiomer alone that is the active form contributing to its analgesic properties. Its inhibition of COX-1 and COX-2 pathways is believed to be responsible for its anti-inflammatory effect by blocking PG synthesis. This inhibition decreases the production of PG and thromboxane precursors from AA [45]. The decrease in PG production and activity is thought to be responsible for the therapeutic and adverse effects of the drug. Relief of pain is probably produced by a peripheral mechanism, in which reduced PG activity decreases the production of pain impulses. Also, inhibiting the synthesis or activity of other compounds that increase the sensitivity of pain receptors to mechanical or chemical stimuli could contribute to the analgesic effect. For ophthalmic administration, ketorolac decreases aqueous humor levels of PGE2 by inhibiting PG biosynthesis [46].

5.3.6 Furanone as Anti-Inflammatory Drugs

The anti-inflammatory activity seen in the aforementioned derivatives has encouraged us to pursue further research for the development of new anti-inflammatory agents. In this research, we describe the synthesis and biological activity of new 2-furanone derivatives. The 2-furanone derivatives were also used to synthesize other heterocyclic and nonheterocyclic compounds, which were evaluated for their ability to act as anti-inflammatory agents, inhibiting COX enzymes, LOX, and tumor necrosis factor α [47].

Furanone derivatives with COX-2 inhibitory activity have been found to exhibit analgesic, anti-inflammatory, and antirheumatic activities. Several quantitative structure–activity relationship (QSAR) studies have been performed to design selective COX-2 inhibitors. QSAR has conventionally been used to determine correlation between changes in chemical structure and changes in biological activity. Several natural products with the 2(5H)-furanone structure have been of interest owing to their noteworthy biological and pharmacological activities [48].

2-Furanones, a popular class of heterocyclic compounds, have attracted huge interest over the last decade because they can ring-open and transform into other heterocycles, including pyrrolones, pyridazinones, pyrazoles, and oxadiazoles. The literature abounds in numerous

2-furanone derivatives that are ring-opened to yield 2-pyrrolone, pyridazinone, and oxadiazole derivatives, all exhibiting excellent anti-inflammatory activity [49].

Rofecoxib (Vioxx) (Figure 5.7) was initially approved by the FDA for use in humans in May 1999, but it was withdrawn from the market in September 2004 because it posed a safety risk. The coxib-class medication was associated with an elevated risk of cardiovascular events. On the other hand, it had anti-inflammatory activity akin to classic NSAIDs, but fewer GI toxins. This decrease in GI complications, including ulcers, made rofecoxib a desirable alternative for patients, although cardiovascular concerns persisted. Rofecoxib eventually acted as a vital reference point during the creation of new NSAIDs that target COX-2 with less risk of side effects [50].

In the present study, we used a ligand-based virtual screening technique to discover new COX-2 inhibitors that were motivated by the structure of rofecoxib. Virtual screening has emerged as a commonly used strategy in the initial phases of drug discovery, accelerating the identification of lead drug candidates and reducing costs of development. We used computational resources and commercial compound libraries to evaluate the similarity and electrostatic characteristics of selected compounds. The outcomes were subsequently filtered according to pharmacokinetic and toxicological information. Moreover, we have also performed docking simulations to determine how tightly these compounds interact with COX-2 and compare their binding affinity to rofecoxib. As a predictive tool for the biological target, Swiss web server was used, whereas bioactivity was determined by the Molinspiration web server. We have also carried out molecular dynamics simulations to check how these potential compounds interact with COX-2 along the way of time [51].

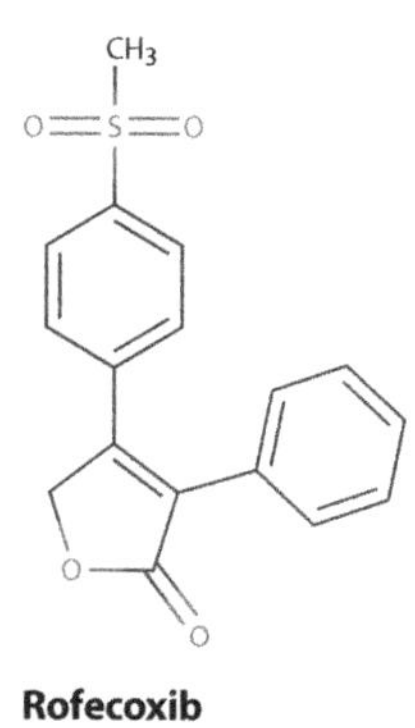

Rofecoxib

Figure 5.7 Furanone as anti-inflammatory drugs [47].

5.3.7 Oxazole and Isoxazole as Anti-Inflammatory Drugs

Oxazole and isoxazole are not only essential structural fragments in medicinal chemistry and organic synthesis but are also highly valued for their chemical properties. Therefore, heterocyclic-containing compounds have become essential in diverse areas of research such as medicinal chemistry, chemical synthesis, supramolecular chemistry, and materials science. Heterocycles are especially critical in being the building blocks for numerous pharmaceuticals and possess vast numbers of anti-inflammatory activities [52].

Valdecoxib, which was approved by the FDA in 2001, is one of the COX-2 selective inhibitors that are used to treat chronic pain due to osteoarthritis, rheumatoid arthritis, and dysmenorrhea (Figure 5.8). Valdecoxib inhibits the COX-2 enzyme that is responsible for converting AA into PGs [53].

Parecoxib is hydrophilic and a prodrug of valdecoxib. Although it was not approved by the FDA in 2005, valdecoxib proved to be more potent in its anti-inflammatory action. Actually, it produced a lower ED_{50} (0.032 ± 0.002 mg/kg per day) than that of CXB, the most effective NSAID, whose ED_{50} was 0.373 ± 0.163 mg/kg per day, in testing in a model of rat adjuvant arthritis. Yet, valdecoxib was recalled in 2005 from Canadian, US, and EU markets because it has been considered by some as an increased risk factor for heart attack and stroke [54].

5.3.8 Thiazole as Anti-Inflammatory Drugs

Thiazoles are heterogeneous units of value, and thiazole derivatives were found to exhibit significant anti-inflammatory activity. Such compounds include pentiazac (2-phenylthiazole), fanetizole (4-phenylthiazole), and meloxicam (5-methylthiazole) [55] (Figure 5.9).

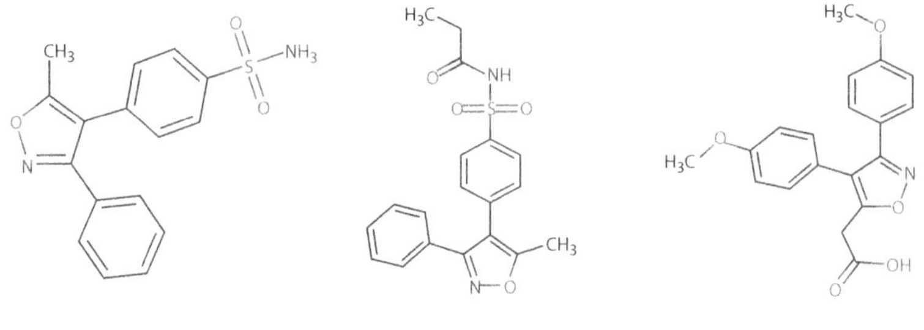

Figure 5.8 Oxazole and isoxazole as anti-inflammatory drugs [54].

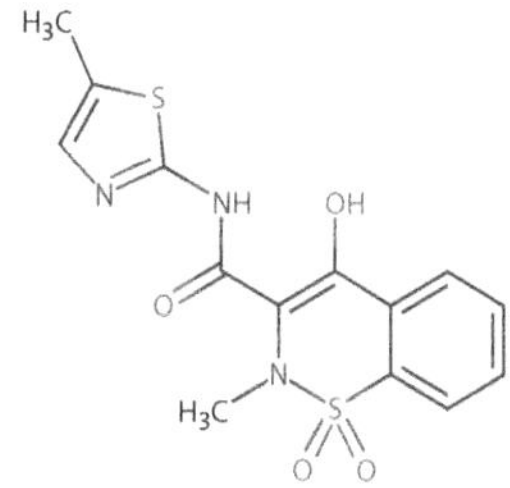

Meloxicam

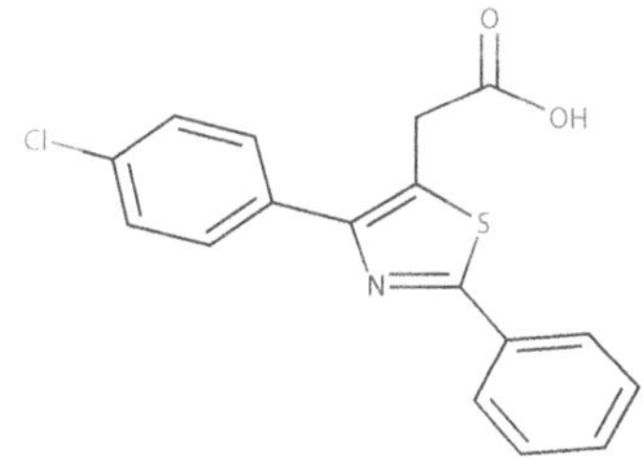

Fentiazac

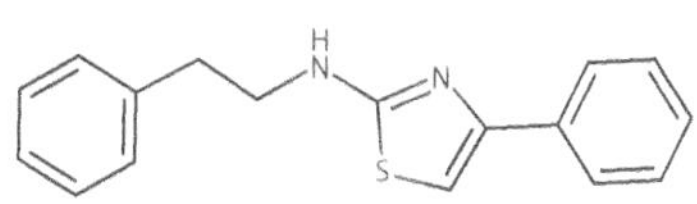

Fanetizole

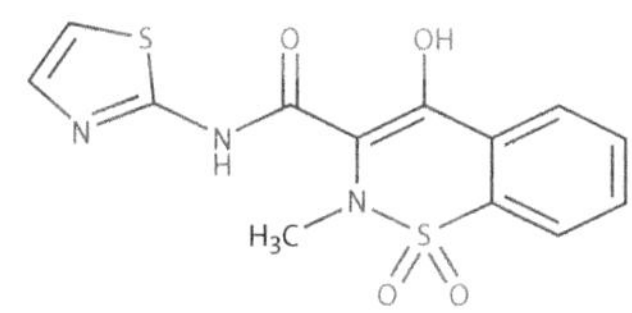

Sudoxicam

Figure 5.9 Thiazole as anti-inflammatory drugs [57].

Thiazole is a five-membered heterocycle with varied pharmacological activities such as anti-inflammatory, antioxidant, anticancer, and anti-HIV activity. Thiazole itself is a key moiety of selective COX-2 inhibitor meloxicam. Recently, 1,2-Benzzothiazole-3(2H)-ON-1,1-dioxide has been screened for COX-1/COX-2 inhibitory activity, and CXB-like activity has been reported [56].

Fanetizole is a 2-aminothiazole derivative and is an anti-inflammatory drug. It has been noted to show some COX inhibitory effect. Interestingly, fanetizole strongly inhibits the generation of superoxide in response to the chemotactic factor formyl-methionyl-leucyl-phenylalanine (f-Met-Leu-Phe). Inhibition of neutrophil-derived toxic oxygen metabolites by this suppression is one way in which fanetizole might exert its antiarthritic effect [57].

In addition, fanetizole has also been found to induce E-rosetting activity in adult thymectomized mice, in addition to improving murine thymic cells *in vitro* proliferation on mitogenic stimuli. Fanetizole synergizes also with the monokine IL-1. It possesses antiarthritic properties when administered *in vivo* in the model of adjuvant arthritis of the rat by inhibiting development of disease within the uninfected foot. Fanetizole induces fewer side effects as compared to levamisole in animal models [58].

Sudoxicam (SDX), a COX inhibitor, was once considered to be a potentially valuable NSAID. Nonetheless, its further development was withdrawn owing to safety issues related to its hepatotoxicity. *In vitro*, SDX was subject to oxidative biotransformation, and as a result, the thiazole ring opened, producing an acylthiourea compound and subsequently a hydroxylated metabolite. Further, the epoxide intermediate underwent hydrolysis, forming a thiazole-4,5-dihydrodiol metabolite. It has been proposed that the generation of secondary electrophilic metabolites, such as S-oxides of the acylthiourea, is associated with SDX-induced liver toxicity [59].

5.3.9 Pyridine as Anti-Inflammatory Drugs

Pyridine, a heterocyclic aromatic compound, has shown promising anti-inflammatory activity in various studies. Its derivatives, such as several pyridine-based compounds (Figure 5.10), control inflammation by inhibiting primary enzymes such as COX and LOX, which are key enzymes involved in the inflammatory process. By suppressing the synthesis of inflammatory mediators such as PGs and LTs, these compounds can effectively alleviate inflammation [29]. Pyridine is also being investigated as a therapeutic agent for diseases such as arthritis, asthma, and inflammatory bowel diseases.

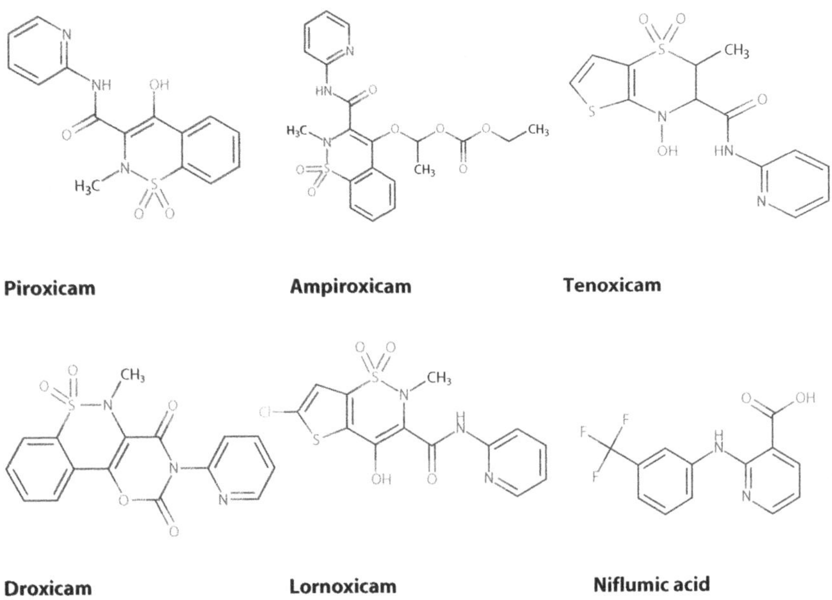

Figure 5.10 Pyridine as anti-inflammatory drugs [62].

Research is currently underway to design more potent and less toxic pyridine derivatives with the goal of making them suitable for therapy [60].

Piroxicam is a new nonsteroidal anti-inflammatory and analgesic that has no chemical similarity to other drugs now available. On a weight-for-weight basis, its effectiveness is similar to indomethacin in nonspecific models of inflammation. In the analgesic mode, piroxicam is more effective (weight-for-weight) than aspirin, fenoprofen, ibuprofen, naproxen, or phenylbutazone against phenylquinone-induced writhing. In human studies, piroxicam 20 mg/d produced less bleeding into the feces than aspirin 3.8 g/d. Piroxicam inhibits the secondary component of platelet aggregation due to aggregation induced by collagen and adenosine diphosphate, as do much other NSAIDs. Piroxicam is a reversible, selective inhibitor of the COX phase of AA metabolism that prevents PG synthesis *in vitro* and *in vivo* [61].

Ampiroxicam is a piroxicam prodrug that is a nonacid ether carbonate. The ampiroxicam does not exhibit visible PG synthesis inhibitory activity *in vitro*, as opposed to piroxicam. Ampiroxicam and piroxicam both exhibited similar *in vivo* potency for inhibition of paw edema in rats with adjuvant arthritis. Ampiroxicam's ED_{50} are 9 and 3.5 times larger than piroxicam's following a single or multiple oral doses per day, respectively, which suggests that piroxicam is superior to ampiroxicam in an acute model of rat paw inflammation. The phenylbenzoquinone stretching test, a measure of acute analgesic efficacy, indicates that ampiroxicam's ED_{50} is nearly three times larger than piroxicam's [62].

A new NSAID belonging to the oxicam group is lornoxicam, which is a tenoxicam congener. In comparison with other NSAIDs, this compound is highly effective as an analgesic and anti-inflammatory. Its analgesic actions are comparable with opiates. At dosages ≥8 mg, when administered, it has been found to be as effective as 10 mg morphine in alleviating pain in oral surgery. Due to its decreased half-life compared to the rest of the oxicams, lornoxicam retains the excellent therapeutic effectiveness of the oxicams along with a favorable GI toxicity profile compared to that of naproxen [63].

5.4 Conclusion and Future Prospects

Heterocyclic compounds have been found to hold a great promise for anti-inflammatory molecules due to the diversity of structural scaffolds and their ability to act on basic molecular mechanisms of inflammation. Their ability to modulate key pathways such as COX, LOX, NF-κB, and cytokines has shown encouraging preclinical and clinical results, making them suitable for the development of new anti-inflammatory drugs. Compared to their conventional counterparts of NSAIDs and corticosteroids, heterocyclic compounds possess the advantages of fewer possible side effects, better selectivity, and more treatment. Greater investigation into heterocyclic compounds has uncovered a vast array of chemical classes—pyrazoles, thiazoles, oxazoles, quinolines, and benzothiazoles—which manifest strong anti-inflammatory activity.

They are selective in inhibiting a range of some enzymes and signaling pathways essential for the inflammatory reaction, providing more specific therapy in the treatment of inflammatory disease. Having the promise of being able to inhibit many pathways at one time, they also provide bright leads for combining drugs against a variety of entities in the inflammatory cascade. Challenges still remain to enhance the pharmacokinetic and pharmacodynamic properties of heterocyclic anti-inflammatory agents. Specific needs for research development are raising the bioavailability, reducing the toxicity, and maximizing selectivity toward target targets.

Heterocyclic compounds have a promising future as anti-inflammatory drugs. The rapid growth of computational approaches, such as molecular modeling and virtual screening, will propel the development of new heterocyclic compounds with higher efficacy and less toxicity. Besides, the combination of heterocyclic compounds with other types of therapies to target multiple pathways of inflammation can provide a more effective therapy for long-term inflammatory conditions. Personalized medicine, wherein the therapy is tailored according to the patient's profile, can also enhance the therapeutic effectiveness of heterocyclic-based anti-inflammatory drugs.

In totality, the area of heterocyclic compounds as anti-inflammatory medications is a growing and intriguing field. With ongoing advancement in research, drug design, and technology, the compounds can revolutionize the treatment of inflammatory diseases through safer and more effective alternatives to existing therapies.

AI disclosure statement: The author acknowledges the use of AI-based tools for minor language editing or grammatical correction. However, all intellectual contributions, analyses, and conclusions presented in this chapter are entirely the author's own.

References

1. Sharma, S., Kumar, D., Singh, G., Monga, V., Kumar, B., Recent advancements in the development of heterocyclic anti-inflammatory agents. *Eur. J. Med. Chem.*, 200, 112438, 2020 Aug 15.
2. Lin, J.Y., Zhang, Y., Bian, Y., Zhang, Y.X., Du, R.Z., Li, M., Tan, Y., Feng, X.S., Non-steroidal anti-inflammatory drugs (NSAIDs) in the environment: Recent updates on the occurrence, fate, hazards and removal technologies. *Sci. Total Environ.*, 904, 166897, 2023 Dec 15.
3. Lisowska, B., Kosson, D., Domaracka, K., Positives and negatives of nonsteroidal anti-inflammatory drugs in bone healing: the effects of these drugs on bone repair. *Drug Des. Dev. Ther.*, 21, 1809–14, 2018 Jun.
4. Abdulkhaleq, L.A., Assi, M.A., Abdullah, R., Zamri-Saad, M., Taufiq-Yap, Y.H., Hezmee, M.N., The crucial roles of inflammatory mediators in inflammation: A review. *Vet. World*, 11, 5, 627, 2018 May 15.
5. Branco, A.C., Yoshikawa, F.S., Pietrobon, A.J., Sato, M.N., Role of histamine in modulating the immune response and inflammation. *Mediators Inflammation*, 2018, 1, 9524075, 2018.
6. Ju, Z., Li, M., Xu, J., Howell, D.C., Li, Z., Chen, F.E., Recent development on COX-2 inhibitors as promising anti-inflammatory agents: The past 10 years. *Acta Pharm. Sin. B*, 12, 6, 2790–807, 2022 Jun 1.
7. Dhiman, P., Arora, N., Thanikachalam, P.V., Monga, V., Recent advances in the synthetic and medicinal perspective of quinolones: A review. *Bioorg. Chem.*, 92, 103291, 2019 Nov 1.
8. Kumar, D., Sharma, S., Kalra, S., Singh, G., Monga, V., Kumar, B., Medicinal perspective of indole derivatives: recent developments and structure-activity relationship studies. *Curr. Drug Targets*, 21, 9, 864–91, 2020 Jul 1.
9. Chauhan, D., George, G., Sridhar, S.N., Bhatia, R., Paul, A.T., Monga, V., Design, synthesis, biological evaluation, and molecular modeling studies of rhodanine derivatives as pancreatic lipase inhibitors. *Arch. Pharm.*, 352, 10, 1900029, 2019 Oct.
10. Paul, A.T., Design, synthesis, biological evaluation, and molecular modeling studies of rhodanine derivatives as pancreatic lipase inhibitors.
11. Mukhopadhyay, N., Shukla, A., Makhal, P.N., Kaki, V.R., Natural product-driven dual COX-LOX inhibitors: Overview of recent studies on the development of novel anti-inflammatory agents. *Heliyon*, 9, 3, 2023 Mar 1.
12. Li, S.M., Tsai, S.E., Chiang, C.Y., Chung, C.Y., Chuang, T.J., Tseng, C.C., Jiang, W.P., Huang, G.J., Lin, C.Y., Yang, Y.C., Fuh, M.T., New methyl 5-(halomethyl)-1-aryl-^{1}H-1, 2, 4-triazole-3-carboxylates as selective COX- 2 inhibitors and anti-inflammatory agents: Design, synthesis, biological evaluation, and docking study. *Bioorg. Chem.*, 104, 104333, 2020 Nov 1.
13. Mohsin, N.U. and Irfan, M., Selective cyclooxygenase-2 inhibitors: A review of recent chemical scaffolds with promising anti-inflammatory and COX-2 inhibitory activities. *Med. Chem. Res.*, 29, 809–30, 2020 May.

14. Bindu, S., Mazumder, S., Bandyopadhyay, U., Non-steroidal anti-inflammatory drugs (NSAIDs) and organ damage: A current perspective. *Biochem. Pharmacol.*, 180, 114147, 2020 Oct 1.
15. Ragab, M.A., Eldehna, W.M., Nocentini, A., Bonardi, A., Okda, H.E., Elgendy, B., Ibrahim, T.S., Abd-Alhaseeb, M.M., Gratteri, P., Supuran, C.T., Al-Karmalawy, A.A., 4-(5-Amino-pyrazol-1-yl) benzenesulfonamide derivatives as novel multi-target anti-inflammatory agents endowed with inhibitory activity against COX-2, 5-LOX and carbonic anhydrase: Design, synthesis, and biological assessments. *Eur. J. Med. Chem.*, 250, 115180, 2023 Mar 15.
16. Mantzanidou, M., Pontiki, E., Hadjipavlou-Litina, D., Pyrazoles and pyrazolines as anti-inflammatory agents. *Molecules*, 26, 11, 3439, 2021 Jun 5.
17. Nehra, B., Rulhania, S., Jaswal, S., Kumar, B., Singh, G., Monga, V., Recent advancements in the development of bioactive pyrazoline derivatives. *Eur. J. Med. Chem.*, 205, 112666, 2020 Nov 1.
18. Raut, D.G., Lawand, A.S., Kadu, V.D., Hublikar, M.G., Patil, S.B., Bhosale, D.G., Bhosale, R.B., Synthesis of asymmetric thiazolyl pyrazolines as a potential antioxidant and anti-inflammatory agents. *Polycyclic Aromat. Compd.*, 42, 1, 70–9, 2021 Dec 31.
19. Priya, D., Gopinath, P., Dhivya, L.S., Vijaybabu, A., Haritha, M., Palaniappan, S., Kathiravan, M.K., Structural Insights into Pyrazoles as Agents against Anti-inflammatory and Related Disorders. *ChemistrySelect*, 7, 5, e202104429, 2022 Feb 4.
20. Khatana, K. and Gupta, A., An update on natural occurrence and biological activity of benzofurans. *Acta. Sci. Med. Sci.*, 4, 114, 2020.
21. Abdellatif, K.R., Abdelall, E.K., Elshemy, H.A., Lamie, P.F., Elnahaas, E., Amin, D.M., Design, synthesis of new anti-inflammatory agents with a pyrazole core: COX-1/COX-2 inhibition assays, anti inflammatory, ulcerogenic, histopathological, molecular Modeling, and ADME studies. *J. Mol. Struct.*, 1240, 130554, 2021 Sep 15.
22. Ayman, R., Abusaif, M.S., Radwan, A.M., Elmetwally, A.M., Ragab, A., Development of novel pyrazole, imidazo [1, 2-b] pyrazole, and pyrazolo [1, 5-a] pyrimidine derivatives as a new class of COX-2 inhibitors with immunomodulatory potential. *Eur. J. Med. Chem.*, 249, 115138, 2023 Mar 5.
23. Naglah, A.M., Almehizia, A.A., Ghazwani, M., Al-Wasidi, A.S., Naglah, A.A., Aboulthana, W.M., Hassan, A.S., *In Vitro* Enzymatic and Computational Assessments of Pyrazole–Isatin and Pyrazole–Indole Conjugates as Anti-Diabetic, Anti-Arthritic, and Anti-Inflammatory Agents. *Pharmaceutics*, 17, 3, 293, 2025 Feb 23.
24. Akhtar, W., Marella, A., Alam, M.M., Khan, M.F., Akhtar, M., Anwer, T., Khan, F., Naematullah, M., Azam, F., Rizvi, M.A., Shaquiquzzaman, M., Design and synthesis of pyrazole–pyrazoline hybrids as cancer-associated selective COX-2 inhibitors. *Arch. Pharm.*, 354, 1, 2000116, 2021 Jan.

25. Reheim, M.A., Rady, H.S., Mohamed, O.A., Hassan, A., Hafiz, I.S., Reffat, H.M., Elsaid, F.G., Eldesoqui, M., Alshaya, D.S., Badawy, A.A., Fayad, E., Synthesis, Anti-Inflammatory, and Molecular Docking Studies of New Heterocyclic Derivatives Comprising Pyrazole, Pyridine, and/or Pyran Moieties. *Pharmaceuticals*, 18, 3, 335, 2025 Feb 26.
26. Osman, E.O., Khalil, N.A., Magdy, A., El-Dash, Y., New pyrazole–pyridazine hybrids as selective COX-2 inhibitors: design, synthesis, molecular docking, in silico studies and investigation of their anti inflammatory potential by evaluation of TNF-α, IL-6, PGE-2 and NO in LPS-induced RAW264. 7 macrophages. *RSC Med. Chem.*, 15, 8, 2692–708, 2024.
27. Najmuldeen, Z.D. and Omar, T.N., Synthesis, characterization and evaluation of new pyrazoline derivatives containing sulfonamide moiety as anti-microbial and anti-inflammatory agents. *J. Res. Med. Dent. Sci.*, 11, 1, 73–81, 2023 Jan.
28. Tylińska, B., Janicka-Kłos, A., Gębarowski, T., Nowotarska, P., Plińska, S., Wiatrak, B., Pyrimidine derivatives as selective COX-2 inhibitors with anti-inflammatory and antioxidant properties. *Int. J. Mol. Sci.*, 25, 20, 11011, 2024 Oct 13.
29. ur Rashid, H., Martines, M.A., Duarte, A.P., Jorge, J., Rasool, S., Muhammad, R., Ahmad, N., Umar, M.N., Research developments in the syntheses, anti-inflammatory activities and structure–activity relationships of pyrimidines. *RSC Adv.*, 11, 11, 6060–98, 2021.
30. Munde, G., Menghani, S., Rarokar, N., Kerzare, D., Chittur, M.A., Khedekar, P., *In-silico* Studies, Synthesis, and Evaluation of Anti-inflammatory Activity of Novel Pyrimidine Scaffold. *Lett. Drug Des. Discovery*, 20, 10, 1621–31, 2023 Oct 1.
31. Bafail, R.S. and Samman, W.A., Anti-parkinsonian, anti-inflammatory, anti-microbial, analgesic, anti hyperglycemic and anticancer activities of poly-fused ring pyrimidine derivatives. *Trop. J. Pharm. Res.*, 23, 1, 67–75, 2024 Feb 5.
32. Siddique, S., Ahmad, K.R., Nawaz, S.K., Raza, A.R., Ahmad, S.N., Ali, R., Inayat, I., Suleman, S., Kanwal, M.A., Usman, M., Evaluation of the anti-inflammatory, analgesic, anti-pyretic and anti-ulcerogenic potentials of synthetic indole derivatives. *Sci. Rep.*, 13, 1, 8639, 2023 May 27.
33. Jin, J., He, H., Zhang, X., Wu, R., Gan, L., Li, D., Lu, Y., Wu, P., Wong, W.L., Zhang, K., The *in vitro* and *in vivo* study of oleanolic acid indole derivatives as novel anti-inflammatory agents: Synthesis, biological evaluation, and mechanistic analysis. *Bioorg. Chem.*, 113, 104981, 2021 Aug 1.
34. Singh, S., Sharma, N., Chandra, R., The indole nucleus as a selective COX-2 inhibitor and anti inflammatory agent (2011–2022). *Org. Chem. Front.*, 9, 13, 3624–39, 2022.
35. Ahmad, A.A., Hussain, K., Shah, M.R., Ashhad Halimi, S.M., Rabbi, F., Ahmad, Z., Khan, I., Rauf, A., Alshammari, A., Alharbi, M., Rasul Suleria, H.A., Molecular insights into the *in vivo* analgesic and anti inflammatory

activity of indomethacin Analogues. *ACS Omega*, 8, 33, 30048–56, 2023 Aug 10.

36. Silva, P., de Almeida, M., Silva, J., Albino, S., Espírito-Santo, R., Lima, M., Villarreal, C., Moura, R., Santos, V., (E)-2-Cyano-3-(1 H-Indol-3-yl)-N-Phenylacrylamide, a Hybrid Compound Derived from Indomethacin and Paracetamol: Design, Synthesis and Evaluation of the Anti-Inflammatory Potential. *Int. J. Mol. Sci.*, 21, 7, 2591, 2020 Apr 8.
37. Dandić, A., Rajkovača, K., Jozanović, M., Pukleš, I., Széchenyi, A., Budetić, M., Samardžić, M., Review of characteristics and analytical methods for determination of indomethacin. *Rev. Anal. Chem*, 41, 1, 34–62, 2022 Jan 13.
38. Sakr, A., Rezq, S., Ibrahim, S.M., Soliman, E., Baraka, M.M., Romero, D.G., Kothayer, H., Design and synthesis of novel quinazolinones conjugated ibuprofen, indole acetamide, or thioacetohydrazide as selective COX-2 inhibitors: anti-inflammatory, analgesic and anticancer activities. *J. Enzyme Inhib. Med. Chem.*, 36, 1, 1810–28, 2021 Jan 1.
39. Tripathy, D.B. and Mishra, A., Microwave synthesis and characterization of waste soybean oil-based gemini imidazolinium surfactants with carbonate linkage. *Surf. Rev. Lett.*, 24.05, 1750062, 2017.
40. Sabale, K.O., Bhor, R.J., Bhosale, M.A., A Review on Biologiv| Cal Activity of "Benzimidazole as a Imidazole Derivatives.". *Int. J. Pharm. Sci. Invent.*, 10, 19–27, 2021.
41. Paprocka, R., Pazderski, L., Mazur, L., Wiese-Szadkowska, M., Kutkowska, J., Nowak, M., Helmin-Basa, A., Synthesis and structural study of Amidrazone derived pyrrole-2, 5-dione derivatives: potential anti-inflammatory agents. *Molecules*, 27, 9, 2891, 2022 Apr 30.
42. Jeelan Basha, N., Basavarajaiah, S.M., Shyamsunder, K., Therapeutic potential of pyrrole and pyrrolidine analogs: an update. *Mol. Diversity*, 26, 5, 2915–37, 2022 Oct.
43. Gogoi, N.G., Rahman, A., Dutta, P., Saikia, J., Baruah, A., Handique, J.G., Design, synthesis, biological evaluation and in silico studies of curcumin pyrrole conjugates. *Chem. Biodiversity*, 21, 5, e202301605, 2024 May.
44. Nambiar MP, A.R.B., Designing of promising Tromethamine-Diflunisal-Pyrrole combinations based on COX binding, drug-properties and safety. *Indian J. Biochem. Biophys. (IJBB)*, 59, 7, 732–40, 2022 Jul 29.
45. Noti, V., Pontiki, E., Hadjipavlou-Litina, D., Development of Novel Pyrrole Derivatives and Their Cinnamic Hybrids as Dual COX-2/LOX Inhibitors. *Molecules*, 28, 24, 7958, 2023 Dec 5.
46. Appetecchia, F., New Pyrolle and Pyrazole based compounds active as anti-inflammatory and antimalarial drugs.
47. Abd El-Hameed, R.H., Mahgoub, S., El-Shanbaky, H.M., Mohamed, M.S., Ali, S.A., Utility of novel 2- furanones in synthesis of other heterocyclic compounds having anti-inflammatory activity with dual COX2/LOX inhibition. *J. Enzyme Inhib. Med. Chem.*, 36, 1, 977–86, 2021 Jan 1.

48. Rani, I., Khokra, S.L., Kaur, G., Sharma, P., Anti-inflammatory, analgesic and antimicrobial activities of some synthetic furanones and their pyrrolone derivatives. *Int. J. Pharm. Sci. Res.*, 12, 12, 3339–48, 2021.
49. Syrova, G., Chalenko, N., Levashova, O., Khaustova, M., Gaichjuk, A., The studying of analgesic activity of new pharmaceutical composition in the experiment.
50. Alizadeh, M., Jalal, M., Hamed, K., Saber, A., Kheirouri, S., Pourteymour Fard Tabrizi, F., Kamari, N., Recent updates on anti-inflammatory and antimicrobial effects of furan natural derivatives. *J. Inflammation Res.*, 19, 451–63, 2020 Aug.
51. Choudhary, D., Rani, I., Monga, J., Goyal, R., Husain, A., Garg, P., Khokra, S.L., Pyrazole based furanone hybrids as novel antimalarial: A combined experimental, pharmacological and computational study. *Cent. Nerv. Syst. Agents Med. ChemistryChemistry-Central Nerv. Syst. Agents*, 22, 1, 39–56, 2022 Apr 1.
52. Gujjarappa, R., Sravani, S., Kabi, A.K., Garg, A., Vodnala, N., Tyagi, U., Kaldhi, D., Singh, V., Gupta, S., Malakar, C.C., An overview on biological activities of oxazole, isoxazoles and 1, 2, 4-oxadiazoles derivatives. *Nanostruct. Biomater. Basic Struct. Appl.*, 8, 379–400, 2022 Mar.
53. Sochacka-Ćwikła, A., Czyżnikowska, Z., Regiec, A., Mączyński, M., Synthesis, structural studies, and molecular docking of isoxazole and oxazole derivatives with potential immunomodulatory and anti-inflammatory activities. *Advances.*, 19, 103, 2023 Oct.
54. Badawi, W.A., Rashed, M., Nocentini, A., Bonardi, A., Abd-Alhaseeb, M.M., Al-Rashood, S.T., Veerakanellore, G.B., Majrashi, T.A., Elkaeed, E.B., Elgendy, B., Gratteri, P., Identification of new 4-(6-oxopyridazin-1-yl) benzenesulfonamides as multi-target anti-inflammatory agents targeting carbonic anhydrase, COX-2 and 5-LOX enzymes: synthesis, biological evaluations and modelling insights. *J. Enzyme Inhib. Med. Chem.*, 38, 1, 2201407, 2023 Dec 31.
55. Hosseininezhad, S. and Ramazani, A., Thiazole ring-the antimicrobial, anti-inflammatory, and anticancer active scaffold. *Arabian J. Chem.*, 16, 11, 105234, 2023 Nov 1.
56. Maghraby, M.T., Abou-Ghadir, O.M., Abdel-Moty, S.G., Ali, A.Y., Salem, O.I., Novel class of benzimidazole thiazole hybrids: The privileged scaffolds of potent anti-inflammatory activity with dual inhibition of cyclooxygenase and 15-lipoxygenase enzymes. *Bioorg. Med. Chem.*, 28, 7, 115403, 2020 Apr 1.
57. Kamat, V., Santosh, R., Poojary, B., Nayak, S.P., Kumar, B.K., Sankaranarayanan, M., Faheem, Khanapure, S., DA, B., Vootla, S.K., Pyridine-and thiazole-based hydrazides with promising anti-inflammatory and antimicrobial activities along with their in silico studies. *ACS Omega*, 5, 39, 25228–39, 2020 Sep 25.

58. Ahmed, A., Molvi, K.I., Patel, H.M., Ullah, R., Bari, A., Synthesis of novel 2, 3, 5-tri-substituted thiazoles with anti-inflammatory and antibacterial effect causing clinical pathogens. *J. Infect. Public Health*, 13, 4, 472–9, 2020 Apr 1.
59. SL M, Novel approach of multi-targeted thiazoles and thiazolidenes toward anti-inflammatory and anticancer therapy—dual inhibition of COX-2 and 5-LOX enzymes. *Med. Chem. Res.*, 30, 1, 236–57, 2021 Jan.
60. Abd El-Lateef, H.M., Abdelhamid, A.A., Khalaf, M.M., Gouda, M., Elkanzi, N.A., El-Shamy, H., Ali, A.M., Green synthesis of novel pyridines *via* one-pot multicomponent reaction and their anti-inflammatory evaluation. *ACS Omega*, 8, 12, 11326–34, 2023 Mar 16.
61. Bilavendran, J.D., Manikandan, A., Thangarasu, P., Sivakumar, K., Synthesis and discovery of pyrazolo pyridine analogs as inflammation medications through pro-and anti-inflammatory cytokine and COX-2 inhibition assessments. *Bioorg. Chem.*, 94, 103484, 2020 Jan 1.
62. Ismael, A.S., Amin, N.H., Elsaadi, M.T., Ali, M.R., Abdel-Rahman, H.M., Design, synthesis and biological evaluation of new imidazo [1, 2-a] pyridine derivatives as selective COX-2 inhibitors. *J. Mol. Struct.*, 1250, 131652, 2022 Feb 15.
63. Abdelgawad, M.A., Elkanzi, N.A., Musa, A., Ghoneim, M.M., Ahmad, W., Elmowafy, M., Ali, A.M., Abdelazeem, A.H., Bukhari, S.N., El-Sherbiny, M., Abourehab, M.A., Optimization of pyrazolo [1, 5-a] pyrimidine based compounds with pyridine scaffold: Synthesis, biological evaluation and molecular m.

6

Heterocyclic Compounds as Antioxidant Agents

Niranjan Kaushik[1]*, Shristi Singh[1], Ajita Paliwal[1] and Sachin Chaudhary[2]

[1]Department of Pharmacy, School of Medical and Allied Sciences, Galgotias University, Greater Noida, India
[2]College of Pharmacy, Department of Medicinal Chemistry, University of Sharjah, Sharjah, United Arab Emirates

Abstract

Oxidative stress occurs when there is a disruption between the production of reactive oxygen species (ROS) and the antioxidant response, which is linked to several illnesses, including cancer, metabolic syndrome, and neurological diseases. Effective antioxidants have been found in heterocyclic compounds, a group of organic compounds that have at least one heteroatom (such as nitrogen [N], oxygen [O], and sulfur [S]), as well as a ring-like structure. They exhibit antioxidant activity through metal chelation, electron donation, and hydrogen transfer mechanisms and are found in both natural items and medicines. Strong antioxidant action has been previously documented for heterocycles containing N, such as pyridine and pyrimidine, and heterocycles containing O, such as furan, benzofuran, and coumarin. Antioxidant capabilities have been found for S heterocycles, such as thiophene, and polyphenolics, such as flavonoids and phenolic acids. Heterocyclic compounds can prevent lipid peroxidation, neutralize ROS, and stop damage caused by oxidation, which is ascribed to their antioxidant properties. This study highlights the antioxidant properties of heterocyclic compounds and discusses their potential therapeutic use in the avoidance and management of conditions linked to the effects of oxidative stress.

***Keywords*:** Oxidative stress, heterocyclic compounds, antioxidant activity, polyphenols, lipid peroxidation, metal chelation

**Corresponding author*: niranjankaushik79@gmail.com

Divya Bajpai Tripathy, Priyanka Chhabra and Anjali Gupta (eds.) Heterocyclic Compounds as Pharmaceutical Ingredients, (207–248) © 2026 Scrivener Publishing LLC

6.1 Introduction: Antioxidants and Oxidative Stress

The inability to detoxify reactive compounds leads to oxidative imbalances, which is known as oxidative stress, produced during metabolism when cells generate reactive oxygen molecules [1, 2]. The rise in free radicals brought on by biological reactions results in modifications to chromosomes, damage to membranes from peroxidation of polyunsaturated fatty acids with lipids, disturbance of within the cell's protein function and structure, and death of cells, as well as oxidative damage to cellular parts and structures such as polysaccharide depolymerization along with carbohydrates breakdown [3]. However, reactive oxygen species (ROS) levels can rise sharply and bring about serious cellular damage in the body during periods of environmental stress and cell malfunction. Therefore, oxidative stress plays a major role in the pathophysiology of aging, Alzheimer disease, cataracts, cardiovascular disease, cancer, diabetes, and inflammatory diseases. To prevent or reduce ROS-induced damage from oxidation, the human body and other species have developed an antioxidant defense mechanism. After these radicals are produced, this system neutralizes them through enzymatic, metal-chelating, and free radical–scavenging actions. Undoubtedly, the subject of antioxidants is experiencing a renaissance, in large part because the progress of many long-term illnesses, such as diabetes and cancer, is closely associated with oxidative damage [4]. A sufficient antioxidant state in the body may also be maintained through dietary antioxidant consumption. The body uses antioxidants to lessen or repair damage and defend against some detrimental consequences of free radicals. It is well recognized that the pathophysiology of numerous diseases, including tumors, metabolic syndrome, coronary artery disease, the parasite malaria, Alzheimer disease, rheumatoid arthritis, neurological diseases, and preeclampsia, is strongly correlated with the mechanisms causing the rise in ROS brought on by oxidative stress [5, 6]. A range of antioxidants have been supplied to both people and animals as nutritional supplements or medications, and they have also been added to foods to maintain their nutritional value [7, 8]. To stop the chain reaction and stop ROS damage, a variety of antioxidants can directly interact with ROS and/or ROS-induced free radical intermediates. There are numerous major categories into which antioxidants can be divided (Figure 6.1): (i) endogenous and exogenous, (ii) natural and synthetic, (iii) polar and non-polar, (iv) enzymatic and nonenzymatic, and (v) by the mechanisms by which they operate [9]. Three primary mechanisms explain the functioning of

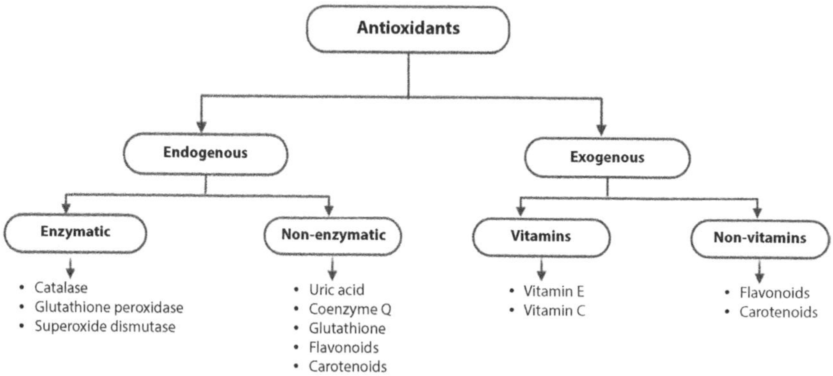

Figure 6.1 Classification of antioxidants.

antioxidants: metal chelation, single electron transfer (SET), and hydrogen atom transfer (HAT) [10].

6.2 Mechanism of Antioxidant Agent

An essential biological function of antioxidants is the scavenging of free radicals, which keeps the organism from suffering from oxidative damage. Numerous illnesses are linked to oxidative damage in the body caused by radicals. Both manufactured and natural antioxidants have the ability to prevent oxidative harm by neutralizing free radicals [11]. The three separate functions that antioxidants display, (a) HAT, (b) metal chelation, and (c) SET, form the basis of their process (as shown in Figure 6.2). According to the HAT process, the antioxidant turns into a radical when one of its hydrogen atoms is removed by the free radical. The bond dissociation enthalpy (BDE) is a crucial metric in this mechanism for assessing antioxidant activity. The inactivation response of free radicals will proceed more easily if the amount of BDE of the potential H-donating molecule antioxidants is smaller [12, 13]. As an example, lipid peroxidation is prevented when vitamin E (α-tocopherol) gives lipid peroxyl radicals a hydrogen atom. A radical intermediate is produced when the nucleophile transfers one electron to the substrate, simulating the SET reaction whose destiny can subsequently influence a variety of events. The antioxidant gives the free radical an electron during SET reactions, after which it transforms into a cation that is radical. The rapid mechanism of SET lessens oxidative damage. In one example, flavonoids (such as quercetin) neutralize ROS and reduce metal ions by donating an electron. The most crucial energy

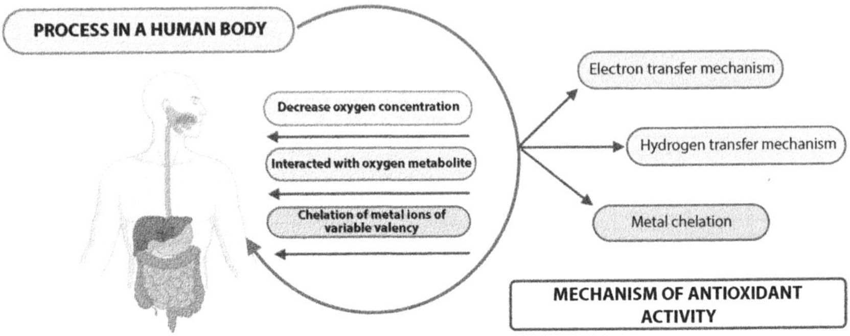

Figure 6.2 Mechanism and effect of antioxidants in the human body.

component in assessing the antioxidant effect in this method is the antioxidant's ionization potential (IP). The electron abstraction process is facilitated by a lower IP [14].

A complicated ring-like structure is formed by the binding of a metal (oxidant), which is known as metal chelation. These metals are bound by antioxidants, which stop ROS from forming. For instance, quinoline compounds and catechins bind Cu^{2+} and Fe^{2+}, lowering oxidative stress. Depending on how they work, antioxidants fall into two major types [15].

(a) Primary antioxidants suppress or reduce the autoxidation's beginning phase and interfere with its propagation stage by stabilizing the free

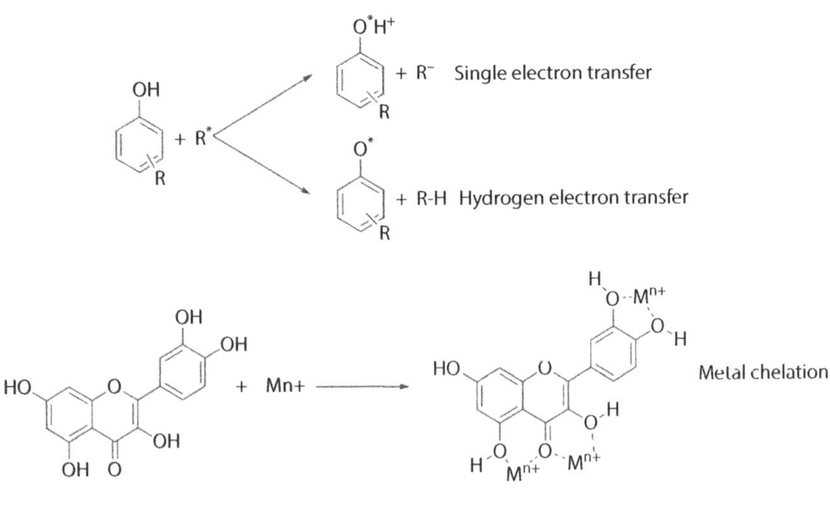

Figure 6.3 Schematic representation of antioxidant's mechanism of action.

radicals by giving the hydroxyl group electrons or hydrogen atoms, therefore disrupting the chain reaction of oxidative free radicals. Before causing cellular harm, these antioxidants directly eliminate reactive oxygen compounds (ROS) and free radicals shown as Figure 6.3. They stop oxidative chain reactions by acting as radical scavengers. These antioxidants involve the HAT, SET, and chain-breaking mechanisms. (b) Antioxidants that are secondary bind metal ions deactivate singlet O, scavenge O, absorb ultraviolet radiation, and aid in the regeneration of the primary antioxidant. Through antioxidant enzymes' activation, they remove damaged molecules for the prevention of ROS generation; these antioxidants strengthen the body's defense mechanisms. They assist other defense mechanisms rather than directly neutralizing radicals. The key mechanism for secondary antioxidants is metal chelation and gene expression modulation. Using electron delocalization and stability, metal chelation characteristics, redox potential and enzyme modulation, lipophilicity, and bioavailability, as indicated in Table 6.1, heterocyclic compounds increase antioxidant activity [16, 17].

Table 6.1 The role of heterocyclic structure enhancing antioxidant activity.

Mechanism	Description
Electron delocalization	Heterocyclic rings, especially those with nitrogen or sulfur atoms, can exhibit electron delocalization throughout their structure. This delocalization, which distributes the electron density over a larger surface area, enhances the molecule's ability to stabilize free radicals.
Metal chelation	Transition metals (Fe^{2+}, Cu^{2+}) can be chelated by a variety of heterocyclic antioxidants, including flavonoids and other polyphenols. By doing this, metals are kept out of redox reactions that result in dangerous ROS.
Redox cycling	Redox cycling allows certain heterocyclic antioxidants, such as those in the polyphenol group, to neutralize several free radicals before going dormant.
Lipophilicity and bioavailability	Antioxidant compounds with heterocyclic structures are frequently more soluble and bioavailable, increasing their potency at lower dosages. For therapeutic uses, this is essential.

6.3 Structures and Classifications of Heterocyclic Compound

Cyclic rings of atoms with at least one heteroatom are known as heterocyclic compounds. [18, 19]. Either cyclic or acyclic is possible. Heterocycles have uses in the agricultural, pharmaceutical, and veterinary industries [20]. Vitamins, hormones, alkaloids, and several other natural goods contain this important class of natural chemicals. Furthermore, they are present in a variety of physiologically active compounds. Most heterocycles are also used within the fields of industry, health, agriculture, and supramolecular chemical research [21]. They are broadly divided into three groups: N, S, and O [22, 23]. One of the reasons heterocyclic compounds are used so widely is that their configurations can be gently changed to create the intended changes in functions. Many heterocycles belong to one of the few major structural groups that vary significantly within the group yet share many traits in general. The variances might include variations in polarity, acidity, or basicity [24]. They can be categorized according to the size and structure of their rings, such as fused rings, rings with five or six members, etc. Two criteria can be used to classify heterocyclic compounds based on their structure or ring size (Figures 6.4. and 6.5) and electronic configuration.

i. **Classification based on ring size and heteroatom**
   a) **Three-membered rings**: Small ring structures with a single heteroatom, such as thiiranes (containing S), oxiranes (carrying O), and aziridines (containing N).
   b) **Four-membered rings**: Azetidines (N), oxetanes (O), and thietanes (S).

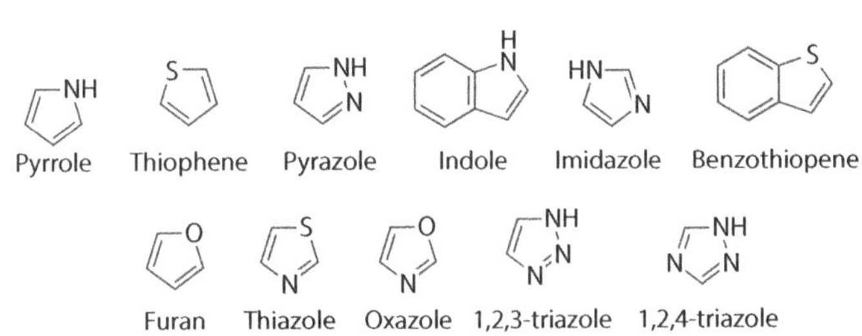

Figure 6.4 Chemical structure of 5-membered heterocyclic compound.

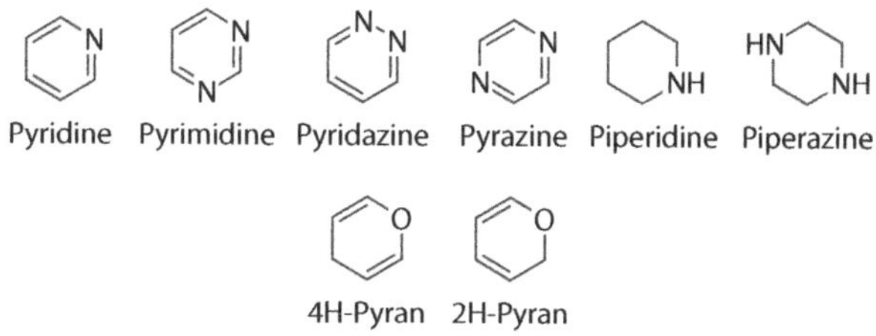

Figure 6.5 Chemical structure of 6-membered heterocyclic compound.

c) **Five-membered ring**: This group includes substances with biological activity and aromatic stability, such as thiophenes (S), furans (O), and pyrroles (N).
d) **Six-membered rings**: Pyridines (N), pyrans (O), and thiopyrans (S) are notable examples.

ii. **Classification based on electronic configuration**
a) Heterocyclic aliphatic compounds are also known as saturated heterocycles because they do not contain any double bonds, for example, cyclic amines, cyclic ethers, cyclic thioethers, etc.
b) Heterocyclic aromatic compound [25]:

6.3.1 Role of Heterocyclic Structure Enhancing Antioxidant Activity

In medical chemistry, heterocyclic molecules are essential, especially when it comes to boosting antioxidant activity. A cyclic structure's at least one heteroatom significantly impacts these compounds' electrical, steric, and redox characteristics. Heterocyclic structures are potent antioxidants because they may scavenge radicals and give electrons. Heterocyclic frameworks and increased antioxidant capability have been linked in recent research, highlighting their importance in medication development, food preservation, and illness prevention. Because of its numerous applications in the biological, chemical, and medical sciences, heterocycles are crucial to organic chemistry [26, 27]. Cyclic organic molecules containing a heteroatom or several, such as S, O, N, etc., are called organic heterocycles [28, 29]. To neutralize ROS and prevent damage from oxidation to proteins, lipids, and genetic material, a variety of naturally occurring heterocycles, including flavonoids, purines, benzofurans, and indoles, donate hydrogen atoms or electrons. Heteroatoms such as N, O, or S enhance electron delocalization, increasing the compounds' ability to stabilize free radicals.

Furthermore, as metal chelators, heterocyclic antioxidants such as pyridine derivatives and quinoline reduce the enzymatic activity of iron and copper ions in processes linked to oxidative stress. Recently, several N-containing heterocycles with various characteristics and applications have become more well-known in medicinal, pharmaceutical, and chemical chemistry [30, 31]. As an anti-inflammatory medication, epirizole has demonstrated a high biological evaluation among heterocyclic N compounds, including pyrazole, pyrimidine, pyrazoline, and isoxazoline [32, 33]. Trimethoprim is used as an antibiotic, whereas dasatinib treats leukemia [34]. The heterocycles that contain O exhibit a variety of pharmacological characteristics, such as antioxidant, antimicrobial, anti-HIV, antimalaria, anticancer, and antitubercular properties [35, 36]. The anti-inflammatory, antifungal, anticancer, and antibacterial qualities of iso-oxazoline are demonstrated (Figure 6.6) [37]. Additionally, S-containing heterocyclic compounds are used in colors, medicines, and agrochemicals and are crucial building components of primary and secondary metabolites [38, 39]. In medicinal chemistry, heterocyclic compounds—characterized by ring structures that have a minimum of one atom other than carbon (C)—are crucial due to their diverse spectrum of biological activity. Their antioxidant capacity is one noteworthy feature that is critical for scavenging free radicals and preventing illnesses linked to oxidative stress. The structural characteristics of heterocyclic molecules are closely related with their antioxidant efficacy.

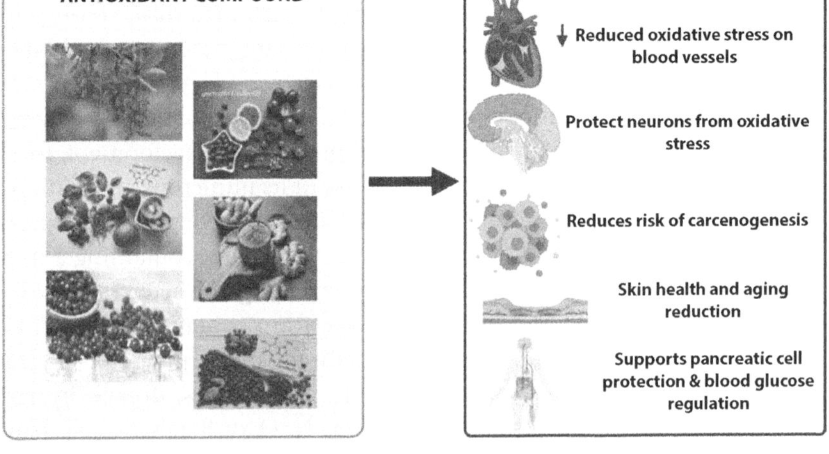

Figure 6.6 Heterocyclic compound containing health benefits.

Electron delocalization: By promoting electron delocalization, heterocyclic rings stabilize and lessen the sensitivity of free radicals.

The existence of heteroatoms in the ring, such as N, O, or S, can improve the ring's capacity to scavenge radicals by donating electron pairs.

Effects of substitutes: Heterocyclic core functional groups can alter electron density, which affects antioxidant potency [40].

6.3.2 Existing Heterocyclic Antioxidant Compounds

Several heterocyclic compounds have strong antioxidant qualities. These substances are distinguished by their capacity to suppress oxidative damage, stop lipid peroxidation, and scavenge free radicals. The presence of heteroatoms such as N, O, or S within the ring structure enhances their ability to interact with ROS. For instance, S-containing heterocycles can bind metal ions and effectively scavenge free radicals, reducing oxidative stress. Similarly, N-containing heterocycles have been found to modify antioxidant defense mechanisms, offering protection against oxidative damage. Major heterocyclic molecules with demonstrable antioxidant activity are presented in Table 6.2, highlighting their structures and functions.

Table 6.2 Examples of heterocyclic antioxidants that are currently in use.

S. no.	Compound name	Structure	Function	References
1.	Flavonoids (quercetin)	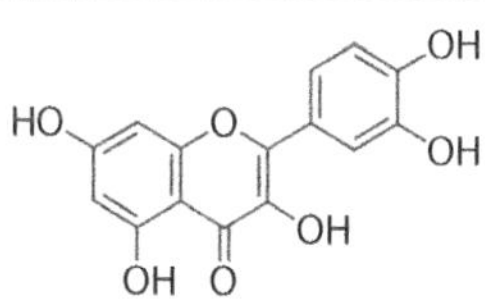	Quercetin reduces oxidative stress and stops lipid peroxidation through the scavenging of free radicals.	[41]
2.	Flavonoids (catechin)	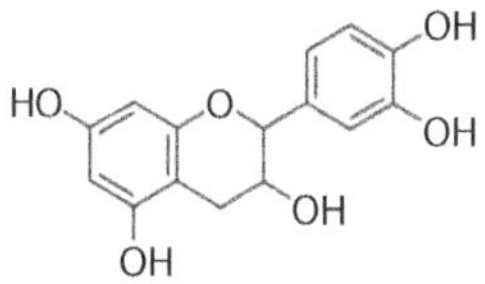	Catechin supplies hydrogen atoms to neutralize and stop cellular damage caused by reactive oxygen species (ROS) and free radicals.	[42]

(*Continued*)

Table 6.2 Examples of heterocyclic antioxidants that are currently in use. *(Continued)*

S. no.	Compound name	Structure	Function	References
3.	Flavonoids (luteolin)	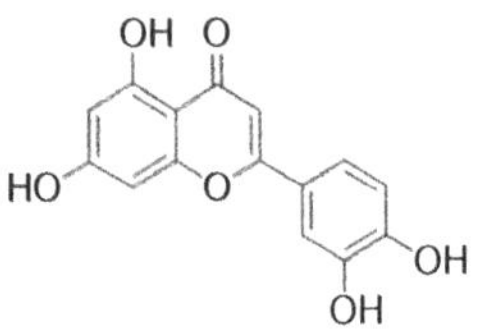	By preventing the production of ROS, luteolin protects DNA, lipids, and proteins against oxidative damage.	[43]
4.	Indoles (melatonin)	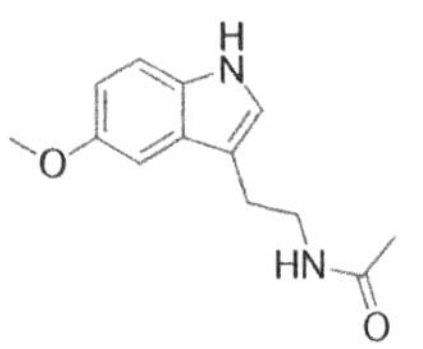	Free radicals are neutralized by melatonin, which lowers inflammation and oxidative stress. Antioxidant enzymes, including SOD and CAT, are expressed in response to melatonin.	[44]
5.	Pyridine (nicotinamide)	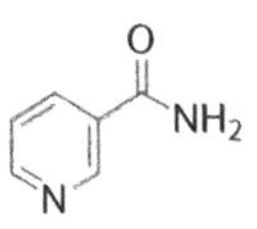	Nicotinamide prevented the ascorbate–Fe^{2+} from causing peroxidation of lipids and the oxidation of proteins (carbonylation).	[45]
6.	Thiazoles (thiamine)	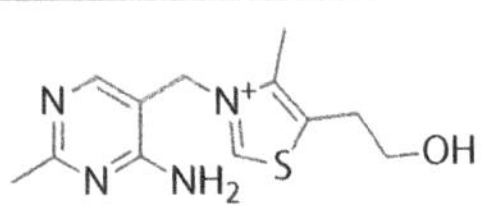	Reduces oxidative damage by functioning as a coenzyme in metabolic pathways.	[46]
7.	Pyranone (curcumin)	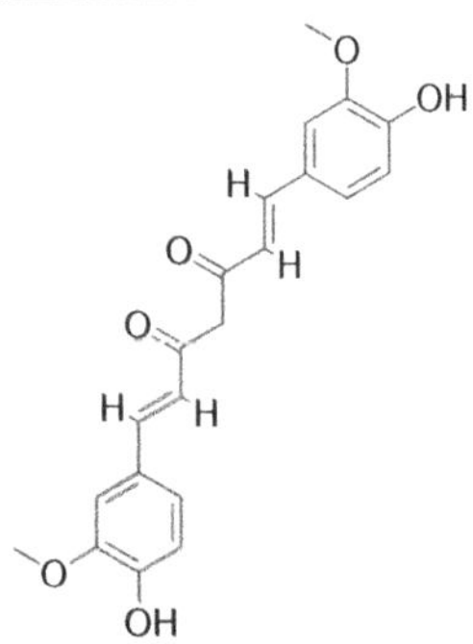	Curcumin reduces oxidative stress and cellular damage by donating electrons to neutralize ROS. It shields cells from oxidative deterioration by stopping the oxidation of lipids in cell membranes.	[47]

(Continued)

Table 6.2 Examples of heterocyclic antioxidants that are currently in use. (*Continued*)

S. no.	Compound name	Structure	Function	References
8.	Purine (theobromine)	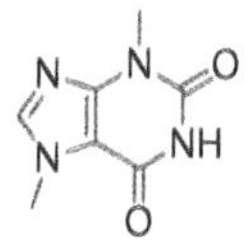	By neutralizing ROS, theobromine scavenges free radicals and prevents oxidative stress. It reduces the contributions of transition metals, including copper and iron, to the generation of ROS by attaching to them.	[20]
9.	Purine (caffeine)	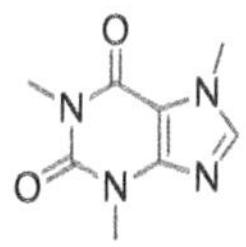	It reduces oxidative stress in cells and ROS are scavenged. Through the prevention of lipid peroxidation, caffeine shields cellular membranes from oxidative stress.	[48]
10.	Stilbene (resveratrol)	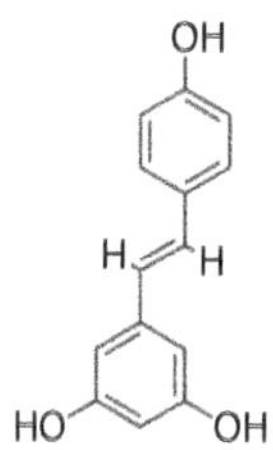	Resveratrol, a polyphenolic stilbene, enhances antioxidant defense by raising the expression of nuclear factor erythroid 2-related factor 2 (Nrf2) activation and antioxidant enzymes.	[49]
11.	Pyridine (pyridoxine)	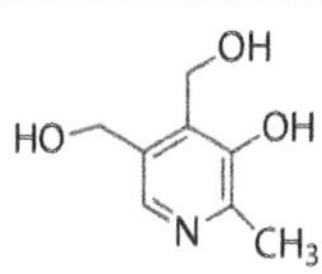	By chelating metal ions as well as eliminating ROS, pyridoxine (vitamin B_6) demonstrates antioxidant action and prevents oxidative stress.	[50]

(*Continued*)

Table 6.2 Examples of heterocyclic antioxidants that are currently in use. (*Continued*)

S. no.	Compound name	Structure	Function	References
12.	Indole (indole-3-carbinol)	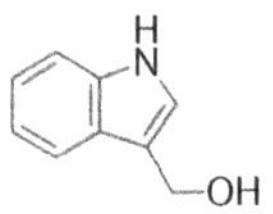	To lessen the effects of oxidation, indole-3-carbinol (I3C) scavenges ROS and upregulates antioxidant enzymes such as superoxide dismutase (SOD) and catalase (CAT). It also improves cellular defense against oxidative damage by modulating the Nrf2/ARE pathway.	[51]

6.4 Types of Heterocyclic Compounds with Antioxidant Properties

Cyclic organic molecules with one or more heteroatoms inside the ring structure are known as heterocyclic compounds. Numerous heterocyclic compounds exhibit potent antioxidant properties because they can:

- Contribute hydrogen atoms or electrons to scavenge free radicals,
- Ions of chelate metal, which aid in preventing oxidative damage.
- Adjust enzymatic antioxidant systems, including glutathione peroxidase and superoxide dismutase.

Due to the distinct structural features that enable them to interact with ROS effectively, heterocyclic antioxidant molecules are crucial in reducing oxidative stress. Heterocyclic compounds fall into three categories: indole, imidazole, piperidine, and pyridine, which are centered on N; thiazole and thiadiazol, which are S; and benzofuran, oxazolidine, and pyran, which are based on O (Figure 6.7).

6.4.1 Nitrogen-Containing Heterocyclic

Nitrogen-containing heterocyclic compounds are molecules that contain both N and C atoms. Examples include pyrrole, imidazole, pyrazole, and

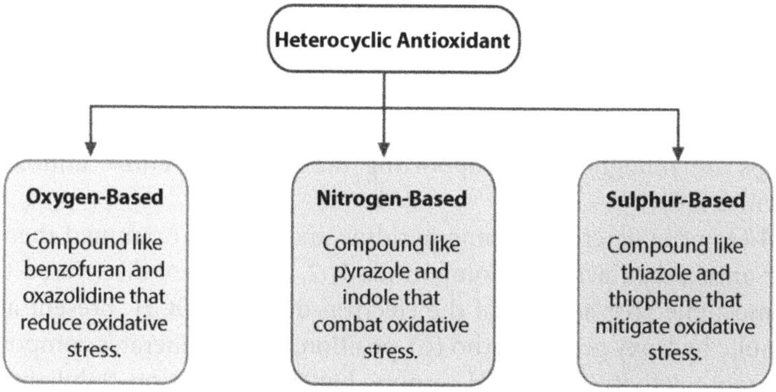

Figure 6.7 Types of heterocyclic antioxidant compound.

pyridine, all of which feature N heteroatoms. These N heterocycles are extensively dispersed throughout nature and found in a range of natural goods and recovery substances. Because of their numerous medical applications, N-based heterocyclic molecules represent an important family of heterocycles in the search for new drugs. This suggests that N heterocycles are crucial for today's medication development. Nowadays, the backbones of these intricate structures are N heterocycles; almost 85% of every physiologically active compound either contains at least one heterocyclic or is heterocycle component. Heterocyclic compounds with N act as powerful antioxidants through various mechanisms, including metal chelation, enzyme activation, HAT, and SET. These mechanisms play a crucial role in preventing oxidative stress and related diseases such as cancer, heart disease, and neurodegeneration. Another name for free radicals is ROS, which are eliminated, and oxidative damage is averted when the antioxidant donates a hydrogen atom (H•). Typically, resonance stabilization results in a less reactive antioxidant radical [52].

6.4.1.1 Pyridine

Pyridine is a prominent primary heterocyclic chemical compound that is a necessary constituent of numerous natural products (C_5H_5N). It is linked to benzene, the most fundamental and well-known aromatic molecule, where one C–H group was swapped out for an N atom [53, 54]. Additionally, certain synthetic analogs have antioxidant chemopreventive effects; they seem to inhibit the activation of carcinogenesis and stop it from starting [55]. Most medications have an enol version of the β-diketone moiety and phenolic hydroxyl groups. In an attempt to create a more potent antioxidant agent, some α, β-unsaturated cyclic ketones were employed to synthesize cyclic-fused nitrogenous pyridine derivatives. This approach preserved the oxidation state of the parent chalcone and, consequently, retained the structural blueprint of curcumin. These functional groups are thought to be supporting the chemopreventive antioxidant activities [56].

Al-Omar *et al.* created some pyridine analogs and examined these for their antioxidant activity. Compounds 1, 2, and 3 were discovered to be the most effective because of the electron-donating OCH_3 present at the phenolic hydroxy group's ortho (o)-position, which generates compounds with more activity. The OCH_3 group's beneficial effect on stabilizing the free radical phenoxy and increasing the scavenging effect of free radicals may be the reason for this improved activity. However, because of the steric hindrance effect, the presence of a large benzyl group reduces activity [57].

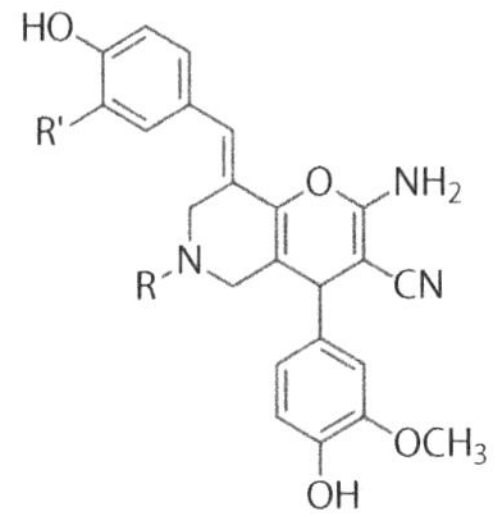

1

Sr. no.	R	R'	Yield [%]
1.	CH_3	OCH_3	99.4
2.	C_2H_5	OCH_3	99.6

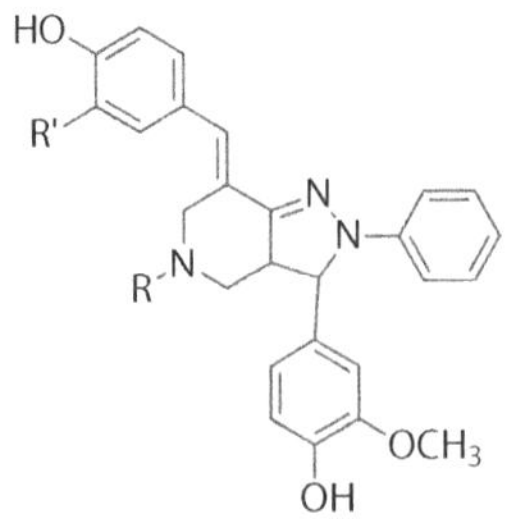

2

Sr. no.	R	R'	Yield [%]
1.	n.C_3H_7	OCH_3	97.3
2.	Bn	OCH_3	84.4

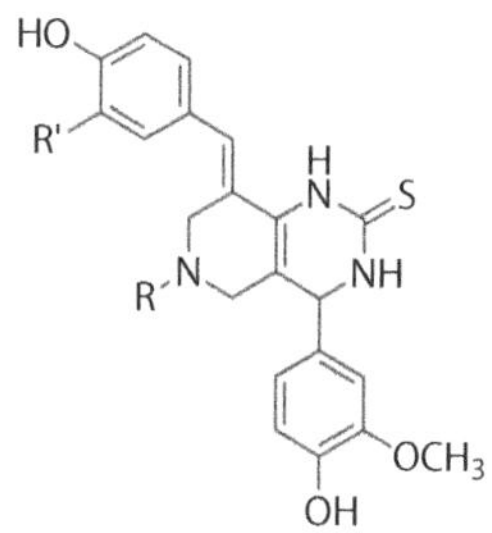

3

Sr. no.	R	R'	Yield [%]
1.	CH_3	OCH_3	99.0
2.	C_2H_5	OCH_3	99.3

6.4.1.2 Pyrimidine

Pyrimidine is extensively distributed in nature and is now recognized as a powerful pharmacophore throughout time. It is a component of nucleic acids as well as many other natural and synthetic chemicals, including pharmaceuticals [58, 59]. Numerous medicinal uses for pyrimidine and its derivatives were demonstrated [60]. Although there are many antioxidants containing pyrimidine nuclei, only a small number are used in therapeutic settings because of the harmful consequences they are linked to. Because the pyrimidine nucleus contains electron-donating substituents [61], several pyrimidine derivatives have shown encouraging antioxidant qualities, which may increase their efficacy and capacity to scavenge radicals [62]. Essentially, the beneficial antioxidant potential of substituted pyrimidines is highlighted by the ability to show a positive correlation between elevated antioxidant capacity as well as decreased density of electrons in the ring system.

Akbas *et al.* synthesized several pyrimidine compounds and examined their antioxidant potential. Because Compound (4) contains a pyrimidine scaffold with a 155.80-μm IC_{50} value, it was determined to be the most potent [63].

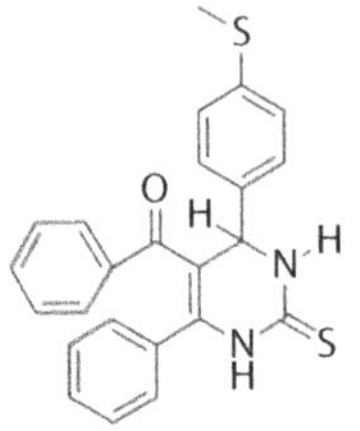

4

Hadi *et al.* synthesized a series of pyrazole-pyrimidine analogs and tested these analogs for antioxidant activity. Among all, Compound (5) was found to be most potent with a % inhibition of 82% because of the existence of hydroxy naphthalene and pyrazole- scaffold of pyrimidines [64].

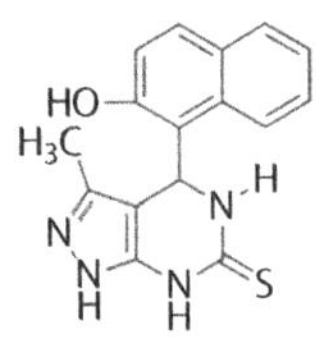

5

Some novel pyrimidine compounds were synthesized, and their antioxidant activity was investigated due to pyran moiety's acetyl and benzoyl substituent groups, Compounds (6) and (7) were determined to be the most powerful among them, with percentage inhibitions of 86.07% and 85.29%, respectively [65].

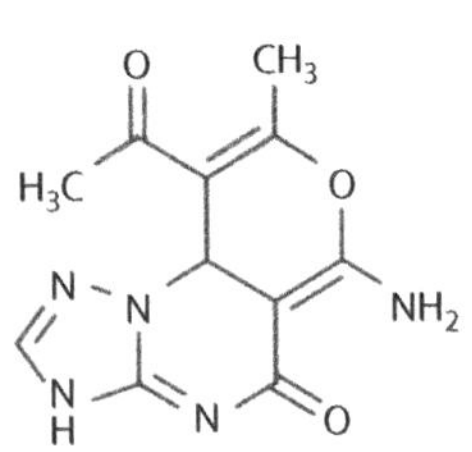

6

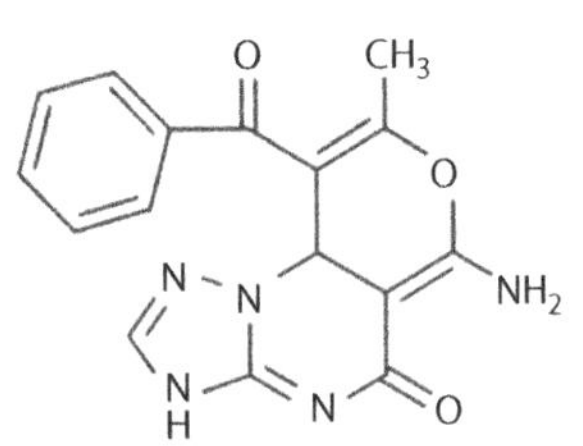

7

6.4.2 Oxygen-Containing Heterocycles

Heterocyclic compounds that contain O have potent antioxidant qualities because they can chelate metal ions, scavenge free radicals, and alter enzymatic antioxidant pathways. These substances, which lower oxidative stress in a variety of ways, include benzopyrans, chromones, and flavonoids. To counteract free radicals, heterocycles that contain O give away a hydrogen atom.

They are more effective at scavenging radicals when hydroxyl (-OH) groups are present. Without rupturing chemical bonds, these substances stabilize free radicals by donating an electron (e^-). The efficiency of electron transfer is increased in O-containing heterocycles when electron-donating groups are present. By binding transition metals (Fe^{2+} and Cu^{2+}), O heterocycles inhibit the Fenton reaction and lower the production of ROS. Lipid peroxidation and oxidative stress are inhibited by chelation. Because of their many biological roles and inherent abundance, O-containing heterocycles are a significant family of molecules in organic chemistry. There are various approaches to categorize O-containing heterocycles, including (a) O atom count, (b) saturation level, (c) aromaticity, or (d) abundance [66]. A variety of therapeutic actions are present in O-based heterocyclic moieties [67]. Heterocycles that include O are unique due to their extensive array of pharmacological as well as biological properties. Their structural resemblance to numerous well-known natural and synthetic chemicals accounts for their significance [68]. Heterocycles that contain O include furans, coumarins, benzofurans, chromones, and others.

6.4.2.1 *Furan*

The heterocyclic organic chemical compound furan (C_4H_4O) is extremely volatile [69]. The synthesis of furan derivatives and the examination of these compounds for various pharmacological properties have attracted a lot of attention lately. A minor change in the pattern of substitution can significantly change the biological activity of furans [70]. Synthesized a series of furan derivatives and tested this derivative for antioxidant activity. Among these compounds, (8) and (9) were found to be most potent due to the presence of electron-donating substituent (OH, NH_2) with IC_{50} value 40 μM [71].

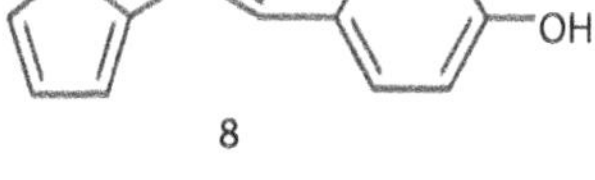

8

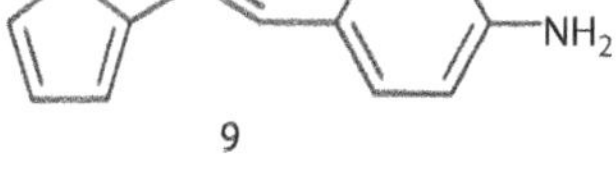

9

6.4.2.2 *Benzofuran*

Heterocycles with a benzene ring joined to one or more furan rings are called benzofurans. It is a member of the phenylethylamine and amphetamine classes of chemicals [72, 73]. The benzofuran ring was originally synthesized by Perkin in 1870. It is one of the most important heterocyclic scaffolds that form the framework of organic molecules with therapeutic value [74]. The structure of benzofuran is the building block of several naturally occurring and artificially produced physiologically active compounds [75].

Produced several compounds of substituted benzofuran and evaluated their antioxidant potential. Because Compound (10) had the OH group in both ortho and para locations, it was determined to be the most effective at lowering intracellular ROS levels [76].

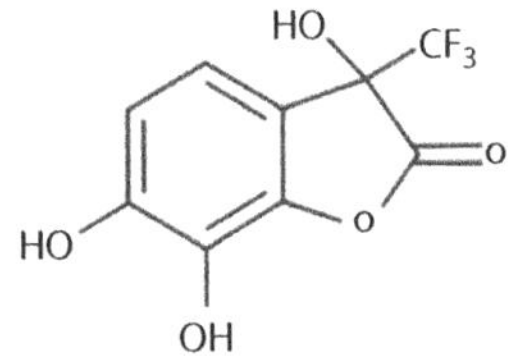

10

A new series of benzofuran derivatives was isolated from *Dalbergia latifolia* and evaluated for antioxidant activity. The DPPH free radical scavenging assay revealed that compounds (11) and (12) exhibited moderate antioxidant activity, with IC_{50} values of 96.7 ± 8.9 μM. In addition, their effect on lipid peroxidation (LPO) was also assessed [77].

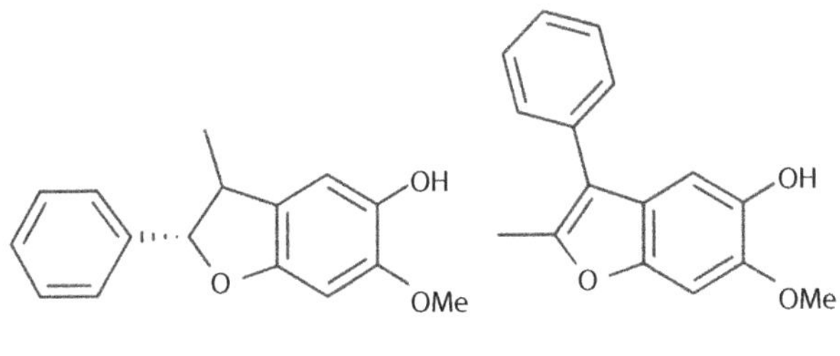

11 12

Synthesized a variety of compounds of substituted benzofuran ester and evaluated their antioxidant potential. Among all, Compound (13) was found to be the most potent and has the highest free radical-scavenging activity [78].

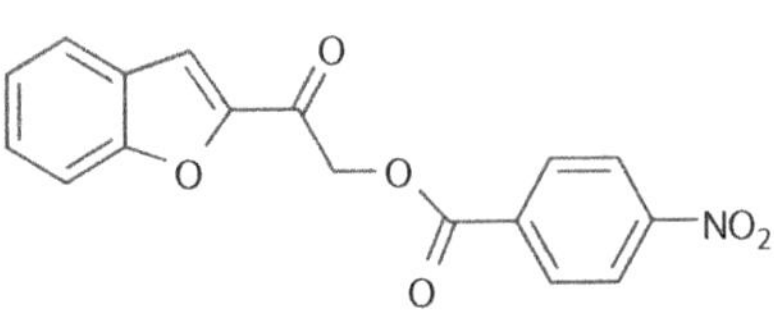

13

Produced many new compounds of benzofuran-2-carboxamide and evaluated their antioxidant potential. Compound (14) is most effective at 100 μM, with a 62% inhibition rate on LPO and a 23.5% inhibition rate on the generation of DPPH radicals [79].

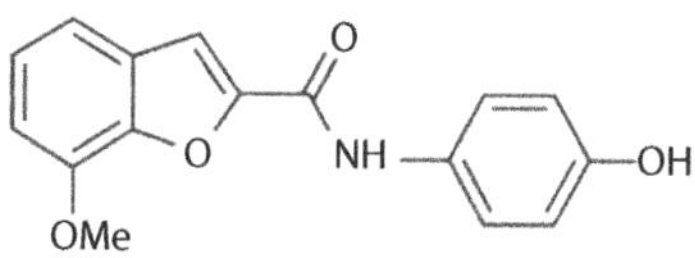

14

6.4.2.3 *Coumarin*

The class of oxygenated, colorless, crystalline, heterocyclic chemicals known as coumarins is rich in polyphenols. When a pyrone ring joins together with benzene, a heterocyclic molecule known as coumarin is produced. Its chemical formula is $C_9H_6O_2$. Another name for it is benzopyrone [80]. Numerous plant groups and essential oils naturally contain coumarin, a secondary metabolite. Their traits and biological processes are essential for developing innovative medicinal therapies. As a result, numerous methods and strategies for producing substituted coumarin have been established [81]. Coumarins have strong antioxidant properties *via* several processes, such as scavenging free radicals, the activity of antioxidant enzyme improvement (catalase and superoxide dismutase), and transcription factor change brought on by oxidative stress. Oxidative stress is a significant factor in the etiology of many diseases, including neurological and cardiovascular sickness, tumor growth, and age-related problems, and is caused by an excess of ROS [82].

Fraxin (15), a coumarin at a high concentration of 0.5-mM derivative, showed a protective scavenging action against free radicals in an *in vitro* model in cells generated with H_2O_2-mediated OS [83]. Esculetin (16) (6,7-dihydroxy coumarin) was another coumarin that showed an antioxidant action inhibiting H_2O_2-induced damage of fibroblasts [84].

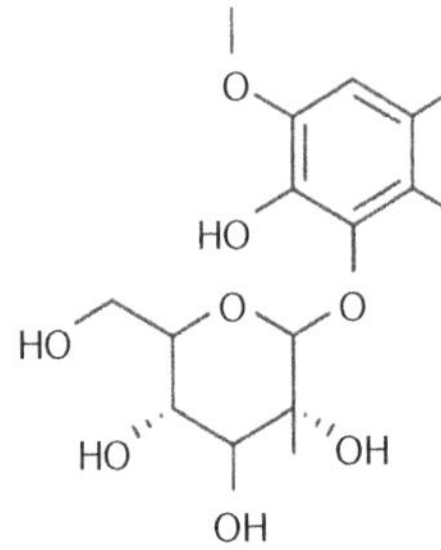

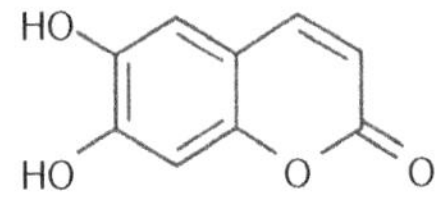

15 16

In 2020, produced a variety of substituted coumarin compounds and evaluated their effectiveness as antioxidants. Compounds (17) and (18) were found to be most potent due to the presence of Cl substitution at position 6, which increases radical-scavenging potential with IC_{50} values of 14.51 and 3.87 μg/mL when compared to ascorbic acid (2.83 μg/mL) [85].

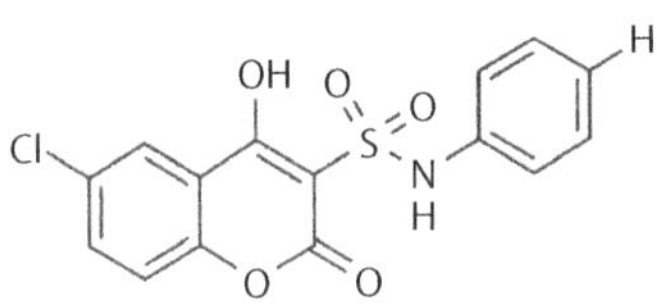

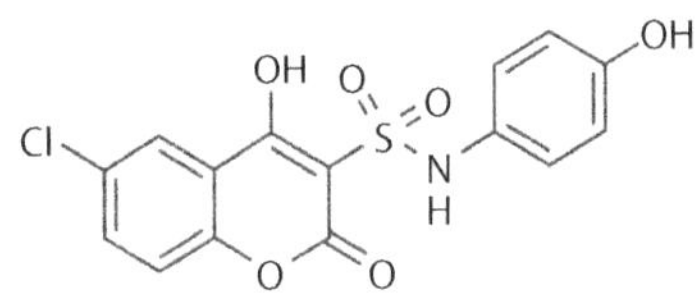

17 18

6.4.3 Sulfur-Containing Heterocycles

S-heterocycles, or heterocycles with S as a heteroatom, are among the most significant kinds of heterocyclic molecules that are constantly used in chemistry [86]. Many drugs and natural compounds that operate as physiologically active compounds in a variety of pathological diseases have several S-containing scaffolds [87]. The main ways that S medications work pharmacologically are by preventing the growth of microorganisms and influencing immunological response [88]. Heterocycles that contain S include thiophenes, benzothiophenes, thiazoles, dithiazoles, and others.

6.4.3.1 Thiophene

Thiophene is a ring with five members and a heteroatom of S. Because it is present in numerous pharmacologically active drugs, it has been recognized as a crucial scaffold. The Greek terms "theion," which means S, and "phaino," which means brilliant, were the sources of the name "thiophene" [89]. Derivatives of thiophene offer beneficial bridges in a variety of scientific and industrial fields, with a broad range of uses and medicinal qualities.

The agrochemical, pharmaceutical, and dye sectors are all very interested in thiophene derivatives, as do many academics [90, 91]. The thiophene moiety activity was tremendously affected by the kind and position of the replacements [92].

Produced several compounds of substituted thiophene and evaluated their antioxidant potential. Compounds (19) and (20) discovered that the electron-withdrawing group was the reason it was most effective (Br) and had 84.26% and 79.08% inhibition; it closely resembles the antioxidant standard L-ascorbic acid (88.44%) [93].

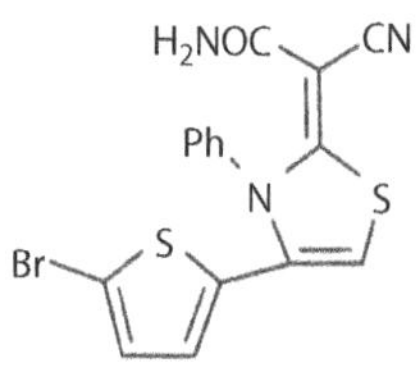

19

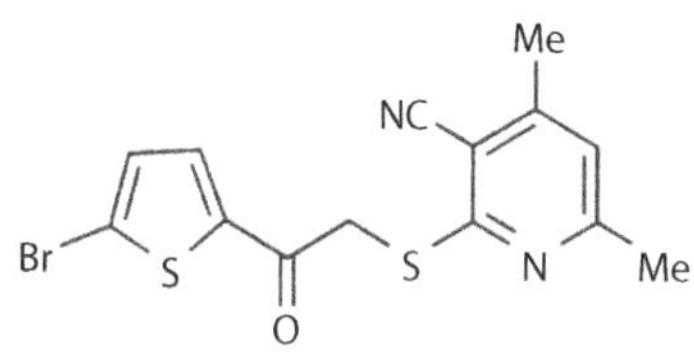

20

Some substitutes for thiophene derivatives and tested these derivatives for antioxidant activity. Compounds (21) and (22) substituted at the third position with hydroxyl and methyl groups were found to be most potent with % inhibition of 87.8% and 87.2% [94].

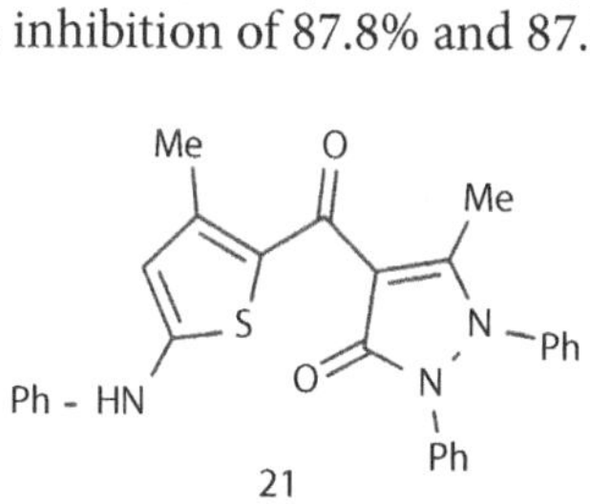

21

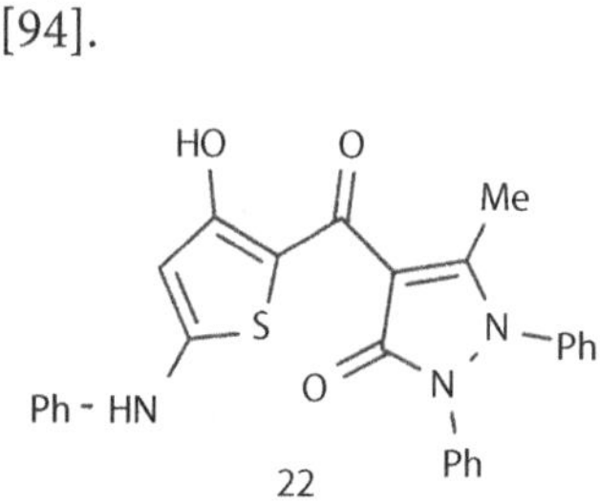

22

Developed some biphenyl-thiophene compounds and assessed how effective they were at preventing antioxidants. Compound (23) containing biphenyl and phenylamino groups exhibited the highest activity, enhancing glutathione S-transferase activity and modulating lipid peroxidation. However, it also reduced catalase activity, indicating an antioxidant response [95].

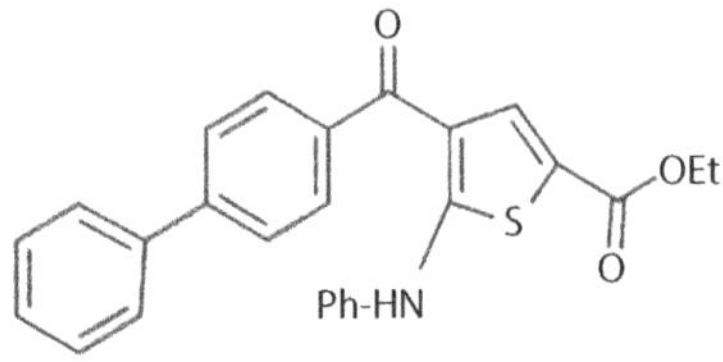

23

6.5 Antioxidant Effect and Mechanism of Action of Polyphenolic Compound

Outstanding biological characteristics of natural antioxidants and polyphenols include their ability to scavenge free radicals, have antiviral effects, and act as antioxidants. They are the most prevalent secondary metabolites in plants and have a polyphenol structure [96]. These substances are distinguished by the presence of two or more hydroxyl (OH) groups on at least one benzene ring [97]. Flavonoids, phenolic acids, and other similar substances are within the broad group of plant compounds known as polyphenols [98]. Many hydroxyl groups are present in them; they can convert free radicals into stable components in the body. This successfully prevents or delays several diseases and stops chain reactions involving free radicals (as illustrated in Figure 6.8) [96, 99]. Commonly termed polyphenols, polyphenolic chemicals are a broad class of naturally occurring molecules widely distributed throughout plants. They are well known for having strong antioxidant qualities, essential for scavenging radicals and shielding the body from damage caused by oxidation. To fully appreciate the potential health advantages of polyphenolic substances, one must comprehend their modes of action and antioxidant properties [97].

Significant antioxidant activity is exhibited by polyphenols, which supports some health advantages:

a) **Reduction of oxidative stress**: Polyphenols help lower oxidative stress, which is connected to aging and some chronic illnesses, by scavenging free radicals.
b) **Neuroprotection**: It has been demonstrated that certain polyphenols shield neurones, which could lessen the chance of neurodegenerative illnesses.
c) **Cardiovascular health**: Probably as a result of their antioxidant and anti-inflammatory qualities, regular consumption of foods high in polyphenols is linked to better heart health [100].

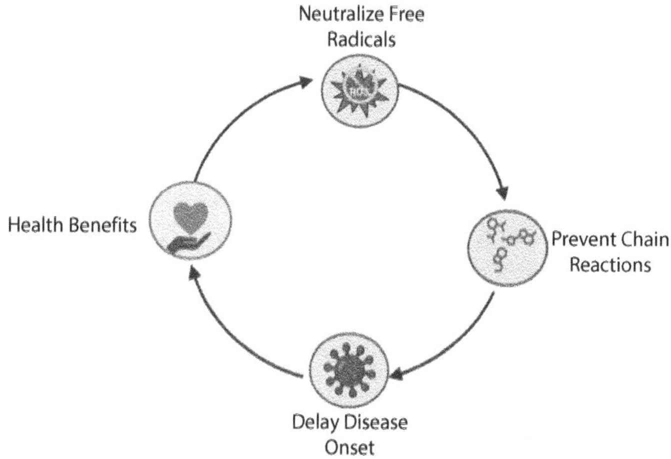

Figure 6.8 Diagrammatic representation of polyphenol-induced health cycle.

6.5.1 Flavonoids

Flavonoids, a large class of polyphenolic compounds, are renowned for their strong antioxidant qualities in fruits, vegetables, and drinks such as wine and tea. These substances are essential for scavenging free radicals, which reduces oxidative stress and helps prevent several chronic illnesses [101].

One family of naturally occurring polyphenolic chemicals that has garnered ever-increasing interest is flavonoids. The phenylpropanoid, shikimate, and flavonoid synthesis routes are the three mechanisms that create flavonoids. They have a benzo-γ-pyrone skeleton in their composition and are secondary metabolites of organisms (Figure 6.9) [102, 103].

Excellent antioxidant qualities are exhibited by flavonoids, which can scavenge ROS and free radicals, chelate metals, and stop low-density lipoproteins from oxidizing (Figure 6.10) [104, 105]. Flavonoids are powerful and effective antioxidants. They shield the body from heart disease by

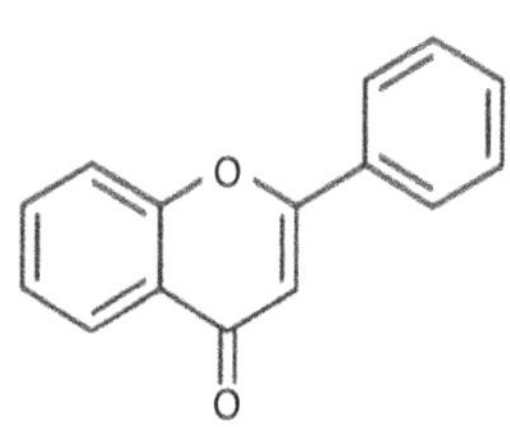

Figure 6.9 Chemical structure of flavonoid.

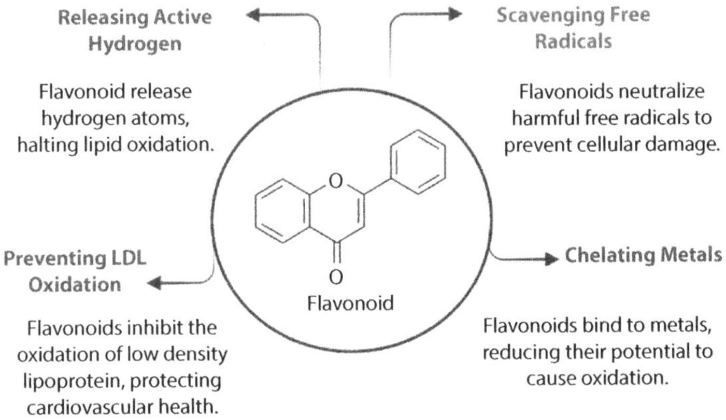

Figure 6.10 Antioxidant mechanism of action of flavonoid.

lowering the oxidation of low-density lipoprotein. Flavonoid molecules' polyphenolic hydroxyl groups release active hydrogen atoms, preventing lipids from automatically oxidizing [96].

6.5.1.1 Catechins

The primary flavanol-based nutraceutical components in teas are catechins (Figure 6.11). The usual C6–C3–C6 flavonoid skeleton serves as the basis for the catechin structure [106]. The structural and property variations between catechins are caused by the ortho-dihydroxyl or vic-tri-hydroxyl B ring groups with the gallate or hydroxyl groups on the C3 position.

In addition to their non-epicatechin forms, for example, (±)-catechin (C), (−)-gallocatechin (GC), (−)-catechin gallate (CG), (−)-gallocatechin gallate (GCG), (−)-epicatechin (EC), (−)-epigallocatechin (EGC), (−)-epicatechin gallate (ECG), and (−)-epicatechin gallate (EGCG) are

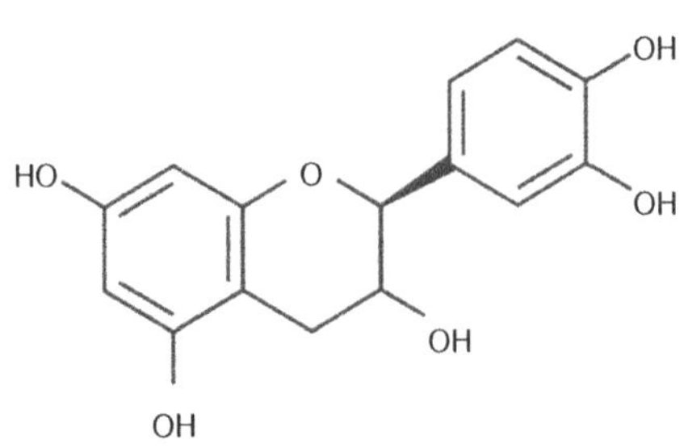

Figure 6.11 Chemical structure of catechins.

common monomeric forms of catechins [107]. Because EGCG is the most prevalent, it is utilized as a quality indicator. Catechins' first-line antioxidant capacity was made possible by the possibility that its B ring's hydroxyl groups may serve as effective electron donors [108]. Studies show that the B ring's hydroxyl groups are superior, allowing EGC and GC to scavenge free radicals more effectively than EC and C. Furthermore, because of the gallate group on the C ring, Compared to EGC and GC, EGCG and GCG exhibited greater antioxidant activity [109]. As a result, the structural aspects of hydroxyl and gallate groups affected the effectiveness of catechins as antioxidants. Below are examples of catechin compounds with demonstrated antioxidant activity (Table 6.3) [110].

6.5.1.2 *Flavonols*

The primary subclass of flavonoids, known as flavonols, exist because the C ring's C3 and C4 locations, respectively, have hydroxyl and carbonyl groups (Figure 6.12). Three phenolic rings (A, B, and C) are present in flavonols.

Table 6.3 Example of catechin containing antioxidant activity.

Compounds	Source	Antioxidant activity
Epigallocatechin-3-gallate (EGCG)	(*Camellia sinensis*) Green tea	EGCG exhibits strong chelating metal ions and scavenging free radicals to produce antioxidant activity, protecting cells from oxidative damage.
Epicatechin (EC)	Green tea, cocoa, berries	It has been demonstrated that EC contains strong antioxidant qualities that help lower inflammation and oxidative stress.
Epicatechin gallate (ECG)	Green tea	By reducing reactive oxygen species and boosting endogenous antioxidant defenses, ECG has significant antioxidant benefits.
Catechin (C)	Berries, cocoa	Catechin exhibits antioxidant effect *via* scavenging free radicals and controlling the actions of antioxidant enzymes.

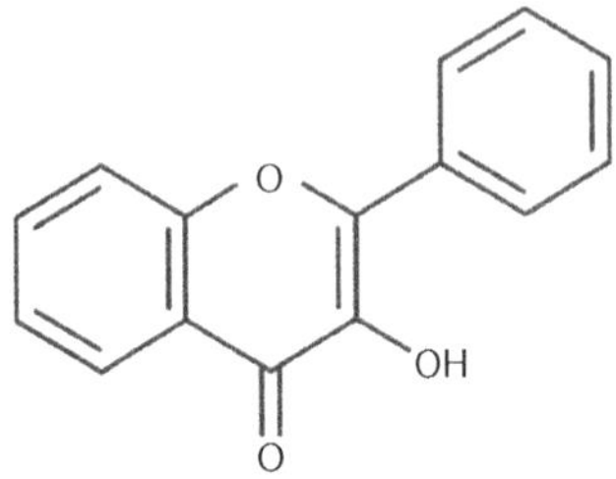

Figure 6.12 Structure of flavonols.

The number of hydroxyl groups on the B ring and their location determine their different kinds [111]. Approximately 900 flavonols have been found in plants, with isorhamnetin, kaempferol, and quercetin being the main types [112].

Flavonols are mostly found in nature in their glycosylated form, with the sugar moiety typically binding with the aglycone's 3, 7, 3′, and 4′ sites. Glucuronic acid, rhamnose, galactose, arabinose, xylose, and glucose are the most prevalent residues. Flavonols' antioxidant potential is dominated by their phenolic hydroxyl groups, which have the ability to provide free radicals hydrogen atoms [113]. The activity is also significantly influenced by the location of the hydroxyl groups [114]. Quercetin's B and C rings contain hydroxyl groups that are more effective at scavenging harmful free radicals compared to its A rings. Additionally, the antioxidant capacity of flavonol aglycones would be inhibited by the substituent glycosides because they reduce the number of hydroxyl groups. Reports on the antioxidant ability of various common flavonols revealed the following decreasing sequences of antioxidants: quercetin, isorhamnetin, quercetin-3-O-glucuronide, and quercetin-3,4′-di-glucoside [115].

6.5.2 Phenolic Acids

Plants, plant meals, and human metabolites are abundant sources of phenolic acids, a class of chemicals created when hydrogen atoms on benzene rings are replaced by a carboxylic acid group and at least one hydroxyl [116]. Phenolic acids (Figure 6.13) are regarded as superior natural antioxidants with potential uses in health foods and medicine. Free phenolic acids, including cinnamic, phenylacetic, and benzoic acids, have good water solubility and high bioavailability in contrast to flavonoids [117].

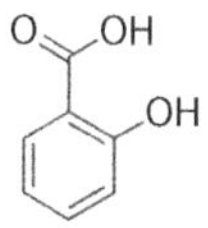

Figure 6.13 Chemical structure of phenolic acid.

Although flavonoids cannot be absorbed in the stomach, they can be passively transferred into the bloodstream through the intestinal wall in trace amounts [118]. Similar to flavonoids, phenolic acids are regarded as superior antioxidants that can prevent chronic illnesses and excessive harm that free radicals inflict to the body. Because phenolic hydroxyls constitute the core of phenolic acids' antioxidant action, their quantity and location are closely correlated with their antioxidant capacity [119, 120]. Examples of phenolic acid compounds with proven antioxidant action are shown in Table 6.4 [121].

6.5.3 Stilbenes

The fundamental building block of stilbene compounds, stilbenes are found in tiny quantities in typical plant tissues. Stilbenes can be divided between *cis* and (more stable) *trans* structures according to their C=C

Table 6.4 Example of phenolic acid compound.

Phenolic acid	Source	Antioxidant activity
Caffeic acid	Coffee, blueberries	Enhances the ability of antioxidants to scavenge free radicals
Ferulic acid	Whole grains, rice bran, oats	Prevents oxidative damage and inhibits lipid peroxidation
Gallic acid	Tea, berries, walnuts	Decreases oxidative stress and has potent radical-scavenging action
Sinapic acid	Mustard seeds, rye, vegetables	Scavenges reactive oxygen species to produce a strong antioxidant effect
Protocatechuic acid	Olives, plums, peanuts	Shows notable antioxidant qualities by scavenging free radicals

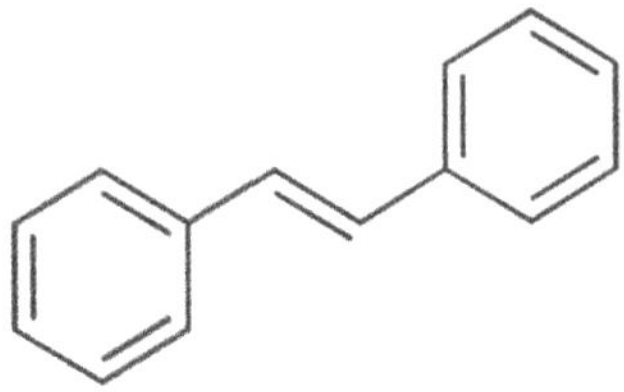

Figure 6.14 Chemical structure of stilbenes.

structures (Figure 6.14). Stilbenes can be either oligomers or monomers. The procedures for methylating, glycosylating, and hydroxylating monomers or the condensation of monomers (such as dimer and trimer) distinguish oligomers, whereas radicals and steric molecular structures distinguish monomers [122]. Stilbenes' structure places them in the category of compounds with intriguing antioxidant qualities [123]. The molecular structure's particular functional groups are in charge of stabilizing the resonance phenomenon and antioxidant action. After stabilizing the double-bond conjugated alkyl chain through resonance, the hydroxy group is essential to antioxidant activity [124]. Typical stilbene representations are piceatannol, pterostilbene, and resveratrol [125].

6.6 Factors Influencing the Antioxidant Capacity of Polyphenols

A vast range of naturally occurring compounds that are extensively present in plants, polyphenols are well renowned for their strong antioxidant capabilities. These substances are essential for scavenging free radicals, which reduces oxidative stress and helps prevent several chronic illnesses. Numerous factors, such as polyphenols' chemical structure, ambient circumstances, interactions with other dietary components, and absorption, affect their antioxidant potential. Optimizing the health advantages of diets high in polyphenols requires an understanding of these elements. Polyphenols' structure, stability, and bioavailability may impact their antioxidant capability (as shown in Figure 6.15).

6.6.1 Structure

Numerous studies have examined the connection between polyphenols' structure and action and recorded them. This relationship depends on

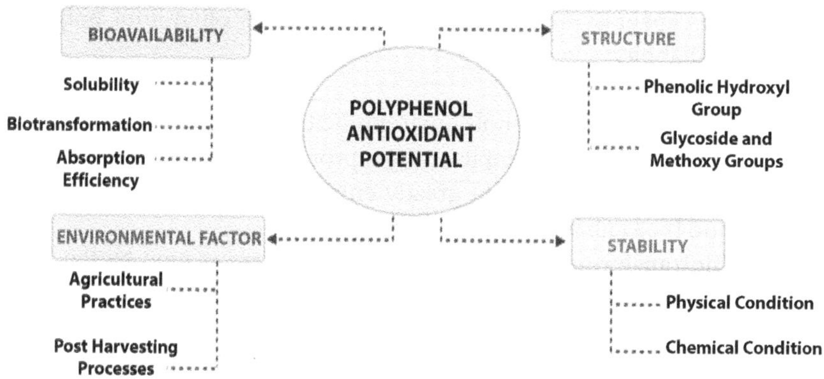

Figure 6.15 Factor influencing polyphenol antioxidant activity.

their nuclear structures, types of substituents, and the positions of these substituents [126]. The phenolic hydroxyl group is the main component that scavenges free radicals, but other groups, as well as the quantity and arrangement of hydroxyl groups, are also important. A double bond at position C2–C3 together contains the carbonyl groups C4 and C3–OH, exhibiting greater opposing radical ability, even though the hydroxyl group on the B ring mostly influences the activity [127]. However, methoxy and glycoside groups typically have adverse effects [128].

6.6.2 Stability

Polyphenols are not very stable, which could rapidly lead to their depletion. Their stability may be impacted by many physical and chemical elements that are frequently present during processing, storage, and digestion, including ascorbic acid, sugar, light, O, pH, temperature, enzymes, and metallic ions [129]. When polyphenols degrade, their antioxidant structure is destroyed, and the nucleus is cleaved [130]. Consequently, a drawback of their limited stability was that it reduced their antioxidant potential in typical situations.

6.6.3 Bioavailability

The percentage of active components that the body absorbs and uses out of their total amount is known as bioavailability. Because the percentage of polyphenols that are not absorbed but are instead eliminated from the body does not contribute, their bioavailability dictates their antioxidant capability *in vivo* [131]. Three variables often limit polyphenols' bioavailability:

1) **Solubility**: Polyphenols' hydrophilicity promotes their bioavailability. For example, using a promoter to make quercetin more soluble could greatly enhance its oral absorption and bioavailability [132].
2) **Biotransformation**: After ingestion and fermentation by the gut bacteria, polyphenols hardly ever exist in their prototype molecular structure; instead, they mostly exist as metabolites. The natural structure of polyphenols would be destroyed by the biotransformation brought on by the gut microbiota, which would affect their antioxidant potential [133].
3) **Absorption**: Polyphenols' bioavailability is significantly impacted by the ability to move from the stomach and intestine into the body [134, 135].

6.6.4 Environmental Factor

Plants' antioxidant capability is impacted by a variety of environmental factors that can change their polyphenolic content:

1) **Agriculture practices**: Organic farming and the application of particular fertilizers can control the production of polyphenols in plants. Stressors such as insect assaults frequently increase polyphenol synthesis as a protective mechanism.
2) **Postharvesting process**: Polyphenolic chemicals may degrade or change due to processes such as fermentation, drying, and storage conditions. For example, foods high in polyphenols may lose some antioxidant properties if processed with too much heat [136–140].

A study that examined how cooking techniques affected the antioxidant activity of polyphenols in edible leaves found that boiling, steaming, and frying all had a substantial impact on this activity.

6.7 Conclusion

Heterocyclic compounds significantly influence the fight against oxidative stress by acting as potent antioxidants. Their diverse shapes, incorporating

N, O, and S, effectively capture free radicals, attach to metal ions, and regulate the body's defense enzymes. These attributes make heterocyclic compounds valuable in preventing and treating illnesses associated with oxidative stress, such as neurological disorders, cardiovascular diseases, and cancer. Research indicates that specific molecules containing N, O, or S possess powerful protective qualities against oxidative stress. This includes pyridine, pyrimidine, benzofuran, coumarin, and thiophene. They function by transferring hydrogen atoms, moving single electrons, or binding to metals, which aids in eliminating harmful molecules known as ROS and preventing cell damage. Additionally, compounds with multiple phenol groups, such as flavonoids, stilbenes, and phenolic acids, help mitigate oxidative stress through their strong antioxidant properties. Despite their promising therapeutic potential, challenges such as bioavailability, stability, and environmental impacts on polyphenol efficacy need to be addressed. Future research should concentrate on adapting these compounds for enhanced medicinal applications, including drug development and dietary supplements. Ongoing exploration of heterocyclic antioxidants could yield innovative therapeutic strategies that improve human health and longevity by minimizing oxidative damage at the molecular level.

AI disclosure statement: The author acknowledges the use of AI-based tools for minor language editing or grammatical correction. However, all intellectual contributions, analyses, and conclusions presented in this chapter are entirely the author's own.

References

1. Pizzino, G., Irrera, N., Cucinotta, M., Pallio, G., Mannino, F., Arcoraci, V., Squadrito, F., Altavilla, D., Bitto, A., Oxidative stress: harms and benefits for human health. *Oxid. Med. Cell. Longevity*, 2017, 1, 8416763, 2017.
2. Kıran, T.R., Otlu, O., Karabulut, A.B., Oxidative stress and antioxidants in health and disease. *J. Lab. Med.*, 47, 1, 1–1, 2023 Feb 23.
3. Juan, C.A., Pérez de la Lastra, J.M., Plou, F.J., Pérez-Lebeña, E., The chemistry of reactive oxygen species (ROS) revisited: outlining their role in biological macromolecules (DNA, lipids and proteins) and induced pathologies. *Int. J. Mol. Sci.*, 22, 9, 4642, 2021 Apr 28.
4. Egea, J., Fabregat, I., Frapart, Y.M., Ghezzi, P., Görlach, A., Kietzmann, T., Kubaichuk, K., Knaus, U.G., Lopez, M.G., Olaso-Gonzalez, G., Petry, A., European contribution to the study of ROS: A summary of the findings and prospects for the future from the COST action BM1203 (EU-ROS). *Redox Biol.*, 13, 94–162, 2017 Oct 1.

5. Arfin, S., Jha, N.K., Jha, S.K., Kesari, K.K., Ruokolainen, J., Roychoudhury, S., Rathi, B., Kumar, D., Oxidative stress in cancer cell metabolism. *Antioxidants.*, 10, 5, 642, 2021 Apr 22.
6. Guler, C., Toy, E., Ozturk, F., Gunes, D., Karabulut, A.B., Otlu, O., Evaluation of salivary total oxidant-antioxidant status and DNA damage of children undergoing fixed orthodontic therapy. *Angle Orthodontist*, 85, 2, 239–44, 2015 Mar 1.
7. Gulcin, I., Antioxidants and antioxidant methods: An updated overview. *Arch. Toxicol.*, 94, 3, 651–715, 2020 Mar.
8. Malki, F., Alouache, A., Krimat, S., Effects of Various Parameters on the Antioxidant Activities of the Synthesized Heterocyclic Pyrimidinium Betaines. *Indones. J. Chemistry*, 23, 1, 90–100, 2023.
9. Mirończuk-Chodakowska, I., Witkowska, A.M., Zujko, M.E., Endogenous non-enzymatic antioxidants in the human body. *Adv. Med. Sci.*, 63, 1, 68–78, 2018 Mar 1.
10. Granato, D., Shahidi, F., Wrolstad, R., Kilmartin, P., Melton, L.D., Hidalgo, F.J., Miyashita, K., van Camp, J., Alasalvar, C., Ismail, A.B., Elmore, S., Antioxidant activity, total phenolics and flavonoids contents: Should we ban *in vitro* screening methods? *Food Chem.*, 264, 471–5, 2018 Oct 30.
11. Patil, S.B., Recent medicinal approaches of novel pyrimidine analogs: A review. *Heliyon*, 9, 6, 2023 Jun 1.
12. Aziz, M. and Karboune, S., Natural antimicrobial/antioxidant agents in meat and poultry products as well as fruits and vegetables: A review. *Crit. Rev. Food Sci. Nutr.*, 58, 3, 486–511, 2018 Feb 11.
13. Beya, M.M., Netzel, M.E., Sultanbawa, Y., Smyth, H., Hoffman, L.C., Plant-based phenolic molecules as natural preservatives in comminuted meats: A review. *Antioxidants.*, 10, 2, 263, 2021 Feb 9.
14. Liang, N. and Kitts, D.D., Antioxidant property of coffee components: assessment of methods that define mechanisms of action. *Molecules*, 19, 11, 19180–208, 2014 Nov 19.
15. Senanayake, S.N., Green tea extract: Chemistry, antioxidant properties and food applications–A review. *J. Funct. Foods*, 5, 4, 1529–41, 2013 Oct 1.
16. Didier, A.J., Stiene, J., Fang, L., Watkins, D., Dworkin, L.D., Creeden, J.F., Antioxidant and anti-tumor effects of dietary vitamins A, C, and E. *Antioxidants*, 12, 3, 632, 2023 Mar 3.
17. Pereira, C.P., Souza, A.C., Vasconcelos, A.R., Prado, P.S., Antioxidant and anti-inflammatory mechanisms of action of astaxanthin in cardiovascular diseases. *Int. J. Mol. Med.*, 47, 1, 37–48, 2021 Jan 1.
18. Alvarez-Builla, J. and Barluenga, J., Heterocyclic compounds: An introduction. *Mod. Heterocycl. Chemistry*, 29, 1–9, 2011.
19. Kabir, E. and Uzzaman, M., A review on biological and medicinal impact of heterocyclic compounds. *Results Chem.*, 4, 100606, 2022 Jan 1.

20. Qadir, T., Amin, A., Sharma, P.K., Jeelani, I., Abe, H., A review on medicinally important heterocyclic compounds. *Open Med. Chem. J.*, 16, 1, 2022 Apr 28.
21. Abdella, A.M., Abdelmoniem, A.M., Abdelhamid, I.A., Elwahy, A.H., Synthesis of heterocyclic compounds *via* Michael and Hantzsch reactions. *J. Heterocycl. Chem.*, 57, 4, 1476–523, 2020 Apr.
22. Mi, Y., Zhang, J., Chen, Y., Sun, X., Tan, W., Li, Q., Guo, Z., New synthetic chitosan derivatives bearing benzenoid/heterocyclic moieties with enhanced antioxidant and antifungal activities. *Carbohydr. Polym.*, 249, 116847, 2020 Dec 1.
23. Hamed, A.A., Abdelhamid, I.A., Saad, G.R., Elkady, N.A., Elsabee, M.Z., Synthesis, characterization and antimicrobial activity of a novel chitosan Schiff bases based on heterocyclic moieties. *Int. J. Biol. Macromol.*, 153, 492–501, 2020 Jun 15.
24. Patnayak, S., Heterocyclic Compounds An Potential Drug And Its Biological Activty: A Review. *J. Nonlinear Anal. Optim.*, 15, 1, 2024.
25. Mahajan, N.D. and Jain, N., Heterocyclic compounds and their applications in the field of biology: A detailed study. *Nveo-Natural Volatiles Essent. Oils J.| NVEO*, 16, 13223–9, 2021 Dec.
26. Mermer, A., Keles, T., Sirin, Y., Recent studies of nitrogen containing heterocyclic compounds as novel antiviral agents: A review. *Bioorg. Chem.*, 114, 105076, 2021 Sep 1.
27. Farag, A.M. and Fahim, A.M., Synthesis, biological evaluation and DFT calculation of novel pyrazole and pyrimidine derivatives. *J. Mol. Struct.*, 1179, 304–14, 2019 Mar 5.
28. Obaid, R.J., Naeem, N., Mughal, E.U., Al-Rooqi, M.M., Sadiq, A., Jassas, R.S., Moussa, Z., Ahmed, S.A., Inhibitory potential of nitrogen, oxygen and sulfur containing heterocyclic scaffolds against acetylcholinesterase and butyrylcholinesterase. *RSC Adv.*, 12, 31, 19764–855, 2022.
29. Fahim, A.M., Shalaby, M.A., Ibrahim, M.A., Microwave-assisted synthesis of novel 5-aminouracil-based compound with DFT calculations. *J. Mol. Struct.*, 1194, 211–26, 2019 Oct 15.
30. Rizk, S.A. and Shaban, S., A facile one-pot synthesis and anticancer evaluation of interesting pyrazole and pyrimidinthione *via* heterocyclic interconversion. *J. Heterocycl. Chem.*, 56, 9, 2379–88, 2019 Sep.
31. Rizk, S.A., El-Sayed, A.A., Mounier, M.M., Synthesis of novel pyrazole derivatives as antineoplastic agent. *J. Heterocycl. Chem.*, 54, 6, 3358–71, 2017 Nov.
32. ur Rashid, H., Martines, M.A., Duarte, A.P., Jorge, J., Rasool, S., Muhammad, R., Ahmad, N., Umar, M.N., Research developments in the syntheses, anti-inflammatory activities and structure–activity relationships of pyrimidines. *RSC Adv.*, 11, 11, 6060–98, 2021.
33. Colom-Fernández, B., Kreutzman, A., Marcos-Jiménez, A., García-Gutiérrez, V., Cuesta-Mateos, C., Portero-Sainz, I., Pérez-García, Y., Casado, L.F., Sánchez-Guijo, F., Martínez-López, J., Ayala, R.M., Immediate effects of

dasatinib on the migration and redistribution of naïve and memory lymphocytes associated with lymphocytosis in chronic myeloid leukemia patients. *Front. Pharmacol.*, 10, 1340, 2019 Nov 25.
34. Smith, D.R., Dolk, F.C., Pouwels, K.B., Christie, M., Robotham, J.V., Smieszek, T., Defining the appropriateness and inappropriateness of antibiotic prescribing in primary care. *J. Antimicrob. Chemother.*, 73, suppl_2, ii11–8, 2018 Feb 1.
35. Aboelnaga, A., Mansour, E., Fahim, A.M., Elsayed, G.H., Synthesis, antiproliferative activity, gene expression, docking and DFT investigation of novel pyrazol-1-yl-thiazol-4 (5H)-one derivatives. *J. Mol. Struct.*, 1251, 131945, 2022 Mar 5.
36. Fahim, A.M., Tolan, H.E., Awad, H., Ismael, E.H., Synthesis, antimicrobial and antiproliferative activities, molecular docking, and computational studies of novel heterocycles. *J. Iran. Chem. Soc.*, 18, 11, 2965–81, 2021 Nov.
37. Wang, X., Hu, Q., Tang, H., Pan, X., Isoxazole/Isoxazoline skeleton in the structural modification of natural products: a review. *Pharmaceuticals.*, 16, 2, 228, 2023 Feb 2.
38. Ashfaq, M., Tahir, M.N., Munawar, K.S., Behjatmanesh-Ardakani, R., Kargar, H., Single crystal exploration, supramolecular behaviour, Hirshfeld surface analysis, linear and non-linear theoretical optical properties of Schiff bases derived from Benzene sulfonamides. *J. Mol. Struct.*, 1261, 132952, 2022 Aug 5.
39. Meyer, B., Elemental sulfur. *Chem. Rev.*, 76, 3, 367–88, 1976 Jun 1.
40. Yamali, C., Nenni, M., Sakarya, M.T., Sakagami, H., Gul, H.I., Biological activities and drug-likeness properties of phenol-based heterocyclic compounds. *Pharm. Chem. J.*, 57, 11, 1754–60, 2024 Feb.
41. Azeem, M., Hanif, M., Mahmood, K., Ameer, N., Chughtai, F.R., Abid, U., An insight into anticancer, antioxidant, antimicrobial, antidiabetic and anti-inflammatory effects of quercetin: A review. *Polym. Bull.*, 80, 1, 241–62, 2023 Jan.
42. Veiko, A.G., Lapshina, E.A., Zavodnik, I.B., Comparative analysis of molecular properties and reactions with oxidants for quercetin, catechin, and naringenin. *Mol. Cell. Biochem.*, 476, 12, 4287–99, 2021 Dec.
43. Lv, J., Song, X., Luo, Z., Huang, D., Xiao, L., Zou, K., Luteolin: Exploring Its Therapeutic Potential and Molecular Mechanisms in Pulmonary Diseases. *Front. Pharmacol.*, 16, 1535555, 2025 Feb 12.
44. Cipolla-Neto, J. and Amaral, F.G., Melatonin as a hormone: new physiological and clinical insights. *Endocr. Rev.*, 39, 6, 990–1028, 2018 Dec.
45. Boo, Y.C., Mechanistic basis and clinical evidence for the applications of nicotinamide (niacinamide) to control skin aging and pigmentation. *Antioxidants.*, 10, 8, 1315, 2021 Aug 21.
46. Piechocka, J., Gramza-Michałowska, A., Szymandera-Buszka, K., The changes in antioxidant activity of selected flavonoids and caffeine depending on the dosage and form of thiamine. *Molecules*, 26, 15, 4702, 2021 Aug 3.

47. Grabarczyk, M., Justyńska, W., Czpakowska, J., Smolińska, E., Bielenin, A., Glabinski, A., Szpakowski, P., Role of plant phytochemicals: resveratrol, curcumin, luteolin and quercetin in demyelination, neurodegeneration, and epilepsy. *Antioxidants*, 13, 11, 1364, 2024 Nov 7.
48. Ősz, B.E., Jîtcă, G., Ştefănescu, R.E., Puşcaş, A., Tero-Vescan, A., Vari, C.E., Caffeine and its antioxidant properties—it is all about dose and source. *Int. J. Mol. Sci.*, 23, 21, 13074, 2022 Oct 28.
49. Meng, T., Xiao, D., Muhammed, A., Deng, J., Chen, L., He, J., Anti-inflammatory action and mechanisms of resveratrol. *Molecules*, 26, 1, 229, 2021 Jan 5.
50. Mutavdzin Krneta, S., Gopcevic, K., Stankovic, S., Jakovljevic Uzelac, J., Todorovic, D., Labudovic Borovic, M., Rakocevic, J., Djuric, D., Insights into the Cardioprotective Effects of Pyridoxine Treatment in Diabetic Rats: A Study on Cardiac Oxidative Stress, Cardiometabolic Status, and Cardiovascular Biomarkers. *Diagnostics*, 14, 14, 1507, 2024 Jul 12.
51. Jasiewicz, B., Kozanecka-Okupnik, W., Przygodzki, M., Warżajtis, B., Rychlewska, U., Pospieszny, T., Mrówczyńska, L., Synthesis, antioxidant and cytoprotective activity evaluation of C-3 substituted indole derivatives. *Sci. Rep.*, 11, 1, 15425, 2021 Jul 29.
52. Smolyaninov, I.V., Burmistrova, D.A., Pomortseva, N.P., Poddel'sky, A.I., Berberova, N.T., Antioxidant Activity of Catechol Thioethers with Heterocyclic Moieties in Reactions with Radical Promoters, in: *Doklady Chemistry*, vol. 504, No. 2, pp. 100–105, Pleiades Publishing, Moscow, 2022 Jun.
53. Alizadeh, S.R. and Ebrahimzadeh, M.A., Antiviral activities of pyridine fused and pyridine containing heterocycles, a review (from 2000 to 2020). *Mini-Rev. Med. Chem.*, 21, 17, 2584–611, 2021 Oct 1.
54. Vessally, E., Hosseinian, A., Edjlali, L., Bekhradnia, A., Esrafili, M.D., New Page to Access Pyridine Derivatives: Synthesis from: N -Propargylamines. *RSC Adv.*, 6, 71662–71675, 2016.
55. Pratt, W.B., *The anticancer drugs*, Oxford University Press, 1994.
56. Youssef, K.M. and El-Sherbeny, M.A., Synthesis and antitumor activity of some curcumin analogs. *Arch. der Pharmazie An Int. J. Pharm. Med. Chem.*, 338, 4, 181–9, 2005 Apr.
57. Al-Omar, M.A., Youssef, K.M., El-Sherbeny, M.A., Awadalla, S.A., El-Subbagh, H.I., Synthesis and *in vitro* antioxidant activity of some new fused pyridine analogs. *Arch. der Pharmazie An Int. J. Pharm. Med. Chem.*, 338, 4, 175–80, 2005 Apr.
58. Kumar, S. and Narasimhan, B., Therapeutic potential of heterocyclic pyrimidine scaffolds. *Chem. Cent. J.*, 12, 1–29, 2018 Dec.
59. Zhuang, J. and Ma, S., Recent development of pyrimidine-containing antimicrobial agents. *ChemMedChem.*, 15, 20, 1875–86, 2020 Oct 19.

60. Wu, W., Lan, W., Wu, C., Fei, Q., Synthesis and antifungal activity of pyrimidine derivatives containing an amide moiety. *Front. Chem.*, 9, 695628, 2021 Jul 12.
61. Ullas, B.J., Rakesh, K.P., Shivakumar, J., Gowda, D.C., Chandrashekara, P.G., Multi-targeted quinazolinone-Schiff's bases as potent bio-therapeutics. *Results Chem.*, 2, 100067, 2020 Jan 1.
62. Nair, N., Majeed, J., Pandey, P.K., Sweety, R., Thakur, R., Antioxidant potential of pyrimidine derivatives against oxidative stress. *Indian J. Pharm. Sci.*, 84, 1, 2022 Jan 1.
63. Akbas, E., Ergan, E., Sahin, E., Ekin, S., Cakir, M., Karakus, Y., Synthesis, characterization, antioxidant properties and DFT calculation of some new pyrimidine derivatives, in: *Phosphorus, Sulfur, and Silicon and the Related Elements*, 2019 Aug 3.
64. Hadi, A.A., Kadhim, Z.Y., Seewan, A.N., Munahi, M.G., Waheeb, M.Q., One-pot synthesis of new pyrazolo [3, 4,-d] pyrimidine derivatives and study of their antioxidant and anticancer activities. *Medico-legal Update*, 20, 2, 327, 2020 Apr 1.
65. Bayazeed, A.A. and Alnoman, R.B., Synthesis of polyheterocyclic ring systems included triazolo [1, 5-a] pyrimidine as antioxidant agents. *Polycyclic Aromat. Compd.*, 42, 3, 735–48, 2022 Apr 22.
66. Suvaiv, Singh, K., Hasan, S.M., Arif, M., Ahmad, F., Sharma, R., Zaidi, S.M., Shahanawaz, M.O., Exploring Novel Benzopyran Derivatives as Antitubercular Agents. *Letters in Drug Design & Discovery,* 21, 18, 4321–9, 2024 Dec.
67. Kumar, D., Sharma, P., Lobe, M.M., Kaur, S., Singh, T.G., Dua, K., Ntie-Kang, F., Emerging Oxygen Based Heterocyclic Scaffolds as Potential Anticancer Candidates.
68. Arce-Ramos, L., Castillo, J.C., Becerra, D., Synthesis and Biological Studies of Benzo [b] furan Derivatives: A Review from 2011 to 2022. *Pharmaceuticals.*, 16, 9, 1265, 2023 Sep 6.
69. Batool, Z., Xu, D., Zhang, X., Li, X., Li, Y., Chen, Z., Li, B., Li, L., A review on furan: Formation, analysis, occurrence, carcinogenicity, genotoxicity and reduction methods. *Crit. Rev. Food Sci. Nutr.*, 61, 3, 395–406, 2021 Feb 4.
70. Saeid, H., Al-sayed, H., Bader, M., A review on biological and medicinal significance of furan. *AlQalam J. Med. Appl. Sci.*, 44–58, 2023 Feb 18.
71. Kumar, N., Gusain, A., Kumar, J., Singh, R., Hota, P.K., Anti-oxidation properties of 2-substituted furan derivatives: A mechanistic study. *J. Lumin.*, 230, 117725, 2021 Feb 1.
72. Lu, B., Huang, S., Cao, J., Hu, Q., Shen, R., Wan, H., Wang, D., Yuan, J., Zhang, L., Zhang, J., Zhang, M., Discovery of EBI-1051: a novel and orally efficacious MEK inhibitor with benzofuran scaffold. *Bioorg. Med. Chemistry,* 26, 3, 581–9, 2018 Feb 1.
73. Nguyen, T.T., Lomberget, T., Tran, N.C., Colomb, E., Nachtergaele, L., Thoret, S., Dubois, J., Guillaume, J., Abdayem, R., Haftek, M., Barret, R., Synthesis

and biological evaluation of novel heterocyclic derivatives of combretastatin A-4. *Bioorg. Med. Chem. Letters*, 22, 23, 7227–31, 2012 Dec 1.

74. Mushtaq, A., Zahoor, A.F., Ahmad, S., Saif, M.J., ulHaq, A., Khan, S.G., Al-Mutairi, A.A., Irfan, A., Al-Hussain, S.A., Zaki, M.E., A Comprehensive Review on Benzofuran Synthesis Featuring Innovative and Catalytic Strategies. *ACS Omega*, 9, 19, 20728–52, 2024 May 6.
75. Baldisserotto, A., Demurtas, M., Lampronti, I., Moi, D., Balboni, G., Vertuani, S., Manfredini, S., Onnis, V., Benzofuran hydrazones as potential scaffold in the development of multifunctional drugs: Synthesis and evaluation of antioxidant, photoprotective and antiproliferative activity. *Eur. J. Med. Chem.*, 156, 118–25, 2018 Aug 5.
76. Scibetta, S., Miceli, M., Iuliano, M., Stefanuto, L., Carbone, E., Piscopo, P., Petrozza, V., Romeo, G., Mangino, G., Calogero, A., Gasperi, T., *In Vitro* Evaluation of the Antioxidant Capacity of 3, 3-Disubstituted-3H-benzofuran-2-one Derivatives in a Cellular Model of Neurodegeneration. *Life.*, 14, 4, 422, 2024 Mar 22.
77. Miao, Y.H., Hu, Y.H., Yang, J., Liu, T., Sun, J., Wang, X.J., Natural source, bioactivity and synthesis of benzofuran derivatives. *RSC Adv.*, 9, 47, 27510–40, 2019.
78. Kumar, C.S., Then, L.Y., Chia, T.S., Chandraju, S., Win, Y.F., Sulaiman, S.F., Hashim, N.S., Ooi, K.L., Quah, C.K., Fun, H.K., Benzofuranyl esters: synthesis, crystal structure determination, antimicrobial and antioxidant activities. *Molecules.*, 20, 9, 16566–81, 2015 Sep 11.
79. Cho, J., Park, C., Lee, Y., Kim, S., Bose, S., Choi, M., Kumar, A.S., Jung, J.K., Lee, H., Neuroprotective and antioxidant effects of novel benzofuran-2-carboxamide derivatives. *Biomol. Therapeutics*, 23, 3, 275, 2015 May.
80. Sahni, T., Sharma, S., Verma, D., Kaur, P., Overview of coumarins and its derivatives: Synthesis and biological activity. *Lett. Org. Chem.*, 18, 11, 880–902, 2021 Nov 1.
81. Bouhaoui, A., Eddahmi, M., Dib, M., Khouili, M., Aires, A., Catto, M., Bouissane, L., Synthesis and biological properties of coumarin derivatives. *A Review ChemistrySelect*, 6, 24, 5848–70, 2021 Jun 28.
82. Katopodi, A., Tsotsou, E., Iliou, T., Deligiannidou, G.E., Pontiki, E., Kontogiorgis, C., Tsopelas, F., Detsi, A., Synthesis, bioactivity, pharmacokinetic and biomimetic properties of multi-substituted coumarin derivatives. *Molecules.*, 26, 19, 5999, 2021 Oct 2.
83. Whang, W.K., Park, H.S., Ham, I., Oh, M., Namkoong, H., Kim, H.K., Hwang, D.W., Hur, S.Y., Kim, T.E., Park, Y.G., Kim, J.R., Natural compounds, fraxin and chemicals structurally related to fraxin protect cells from oxidative stress. *Exp. Mol. Med.*, 37, 5, 436–46, 2005 Oct.
84. Kim, S.H., Kang, K.A., Zhang, R., Piao, M.J., Ko, D.O., Wang, Z.H., Chae, S.W., Kang, S.S., Lee, K.H., Kang, H.K., Kang, H.W., Protective effect of esculetin against oxidative stress-induced cell damage *via* scavenging reactive oxygen species. *Acta Pharmacol. Sin.*, 29, 11, 1319–26, 2008 Nov.

85. Alshibl, H.M., Al-Abdullah, E.S., Haiba, M.E., Alkahtani, H.M., Awad, G.E., Mahmoud, A.H., Ibrahim, B.M., Bari, A., Villinger, A., Synthesis and evaluation of new coumarin derivatives as antioxidant, antimicrobial, and anti-inflammatory agents. *Molecules.*, 25, 14, 3251, 2020 Jul 16.
86. Dua, R., Shrivastava, S., Sonwane, S.K., Srivastava, S.K., Pharmacological significance of synthetic heterocycles scaffold: a review. *Adv. Biol. Res.*, 5, 3, 120–44, 2011.
87. Feng, M., Tang B, H., Liang, S., Jiang, X., Sulfur containing scaffolds in drugs: synthesis and application in medicinal chemistry. *Curr. Top. Med. Chem.*, 16, 11, 1200–16, 2016 May 1.
88. Kapoor, K., Kaur, N., Sohal, H.S., Kaur, M., Singh, K., Kumar, A., Drugs and Their Mode of Action: A Review on Sulfur-Containing Heterocyclic Compounds. *Polycyclic Aromat. Compd.*, 26, 1–40, 2024 Aug.
89. Pathania, S. and Chawla, P.A., Thiophene-based derivatives as anticancer agents: An overview on decade's work. *Bioorg. Chem.*, 101, 104026, 2020 Aug 1.
90. Lahsasni, S., Al-Hemyari, D.A., Ghabbour, H.A., Mabkhoot, Y.N., Aleanizy, F.S., Alothman, A.A., Almarhoon, Z.M., Synthesis, Characterization, and Antibacterial and Anti-Inflammatory Activities of New Pyrimidine and Thiophene Derivatives. *J. Chem.*, 2018, 1, 8536063, 2018.
91. Keri, R.S., Chand, K., Budagumpi, S., Somappa, S.B., Patil, S.A., Nagaraja, B.M., An overview of benzo [b] thiophene-based medicinal chemistry. *Eur. J. Med. Chem.*, 138, 1002–33, 2017 Sep 29.
92. Mishra, R., Kumar, N., Mishra, I., Sachan, N., A review on anticancer activities of thiophene and its analogs. *Mini-Rev. Med. Chem.*, 20, 19, 1944–65, 2020 Nov 1.
93. Abed, N.A., Hammouda, M.M., Ismail, M.A., Abdel-Latif, E., Synthesis of new heterocycles festooned with thiophene and evaluating their antioxidant activity. *J. Heterocycl. Chem.*, 57, 12, 4153–63, 2020 Dec.
94. Saleh, A., Saeed, A., Abdel-Ghani, G.E., El-Rayyes, A., Abdel-Latif, E., Synthesis of some new antipyrine-thiophene hybrids and their evaluations as antioxidant and antibacterial agents. *Bull. Chem. Soc. Ethiop.*, 37, 1, 123–40, 2023.
95. Eldebss, T.M., Shaurub, E.S., Mekky, A.E., Shaaban, M.R., Farag, A.M., Synthesis of novel biphenyls containing thiophene, pyridazinimine and 1, 3, 4-thiadiazole moieties, and their effects on the mortality, oxidative stress and antioxidant enzymes in the desert locust, Schistocerca gregaria (Forskål) (Orthoptera: Acridiidae).
96. Lv, Q.Z., Long, J.T., Gong, Z.F., Nong, K.Y., Liang, X.M., Qin, T., Huang, W., Yang, L., Current state of knowledge on the antioxidant effects and mechanisms of action of polyphenolic compounds. *Nat. Prod. Commun.*, 16, 7, 1934578X211027745, 2021 Aug.
97. Andrés, C.M., Pérez de la Lastra, J.M., Juan, C.A., Plou, F.J., Pérez-Lebeña, E., Polyphenols as antioxidant/pro-oxidant compounds and donors of reducing

species: relationship with human antioxidant metabolism. *Processes.*, 11, 9, 2771, 2023 Sep 16.

98. Staveckienė, J., Kulaitienė, J., Levickienė, D., Vaitkevičienė, N., Vaštakaitė-Kairienė, V., The effect of ripening stages on the accumulation of polyphenols and antioxidant activity of the fruit extracts of Solanum species. *Plants.*, 12, 14, 2672, 2023 Jul 17.
99. Xin-yue, H.O., Jian-mei, W.U., Xiao-qiao, L.U., Kai, L.I., Wei-jie, Z.H., Guo-wen, Z.H., Research progress of inhibitory effects of polyphenolic compounds on xanthine oxidase. *Food Mach.*, 37, 2, 1–8, 2021.
100. Rudrapal, M., Khairnar, S.J., Khan, J., Dukhyil, A.B., Ansari, M.A., Alomary, M.N., Alshabrmi, F.M., Palai, S., Deb, P.K., Devi, R., Dietary polyphenols and their role in oxidative stress-induced human diseases: Insights into protective effects, antioxidant potentials and mechanism (s) of action. *Front. Pharmacol.*, 13, 806470, 2022 Feb 14.
101. Liga, S., Paul, C., Péter, F., Flavonoids: Overview of biosynthesis, biological activity, and current extraction techniques. *Plants.*, 12, 14, 2732, 2023 Jul 23.
102. Rehan, M., Biosynthesis of diverse class flavonoids *via* shikimate and phenylpropanoid pathway. *Bioact. Compounds-biosynthesis Characterization Appl.*, 75392, 2021 May 18.
103. Tariq, H., Asif, S., Andleeb, A., Hano, C., Abbasi, B.H., Flavonoid production: current trends in plant metabolic engineering and de novo microbial production. *Metabolites*, 13, 1, 124, 2023 Jan 13.
104. Chu, Y.H., Chang, C.L., Hsu, H.F., Flavonoid content of several vegetables and their antioxidant activity. *J. Sci. Food Agric.*, 80, 5, 561–6, 2000 Apr.
105. Hassanpour, S.H. and Doroudi, A., Review of the antioxidant potential of flavonoids as a subgroup of polyphenols and partial substitute for synthetic antioxidants. *Avicenna J. Phytomedicine*, 13, 4, 354, 2023 Jul.
106. Liu, W., Lu, H., Chu, X., Lou, T., Zhang, N., Zhang, B., Chu, W., Tea polyphenols inhibits biofilm formation, attenuates the quorum sensing-controlled virulence and enhances resistance to Klebsiella pneumoniae infection in Caenorhabditis elegans model. *Microb. Pathogen.*, 147, 104266, 2020 Oct 1.
107. Baranwal, A., Aggarwal, P., Rai, A., Kumar, N., Pharmacological actions and underlying mechanisms of catechin: A review. *Mini-Rev. Med. Chem.*, 22, 5, 821–33, 2022 Mar 1.
108. Lang, Y., Gao, N., Zang, Z., Meng, X., Lin, Y., Yang, S., Yang, Y., Jin, Z., Li, B., Classification and antioxidant assays of polyphenols: A review. *J. Future Foods*, 4, 3, 193–204, 2024 Sep 1.
109. Braicu, C., Ladomery, M.R., Chedea, V.S., Irimie, A., Berindan-Neagoe, I., The relationship between the structure and biological actions of green tea catechins. *Food Chem.*, 141, 3, 3282–9, 2013 Dec 1.
110. Bawono, L.C., Khairinisa, M.A., Jiranusornkul, S., Levita, J., The role of catechins of Camellia sinensis leaves in modulating antioxidant enzymes: A review and case study. *J. Appl. Pharm. Science*, 13, 12, 052–65, 2023 Dec 5.

111. Park, M.J., Ryu, D.H., Cho, J.Y., Ha, I.J., Moon, J.S., Kang, Y.H., Comparison of the antioxidant properties and flavonols in various parts of Korean red onions by multivariate data analysis. *Hortic. Environ. Biotechnol.*, 59, 919–27, 2018 Dec.
112. Barreca, D., Trombetta, D., Smeriglio, A., Mandalari, G., Romeo, O., Felice, M.R., Gattuso, G., Nabavi, S.M., Food flavonols: Nutraceuticals with complex health benefits and functionalities. *Trends Food Sci. Technol.*, 117, 194–204, 2021 Nov 1.
113. Jakobek, L., Drenjančević, M., Jukić, V., Šeruga, M., Phenolic acids, flavonols, anthocyanins and antiradical activity of "Nero","Viking","Galicianka" and wild chokeberries. *Sci. Hortic.*, 147, 56–63, 2012 Nov 12.
114. Song, X., Wang, Y., Gao, L., Mechanism of antioxidant properties of quercetin and quercetin-DNA complex. *J. Mol. Model.*, 26, 1–8, 2020 Jun.
115. Oh, W.Y., Ambigaipalan, P., Shahidi, F., Quercetin and its ester derivatives inhibit oxidation of food, LDL and DNA. *Food Chem.*, 364, 130394, 2021 Dec 1.
116. Sozen, E., Karademir, B., Ozer, N.K., Basic mechanisms in endoplasmic reticulum stress and relation to cardiovascular diseases. *Free Radical Biol. Med.*, 78, 30–41, 2015 Jan 1.
117. Middleton, E., Kandaswami, C., Theoharides, T.C., The effects of plant flavonoids on mammalian cells: implications for inflammation, heart disease, and cancer. *Pharmacol. Rev.*, 52, 4, 673–751, 2000 Dec 1.
118. Wang, W., Sun, C., Mao, L., Ma, P., Liu, F., Yang, J., Gao, Y., The biological activities, chemical stability, metabolism and delivery systems of quercetin: A review. *Trends Food Sci. Technology*, 56, 21–38, 2016 Oct 1.
119. Farhoosh, R., Johnny, S., Asnaashari, M., Molaahmadibahraseman, N., Sharif, A., Structure–antioxidant activity relationships of o-hydroxyl, o-methoxy, and alkyl ester derivatives of p-hydroxybenzoic acid. *Food Chem.*, 194, 128–34, 2016 Mar 1.
120. Siquet, C., Paiva-Martins, F., Lima, J.L., Reis, S., Borges, F., Antioxidant profile of dihydroxy-and trihydroxyphenolic acids-A structure–activity relationship study. *Free Radical Res.*, 40, 4, 433–42, 2006 Jan 1.
121. Spiegel, M., Kapusta, K., Kołodziejczyk, W., Saloni, J., Żbikowska, B., GA, H., Sroka, Z., Antioxidant activity of selected phenolic acids–ferric reducing antioxidant power assay and QSAR analysis of the structural features. *Molecules.*, 25, 13, 3088, 2020 Jan.
122. Aliaño-González, M.J., Richard, T., Cantos-Villar, E., Grapevine cane extracts: Raw plant material, extraction methods, quantification, and applications. *Biomolecules.*, 10, 8, 1195, 2020 Aug 17.
123. Hamadouche, S., Ounissi, A., Baira, K., Ouddai, N., Balsamo, M., Erto, A., Benguerba, Y., Theoretical evaluation of the antioxidant activity of some stilbenes using the Density Functional Theory. *J. Mol. Struct.*, 1229, 129496, 2021 Apr 5.

124. Murtaza, G., Sajjad, A., Mehmood, Z., Shah, S.H., Siddiqi, A.R., Possible molecular targets for therapeutic applications of caffeic acid phenethyl ester in inflammation and cancer. *J. Food Drug Anal.*, 23, 1, 11–8, 2015 Mar 1.
125. El Khawand, T., Courtois, A., Valls, J., Richard, T., Krisa, S., A review of dietary stilbenes: Sources and bioavailability. *Phytochem. Rev.*, 17, 5, 1007–29, 2018 Oct.
126. Latos-Brozio, M. and Masek, A., Structure-activity relationships analysis of monomeric and polymeric polyphenols (Quercetin, Rutin and Catechin) obtained by various polymerization methods. *Chem. Biodiversity*, 16, 12, e1900426, 2019 Dec.
127. Bors, W., Michel, C., Stettmaier, K., Structure-activity relationships governing antioxidant capacities of plant polyphenols, in: *Methods in enzymology*, vol. 335, pp. 166–180, Academic Press, 2001 Jan 1.
128. Truong, V.L. and Jeong, W.S., Cellular defensive mechanisms of tea polyphenols: Structure-activity relationship. *Int. J. Mol. Sci.*, 22, 17, 9109, 2021 Aug 24.
129. Khoo, H.E., Azlan, A., Tang, S.T., Lim, S.M., Anthocyanidins and anthocyanins: Colored pigments as food, pharmaceutical ingredients, and the potential health benefits. *Food Nutr. Res.*, 2017 Aug 13.
130. Hanuka-Katz, I., Okun, Z., Parvari, G., Shpigelman, A., Structure dependent stability and antioxidant capacity of strawberry polyphenols in the presence of canola protein. *Food Chem.*, 385, 132630, 2022 Aug 15.
131. Chen, L., Cao, H., Huang, Q., Xiao, J., Teng, H., Absorption, metabolism and bioavailability of flavonoids: A review. *Crit. Rev. Food Sci. Nutr.*, 62, 28, 7730–42, 2022 Oct 3.
132. Kandemir, K., Tomas, M., McClements, D.J., Capanoglu, E., Recent advances on the improvement of quercetin bioavailability. *Trends Food Sci. Technol.*, 119, 192–200, 2022 Jan 1.
133. Lang, Y., Tian, J., Meng, X., Si, X., Tan, H., Wang, Y., Shu, C., Chen, Y., Zang, Z., Zhang, Y., Wang, J., Effects of α-casein on the absorption of blueberry anthocyanins and metabolites in rat plasma based on pharmacokinetic analysis. *J. Agric. Food. Chem.*, 69, 22, 6200–13, 2021 May 28.
134. Hahm, T.H., Tanaka, M., Matsui, T., Current knowledge on intestinal absorption of anthocyanins. *J. Agric. Food. Chem.*, 70, 8, 2501–9, 2022 Feb 18.
135. Matsui, T., Polyphenols-absorption and occurrence in the body system. *Food Sci. Technol. Res.*, 28, 1, 13–33, 2022.
136. Stagos, D., Antioxidant activity of polyphenolic plant extracts. *Antioxidants.*, 9, 1, 19, 2019 Dec 24.
137. Rai, T., Kaushik, N., Malviya, R., Sharma, P.K., A review on marine source as anticancer agents. *J. Asian Nat. Prod. Res.*, 26, 4, 415–51, 2024 Apr 2.
138. Chauhan, A., Sharma, P.K., Kaushik, N., Kumar, N., Synthesis of novel pyrazole analogues as efficacious antimicrobial agents. *Int. J. Pharm. Pharmaceutical Sci.*, 3, 5, 166–76, 2011.

139. Kumar, S., Kaushik, N., Sahu, J.K., Jatav, S., Quinoline analogues and nanocarrier systems: A dual approach to anti-tubercular therapy. *Eur. J. Med. Chem. Rep.*, 12, 100212, 2024 Dec 1.
140. Kumar, S., Kaushik, N., Design, Synthesis and Evaluation of Novel Quinoline-Thiazolidinone-Isonicotinamide Hybrids as Potent InhA Inhibitors for Antitubercular Therapy. *ajc* [Internet]. 37, 12, 3158–66, 2025 Nov. 30 [cited 2026 Jan. 10]. Available from: https://asianpubs.org/index.php/ajchem/article/view/37_12_28.

7

Heterocyclic Compounds as Anticonvulsants, Antipyretics, Antiallergic Agents, and Antileprosy Agents

Anushka Sharma[1], Awaneet Kaur[1*], Pranav Gupta[1], Debashis Paramanick[2] and Abdulsalam Alhalmi[3]

[1]Department of Pharmacy, School of Medical and Allied Sciences, Galgotias University, Greater Noida, Uttar Pradesh, India
[2]School of Medical and Allied Sciences, K. R. Mangalam University, Gurugram, India
[3]Department of Pharmaceutics, College of Pharmacy, University of Aden, Aden, Yemen

Abstract

Heterocyclic compounds, defined by rings that include at least one noncarbon element (usually nitrogen, oxygen, or sulfur), are the foundation of many pharmacologically active substances in contemporary medical chemistry. Their structural variety and chemical reactivity render them essential in formulating pharmaceuticals for various ailments. This chapter emphasizes the therapeutic importance of heterocycles, concentrating on their functions as anticonvulsants, antipyretics, antiallergic drugs, and in the management of leprosy. In anticonvulsant medicine, heterocycles influence neurotransmitter pathways, with several Food and Drug Administration–approved medications and ongoing structure–activity relationship research highlighting their significance. Heterocycles function as antipyretic drugs by suppressing prostaglandin production, hence facilitating efficient fever treatment. Their antiallergic characteristics arise from their capacity to inhibit histamine receptors, providing alleviation in allergic situations. Additionally, in the treatment of leprosy, heterocyclic scaffolds have significant antimycobacterial efficacy, combating medication resistance and enhancing therapeutic results.

**Corresponding author*: kaurawaneet@gmail.com

Divya Bajpai Tripathy, Priyanka Chhabra and Anjali Gupta (eds.) Heterocyclic Compounds as Pharmaceutical Ingredients, (249–268) © 2026 Scrivener Publishing LLC

Recent developments in synthetic methods and molecular modeling have expedited the identification of new heterocyclic medication candidates. Notwithstanding difficulties in toxicity and bioavailability, current research presents encouraging prospects for enhanced medication design and therapeutic effectiveness. This chapter summarizes the diverse applications of heterocyclic compounds and examines prospective advancements in their development for targeted medication therapy.

Keywords: Heterocyclic compounds, anticonvulsants, antipyretics, antiallergic agents, antileprosy agents

7.1 Introduction

Heterocycles are a kind of chemical compound with at least one heteroatom in their ring structure. The two primary kinds are cyclic heterocycles and acyclic heterocycles. Heterocycles' fundamental structure resembles that of circular organic molecules composed only of carbon atoms. Replacing carbon in a heterocycle with one or more heteroatoms alters its chemical and physical characteristics, hence distinguishing it from a ring composed only of carbon [1–5]. The figures indicate that about 85% of chemical compounds with physiological effect include heterocycles. This demonstrates how crucial heterocycles are for developing novel medications nowadays [6–8]. Every heterocycle has a therapeutic impact whether it is natural or manmade. Heterocyclic compounds are growing more significant in medical research because they function in the body and in pharmacology. Many medically significant compounds, including vitamins, hormones, and antibiotics, include heterocyclic molecules. Heterocyclic compounds with nitrogen atoms comprise many different kinds of molecules, including natural products, compounds frequently used in medical chemistry, and complexes with physiological effects [9, 10].

Most of them are heterocycles, which consist of five or six components formed from various atoms of nitrogen (N), oxygen (O), or sulfur (S). Among the most well-known simple heterocyclic compounds are thiophene, furan, pyridine, and pyrrole. Five carbon atoms and one nitrogen atom make up pyridine's ring form. The five-membered rings of pyrrole, furan, and thiophene, in that sequence, contain four carbon atoms and one N, O, or S atom [11]. Many objects use heterocyclic compounds because they can be modified little to alter their function. Although they differ greatly internally in terms of acidity or polarity, they can be grouped into large categories with comparable characteristics. These modifications could require moving the same heteroatoms around in the ring structure or replacing one heteroatom with another inside the ring. Heterocyclic compounds can have functional groups as substituents or as fundamental components of the ring, hence increasing the number of methods to provide or imitate them. One example of tetrazole is its ring structure allowing basic N atoms to function as both amino substituents and ring components [12].

Many other natural compounds, including DNA, RNA, chlorophyll, hemoglobin, and vitamins, rely on the heterocyclic ring in their structure. The O-transporting pigment hemoglobin, the photosynthetic pigment chlorophyll, and the necessary amino acid proline are just a few examples of the many proteins and other compounds that comprise the body's structure that rely on heterocycles as building blocks [13–15]. Many heterocyclic compounds, including analogs of triazine, have been used as anti-inflammatory drugs, urinary antiseptics, and therapeutic agents to treat a variety of ailments. Derivatives of benzimidazole have demonstrated anthelmintic, antiviral, antifungal, and antibacterial qualities [16, 17]. A large number of pharmacological drugs are heterocyclic compounds. The most frequent types of heterocycles are amino acids such as histidine, tryptophan, and proline as well as vitamins and synthetase precursors such as pyridoxine, riboflavin, thiamine, biotin, folic acid, and B_{12}. Among the many practical applications of synthetic heterocyclic compounds are those of fungicides, photostabilizers, agrochemicals, colorants, copolymers, flavorings, fluorescents, and Kodak developer, among many more [18–21].

This chapter examines the application of heterocyclic compounds across four primary therapeutic domains: anticonvulsants, antipyretics, antiallergic agents, and antileprosy agents. Epilepsy and seizure-related disorders present considerable therapeutic challenges, highlighting the need for the development of new anticonvulsant agents that offer enhanced safety profiles. Heterocyclic scaffolds, comprising five-membered, six-membered, and fused ring systems, have demonstrated significant potential in both preclinical and clinical contexts. Recent advancements in synthetic methodologies, computational drug design, and structure–activity relationship (SAR) studies have broadened the exploration of novel heterocyclic candidates exhibiting improved bioactivity and reduced side effects.

7.2 Heterocyclic Compounds as Anticonvulsant Agents

Epilepsy stands out as a prevalent neurological disorder in humans, marked by excessive temporary neuronal discharges that lead to uncontrolled convulsions. This condition impacts more than 2 million individuals in the United States and approximately 60 million people globally [22]. Anticonvulsant medications such as phenobarbital, primidone, phenytoin, carbamazepine, ethosuximide, and benzodiazepines are commonly utilized. Nonetheless, these substances exhibit toxicity that is related to dosage and can also present unique side effects in certain individuals [23–26].

The five-membered heterocyclic chemical class known as imidazole and its derivatives is distinguished by the presence of two N atoms that are not next to one another. According to recent research, substituted imidazole derivatives have attracted a lot of attention because of their diverse pharmacological properties, which include analgesic and anti-inflammatory actions [27, 28]. Imidazole–heterocyclic compounds may represent new families of anticonvulsant medicines because of their potential anticonvulsant effects, according to a study of the literature. A group of compounds from 2-acetylnaphthalene was described by Karakurt and colleagues. The intraperitoneal route (ED_{50} and TD_{50}) was used to evaluate the anticonvulsant protection of the most promising Compound 1 (Figure 7.1). In the MES model, mice showed protection of 38.46 and 123.83 mg/kg, whereas rats showed protection of 20.44 and 56.36 mg/kg, respectively [29].

Numerous 1,2,4-trisubstituted-1H-imidazole derivatives were synthesized by Husain *et al.* The majority of the compounds exhibited anticonvulsant effects in the MES and sc-PTZ screens when administered intraperitoneally to mice. Only Compound 2, which was screened for anticonvulsants, showed strong action that was on par with the common medications carbamazepine and phenytoin (Figure 7.1) [30]. Amir and colleagues demonstrated the synthesis of many novel semicarbazones with imidazole incorporation. Compound 3, which showed no neurotoxic or depressive effects on the central nervous system, was the most active of the produced compounds (Figure 7.1). There were no discernible changes

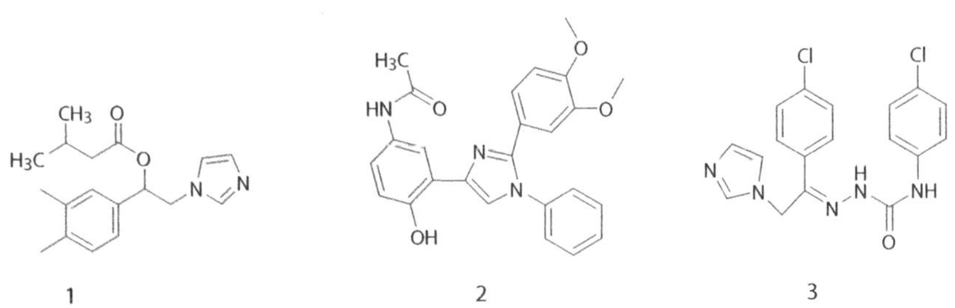

Figure 7.1 Structure of Compounds 1, 2, and 3.

in the compound's enzyme levels, according to measurements of liver enzymes such as serum glutamate oxaloacetate transaminase, serum glutamate pyruvate transaminase, and alkaline phosphatase [31].

The oxadiazole scaffold has been extensively studied in recent years and has amazing adaptability. The biological actions of compounds with oxadiazole rings have been studied [32]. Bhat *et al.* [33] researched a variety of 3-(4-acetyl-5H/methyl-5-substituted phenyl-4,5-dihydro-1,3,4-oxadiazol-2-yl)-2H-chromene-2-ones for their anticonvulsant and potentially neurotoxic properties. Figure 7.2 shows that Compound 4, which exhibited effect at a lower dosage of 30 mg/kg in the MES-test, was quite potent. All of the compounds exhibited reduced toxicity when contrasted with the widely used medicine phenytoin [33]. Several 2-(2-phenoxy)phenyl-1,3,4-oxadiazole derivatives were produced by Tabatabai *et al.* Figure 7.2 shows that Compound 5's anticonvulsant effectiveness was lower than diazepam's. However, Compound 5 had a far better margin of safety and an LD_{50} that was 15 times higher than its ED_{50} [34].

Zhang *et al.* [35] created 7-alkoxy-2H-1,4-benzothiazin-3(4H)-ones and 4H-[1,2,4]triazolo[4,3-d]benzo[b][1,4]thiazine derivatives. After being administered intraperitoneally to KunMing mice, the anticonvulsant activity of these compounds was assessed using the MES and tarod tests. Among the compounds, one was determined to be the most active, with values of 14.3 for PI, 243.9 for TD_{50}, and 17.0 for ED_{50}. The manufactured compounds exhibited the lowest levels of neurotoxicity. On top of that, it was significantly lower than the reference medicine, carbamazepine [35]. More than that, Piao *et al.* created a series of 7-benzylamino-2H-1,4-benzoxazin-3(4H)-ones. Neurotoxicity was determined using the rotarod test, whereas anticonvulsant characteristics were assessed using the MES test. Their molecule had the maximum potency, according to the MES test, with an ED_{50} value of 31.7 mg/kg and a PI value of 7.2 [36].

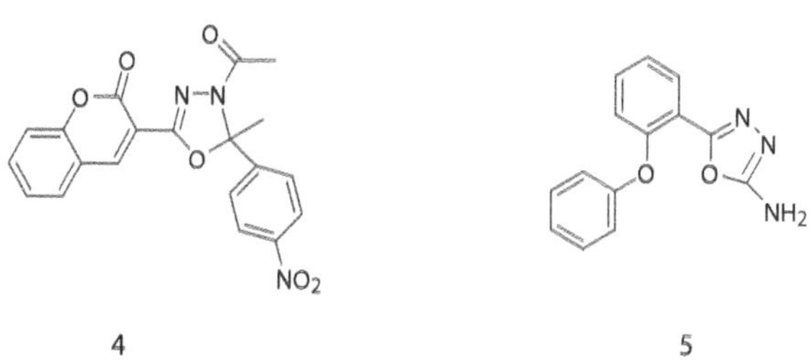

Figure 7.2 Structure of Compounds 4 and 5.

7.3 Heterocyclic Compounds as Antipyretic Agents

Drugs that reduce fevers are known as antipyretics. The majority of antipyretic drugs function by inhibiting the enzyme cyclooxygenase, which in turn stops the body from making prostaglandins, which are inflammatory. Some examples of nonsteroidal anti-inflammatory drugs that are sold as over-the-counter antipyretics are acetaminophen, aspirin, ibuprofen, and naproxen. do Carmo Malvar *et al.* created a novel heterocyclic pyrazol derivative, 1,5-diphenyl-1H-pyrazole-3-carbohydrazide (1.5-DHP), to study the compound's antipyretic effectiveness on fever produced by lipopolysaccharide (LPS). Intravenous injections of 5 g/kg LPS plus pretreatment with 1.5-DHP (30–60 mg/kg, p.o.) reduced LPS-induced flu in rats in a dose-dependent manner [37].

Pyrazole compounds, including ampyrone, phenazone, and propylphenazone, are well known for their analgesic and antipyretic effects. Flora F. Barsoum *et al.* developed derivatives of pyrazole-1-thio carboxamide. In a carrageenan-induced paw edema test on rats, the substituted derivative with the formula A = 4-O $(CH_2)_2O_4'$ and R = Ph exhibited strong anti-inflammatory effects [38]. By administering 50 mg/kg of pyrazoline derivatives orally to albino rats, researcher tested their anti-inflammatory properties against edema caused by carrageenan. There is encouraging evidence that all of the compounds belonging to this class can reduce inflammation. With a greater percentage of edema inhibition, less ulcerogenic liability, and less acute toxicity, phenylbutazone was the most powerful chemical in the series, outperforming the traditional medication, 3-[1-acetyl-5-(p-hydroxyphenyl)-2-pyrazolin-3-yl] indole (Compound 6) (Figure 7.3) [39] .

The anti-inflammatory properties of a new series of 1-thiocarbamoyl-3-substituted phenyl-5-(2-pyrrolyl)-4,5-dihydro (1H)-pyrazole derivatives were evaluated *in vivo* in a separate study by Gökhan-Kelekçi *et al.* [40],

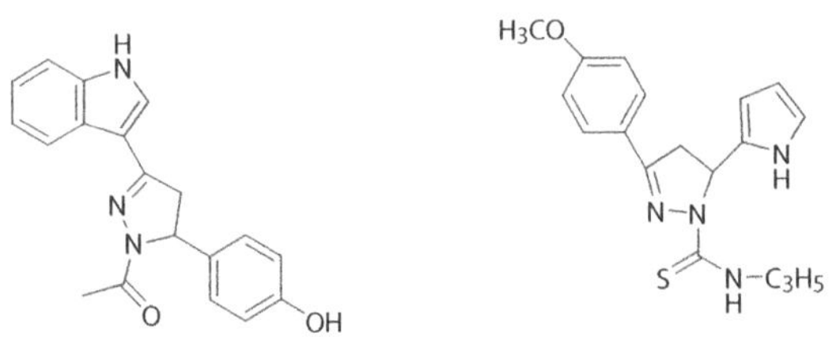

6 7

Figure 7.3 Structure of Compounds 6 and 7.

who used two bioassays to measure the effects on carrageenan-induced edema and acetic acid–induced capillary permeability in mice. Additionally, the evaluation of ulcerogenic and analgesic properties was carried out. Figure 7.3 shows that Compound 7, when paired with the anti-inflammatory results from *in vivo* animal research [40], had an activity similar to indomethacin's but without the ulcerogenic consequences. Additionally, Khode *et al.* [41] evaluated the analgesic and anti-inflammatory effects of a new series of 5-(substituted) aryl-3-(3-coumarinyl)-1-phenyl-2-pyrazolines *in vivo* at a dosage of 200 mg/kg of body weight. Of the 12 compounds he tested, Compound 44 had the most anti-inflammatory impact in the acute inflammation model, especially in the carrageenan-induced rat edema paw model [41].

7.4 Heterocyclic Compounds as Antiallergic Agents

Probiotics can reduce the likelihood of allergic reactions by controlling proinflammatory mediators, stabilizing aberrant microbiota, enhancing gut–barrier function, and eliminating or structurally changing enteral antigens. To study their potential antiallergic effects, Putta *et al.* [42] created bis-heteroaryl hydrazines. Immunoglobulin E/silver accelerated RBL-2H3 cells showed variable degrees of antiallergic efficacy at 50 and 1 dosages for all compounds tested. The development of antiallergy drugs has traditionally relied heavily on heterocyclic moieties. The methyl-substituted molecules were weaker in comparison to the electron-deficient and chlorosubstituted ones [42].

Heterocyclic compounds have been investigated for their possible antiallergic effects, especially regarding allergic conditions such as asthma, allergic rhinitis, and atopic dermatitis [43–50]. It is important to understand that although these heterocyclic compounds have antiallergic qualities, the specific mechanisms of action and efficacy might vary depending on the chemical, the kind of allergy, and the patient's unique features. Further research is needed to fully understand how heterocyclic compounds can help with allergies and to find new ones with better antiallergic properties. Here are some examples of heterocyclic compounds that show promise for antiallergic effects; nonetheless, research in this area is ongoing: There is evidence of the creation of several pyridothienopyrimidines (Figure 7.4) and their assessment as histamine release inducers or inhibitors in rat mast cells. Evaluation of activity followed chemical stimulation with polymer 48/80 and the medicines vinorelbine and adriamycin, as well as immunological stimulation with ovalbumin [51].

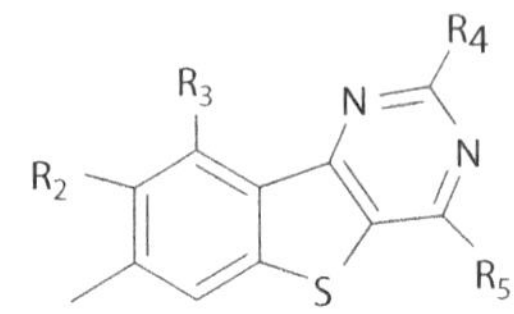

8

Figure 7.4 Structure of pyridothienopyrimidines.

In this study, many compounds from the furo[2, 3-b][1,8]naphthyridine-3,4(2H,9H)-dione and 4H-furo[2, 3-d] families were synthesized. Through the use of rat pas- sive cutaneous anaphylaxis testing, the antiallergic potential of these drugs was evaluated. The majority of the compounds demonstrated oral activity, suggesting that these drugs are best delivered orally. The compound [1,8]naphthyridine-3,4(2H,9H)-dione was selected for further study and modification because it was the most powerful one identified. These findings point to the compound's and its derivatives' potential for use in the development of new antiallergenic medications [52, 53].

7.5 Heterocyclic Compounds as Antileprosy Agents

Heterocyclic compounds are integral to modern leprosy treatment, particularly through their inclusion in multidrug therapy (MDT). Clofazimine, a phenazine-based heterocycle, has been a cornerstone since the 1960s, combining bactericidal and anti-inflammatory actions. It disrupts *Mycobacterium leprae* membrane function, intercalates bacterial DNA [54–58], and suppresses T-cell activity to mitigate inflammatory reactions such as erythema nodosum leprosum (ENL) [59]. Synthetic heterocyclic analogs, such as fluoroquinolones (e.g., ofloxacin) and tetracyclines (e.g., minocycline), have expanded therapeutic options. These agents inhibit DNA gyrase and ribosomal protein synthesis, respectively, and are used in combination regimens such as ROM (rifampicin, ofloxacin, minocycline) to reduce treatment duration and resistance risks. Clarithromycin, a macrolide heterocycle, further enhances efficacy by targeting bacterial ribosomes [60–65]. Recent research emphasizes structural optimization of heterocyclic scaffolds to improve target affinity and overcome drug resistance, underscoring their enduring relevance in leprosy management [66, 67].

A long-term infectious disease predominantly affecting the eyes, skin, peripheral nerves, and the lining of the upper respiratory tract, leprosy is brought on by the obligate intracellular bacterium *M. leprae* [68–74]. Although much effort has been made to reduce the number of cases globally, more than 200,000 fresh cases are still recorded year, particularly in nations where the illness is prevalent, including India, Brazil, and Indonesia. From paucibacillary (PB) forms with few bacteria to multibacillary (MB) forms with a lot of bacteria, the disease manifests in the body in a wide spectrum. To efficiently cure leprosy and prevent its spread, the World Health Organization (WHO) advises MDT using dapsone, rifampicin, and clofazimine. Some issues make it more difficult to obtain therapy results. Standard MDT regimens become less successful with the rise of drug-resistant *M. leprae* strains, particularly those resistant to rifampicin and dapsone. This also increases the likelihood that the illness may return and spread [75]. Often, the lengthy treatment time—6 months for PB cases and 12 months or more for MB cases—causes patients to not complete their therapy, which accelerates resistance development. *M. leprae*'s existence inside macrophages and Schwann cells complicates drug access to such cells. Moreover, inflammatory complications such as ENL require adjunct immunomodulatory therapies, adding complexity to management [76]. These challenges underscore the urgent need for novel therapeutic agents and regimens with improved efficacy, reduced toxicity, and shorter duration.

An extensive synthesis of dialkyldithiocarbamates was carried out by Makarov *et al.* [77]. The capacity of all dithiocarbamate derivatives that were produced to suppress the development of *M. leprae* on mice footpads was tested. Depending on the dosage, the efficacy against leprosy and its toxicity might vary greatly. Three compounds were identified as the most promising candidates for antileprosy drugs, showing actions similar to dapsone, because of the enhanced survivorship and significant decrease in *M. leprae* populations. These medicines were well-tolerated for the whole 7.5-month treatment duration. On a typical basis, their minimal inhibitory values vary from 0.45 to 6.25 mg/L [77].

Clofazimine, a phenazine heterocyclic compound, is a key MDT drug with both bactericidal and anti-inflammatory effects. It accumulates in macrophages and skin tissues, where it generates reactive oxygen species that disrupt *M. leprae* membrane integrity and intercalate bacterial DNA, leading to bacterial death. Additionally, clofazimine modulates host immune responses by suppressing T-cell activity, thereby reducing inflammatory lepra reactions such as ENL [78]. Rifampicin, a rifamycin heterocyclic antibiotic, inhibits bacterial DNA-dependent RNA polymerase, resulting in potent bactericidal activity. Dapsone, a sulfone heterocyclic compound, inhibits dihydropteroate synthase, interfering with folate synthesis and exerting a bacteriostatic effect. Fluoroquinolones such as ofloxacin, containing a quinolone heterocyclic core, inhibit bacterial DNA gyrase and topoisomerase IV, enzymes critical for DNA replication and transcription, leading to bactericidal effects against *M. leprae* [79]. Tetracyclines such as minocycline, with a fused heterocyclic ring system, inhibit protein synthesis by binding to the 30S ribosomal subunit, enhancing intracellular penetration and efficacy. Macrolide antibiotics such as clarithromycin also contribute by targeting bacterial ribosomes and inhibiting protein synthesis. More recently, novel heterocyclic compounds, such as bedaquiline, a diarylquinoline, have shown promise by targeting the mycobacterial ATP synthase proton pump, disrupting energy metabolism and killing drug-resistant *M. leprae* strains *in vitro* and in animal models [80]. Research continues to focus on optimizing heterocyclic scaffolds, including oxazole and imidazole derivatives, to improve antimycobacterial potency and pharmacokinetic profiles [81].

The clinical management of leprosy relies heavily on MDT regimens incorporating heterocyclic compounds. For PB cases, WHO recommends a 6-month regimen of rifampicin and dapsone, while MB cases require a 12-month or longer combination of rifampicin, dapsone, and clofazimine. Clofazimine's dual bactericidal and anti-inflammatory properties make it indispensable in controlling bacterial load and managing lepra reactions. In cases of drug resistance or intolerance, alternative regimens such as the ROM combination (rifampicin, ofloxacin, minocycline) are used, with ofloxacin and minocycline providing potent bactericidal activity through inhibition of DNA gyrase and protein synthesis, respectively. This regimen has demonstrated efficacy in shortening treatment duration and improving patient adherence, especially in PB leprosy. Clarithromycin is also used as an adjunct in resistant cases due to its macrolide heterocyclic structure and ribosomal targeting. Emerging therapies include bedaquiline, currently under investigation in clinical trials for leprosy, which offers potential for treating multidrug-resistant infections by targeting ATP synthase [82]. Immunomodulatory agents such as thalidomide, a glutamic acid derivative heterocycle, are utilized to manage severe ENL reactions. Additionally, immunotherapeutic approaches such as the *Mycobacterium indicus pranii* vaccine enhance host immune responses, complementing chemotherapy and reducing relapse risk [83]. Scientists want to create drugs that function better, are safer, and need shorter treatment times by means of novel heterocyclic compounds. This would address present issues with constraints and opposition.

7.6 Recent Advances and Future Prospects

7.6.1 Novel Synthetic Approaches

Recent advances in leprosy research have mostly concentrated on creating novel synthetic compounds to enhance the efficacy of treatment and combat medication resistance. New dapsone compounds targeting the folic acid production pathway—a key metabolic mechanism in *M. leprae*—have been created using tools from bioinformatics and medicinal chemistry. Molecular docking and molecular dynamics simulations helped us to predict how these dapsone derivatives would bind to and remain stable with wild-type and mutant dihydropteroate synthase enzymes. This made us discover encouraging options such as DD3 with lower binding free energy than normal dapsone [84–87]. Synthetic peptides were created to improve serological testing; they can also hasten early diagnosis, which helps to respond swiftly and prevent the spread of disease. By means of molecular designs, these synthetic drugs seek to enhance target selectivity and pharmacokinetics, hence addressing issues with current medications such as their toxicity and resistance.

7.6.2 Drug Design and Molecular Modeling

The discovery of antileprosy medications has been altered by combining molecular models with computational drug design. Heterocyclic compounds that bind strongly to *M. leprae* targets can be found and improved using molecular docking, molecular dynamics simulations, and binding free energy studies all utilized together. Using these computer-based techniques to logically design dapsone derivatives, for example, has provided us with knowledge of their interaction with dihydropteroate synthase, which has guided us in the synthesis and confirmation of our findings [88]. Furthermore, molecular modeling supports the development of novel drug delivery systems, such as pH-sensitive polymeric nanoparticles and hydrogels, which enhance oral bioavailability and reduce drug toxicity by controlled release of agents such as dapsone and clofazimine [89]. These computational approaches significantly reduce time and resources required for drug development and hold promise for generating next-generation therapeutics with improved efficacy and safety profiles.

Despite these advances, several challenges remain in leprosy treatment. The emergence of resistance to first-line drugs such as dapsone and rifampicin continues to threaten disease control efforts, necessitating ongoing surveillance and development of alternative regimens [90]. Adverse effects associated with current MDT drugs, including skin discoloration from clofazimine and hematologic toxicity from dapsone, impact patient adherence and quality of life [91]. Moreover, the intracellular persistence of *M. leprae* and the complexity of host immune responses complicate eradication efforts. However, these challenges also present opportunities: the incorporation of fluoroquinolones such as moxifloxacin and ofloxacin into treatment regimens has demonstrated rapid bacterial clearance with fewer side effects, improving compliance [92]. Ethnomedicinal plants and bioactive natural compounds offer numerous potential antileprosy medications. Nanotechnology-based drug delivery systems, on the other hand, provide precise distribution and reduced toxicity. Ongoing research combining synthetic chemistry, computer modeling, and innovative approaches to deliver drugs may help to find better, safer methods to treat leprosy and finally eradicate it [93].

7.7 Conclusion

Right now, medicine development greatly depends on heterocyclic compounds because they have many varied chemical structures and biological activities. Their efficacy as anticonvulsants, fever reducers, allergy medications, and leprosy treatments highlights their significance in managing both frequent and less common health concerns. By inhibiting histamine receptors, halting the synthesis of prostaglandins, regulating neurotransmission, and killing bacteria, these compounds have significant therapeutic properties. Drugs such as dapsone, loratadine, paracetamol, and phenytoin help us all know that heterocyclic scaffolds can be effectively used in therapeutic environments. Recent advancements in synthetic techniques, SAR research, and computer models have accelerated and simplified the creation of novel heterocyclic compounds that are safer and more effective. Although issues such as side effects and drug resistance remain, heterocyclic drug research has a promising future. Ongoing research and development in this field will probably lead to more individualized and successful therapies for a great spectrum of disorders.

AI disclosure statement: The author acknowledges the use of AI-based tools for minor language editing or grammatical correction. However, all intellectual contributions, analyses, and conclusions presented in this chapter are entirely the author's own.

References

1. Alvarez-Builla, J. and Barluenga, J., Heterocyclic compounds: An introduction. *Mod. Heterocyclic Chem.*, 29, 1–9, 2011.
2. Jeelani, I., Abe, H., Nawaz, A., Bhosale, M., Ahmad, S., Jamadar, A., Ahmed, K., Qadir, T., Amin, A., Sharma, P.K., Abidi, S., Anti-cancer potential of natural products containing (6Hdibenzo [b, d] pyran-6-one) framework using docking tools. *Pak. J. Pharm. Sci.*, 34, 2021 Sep 2.
3. Sharma, P.K., Antifungal, antibacterial and antioxidant activities of substituted morpholinylbenzothiazine. *Pharm. Lett.*, 8, 11, 140–2, 2016.
4. Sharma, P.K., Makkar, R., Singh, S., Antibacterial, antifungal and antioxidant activities of substituted 4H-1, 4-benzothiazines. *Der Pharma Chem.*, 8, 11, 156, 2016.
5. Ahmed, K. and Jeelani, I., Synthesis and *in vitro* antimicrobial screening of 3-acetyl-4-hydroxycoumarin hydrazones. *Int. J. Pharm. Biol. Sci.*, 9, 1, 1000–5, 2019.
6. Tietze, L.F. and Rackelmann, N., Domino reactions in the synthesis of heterocyclic natural products and analogs. *Pure Appl. Chem.*, 76, 11, 1967–83, 2004 Jan 1.
7. Zhang, B. and Studer, A., Recent advances in the synthesis of nitrogen heterocycles *via* radical cascade reactions using isonitriles as radical acceptors. *Chem. Soc. Rev.*, 44, 11, 3505–21, 2015.
8. Mahmood, R.M. and Aljamali, N.M., Synthesis, spectral investigation, and microbial studying of pyridine-heterocyclic compounds. *Eur. J. Mol. Clin. Med.*, 7, 11, 4444–53, 2020.
9. Panchal, N.B., Patel, P.H., Chhipa, N.M., Parmar, R.S., Acridine a versatile heterocyclic moiety as anticancer agent.
10. Marín-Ocampo, L., Veloza, L.A., Abonia, R., Sepúlveda-Arias, J.C., Anti-inflammatory activity of triazine derivatives: A systematic review. *Eur. J. Med. Chem.*, 162, 435–47, 2019 Jan 15.
11. Patnayak, S., Heterocyclic Compounds An Potential Drug And Its Biological Activty: A Review. *J. Nonlinear Anal. Optim.*, 15, 1, 2024.
12. Arora, P., Arora, V., Lamba, H.S., Wadhwa, D., Importance of heterocyclic chemistry: A review. *Int. J. Pharm. Sci. Res.*, 3, 9, 2947, 2012 Sep 1.
13. Dua, R., Shrivastava, S., Sonwane, S.K., Srivastava, S.K., Pharmacological significance of synthetic heterocycles scaffold: A review. *Adv. Biol. Res.*, 5, 3, 120–44, 2011.
14. Fürst, P. and Stehle, P., What are the essential elements needed for the determination of amino acid requirements in humans? *J. Nutr.*, 134, 6, 1558S–65S, 2004 Jun 1.
15. Kabir, E. and Uzzaman, M., A review on biological and medicinal impact of heterocyclic compounds. *Results Chem.*, 4, 100606, 2022 Jan 1.
16. Al-Mulla, A., A review: Biological importance of heterocyclic compounds. *Der Pharma Chem.*, 9, 13, 141–7, 2017.

17. Shaikh, A.R., Farooqui, M., Satpute, R.H., Abed, S., Overview on nitrogen containing compounds and their assessment based on 'International Regulatory Standards'. *J. Drug Delivery Ther.*, 8, 6-s, 424–8, 2018 Nov 2.
18. Saini, M.S., Kumar, A., Dwivedi, J., Singh, R., A review: Biological significances of heterocyclic compounds. *Int. J. Pharm. Sci. Res.*, 4, 3, 66–77, 2013.
19. Chen, Q., Zhu, X.L., Jiang, L.L., Liu, Z.M., Yang, G.F., Synthesis, antifungal activity and CoMFA analysis of novel 1, 2, 4-triazolo [1, 5-a] pyrimidine derivatives. *Eur. J. Med. Chem.*, 43, 3, 595–603, 2008 Mar 1.
20. Ahmad, A., Husain, A., Khan, S.A., Mujeeb, M., Bhandari, A., Synthesis, antimicrobial and antitubercular activities of some novel pyrazoline derivatives. *J. Saudi Chem. Soc.*, 20, 5, 577–84, 2016 Sep 1.
21. Ziajahromi, S., Neale, P.A., Leusch, F.D., Wastewater treatment plant effluent as a source of microplastics: Review of the fate, chemical interactions and potential risks to aquatic organisms. *Water Sci. Technol.*, 74, 10, 2253–69, 2016 Nov 18.
22. Beghi, E. and Hesdorffer, D., Prevalence of epilepsy—an unknown quantity. *Epilepsia*, 55, 7, 963–7, 2014 Jul.
23. Al-Soud, Y.A., Al-Masoudi, N.A., Ferwanah, A.E., Synthesis and properties of new substituted 1, 2, 4-triazoles: Potential antitumor agents. *Bioorg. Medicinal Chem.*, 11, 8, 1701–8, 2003 Apr 1.
24. Meador, K.J., Newer anticonvulsants: dosing strategies and cognition in treating patients with mood disorders and epilepsy. *J. Clin. Psychiatry*, 64, 30–4, 2003 Jan 1.
25. Penovich, P.E. and James Willmore, L., Use of a new antiepileptic drug or an old one as first drug for treatment of absence epilepsy. *Epilepsia*, 50, 37–41, 2009 Sep.
26. Belcastro, V., Striano, P., Gorgone, G., Costa, C., Ciampa, C., Caccamo, D., Pisani, L.R., Oteri, G., Marciani, M.G., Aguglia, U., Striano, S., Hyperhomocysteinemia in epileptic patients on new antiepileptic drugs. *Epilepsia*, 51, 2, 274–9, 2010 Feb.
27. Tripathy, D.B. and Mishra, A., Microwave synthesis and characterization of waste soybean oil-based gemini imidazolinium surfactants with carbonate linkage. *Surf. Rev. Lett.*, 24.05, 1750062, 2017.
28. Thenmozhiyal, J.C., Wong, P.T., Chui, W.K., Anticonvulsant activity of phenylmethylenehydantoins: A structure–activity relationship study. *J. Med. Chem.*, 47, 6, 1527–35, 2004 Mar 11.
29. Karakurt, A., Özalp, M., Işık, S., Stables, J.P., Dalkara, S., Synthesis, anticonvulsant and antimicrobial activities of some new 2-acetylnaphthalene derivatives. *Bioorg. Medicinal Chem.*, 18, 8, 2902–11, 2010 Apr 15.
30. Husain, A., Siddiqui, N., Sarafroz, M.D., Khatoon, Y., Rasid, M., Ahmad, N., Synthesis, anticonvulsant and neurotoxicity screening of some novel 1, 2, 4-trisubstituted-1H-imidazole derivatives. *Acta Pol. Pharm.*, 68, 5, 657–63, 2011 Sep 1.

31. Amir, M., Ali, I., Zaheen Hassan, M., Imidazole Incorporated Semicarbazone Derivatives as a New class of anticonvulsants: design, synthesis and *in-vivo* screening. *Med. Chem.*, 9, 4, 571–80, 2013 Jun 1.
32. Zarghi, A. and Hajimahdi, Z., Substituted oxadiazoles: A patent review (2010–2012). *Expert Opin. Ther. Pat.*, 23, 9, 1209–32, 2013 Sep 1.
33. Bhat, M.A., Siddiqui, N., Khan, S.A., Synthesis of novel 3-(4-acetyl-5H/methyl-5-substituted phenyl-4, 5-dihydro-1, 3, 4-oxadiazol-2-yl)-2H-chromen-2-ones as potential anticonvulsant agents. *Acta Pol. Pharm.*, 65, 2, 235–9, 2008 Mar 1.
34. Tabatabai, S.A., Lashkari, S.B., Zarrindast, M.R., Gholibeikian, M., Shafiee, A., Design, synthesis and anticonvulsant activity of 2-(2-phenoxy) phenyl-1, 3, 4-oxadiazole derivatives. *Iran. J. Pharmaceutical Research IJPR*, 12, Suppl, 105, 2013.
35. Zhang, L.Q., Guan, L.P., Wei, C.X., Deng, X.Q., Quan, Z.S., Synthesis and anticonvulsant activity of some 7-alkoxy-2H-1, 4-benzothiazin-3 (4H)-ones and 7-alkoxy-4H-[1,2,4]triazolo[4,3-d]benzo [b][1,4]thiazines. *Chem. Pharm. Bull.*, 58, 3, 326–31, 2010 Mar 1.
36. Piao, Z.T., Guan, L.P., Zhao, L.M., Piao, H.R., Quan, Z.S., Synthesis of novel 7-benzylamino-2H-1, 4-benzoxazin-3 (4H)-ones as anticonvulsant agents. *Eur. J. Med. Chem.*, 43, 6, 1216–21, 2008 Jun 1.
37. do Carmo Malvar, D., Ferreira, R.T., de Castro, R.A., de Castro, L.L., Freitas, A.C., Costa, E.A., Florentino, I.F., Mafra, J.C., de Souza, G.E., Vanderlinde, F.A., Antinociceptive, anti-inflammatory and antipyretic effects of 1.5-diphenyl-1H-Pyrazole-3-carbohydrazide, a new heterocyclic pyrazole derivative. *Life Sci.*, 95, 2, 81–8, 2014 Jan 30.
38. Debnath, S., Design, Synthesis, Characterization of Some New Pyrazoline Derivatives and Evaluation for Their Antimicrobial Profiles. Master's thesis, Rajiv Gandhi University of Health Sciences (India), 2017.
39. Bajaj, K. and Kumar, A., Synthesis and pharmacological evaluation of newer substituted benzoxazepine derivatives as potent anticonvulsant agents. *Eur. J. Med. Chem.*, 39, 4, 369–76, 2004 Apr 1.
40. Gökhan-Kelekçi, N., Yabanoğlu, S., Küpeli, E., Salgın, U., Özgen, O., Uçar, G., Yeşilada, E., Kendi, E., Yeşilada, A., Bilgin, A.A., A new therapeutic approach in Alzheimer disease: some novel pyrazole derivatives as dual MAO-B inhibitors and antiinflammatory analgesics. *Bioorg. Medicinal Chem.*, 15, 17, 5775–86, 2007 Sep 1.
41. Khode, S., Maddi, V., Aragade, P., Palkar, M., Ronad, P.K., Mamledesai, S., Thippeswamy, A.H., Satyanarayana, D., Synthesis and pharmacological evaluation of a novel series of 5-(substituted) aryl-3-(3-coumarinyl)-1-phenyl-2-pyrazolines as novel anti-inflammatory and analgesic agents. *Eur. J. Med. Chem.*, 44, 4, 1682–8, 2009 Apr 1.
42. Putta, R.R., Donthamsetty, V.S., Guda, D.R., Adivireddy, P., Venkatapuram, P., Synthesis and anti-allergic activity of bis-heteroaryl hydrazines. *J. Heterocycl. Chem.*, 54, 4, 2216–22, 2017 Jul.

43. Kaur, R., Kaur, K., Arora, R., Saini, B., Arora, S., Natural fused heterocyclic flavonoids: Potent candidates as anti-inflammatory and anti-allergic agents in the treatment of asthma. *Curr. Bioact. Compd.*, 17, 1, 28–40, 2021 Jan 1.
44. Kapoor, Y. and Kumar, K., Structural and clinical impact of anti-allergy agents: An overview. *Bioorg. Chem.*, 94, 103351, 2020 Jan 1.
45. Zhang, S., Liu, Y., Javeed, A., Jian, C., Sun, J., Wu, S., Han, B., Treatment of allergy: Overview of synthetic anti-allergy small molecules in medicinal chemistry. *Eur. J. Med. Chem.*, 249, 115151, 2023 Mar 5.
46. Naik S, R. and Wala S, M., Inflammation, allergy and asthma, complex immune origin diseases: Mechanisms and therapeutic agents. *Recent Patents Inflammation Allergy Drug Discovery*, 7, 1, 62–95, 2013 Jan 1.
47. Tilkat, E. and Sökmen, A., Antiallergic Metabolite Production from Plants *via* Biotechnological Approaches, in: *Biotechnology of Medicinal Plants with Antiallergy Properties: Research Trends and Prospects*, pp. 433–485, Springer Nature Singapore, Singapore, 2024 May 31.
48. Adamu, R.M., Majumdhar, R.S., Uba, A.I., Structure-based virtual screening of natural compounds as potential anti-allergy agents against cytokine alarmins (TSLP and IL-33). *Lett. Drug Des. Discovery*, 18, 9, 932–42, 2021 Sep 1.
49. El-Qutob Lopez, D., New methods of prevention and treatment of allergic diseases. *Recent Pat. Inflammation Allergy Drug Discovery*, 6, 1, 46–64, 2012 Jan 1.
50. Mishra G, P., Tamboli, V., Jwala, J., K. Mitra, A., Recent patents and emerging therapeutics in the treatment of allergic conjunctivitis. *Recent Patents Inflammation Allergy Drug Discovery*, 5, 1, 26–36, 2011 Jan 1.
51. Quintela, J.M., Peinador, C., Veiga, C., González, L., Botana, L.M., Alfonso, A., Riguera, R., Synthesis and antiallergic activity of pyridothienopyrimidines. *Bioorg. Medicinal Chem.*, 6, 10, 1911–25, 1998 Oct 1.
52. Kuo, S.C., Tsai, S.Y., Li, H.T., Wu, C.H., Ishii, K., Nakamura, H., Studies on Heterocyclic Compounds. IX. Synthesis and Antiallergic Activity of Furo [2,3-b][1,8]naphthyridine-3, 4 (2H, 9H)-diones and 4H-Furo [2,3-d]pyrido[1,2-a]-pyrimidine-3, 4 (2H)-diones. *Chem. Pharm. Bull.*, 36, 11, 4403–7, 1988 Nov 25.
53. Thakur, D.A., Kumar, V., Tyagi, S.K., Parmar, R.S., Review on the Medicinal Significance of Heterocyclic Compounds. *Int. J. Sci. Technol. Eng.*, 2023.
54. Cruz, R.C., Bührer-Sékula, S., Penna, M.L., Penna, G.O., Talhari, S., Leprosy: Current situation, clinical and laboratory aspects, treatment history and perspective of the uniform multidrug therapy for all patients. *Anais Brasileiros de Dermatologia*, 92, 6, 761–73, 2017.
55. Lockwood, D.N., Penna, G.O., Lambert, S., Pai, V.V., Walker, S.L., Safer and newer antimicrobial drugs for leprosy–time to test monthly ROM in an adequately powered randomised trial? *Leprosy Rev.*, 93, 2, 96–101, 2022 Jun 1.

56. Laursen, J.B. and Nielsen, J., Phenazine natural products: biosynthesis, synthetic analogues, and biological activity. *Chem. Rev.*, 104, 3, 1663–86, 2004 Mar 10.
57. Yan, J., Liu, W., Cai, J., Wang, Y., Li, D., Hua, H., Cao, H., Advances in phenazines over the past decade: Review of their pharmacological activities, mechanisms of action, biosynthetic pathways and synthetic strategies. *Mar. Drugs*, 19, 11, 610, 2021 Oct 27.
58. Chaudhary, A. and Khurana, J.M., Synthetic routes for phenazines: an overview. *Res. Chem. Intermed.*, 44, 1045–83, 2018 Feb.
59. Pai, V.V., Role of clofazimine in management of reactions in leprosy: A brief overview. *Indian J. Drugs Dermatol.*, 1, 1, 12–5, 2015 Jul 1.
60. Rusu, A., Moga, I.M., Uncu, L., Hancu, G., The role of five-membered heterocycles in the molecular structure of antibacterial drugs used in therapy. *Pharmaceutics*, 15, 11, 2554, 2023 Nov.
61. Charushin, V.N., Nosova, E.V., Lipunova, G.N., Chupakhin, O.N., Fluoroquinolones: Synthesis and application, in: *Fluorine in Heterocyclic Chemistry Volume 2: 6-Membered Heterocycles*, pp. 111–79, 2014.
62. Lungu, I.A., Moldovan, O.L., Biriş, V., Rusu, A., Fluoroquinolones hybrid molecules as promising antibacterial agents in the fight against antibacterial resistance. *Pharmaceutics*, 14, 8, 1749, 2022 Aug 22.
63. Petersen, U., Quinolone antibiotics: the development of moxifloxacin. *Analogue-based Drug Discovery*, 20, 315–70, 2006.
64. Hryhoriv, H., Kovalenko, S.M., Georgiyants, M., Sidorenko, L., Georgiyants, V., A comprehensive review on chemical synthesis and chemotherapeutic potential of 3-heteroaryl fluoroquinolone hybrids. *Antibiotics*, 12, 3, 625, 2023 Mar 21.
65. Williams, D.L. and Gillis, T.P., Drug-resistant leprosy: Monitoring and current status. *Leprosy Rev.*, 83, 3, 269–81, 2012 Sep 1.
66. Bera, S. and Mondal, D., Insights of synthetic analogues of anti-leprosy agents. *Bioorg. Med. Chem.*, 27, 13, 2689–717, 2019 Jul 1.
67. Maheshwari, K.K. and Bandyopadhyay, D., Heterocycles in the treatment of neglected tropical diseases. *Curr. Med. Chem.*, 28, 3, 472–95, 2021 Jan 1.
68. Hussain, T., Leprosy and tuberculosis: an insight-review. *Crit. Rev. Microbiol.*, 33, 1, 15–66, 2007 Jan 1.
69. Mungroo, M.R., Khan, N.A., Siddiqui, R., Mycobacterium leprae: Pathogenesis, diagnosis, and treatment options. *Microb. Pathogen.*, 149, 104475, 2020 Dec 1.
70. Gillis, T.P., Mycobacterium leprae, in: *Molecular medical microbiology*, pp. 1655–1668, Academic Press, 2015 Jan 1.
71. Makhakhe, L., Leprosy review. *South Afr. Fam. Pract.*, 63, 4, 2021 Oct 29.
72. Ooi, W.W. and Srinivasan, J., Leprosy and the peripheral nervous system: Basic and clinical aspects. *Muscle Nerve*, 30, 4, 393–409, 2004 Oct.
73. Karunarathna, I., Rajapaksha, S., Ekanayake, U., Gunawardana, K., Aluthge, P., Gunasena, P., T Gunathilake, S., Hapuarachchi, S.B., Jayawardana, A.,

De Alvis, K., The Pathophysiology of Leprosy: How Mycobacterium leprae Affects the Nervous System.

74. Sugawara-Mikami, M., Tanigawa, K., Kawashima, A., Kiriya, M., Nakamura, Y., Fujiwara, Y., Suzuki, K., Pathogenicity and virulence of Mycobacterium leprae. *Virulence*, 13, 1, 1985–2011, 2022 Dec 31.
75. Celestino, I.C., Antunes, D.E., Santos, D.F., Gimenes, V.L., de Souza, F.M., Goulart, I.M., Adverse reactions induced by MDT/WHO (Rifampicin+ Clofazimine+ Dapsone) and ROM (Rifampicin+ Ofloxacin+ Minocycline) regimens used in the treatment of leprosy: A cohort study in a National Reference Center in Brazil. *Front. Pharmacol.*, 15, 1346169, 2024 Mar 7.
76. Dutra, F.A., Araújo, M.G., Farah, K.D., Maciel, M.M., Lucas Junior, F.D., Araújo, S.D., Guedes, A.C., Hanseníase multibacilar em paciente transplantado renal: Relato de caso. *Braz. J. Nephrol.*, 37, 131–4, 2015 Jan.
77. Makarov, V., Riabova, O.B., Yuschenko, A., Urlyapova, N., Daudova, A., Zipfel, P.F., Möllmann, U., Synthesis and antileprosy activity of some dialkyldithiocarbamates. *J. Antimicrob. Chemother.*, 57, 6, 1134–8, 2006 Jun 1.
78. Jardim, G.A., Cruz, E.H., Valença, W.O., Resende, J.M., Rodrigues, B.L., Ramos, D.F., Oliveira, R.N., Silva, P.E., Silva, E.N., On the search for potential antimycobacterial drugs: Synthesis of naphthoquinoidal, phenazinic and 1, 2, 3-triazolic compounds and evaluation against Mycobacterium tuberculosis. *J. Braz. Chem. Soc.*, 26, 5, 1013–27, 2015 May.
79. Surette, M.D., Spanogiannopoulos, P., Wright, G.D., The enzymes of the rifamycin antibiotic resistome. *Acc. Chem. Res.*, 54, 9, 2065–75, 2021 Apr 20.
80. Andries, K., Verhasselt, P., Guillemont, J., Göhlmann, H.W., Neefs, J.M., Winkler, H., Van Gestel, J., Timmerman, P., Zhu, M., Lee, E., Williams, P., A diarylquinoline drug active on the ATP synthase of Mycobacterium tuberculosis. *Science*, 307, 5707, 223–7, 2005 Jan 14.
81. Ghobadi, E., Saednia, S., Emami, S., Synthetic approaches and structural diversity of triazolylbutanols derived from voriconazole in the antifungal drug development. *Eur. J. Med. Chem.*, 231, 114161, 2022 Mar 5.
82. Heller, M., Henrici, C., Büttner, J., Leube, S., Treske, I., Pospischil, P., Doll, M., Schanz, I., Hallier, A., Herrmann, E., Schmidt, M., SARS-CoV-2 neutralizing antibody therapies: An early retrospective cohort study of 26 hospitalized patients treated with bamlanivimab or casirivimab/imdevimab. *Int. J. Infect. Dis.*, 129, 260–5, 2023 Apr 1.
83. Weder, B., Mozaffari, M., Biedermann, L., Mamie, C., Moncsek, A., Wang, L., Clarke, S.H., Rogler, G., McRae, B.L., Graff, C.L., Ruiz, P.A., BCL-2 levels do not predict azathioprine treatment response in inflammatory bowel disease, but inhibition induces lymphocyte apoptosis and ameliorates colitis in mice. *Clin. Exp. Immunol.*, 193, 3, 346–60, 2018 Sep.
84. Potshangbam, A.M., Rathore, R.S., Nongdam, P., Discovery of sulfone-resistant dihydropteroate synthase (DHPS) as a target enzyme for kaempferol, a natural flavanoid. *Heliyon*, 6, 2, 2020 Feb 1.

85. Chahal, V., Nirwan, S., Kakkar, R., Combined approach of homology modeling, molecular dynamics, and docking: Computer-aided drug discovery. *Phys. Sci. Rev.*, 4, 10, 20190066, 2019 Oct 25.
86. Aamir, M., Singh, V.K., Dubey, M.K., Meena, M., Kashyap, S.P., Katari, S.K., Upadhyay, R.S., Umamaheswari, A., Singh, S., *In silico* prediction, characterization, molecular docking, and dynamic studies on fungal SDRs as novel targets for searching potential fungicides against Fusarium wilt in tomato. *Front. Pharmacol.*, 9, 1038, 2018 Oct 22.
87. Swain, S.S., Paidesetty, S.K., Dehury, B., Sahoo, J., Vedithi, S.C., Mahapatra, N., Hussain, T., Padhy, R.N., Molecular docking and simulation study for synthesis of alternative dapsone derivative as a newer antileprosy drug in multidrug therapy. *J. Cell. Biochem.*, 119, 12, 9838–52, 2018 Dec.
88. Bhardwaj, R. and Thounaojam, A.S., Employing advanced computational drug discovery techniques to identify novel inhibitors against ML2640c protein: A potential therapeutic approach for combatting leprosy. *Mol. Diversity*, 29, 2, 1229–43, 2025 Apr.
89. da Rocha, N.P., de Souza, A., Yukuyama, M.N., Barreto, T.L., Macedo, L.D., Löbenberg, R., de Araújo, G.L., Ishida, K., Bou-Chacra, N.A., Highly water-soluble dapsone nanocrystals: Towards innovative preparations for an undermined drug. *Int. J. Pharm.*, 630, 122428, 2023 Jan 5.
90. Sgaragli, G., Frosini, M., Saponara, S., Corelli, F., Human tuberculosis. III. Current and prospective approaches in anti-tubercular therapy. *Curr. Med. Chem.*, 23, 21, 2245–74, 2016 Jun 1.
91. Krismawati, H., Harianja, M., Oktavian, A., Bøgh, C., Ataupah, M.R., Laiskodat, R.D., Pongtiku, A., Geluk, A., Baird, J.K., Hamers, R.L., Soebono, H., Challenges associated with dapsone for leprosy treatment in Indonesia-urgent need for access to alternative antimicrobial drugs. *Lancet Reg. Health-Southeast Asia*, 34, 2025 Mar 1.
92. Pardillo, F.E., Burgos, J., Fajardo, T.T., Crux, E.D., Abalos, R.M., Paredes, R.M., Andaya, C.E., Gelber, R.H., Rapid killing of M. leprae by moxifloxacin in two patients with lepromatous leprosy. *Leprosy Rev.*, 80, 2, 205–9, 2009 Jun 1.
93. Kimta, N., Majdalawieh, A.F., Nasrallah, G.K., Puri, S., Nepovimova, E., Jomova, K., Kuča, K., Leprosy: Comprehensive insights into pathology, immunology, and cutting-edge treatment strategies, integrating nanoparticles and ethnomedicinal plants. *Front. Pharmacol.*, 15, 1361641, 2024 May 16.

8

Heterocyclic Compounds as Antihistamine Agents

Avinash Jha[1], Saurabh Sharma[1], Deepika Paliwal[1*], Aman Thakur[2] and Saurabh Srivasatva[3]

[1]School of Medical & Allied Sciences, Galgotias University, Greater Noida, India
[2]Adarsh Vijendra Institute of Pharmaceutical Sciences Gangoh Saharanpur, Uttar Pradesh, India
[3]Department of Pharmaceutics, KPJ Healthcare University, Nilai, Malaysia

Abstract

Histamine is essential for brain activity, immunological responses, and allergic reactions, which is a naturally occurring substance in the human body. But an excess of histamine can cause a number of illnesses, such as inflammation and allergies. This chapter studies the function of histamine in the human body and how antihistamines help in the treatment of histamine-related disorders. It highlights the efficacy and adverse effects of first- and second-generation H1 antihistamines and also explains the differences between them. Histamine receptors (H1, H2, H3, and H4) and their function in drug development are also provided in this chapter. The importance of heterocyclic compounds is focused in this chapter and the development of improved medications as they form the basis for many effective antihistamines. As they are important building blocks for antihistamine medicines, important chemical structures such as imidazole, piperazine, quinazoline, piperidine, and ethanolamines are investigated. By understanding these molecules, Scientists can develop safer and more effective drugs to treat histamine-related illnesses. This chapter offers a comprehensive and lucid explanation of histamine, its receptors, and the chemistry underlying antihistamine medications for investigators, learners, and practitioners in medicinal chemistry and pharmacology. It provides insight into the ongoing efforts to improve antihistamine treatments by integrating biological and pharmacological perspectives.

***Keywords*:** Histamine, antihistamine agents, heterocyclic, imidazole, pyrazole

**Corresponding author*: deepika14paliwal@gmail.com

Divya Bajpai Tripathy, Priyanka Chhabra and Anjali Gupta (eds.) Heterocyclic Compounds as Pharmaceutical Ingredients, (269–306) © 2026 Scrivener Publishing LLC

8.1 Introduction

8.1.1 Overview of Histamine

Histamine is an important biogenic amine and is mostly found in mast cells (MCs), basophils, B cells of the gastrointestinal mucosa, and neurons within the central nervous system (CNS) [1]. Many human diseases are significantly impacted by it. Urgent-type (type I) allergies are characterized by enormous IgE-induced MC that releases histamines, which triggers the activation of the H1 receptor (H1R) and results in conjunctivitis and rhinitis with the main symptoms of swelling (accumulation of fluid in the skin), erythema (reddening of the skin), and pruritus (itching) [2]. L-Histidine is changed into histamine by an enzyme known as histidine decarboxylase. The physiological effects of histamine as a neurotransmitter and local mediator are mediated by four distinct kinds of histamine receptors (HRs), which belong to the group of G protein–coupled receptors (GPCRs) [3]. Histamine is an essential mediator in the decoding of nociceptive information, having a nociceptive impact in the peripheral nervous system and an antinociceptive impact in the CNS [4].

Histamine is released by MCs and basophiles as part of the localized immune response to invasive infections shown in Figure 8.1. Most histaminergic neurons are located in the rear hypothalamus, the part of the brain that regulates wakefulness, and hypersomnia that is caused by injuries in

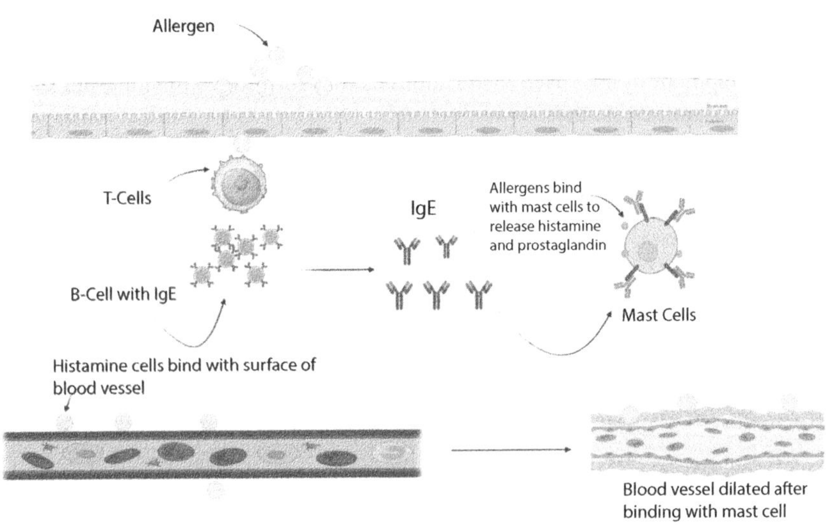

Figure 8.1 Release of histamine in response to an allergen, involving B cells, IgE antibodies, mast cells, and blood vessel dilation.

this area. Scientists have shown that the acidification activates histaminergic neurons, which are critical regulators in the awake state because they are responsive to carbon dioxide. Therefore, a range of hypothalamic functions, including hormone release, energy balance, and sleep/wake cycles, can be impacted by either normal or pathological changes in pH [5].

Histaminergic neurons, enterochromaffin-like cells, basophils, and MCs all exhibit higher rates of histamine production, which is stored in specific granules and released upon degranulation. Other cell types (including dendritic cells [DCs], T cells, and certain nonimmune cells) do not retain histamine, although they do produce it more slowly when needed. Nobody is free of histamine or MCs at the moment [6]. In the CNS, histamine is metabolized by histamine N-methyltransferase (HNMT), but not by diamine oxidase (DAO), which is not present in the CNS. HNMT is often found inside glia, neurons, hematopoietic cells, and epithelial cells [7, 8]. Histamine in the environment is broken down by the enzyme histamine-*N*-methyltransferase, which is expressed in the cells lining the respiratory system as shown in Figure 8.2. Research shows that histamine-*N*-methyltransferase blockers promote bronchoconstriction both *in vitro* and *in vivo* [9].

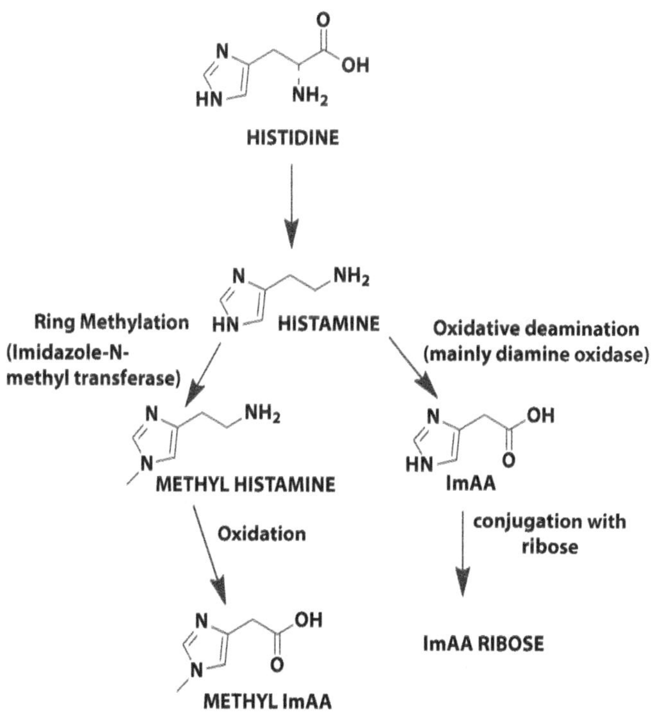

Figure 8.2 Metabolic pathway of histamine.

8.1.2 Histamine Role in Disease

Histamine is involved in the pathogenesis of several allergic diseases, such as atopic dermatitis, allergic rhinitis, and allergic asthma, by variably controlling T helper cells. Histamine can suppress the synthesis of T-helper 1 (T_H1) cytokines, such as interferon γ (IFN-γ), the monokine interleukin 12 (IL-12), and IL-2, while increasing the release of T-helper 2 (T_H2) cytokines, such as IL-5, IL-4, IL-10, and IL-13. The effective imbalance between T_H2 and T_H1 cells is thus regulated by histamine, which promotes a shift to T_H2 cells [10]. Stimulation of MCs by histamine is essential in many allergic disorders. Leukotrienes, cytokines, and chemokines may be released by histamine through H4 receptor (H4R) in human MCs generated from $CD34^+$ cord blood [11]. LPS stimulation amplifies the impact of histamine and 4-methylhistamine, which both independently stimulate the synthesis of IL-6 in mouse MCs. H4R antagonists can prevent this reaction, and H4R-deficient allergic mice do not experience it [12]. Recent studies have shown that histamine-induced activation of the H4R stimulates the generation of tumor necrosis factor α (TNF-α) in bone marrow–derived murine MCs and IL-4 and IL-5 in human cord blood MCs, which may all be involved in allergic inflammation [13].

8.1.3 Antihistamine Agents

A substance known as histamine is created when our bodies come into touch with allergens such as dust mites, pollen, pet dander, etc. Antihistamines shield us against allergies by preventing histamines from doing their job. They combat the manifestations of a variety of allergies, such as food allergies and hay fever. Blocking H1R-mediated responses such vasodilatation, sneezing, and itching, antihistamines function as competitive antagonists of histamine at H1 HRs [14].

8.1.3.1 H1 Antihistamine

Previously referred to as H1R antagonists or H1R blockers, H1-As are a class of pharmaceuticals that prevent the body's H1Rs from suppressing proinflammatory pathways. In summary, H1-As are classified into two functional groups [15], namely, first generation (G1) and second generation. The main distinction between the two is that G1 H1-As, which consist of triprolidine, hydroxyzine, and ketotifen, are able to traverse the blood–brain barrier (BBB) and prevent the expulsion of histamine-related neurotransmitters, which causes sedation and impacts on the CNS, including cognitive, memory, and psychomotor functions. In addition, G1 H1-As have been linked to general adverse reactions such as dry mouth, urine retention, peripheral vasodilatation, and greater hunger, as well as cardiac adverse reactions, such as a higher chance of QT interval, a longer duration, tachycardia, and cardiac arrhythmia, and even hazards related to an excessive dose or assault such as severe CNS depression, delirium, and coma [16]. Conversely, G2 H1-As (usually G1 H1-As metabolites, including acrivastine, cetirizine, fexofenadine, and loratadine) show only little recorded heart toxicity and overdose-related adverse effects, along with mild BB crossing and the ensuing CNS-related consequences.

8.1.3.1.1 First-Generation H1 Antihistamine

Unlike second-generation antihistamines, which only bind with peripheral receptors for histamine, first-generation antihistamines easily cross the BBB through the CNS and block H1Rs, which produces a unique pharmacological and unfavorable influence pattern [17].

Probably, the category of drugs to relieve urticaria and allergic rhinoconjunctivitis the majority often used is H1 antihistamines [18]. Considering 1946, diphenhydramine (Benadryl), the most renowned first-generation AH, has been on the market. This medication does not meet the strict precautions and efficacy requirements nowadays because it was released prior to existing certification criteria [19].

Skin, intranasal, and ocular formulations of first-generation anti-H1 antihistamines, such as promethazine, diphenhydramine, and clemastine, can be used topically, orally, and also by injection [20]. Following oral treatment, the majority are readily ingested from the digestive system; they attach to 70% to 97% of plasma proteins, undergo liver metabolism, and are mostly eliminated in the urine within 24 hours of ingestion. Within 30 to 60 minutes, the therapeutic effect starts to manifest, peaks in 1 to 3 hours, and typically lasts for 4 to 6 hours. Chlorpheniramine and hydroxyzine, for example, have a half-life of more than 20 hours in adults and shorter time in children, who metabolize these medications more rapidly [21]. The drug's primary, if not exclusive, function in first-generation molecules is binding to the H1R; in second-generation molecules, this method is refined and extended.

A number of first-generation compounds' therapeutic properties now severely restrict their therapeutic use. The first-generation anti-H1 antihistamines' most frequent negative consequences are sleepiness, sedation, and impaired focus and response moments, which are brought on by their elevated liposolubility and ease of way across the BBB [22].

Accordingly, the most permeable antihistamines in the brain are hydroxyzine, chlorphenamine, and ketotifen, according to a research on cerebral histamine H1R occupancy using positron emission tomography [23].

Anti-H1 antihistamines are mainly used for the treatment of hypersensitive indicators with prevalent exudative, as well as irritative neurogenic characteristics; however, depending on the significance of histamine's function in clinical manifestations as well local health authority regulations, antihistamine efficacy varies among different pathologies [24, 25]. Academic performance was unaffected when adolescents with allergic rhinitis are managed with a second-generation anti-H1 antihistamine (such as loratadine) as opposed to a first-generation compound (such as diphenhydramine) [26]. Several clinical studies back the application of anti-H1 antihistamines to treat cutaneous sores and pruritus in mastocytosis [27], contact dermatitis, allergic responses to insect stings, and venom allergies in hymenoptera [28]. Antihistamines have not been shown to be effective in managing the ordinary chill in nonatopic patients or in preventing recurring pulmonary illnesses [29].

Lastly, several first-generation anti-H1 antihistamines (Figure 8.3) are applied at modest quantities to avoid or alleviate sleep deprivation, whereas others (promethazine, hydroxyzine, and diphenhydramine) are utilized in mixture with additional medications to provide relaxation and pain relief, as well as to prevent motion sickness [30].

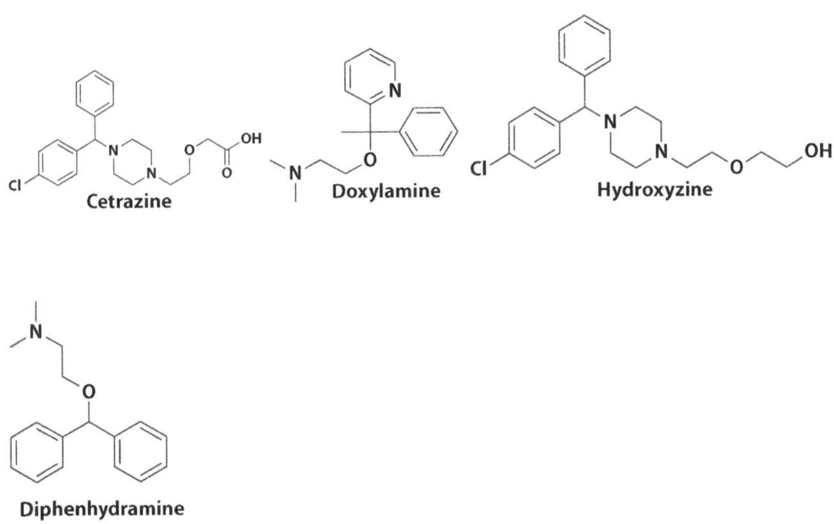

Figure 8.3 First-generation H1 antihistamine.

8.1.3.1.2 Second-Generation H1 Antihistamine

Second-generation H1 antihistamines (Figure 8.4), such as desloratadine, mizolastine, fexofenadine, cetirizine, loratadine, and levocetirizine, have a much lower affinity for muscarinic cholinergic, α-adrenergic, and 5-hydroxytryptaminergic receptors and have poorer CNS penetration because of their poor lipid solubility, relatively high molecular weight, and similarity to the P-glycoprotein escape pump. Fexofenadine does not tend to occupy CNS H1Rs, whereas cetirizine has a 30% chance of doing so.

Because second-generation H1 antihistamines do not cause sedative, antimuscarinic, anticholinergic, or cognitive and psychomotor performance deficits, they should be the preferred medication for treating these illnesses in place of first-generation H1 antihistamines. Second-generation H1 antihistamines were far less likely to cause adverse responses than first-generation drugs. At the manufacturer-recommended doses, second-generation H1 antihistamines significantly reduce CNS activity compared to first-generation H1 antihistamines. Furthermore, they have no anticholinergic or antimuscarinic properties. Upon consuming cetirizine or loratadine, there have been isolated cases of hepatitis, exacerbations of preexisting urticaria, and fixed-drug eruptions. The fast element of awaited reversal potassium current (IKr) is blocked by two early second-generation H1 antihistamines that are not anymore regulated: astemizole and terfenadine. By extending the QT interval, each of these substances has the

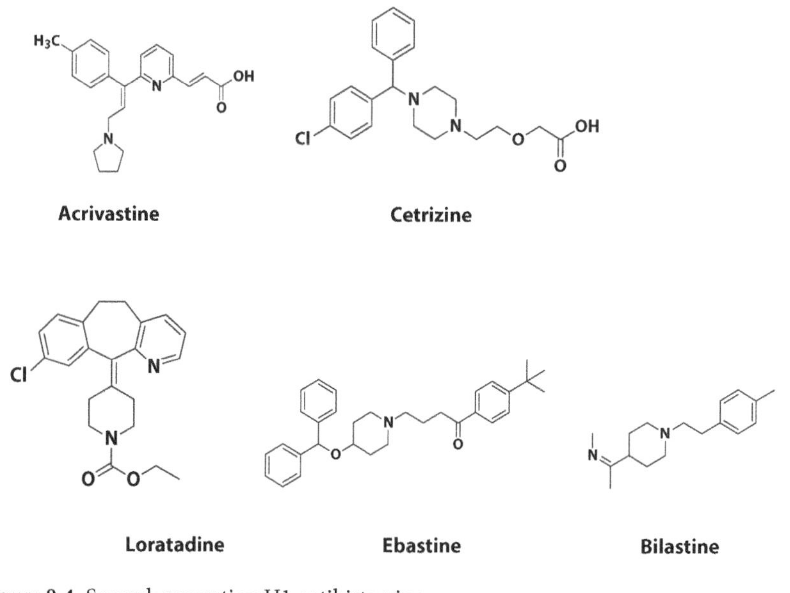

Figure 8.4 Second-generation H1 antihistamine.

potential to result in cardiac arrhythmias, including atypical ventricular tachycardias with a torsade de pointes and ventricular fibrillation.

At physiological and supratherapeutic dosages, new second-generation H1 antihistamines, including cetirizine, desloratadine, loratadine, and fexofenadine, are devoid of possible cardiac adverse effects and have a 1:1000 potency in inhibiting the IKr current. The evidence that is now available suggests that the class of second-generation H1 antihistamines does not affect their capacity to cause ventricular arrhythmias [31]. The second-generation antihistamines work quickly and are quite effective and nonsedative when taken as directed [32].

8.1.4 Role of Heterocyclic Compound in Drug Development

Heteroatoms are an integral part of many active pharmaceutical ingredients and excipients; from a significance perspective, it makes no difference if these are genuine heterocycles or isosterically or bioisosterically substituted carbons or carbon substructures in aliphatic structures. Research shows that heterocycles are present in almost 85% of all chemical substances that are physiologically relevant. Developments in artificial techniques, such as metal-catalyzed cross-coupling and hetero-coupling procedures, which provide quick admission to a large range of functionalized heterocyclic ring, are linked to the growing occurrence of different heterocycles in medications [33].

The scope and complexity of this issue are illustrated in two academic publications. The first section provides a comprehensive and up-to-date overview of cinnoline derivatives with a variety of pharmacological characteristics, such as antifungal, antibacterial, anti-inflammatory, analgesic, antimalarial, anticancer, and anxiolytic effects [34]. The study and applications of nucleotide analogs, especially second-generation bridging nucleic acids, are covered in the second section [35].

Numerous studies have examined and identified benzoimidazoles as bioactive substances with significant therapeutic promise; one such instance involves bis-benzimidazole derivatives, which are novel and specific antagonists of *Leishmania mexicana*'s arginase [36]. Analyzed as anticancer chemicals, pyrazoline derivatives are one of the most intriguing scaffolds identified as therapeutic substances [37].

Thiirane, sometimes referred to as ethylene sulfide or this cyclopropane, is an unstable, unsaturated three-membered sulfur heterocycle. Thiirane finds application in the manufacturing of adhesives, polymers, and other intermediates, such as 2-mercaptoethylamines. Furthermore, thiirane is a useful heterocyclic molecule with antibacterial and anticancer properties [38].

The four-membered heterocycles, sometimes referred to as β-lactams, are essential among various pharmacophores because of their significant medicinal qualities. Because β-lactamase is resistant to bacteria and has a diverse core structure, its creation is the main objective in medication research. Apart from its anti-infective qualities, β-lactam derivatives also exhibit a number of therapeutic actions, including antidiabetic and anti-inflammatory effects [39].

π-Rich or π-excessive heterocycles are five-membered heterocyclic ring that include heteroatoms (nitrogen, sulfur, and oxygen) and are very reactive in electrophilic substitution processes. Benzene-attached rings are referred to as pentathiophene, benzofuran, and indole, respectively [40]. Important natural compounds with indole rings include tryptophan, an important amino acid found in proteins, and indole 3-acetic acid, a plant hormone that controls plant growth.

Two nitrogen atoms make up quinoxalines, a significant six-membered heterocycle that starts the main structural patterns in a variety of organic chemical categories. Quinoxalines are an enzyme inhibitor, according to several investigations, and are also present in a number of artificial drugs, such as antifungal, anti-inflammatory, antiviral, and antibacterial ones [41].

8.2 Histamine Receptor

Histamine is a bioactive amine that functions as a neurotransmitter and signaling molecule. It uses its diverse biological effects by activating four types of membrane-bound receptors belonging to the aminergic GPCR family: H1, H2, H3, and also H4 as illustrated in Table 8.1. Histamine receptors were first differentiated into H1 and H2 in 1966 [42], and it was reported that some histamine-induced responses were unsympathetic, whereas others were inhibited by low doses of mepyramine (pyrilamine). A third HR subtype was cloned in 1999 and named H3 [43], and the fourth HR subtype was then reported in 2000, which was named H4 [44]. This marked a significant chapter in the history of histamine effects.

Human chronic conditions undoubtedly require an understanding of receptor regulation; HRs, including H1R, H2R, H3R, and H4R and their cellular dispersion, functional evaluation, and structural biology, along with messenger mechanisms; nonclassical histamine-binding sites, such as cytochrome P450; and histamine transporters; immune oversight by histamine *via* immunomodulation along with allergic inflammation; impacts of histamine in immune cells with regard to allergic diseases; significance of histamine in

Table 8.1 Types of histamine receptor [45–48].

Receptor type	Location	Functions	Signaling pathways	Antagonists	Used
H1 receptor	Central nervous system, smooth muscle, endothelium	Mediates allergic reactions, promotes wakefulness, induces bronchoconstriction and vasodilation	Gq/11 → PLC → IP3/DAG → Ca^{2+} release	• Cetirizine • Fexofenadine • Loratadine • Clemastine • Rupatadine • Azelastine • Diphenhydramine	• Sleep • Nausea disturbances • Allergic reactions
H2 receptor	Gastric parietal cells, heart, immune cells	Stimulates gastric acid secretion, increases heart rate, promotes vasodilation	Gs → AC → cAMP increase	• Cimetidine • Nizatidine • Famotidine • Ranitidine	• Peptic ulcer condition • Aspiration pneumonia • Ulcers o Gastroesophageal reflux disorder o Stress-related
H3 Receptor	Central nervous system, peripheral nervous system	Modulates neurotransmitter release, regulates appetite and cognition	Gi/o → inhibition of AC → decrease in cAMP	• Ciproxifan • Thioperamide • ABT-239 • Clobenpropit	• Narcolepsy • Schizophrenia • Attention-deficit/ hyperactivity disorder (ADHD) • Alzheimer's condition
H4 receptor	Bone marrow, immune cells, Gastrointestinal tract	Involved in immune responses, chemotaxis of eosinophils and mast cells	Gi/o → inhibition of AC → Decrease in cAMP	• Thioperamide • JNJ 7777120.	• Rheumatoid arthritis • Asthma • Allergic rhinitis—dermatitis • Itching

autoimmunity and allergic diseases, as well as malignancies; and, lastly, the relationship between histamine and cytokines during hematopoiesis.

8.2.1 H1 Receptor

The H1R is widely distributed throughout the body, being expressed in the heart, adrenal medulla, smooth muscle, sensory nerves, immune cells, hepatocytes, chondrocytes, epithelial cells, neutrophils, monocytes, DCs, eosinophils, B cells, and T cells. It is essential for mediating most of the postsynaptic effects of histamine in the CNS, where histamine binds to the transmembrane domains 3 and 5 of the H1R, causing smooth muscle contraction in the respiratory and gastrointestinal systems, as well as the activation of sensory nerves that causes itching and sneezing. The presence of H1Rs in the alive human head has been confirmed through the use of specific ligands, namely, [^{11}C]mepyramine and [^{11}C]doxepin [49, 50]. Extensive research has been conducted on H1Rs in blood vessels, as well as in smooth muscle tissues [51, 52]. Low histamine concentrations modulate action-potential discharge, which is essential for raising tension in smooth muscles that can produce muscular action potentials on their own, such the guinea pig ileum [53]. Evidence proposes that the contractile retort to histamine is facilitated by the mobilization of intracellular calcium (Ca^{2+}) through inositol 1, 4, 5-triphosphate pathways. Additionally, the effects of H1R stimulation have been observed in not-excitable smooth muscles, including those in the airways and vascular systems. Calcium (Ca^{2+}) is released from intracellular reserves as part of the first muscular response to H1R activation, for example, by hydrolyzing inositol phospholipids [54, 55]. In addition, stimulation of H1Rs causes vascular endothelial cells to undergo a number of cellular reactions, such as muscle contraction of the smooth muscle cells, changes in vascular permeability *via* prostacyclin-mediated mechanisms, platelet factor activation, producing nitric oxide, arachidonic acid along with derivatives, thromboxane, and more. Moreover, H1R activation boosts the ability of antigen-presenting cells to function, activates T_H1 lymphocytes, decreases humoral immunity, and increases the generation of IgE. It also improves the chemotaxis of neutrophils and eosinophils to inflammatory areas [56]. Therefore, to reduce allergy symptoms, H1R antagonists such promethazine, pyrilamine, fexofenadine, and diphenhydramine are commonly used. The investigation of H1Rs in human T lymphocytes has been conducted using [^{125}I]iodobolpyramine and has been found to elevate intracellular calcium levels ([Ca^{2+}]i) [57, 58]. Mice with no H1Rs react to antigens with strong systemic T-cell activation as well as effective B-cell responses [59].

Several years ago, it was discovered that the central medulla, which causes catecholamines to be released, is connected to H1Rs. As a result, histamine can stimulate the release of noradrenaline and adrenaline [60] and phosphorylate tyrosine hydroxylase, an enzyme essential to catecholamine biosynthesis, *via* a particular mechanism that facilitates the intracellular calcium release from developed bovine adrenal chromaffin cells [61].

A number of intracellular transcription factors, such as phospholipase C (PLC), inositol triphosphate (IP3), diacylglycerol (DAG), protein kinase C (PKC), calcium ions (Ca^{2+}), cGMP, phospholipase D, phospholipase A2, and nuclear factir κB, are also activated by H1R signaling as shown in Figure 8.5. When histamine simultaneously activates both H1 (H1R) and H2 receptors (H2R), it can cause clinical symptoms of anaphylaxis, such as headaches, flushing, tachycardia, hypotension, and flushing. Recent studies have linked H1R and H4R signaling to mitogen-activated protein kinase (MAPK) signaling pathways and cAMP accumulation, which contribute to the upregulation of proinflammatory gene expression [62]. Additionally, H1R activation is crucial for the development of T_H1 responses, whereas H2R is involved in regulating T_H2 responses. Mice that are genetically deficient in H1R ($H1R^{-/-}$) exhibit an enhanced T_H2 profile due to diminished T_H1 responses [63]. Moreover, in an experimental allergy model, H1R has been shown to play a vital role alongside histamine in facilitating the recruitment of T_H2 cells to areas of allergic lung inflammation. The H1R is implicated in the pathogenesis of allergic rhinitis, asthma, atopic dermatitis, conjunctivitis, and urticarial [64].

The functional analysis of H1Rs has greatly profited from the application of numerous effective and specific antagonists [65, 66]. H1R antagonists represent some of the earliest therapeutic agents in modern medicine, primarily due to their sedative properties. Initially developed as antiallergic medications, many of these drugs have since fallen out of favor. Histamine-induced inhibition of exploration behavior in the brain and the disturbance of the circadian cycle and locomotor activities are all linked to the H1R. As a result, "nonsedating" H1 blockers have been developed to prevent BBB bridging. Because they often make allergy symptoms worse rather than better, H1R stimulants are not widely accessible. Hisataprodifens (Histaprodifen is a potent and selective histamine H1-receptor agonist, recognized for its high activity and utility in pharmacological research.) are highly effective H1R agonists, surpassing histamine in their ability to activate H1R [67]. Additionally, the anti-inflammatory effects observed with high doses of H1R antagonists may be nonspecific, arising from the expulsion of histamine along with additional inflammatory mediators, including leukotrienes and platelet-activating factors, *via* basophils in response to specific H1R antagonists [68, 69].

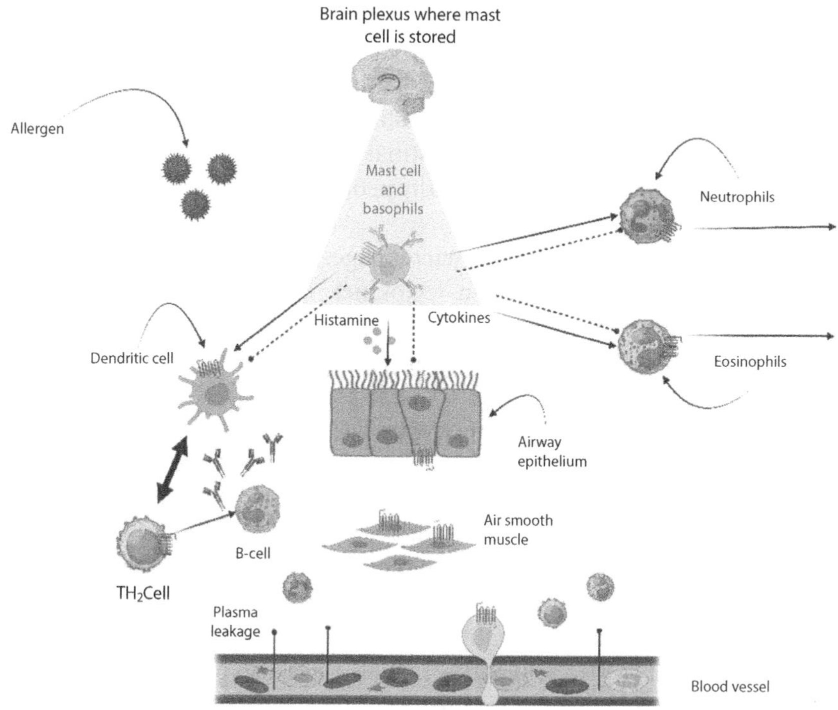

Figure 8.5 Mechanism of H1 receptor.

8.2.2 H2 Receptor

In addition to being found in the heart, the CNS, immunological cells, cardiomyocytes, DCs, and T and B cells, including the smooth muscles of the airways, the H2R is widely dispersed throughout the body. It has a long C-terminal tail and a distinctively short intracellular loop. Postsynaptic H2R receptors primarily use cyclic adenosine monophosphate (cAMP) and Gαs coupling to transfer signals [70]. Both cAMP-dependent and cAMP-autonomous signaling pathways, such as adenylate cyclase, c-Fos, c-Jun, PKC, and p70S6K, are activated by the H2R. Differences in cellular expression levels lead to multiple physiological impacts, such as the activation of hydrochloric acid secretion *via* parietal cells in the gastric mucosa, tranquility of smooth muscles in the vasculature along with airways, raised heart rates and elasticity, and immune-modulating actions *via* the suppression of basophils [71]. Furthermore, H2Rs are implicated in the inhibition of T_H1 and T_H2 immunological responses [72]. They also play a vital role in maintaining gut immunological homeostasis by modulating innate immune reactions to pathogen components. H2R deletion mice show iNKT-dependent lung inflammation because H2Rs promote invariant natural killer (NK) T-cell–reliant lung inflammation in a mouse model of asthma [73]. A possible reduction of natural immune responses controlled by H2R activation in the gastrointestinal tract is suggested by a study of H2R gene transcription among healthy people and those who have inflammatory bowel disease [74].

By preventing monocytes, macrophages, and MCs from producing IL-12, IFN-γ, and TNF-α cytokines, H2R activation also has anti-inflammatory effects by lowering T_H1 proliferation and antibody production [75]. Histamine's requirement for the H2R causes a rise in IL-10 secretion and a fall in IL-12 levels. Consequently, naive $CD4^+$ T cells were polarized toward the T_H2 phenotype by DCs with histamine maturation. According to this study, T_H2 cells promote the synthesis of IgE, which may lead to enhanced histamine secretion by MCs. Atopic diseases may worsen as a result of this effect, which could create a positive feedback loop [76]. Histamine (endogenous and exogenous) dramatically alters the immune system's natural reaction to pathogens through H2R [77]. Knockdown of $H2R^{-/-}$ mice induced abnormalities of immunity as well as stomach problems (decreased gastric acid output). There have also been reports of cognitive deterioration and aberrant nociception [78–80].

In the brain of guinea pigs, a comparable delivery has been noted. Irretrievable cataloguing has also been effectively demonstrated using [^{125}I] iodoazidopotentidine [81, 82]. The stimulation of H2Rs has been associated with the accumulation of cAMP or the activity of adenylyl cyclase in numerous tissues, including stomach cells, cardiac tissue, and the brain [83]. The physiological significance of histamine in stomach acid secretion is supported by studies showing that H2R antagonists impede this secretory process, demonstrating the substantial impact of H2Rs on the production of gastric acid as shown in Figure 8.6 [84, 85]. Elevated histamine levels in the heart tissues of many animal species can activate H2R and produce favorable chronotropic and inotropic impacts on atrial and ventricular tissues [86, 87]. Moreover, uterine muscle, vascular smooth muscle, and the airways have all been shown to exhibit H2R-mediated smooth muscle relaxation. H2Rs have the ability to block a number of immune system processes. Histamine release from MCs and basophils has been shown to be adversely regulated by H2Rs [88, 89]. Furthermore, H2Rs have been shown to play a part in preventing T-cell proliferation, antibody formation, cell-mediated cytolysis, and lymphocyte cytokine production [90].

H2Rs are linked to a number of other signaling pathways along with Gs coupling with adenylyl cyclase. In stomach parietal cells, for example, it has been demonstrated that activating H2Rs raises the intracellular level of calcium (Ca^{2+}) ions [91, 92]. In several cell systems, there is evidence of Gq linking to the production of PLC and consequent intracellular Ca^{2+} increase. Hepatoma-derived cells transfected with canine H2R cDNA and HL-60 cells have both shown a similar Ca^{2+} response to H2R activation [93]. Moreover, H2R activation causes CHO cells transfected with the rodent H2R to accumulate more cAMP and block P2U receptor-mediated arachidonic acids secretion [94]. However, it was shown that forskolin, PGE1, and 8-bromo-cAMP did not reproduce their impact on phospholipase A2 action, which is in charge of releasing arachidonic acid.

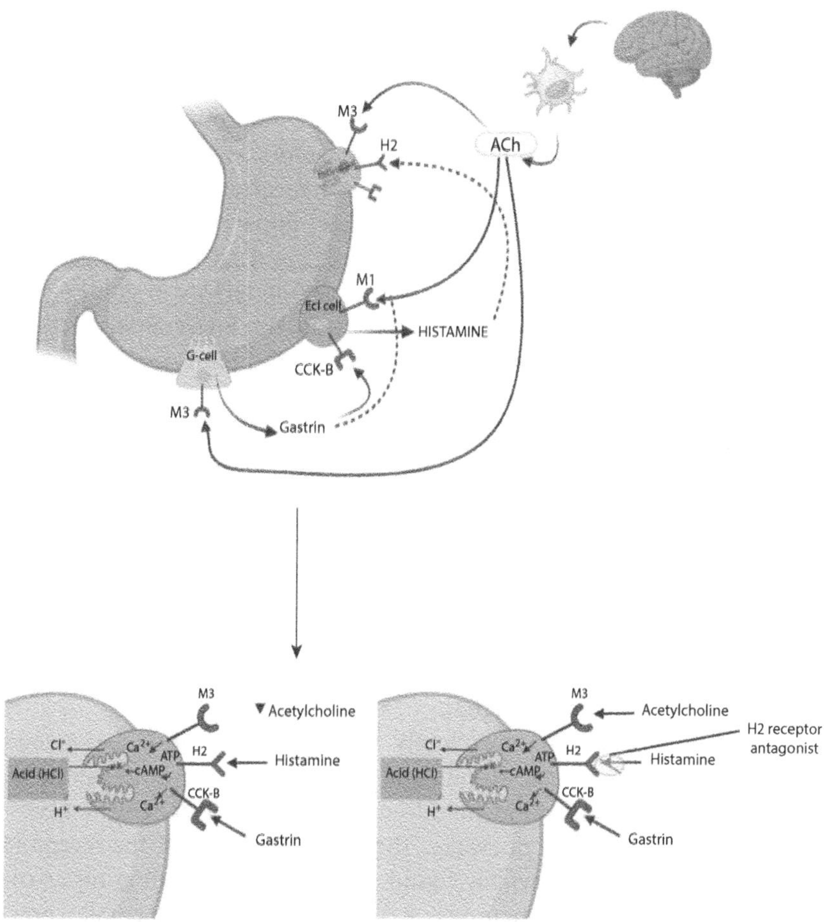

Figure 8.6 H2 receptor mechanism in gastric acid secretion and its inhibition by H2 receptor antagonists.

This suggests that the mechanism of action is independent of cAMP-mediated protein kinase A activities [95]. However, CHO cells transfected with a human H2R showed inhibitory effects of H2R activation on phospholipase A2 activities [96]. Therefore, the degree of receptor production or minute variations among clonal cell lines may influence these cAMP-independent actions.

8.2.3 H3 Receptor

Although it is also present in the peripheral nervous system, which includes the airways, cardiovascular system, gastrointestinal tract, and MCs, the H3 receptor is primarily expressed in the CNS (basal ganglia, hippocampus, and cortical areas). Numerous isoforms have been associated with distinct signaling pathways [97], which may indicate that different brain functions are selectively regulated [98]. H3 receptor (H3R) has been shown to regulate the emission of histamine along with other neurotransmitters in the CNS as well as peripheral and presynaptic receptors. Histamine's H3R controls both its own release and that of other neurotransmitters, including γ-aminobutyric acid (GABA), noradrenaline, dopamine, serotonin, and acetylcholine. By stimulating phospholipase A2 and subsequently increasing the secretion of endothelium-derived relaxing factors in guinea pigs, H3Rs on postganglionic cholinergic nerves in the lower airways counteract bronchoconstriction [99, 100]. The acetylcholine produced in a mouse's cortex, which regulates neurogenic inflammation by preventing cAMP production and Ca^{2+} buildup, is inhibited by the asymmetry of histamine through H3R [101]. The obese phenotype of H3R sensation mice indicates that H3R controls leptin release and insulin resistance in addition to a reduction in homeostatic energy, which is the cellular mechanism for coordinating the homeostatic regulation of energy expenditure (energy outflow) and food intake (energy inflow), as linked to the UCP1 and UCP3 genes [102]. Increased proinflammatory activity, pruritus (without MC participation), bronchoconstriction, and antigen-presentation ability are all linked to H3R expression [103].

Histaminergic neurons are closely related to neuromodulation and wakefulness. The constant stimulation of aminergic neurotransmitters, such as histamine, dopamine, noradrenaline, and acetylcholine, is necessary for the maintenance of awake. Three distinct kinds of the receptors for histamine (HRs) are widely found in the brain, including on astrocytes, blood vessels, and neurons. A natural state that resembles non-rapid eye movement sleep can be induced by positive allosteric inhibitors of receptors for GABA-A that act on histamine neuron in the posterior hypothalamus. By concentrating on the histaminergic and noradrenergic systems, we might better comprehend the biological sleep–wake circuitry and potentially produce more specialized sedatives [104]. In addition, histamine H3R inverse agonists and antagonists have potential for treating illnesses of the CNS. Histamine and/or HR agonists have the ability to downregulate the adverse impacts of microglia-mediated inflammation on neurons that produce dopamine when they are subjected to lipopolysaccharide challenges, according to an experimental investigation. Innovative therapeutic approaches for the management of neurodegenerative illnesses may be made possible by this research [105].

It has been investigated in a variety of tissues, including the heart, skin, meninges, and airways, how activation of the H3R significantly inhibits the release of neuropeptides from sensory C fibers, including calcitonin gene-related peptide and tachykinins [106]. Furthermore, it has been reported that H3Rs affect the effects of substance P, capsaicin, and acetylcholine in separated perfused rabbit lungs [107]. There is evidence that H3R activation can inhibit the release of neurotransmitter from guinea pig ileum and bronchioles' nonadrenergic–noncholinergic neurons. Additionally, compared to their efficacy in inhibiting the release of histamine in the rat cerebral cortex, the H3R antagonists betahistine along with phenylbutanoylhistamine showed noticeably less potency in the guinea pig ileum as inhibitions of H3R-mediated impacts on nonadrenergic–noncholinergic spread [108].

Although the H3R's signaling mechanisms are still primarily theoretical, there is mounting evidence that this receptor belongs to the GPCR family (Gi/o). It appears that when H3R is activated, the synthesis of cAMP is suppressed, Ca^{2+} levels rise, and the MAPK pathway is activated [109]. This evidence comes from ligand-binding tests that look at how guanine nucleotides affect H3R-agonist binding [110] in addition to how H3R agonists hinder H3R-antagonist binding [111, 112]. Studies using [35S] GTPgS coupling to rat central cortical membranes provided direct evidence for a functional interaction among H3R and G proteins [113]. R-(α)-methylhistamine and *N*-(α)-methylhistamine, as well as H1R- and H2R-antagonists (0.1 mM mepyramine or 10 mM titotidine), increased [35S]GTPgS binding in these membranes in a concentration-dependent manner (EC50 = 0.4 and 0.2 nM) [114]. Crucially, conditioning these membranes with the toxin from pertussis prevented this response, indicating an explicit coupling to either the Gi or Go proteins. Research on H3R signaling in human and guinea pig hearts has shown proof of pertussis toxin-sensitive G proteins in reaction to H3R activation [115, 116]. N-type calcium channels, which are essential for the voltage-dependent production of noradrenaline in the hearts of both humans and guinea pigs, appear to be inhibited by H3R activation. H3Rs may be linked to Go proteins, although several investigations have not shown a suppression of adenylyl cyclase function across different tissues and cell types.

8.2.4 H4 Receptor

Primarily expressed in the marrow of the bone and bloodstream leukocytes, the H4R is also found in the liver, thymus, the lungs, digestive system, nerves in the extremities, and central neurons in the hippocampus and cerebellum [117]. The majority of hematopoietic cells, such as MCs, eosinophils, basophils, DC-like monocytes, NK, and T cells (T_H17), have this receptor. The H4R's activation causes MCs to mobilize calcium through cAMP, which promotes the cells' movement in the direction of histamine. The receptor is also essential for regulating T and DC activity. Numerous nonhematopoietic systems are also regulated by the H4R, which is more robustly expressed in the dermis and epidermis, specifically in keratinocytes, chondrocytes, and intra-articular synovial cells.

In diseases such as atopic dermatitis, bronchial asthma, and chronic arthritis, the H4R is thought to be a prospective target for therapeutic approaches. This is ascribed to both its effect on regulatory T cells and the direct stimulation of effector cells inside the hematopoietic system. H4R agonists have been shown to inhibit the production of CCL2 mRNA into monocyte-derived Langerhans cell types and encourage their migration away from the epidermis [118]. H4R antagonist studies have demonstrated a substantial decrease in CCL17 and CCL22 production by atopic dermatitis patients' monocytes, with no effects seen in healthy subjects [119]. Additionally, the H4R has been linked to the development of pruritus, as evidenced by the utilization of H1 and H4R antagonists, which have been linked to a drop in pruritus in atopic dermatitis [120–122].

Mostly present on immune system cells, such as T cells, NK cells, eosinophils, MCs, DCs, and monocytes, the histamine H4R is a GPCR that is sensitive to pertussis toxin. Peripheral organs, such as the thymus, bone marrow, colon, spleen, and blood leukocytes, are notable for their expression, which is altered in response to inflammatory signals [123, 124]. The H4R has more affinity with histamine over the H1R, and when it is activated, intracellular calcium concentration rises and leukocyte chemotaxis to inflammation areas is facilitated by Gαi/o proteins [125]. Moreover, it has been discovered that the liver-expressed chemokine LEC/CCL16 is a nonhistamine endogenous activator of the H4R, which functions in concert with histamine and contributes to eosinophil migration [126].

The occurrence of the H4R in various human immune cell types, along with its chemotactic capabilities, underscores its significance in immunomodulation. Although there are interspecies variations in amino acid sequences, countenance levels, ligand interactions, and receptor stimulation, the similar dispersal across tissues indicates that this receptor likely serves analogous physiological functions in different species [127].

The human H4R gene is located on chromosome 18q11.2 and is linked to a seven-transmembrane G protein–related receptor. It encodes a 390-amino-acid protein. This gene and the H3R share a similar genomic architecture and 37% to 43% similarity, with a noteworthy 58% homology across the transmembrane domains. The three exons of the 21-kbp-long H4R gene are divided by two sizable introns that are more than 7 kb long. Significant differences exist between species, with sequence homology varying between 65% and 72%. The normal TATA and CAAT boxes were not found when the 5′ flanking domain was examined. Nonetheless, a number of potential regulatory elements that are connected to proinflammatory cytokine signaling networks are present in the promoter region. H4Rs are linked to Gi/o proteins, which trigger a number of signaling cascades, including as MAPK activation, enhanced calcium influx, and the suppression of forskolin-induced cAMP synthesis. Numerous H3R antagonists and agonists have been found to interact with the H4R, albeit with differing affinities, because of the similarities between the H3 and H4Rs. Notably, it has been demonstrated that R-α-methyl histamine, an H3R agonist, acts on H4R with hundreds of times less efficacy. Similar findings have been reported with clobenpropit, other H3R antagonist that exhibits agonistic activity on H4R, and thioperamide, a conventional H3R antagonist that also acts as an H4R antagonist, albeit with a much lower affinity [128–132].

8.3 Imidazole as Antihistamine Agents

Imidazoles are significant heterocyclic compounds [133] with key properties of many different types of medications. Imidazole is a chemical with a five-membered planar ring that dissolves in water and other polar solvents. Being able to place the hydrogen atom on either of the two nitrogen atoms allows it to exist in two conventional tautomeric forms. In order to increase its activity, the synthesis of the imidazole moiety and its functionalization at different positions are ongoing.

Dharmishtha *et al.* synthesized and designed a novel imidazole derivatives Compound 1 and Compound 2. All synthesized compounds were pharmacologically screened for H1-antihistamine ability *in vitro* using the chicken ileum. The potent medications' sedative effects were evaluated *in vivo* [134].

Iemura *et al.* [135] synthesized a Compound **3** and tested for H1-antihistaminic activity with the help of *in vitro* and *in vivo* models. Compound **3** exhibited greater potency and effectiveness BY demonstrated 39 times the potency toward chlorpheniramine maleate in HI-antihistaminic activity *in vivo*, it was selected for clinical testing. In contrast to most Hl-antihistamine medicines, which have structures that feature a double-aromatic unit coupled to an essential tertiary amino group with a chain, the compound's structure is notable because it provides only one aromatic unit bound to a basic nitrogen by a chain [135].

In their review, Zhang *et al.* [136] declare that there has been a lot of research done on the structurally basic imidazole ring (Compounds **4** and **5**) and that it is a crucial component of medicinal chemistry. Significant progress has been made in medicinal chemistry based on imidazoles [136].

Imidazole derivative compound 1-5 shown in Figure 8.7. With the exception of cimetidine, imidazole rings, a crucial biological building component found in histidine, are absent from the most widely used H1R and H2R antagonists, most likely because of their metabolic susceptibility. Numerous H3R ligands, such as imetit, thioperamide, and clobenpropit, also exhibit a strong affinity for the H4R because of the structural similarities between the two molecules. The potential therapeutic uses of specific H3R and H4R ligands have drawn increased interest in recent years [137, 138].

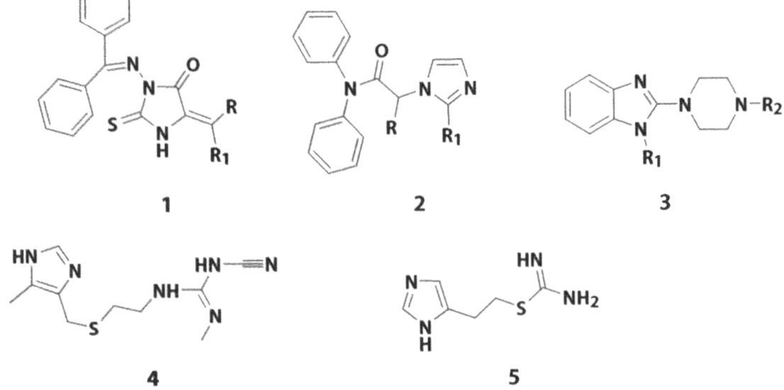

Figure 8.7 Imidazole derivatives (1–5) as antihistamine agents.

8.4 Piperazine as Antihistamine Agents

In medicinal chemistry, piperazine—also known as 1,4-diazacyclohexane—is an important heterocyclic scaffold. Two nitrogen atoms are positioned at opposite ends of a ring of six atoms that makes up this compound [139]. Piperazine, a heterocyclic nucleus, forms the core structure in a wide range of organic compounds known as piperazine derivatives, many of which have significant pharmacological characteristics. A membrane receptor protein that is activated by a neurotransmitter is known as a neurotransmitter receptor [140]. Neurotransmitters can cause actions at presynaptic and postsynaptic locations on target neurons and elicit whether a stimulating or a suppressive synaptic membrane potential [141].

A number of new piperazine derivatives of cyclizine (Compounds **6** and 7) were created and tested by Ahmadi *et al.* [142] for their ability to reduce inflammation in response to formalin-induced, histamine-induced rat paw edema (acute inflammation), and cotton pellet-induced granuloma development (chronic inflammation) [142].

A new nonreversible and specific long-acting histamine H1R antagonist called KAA-276 (Compound **8**) was discovered from guinea pigs and causes constriction of the ileum and trachea. Antihistaminic activity was displayed by 4-(diphenyl methyl)-1-piperazine derivatives (Compound **9**) with a terminal heteroaryl or cycloalkyl amide fragment [143, 144].

A novel series of derivatives of xanthone containing a moiety of piperazine (**10**) has been created by Waszkielewicz *et al.*, and their biological activity has been assessed. They underwent binding tests for 5-HT6, 5-HT7b, and 5-HT1A serotoninergic receptors in addition to α1 and β1 adrenergic receptors. For serotoninergic 5-HT1A receptors, 3-methoxy-5-{[4-(2-methoxyphenyl) piperazin-1-yl] methyl}-9-xanthen-9-one hydrochloride showed a substantially greater affinity than other compounds (Ki = 24 nM). Following oral administration, molecules with a higher bioavailability were obtained, thanks to the piperazine moiety [145].

Szczepańska *et al.* designed and synthesized phenoxyalkyl piperazine derivatives (**11**) and screened for their binding affinity toward human H3 HR. This study's results show that the 4-pyridyl-piperazino group can replace piperidine in H3R ligands as a new bioisosteric substitute [146].

Wang *et al.* in 2010 synthesized and designed a novel piperazine drug derivative named as (R)(-)-1-[(4-chlorophenyl) phenylmethyl] and investigated for antiallergic by choosing an *in vivo* model. The majority of these compounds showed noteworthy impacts on allergic irritation and asthma.

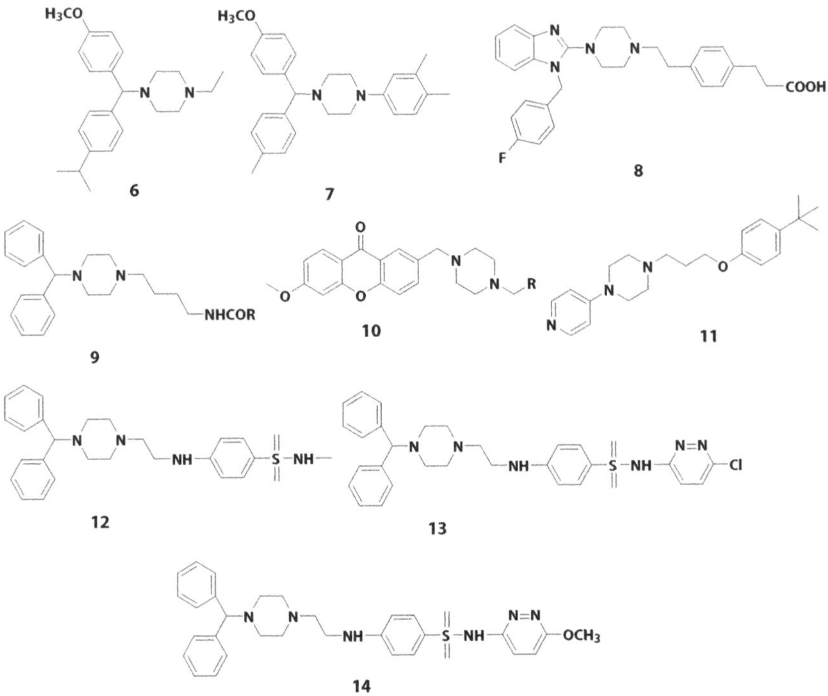

Figure 8.8 Piperazine derivatives (6–14) as antihistamine agents.

The positive control medication, levocetirizine, is less effective against allergic asthma than 3 of the piperazine derivatives (Compounds **12–14**) [147]. Piperazine derivative compound 6-14 shown in Figure 8.8.

8.5 Quinazoline as Antihistamine Agents

Due to their diverse spectrum of biological properties, the fused heterocycle classes quinazolines and quinazolinones are of great interest [148]. Numerous derivatives of substituted quinazoline and quinazolinone have a variety of biological activities, including weedicide, muscle relaxant, antitubercular, antimalarial, antidepressant, anticonvulsant, anticancer, antifungal, anti-inflammatory, antiviral, antiprotozoan, diuretic, antimicrobial, and many more [149, 150].

Alagarsamy *et al.* assessed a range of new 1,2,4-triazolo-[4,3-a]-quinazolin-5-(4H)-one derivatives **(15)** to evaluate their antihistaminic properties. The compound with a methyl substitution demonstrated better

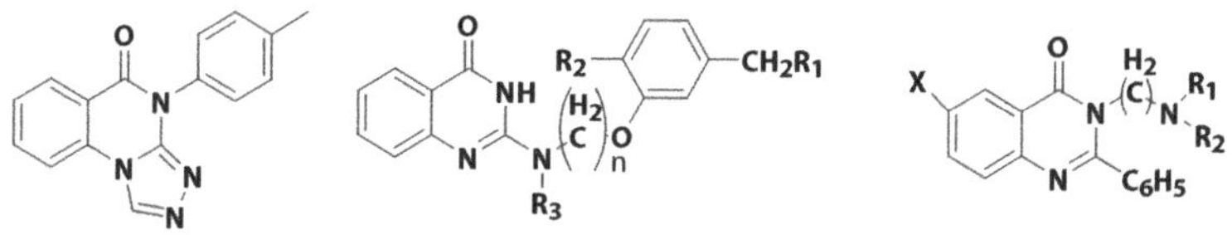

Figure 8.9 Quinazoline derivatives (15–17) as antihistamine agents.

activity than its unsubstituted counterpart, whereas the ethyl-substituted version maintained its effectiveness with enhanced lipophilicity. On the other hand, the propyl-substituted analog showed a decrease in activity because of its excessive lipophilicity. Adding chlorine to the methyl group led to a further reduction in activity [151].

Ogawa *et al.* screened new 4-quinazolinone compounds (**16**) against H_2 receptors. They used guinea pig heart tissue and the Furchgott technique to measure this. The compound with a phenyl group attached worked best and even better than ranitidine, a well-known H_2 blocker. Compounds with thiazole and furan groups worked okay, but not as well. The one with an imidazole group did not work very well at all. This tells us that adding a phenyl ring to the quinazolinone structure seems to make the best H_2 blocker. So, researchers decided to focus on making more compounds with phenyl groups to try to create even stronger H_2 blockers [152].

Raju *et al.* screened 3-(N,N-dialkylamino)alkyl derivatives of dihydroquinazolinone (17) for H_1-antihistaminic activity. Results revealed that most of the compounds inhibited histamine-induced contractions, whereas some compounds inhibit partial inhibition The surmountable nature of this antagonism suggests reversible binding to the H_1 receptor [153]. Quinazoline derivatives compound 15-17 shown in Figure 8.9.

8.6 Piperadine as Antihistamine Agents

A common molecular fragment seen in both natural goods and pharmaceutically active substances is the *N*-heterocyclic piperidine ring. Over the course of a relatively recent decade, thousands of piperidine substances have been identified in preclinical and clinical research [154]. As a result, the synthesis of functionalized piperidines has been thoroughly investigated, with chiral and multifunctionalized derivatives currently of special interest due to synthetic approaches [155]. Creation of disubstituted piperidines with pharmacological activity, a particular subclass of functionalized piperidines, has attracted significant interest due to their diverse therapeutic potential and structural versatility. In addition to being present in other plants, piperidines may also be produced from black pepper [156]. Numerous medications, including opioids, neuroleptics, antipsychotics, and vasodilators, belong to this class of compounds.

A drug called loratadine a piperidine derivative shown in Figure 8.10 (**18**) is used to treat and control urticaria and allergic rhinitis. It belongs to the class of second-generation antihistamines. As a useful treatment and management tool for urticaria and allergic rhinitis, loratadine's indications, activities, and contraindications are reviewed in this exercise. A piperidine histamine H1R antagonist, loratadine has antiallergic qualities without causing drowsiness [157]. In 1998, the Food and Drug Administration approved an appeal to make the medication available over-the-counter. The product was made accessible over-the-counter by the original manufacturer in 2002.

A second-generation, nonsedating, long-acting tricyclic antihistamine (piperidine derivative), loratadine selectively inhibits peripheral histamine H1Rs. Loratadine preferentially inhibits H1Rs, which are mostly found on immunological cells, gastrointestinal tract cells, vascular endothelial cells, and respiratory smooth muscle cells. Loratadine is a competitively histamine inhibitor that does not penetrate the BBB, in contrast to first-generation antihistamines such as diphenhydramine. As a result, it has no effect on CNS neurons, which checks for sedation or somnolence during the day [158].

By binding to H1Rs in many cells, loratadine reduces smooth muscle tone (bronchodilation), vascular permeability (preventing edema and flushing), and peripheral nociceptive receptor stimulation (reducing pain and itch). Second-generation antihistamines, such as loratadine, can suppress MC and basophil histamine release at high concentrations.

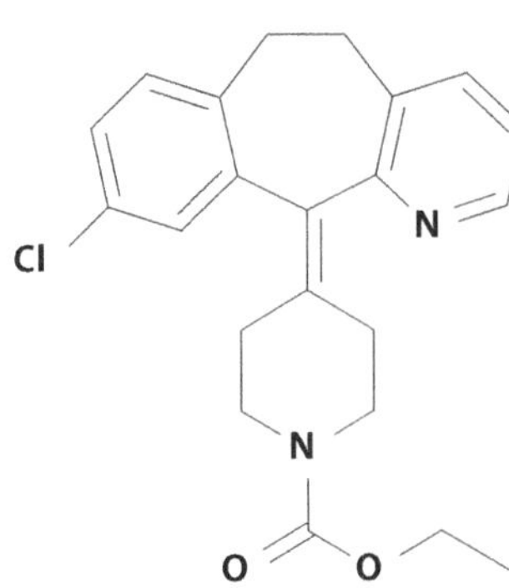

Figure 8.10 Piperidine derivative (18) as antihistamine agent.

This reduces ICAM-1 expression in epithelial cells and prevents type 1 hypersensitivity reactions, such as hay fever, urticaria, pruritus, and edema [159]. Second-generation antihistamines have also been shown to have anti-inflammatory qualities.

8.7 Ethanolamines as Antihistamine Agents

Certain bacteria can use ethanolamine, a substance that is easily extracted from cell membranes, as a source of carbon and/or nitrogen. It is believed that the development of a microcompartment aids in the retention of the volatile precursor acetaldehyde in ethanolamine use, preventing the loss of this carbon source and shielding the cell from any potential harmful effects [160, 161]. As an over-the-counter drug, diphenhydramine a ethanolamine derivative shown in Figure 8.11. (19) is a first-generation antihistamine used to treat and prevent a number of ailments, including vertigo, urticaria, sleeplessness, motion sickness, pruritus, and dystonias. Although this is an off-label usage of the drug, it also has local anesthetic qualities for patients who are allergic to other, more widely used local anesthetics. It can also be used off-label to treat oral mucositis.

Although it has different modes of action, diphenhydramine primarily functions by antagonistically binding to the H1R [162]. Increased vascular permeability, increased vasodilation leading to flushing, reduced atrioventricular node conduction time, activation of airway sensory nerves causing coughing, smooth muscle contraction of the bronchi and gastrointestinal tract, and eosinophilic chemotaxis fostering the allergic immune response are just a few of the many actions that occur when the H1R activated in these tissues. As a first-generation antihistamine, diphenhydramine easily penetrates the BBB and inversely frets the H1 CNS receptors, which suppresses the medullary cough center and causes drowsiness [163]. Finally, diphenhydramine produces local anesthetic effects by blocking intracellular sodium channels.

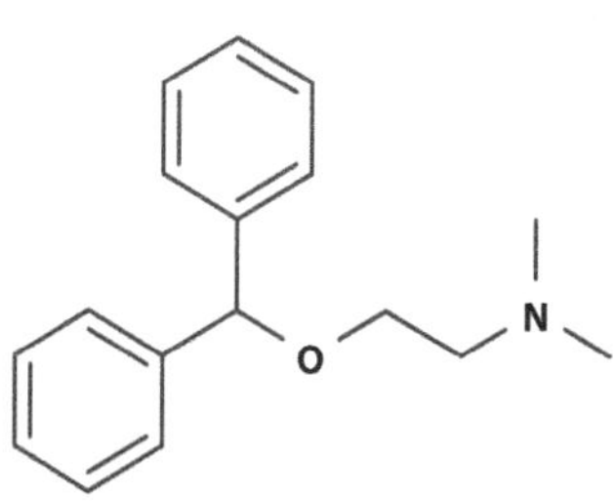

Figure 8.11 Ethanolamine derivative (19) as antihistamine agent.

8.8 Conclusion

Histamine is a naturally occurring chemical that is important for various bodily processes, including brain activity, stomach acid secretion, and allergic reactions. It functions by binding to four distinct HR types (H1, H2, H3, and H4); each of these HRs has a unique physiological role. When it functions excessively or improperly, histamine can cause several diseases, including allergies, asthma, ulcers, and even certain issues with the brain. Scientists have discovered medications known as antihistamines to inhibit the effects of histamine. These medicines work by preventing histamine from binding to its receptors. H1 antihistamines are the most often utilized, which are separated into first and second generation. The more recent second-generation ones are more selective, and they do not make individuals drowsy, whereas the first-generation ones can. The significance of the creation of antihistamine medications from heterocyclic compounds is also covered in this chapter. Imidazole, piperazine, quinazoline, piperidine, and ethanolamines are the compounds that have shown capability in the development of strong medications that can target particular HRs. These structures help in improving the medication's potency and safety for the body. Researchers are using frontline methods and improving chemical structures to produce antihistamines that are even safer and more potent. There is an expectation that many histamine-related disorders may be treated more effectively in the near future, thanks to the thorough understanding of each HR and the ongoing study for better compounds. This chapter provides an upfront but detailed review of histamine, its receptors, and associated illnesses and how various heterocyclic compounds are assisting in the development of potent antihistamine medications that might help in improving human health.

AI disclosure statement: The author acknowledges the use of AI-based tools for minor language editing or grammatical correction. However, all intellectual contributions, analyses, and conclusions presented in this chapter are entirely the author's own.

References

1. Li, L., Liu, R., Peng, C., Chen, X., Li, J., Pharmacogenomics for the efficacy and side effects of antihistamines. *Exp. Dermatol.*, 31, 7, 993–1004, 2022 Jul, doi: 10.1111/exd.14602. Epub 2022 May 17. PMID: 35538735.

2. Seifert, R., Strasser, A., Schneider, E.H., Neumann, D., Dove, S., Buschauer, A., Molecular and cellular analysis of human histamine receptor subtypes. *Trends Pharmacol. Sci.*, 34, 1, 33–58, 2013 Jan, doi: 10.1016/j.tips.2012.11.001. Epub 2012 Dec 17. PMID: 23254267; PMCID: PMC3869951.
3. Hirasawa, N., Expression of histidine decarboxylase and its roles in inflammation. *Int. J. Mol. Sci.*, 20, 2, 376, 2019 Jan 16.
4. Khalilzadeh, E., Azarpey, F., Hazrati, R., Vafaei Saiah, G., Evaluation of different classes of histamine H1 and H2 receptor antagonist effects on neuropathic nociceptive behavior following tibial nerve transection in rats. *Eur. J. Pharmacol.*, 834, 221–229, 2018, https://doi.org/10.1016/j.ejphar.2018.07.011.
5. Sharma, A., Menon, P.K., Patnaik, R., Muresanu, D.F., Lafuente, J.V., Tian, Z.R., Sharma, H.S., Novel Treatment Strategies Using TiO 2 -Nanowired Delivery of Histaminergic Drugs and Antibodies to Tau With Cerebrolysin for Superior Neuroprotection in the Pathophysiology of Alzheimer's Disease. *Nanomed. Cent. Nerv. Syst. Inj. Repair*, 123–165, 2017, doi:10.1016/bs.irn.2017.09.002.
6. Wong, G.W., Zhuo, L., Kimata, K., Lam, B.K., Satoh, N., Stevens, R.L., Ancient origin of mast cells. *Biochem. Biophys. Res. Commun.*, 451, 314–318, 2014.
7. Yoshikawa, T., Naganuma, F., Iida, T., *et al.*, Molecular mechanismof histamine clearance by primary human astrocytes. *Glia*, 61, 905–916, 2013.
8. Schwelberger, H.G., Feurle, J., Houen, G., Monoclonal antibodies for human and porcine histamine N-methyltransferase (HMT)facilitate protein expression and localization studies. *Inflamm. Res.*, 66, 67–77, 2017.
9. Patel, R.H. and Mohiuddin, S.S., Biochemistry, Histamine, in: *StatPearls*, StatPearls Publishing, Treasure Island (FL), 2023, PMID: 32491722.
10. Jutel, M. and Akdis, C.A., Histamine as an immune modulator in chronic inflammatory responses. *Clin. Exp. Allergy*, 37, 308–10, 2007, 10.1111/j.1365-2222.2007.02666.x.
11. Jemima, E.A., Prema, A., Thangam, E.B., Functional characterization of histamine H4 receptor on human mast cells. *Mol. Immunol.*, 62, 19–28, 2014, 10.1016/j.molimm.2014.05.007.
12. Desai, P. and Thurmond, R.L., Histamine H(4) receptor activation enhances LPS-induced IL-6 production in mast cells *via* ERK and PI3K activation. *Eur. J. Immunol.*, 41, 1764–73, 10.1002/eji.201040932, 2011.
13. Zhang, B., Alysandratos, K.D., Angelidou, A., Asadi, S., Sismanopoulos, N., Delivanis, D.A., *et al.*, Human mast cell degranulation and preformed TNF secretion require mitochondrial translocation to exocytosis sites: Relevance to atopic dermatitis. *J. Allergy Clin. Immunol.*, 127, 1522–31.e8, 2011, 10.1016/j.jaci.2011.02.005 [Europe PMC free article] [Abstract] [CrossRef] [Google Scholar].
14. Antihistamines, Meyler's Side Effects of Drugs, pp. 606–618, 2016, doi:10.1016/b978-0-444-53717-1.00314-0.

15. Simon, F.E., Estelle, R., Simons Keith, J., H1antihistamines: Current status and future directions. *World Allergy Organ. J.*, 1, 145–55, 2008.
16. Mantha, K.B., An opinion on H1-antihistamines as a potential avenue for endometriosis management, https://doi.org/10.1016/j.xagr.2023.100274.
17. Farzam, K., Sabir, S., O'Rourke, M.C., Antihistamines. [Updated 2023 Jul 10], in: *StatPearls [Internet]*, StatPearls Publishing, Treasure Island (FL), 2024 Jan, Available from: https://www.ncbi.nlm.nih.gov/books/NBK538188/.
18. Simons, F.E. and Simons, K.J., Histamine and H1 antihistamines, in: *Middleton's allergy principles and practice*, J.N. Adkinson, *et al.*, (Ed.), pp. 1503–33, Elsevier Saunders, Philadelphia, 2014.
19. Sanchez-Borges, M., Ignacio Ansotguei. Second generation antihistamines: An update. *Curr. Opin. Allergy Clin. Immunol.*, 19, 4, 358–64, 2019.
20. Parisi, G.F., Licari, A., Papale, M., Manti, S., Salpietro, C., Marseglia, G.L., Leonardi, S., Antihistamines: ABC for the pediatricians. *Pediatr. Allergy Immunol.*, 31, Suppl. 24, 34—6, 2020.
21. Tiligada, E. and Ennis, M., Histamine pharmacology: from Sir Henry Dale to the 21st century. *Br. J. Pharmacol.*, 2018, http://dx.doi.org/10.1111/bph.14524.
22. Church, M.K., Maurer, M., Simons, F.E., Bindslev-Jensen, C., van Cauwenberge, P., Bousquet, J., *et al.*, Risk of first-generation H(1)-antihistamines: a GA(2)LEN position paper. *Allergy*, 65, 459—66, 2010.
23. Yanai, K., Yoshikawa, T., Yanai, A., Nakamura, T., Iida, T., Leurs, R., *et al.*, The clinical pharmacology of non-sedating antihistamines. *Pharmacol. Ther.*, 178, 148—56, 2017.
24. Simons, F.E. and Simons, K.J., Histamine and H1-antihistamines: Celebrating a century of progress. *J. Allergy Clin. Immunol.*, 128, 1139—50.e4, 2011.
25. Hu Y, Sieck, D.E., Hsu, W.H., Why are second-generation H1-antihistamines minimally sedating? *Eur. J. Pharmacol.*, 765, 100—6, 2015.
26. Giallongo A, Parisi, G.F., Licari, A., Pulvirenti, G., Cuppari, C., Salpietro, C., *et al.*, Novel therapeutic targets for allergic airway disease in children. *Drugs Context*, 8, 212590, 2019.
27. Nurmatov, U.B., Rhatigan, E., Simons, F.E., Sheikh, A., H1- antihistamines for primary mast cell activation syndromes: a systematic review. *Allergy*, 70, 9, 1052—61, 2015, http://dx.doi.org/10.1111/all.12672, sep; Epub 2015 Jul 6.
28. Juckett, G., Arthropod bites. *Am. Fam. Physician*, 88, 841—7, 2013.
29. De Sutter AI, Saraswat, A., van Driel, M.L., Antihistamines for the common cold. *Cochrane Database Syst. Rev.*, 11, CD009345, 2015.
30. Motola, D., Donati, M., Biagi, C., Calamelli, E., Cipriani, F., Melis, M., *et al.*, Safety profile of H1-antihistamines in paediatrics: An analysis based on data from VigiBase. *Pharmacoepidemiol. Drug Saf.*, 26, 1164—71, 2017.
31. Mahdy, A.M. and Webster, N.R., Histamine and antihistamines. *Anaesth. Intensive Care Med.*, *18*, 4, 210–215, 2017, doi:10.1016/j.mpaic.2017.01.007.
32. Paśko, P., Rodacki, T., Domagała-Rodacka, R., Palimonka, K., Marcinkowska, M., Owczarek, D., Second generation H1 - antihistamines interaction with

food and alcohol—A systematic review. *Biomed. Pharmacother.*, 93, 27–39, 2017, doi:10.1016/j.biopha.2017.06.008.
33. Jampilek, J., Heterocycles in Medicinal Chemistry. *Molecules*, 24, 21, 3839, 2019, https://doi.org/10.3390/molecules24213839.
34. Szumilak, M. and Stanczak, A., Cinnoline scaffold-A molecular heart of medicinal chemistry? *Molecules*, 24, 2271, 2019, [CrossRef] [PubMed].
35. Soler-Bistue, A., Zorreguieta, A., Tolmasky, M.E., Bridged nucleic acids reloaded. *Molecules*, 24, 2297, 2019, [CrossRef] [PubMed].
36. Betancourt-Conde, I., Avitia-Domínguez, C., Hernández-Campos, A., Castillo, R., Yépez-Mulia, L., Oria-Hernández, J., Méndez, S.T., Sierra-Campos, E., Valdez-Solana, M., Martínez-Caballero, S., *et al.*, Benzimidazole Derivatives as New and Selective Inhibitors of Arginase from Leishmania mexicana with Biological Activity against Promastigotes and Amastigotes. *Int. J. Mol. Sci.*, 22, 13613, 2021, [CrossRef] [PubMed].
37. Al-Wahaibi, L.H., Abou-Zied, H.A., Beshr, E.A.M., Youssif, B.G.M., Hayallah, A.M., Abdel-Aziz, M., Design, Synthesis, Antiproliferative Actions, and DFT Studies of New Bis–Pyrazoline Derivatives as Dual EGFR/BRAFV600E Inhibitors. *Int. J. Mol. Sci.*, 24, 9104, 2023, [CrossRef] [PubMed].
38. Bandyopadhyay, D., Rhodes, E., Banik, B.K., A green, chemoselective, and practical approach toward N-(2-azetidinonyl) 2, 5-disubstituted pyrroles. *RSC Adv.*, 3, 37, 16756–64, 2013.
39. Palaniappan, S.P., Pharmacological Role of Heterocyclic Compounds in the Treatment of Alzheimer's Disease: A Review.
40. Khatana, K. and Gupta, A., An update on natural occurrence and biological activity of benzofurans. *Acta Sci. Med. Sci.*, 4, 114, 2020.
41. Hiremathad, A. and Piemontese, L., Heterocyclic compounds as key structures for the interaction with old and new targets in Alzheimer's disease therapy. *Neural Regener. Res.*, 12, 8, 1256, 2017.
42. Ash, A.S.F. and Schild, H.O., Receptors mediating some actions of histamine. *Br. J. Pharmacol.*, 27, 427–3, 1966.
43. Lovenberg, T.W., Roland, B.L., Wilson, S.J., *et al.*, Cloning and functional expression of the human histamine H3 receptor. *Mol. Pharmacol.*, 55, 1101–7, 1999.
44. Oda, T., Morikawa, N., Saito, Y., Masuho, Y., Matsumoto, S., Molecular cloning and characterization of a novel type of histamine receptor preferentially expressed in leukocytes. *J. Biol. Chem.*, 275, 36781–6, 2000.
45. Bakker, R.A., Timmerman, H., Leurs, R., Histamine receptors: Specific ligands, receptor biochemistry, and signal transduction. *Clin. Allergy Immunol.*, 17, 27–64, 2002.
46. de Esch, I.J.P., Thurmond, R.L., Jongejan, A., Leurs, R., The histamine H4 receptor as a new therapeutic target for inflammation. *Trends Pharmacol. Sci.*, 26, 462–469, 2005. Leurs, R., *et al.*, Histamine receptors: pharmacology and functions. *Pharmacol. Ther.*, 94, 3, 305–331, 2002.

47. Wang, J., *et al.*, Histamine H2-receptor antagonists: pharmacology and clinical use. *Expert Opin. Pharmacother.*, 5, 3, 501–510, 2004.
48. Arrang, J.M., *et al.*, A novel brain histamine receptor. *Nature*, 302, 5908, 832–837, 1983.
49. Tliba, O., *et al.*, The role of H4 receptor in mast cell functions. *Curr. Opin. Pharmacol.*, 9, 3, 318–324, 2009.
50. Histamine H4 Receptor Antagonist - an overview | ScienceDirect Topics, www.sciencedirect.com. Retrieved 2023-10-03.
51. Villemagne, V.L., Dannals, R.F., Sánchez-Roa, P.M., *et al.*, Imaging histamine H1 receptors in the living human brain with carbon-11- pyrilamine. *J. Nucl. Med.*, 32, 308–11, 1991.
52. Yanai, K., Ryu, J.H., Watanabe, T., *et al.*, Histamine H1-receptor occupancy in human brains after single oral doses of histamine H1 antagonists measured by positron emission tomography. *Br. J. Pharmacol.*, 116, 1649–55, 1995.
53. Barger, G. and Dale, H.H., Chemical structure and sympathomimetic action of amines. *J. Physiol. (Lond)*, 41, 19–59, 1910.
54. Black, J.W., Duncan, W.A.M., Durant, C.J., Ganellin, C.R., Parsons, E.M., Definition and antagonism of histamine H2-receptors. *Nat. (Lond)*, 236, 385–90, 1972.
55. Bolton, T.B., Clark, J.P., Kitamura, K., Lang, R.J., Evidence that histamine and carbachol may open the same ion channels in longitudinal smooth muscle of guinea-pig ileum. *J. Physiol. (Lond)*, 320, 363–79, 1981.
56. Bolton, T.B. and Lim, S.P., Properties of calcium stores and transient outward currents in single smooth muscle cells of rabbit intestine. *J. Physiol. (Lond)*, 409, 385–401, 1989.
57. Donaldson, J. and Hill, S.J., Histamine-induced hydrolysis of polyphosphoinositides in guinea-pig ileum and brain. *Eur. J. Pharmacol.*, 124, 255–65, 1986.
58. Jutel, M., Blaser, K., Akdis, C.A., The role of histamine in regulation of immune responses. *Chem. Immunol. Allergy*, 91, 174–187, 2006, doi: 10.1159/000090280.
59. Villemain, F.M., Bach, J.F., Chatenoud, L.M., Characterization of histamine H1 binding sites on human T lymphocytes by means of 125Iiodobolpyramine. *J. Immunol.*, 144, 1449–54, 1990.
60. Kitamura, Y., Arima, T., Kitayama, Y., Nomura, Y., Regulation of [Ca21]i rise activated by doxepin-sensitive H1-histamine receptors in jurkat cells, cloned human T lymphocytes. *Gen. Pharmacol.*, 27, 289–91, 1996.
61. Bryce, P.J., Mathias, C.B., Harrison, K.L., Watanabe, T., Geha, R.S., Oettgen, H.C., The H1 histamine receptor regulates allergic lung responses. *J. Clin. Invest.*, 116, 1624–32, 2006.
62. Livett, B.G. and Marley, P.D., Effects of opioid peptides and morphine on histamine-induced catecholamine secretion from cultured, bovine adrenal chromaffin cells. *Br. J. Pharmacol.*, 89, 327–34, 1986.

63. Bunn, S.J., Sim, A.T.R., Herd, L.M., Dunkling, P.R., Tyrosinehydroxylase phosphorylation in bovine adrenal chromaffin cells: The role of intracellular Ca21 in the histamine H1-receptorstimulated phosphorylation of Ser (8), Ser (19), Ser (31) and Ser (40). *J. Neurochem.*, 64, 1370–8, 1995.
64. Beermann, S., Bernhardt, G., Seifert, R., Buschauer, A., Neumann, D., Histamine H1- and H4-receptor signaling cooperatively regulate MAPK activation. *Biochem. Pharmacol.*, 98, 3, 432–439, 2015, doi: 10.1016/j.bcp.2015.09.011.
65. Jutel, M., Watanabe, T., Klunker, S., *et al.*, Histamine regulates T-cell and antibody responses by differential expression of H1 and H2 receptors. *Nature*, 413, 6854, 420–425, 2001, doi: 10.1038/35096564.
66. Tanaka, S., Multifaceted Roles of Mast Cell-Derived Histamine, in: *Histamine Receptors: Preclinical and Clinical Aspects,* pp. 1–14, Springer Nature Switzerland, Cham, 2025 Oct 1.
67. Leurs, R., Smit, M.J., Timmerman, H., Molecular pharmacological aspects of histamine receptors. *Pharmacol. Ther.*, 66, 413–63, 1995.
68. Sharma, A. and Hamelin, B.A., Classic histamine H1 receptor antagonists: A critical review of their metabolic and pharmacokinetic fate from a bird's eye view. *Curr. Drug Metab.*, 4, 105–29, 2003.
69. Elz, S., Kramer, K., Pertz, H.H., *et al.*, Histaprodifens: Synthesis, pharmacological *in vitro* evaluation and molecular modeling of a new class of highly active and selective histamine (H1) receptor agonists. *J. Med. Chem.*, 43, 1071–84, 2000.
70. MacGlashan, D., Jr., Histamine: a mediator of inflammation. *J. Allergy Clin. Immunol.*, 112, S53–S, 2003.
71. Dy, M. and Schneider, E., Histamine-cytokine connection in immunity and hematopoiesis. *Cytokine Growth Factor Rev.*, 15, 393– 410, 2004.
72. Tanaka, S., Takasu, Y., Mikura, S., Satoh, N., Ichikawa, A., Antigenindependent induction of histamine synthesis by immunoglobulin E in mouse bone marrow derived mast cells. *J. Exp. Med.*, 196, 229–35, 2002.
73. Novak, N., Mete, N., Bussmann, C., Maintz, L., Bieber, T., Akdis, M., Zumkehr, J., Jutel, M., Akdis, C., Early suppression of basophil activation during allergen-specific immunotherapy by histamine receptor 2. *J. Allergy Clin. Immunol.*, 130, 5, 1153–1158.e2, 2012.
74. Palestra, F., Memoli, G., Secondo, A., Cristinziano, L., Tedeschi, V., Ferrara, A.L., Poto, R., Galdiero, M.R., Rocca, A.L., Illiano, A., Varricchi, G., Histamine as a mediator of cross-talk between human lung mast cells and macrophages. *Sci. Rep.*, 15, 1, 31969, 2025 Aug 30.
75. Ferstl, R., Frei, R., Barcik, W., Schiavi, E., Wanke, K., Ziegler, M., Rodriguez-Perez, N., Groeger, D., Konieczna, P., Zeiter, S., Nehrbass, D., Lauener, R., Akdis, C.A., O'Mahony, L., Histamine receptor 2 modifies iNKT cell activity within the inflamed lung. *Allergy.*, 72, 12, 1925–1935, 2017.
76. Smolinska, S., Groeger, D., Perez, N.R., Schiavi, E., Ferstl, R., Frei, R., Konieczna, P., Akdis, C.A., Jutel, M., O'Mahony, L., Histamine receptor 2 is

required to suppress innate immune responses to bacterial ligands in patients with inflammatory bowel disease. *Inflamm. Bowel Dis.*, 22, 7, 1575–1586, 2016.
77. Shi, Z., Fultz, R., Engevik, M.A., Gao, C., Hall, A., Major, A., Mori-Akiyama, Y., Versalovic, J., Distinct roles of histamine H1- and H2-receptor signaling pathways in inflammation-associated colonic tumorigenesis. *Am. J. Physiol. Gastrointest. Liver Physiol.*, 316, 205–216, 2019, doi: 10.1152/ajpgi.00212.2018.
78. Mazzoni, A., Young, H.A., Spitzer, J.H., Visintin, A., Segal, D.M., Histamine regulates cytokine production in maturing dendritic cells, resulting in altered T cell polarization. *J. Clin. Investig.*, 108, 1865–1873, 2001, doi: 10.1172/JCI200113930.
79. Frei, R., Ferstl, R., Konieczna, P., Ziegler, M., Simon, T., Rugeles, T.M., Mailand, S., Watanabe, T., Lauener, R., Akdis, C.A., *et al.*, Histamine receptor 2 modifies dendritic cell responses to microbial ligands. *J. Allergy Clin. Immunol.*, 132, 194–204.e12, 2013, doi: 10.1016/j.jaci.2013.01.013.
80. Teuscher, C., Poynter, M., Offner, H., Zamora, A., Watanabe, T., Fillmore, P.D., Zachary, J.F., Blankenhorn, E.P., Attenuation of Th1 Effector Cell Responses and Susceptibility to Experimental Allergic Encephalomyelitis in Histamine H2 Receptor Knockout Mice Is Due to Dysregulation of Cytokine Production by Antigen-Presenting Cells. *Am. J. Pathol.*, 164, 883–892, 2004, doi: 10.1016/S0002-9440(10)63176-8.
81. Dai, H., Kaneko, K., Kato, H., Fujii, S., Jing, Y., Xu, A., Sakurai, E., Kato, M., Okamura, N., Kuramasu, A., *et al.*, Selective cognitive dysfunction in mice lacking histamine H1 and H2 receptors. *Neurosci. Res.*, 57, 306–313, 2007.
82. Mobarakeh, J.I., Takahashi, K., Sakurada, S., Kuramasu, A., Yanai, K., Enhanced antinociceptive effects of morphine in histamine H2 receptor gene knockout mice. *Neuropharmacology*, 51, 612–622, 2006.
83. Ruat, M., Bouthenet, M.L., Schwartz, J.C., Ganellin, C.R., Histamine H1-receptor in heart: unique electrophoretic mobility and autoradiographic localization. *J. Neurochem.*, 55, 379–85, 1990.
84. Hirschfeld, J., Buschauer, A., Elz, S., *et al.*, Iodoaminopotentidine and related compounds: a new class of ligands with high affinity and selectivity for the histamine H2 receptor. *J. Med. Chem.*, 35, 2231–8, 1992.
85. Johnson, C.L., Histamine receptors and cyclic nucleotides, in: *Pharmacology of Histamine Receptors*, R. Ganellin and M. Parsons (Eds.), pp. 146–216, Wright & Bristol, England, 1982.
86. Black, J.W. and Shankley, N.P., The isolated stomach preparation of the iuphar classification of histamine receptors 271 mouse: a physiological unit for pharmacological analysis. *Br. J. Pharmacol.*, 86, 571–79, 1985.
87. Soll, A.H. and Berglindh, T., Physiology of isolated gastric glands and parietal cells: Receptors and effectors regulating function, in: *Physiology of the gastrointestinal tract*, L.R. Johnson (Ed.), pp. 883–909, 1987.

88. Hescheler, J., Tang, M., Jastorff, B., Trauteein, W., On the mechanism of histamine induced enhancement of the cardiac Ca21 current. *Pflü gers Arch.*, 410, 23–9, 1987.
89. Levi, R.C. and Alloatti, G., Histamine modulates calcium current in guineapig ventricular myocytes. *J. Pharmacol. Exp. Ther.*, 246, 377–83, 1988.
90. Ting, S., Dunsky, E.H., Zweiman, B., Histamine suppression of eosinophilotaxis and histamine release *in vivo*. *J. Allergy Clin. Immunol.*, 65, 196–7, 1980.
91. Plaut, M. and Lichtenstein, L.M., Histamine and immune responses, in: *The Pharmacology of Histamine Receptors*, C.R. Ganellin and M.E. Parsons (Eds.), pp. 392–435, Wright & Bristol, England, 1982.
92. Zheng, Y., Gao, M., Wijtmans, M., Vischer, H.F., Leurs, R., Synthesis and Pharmacological Characterization of New Photocaged Agonists for Histamine H3 and H4 Receptors. *Pharmaceuticals*, 17, 4, 536, 2024 Apr 21.
93. Delvalle, J., Tsunoda, Y., Williams, J.A., Yamada, T., Regulation of [Ca21]i by secretagogue stimulation of canine gastric parietal cells. *Am. J. Physiol.*, 262, G420–G6, 1992.
94. Seifert, R., Hoer, A., Schwaner, I., Bschauer, A., Histamine increases cytosolic Ca21 in HL-60 promyelocytes predominantly *via* H2 receptors with an unique agonist/antagonist profile and induces functional differentiation. *Mol. Pharmacol.*, 42, 2, 1992.
95. Delvalle, J., Wang, L., Gantz, I., Yamada, T., Characterization of H2 histamine receptor: Linkage to both adenylate cyclase and [Ca^{2+}] signaling systems. *Am. J. Physiol.*, 263, G967–G72, 1992.
96. Leurs, R., Smit, M.J., Tensen, C.P., Ter Laak, A.M., Timmerman, H., Sitedirected mutagenesis of the histamine H1-receptor reveals a selective interaction of asparagine207 with subclasses of H1-receptor agonists. *Biochem. Biophys. Res. Commun.*, 201, 295–301, 1994.
97. Traiffort, E., Ruat, M., Arrang, J.M., Leurs, R., Piomelli, D., Schwartz, J.C., Expression of a cloned rat histamine H2 receptor mediating inhibition of arachidonate release and activation of cAMP accumulation. *Proc. Natl. Acad. Sci. U.S.A.*, 89, 2649–53, 1992.
98. Ishikawa, M., Watanabe, T., Kudo, T., Yokoyama, F., Yamauchi, M., Kato, K., Kakui, N., Sato, Y., Investigation of the histamine H3 receptor binding site. Design and synthesis of hybrid agonists with a lipophilic side chain. *J. Med. Chem.*, 53, 17, 6445–6456, 2010.
99. García-Gálvez, A.-M., Escamilla-Sanchez, J., Flores-Maldonado, C., Contreras, R.-G., Arias, J.-M., Arias-Montano, J.-A., Differential homologous desensitization of the human histamine H3 receptors of 445 and 365 amino acids expressed in CHO-K1 cells. *Neurochem. Int.*, 112, 114–123, 2018.
100. Barnes, P.J., Histamine receptors in the lung. *Agents Actions Suppl.*, 33, 103–122, 1991.

101. Burgaud, J.L. and Oudart, N., Bronchodilatation of guinea-pig perfused bronchioles induced by the H3-receptor for histamine: Role of epithelium. *Br. J. Pharmacol.*, 109, 4, 960–966, 1993.
102. Nieto-Alamilla, G., Márquez-Gómez, R., García-Gálvez, A.M., Morales-Figueroa, G.E., Arias-Montaño, J.A, The histamine H3 receptor: Structure, pharmacology, and function. *Mol. Pharmacol.*, 90, 5, 649–673, 2016.
103. Dimitriadou, V., Rouleau, A., Tuong, M.D.T., *et al.*, Functional relationship between mast cells and C-sensitive nerve fibres evidenced by histamine H3-receptor modulation in rat lung and spleen. *Clin. Sci.*, *87*, 2, 151–163, 1994, doi: 10.1042/cs0870151.
104. Takahashi, N., Patel, H.R., Qi, Y., Dushay, J., Ahima, R.S., Divergent effects of leptin in mice susceptible or resistant to obesity. *Hormone Metab. Res.*, 34, 11/12, 691–697, 2002, doi: 10.1055/s-2002-38251.
105. Akdis, C.A. and Simons, F.E.R., Histamine receptors are hot in immunopharmacology. *Eur. J. Pharmacol.*, 533, 1–3, 69–76, 2006, doi: 10.1016/j.ejphar.2005.12.044.
106. Yu, X., Franks, N.P., Wisden, W., Sleep and sedative states induced by targeting the histamine and noradrenergic systems. *Front. Neural Circuits*, 12, 4, 2018, doi: 10.3389/fncir.2018.00004.
107. Barata-Antunes, S., Cristóvão, A.C., Pires, J., Rocha, S.M., Bernardino, L., Dual role of histamine on microglia-induced neurodegeneration. *Biochim. Biophys. Acta (BBA) Mol. Basis Dis.*, 1863, 3, 764–769, 2017, doi: 10.1016/j.bbadis.2016.12.016.
108. Imamura, M., Smith, N.C.E., Garbarg, M., Levi, R., Histamine H3- receptor-mediated inhibition of calcitonin gene-related peptide release from cardiac C fibers: a regulatory negative feedback loop. *Circ. Res.*, 78, 863–9, 1996.
109. Delaunois, A., Gustin, P., Garbarg, M., Ansay, M., Modulation of acetylcholine, capsaicin and substance P effects by histamine H3 receptors in isolated perfused rabbit lungs. *Eur. J. Pharmacol.*, 277, 243–50, 1995.
110. Taylor, S.J. and Kilpatrick, G.J., Characterization of histamine-H3 receptors controlling non-adrenergic non-cholinergic contractions of the guinea pig isolated ileum. *Br. J. Pharmacol.*, 105, 667–74, 1992.
111. Dimitriadou, V., Rouleau, A., Dam Trung Tuong, M., *et al.*, Functional relationship between mast cells and C-sensitive nerve fibres evidenced by H3-receptor modulation in rat lung and spleen. *Clin. Sci.*, 87, 151–63, 1994.
112. West, R.E., Jr, Zweig, A., Shih, N.Y., Siegel, M.I., Egan, R.W., Clark, M.A., Identification of two H3-histamine receptor subtypes. *Mol. Pharmacol.*, 38, 610, 1990.
113. Brown, J.D., O'Shaughnessy, C.T., Kilpatrick, G.J., *et al.*, Characterisation of the specific binding of the histamine H3-receptor antagonist radioligand 3HGR16820. *Br. J. Pharmacol.*, 114, 344, 1994.
114. Clark, M.A., Korte, A., Myers, J., Egan, R.W., High affinity histamine H3 receptors regulate ACTH release by AtT-20 cells. *Eur. J. Pharmacol.*, 210, 31–5, 1992.

115. Clark, E.A. and Hill, S.J., Sensitivity of histamine H3 receptor agoniststimulated [35S]-GTPgS binding to pertussis toxin. *Eur. J. Pharmacol.*, 296, 223–5, 1996.
116. Endou, M., Poli, E., Levi, R., Histamine H3-receptor signalling in the heart: possible involvement of Gi/Go proteins and N-type Ca21 channels. *J. Pharmacol. Exp. Ther.*, 269, 221–9, 1994.
117. Imamura, M., Seyedi, N., Lander, H.M., Levi, R., Functional identification of histamine H3-receptors in the human heart. *Circ. Res.*, 77, 206–10, 1995.
118. Cogé, F., Guénin, S.P., Rique, H., Boutin, J.A., Galizzi, J.P., Structure and expression of the human histamine H4-receptor gene. *Biochem. Biophys. Res. Commun.*, 284, 2, 301–309, 2001.
119. Gschwandtner, M., Rossbach, K., Dijkstra, D., Bäumer, W., Kietzmann, M., Stark, H., Werfel, T., Gutzmer, R., Murine and human Langerhans cells express a functional histamine H4 receptor: modulation of cell migration and function. *Allergy*, 65, 840–849, 2010.
120. Miyano, K., Matsushita, S., Tsuchida, T., Nakamura, K., Inhibitory effect of a histamine 4 receptor antagonist on CCL17 and CCL22 production by monocyte-derived Langerhans cells in patients with atopic dermatitis. *J. Dermatol.*, 43, 9, 1024–1029, 2016.
121. Köchling, H., Schaper, K., Wilzopolski, J., Gutzmer, R., Werfel, T., Bäumer, W., Kietzmann, M., Rossbach, K., Combined treatment with H1 and H4 receptor antagonists reduces inflammation in a mouse model of atopic dermatitis. *J. Dermatol. Sci.*, 87, 2, 130–137, 2017.
122. Ohsawa, Y. and Hirasawa, N., The role of histamine H1 and H4 receptors in atopic dermatitis: from basic research to clinical study. *Allergol. Int.*, 63, 4, 533–42, 2014.
123. Morse, K.L., Behan, J., Laz, T.M., West, R.E., Jr, Greenfeder, S.A., Anthes, J.C., *et al.*, Cloning and characterization of a novel human histamine receptor. *J. Pharmacol. Exp. Ther.*, 296, 1058–1066, 2001.
124. Dijkstra, D., Leurs, R., Chazot, P., Shenton, F.C., Stark, H., Werfel, T., *et al.*, Histamine downregulates monocyte CCL2 production through the histamine H4 receptor. *J. Allergy Clin. Immunol.*, 120, 300–307, 2007, doi: 10.1016/j.jaci.2007.03.024.
125. Hofstra, C.L., Desai, P.J., Thurmond, R.L., Fung-Leung, W.P., Histamine H4 receptor mediates chemotaxis and calcium mobilization of mast cells. *J. Pharmacol. Exp. Ther.*, 305, 1212–1221, 2003, doi: 10.1124/jpet.102.046581.
126. Nakayama, T., Kato, Y., Hieshima, K., Nagakubo, D., Kunori, Y., Fujisawa, T., *et al.*, Liver-expressed chemokine/CC chemokine ligand 16 attracts eosinophils by interacting with histamine H4 receptor. *J. Immunol.*, 173, 2078–2083, 2004, doi: 10.4049/jimmunol.173.3.2078.
127. Liu, C., Wilson, S.J., Kuel, C., Lovenberg, T.W., Comparison of human, mouse, rat, and guinea pig histamine H4 receptors reveals substantial pharmacological species variation. *J. Pharmacol. Exp. Ther.*, 299, 121–130, 2001b.

128. Tiligada, E., Stefanaki, C., Ennis, M., Neumann, D., Opportunities and challenges in the therapeutic exploitation of histamine and histamine receptor pharmacology in inflammation-driven disorders. *Pharmacol. Ther.*, 263, 108722, 2024 Nov 1.
129. Nakamura, T., Itadani, H., Hidaka, Y., Ohta, M., Tanaka, K., Molecular cloning and characterization of a new human histamine receptor, H4R. *Biochem. Biophys. Res. Commun.*, 279, 615–20, 2000.
130. Zhu, Y., Michalovich, D., Wu, H., *et al.*, Cloning, expression, and pharmacological characterization of a novel human histamine receptor. *Mol. Pharmacol.*, 59, 434, 2001.
131. Ottosson, A., Jansen, I., Edvinsson, L., Pharmacological characterization of histamine receptors in human temporal artery. *Br. J. Clin. Pharmacol.*, 27, 139–45, 1989.
132. Melmon, K.L., Rocklin, R.E., Rosenkranz, R.P., Autacoids as modulators of the inflammatory and immune response. *Am. J. Med.*, 71, 100–6, 1981.
133. Tripathy, D.B. and Mishra, A., Microwave synthesis and characterization of waste soybean oil-based gemini imidazolinium surfactants with carbonate linkage. *Surf. Rev. Lett.*, 24.05, 1750062, 2017.
134. Parmar, D.R., Vala, B.S., Design, synthesis and biological screening of some novel imidazole derivatives as H1-receptor antagonists. *Pharma Sci. Mon.*, 8, 2, 461–472, 2017.
135. Iemura, R., Kawashima, T., Fukuda, T., Ito, K., Tsukamoto, G., Synthesis of 2-(4-substituted-1-piperazinyl)benzimidazoles as H1-antihistaminic agents. *J. Med. Chem.*, 29, 7, 1178–1183, 1986, doi:10.1021/jm00157a010.
136. Zhang, L., Peng, X.-M., Damu, G.L.V., Geng, R.-X., Zhou, C.-H., Comprehensive Review in Current Developments of Imidazole-Based Medicinal Chemistry. *Med. Res. Rev.*, 34, 2, 340–437, 2013, doi:10.1002/med.21290.
137. Jafarzadeh, A., Nemati, M., Khorramdelazad, H., Hassan, Z.M., Immunomodulatory properties of cimetidine: Its therapeutic potentials for treatment of immune-related diseases. *Int. Immunopharmacol.*, 70, 156–166, 2019 May.
138. Al-Ghorbani, M., Bushra, B.A., Zabiulla, M.S.V., Shaukath, A.K., Piperazine and morpholine: Synthetic preview and pharmaceutical applications. *J. Chem. Pharm. Res.*, 7, 5, 281–301, 2015.
139. Rathi, A.K., Syed, R., Shin, H.S., Patel, R.V., Piperazine derivatives for therapeutic use: A patent review (2010-present). *Expert Opin. Ther. Pat., 26,* 7, 777–797, 2016.
140. Beliyaiah, L., Dilkalal, A., Vinaya, Basavaiah, U., Parkin, S., Butcher, R.J., Yathirajan, H.S., Basavaraju, Y.B., Disulfonamide Piperazine Derivatives: Synthesis, Characterization, Crystal Structures Studies, and Evaluation of Cell Proliferation Activity. *Russ. J. Gen. Chem.*, 94, 8, 2167–79, 2024 Aug.
141. Rasheed, A.M., Shetty, K.K., Roy, L.D., Kumar, J., Novel Piperazine Derivatives as Potent Antihistamine, Anti-Inflammatory, and Anticancer Agents, their

Synthesis and Characterization. *Anti-Cancer Agents Med. Chemistry-Anti-Cancer Agents, 24,* 14, 1063–1073, 2024.

142. Ahmadi, A., Khalili, M., Nafarie, A., Yazdani, A., Nahri-Niknafs, B., Synthesis and anti-inflammatory effects of new piperazine and ethanolamine derivatives of H1-antihistaminic drugs. *Mini-Rev. Med. Chem.*, 12, 12, 1282–1292, 2012, doi:10.2174/138955712802762013.
143. Perli, M. and Govindarajan, R., Piperazine derivatives: A review of biological activities. *World J. Pharm. Res.*, *9*, 14, 194–204, 2020.
144. Carrière, I., Fourrier-Reglat, A., Dartigues, J.-F., Rouaud, O., Pasquier, F., Ritchie, K., Ancelin, M.-L., Drugs with anticholinergic properties, cognitive decline, and dementia in an elderly general population: The 3-city study. *Arch. Intern. Med.*, 169, 14, 1317–1324, July 2009.
145. Resende, D.I., Durães, F., Maia, M., Sousa, E., Pinto, M.M., Recent advances in the synthesis of xanthones and azaxanthones. *Org. Chem. Front.*, 7, 19, 3027–66, 2020.
146. Szczepańska, K., Karcz, T., Mogilski, S., Siwek, A., Kuder, K.J., Latacz, G., Kubacka, M., Hagenow, S., Lubelska, A., Olejarz, A., Kotańska, M., Synthesis and biological activity of novel tert-butyl and tert-pentylphenoxyalkyl piperazine derivatives as histamine H3R ligands. *Eur. J. Med. Chem.*, 152, 223–34, 2018.
147. Wang, L., Wang, T., Yang, B., Chen, Z., Yang, H., Design, synthesis, and anti-allergic activities of novel (R)(-)-1-[(4-chlorophenyl) phenyl methyl] piperazine derivatives. *Med. Chem. Res.*, 21, 124–32, 2012 Jan.
148. Selvam, T.P. and Kumar, P.V., Quinazoline marketed drugs. *Res. Pharm.*, *1*, 1, 2015.
149. Suthakaran, R., Kavimani, S., Venkappayya, D., Sakina, F., Suganthi, K., Inflammation: Synthesis And Pharmacological Investigation Of Some New 4 (3h)-Quinazolinone Analogs As Antioxidant, Antihistaminic, Anti-Inflammatory And Antitumor Agents. *Chem*, *1*, 2, 2008.
150. Gobinath, M., Subramanian, N., Alagarsamy, V., Design, synthesis and H1-antihistaminic activity of novel 1-substituted-4-(3-chlorophenyl)-[1, 2, 4] triazolo [4, 3-a] quinazolin-5 (4H)-ones. *J. Saudi Chem. Soc.*, *19*, 3, 282–286, 2015.
151. Alagarsamy, V., Giridhar, R., Yadav, M.R., Synthesis and H1-Antihistaminic Activity of Some Novel 1-Substituted-4-(3-methylphenyl)-1, 2, 4-triazolo [4, 3-a] quinazolin-5 (4H)-ones. *Biol. Pharm. Bull.*, 28, 8, 1531–34, 2005.
152. Ogawa, N., Yoshida, T., Aratani, T., Koshinaka, E., Kato, H., Ito, Y., Synthesis and histamine H2-antagonist activity of 4-quinazolinone derivatives. *Chem. Pharm. Bull.*, 36, 8, 2955–67, 1988 Aug 25.
153. Kumar Raju, V.S., Raju, M.B., Bahekar, R.H., Rajan, K.S., Ram Rao, A.R., New Antihistaminic Agents-5: Synthesis and H~ 1-Antihistaminic Evaluation of 3-(N, N-Dialkylamino) alkyl derivatives of 2-phenyl-3, 4-dihydroquinazolin-4 (3H)-ones. *Indian Drugs-Bombay*, 36, 12, 759, 1999.

154. Saxena, M., Gaur, S., Prathipati, P., Saxena, A.K., Synthesis of some substituted pyrazinopyridoindoles and 3D QSAR studies along with related compounds: piperazines, piperidines, pyrazinoisoquinolines, and diphenhydramine, and its semi-rigid analogs as antihistamines (H1). *Bioorg. Med. Chem.*, *14*, 24, 8249–8258, 2006.
155. Schon, U., Messinger, J., Buckendahl, M., Prabhu, M.S., Konda, A., Animproved synthesis of N-aryl and N-heteroaryl substituted homopiperazines fromconventional thermal conditions to scaling-up using microwave heating. *Tetrahedron*, 65, 8125–8131, 2009.
156. Orjales, A., Gil-Stichez, J.C., Alonso-Cires, L., Labeaga, L., Mosquera, R., Berisa, A., Synthesis and histamine HI-receptor antagonist activity of 4-(Diphenylmethyl)-1-piperazine derivatives with a terminal heteroaryl or cycloalkyl amidefragment. *Eur. J. Med. Chem.*, 31, 813–818, 1996.
157. Waszkielewicz, A.M., Gunia, A., Szkaradek, N., Pytka, K., Siwek, A., Satała, G., Marona, H., Synthesis and evaluation of pharmacological properties of some new xanthone derivatives with piperazine moiety. *Bioorg. Med. Chem. Lett.*, 23, 15, 4419–4423, 2013, doi:10.1016/j.bmcl.2013.05.062.
158. Kuder, K.J., Łażewska, D., Kaleta, M., Latacz, G., Kottke, T., Olejarz, A., Kieć-Kononowicz, K., Synthesis and biological activity of novel tert-amylphenoxyalkyl (homo) piperidine derivatives as histamine H3R ligands. *Bioorg. Med. Chem.*, *25*, 10, 2701–2712, 2017.
159. Lavrador-Erb, K., Ravula, S.B., Yu, J., Zamani-Kord, S., Moree, W.J., Petroski, R.E., Beaton, G., The discovery and structure–activity relationships of 2-(piperidin-3-yl)-1H-benzimidazoles as selective, CNS penetrating H1-antihistamines for insomnia. *Bioorg. Medicinal Chem. Lett.*, *20*, 9, 2916–2919, 2010.
160. Belsito, M.D., Hill, R.A., Klaassen, C.D., Liebler, D., Marks Jr., J.G., Ronald, C., On the Safety Assessment of Ethanolamine and Ethanolamine Salts as Used in Cosmetics, 2012.
161. Fiume, M.M., Heldreth, B.A., Bergfeld, W.F., Belsito, D.V., Hill, R.A., Klaassen, C.D., Andersen, F.A., Safety assessment of ethanolamine and ethanolamine salts as used in cosmetics. *Int. J. Toxicol.*, *34*, 2_suppl, 84S–98S, 2015.
162. Akimoto, H., Uesawa, Y., Hishinuma, S., Molecular determinants of the kinetic binding properties of antihistamines at the histamine H1 receptors. *Int. J. Mol. Sci.*, 22, 5, 2400, 2021 Feb 27.
163. Mahdy, A.M. and Webster, N.R., Histamine and antihistamines. *Anaesth. Intensive Care Med.*, *12*, 7, 324–329, 2011.

9

Heterocyclic Compound Having Herbicides Activity

Niranjan Kaushik[1]*, Mridul Singh Sengar[1], Deepika Paliwal[1] and Nadeem Ahmad Siddique[2]

[1]Department of Pharmacy, School of Medical and Allied Sciences, Galgotias University, Greater Noida, India
[2]Department of Pharmaceutical Chemistry, University of Hafr Al Batin, Hafar Al Batin, Kingdom of Saudi Arabia

Abstract
Heterocyclic compounds are essential to agricultural chemistry because of their herbicidal activity, successfully preventing weeds from competing with crops for essential resources. Triazines, pyridines, and imidazolines are among the key compounds that act through various mechanisms, such as inhibiting photosynthesis, disrupting amino acid synthesis, and affecting cell membrane integrity. Advances in herbicide research have led to the development of new heterocyclic derivatives with enhanced selectivity and a lower environmental impact. However, the growing challenge of herbicide resistance necessitates innovative approaches, including dual-mode herbicides and genetically modified crops. This study explores the significance, classification, and mode of action of heterocyclic herbicides in the context of increasing resistance and environmental sustainability to enhance weed control management.

***Keywords*:** Heterocyclic compounds, herbicides, weed management, resistance, photosynthesis inhibitors, agricultural chemistry

**Corresponding author*: niranjankaushik79@gmail.com

Divya Bajpai Tripathy, Priyanka Chhabra and Anjali Gupta (eds.) Heterocyclic Compounds as Pharmaceutical Ingredients, (307–348) © 2026 Scrivener Publishing LLC

9.1 Introduction

Heterocycles are cyclic atomic rings containing one or more heteroatoms [1]. They represent a fundamental field in organic chemistry and are gaining increasing research attention due to their diverse applications in industry, antimicrobial activity, and medicine [2–4]. Many essential biomolecules, including DNA, RNA, chlorophyll, hemoglobin, and vitamins, are largely synthesized through heterocyclic rings [5]. Heterocycles not only contribute to the formation of vital amino acids, such as proline, but also play a crucial role in synthesizing chlorophyll and hemoglobin, which transport oxygen. Additionally, they serve as key substrates for numerous compounds that contribute to biological structures [6]. Due to their diverse variety of biological functions, heterocyclic compounds are crucial scaffolding for both natural and synthesized derivatives [7–11]. Many heterocyclic compounds, including triazine analogs, have been used as medicinal agents, urinary antiseptics, and anti-inflammatory drugs to treat a variety of diseases [12]. Humans, animals, plants, and microbes all contain uracils, a family of naturally occurring bioactive substances with a variety of biological functions [13–17]. Because of their exceptional insecticidal, fungicidal, and herbicidal bioactivities, uracils are significant nitrogen-linked heterocyclic compounds that have drawn a lot of interest in the field of agricultural chemistry [18–22]. A crucial active group in both natural and manufactured drugs is the uracil skeleton. For instance, benzfendizone (Figure 9.1) [23] and tiafenacil [24] are protoporphyrinogen oxidase (PPO) inhibitors that have been effectively used to suppress weeds in cereal, corn, and cotton crops. The glyphosate-resistant weeds and broadleaf weeds are managed using saflufenacil (Figure 9.1) [25], another PPO inhibitor. There is a uracil motif in these PPO inhibitors. These PPO inhibitors' connections between structure and action [26] indicate that compounds with a replaced aryl group at number 1, an amino and methyl group at position 3, and a trifluormethyl group at position 6 have high herbicidal action. Most recently, studies [27, 28] said that substantial herbicidal activity was discovered in Compounds A and B (Figure 9.1). Likewise, the pyridine ring is a significant active heterocyclic that finds extensive application in agricultural chemistry [29–31]. Although its structure is similar to that of a benzene ring, it differs from the benzene ring in certain ways, such as its features related to the oil–water partition coefficient [32].

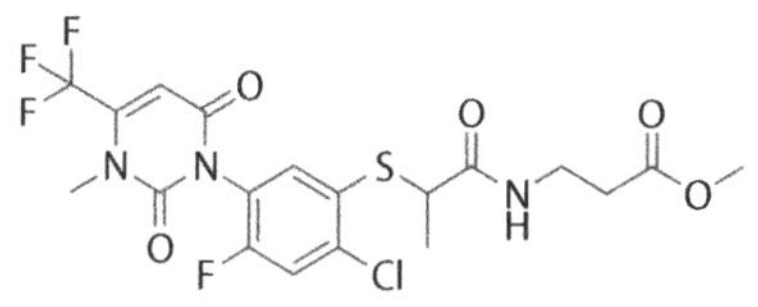

Tiafenacil

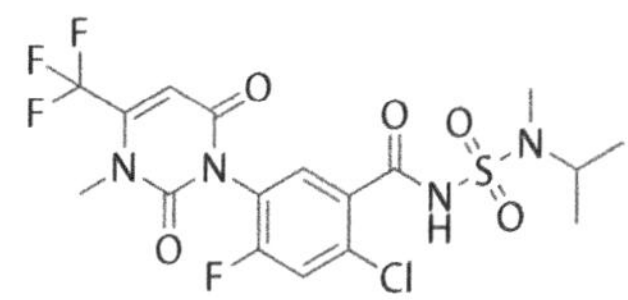

Salfufenacil

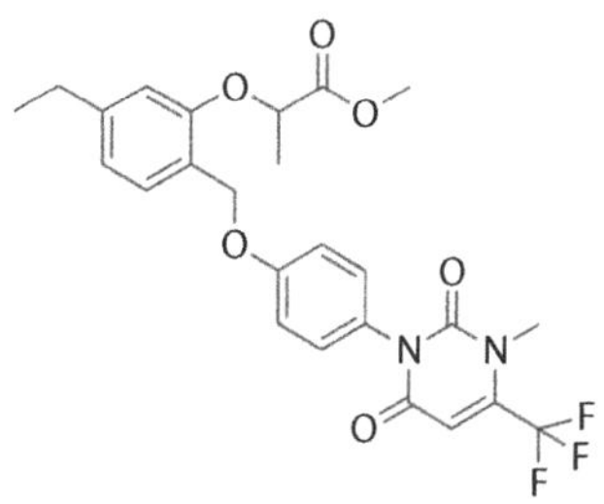

Benzendizone

Figure 9.1 Uracil-skeletoned bioactive compounds.

9.1.1 Historical Background

"The field of chemical study remained unbroken two centuries ago, but by 1900, it was essential to separate it into inorganic, organic, and physical chemistry. Growing amounts of factual information compelled a gradual division into subfields. The largest and most diverse family of organic substances is made up of heterocyclic compounds." Heterocyclic chemistry developed alongside the advancement of organic chemistry in the early 1800s. Several landmark discoveries in 1818, including the identification of heterocyclic ring systems, played a crucial role in shaping the modern understanding of heterocyclic compounds. Alloxan was separated from uric acid 1832 by Brugnatelli. Dobereiner uses sulfuric acid to process starch to make furfural (afuran). In 1834, Runge extracted pyrrole (also known as "fiery oil") from bones using dry distillation. When Friedlander creates indigo dye, a significant portion of the agricultural sector is displaced by synthetic chemistry. In 1936, the fundamental principles governing genetic inheritance were further elucidated, highlighting the crucial role of heterocyclic compounds such as purines and pyrimidines in the genetic code. In the same period, Treibs successfully separated crude oil derivatives related to chlorophyll, providing insight into the biological origin of petroleum.

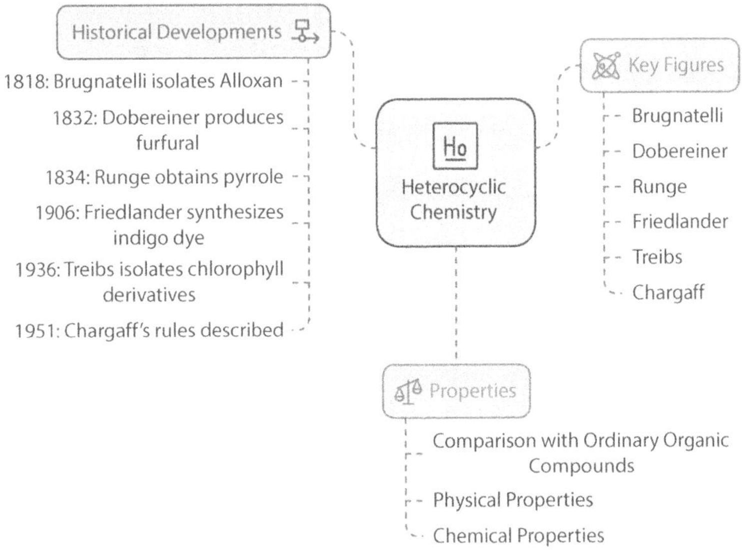

Figure 9.2 Flowchart for history of heterocyclic compound.

A flowchart of history is shown in Figure 9.2. Notable advancements in heterocycles include the development of systematic nomenclature, understanding of structure–reactivity relationships, synthesis of biologically important heterocyclic systems. The easiest way to understand the physical and chemical characteristics of heterocyclic compounds is generally to compare them to regular organic molecules without heteroatoms.

Two criteria can be used to classify chemicals that are heterocyclic according to their structure and electrical configuration.

i. Heterocyclic aliphatic compounds
ii. Heterocyclic aromatic compounds
iii. Aliphatic heterocyclic compounds include cyclic amines, cyclic amides, which cyclic ethers, or cyclic thioethers. Aliphatic heterocycles without double bonds are known as saturated heterocycles [33].

9.1.2 Herbicides

It is necessary to comprehend the significance of weed control to comprehend herbicides. Even before the invention of herbicides, weed control was an essential component of agriculture. When cultivating crops, weeds—unwanted plants—may also grow next to the crops. These undesirable plants can have a detrimental effect on the development and output of your intended crops because they eat the same resources such as water, nutrients, and sunlight and harbor insects and fungi. Herbicidal compound and heterocyclic chemicals are classified in Figures 9.3 and 9.4.

Herbicides are mostly used in agriculture, but they are also utilized in other industries where it is required to remove unwanted plants. Herbicides are helpful in agriculture because they may suppress weeds at nearly every stage of growth, giving crop producers more freedom. Because different types of weeds affect different crops, different herbicides will be required. It is possible to use some herbicides prior to seeding to guarantee that no

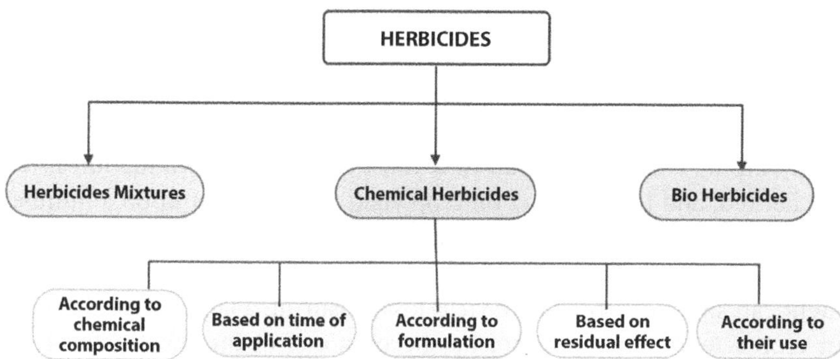

Figure 9.3 Classification of heterocyclic compound.

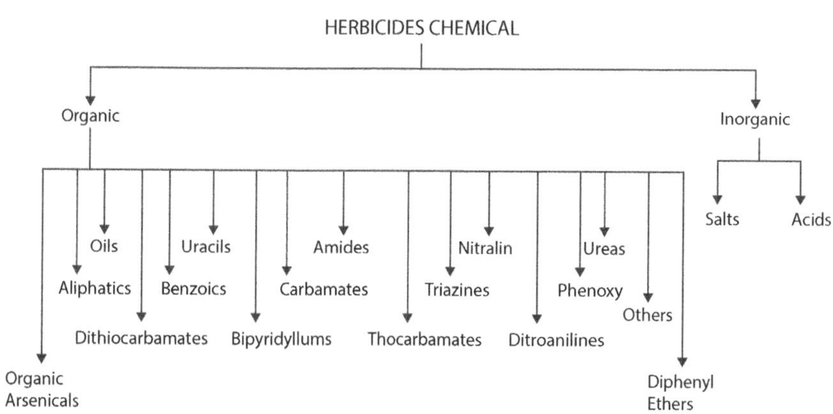

Figure 9.4 Classification of heterocyclic chemical.

herbicide residue remains. When weeds are present in addition to developing crops, a herbicide that is selective may be used to manage the weeds without harming the crops. Because they may significantly impact production and stop dangerous weeds from growing and processing alongside crops, pesticide usage will continue to be crucial [34].

The rapid growth of the global population is increasing the demand for food crops. However, several factors, including diseases, pests, and weeds, hinder crop production. Among these, weeds are a major cause of productivity loss in field crops. A weed is any plant that grows in an unfavorable location, competing with crops for essential resources such as nutrients, sunlight, and moisture. In general, any plant growing in an unintended location is considered a weed [35], weeds are known to lower output more than pests, diseases, and insects, which are believed to cost ten major crops an estimated US $11 billion. Of these, soybeans lose 31.4%, green grams 30.8%, maize 25.3%, mustard 21.4%, wheat 18.6%, and 21.4% direct-seeded rice of their production. Weeds and crops compete with one another for space and other growth factors, which lowers crop output. Weeds can result in a 100% loss in crop production if left unchecked, and their impact on crop yield is contingent upon a number of parameters, including weed emergence time, weed density, weed species, and crop competitive ability. Effective and targeted weed management can significantly reduce yield loss, ensuring a larger grain harvest. In crop science, weed control has become one of the most extensively studied topics. On small farms, weeds can be controlled by automated or manual weeding tools such as weed rollers and cono-weeders. However, controlling weeds on large farms presents greater challenges. In commercial cultivation, various pre-emergence and post-emergence herbicides are commonly used to manage weed species. With the increasing resistance of weeds to herbicides, research in weed science must focus on resistance to herbicides and modes of action. There have been 511 reported instances of herbicide resistance worldwide, involving 133 monocots and 153 dicots, making up the 266 weed species, most commonly herbicide-resistant, the following crops had weed species: wheat, maize, rice, soybean, summer barley, and cotton. Due to natural selection and the adaptation of weeds to evade herbicidal effects, the following crops had weed species: wheat, maize, rice, soybean, summer barley, and cotton. The number of reported herbicide resistance cases has increased sharply, rising 509 instances compared to seven cases on 1975 by the conclusion of 2020 [36]. Herbicide-resistant weeds are predominantly members of the Poaceae family. Together with Asteraceae, Brassicaceae, Amaranthaceae, and Chenopodiaceae, these five families represent only about 50% of major weed species, yet they account for approximately 70% of all reported herbicide-resistance cases [37]. With an estimated 160 weed species resistant to photosystem II (PSII) inhibitors, ALS inhibitors rank next in prevalence. Glyphosate, one of the most widely used post-emergence herbicides for broad-spectrum weed management, has become less effective due to extensive use, accelerating the emergence of glyphosate-resistant biotypes [38]. Advancements in genetics, plant physiology, chemistry, and plant science have enabled researchers to better understand the underlying causes and mechanisms of resistance to herbicides. Resistance to herbicides can manifest through various mechanisms, including multiple resistance, cross-resistance, and both non–target-site and target-site resistance [39]. Resistance to non–target-site herbicides occurs when the effectiveness of a herbicide is reduced due to decreased absorption or translocation of its active ingredients. In contrast, target-site resistance arises from mutations in the genes that encode herbicide-targeted enzymes. Cross-resistance occurs when weeds develop resistance to multiple herbicides with the same manner of action due to repeated application. In contrast, multiple resistance arises when weeds gain resistance to herbicides that work in different ways through successive selection involving two or more resistance mechanisms. The herbicide's method of action defines how its active ingredients affect plant growth and survival. Herbicide efficacy varies depending on the targeted weed species and the chemical composition of its active ingredients. Some herbicides primarily target the root system, whereas others inhibit photosynthesis, disrupt enzymatic pathways, or compromise cell membrane integrity. Understanding herbicide modes of action is essential not only for proper application, classification, and organization but also for managing herbicide resistance—an increasingly critical issue in sustainable agricultural management [40]. Research on herbicide modes of action continues to progress, shown in Table 9.1, leading to the identification of novel mechanisms that support the development of new herbicides. Currently, 22 distinct herbicide action modes have been classified based on their site and mechanism of action. Understanding these mechanisms is crucial for managing herbicide resistance and developing innovative weed control strategies. This chapter explores various herbicide modes of action, focusing on their mechanisms and chemical classifications.

Table 9.1 Mode of action and field application of heterocyclic herbicides.

Heterocyclic core	Example compound	Mode of action	Target weeds	Common application
Triazine	Simazine	Inhibits photosystem II (PSII)	Broadleaf and grassy weeds	Corn, orchards, vineyards
Triazinone	Hexazinone	Inhibits photosystem II (PSII)	Woody and broadleaf weeds	Forestry, pastures
Pyridine	Aminopyralid	Synthetic auxin, disrupts growth	Broadleaf weeds	Pastures, rangelands
Isoxazole	Tembotrione	Inhibits 4-hydroxyphenylpyruvate dioxygenase (HPPD)	Broadleaf and grass weeds	Corn
Pyrazole	Pyrazolate	Inhibits photosystem II (PSII)	Grassy weeds	Rice
Pyrimidinyl-thiobenzoate	Pyrithiobac-sodium	Inhibits acetolactate synthase (ALS)	Broadleaf weeds	Cotton
Thiophene	Fomesafen	Inhibits protoporphyrinogen oxidase (PPO)	Broadleaf weeds	Soybeans, cotton
Imidazolinone	Imazamox	Inhibits acetolactate synthase (ALS)	Broad-spectrum weeds	Clearfield crops (wheat, rice)
Quinoline	Halauxifen-methyl	Synthetic auxin, disrupts cell elongation	Broadleaf weeds	Cereals
Isoxazolidinone	Tiafenacil	Inhibits protoporphyrinogen oxidase (PPO)	Broadleaf weeds	Pre-emergence applications
Thiadiazole	Thidiazuron	Inhibits ethylene production (defoliant)		

9.2 Mechanisms of Herbicide Action and Their Role in Plant Growth Inhibition

9.2.1 Inhibitors for Lipid Synthesis

These herbicides inhibit lipid synthesis, halting plant development by destroying cell walls and causing cell leakage. Among them are Group 1 herbicides, which target acetyl CoA carboxylase (ACCase) enzymes. ACCase plays a crucial role in the production of fatty acids, essential components of the phospholipid bilayer in plant cell membranes. Inhibiting ACCase disrupts meristematic growth by preventing the formation of new cell walls, ultimately leading to plant cell death. ACCase inhibitors are herbicides specifically designed to target plant species that have developed resistance to glyphosate [41]. Once applied, the herbicide is absorbed by the plant and translocated through the phloem to meristematic regions, where it disrupts growth by inhibiting meristematic activity. Within a week, this leads to necrotic symptoms in developing tissues.

The chemical families of ACCase inhibitors include aryloxyphenoxypropionates, cyclohexanediones, and phenylpyrazoles. These post-emergence herbicides are widely used to manage broadleaf crops' grassy weeds [42]. Typical herbicides in this class include fluazifop, diclofop, quizalofop, clethodim, sethoxydim, fenoxaprop, and pinoxaden. These herbicides are applied to leaves and translocate through the phloem to meristematic regions. Three to 4 days after application, affected plants exhibit browning, chlorotic symptoms on the leaves, and purpling or browning of the veins. ACCase-inhibiting herbicides have a short half-life in soil, are infrequently used, and have low soil solubility. Due to their reduced leaching potential, they are considered to have minimal environmental impact. However, resistance to ACCase inhibitors has been reported in 43 grass weed species [43].

9.2.2 Inhibitors of Amino Acid Synthesis

These herbicides have been found to act at two different target sites. Group 2 herbicides inhibit ALS, or acetolactate synthase, an enzyme that is in charge of catalyzing the synthesis of amino acids with branching chains, including valine, isoleucine, and leucine. These amino acids are necessary components of proteins required for the development of cell membranes, and their biosynthesis is disrupted when ALS is inhibited. In some plant species, ALS inhibitors have been reported to affect reproduction by inducing male sterility. Their high efficacy at low doses has contributed to their widespread use, but it has also caused resistance to grow quickly in certain weed species [44]. ALS-inhibiting herbicides are the most widely used class of post-emergence selective herbicides. After being absorbed through the roots and leaves, their chemical components are translocated through the xylem and phloem. The primary symptoms of damage caused by these herbicides include root trimming, purpling, and interveinal chlorosis. The main chemical families in this group include sulfonylurea, imidazolinone, and triazolopyrimidine herbicides [45]. Common herbicides include flumetsulam, pyroxsulam, diclosulam, nicosulfuron, chlorsulfuron, rimsulfuron, prosulfuron, and triasulfuron, imazamox, imazapic, imazaquin, and imazethapyr. In Dor and colleagues' studies [46], Group 9 herbicides belong to the second class of herbicides that inhibit synthesis of amino acids by blocking enzyme production in the 5-enolpyruvylshikimate-3-phosphate (EPSP) pathway. These enzymes are essential for the production of amino acids that are aromatic, such as tryptophan, tyrosine, and phenylalanine. The inhibition of this pathway disrupts protein synthesis, leading to cell membrane breakdown and, ultimately, plant death. The chemical family that inhibits the EPSP pathway includes glyphosate, a broad-spectrum herbicide [47]. This method of action involves herbicides that are absorbed through phloem tissues and are nonselective. They cause significant damage to foliage, eventually leading to discoloration, stunted growth, and plant death. Although plant growth and development are monitored immediately after herbicide application, the primary effects typically do not become apparent until several days later.

9.2.3 Photosynthesis Inhibitors

Inhibitors of photosynthesis disrupt the procedure by attaching to particular locations in PSII within plant cell chloroplasts. When photosynthesis is inhibited, plants may slowly starve due to impaired starch translocation or die rapidly from the accumulation of secondary toxic compounds. PSII herbicides are categorized into various chemical classes, including triazinones, s-triazines, uracils, ureas, phenyl carbamates, anilides, cyanophenols, and dinitrophenols. Considering their location of the action, these herbicides are categorized into three groups: 5, 6, and 7 [48]. Frequently applied herbicides with this mechanism of action include pyrazone, bentazon, diuron, linuron, terbacil, bromoxynil, bromacil, atrazine, simazine, the substance metribuzin, hexazinone, and diuron. Group 5 herbicides function as post-emergence contact herbicides, inhibiting photosynthesis by attaching itself to PSII's serine. These herbicides are taken up by the roots and branches and transported *via* the Group 6: Phloem and xylem herbicides also act as post-emergence contact herbicides, making foliar spraying the recommended application method. Herbicides belonging to Group 7 attach to thylakoid membrane protein complexes, disrupting electron movement within the chain of electron transport. The blockage of electron transport leads to decreased fixation of carbon dioxide and decreased synthesis NADPH2 and ATP, which serve as the energy carriers in respiration. These herbicides effectively control both broadleaf and narrowleaf weeds. They perform optimally during daylight hours when sunlight directly activates their mechanism of action. Within a few hours of application, symptoms such as chlorosis and necrosis appear along the leaf edges and gradually extended toward the base of the leaves.

9.2.4 Growth Regulators

Herbicides that regulate plant development typically function by disrupting the balance of regulators and growth hormones in the plant body. This mechanism of action is represented by two distinct site-of-action groups, including Group 4 herbicides such as 3,4-D, which are synthetic auxins. These synthetic auxins cause growth irregularities by altering indole acetic acid levels, ultimately leading to plant mortality. The primary chemical families include Benzoic, carboxylic, and phenoxylic acids. Commonly used herbicides in this category include dichlorprop, MCPA, MCPB, 2,4-D, 2,4-DB, dicamba, clopyralid, and picloram. These post-emergence herbicides are primarily used to control broadleaf weeds in narrow-leaved plant species [49]. Their activity is regulated by multiple processes, including the disruption of cell wall integrity and nucleic acid metabolism. However, due to herbicide resistance and the development of alternative herbicides, the effectiveness and usage of synthetic auxin herbicides have declined in recent years. Herbicides that inhibit auxin transport in meristematic regions fall under the growth regulator mode of action. Most of these herbicide groups are known to disrupt vesicle trafficking in plants, thereby preventing the two-way movement of auxin, although their exact mechanism of action remains unclear [50]. The primary chemical family, semicarbazone, is effective in controlling broadleaf weeds in grass crops. These pesticides are absorbed by both the roots and leaves of plants and are subsequently translocated through the xylem and phloem. These herbicides are more effective in controlling broadleaf weeds when applied either before or after emergence. They disrupt normal growth and reproduction, leading to stem and stalk brittleness, leaf distortion, cupping, and abnormal tissue outgrowths.

9.2.5 Nitrogen Metabolism Inhibitors

This mode of action is associated with Group 10 herbicides, which target the glutamate production pathway. These herbicides inhibit the synthesis of glutamate synthase enzymes, which are necessary for converting ammonia into additional nitrogenous substances [51]. As a result of this blockage, ammonia ions accumulate within the plant, increasing the pH of surrounding tissues. This leads to cellular rupture, fatty acid degradation, protein breakdown, and overall ion imbalance within the cell sap.

Glufosinate, a commonly marketed herbicide, belongs to the phosphorylated amino acid chemical family. These herbicides function as post-emergence, broad-spectrum weed control agents with limited translocation within plant systems, making spraying the preferred application method for optimal effectiveness. The primary damage caused by these herbicides is foliar injury, with symptoms more pronounced in younger leaves compared to typical deficiency or plant stress indicators.

9.2.6 Pigment Inhibitors

Pigment-inhibiting herbicides disrupt the synthesis of essential pigments such as chlorophyll, retinol, carotene, and anthocyanins. These pesticides fall under Group 12 in the location of the action classification, targeting the enzyme 4-hydroxyphenylpyruvate dioxygenase (HPPD) (dehydrogenase), which has a crucial part in pigment biosynthesis. Group 13 herbicides, on the other hand, inhibit retinol production and degrade phytin pigments, further affecting plant development [52]. Herbicides from three main chemical families—pyrazole, pyrazolone, and pyridazinone—are widely used for their pigment-inhibiting properties. Commonly used herbicides in this category include amitrole, clomazone, isoxaflutole, and mesotrione. These herbicides significantly reduce pigment levels, resulting in the existence of free lipid radicals that contribute to lipid oxidation, protein malfunction, and cell membrane rupture. The resulting damage is highly visible, with immediate bleaching or whitening of treated plants. Over time, symptoms progress to transparent leaf expression as well as quick withering of weed species in the affected area.

9.2.7 Cell Membrane Disrupter

These pesticides disrupt cell structure and function by interfering with cell membrane activity. Group 14 herbicides PPO inhibition, an enzyme that facilitates the transformation of protoporphyrinogen IX (PPGIX) into protoporphyrin. The build-up of PPGIX leads to the formation of PPGIX triplet, which, when exposed to light, can break fatty acid bonds, disrupt hydrogen bonding, and degrade protein structures. Additionally, triplet PPGIX may hinder the synthesis of heme and chlorophyll pigments.

The substance families associated with these herbicides include triazolinone, thiadiazole, diphenyl ether, and trifluoromethyl uracil. Common herbicides within these groups include sulfentrazone, lactofen, acifluorfen, fomesafen, and flumiclorac. Another class of herbicides that disrupt cell membranes belongs to Unit 22, which inhibits PSI, while photosynthesis is occurring. The primary bipyridylium is a chemical species in this category that includes widely used commercial herbicides such as paraquat and diquat. These compounds divert electrons from PSI, generating herbicide radicals that react with oxygen to form hydroxyl radicals and hydrogen peroxide. These ROS, or reactive oxygen species, break down proteins, lipids, unsaturated fatty acids, and chlorophyll in the cell membrane, ultimately leading to cell destruction. These herbicides function through a contact mode of action and are pesticides used after emergence that become active under strong light [53]. They have been found to effectively control weeds even before they reach maturity. Within 1 to 2 hours of application, visible injury symptoms appear, including necrosis, browning, and water-soaked foliage.

9.2.8 Seedling Shoot Growth Inhibitors

Herbicides that act on Group 8 location that block lipid synthesis by using the chemical thiocarbamate are known as seedling shoot growth inhibitors. These substances prevent the formation of gibberellins, fatty acids, proteins, and flavonoids. Long-chain fatty acid inhibitors are another class of herbicides that limit the development of seedling shoots. These herbicides impede the formation of long-chain fatty acids in seedlings by conjugating with acetyl CoA to produce thiocarbamate sulfoxide [54]. Herbicides with this mechanism of action belong to the chloroacetamide chemical family, which includes alachlor, butachlor, and metolachlor. These compounds are primarily utilized as herbicides before emergence, with their effectiveness enhanced when incorporated into the soil. Due to their volatility, the absorption of these herbicides by plant roots and emerging shoots is then moved through the xylem. The primary phytotoxic effects of this chemical group include stunted growth and enlarged cotyledons.

9.2.9 Seedling Root Growth Inhibitors

These herbicides are within the site-of-action's Group 3 classification, which prevents early seedling growth of roots by disrupting microtubules in cell walls. Due to this mechanism, they are often referred to as microtubule inhibitors. These compounds hinder cell division by preventing The formation of the tubulin–herbicide complex within microtubules ultimately blocks microtubule polymerization and disrupts development of the root cell wall [55]. These pesticides are applied directly to the soil for optimal effectiveness, making them pre-emergence herbicides. They belong to the dinitroaniline chemical family, with pendimethalin being a commonly marketed example. Other frequently used herbicides in this group include cycloate, ethafluralin, and trifluralin. The primary adverse effects of these herbicides include swelling of the coleoptile and brittle, callus-forming, and hypocotyl stems and the development of brief secondary roots.

9.2.10 Several Innovative Ways That Herbicides Work

Due to the growing challenge of herbicide resistance, researchers are actively exploring novel mechanisms of herbicide activity. Investigating herbicides that have unique modes of action could provide effective solutions for weed species that have developed resistance to conventional herbicides. Various techniques are being used to identify herbicides that target new sites of action. One approach involves investigating phytotoxic primary and secondary metabolites, including protoporphyrin IX and sphingoid bases, which have attracted considerable scientific interest. Another widely used method in research is identifying potential sites of action by analyzing enzyme activity at extremely low levels [56]. When applied to *Arabidopsis thaliana*, the herbicide targeting dihydrodipiconitase synthase (DHDPS)—an enzyme that catalyzes the first and rate-limiting step in lysine biosynthesis—has demonstrated effectiveness. Inhibitors from this herbicide family bind to newly identified, previously unknown pockets within DHDPS, leading to symptoms such as delayed germination and stunted growth [57]. Tetflupyrolimet, a novel herbicide classified under Group 28 of the Weed Research Society of the United States, has demonstrated effectiveness in controlling long-season grass weeds in rice fields. This herbicide exerts its action by inhibiting dihydroorotate dehydrogenase and homogentisate solanesyltransferase [58].

9.3 Heterocyclic Having Herbicides Activity

One of the largest risks to the world's food supply, weeds deprive crops of nutrients, light, or water, costing the economy billions of dollars annually [59, 60]. Figure 9.5 demonstrates the several types of herbicidal heterocyclic compounds used to control weeds.

Weed control is therefore essential to contemporary sustainable agricultural output [61, 62]. Herbicides are the most cost-effective and efficient technique to eradicate weeds in agricultural fields, although there are other methods as well [63].

Therefore, the development of novel herbicides is crucial for crop protection [64–66]. When developing herbicides, phytoene desaturase (PDS)

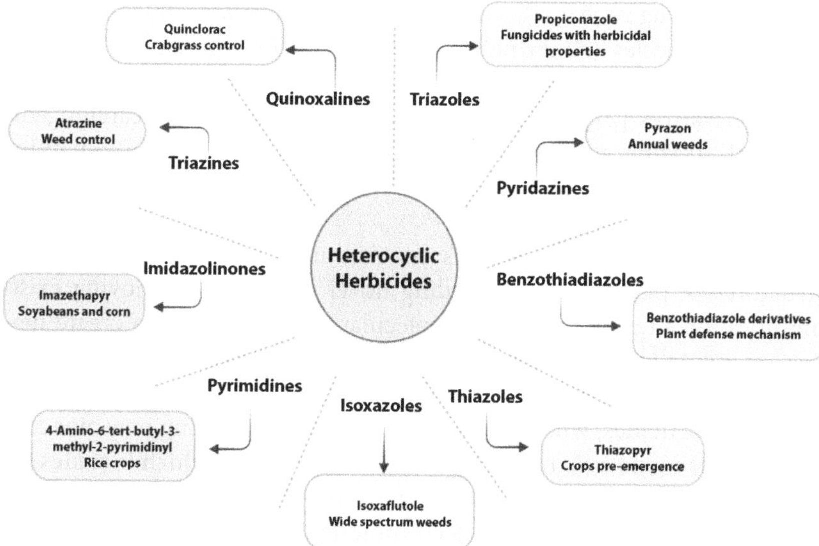

Figure 9.5 Classification of heterocyclic herbicides.

was a key target [67]. In the synthesis of carotenoids, it is a rate-limiting protein that catalyzes the symmetric desaturation of phytoene to carotene. Inhibition of PDS within plants could block the biosynthesis of carotene and lead to the accumulation of phytoene, which, in turn, would result in the photo-oxidation of the plant cell. Then, the treated new leaves would first exhibit a distinct whitening symptom, then the entire leaf would begin to whiten, and finally the plant would die [68, 69].

9.3.1 Triazines

Broadleaf and grassy weeds are frequently controlled both before and after emergence. To meet one of the current top priorities—supplying adequate food for the expanding global population—synthetic pesticides are still often used in agriculture across the world. Triazine derivatives are a class of nitrogen-containing chemicals that are frequently used in agricultural cultivation to suppress weeds [70, 71]. A variety of fungi, such as *Aspergillus niger*, *Candida albicans*, *Colletotrichum capsici*, *Cryptococcus neoformans*, and others, are susceptible to the antifungal activity of some triazine derivatives [72–76]. However, both biotic and abiotic mechanisms have the ability to break down triazines, producing potentially harmful metabolic products. *Aspergillus* species (atrazine, simazine, propazine), *Penicillium* species (the drug simazine), *Arthrobacter nicotinovorans* (atrazine, which), *Variovorax* species (simazine), *Phanerochaete chrysosporium* (atrazine), and other microbes have a significant capacity for biodegradation of triazine derivatives [77]. Because of its strong mutagenic and carcinogenic potential, atrazine, among the most commonly used herbicides containing triazine, can accumulate in the surroundings and pose a harm to survive things [78].

Numerous current research is concentrated on altering the current structures of pesticides by including novel substitutes, removing existing ones, or altering important molecular characteristics to create new environmentally friendly, effective, affordable, and safer insecticides. This is due to the numerous issues (health and environmental risks) with the current insecticides, whose use is allowed globally. Finding new bioactive chemicals is made easier by computational chemistry, which includes the link between quantitative structure and property (QSPR) and quantitative structure and activity (QSAR) techniques. When used in conjunction with regression and chemometric pattern recognition techniques, QSPR and QSAR methodologies help researchers anticipate the physicochemical characteristics and biological activity of new compounds by extracting important information from the data pool. Four hybrids of anilinoquinoline and triazine were modeled as inhibitors of pf-DHFR using the QSAR approach and molecular docking [79]. In addition, a group of triazine herbicides' antigen–antibody interaction processes was modeled using computer-aided molecular modeling [80]. A collection of triazine chemicals' toxicity was effectively predicted using QSAR in conjunction with chemometric approaches [81]. Several studies have focused on QSAR analyses of the binding interactions between monoclonal and polyclonal antibodies and triazine compounds [82] and the examination derived from triazine as potential inhibitors of h-DAAO using molecular docking, molecular dynamics models, and 3D-QSAR [83].

Compounds 1–3 (Figure 9.6) have a greater lipophilicity than commercial ones, making them possible contact pre-emergence herbicides.

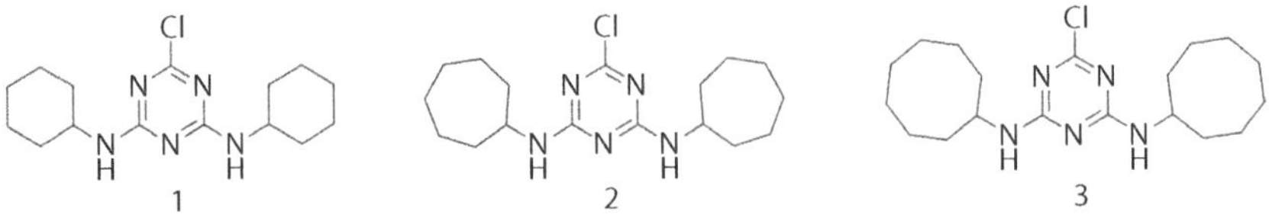

Figure 9.6 Triazine derivatives as potential h-DAAO inhibitors.

Because alkyl derivatives are more polarizable than cycloalkyl compounds, they are more prone to contaminate the air. Cycloalkyl derivatives have a greater potential to inhibit the synthesis of aflatoxins found in *Aspergillus flavus* and *Aspergillus parasiticus*, according to the examination of docking, which additionally showed that they had a greater capacity to block the biosynthetic enzyme 6IX8, which was overlaid with the Omt-1 protein, compared to alkyl derivatives.

The results of the molecular docking research also suggested that cycloalkyl derivatives are probably not the best herbicides to use on popular crops such as maize, rice, barley, and wheat. The nonselective activity of the chemicals toward certain protein groups and grains were noted because hierarchical clustering did not categorize the cereals or enzymes according to their activities and kinds [84].

General procedure for the synthesis of symmetrical 2,4,6-trisubstitued1,3,5-triazine

A solution was created by combining benzaldehyde (0.5 mmol, 1.0 equivalent), NH_4OAc (0.5 mmol, 1.0 equivalent), and CuCl2 (10 mol%) in toluene in a 25 mL RBF that held 2 mL. For 15 hours, at reflux, the resulting mixture was agitated, shown in Scheme 9.1. Once the reaction was finished, the mixture was given time to cool to room temperature before being diluted with 10 mL of ethyl acetate, washed with 5 mL of water, and then extracted with 3 to 5 mL of ethyl acetate. The solution dried on anhydrous Na_2SO_4 and then focused with less pressure.

9.3.2 Triazinone

Herbicides in the triazinone class primarily act by disrupting essential biochemical pathways in target weed species. Because protoporphyrinogen IX oxidase (PPO) plays a crucial role in chlorophyll production, these

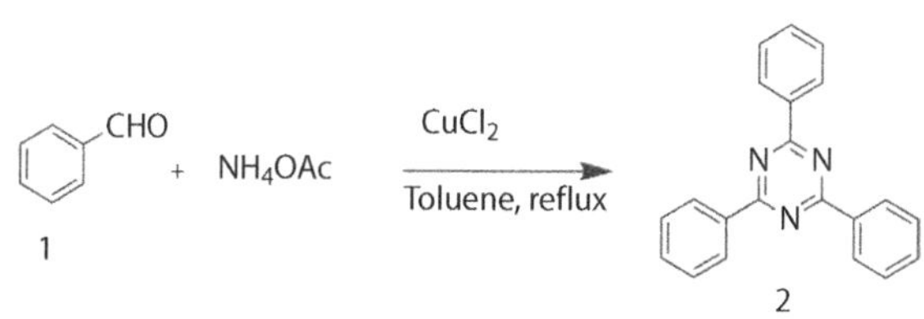

Scheme 9.1 NH4OAc-mediated synthesis of triazine.

herbicides are widely recognized for their effectiveness in controlling resistant weed populations. The primary mechanisms of action of triazinone herbicides are detailed in the following sections.

i) **Inhibition of protoporphyrinogen IX oxidase**
   - Triazinones, including trifludimoxazin, function as PPO inhibitors, leading to the accumulation of harmful porphyrin molecules in plants, ultimately resulting in cell death.
   - This mode of action is highly beneficial for crop management, particularly in controlling resistant weed species.

ii) **Absorption and translocation**
   - After being absorbed through plant surfaces, these herbicides are transported to specific target sites, where they disrupt normal physiological functions [85].
   - Many triazinone compounds demonstrate high potency, with effective concentrations ranging from 37.5 to 150 g ai/ha.

iii) **Crop safety and efficacy**
   - New formulations of triazinone derivatives have improved crop safety, allowing for higher application rates without harming crops such as wheat and rice [86].
   - Achieving this balance between safety and effectiveness is essential for sustainable agriculture.

Despite their effectiveness in weed management, ongoing research focuses on developing herbicides with dual modes of action to enhance efficacy and mitigate resistance. This approach could provide a more comprehensive solution to agricultural weed management challenges [87].

9.3.3 Pyridine

Aminopyralid, a pyridine herbicide, is highly effective in controlling various invasive plant species, particularly in agricultural and noncrop environments. Its application has significantly reduced the production of viable seeds in invasive species, making it a valuable tool for land management. The following sections provide a detailed overview of aminopyralid's efficacy, application methods, and potential synergies with other herbicides.

i. **Efficacy in weed control**
   - When applied at the flowering stage, aminopyralid reduces seed viability in invasive grasses such as barb goatgrass by 65% to 95% [88].
   - It effectively controls invasive winter annual grasses and broadleaf weeds in treated fields, thereby enhancing forage production [89].

ii. **Application methods**
   - To expand its control spectrum and minimize environmental impact, aminopyralid is formulated in various mixtures, including combinations with clomazone and mesotrione [90].
   - It has demonstrated high efficacy in managing annual weeds in maize and sugarcane crops.

iii. **Synergistic effects**
   - Combining aminopyralid with other herbicides, such as triclopyr, enhances weed control efficiency while reducing the required treatment rates [91].
   - Such combinations provide broader weed management with lower environmental risks [92].

Although highly effective, aminopyralid must be applied with caution due to its phytotoxicity, which poses a risk to desirable broadleaf plants. This underscores the importance of an integrated weed management approach to balance efficacy with ecological sustainability.

9.3.4 Pyrazole Benzophenone Derivatives

In a single catalytic cycle, HPPD catalyzes the decarboxylation, substituent migration, and aromatic oxygenation processes necessary to convert 4-hydroxyphenylpyruvic alcohol to homogentisate (HGA) [93–95]. The manufacture of tocopherol and plastoquinone, which are essential for plants to develop normally, depends on HGA as a precursor. When HPPD is suppressed, the plant will suffer significant harm; the meristem will bleach in the sun, resulting in necrosis and eventual death [96, 97]. Thus, bleaching herbicides are another name for HPPD inhibitors [98–100]. These herbicides' broad range of weed-control properties, adjustable treatment schedules, and compatibility with tank mixes with other herbicides are their main advantages [101, 102].

With its strong activity, minimal residue, and application safety, the HPPD inhibitor may also eradicate perennial broadleaf weeds in broadleaf plant fields. In the fields of biology and pharmacology, pyrazole and its derivatives are attracting a lot of interest [103]. An essential chemical scaffold is N-substituted pyrazole, and several pyrazole derivatives with notable bioactivity have been created. N-substituted pyrazoles, for instance, are effective p38 mitogen-activated protein kinase inhibitors that can be used to treat cancer cells [104] Pyrazole–benzimidazolone has been suggested as a possible inhibitor of type I tyrosinemia [105]. Additionally, it was confirmed that pyrazolone quinazolone hybrids were new inhibitors of HPPD in humans [106]. These pyrazole inhibitors frequently contain the organic compound 1-ol 2-benzoylethen. To continue developing and creating new nitrogen-containing heterocycle insecticides and to look for potential analogs of HPPD inhibitors, researchers offer the combination of a number of new pyrazole compounds through cyclization along with acylation without the need of any pricy catalyst or reagent; 0.05 mol/m^2 is the rate at which some of them demonstrated promoting herbicidal effects against field grass in greenhouse testing.

9.3.5 Imidazolines

The literature has detailed the 5-imidazolinones' substantial biological significance in the realm of medications and medicines [107]. The 5-imidazolinone central moiety is found in several heterocyclic compounds that are utilized as antitubercular, antihistamine, antibacterial, antiviral medications, antihelminthic, sedatives, hypnotics, and local anesthetics. Furthermore, it was discovered that oxo imidazolines, sometimes referred to as imidazolines, were highly potent anticancer, fungicidal, herbicidal, and vasodilator agents. The six commercial imidazoline herbicides (Figure 9.7) have different levels of herbicidal action and crop selectivity based on their chemistry.

All of these herbicides include nicotinic acid 2-(4-isopropyl-4-methyl-5-oxo-4,5-dihydro-1H-imidazol-2-yl) as its main derivative (or its benzoic acid analog, which occurs as a mixture of the m- and p-isomers).

These herbicides are beneficial for a variety of crops used in the production of food and fiber because of their adaptability, low toxicity, and environmental safety. The effectiveness and prospective applications of these imidazolinones for weed control have been extensively studied due to their crucial functions in agriculture. In view of these facts, we present here our experiment to investigate the synthesis of certain novel 5-imidazolinone derivatives as a safe substitute for toxic pesticides against cotton leaf worms.

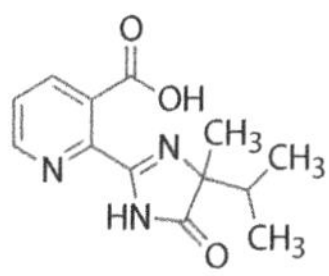
4

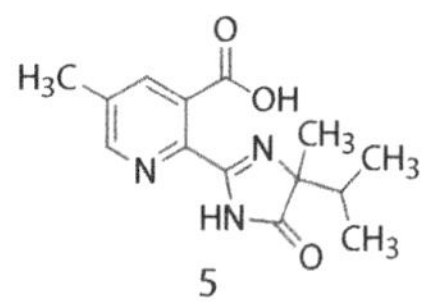
5

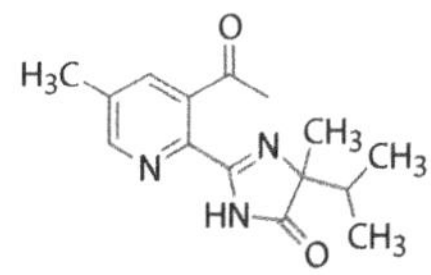
6

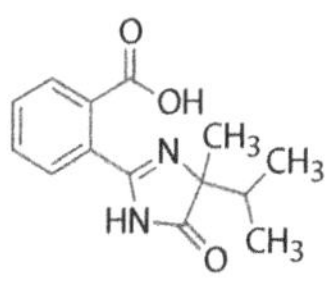
7

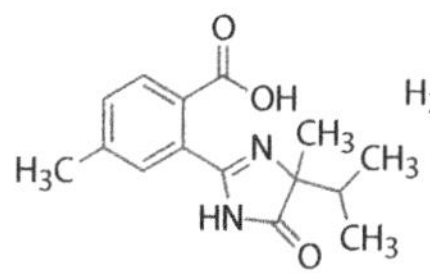
8

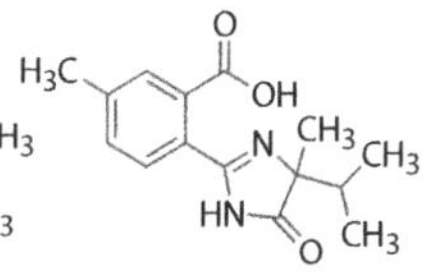
9

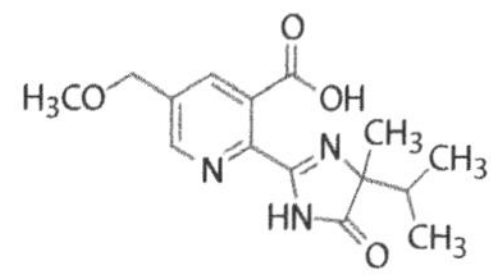
10

Figure 9.7 Six commercial imidazoline herbicides.

Until new biologically based management systems can be created, insecticides will remain the mainstay of Egypt's strategy for managing Egyptian cotton leaf worms.

9.4 Environmental Impact and Resistance Management

9.4.1 Environmental Considerations

Due to historical and ongoing greenhouse gas emissions, there will probably be a rise in global temperature, irregular rainfall distribution, and altered wind patterns, as well as extreme weather events such as severe hurricanes, droughts, and flooding [108–110].

The quality and quantity of food and feed crops are known to be significantly impacted by climate change, which has major ramifications for agriculture; altered cropping systems and livestock production; and crop management practices for sustaining crop output under scenarios of anticipated climate change [111]. Along with these effects, weed control is made more difficult by a changing climate [112]. Shown as Table 9.2 of weeds are the most dangerous of the various biotic factors that lower crop output; they can cause yield losses of 34% to 37%, compared to pests (18%–29%), diseases (16%–22%), and other factors (12%) [113, 114].

Furthermore, unchecked weeds harm crops in a number of ways, both directly and indirectly. Weeds can adjust to shifting climatic conditions because of their enormous genetic variety and innate phenotypic flexibility.

Table 9.2 Herbicide mode of action and its impact on plant physiology.

Mode of action	Description	References
Inhibition of photosynthetic electron transport	Interferes with electron transport in thylakoid membranes, disrupting ATP and NADPH synthesis	[115]
Inhibition of protoporphyrinogen IX oxidase (PPO)	Prevents heme and chlorophyll biosynthesis, leading to oxidative damage	[116]
Disruption of microtubule formation	Inhibits cell division and elongation, causing root tip swelling	[117]
Inhibition of acetolactate synthase (ALS)	Reduces branched-chain amino acid synthesis, essential for protein production	[118]
Enantioselective effects	Different enantiomers exhibit distinct modes of action	[119]
Enhanced crop safety	Pharmacokinetic properties ensure lower toxicity to crops	[120]
Proteomics-based insights	Molecular-level understanding of target protein variations	[121]
Dual-action mechanisms	Combines multiple modes of action for enhanced herbicidal activity	[122]
Environmental and toxicological considerations	Generally low toxicity to animals, but resistance management is crucial	[123]

A change in climate can result in a change in the floral species diversity of different ecosystems because climate factors affect plant flowering, fruiting, or seed dormancy [124] shown as Table 9.3. By migrating, modifying their phenology as well as physiology, and changing their development patterns, weed species will adjust to the changing climate. In addition to influencing plant development, higher temperatures and changed soil moisture levels can have an impact on how well pesticides work to control weeds. Because farmers all over the world mainly depend on herbicides to control weeds in their crops, rangelands, or fallow lands, declining herbicide efficacy is an issue [125]. Herbicide efficiency is influenced by environmental conditions

Table 9.3 Classification of heterocyclic herbicides by persistence and selectivity.

Heterocyclic class	Example compound	Soil persistence	Selectivity
Triazine	Atrazine	Moderate (30–60 days)	Selective (corn, sorghum)
Triazinone	Metribuzin	Moderate (30–60 days)	Selective (soybeans, potatoes)
Pyridine	Fluroxypyr	Low (10–30 days)	Selective (grasses, cereals)
Isoxazole	Isoxaflutole	High (>90 days)	Selective (corn)
Pyrazole	Pyrazosulfuron-ethyl	Low (10–30 days)	Selective (rice)
Imidazolinone	Imazapic	High (>90 days)	Selective (soybeans, peanuts)
Quinoline	Quinclorac	Moderate (30–60 days)	Selective (rice)
Thiadiazole	Flumioxazin	Low (10–30 days)	Selective (soybeans, peanuts)
Benzothiadiazole	Bentazon	Low (10–30 days)	Selective (soybeans, peanuts)
Diketopiperazine	Asulam	Moderate (30–60 days)	Selective (sugarcane, pastures)
Pyrimidinedione	Saflufenacil	Moderate (30–60 days)	Nonselective (broadleaf weeds)
Oxazolidinone	Clomazone	Moderate (30–60 days)	Selective (rice, soybeans)

in addition to plant characteristics and chemical characteristics. Herbicide absorption in weeds depends on how the herbicide interacts with the soil, plant leaves, atmosphere, and the interface between the soil, plant leaves, and atmosphere [126]. Therefore, the growth of weed plants or the effectiveness of herbicides can be greatly impacted by changes in the climate brought on by rising CO_2 levels, as well as their effects on the Earth's average surface temperature or changing rainfall patterns.

9.4.2 Fundamentals of How Weeds React to a Climate with Higher CO_2 and Temperature

Weed species must be able to adjust to shifting climatic circumstances to continue to survive. Three distinct adaptation strategies are available to weeds for survival [127].

- **Migration**: Also known as range shift, this process involves weed species moving or shifting to a more advantageous location. Weed seeds are also spread by a number of human actions, including the use of organic manures and crop seeds contaminated with weed seeds, the transportation of agricultural apparatus, and the transportation of crop products contaminated with weed seeds [128, 129].
- **Acclimation**: This process, also known as a niche shift, describes how plant species respond to changing environmental conditions by changing their phenotypes; nevertheless, these changes are neither heritable nor evolutionary [130].

9.4.3 Resistance Develop in Transgenic Plants

The fact that herbicides successfully suppress weeds contributes to excellent agricultural yields and high-quality crops is undeniable. Effective herbicides that are widely used across the world are often nonselective and broad range. Their application in arable fields is thus limited. Overuse of herbicides, which promotes the selection of resistant weeds, is another contributing factor. The creation of herbicide-resistant transgenic plants (HRTPs) was required by the aforementioned facts. Typically, herbicides target vital plant metabolic processes including photosynthetic, mitotic, or amino acid production. Both weeds and crops go through these processes. Crops that are herbicide-tolerant have been engineered using biotechnological techniques. The commercial potential of genetically modified crops has motivated several scientists and businesses worldwide to create HRTP [131].

According to one study, at an annual growth rate of 6%, the area under biotechnology-produced crops reached 170.3 million hectares. In 2012, genetically modified crops were cultivated worldwide, covering an additional 10.3 million hectares compared to 2011, bringing the total global area to 170.3 million hectares. The amount of acreage covered by transgenic crops has increased by an impressive 100 times when compared to statistics from 1996. In 2010, 70% of all types of maize, 78% all cotton, and 93% of all soybeans in the United States were resistant to herbicides. Additionally, since 1996, the number of transgenic herbicide-resistant crops has significantly increased in comparison to the increase in transgenic crops that are resistant to insects (Bt crops). Resistance status is shown in Table 9.4. Globally, HRTP covers more than 100 million hectares of land, outnumbering insect-resistant transgenic crops, or Bt crops [132].

9.5 Field Applications

Due to their versatility, herbicides provide several benefits over other weed-control techniques. The following benefits emphasize the significance of herbicides:

- Herbicides can be sprayed prior to weed growth.

Applying herbicides during the pre-plant as well as pre-emergence phases is feasible, in contrast to other weed management techniques. This enables the crops to develop in an environment free of weeds and free from competition for nutrients, even in their early stages.

- Herbicides work well in a variety of crop configurations.

Although mechanical weed control methods can be useful in some situations, they are unable to completely eradicate all weeds in the field, something herbicides can readily accomplish with broadcast-sown and narrow-spaced crops. Intra-row weeds can also be efficiently targeted with herbicides.

- Deep-rooted weeds can be eliminated with herbicides.

Deeply rooted weeds are sometimes impossible to eradicate mechanically because, even if the top portion of the plant may be removed, the roots may allow the weed to regrow. For these, successful weed control may require a mix of chemical and mechanical herbicides.

Table 9.4 Commercial herbicides and their resistance status in agriculture.

Herbicide showing resistance	Resistance status	Commercial brand(s)	Primary crops
Atrazine	Widespread resistance in weeds	Aatrex, Gesaprim	Corn, sorghum, sugarcane
Metribuzin	Moderate resistance reported	Sencor, Lexone	Potatoes, soybeans, tomatoes
Clopyralid	Low resistance reported	Stinger, Transline	Wheat, barley, pastures
Isoxaflutole	Emerging resistance cases	Balance Flexx	Corn
Pyrazosulfuron-ethyl	Widespread resistance in grasses	Sirius, Liquis	Rice
Imazethapyr	High resistance in weeds	Pursuit	Soybeans, peanuts, alfalfa
Quinclorac	Moderate resistance reported	Facet, Drive	Rice, turfgrass
Flumioxazin	Moderate resistance cases	Valor, SureGuard	Soybeans, cotton, grapes
Bentazon	Limited resistance cases	Basagran	Peanuts, beans, corn
Asulam	No significant resistance	Asulox, Asulam 80	Sugarcane, turf, pastures
Saflufenacil	Emerging resistance cases	Sharpen, Detail	Cereals, soybeans
Clomazone	No major resistance reported	Command	Rice, soybeans

• For a long period, herbicides stop weeds from growing.
Herbicides have a long-lasting impact and stop weeds from coming back quickly after they are treated.

• Weeds with structural similarities can be eliminated by herbicides.
Because it is difficult to tell the difference between weeds and crops, weeds that seem identical to the crops may be overlooked when they are removed mechanically. Herbicides, on the other hand, can be used to eradicate these weeds—which may have the same appearance but vary biologically—without harming the crops.

Advantages of herbicides are shown in Figure 9.8.

Crop application timing, crop safety parameters, and application methods are crucial factors in ensuring effective and safe agricultural practices (shown in Table 9.5). Proper application timing is essential to maximize crop uptake and minimize environmental impact, as applying fertilizers or pesticides at the wrong stage can lead to reduced efficacy or potential crop damage. Crop safety parameters, such as tolerance levels and phytotoxicity risks, must be considered to prevent adverse effects on plant health. Additionally, the choice of application method—whether foliar spraying, soil incorporation, or seed treatment—plays a significant role in determining the effectiveness of the treatment. Selecting the appropriate method based on crop type, growth stage, and environmental conditions enhances productivity while minimizing risks to both the crop and the surrounding ecosystem.

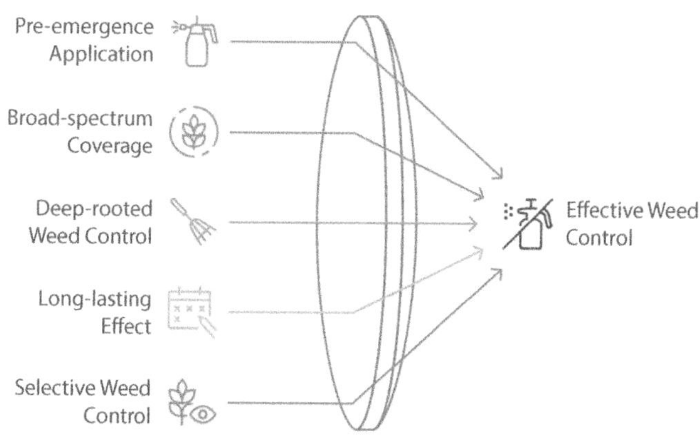

Figure 9.8 Advantages of herbicides.

Table 9.5 Application strategies of heterocyclic herbicides in agriculture.

Heterocyclic class	Application timing	Crop safety	Application method
Triazine	Pre- and post-emergence	Selective (corn, sorghum)	Soil or foliar spray
Triazinone	Pre- and early post-emergence	Selective (potatoes, soybeans)	Broadcast or band spray
Pyridine	Post-emergence	Selective (cereals, pastures)	Foliar spray
Isoxazole	Pre-emergence	Selective (corn)	Soil incorporation
Pyrazole	Post-emergence	Selective (rice)	Foliar spray
Imidazolinone	Pre- and post-emergence	Nonselective (affects most crops)	Soil and foliar application
Quinoline	Post-emergence	Selective (rice, turfgrass)	Foliar spray
Thiadiazole	Pre-emergence	Selective (soybeans, cotton)	Soil-applied
Benzothiadiazole	Post-emergence	Selective (soybeans, peanuts)	Foliar spray
Diketopiperazine	Post-emergence	Selective (sugarcane, pastures)	Foliar spray
Pyrimidinedione	Pre- and post-emergence	Nonselective (desiccation use)	Soil or foliar application
Oxazolidinone	Pre-emergence	Selective (rice, soybeans)	Soil-applied

9.6 Future Prospects of Heterocyclic Herbicides

The future of heterocyclic herbicides lies in innovation and sustainability. Research is increasingly focused on developing herbicides with novel targets to reduce resistance risks and enhance weed control efficiency. Computational chemistry, including molecular docking and measurable structure–activity relationship mapping experiments, has played a crucial role in designing new herbicidal molecules with high selectivity and activity.

Beyond chemical advancements, precision agriculture is transforming weed management. The integration of artificial intelligence and remote sensing technologies enables site-specific herbicide application, reducing overall chemical usage while maintaining effective weed control. Additionally, herbicide delivery through drones and weeding robots is being explored as an efficient and sustainable alternative to conventional spraying methods.

Another promising research avenue is the development of bioherbicides derived from biological sources. Various microorganisms and plants produce metabolites with herbicidal properties that can serve as environmentally friendly weed control agents. By incorporating bioherbicides alongside synthetic herbicides, researchers aim to create synergistic formulations that maximize effectiveness while minimizing environmental impact.

Heterocyclic herbicides have long played a vital role in modern agriculture by providing cost-effective and efficient weed control solutions. Their chemical diversity and diverse modes of action make them indispensable tools for managing invasive plant species and optimizing crop yields. However, the rise of herbicide-resistant weeds and increasing environmental concerns necessitate continuous innovation and responsible application.

To sustain the benefits of heterocyclic herbicides, a balanced, integrated approach is essential—combining chemical and nonchemical weed management strategies while adhering to environmentally conscious principles. The convergence of advances in chemistry, biotechnology, and precision agriculture will shape the future of weed control, guaranteeing food safety for a world population that is expanding while preserving ecological integrity.

As the agricultural sector continues to evolve, heterocyclic herbicides will remain fundamental to weed management. However, their use must align with sustainability principles to safeguard future generations' ability to maintain efficient weed control without compromising environmental health. The success of herbicide development will ultimately depend on the collaboration of scientific research, technological innovation, and ethical farming practices.

9.7 Conclusion

Heterocyclic compounds have emerged as a cornerstone of modern agricultural chemistry due to their significant herbicidal activities. These compounds, which include triazines, pyridines, imidazolines, pyrazoles, and thiadiazoles, have been extensively used in weed management strategies to enhance agricultural productivity. The diverse modes of action of heterocyclic herbicides—such as inhibiting photosynthesis, disrupting amino acid synthesis, interfering with cell membrane integrity, and preventing lipid synthesis—have made them indispensable tools in crop protection.

The increasing global population and rising food demand necessitate sustainable farming practices that maximize crop production while minimizing losses caused by weeds. Weeds compete with crops for essential resources such as nutrients, sunlight, and water, ultimately reducing agricultural productivity. Heterocyclic compound-based herbicides provide an effective and cost-efficient method of weed control, offering both pre-emergence and post-emergence applications based on crop type and weed species.

A key advantage of heterocyclic herbicides is their ability to target weeds selectively or nonselectively, depending on their chemical composition and mode of action. For instance, triazine herbicides, such as atrazine and simazine, inhibit PSII by blocking electron transport and ATP synthesis, ultimately leading to plant death. Similarly, imidazolinone herbicides such as imazamox inhibit ALS (acetolactate synthase), a crucial enzyme involved in the biosynthesis of branched-chain amino acids, thereby disrupting protein formation and halting plant growth.

Another significant category includes pyridine-based herbicides, such as aminopyralid, which function as synthetic auxins. These compounds mimic natural plant hormones, inducing uncontrolled growth that eventually proves fatal to the plant. Pyrazole-derived herbicides, such as pyrazolate, act by interfering with PSII and are frequently used to regulate grassy weeds in rice crops. Additionally, triazinone compounds, including metribuzin and hexazinone, exhibit dual functionality by inhibiting both photosynthetic electron transport and protoporphyrinogen PPO (oxidase), a crucial enzyme in chlorophyll biosynthesis.

These diverse mechanisms make heterocyclic herbicides indispensable for effective weed management, ensuring higher crop yields while preserving soil health. Continuous advancements in herbicide formulations are improving their efficacy, selectivity, and environmental compatibility.

However, the prolonged and widespread use of herbicides has led to an emerging challenge: herbicide resistance. Over time, repeated application of herbicides with similar mechanisms of action leads to the selection of resistant weed populations. For instance, resistance to ALS- herbicide inhibition has been reported in numerous weed species, complicating weed management in croplands. The growing resistance to glyphosate, one of the most widely used herbicides, further underscores the urgent need for alternative strategies in weed control.

To combat herbicide resistance, researchers are investigating multiple approaches, including the development of new herbicides with dual or multiple modes of action. By combining different herbicidal mechanisms, it is possible to slow the development of resistant weed biotypes. Additionally, rotating herbicides with different targets and incorporating nonchemical weed control strategies, such as mechanical weeding and cover cropping, can help mitigate resistance issues.

Although heterocyclic herbicides have proven to be highly effective, their environmental impact remains a concern. Some herbicides, particularly triazines, have been found to persist in soil and water, leading to contamination and potential toxicity to nontarget organisms. As a result, regulatory agencies have imposed restrictions on certain herbicides, prompting researchers to develop eco-friendly alternatives.

Soil persistence, biodegradability, and water solubility are key factors in assessing the environmental safety of herbicides. Herbicides with high soil mobility can leach into groundwater, posing risks to aquatic ecosystems. To mitigate these effects, advancements in herbicide formulation focus on reducing environmental persistence while maintaining weed control efficacy. Modern herbicides are being designed with enhanced selectivity to minimize their impact on nontarget plant species and soil microorganisms.

Additionally, integrated weed management promotes a holistic approach to sustainable agriculture. By combining chemical, biological, and mechanical weed control methods, farmers can achieve effective weed suppression while reducing the environmental risks associated with excessive herbicide use. Genetic engineering has further contributed to weed management by enabling the development of herbicide-resistant transgenic crops, allowing for the targeted application of herbicides without harming crops. However, concerns persist regarding the long-term environmental impact of genetically modified organisms.

AI disclosure statement: The author acknowledges the use of AI-based tools for minor language editing or grammatical correction. However, all intellectual contributions, analyses, and conclusions presented in this chapter are entirely the author's own.

References

1. Alvarez-Builla, J. and Barluenga, J., Heterocyclic compounds: An introduction. *Mod. Heterocycl. Chem.*, 29, 1–9, 2011 Jun.
2. Mahmood, R.M. and Aljamali, N.M., Synthesis, spectral investigation, and microbial studying of pyridine-heterocyclic compounds. *Eur. J. Mol. Clin. Med.*, 7, 11, 4444–53, 2020.
3. Ogawa, Y., Tokunaga, E., Kobayashi, O., Hirai, K., Shibata, N., Current contributions of organofluorine compounds to the agrochemical industry. *Iscience*, 23, 9, 2020 Sep 25.
4. Pozharskii, A.F., Soldatenkov, A.T., Katritzky, A.R., *Heterocycles in life and society: an introduction to heterocyclic chemistry, biochemistry and applications*, John Wiley & Sons, 2011 Jun 7.
5. Dua, R., Shrivastava, S., Sonwane, S.K., Srivastava, S.K., Pharmacological significance of synthetic heterocycles scaffold: A review. *Adv. Biol. Res.*, 5, 3, 120–44, 2011.
6. Fürst, P. and Stehle, P., What are the essential elements needed for the determination of amino acid requirements in humans? *J. Nutr.*, 134, 6, 1558S–65S, 2004 Jun 1.
7. Zhang, S., Liu, H., Yang, N., Xiong, L., Wang, B., Synthesis and evaluation of novel xanthine-acrylamides and xanthine-acrylates as insecticidal agents. *Pest. Manage. Sci.*, 78, 5, 2086–95, 2022 May.
8. Zhong, L.K., Sun, X.P., Han, L., Tan, C.X., Weng, J.Q., Xu, T.M., Shi, J.J., Liu, X.H., Design, Synthesis, Insecticidal Activity, and SAR of Aryl Isoxazoline Derivatives Containing Pyrazole-5-carboxamide Motif. *J. Agric. Food. Chem.*, 71, 40, 14458–70, 2023 Oct 2.
9. Zhang, Z., Sun, P., Zhao, J., Zhang, H., Wang, X., Li, L., Xiong, L., Yang, N., Li, Y., Yuchi, Z., Li, Z., Design, synthesis and biological activity of diamide compounds based on 3-substituent of the pyrazole ring. *Pest. Manage. Sci.*, 78, 5, 2022–33, 2022 May.
10. Zhao, Y.Q., Li, X., Guo, H.Y., Shen, Q.K., Quan, Z.S., Luan, T., Application of quinoline ring in structural modification of natural products. *Molecules*, 28, 18, 6478, 2023 Sep 6.
11. Frühauf, A., Behringer, M., Meyer-Almes, F.J., Significance of five-membered heterocycles in human histone deacetylase inhibitors. *Molecules*, 28, 15, 5686, 2023 Jul 27.

12. Singla, P., Luxami, V., Paul, K., Triazine as a promising scaffold for its versatile biological behavior. *Eur. J. Med. Chem.*, 102, 39–57, 2015 Sep 18.
13. Tang, X., Fan, L., Zhang, Z., Yang, D., Design, Synthesis and Biological Activity of Dipeptide Derivatives Bearing Uracil Unit. *Chin. J. Org. Chem.*, 39, 5, 1460, 2019 May 25.
14. Diab, H.M., Salem, M.E., Abdelhamid, I.A., Elwahy, A.H., Aminouracil and aminothiouracil as versatile precursors for a variety of heterocyclic systems. *ARKIVOC Online J. Org. Chem.*, 2021, 2021 Jan 1.
15. Reddy, N.N., Hung, S.J., Swamy, M.K., Sanjeev, A., Rao, V.S., Rohini, R., Raju, A.K., Bhaskar, K., Hu, A., Reddy, P.M., Synthesis and rational design of new appended 1, 2, 3-triazole-uracil ensembles as promising anti-tumor agents *via in silico* VEGFR-2 transferase inhibition. *Molecules*, 26, 7, 1952, 2021 Mar 30.
16. Semenov, V.E., Zueva, I.V., Lushchekina, S.V., Suleimanov, E.G., Gubaidullina, L.M., Shulaeva, M.M., Lenina, O.A., Petrov, K.A., Novel Uracil-Based Inhibitors of Acetylcholinesterase with Potency for Treating Memory Impairment in an Animal Model of Alzheimer's Disease. *Molecules*, 27, 22, 7855, 2022 Nov 14.
17. Tatarinov, D.A., Garifullin, B.F., Belenok, M.G., Andreeva, O.V., Strobykina, I.Y., Shepelina, A.V., Zarubaev, V.V., Slita, A.V., Volobueva, A.S., Saifina, L.F., Shulaeva, M.M., The first 5′-phosphorylated 1, 2, 3-triazolyl nucleoside analogues with uracil and quinazoline-2, 4-dione moieties: A synthesis and antiviral evaluation. *Molecules*, 27, 19, 6214, 2022 Sep 21.
18. Yang, J., Guan, A., Wu, Q., Cui, D., Liu, C., Design, synthesis and herbicidal evaluation of novel uracil derivatives containing an isoxazoline moiety. *Pest. Manage. Sci.*, 76, 10, 3395–402, 2020 Oct.
19. El-Kalyoubi, S., Agili, F., Zordok, W.A., El-Sayed, A.S., Synthesis, *in silico* prediction and *in vitro* evaluation of antimicrobial activity, dft calculation and theoretical investigation of novel xanthines and uracil containing imidazolone derivatives. *Int. J. Mol. Sci.*, 22, 20, 10979, 2021 Oct 12.
20. Kezin, V.A., Matyugina, E.S., Novikov, M.S., Chizhov, A.O., Snoeck, R., Andrei, G., Kochetkov, S.N., Khandazhinskaya, A.L., New derivatives of 5-substituted uracils: Potential agents with a wide spectrum of biological activity. *Molecules*, 27, 9, 2866, 2022 Apr 30.
21. Zhou, H., Li, Q.X., Zeng, L., Cao, C., Zhang, T., Zhou, Y., He, H., Uracil hydrazones: Design, synthesis, antimicrobial activities, and putative mode of action. *Pest. Manage. Sci.*, 80, 2, 414–25, 2024 Feb.
22. Mattison, R.L., Beffa, R., Bojack, G., Bollenbach-Wahl, B., Dörnbrack, C., Dorn, N., Freigang, J., Gatzweiler, E., Getachew, R., Hartfiel, C., Heinemann, I., Design, synthesis and screening of herbicidal activity for new phenyl pyrazole-based protoporphyrinogen oxidase-inhibitors (PPO) overcoming resistance issues. *Pest. Manage. Sci.*, 79, 6, 2264–80, 2023 Jun.
23. Park, J., Ahn, Y.O., Nam, J.W., Hong, M.K., Song, N., Kim, T., Yu, G.H., Sung, S.K., Biochemical and physiological mode of action of tiafenacil, a

new protoporphyrinogen IX oxidase-inhibiting herbicide. *Pestic. Biochem. Physiol.*, 152, 38–44, 2018 Nov 1.
24. Min, L., Liang, W., Bajsa-Hirschel, J., Ye, P., Wang, Q., Sun, X., Cantrell, C.L., Han, L., Sun, N., Duke, S.O., Liu, X., Synthesis, Herbicidal Activity, Mode of Action, and *In Silico* Analysis of Novel Pyrido [2, 3-d] pyrimidine Compounds. *Molecules*, 28, 21, 7363, 2023 Oct 31.
25. Silva, J.N., Jakelaitis, A., Zuchi, J., Pereira, L.S., Costa, E.M., Silva, C.H., Production and quality of cowpea seeds desiccated with saflufenacil and flumioxazin. *Rev. Bras. De Engenharia Agrícola e Ambiental*, 25, 3, 209–15, 2021 Jan 22.
26. Wang, D.W., Li, Q., Wen, K., Ismail, I., Liu, D.D., Niu, C.W., Wen, X., Yang, G.F., Xi, Z., Synthesis and herbicidal activity of pyrido [2, 3-d] pyrimidine-2, 4-dione–benzoxazinone hybrids as protoporphyrinogen oxidase inhibitors. *J. Agric. Food. Chem.*, 65, 26, 5278–86, 2017 Jul 5.
27. He, B., Wu, F.X., Yu, L.K., Wu, L., Chen, Q., Hao, G.F., Yang, W.C., Lin, H.Y., Yang, G.F., Discovery of novel pyrazole–quinazoline-2, 4-dione hybrids as 4-hydroxyphenylpyruvate dioxygenase inhibitors. *J. Agric. Food. Chem.*, 68, 18, 5059–67, 2020 Apr 14.
28. Yang, L., Demares, F., Norris, E.J., Jiang, S., Bernier, U.R., Bloomquist, J.R., Bioactivities and modes of action of VUAA1. *Pest. Manage. Sci.*, 77, 8, 3685–92, 2021 Aug.
29. Sivakumar, C., Revathi, B., Balachandran, V., Narayana, B., Salian, V.V., Shanmugapriya, N., Vanasundari, K., Molecular structure, spectroscopic, quantum chemical, topological, molecular docking and antimicrobial activity of 3-(4-Chlorophenyl)-5-[4-propan-2-yl) phenyl-4, 5-dihydro-1H-pyrazol-1-yl] (pyridin-4-yl) methanone. *J. Mol. Struct.*, 1224, 129286, 2021 Jan 15.
30. Kappenberg, Y.G., Nogara, P.A., Stefanello, F.S., Moretto, L.H., Santos, P.M., Rocha, J.B., Zanatta, N., Martins, M.A., Bonacorso, H.G., Design, synthesis, AChE/BChE inhibitory activity, and molecular docking of spiro [chromeno [4, 3-b] thieno [3, 2-e] pyridine]-7-amine tacrine hybrids. *J. Mol. Struct.*, 1266, 133485, 2022 Oct 15.
31. El-Naggar, M.A., Abu-Youssef, M.A., Soliman, S.M., Haukka, M., Al-Majid, A.M., Barakat, A., Badr, A.M., Synthesis, X-ray structure, Hirshfeld, and antimicrobial studies of new Ag (I) complexes based on pyridine-type ligands. *J. Mol. Struct.*, 1264, 133210, 2022 Sep 15.
32. Zhu, K., Lu, X., Li, X., Han, Q., Zou, R., Yang, Z., Li, H., Duan, H., Design and structure optimization of novel butenolide derivatives as low bee-toxicity candidates. *J. Mol. Struct.*, 1282, 135257, 2023 Jun 15.
33. Mahajan, N.D. and Jain, N., Heterocyclic compounds and their applications in the field of biology: A detailed study. *NVEO-Natural Volatiles Essent. Oils J. | NVEO*, 16, 13223–9, 2021 Dec.
34. https://www.google.com/url?sa=t&source=web&rct=j&opi=89978449&url=http://www.jnkvv.org/PDF/04042020123923weed_managementLecture_classification_of_herbicide.pdf&ved=2ahUKEwiJrOn-

h8Y-KAxVik68BHYuFMY0Q-tANegQIGxAc&usg=AOvVaw0t55SfkkkFY-c0xDl3YhdJk
35. Gharde, Y., Singh, P.K., Dubey, R.P., Gupta, P.K., Assessment of yield and economic losses in agriculture due to weeds in India. *Crop Prot.*, 107, 12–8, 2018 May 1.
36. Chauhan, B.S., Grand challenges in weed management. *Front. Agron.*, 1, 3, 2020 Jan 22.
37. Heap, I., The international survey of herbicide resistant weeds, 2010. http://www.weedscience.com.
38. Mendes, K.F., de Sousa, R.N., Laube, A.F., Current approaches to pesticide use and glyphosate-resistant weeds in Brazilian agriculture, in: *Multifunctionality and Impacts of Organic and Conventional Agriculture*, IntechOpen, London, UK, 2020 Mar 26.
39. Vrbničanin, S., Pavlović, D., Božić, D., Weed resistance to herbicides, in: *Herbicide resistance in weeds and crops*, vol. 4, pp. 7–35, 2017 Oct.
40. Sherwani, S.I., Arif, I.A., Khan, H.A., Modes of action of different classes of herbicides. *Herbicides Physiol. Action Saf.*, 10, 61779, 2015 Dec 2.
41. Takano, H.K., Ovejero, R.F., Belchior, G.G., Maymone, G.P., Dayan, F.E., ACCase-inhibiting herbicides: mechanism of action, resistance evolution and stewardship. *Sci. Agric.*, 78, 1, e20190102, 2020 Mar 13.
42. Kukorelli, G., Reisinger, P., Pinke, G., ACCase inhibitor herbicides–selectivity, weed resistance and fitness cost: a review. *Int. J. Pest Manage.*, 59, 3, 165–73, 2013 Jul 1.
43. Kaundun, S.S., Resistance to acetyl-CoA carboxylase-inhibiting herbicides. *Pest. Manage. Sci.*, 70, 9, 1405–17, 2014 Sep.
44. Liu, X.Q., Yu, C.Y., Dong, J.G., Hu, S.W., Xu, A.X., Acetolactate synthase-inhibiting gametocide amidosulfuron causes chloroplast destruction, tissue autophagy, and elevation of ethylene release in rapeseed. *Front. Plant Sci.*, 8, 1625, 2017 Sep 21.
45. Gaines, T.A., Duke, S.O., Morran, S., Rigon, C.A., Tranel, P.J., Küpper, A., Dayan, F.E., Mechanisms of evolved herbicide resistance. *J. Biol. Chem.*, 295, 30, 10307–30, 2020 Jul 24.
46. Dor, E., Galili, S., Smirnov, E., Hacham, Y., Amir, R., Hershenhorn, J., The effects of herbicides targeting aromatic and branched chain amino acid biosynthesis support the presence of functional pathways in broomrape. *Front. Plant Sci.*, 8, 707, 2017 May 4.
47. Dong, Y., Ng, E., Lu, J., Fenwick, T., Tao, Y., Bertain, S., Sandoval, M., Bermudez, E., Hou, Z., Patten, P., Lassner, M., Desensitizing plant EPSP synthase to glyphosate: Optimized global sequence context accommodates a glycine-to-alanine change in the active site. *J. Biol. Chem.*, 294, 2, 716–25, 2019 Jan 11.
48. Oettmeier, W., Herbicides, Inhibitors of Photosynthesis at Photosystem II, in: *Encycl Agrochem*, 2013, Available from: https://onlinelibrary.wiley.com/doi/ full/10.1002/047126363X.agr129.

49. Busi, R., Goggin, D.E., Heap, I.M., Horak, M.J., Jugulam, M., Masters, R.A., Napier, R.M., Riar, D.S., Satchivi, N.M., Torra, J., Westra, P., Weed resistance to synthetic auxin herbicides. *Pest. Manage. Sci.*, 74, 10, 2265–76, 2018 Oct.
50. Dhonukshe, P., Grigoriev, I., Fischer, R., Tominaga, M., Robinson, D.G., Hašek, J., Paciorek, T., Petrášek, J., Seifertová, D., Tejos, R., Meisel, L.A., Auxin transport inhibitors impair vesicle motility and actin cytoskeleton dynamics in diverse eukaryotes. *Proc. Natl. Acad. Sci.*, 105, 11, 4489–94, 2008 Mar 18.
51. Latorre, D.D., Herbicide inhibitors glutamate synthetase. *Rev. Brasileria de Engenharia de Biossistemas*, 7, 3, 134–41, 2013.
52. Kraehmer, H., Laber, B., Rosinger, C., Schulz, A., Herbicides as weed control agents: State of the art: I. Weed control research and safener technology: The path to modern agriculture. *Plant Physiol.*, 166, 3, 1119–31, 2014 Nov 1.
53. Togo, T., Cell membrane disruption stimulates cAMP and Ca2+ signaling to potentiate cell membrane resealing in neighboring cells. *Biol. Open*, 6, 12, 1814–9, 2017 Dec 15.
54. Tuyen, P.T., Xuan, T.D., Tu Anh, T.T., Mai Van, T., Ahmad, A., Elzaawely, A.A., Khanh, T.D., Weed suppressing potential and isolation of potent plant growth inhibitors from Castanea crenata Sieb. et Zucc. *Molecules*, 23, 2, 345, 2018 Feb 7.
55. Wang, H., Liu, W., Zhao, K., Yu, H., Zhang, J., Wang, J., Evaluation of weed control efficacy and crop safety of the new HPPD-inhibiting herbicide-QYR301. *Sci. Rep.*, 8, 1, 7910, 2018 May 21.
56. Dayan, F.E. and Duke, S.O., Discovery for new herbicide sites of action by quantification of plant primary metabolite and enzyme pools. *Engineering*, 6, 5, 509–14, 2020 May 1.
57. da Costa, T.P., Hall, C.J., Panjikar, S., Wyllie, J.A., Christoff, R.M., Bayat, S., Hulett, M.D., Abbott, B.M., Gendall, A.R., Perugini, M.A., Towards novel herbicide modes of action by inhibiting lysine biosynthesis in plants. *Elife*, 10, e69444, 2021 Jul 27.
58. Umetsu, N. and Shirai, Y., Development of novel pesticides in the 21st century. *J. Pestic. Sci.*, 45, 2, 54–74, 2020 May 20.
59. Abouziena, H.F. and Haggag, W.M., Weed control in clean agriculture: A review1. *Planta Daninha*, 34, 2, 377–92, 2016.
60. Liu, J., Hua, R., Lv, P., Tang, J., Wang, Y., Cao, H., Wu, X., Li, Q.X., Novel hydrolytic de-methylthiolation of the s-triazine herbicide prometryn by Leucobacter sp. JW-1. *Sci. Total Environ.*, 579, 115–23, 2017 Feb 1.
61. Baćmaga, M., Wyszkowska, J., Kucharsk,i J., Environmental Implication of Herbicide Use. *Molecules*, 29, 24, 5965, 2024 Dec 18.
62. Lin, H.Y., Chen, X., Dong, J., Yang, J.F., Xiao, H., Ye, Y., Li, L.H., Zhan, C.G., Yang, W.C., Yang, G.F., Rational redesign of enzyme *via* the combination of quantum mechanics/molecular mechanics, molecular dynamics, and structural biology study. *J. Am. Chem. Soc.*, 143, 38, 15674–87, 2021 Sep 20.

63. Duke, S.O., The history and current status of glyphosate. *Pest. Manage. Sci.*, 74, 5, 1027–34, 2018 May.
64. Zhao, L.X., Jiang, M.J., Hu, J.J., Zou, Y.L., Cheng, Y., Ren, T., Gao, S., Fu, Y., Ye, F., Design, synthesis, and herbicidal activity of novel diphenyl ether derivatives containing fast degrading tetrahydrophthalimide. *J. Agric. Food. Chem.*, 68, 12, 3729–41, 2020 Mar 3.
65. Zuo, Y., Wu, Q., Su, S.W., Niu, C.W., Xi, Z., Yang, G.F., Synthesis, herbicidal activity, and QSAR of novel N-benzothiazolyl-pyrimidine-2, 4-diones as protoporphyrinogen oxidase inhibitors. *J. Agric. Food. Chem.*, 64, 3, 552–62, 2016 Jan 27.
66. Zakharychev, V.V., Kuzenkov, A.V., Martsynkevich, A.M., Good pyridine hunting: A biomimic compound, a modifier and a unique pharmacophore in agrochemicals. *Chem. Heterocycl. Compd.*, 56, 12, 1491–516, 2020 Dec.
67. Xu, H., Zou, X.M., Zhu, Y.Q., Liu, B., Tao, H.L., Hu, X.H., Song, H.B., Hu, F.Z., Wang, Y., Yang, H.Z., Synthesis and herbicidal activity of novel α, α, α-trifluoro-m-tolyl pyridazinone derivatives. *Pest Manage. Science formerly Pestic. Sci.*, 62, 6, 522–30, 2006 Jun.
68. Matthias, W. and Gerhard, H., Herbicides with Bleaching Properties, in: *Modern Crop Protection Compounds*, pp. 213–302, John Wiley & Sons Ltd, Chichester, UK, 2019.
69. Zhang, H., Wang, J., Ji, Z., Sun, X., Tian, Q., Wei, S., Ji, Z., Discovery, SAR, and putative mode of action of N-benzyl-2-methoxybenzamides as potential bleaching herbicides. *Pest. Manage. Sci.*, 77, 6, 2804–11, 2021 Jun.
70. Klementova, S. and Keltnerova, L., Triazine herbicides in the environment. *Herbicides Physiol. Action Saf.*, 2, 71–96, 2015.
71. Forgacs, A.L., D'Souza, M.L., Huhtaniemi, I.T., Rahman, N.A., Zacharewski, T.R., Triazine herbicides and their chlorometabolites alter steroidogenesis in BLTK1 murine Leydig cells. *Toxicol. Sci.*, 134, 1, 155–67, 2013 Jul 1.
72. Patil, V., Noonikara-Poyil, A., Joshi, S.D., Patil, S.A., Patil, S.A., Lewis, A.M., Bugarin, A., Synthesis, molecular docking studies, and *in vitro* evaluation of 1, 3, 5-triazine derivatives as promising antimicrobial agents. *J. Mol. Struct.*, 1220, 128687, 2020 Nov 15.
73. Liu, H., Long, S., Rakesh, K.P., Zha, G.F., Structure-activity relationships (SAR) of triazine derivatives: Promising antimicrobial agents. *Eur. J. Med. Chem.*, 185, 111804, 2020 Jan 1.
74. Sharma, A., Sheyi, R., de la Torre, B.G., El-Faham, A., Albericio, F., s-Triazine: A privileged structure for drug discovery and bioconjugation. *Molecules*, 26, 4, 864, 2021 Feb 6.
75. Singh, U.P., Bhat, H.R., Gahtori, P., Antifungal activity, SAR and physicochemical correlation of some thiazole-1, 3, 5-triazine derivatives. *J. de Mycologie Médicale*, 22, 2, 134–41, 2012 Jun 1.
76. Li, L.X., Jiao, J., Wang, X.B., Chen, M., Fu, X.C., Si, W.J., Yang, C.L., Synthesis, characterization, and antifungal activity of novel benzo [4, 5] imidazo [1, 2-d][1, 2, 4] triazine derivatives. *Molecules*, 23, 4, 746, 2018 Mar 23.

77. Esquirol, L., Peat, T.S., Sugrue, E., Balotra, S., Rottet, S., Warden, A.C., Wilding, M., Hartley, C.J., Jackson, C.J., Newman, J., Scott, C., Bacterial catabolism of s-triazine herbicides: biochemistry, evolution and application. *Adv. Microbial Physiol.*, 76, 129–86, 2020 Jan 1.
78. Larif, M., Adad, A., Hmammouchi, R., Taghki, A.I., Soulaymani, A., Elmidaoui, A., Bouachrine, M., Lakhlifi, T., Biological activities of triazine derivatives. Combining DFT and QSAR results. *Arabian J. Chem.*, 10, S946–55, 2017 Feb 1.
79. Hadni, H., Mazigh, M., El Hallaoui, M., QSAR and Molecular docking studies of 4-anilinoquinoline-triazine hybrids as pf-DHFR inhibitors. *Mediterr. J. Chem.*, 8, 2, 84–93, 2019 Apr 4.
80. Chauhan, A., Sharma, P.K., Kaushik, N., Kumar, N., Synthesis of novel pyrazole analogues as efficacious antimicrobial agents. *Int. J. Pharm. Pharm. Sci.*, 3, 5, 166–76, 2011.
81. Ledenyova, I.V., Kartavtsev, P.A., Vandyshev, D.Y., Denisov, G.L., Shikhaliev, K.S., Sulfonyl derivatives of 1, 4-dihydropyrazolo [5, 1-c][1, 2, 4] triazines in reactions with acyl/alkyl halides. *Mendeleev Commun.*, 36, 1, 66–8, 2026.
82. Buglak, A.A., Zherdev, A.V., Lei, H.T., Dzantiev, B.B., QSAR analysis of immune recognition for triazine herbicides based on immunoassay data for polyclonal and monoclonal antibodies. *PLoS One*, 14, 4, e0214879, 2019 Apr 3.
83. Qian, P.P., Wang, S., Feng, K.R., Ren, Y.J., Molecular modeling studies of 1, 2, 4-triazine derivatives as novel h-DAAO inhibitors by 3D-QSAR, docking and dynamics simulations. *RSC Adv.*, 8, 26, 14311–27, 2018.
84. Salaković, B., Kovačević, S., Karadžić Banjac, M., Podunavac-Kuzmanović, S., Jevrić, L., Pajčin, I., Grahovac, J., New Perspective on Comparative Chemometric and Molecular Modeling of Antifungal Activity and Herbicidal Potential of Alkyl and Cycloalkyl s-Triazine Derivatives. *Processes*, 11, 2, 358, 2023 Jan 22.
85. Bamal, D., Duhan, A., Pal, A., Beniwal, R.K., Kumawat P., Dhanda, S., Goyat, A., Hooda, V.S., Yadav, R., Herbicide risks to non-target species and the environment: A review. *Environ. Chem. Lett.*, 22, 6, 2977–3032, 2024 Dec.
86. Zhang, H., Song, C., Shen, Z., Wang, X., Zhu, Y., Lou, M., Wu, Z., Song, R., Song, B., Design, Synthesis and Proteomics-Based Analysis of Novel Triazinone Derivatives Containing Amide Structures as Safer Protoporphyrinogen IX Oxidase Inhibitors. *J. Agric. Food. Chem.*, 72, 33, 18378–90, 2024 Aug 7.
87. Nogueira, A., Marques, R.T., Ortega, Y., Tubaldini, D., Welang, R., inventeur. Herbicidal formulation with double action mechanism for controlling weeds, method for controlling undesirable weeds and method for increasing harvest yield [En ligne], 28 juill 2015. Disponible: https://patentscope.wipo.int/search/en/detail.jsf?docId=WO2016015113.
88. Rinella, M.J., Bellows, S.E., Beitz, P.A., Low rate of aminopyralid nearly eliminates viable seed production in barb goatgrass (Aegilops triuncialis) in the greenhouse. *Invasive Plant Sci. Manage.*, 15, 1, 57–60, 2022 Mar.

89. Sebastian, D.J., Nissen, S.J., Sebastian, J.R., Meiman, P.J., Beck, K.G., Preemergence Control of Nine Invasive Weeds with Aminocyclopyrachlor, Aminopyralid, and Indaziflam. *Invasive Plant Sci. Manage.*, 10, 1, 99–109, 2017. doi:10.1017/inp.2017.7.
90. Wang, J., Peng, Y., Gan, H., Shi, Z., Xing, G., inventeur. Aminopyralid-clomazone-containing mixed herbicide, 10 juin 2015.
91. Lourdet, Y., inventeur. Weed control from applications of aminopyralid, triclopyr, and an organosilicone surfactant [En ligne], 19 déc 2013. Disponible: https://patents.google.com/patent/US20140179530A1/en.
92. Wang, X., Wang, X., Zhang, Y., Zhang, J., Li, S., Wu, L., inventeur. Herbicide composition containing aminopyralid and mesotrione, and application thereof, 19 janv 2018.
93. Moran, G.R., 4-Hydroxyphenylpyruvate dioxygenase. *Arch. Biochem. Biophys.*, 433, 1, 117–28, 2005 Jan 1.
94. Wang, D.W., Lin, H.Y., Cao, R.J., Ming, Z.Z., Chen, T., Hao, G.F., Yang, W.C., Yang, G.F., Design, synthesis and herbicidal activity of novel quinazoline-2, 4-diones as 4-hydroxyphenylpyruvate dioxygenase inhibitors. *Pest. Manage. Sci.*, 71, 8, 1122–32, 2015 Aug.
95. Raspail, C., Graindorge, M., Moreau, Y., Crouzy, S., Lefèbvre, B., Robin, A.Y., Dumas, R., Matringe, M., 4-hydroxyphenylpyruvate dioxygenase catalysis: identification of catalytic residues and production of a hydroxylated intermediate shared with a structurally unrelated enzyme. *J. Biol. Chem.*, 286, 29, 26061–70, 2011 Jul 22.
96. Santucci, A., Bernardini, G., Braconi, D., Petricci, E., Manetti, F., 4-Hydroxyphenylpyruvate dioxygenase and its inhibition in plants and animals: small molecules as herbicides and agents for the treatment of human inherited diseases. *J. Med. Chem.*, 60, 10, 4101–25, 2017 May 25.
97. Maeda, H. and DellaPenna, D., Tocopherol functions in photosynthetic organisms. *Curr. Opin. Plant Biol.*, 10, 3, 260–5, 2007 Jun 1.
98. Dayan, F.E., Duke, S.O., Sauldubois, A., Singh, N., McCurdy, C., Cantrell, C., p-Hydroxyphenylpyruvate dioxygenase is a herbicidal target site for β-triketones from Leptospermum scoparium. *Phytochemistry*, 68, 14, 2004–14, 2007 Jul 1.
99. Godar, A.S., Varanasi, V.K., Nakka, S., Prasad, P.V., Thompson, C.R., Mithila, J., Physiological and molecular mechanisms of differential sensitivity of Palmer amaranth (Amaranthus palmeri) to mesotrione at varying growth temperatures. *PLoS One*, 10, 5, e0126731, 2015 May 19.
100. Jhala, A.J., Sandell, L.D., Rana, N., Kruger, G.R., Knezevic, S.Z., Confirmation and control of triazine and 4-hydroxyphenylpyruvate dioxygenase-inhibiting herbicide-resistant Palmer amaranth (Amaranthus palmeri) in Nebraska. *Weed Technol.*, 28, 1, 28–38, 2014 Mar.
101. Walsh, M.J., Stratford, K., Stone, K., Powles, S.B., Synergistic effects of atrazine and mesotrione on susceptible and resistant wild radish (Raphanus

raphanistrum) populations and the potential for overcoming to triazine herbicides. *Weed Technol.*, 26, 2, 341–7, 2012 Jun.
102. Williams II, M.M. and Pataky, J.K., Factors affecting differential sensitivity of sweet corn to HPPD-inhibiting herbicides. *Weed Sci.*, 58, 3, 289–94, 2010 Sep.
103. Cheekatla, S,R., Azetidines in medicinal chemistry: Emerging applications and approved drugs. *Future Med. Chem.*, 1–35, 2026 Jan 7.
104. Shaw, A.Y., Liau, H.H., Lu, P.J., Yang, C.N., Lee, C.H., Chen, J.Y., Xu, Z., Flynn, G., 3, 5-Diaryl-1H-pyrazole as a molecular scaffold for the synthesis of apoptosis-inducing agents. *Bioorg. Medicinal Chem.*, 18, 9, 3270–8, 2010 May 1.
105. Xu, Y.L., Lin, H.Y., Ruan, X., Yang, S.G., Hao, G.F., Yang, W.C., Yang, G.F., Synthesis and bioevaluation of pyrazole-benzimidazolone hybrids as novel human 4-Hydroxyphenylpyruvate dioxygenase inhibitors. *Eur. J. Med. Chem.*, 92, 427–38, 2015 Mar 6.
106. Kumar, S., Kaushik, N., Sahu, J.K., Jatav, S., Quinoline analogues and nanocarrier systems: A dual approach to anti-tubercular therapy. *Eur. J. Med. Chem. Rep.*, 12, 100212, 2024 Dec 1.
107. Tripathy, D.B. and Mishra, A., Microwave synthesis and characterization of waste soybean oil-based gemini imidazolinium surfactants with carbonate linkage. *Surf. Rev. Lett.*, 24, 05, 1750062, 2017.
108. Gillett, N.P., Arora, V.K., Zickfeld, K., Marshall, S.J., Merryfield, W.J., Ongoing climate change following a complete cessation of carbon dioxide emissions. *Nat. Geosci.*, 4, 2, 83–7, 2011 Feb.
109. Robinson, T.M. and Gross, K.L., The impact of altered precipitation variability on annual weed species. *Am. J. Bot.*, 97, 10, 1625–9, 2010 Oct.
110. Tubiello, F.N., Soussana, J.F., Howden, S.M., Crop and pasture response to climate change. *Proc. Natl. Acad. Sci.*, 104, 50, 19686–90, 2007 Dec 11.
111. Chauhan, B.S., Mahajan, G., Randhawa, R.K., Singh, H., Kang, M.S., Global warming and its possible impact on agriculture in India. *Adv. Agron.*, 123, 65–121, 2014 Jan 1.
112. Ziska, L.H., Blumenthal, D.M., Franks, S.J., Understanding the nexus of rising CO2, climate change, and evolution in weed biology. *Invasive Plant Sci. Manage.*, 12, 2, 79–88, 2019 Jun.
113. Oerke, E.C., Crop losses to pests. *J. Agric. Sci.*, 144, 1, 31–43, 2006 Feb.
114. Rana, K.S., Choudhary, A.K., Sepat, S., Bana, R.S., *Advances in field crop production*, p. 475, Post Graduate School, Indian Agricultural Research Institute, New Delhi, India, 2014.
115. Ohki, A., Kuboyama, N., Koizumi, K., Tanaka, A., Sato, Y., Kohno, H., Böger, P., Wakabayashi, K., Mode of action of novel 2-(benzylamino)-4-methyl-6-(trifluoro-methyl)-1, 3, 5-triazine herbicides: Inhibition of photosynthetic electron transport and binding studies. *J. Agric. Food. Chem.*, 47, 10, 4398–402, 1999 Oct 18.

116. Khan, B.A., Nadeem, M.A., Nawaz, H., Amin, M.M., Abbasi, G.H., Nadeem, M., Ali, M., Ameen, M., Javaid, M.M., Maqbool, R., Ikram, M., Pesticides: impacts on agriculture productivity, environment, and management strategies, in: *Emerging contaminants and plants: Interactions, adaptations and remediation technologies*, pp. 109–134, Springer International Publishing, Cham, 2023 Feb 17.
117. Grossmann, K., Tresch, S., Plath, P., Triaziflam and diaminotriazine derivatives affect enantioselectively multiple herbicide target sites. *Z. für Naturforschung C*, 56, 7-8, 559–69, 2001 Aug 1.
118. Bristow, J.T., inventeur. Herbicidal compositions comprising triazinone herbicide and als inhibitor herbicide and uses thereof [En ligne], 16 oct 2014. Disponible: https://patents.google.com/patent/WO2015078243A1/en.
119. Rai, T., Kaushik, N., Malviya, R., Sharma, P.K., A review on marine source as anticancer agents. *J. Asian Nat. Prod. Res.*, 26, 4, 415–51, 2024 Apr 2.
120. Anderson, R.L., Managing weeds with a dualistic approach of prevention and control. A review. *Agron. Sustain. Dev.*, 27, 1, 13–8, Mar 2007.
121. Muchhadiya, R.M., Kumawat, P.D., Sakarvadia, H.L., Muchhadiya, P.M., Weed management with the use of nano-encapsulated herbicide formulations: A review. *J. Pharm. Innov.*, 11, 2068–75, 2022.
122. Kumar, S., Kumar, A., Agrawal, A., Sahu, J.K., Synthesis, in vivo biological assessment and molecular docking study of some newer indole derivatives as cox 1/2 inhibitors. *J. Mol. Struct.*, 1230, 129831, 2021 Apr 15.
123. Stevens, J.T., Breckenridge, C.B., Simpkins, J., Eldridge, J.C., Symmetrical and asymmetrical triazine herbicides, in: *Handbook of Pesticide Toxicology*, pp. 1511–1519, Academic Press, 2001 Jan 1.
124. Singh, R.P., Singh, R.K., Singh, M.K., Impact of climate and carbon dioxide change on weeds and their management-a review.
125. Gianessi, L.P., The increasing importance of herbicides in worldwide crop production. *Pest. Manage. Sci.*, 69, 10, 1099–105, 2013 Oct.
126. Varanasi, A., Prasad, P.V., Jugulam, M., Impact of climate change factors on weeds and herbicide efficacy. *Adv. Agron.*, 135, 107–46, 2016 Jan 1.
127. Pautasso, M., Dehnen-Schmutz, K., Holdenrieder, O., Pietravalle, S., Salama, N., Jeger, M.J., Lange, E., Hehl-Lange, S., Plant health and global change–some implications for landscape management. *Biol. Rev.*, 85, 4, 729–55, 2010 Nov.
128. Kubisch, A., Degen, T., Hovestadt, T., Poethke, H.J., Predicting range shifts under global change: The balance between local adaptation and dispersal. *Ecography*, 36, 8, 873–82, 2013 Aug.
129. Kumar, V., Bana, R.S., Singh, T., Louhar, G., Ecological weed management approaches for wheat under rice–wheat cropping system. *Environ. Sustainability*, 4, 51–61, 2021 Mar.
130. Pearman, P.B., Guisan, A., Broennimann, O., Randin, C.F., Niche dynamics in space and time. *Trends Ecol. Evol.*, 23, 3, 149–58, 2008 Mar 1.

131. Vats, S., Herbicides: History, classification and genetic manipulation of plants for herbicide resistance. *Sustainable Agric. Rev.*, 15, 153–92, 2015.
132. James, C., *Global status of commercialized biotech/GM crops: 2012*, ISAAA, Ithaca, NY, 2012.

10

Heterocyclic Compounds as Antihypertensive Agents

Manjot Singh[1], Deepika Paliwal[1*], Aman Thakur[2], Manvi Karayat[1] and Saurabh Srivasatva[3]

[1]Department of Pharmacy, School of Medical & Allied Sciences, Galgotias University, Greater Noida, India
[2]Department of Pharmacy, Adarsh Vijendra Institute of Pharmaceutical Sciences Gangoh Saharanpur, Uttar Pradesh, India
[3]KPJ Healthcare University, Nilai, Malaysia

Abstract

Heterocyclic compounds are important in medicinal chemistry because of their various biological activity and structural plasticity. Among their numerous pharmacological applications, their possibility as antihypertensive medicines has received considerable attention. Hypertension, a major risk indicator for cardiovascular disease, needs the development of new and efficient therapies. This chapter provides a thorough summary of the numerous heterocyclic scaffolds that have shown antihypertensive activity, such as imidazole, pyridines, quinazolines, and thiazoles. The emphasis is on structure–activity relationship, the mechanisms of action, and current advances in the synthesis of these molecules. The chapter also discusses the therapeutic significance and future prospects toward heterocyclic medication development targeting hypertension.

Keywords: Heterocyclic compounds, antihypertensive medicines, hypertension, imidazole, pyridines, quinazolines, thiazoles

**Corresponding author*: deepika14paliwal@gmail.com

Divya Bajpai Tripathy, Priyanka Chhabra and Anjali Gupta (eds.) Heterocyclic Compounds as Pharmaceutical Ingredients, (349–388) © 2026 Scrivener Publishing LLC

10.1 Introduction

Cardiovascular disease (CVD) remains the leading cause of disability and early death rate, therefore representing a significant public health issue. One of the main risk factors, which includes diabetes, dyslipidemia, and smoking, unquestionably, the most common cause of CVDs is hypertension, and its association with additional danger factors is a lot stronger. It is the primary contributor to CVD-related morbidity and mortality, accounting for approximately 16.5% of all annual deaths worldwide [1–6]. A systolic arterial pressure (SBP) of 150 mm Hg or a diastolic arterial pressure (DBP) of 90 mm Hg is the recommended starting point for pharmacologic treatment in the general population, according to the Eighth Joint National Committee. However, treatment should start for those with chronic renal disease when their SBP and DBP readings are at least 140 or 90 mm Hg, respectively [7, 8]. Secondary hypertension and primary (essential) hypertension are the two categories of elevated blood pressure. Five percent to 10% of hypertensive patients have secondary hypertension, which has a comparatively greater chance of being well treated because it may be linked to recognizable causes including diabetes and renal impairment [9, 10].

Pharmacological agents such as diuretics (indapamide, furosemide, Amiloride), sympatholytics (clonidine, reserpine), renin inhibitors (aliskiren), angiotensin-converting enzyme (ACE) inhibitors (enalapril, captopril, quinapril), angiotensin receptor blockers (ARBs—losartan, irbesartan, olmesartan), and calcium-channel blockers (nifedipine, verapamil, diltiazem) are among the pharmacological agents used to control blood pressure in patients with hypertension. Nonetheless, a notable concern for physicians and healthcare practitioners is the disturbing and unpleasant reality that only 34% of hypertension patients have their high blood pressure well treated [11–18]. The primary concerns that frequently impede treatment pertain to the elevated costs of antihypertensive medications, their availability and accessibility [19], the undesirable side effects associated with these drugs, and and the diminished patient compliance required; alternative methods are pursued such as herbal therapies for the treatment of hypertension. During this timeframe, numerous significant antihypertensive medications were discovered: Hydroplane is a vasodilator with a potent antihypertensive effect; thiazides promote sodium excretion, and diuretics decrease cardiac output and peripheral vascular resistance, utilized either independently or in conjunction with other antihypertensive medications, and continue to be fundamental agents in hypertension management; guanethidine and rifampicin, both classified as norepinephrine-releasing nerve ending blockers, operate by affecting the release and storage of norepinephrine [20, 21]. Guanethidine and rifampicin are both agents that inhibit norepinephrine nerve endings. Their mechanism of action involves altering the storage and release of norepinephrine, resulting in the depletion of vesicles containing norepinephrine and consequently lowering blood pressure. However, due to the prevalence of adverse effects on the neurological and gastrointestinal systems, these drugs are being supplanted by newer alternatives with fewer side effects. Currently, they are primarily utilized as tools for investigating sympathetic activity. Rifampicin has diminished efficacy and increased adverse responses; it is no longer utilized as a monotherapy but remains a fundamental component in combination formulations of antihypertensive medications. The antihypertensive medications introduced in the 1960s encompass central antihypertensives (methyldopa, clonidine), vasodilators (diazoxide), β-receptor antagonists (propranolol, among others), and calcium-channel blockers (nifedipine, among others). Since then, selective α1-receptor antagonists (such as prazosin), potassium-channel openers (such as minoxidil), and selective imidazoline receptor agonists (such as moxonidine and rilmenidine) have been introduced, significantly enhancing the diversity of antihypertensive medications. In the 1980s, the advent of ACE inhibitors and ARBs (such as chlorsartan) marked a significant advancement in the pharmaceutical management of hypertension, effectively lowering blood pressure and preventing and controlling the condition. These medications are helpful not only in reducing blood pressure but also in avoiding and reversing the structural alterations of the cardiovascular system induced by hypertension. Alongside the aforementioned antihypertensive medications, other novel antihypertensive treatments have been created in recent years. For instance, dual inhibitors of neutral endopeptidase and ACE, including omadacycline, fasidotril, and sambazolol, concurrently inhibit ACE and neutral endopeptidase activity, diminish the renin–angiotensin system's activity, and elevate bradykinin (BK) and cardiac natriuretic peptide levels, thereby exerting antihypertensive effects; the prostaglandin synthesis–enhancing agent, sicloxacillin, facilitates prostaglandin synthesis to yield antihypertensive effects [22–28], the 5-hydroxytryptamine receptor agonist, furaldehyde, promotes serotonergic vasodilatory signalling *via* 5-HT receptor activation, hence boosting peripheral vasodilation and adding to its overall antihypertensive action. The 5-hydroxytryptamine receptor agonist uradil activates the central 5-HT_{1A} receptor, diminishes peripheral sympathetic nerve activity, and elicits antihypertensive effects; the 5-HT receptor antagonist ketanserin inhibits the 5-HT_{2A} receptor and mildly antagonizes the α1-receptor, thereby reducing peripheral vascular resistance and producing antihypertensive effects; additionally, the endothelin receptor antagonists bosentan, ketanserin, and enrasentan obstruct the binding of endothelin to its receptor, resulting in a potent antihypertensive effect [29]. The impact of these medications on cardiocerebral canal incidents has not been assessed in extensive clinical trials. With an enhanced comprehension of the pathogenesis of hypertension, numerous novel antihypertensive agents are under investigation. These include vasoactive peptidase inhibitors, natriuretic peptide A agonists, aldosterone synthase inhibitors, soluble epoxide hydrolase inhibitors, vasoactive intestinal peptide Type 2 receptor agonists, salicylated corticosteroid receptor antagonists, aminopeptidase A inhibitors, dopamine β-hydroxylase inhibitors, small intestinal Na^+/H^+ exchanger 3 inhibitors, ACE II/angiotensin (1–7)/Mas receptor axis component agonists, enkephalinase inhibitors, and purinergic P2X3 receptor inhibitors (Figure 10.1) [30–32].

By showing how different organs and hormonal systems interact to regulate blood pressure and contribute to damage related to hypertension, this diagram illustrates the pathophysiology of hypertension or high blood pressure.

Key mechanisms in hypertension are as follows:

1. **Sympathetic nervous system and brain**: The brain uses sympathetic nerve activity to manage blood pressure. Whenever these nerves are overactivated, the heart rate rises,

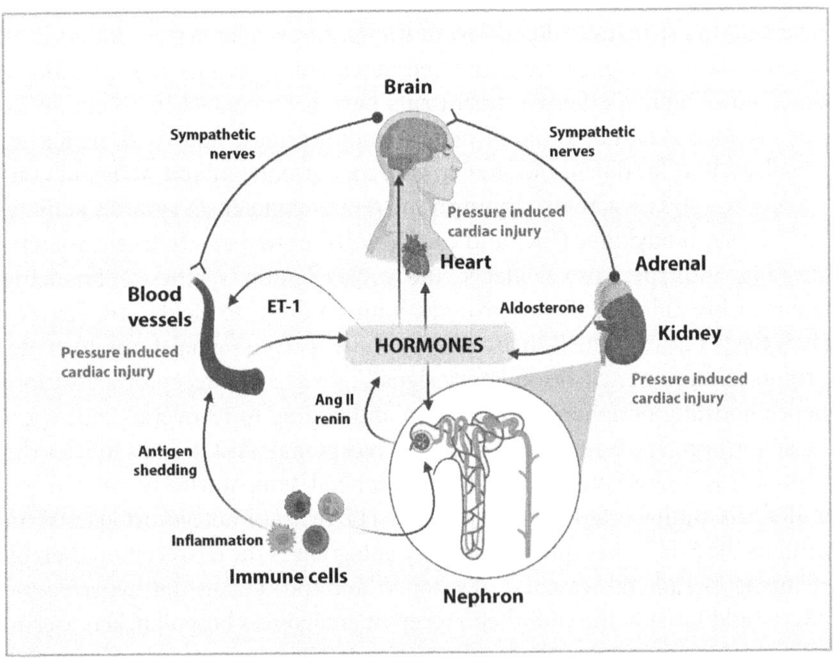

Figure 10.1 Pathophysiology of hypertension.

blood vessels constrict, and renal hormones are released, all of which lead to hypertension.

Pressure-induced cardiac injury, or heart damage, can result from persistently high blood pressure.

2. **Heart**: Hypertrophy, or the thickening of the heart's muscles, and ultimately heart failure arise from the heart having to work harder due to elevated blood pressure. This process is aided by hormones, such as endothelin-1 (ET-1) and aldosterone.
3. **Kidney and adrenal gland**: The secretion of aldosterone by the adrenal gland raises blood pressure *via* increasing water and sodium retention. Through the renin–angiotensin system, the kidney is a key player in hypertension. Vasoconstriction, or the narrowing of blood arteries, is brought on by an excess of renin and angiotensin II (Ang II), which can harm the heart and kidneys.
4. **The kidney's functional unit, or nephron**: Nephron damage from high blood pressure exacerbates hypertension and causes kidney disease. Damage causes the kidney to release more renin, which feeds the vicious cycle of rising blood pressure.
5. **ET-1 and blood vessels**: Vasoconstrictor ET-1 restricts blood arteries, which raises blood pressure even further. Vascular damage from chronic hypertension results in arteriosclerosis, or the hardening of arteries, which lowers blood flow to organs.
6. **Inflammation and immune system**: An immunological reaction brought on by high blood pressure results in inflammation. Damage to blood vessels is exacerbated by immune cells being further activated by antigen spilling from injured cells.

Overall, a multiorgan disease, hypertension affects the heart, brain, kidneys, blood arteries, and immune system. Sustained high blood pressure is caused by immunological activation, overactive sympathetic neurons, and excessive hormone release (aldosterone, Ang II, ET-1). Heart failure, renal damage, and vascular problems result with untreated hypertension.

10.1.1 Marketed Drugs for Hypertension

The commercially available antihypertensive medications and their corresponding pharmacological activities relevant to blood-pressure control are shown in Table 10.1.

Table 10.1 Commercially available drugs for hypertension.

S. no.	Name	Structure	Activity	Side effect	Brand name
1.	Captopril	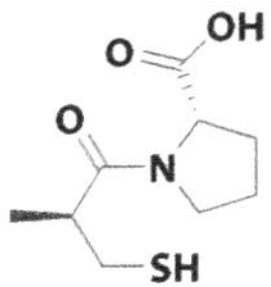	ACE inhibitor	Hyperkalemia, angioedema, renal impairment	Capoten
2.	Lisinopril	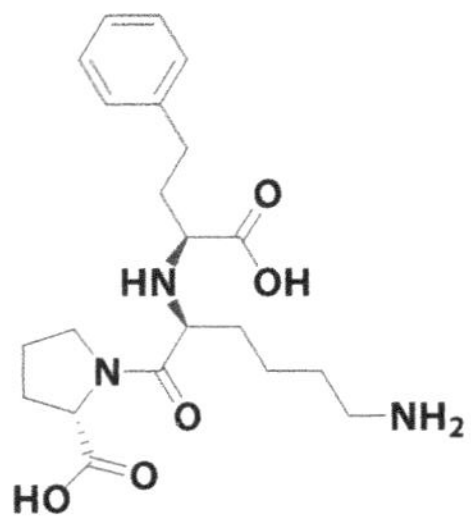	ACE inhibitor	Dry cough, renal dysfunction, angioedema	Prinivil, Zestril
3.	Candesartan	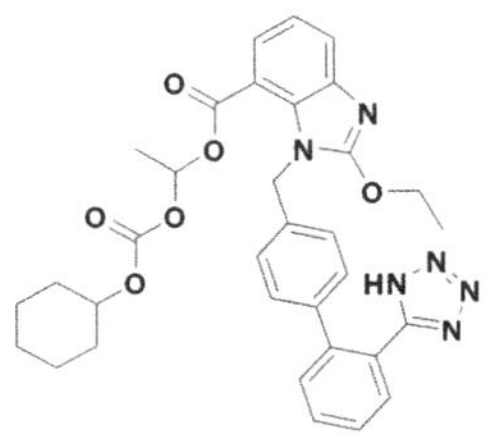	Angiotensin II receptor blocker (ARB)	Back pain, dizziness, renal impairment	Atacand

(*Continued*)

Table 10.1 Commercially available drugs for hypertension. (*Continued*)

S. no.	Name	Structure	Activity	Side effect	Brand name
4.	Losartan	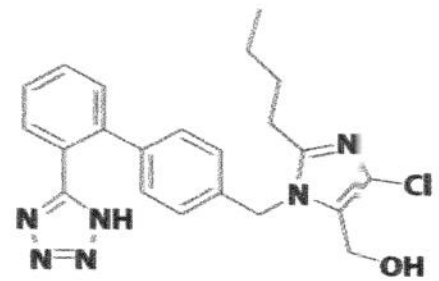	Angiotensin II receptor blocker (ARB)	Nasal congestion, back pain, hyperkalemia	Cozaar
5.	Valsartan	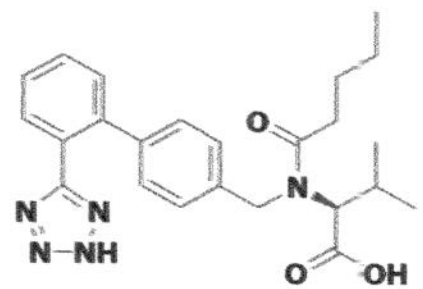	Angiotensin II receptor blocker (ARB)	Renal dysfunction, Fatigue, Headache	Diovan
6.	Aliskiren	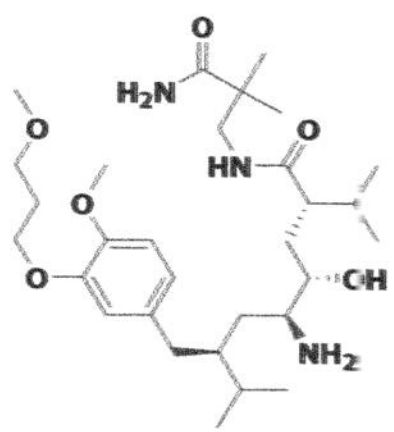	Direct renin inhibitor	Diarrhea, cough, renal impairment	Tekturna, Rasilez
7.	Atenolol	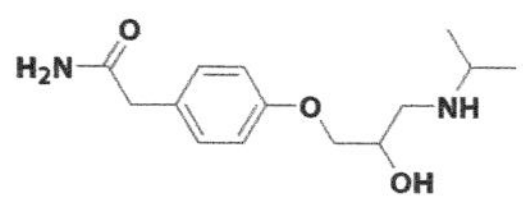	β1-adrenergic blocker (β-blocker)	Bradycardia, sexual dysfunction, depression	Tenormin

(*Continued*)

Table 10.1 Commercially available drugs for hypertension. (*Continued*)

S. no.	Name	Structure	Activity	Side effect	Brand name
8.	Metoprolol	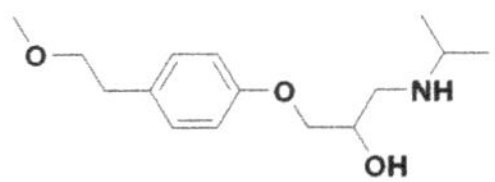	β1-adrenergic blocker (β blocker)	Bradycardia, sexual dysfunction, depression	Lopressaor, Toprol XL
9.	Propanolol	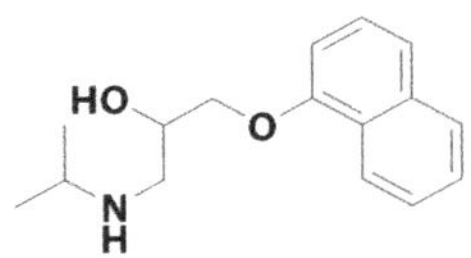	Nonselective β-blocker (blocks both β1 and β2)	Bronchospasm, sleep disturbance	Inderal, Inderal LA (long acting)
10.	Doxazosin	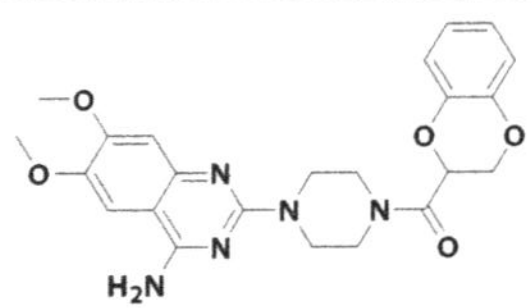	α1-Adrenergic receptor blocker	Orthostatic hypotension, nasal congestion, Edema	Cardura
11.	Verapamil	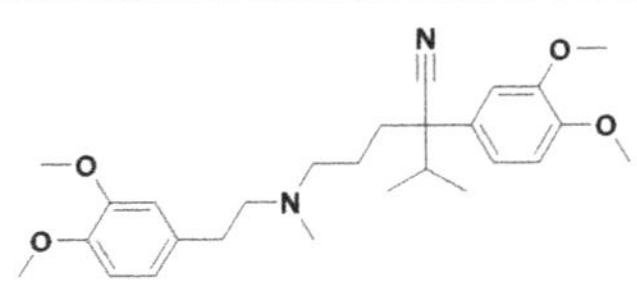	Calcium-channel blocker	Bradycardia, heart block, constipation	Calan, Isoppin, Verelan, Covera-HS

10.1.2 Significance of Heterocyclic Compounds

Heterocyclic compounds are fundamental to contemporary medicinal chemistry and have demonstrated considerable importance in the management of several disorders, including hypertension. Hypertension, or high blood pressure, is a persistent illness that may result in severe cardiovascular problems, including heart disease, stroke, and renal failure. The quest for efficacious antihypertensive medicines has resulted in the identification of various types of pharmaceuticals, many of which are derived from heterocyclic compounds. These compounds possess rings that include at least one atom other than carbon, such as nitrogen, oxygen, or sulfur, and they demonstrate distinctive features that render them suitable for medicinal applications. This discussion will investigate the importance of heterocyclic compounds in the creation of antihypertensive medications, focusing on their structural, pharmacological, and therapeutic benefits.

10.1.3 Structural Versatility of Heterocyclic Compounds

The structural plasticity of heterocyclic compounds is a principal reason for their significance in the design of antihypertensive drugs. The inclusion of noncarbon atoms, such as nitrogen, oxygen, or sulfur, within the ring structure generates opportunities for hydrogen bonding, pi-stacking, and other interactions, hence augmenting binding affinity to particular biological targets. This structural diversity enables chemists to engineer molecules that accurately align with the active areas of enzymes and receptors implicated in blood pressure regulation [33–37]. Pyridines, imidazoles, and quinolines are heterocyclic compounds frequently present in antihypertensive medications. The varying electronic characteristics of these rings can affect the drug's lipophilicity, electron density, and stability, which are essential for its efficacy and bioavailability. Structural alterations of these heterocyclic rings provide the precise adjustment of pharmacological attributes, including enhanced potency, selectivity, and less toxicity. This versatility renders heterocycles a cornerstone for the development of novel pharmacological candidates [38–40].

10.1.4 Pharmacophoric Potential of Heterocyclic Rings

A pharmacophore is the molecular component that mediates its biological activity. Heterocyclic rings, due to their three-dimensional configuration and capacity to accommodate diverse substituents, serve as crucial pharmacophores in antihypertensive medications. They frequently delineate the critical interaction locations with biological receptors, thus augmenting the drug's specificity for its target and diminishing the probability of off-target effects [41–43].

Imidazole and pyridine rings are present in various ARBs, such as losartan and irbesartan. The heterocyclic rings are essential for binding to the Ang II receptor, a protein that is pivotal in the control of blood pressure. By binding to this receptor, ARBs obstruct the actions of Ang II, a powerful vasoconstrictor, resulting in vasodilation and decreased blood pressure. This selectivity is attained through the meticulous positioning of the heterocyclic rings within the receptor binding site, underscoring the significance of heterocyclic structures as pharmacophores [44–49].

10.1.5 Enhancement of Bioavailability and Stability

Heterocyclic compounds are recognized for improving the solubility, stability, and bioavailability of pharmaceuticals, which are crucial for sustaining therapeutic levels in the bloodstream. These attributes are especially crucial in the management of hypertension, which necessitates stable medication concentrations to regulate blood pressure over extended durations. The hydrophilic–lipophilic balance of heterocyclic compounds can be modified to enhance absorption across biological membranes while maintaining enough aqueous solubility for systemic circulation. Thiazide diuretics, a category of antihypertensive medications, possess a thiazide ring that increases water solubility and promotes renal excretion, hence augmenting their diuretic efficacy. This characteristic is particularly significant in those with hypertension, because it aids in decreasing blood volume and thereby reduces blood pressure. Furthermore, the stability provided by heterocyclic rings guarantees that the medications maintain efficacy over prolonged periods without premature degradation, which is crucial for the therapy of chronic conditions such as hypertension [50, 51].

10.1.6 Specific Mechanistic Action

Heterocyclic compounds exhibit significant versatility in interacting with various biological targets, many of which are critically involved in the regulation of blood pressure. Antihypertensive medications often function by obstructing particular enzymes or receptors within the renin–angiotensin–aldosterone system (RAAS), a hormonal system that governs blood pressure and fluid equilibrium. Heterocyclic compounds have been engineered to target elements of the RAAS, including the ACE and Ang II receptors, thereby efficiently reducing blood pressure by disrupting pathways that induce vasoconstriction and fluid retention.

10.2 Heterocyclic Moieties against Hypertension

10.2.1 Indole and Indazole Moiety

Kreighbaum *et al.* [52] synthesized 25 indole analogs containing phenoxypropanolamine in three steps. Based on their vasodilatory effects at an oral dose of 100 mL/kg and an intravenous dosage of 2 mL/kg in rats with ganglion block and spontaneously hypertensive rats (SHRs), respectively, the synthesized analogs were assessed for antihypertensive effectiveness *in vivo*. Whereas Compounds 1b and 1c, which were substituted with 2-methyl and 2-cyano groups on the benzene ring of the phenylpropanolamine moiety, accordingly, along with indoles featuring 5-methoxy substitution, demonstrated superior vasodilatory properties in comparison to the other synthesized compounds, Compound 1a demonstrated significant antihypertensive activity in SHR [52].

Monge Vega *et al.* [53] synthesized a series of compounds, specifically **2a–2d** and **3a–3d** as shown in Figure 10.2, and evaluated their antihypertensive characteristics. The alterations in arterial pressure in SHR were assessed following intraperitoneal (i.p.) injection of 1, 2, and 5 mL/kg, and oral administration of Compounds **2a** and **2b** led to a significant reduction in arterial pressure. The desoxycorticosterone-saline rats were used to test the most active compounds, 2c and 2d, at doses of 2 and 5 mL/kg, intraperitoneally. According to the findings, 2c and 2d considerably lowered arterial blood pressure. It is possible that compounds with antihypertensive properties may not always have a hydrazine group, indicating that hydrazine is not necessary for antihypertensive action [53].

Compound 4 was produced by Kim *et al.*, who then assessed its antihypertensive effectiveness. The substituted ethyl 1-acryloyl-indoline-2-carboxylate derivatives 15 were obtained by acylating Compound 14 with methacryloyl chloride. These derivatives were then hydrolyzed under basic conditions to generate 1-acryloyl-indoline-2-carboxylic acids. Among the compounds, **4b** (S, S) exhibited the highest activity with an IC_{50} of 3.7×10^{-9} M, which is thrice that of captopril. In comparison to captopril, the hydrophobicity of **4b** (S, S) was enhanced, suggesting the presence of a hydrophobic group in the active site of ACE. Compound **4b** (S, S) exhibited oral antihypertensive efficacy in SHRs that was 27-fold superior to that

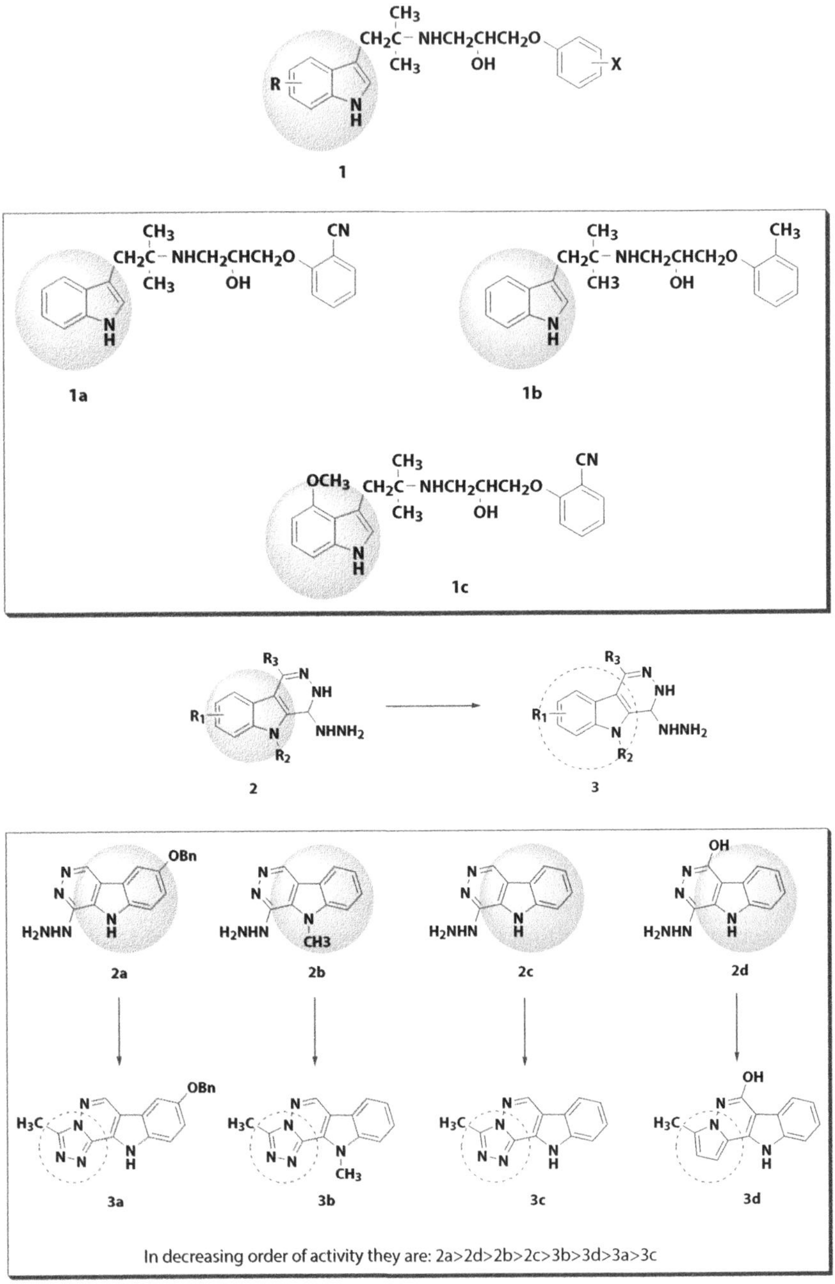

Figure 10.2 Indole derivatives (1–3) as antihypertensive agents.

of captopril. Benzoyl **4a** (S, S) had a potency 24 times greater than that of captopril. The significant antihypertensive efficacy demonstrated by **4a** and **4b** suggests that, in addition to peripheral ACE inhibitory action, other antihypertensive pathways may be implicated [54].

Compound 5's antihypertensive effectiveness was assessed by Asselin *et al.* in 1986 as indicated in Figure 10.3. Tissue testing of Compounds 5a and 5b showed that they act as nonselective β-adrenergic receptor antagonists. Additionally, there were noticeable intrinsic sympathomimetic effects on β-adrenoceptors (ARs) in 5a and 5b. By forming a hydrogen bond with the receptor, 5a and 5b were able to induce their β-adrenergic receptor antagonist as well as agonist actions.

Monge *et al.* [55] reported the synthesis of 5H-pyridazino[4,6-b]indole derivatives in 1987 in their study on selective thromboxane synthetase inhibitors and related antihypertensive agents. Compounds show how intravenous administration of Compounds **6a–6c** as shown in Figure 10.3 at a dosage of 1 mg/kg affects rats' systolic blood pressure over time. The following is the order in which the substances lowered blood pressure: 6a > 6b = 6c > hydralazine. At an increased dose, 6b = 6a < 6c at a dosage of 1 mg/kg i.p., and hydralazine = 6a = 6b < 6c, the stimulation was found in progressive order. Derivatives, however, had little effect. Compared to hydralazine, the LD_{50} values for 6a, 6b, and 6c are roughly 2.2 to 2.5 times less hazardous. In mice, 6a has an acute toxicity of 225 ± 12 mg/kg. According to the previously cited information, 6a might be an appealing antihypertensive drug [56].

In 1990, Squadrito *et al.* investigated the impact of Compound **8** on systolic blood pressure in rats with spontaneous hypertension and Grollman hypertension. Administration of Compound **8** significantly decreased systolic blood pressure in a hypertension model at a dosage of 100 mL/kg *via* oral gavage for 10 days. Furthermore, Compound **8** partly suppressed the progression of hypertension when administered daily to 5-week-old SHR for a duration of 7 weeks. The administration of Compound **8** also elevated the concentrations of tryptophan and 5-HIAA in the diencephalon and cortex. The serotonin concentration in the brain stem of SHR was elevated following treatment with Compound **8**. The results suggested that Compound 8 can lower blood pressure in many rat models of hypertension [57].

In 1991, Todd and Fitton revealed that perindopril (Compound **9**), a prodrug, was a long-acting ACE inhibitor. Compound **9** is a lipophilic small molecule featuring an indole ring. Compound **9** is administered to regulate blood pressure in individuals with mild to moderate essential

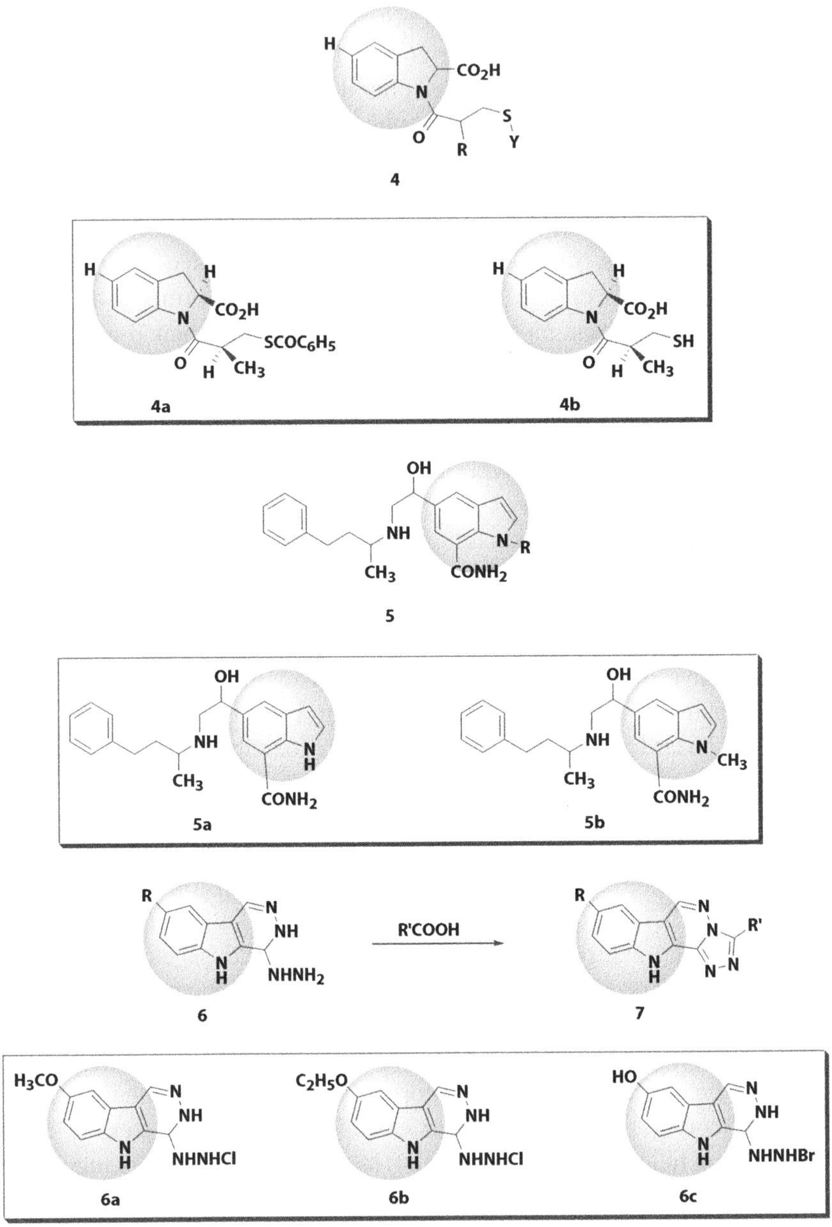

Figure 10.3 Indole derivatives (4–7) as antihypertensive agents.

hypertension, at a dosage of 4 to 8 mg once daily. Furthermore, the study indicated that Compound **9** might serve as a viable option to existing ACE inhibitors for the management of all forms of hypertension [58]. In 2001, El-Gendy and El-Banna synthesized two series of Mannich base derivatives, **10** and **11** as highlighted in Figure 10.4, and evaluated their antihypertensive effects. Compounds **10a, 10b, 10c, and 10d** markedly decreased blood pressure. Nevertheless, derivatives did not exhibit a reduction in blood pressure. Three anaesthetized dogs, each weighing between 12 and 16 kg, were injected with 35 mg/kg of pentobarbital sodium and subsequently administered Compound **10b**. Compound **10b**, administered at a dosage of 40 mg/kg, dramatically reduced systolic blood pressure from 70 ± 10 to 120 ± 10 mm Hg, with an LD_{50} value of 180 mg/kg, demonstrating antihypertensive action. The majority of the compounds exhibited selective affinity [59].

Compounds **13a** and **13b** (Figure 10.4) exhibited considerable affinities for the α1D-adrenergic receptor (pKi = 7.45 and 7.70, respectively), with no detectable affinity for the α1A-adrenergic and α1B-adrenergic receptor subtypes during binding assays conducted on the three human cloned receptors. Figure 10.4 outlines that among Compounds **12a–12g, 12a** and **12b** exhibited the highest affinity (pKi = 9.14 and 9.39), with a little selectivity for the α1D-AR. The structure–activity relationship (SAR) derived from the pharmacophore model fitting for the α1-adrenergic receptor of the antihypertensive is examined [60].

In 2006, a series of novel pyramido[5,4-b]indole and benzothieno[3,2-d] pyrimidine derivatives were synthesized. The impacts of Compounds **14a** and **14b** (Figure 10.5) were examined on a1-AR subtypes from isolated rat spleen (a1B-AR), prostate vas deferens (a1A-AR), and thoracic aorta (a1D-AR). Moreover, Compounds **14a** and **14b** exhibited significant affinity and modest selectivity for the a1D-AR in binding assays utilizing human cloned receptors. The SAR of indole and pyrimidine derivatives for the α1D-adrenergic receptor in antihypertensive applications is examined [61].

Compound **15** as observed in Figure 10.5 was synthesized by a five-step process [62]. Compound **15** demonstrated efficacy as a selective and orally bioavailable MR antagonist. Compound **8** exhibited high affinity (hMR Ki) of 0.494 (0.23 nM) and acted as a functional antagonist (hMR Kb) with a value of 19 (12.8 nM), demonstrating moderate activity *in vivo*. In salt-containing, mononephrectomized Sprague–Dawley rats, following 14 days of oral administration of Compound **15** at a dosage of 3 mg/kg, a reduction in blood pressure was seen, which further lowered to 1 at 30 mg/kg. The introduction of ortho-F in Compound **15** produced a significant effect. Zhou *et al.* [63] identified the chemicals rhynchophylline (Compound **16**)

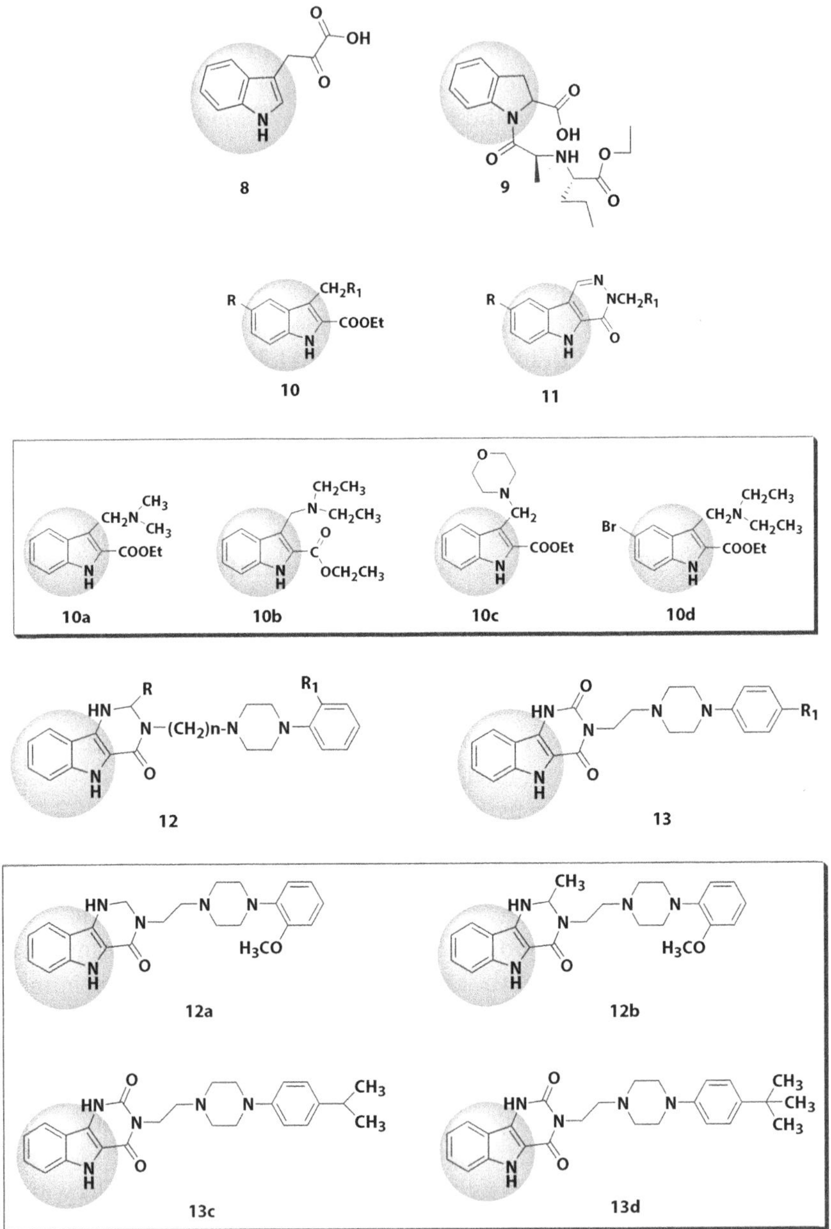

Figure 10.4 Indole derivatives (8–13) as antihypertensive agents.

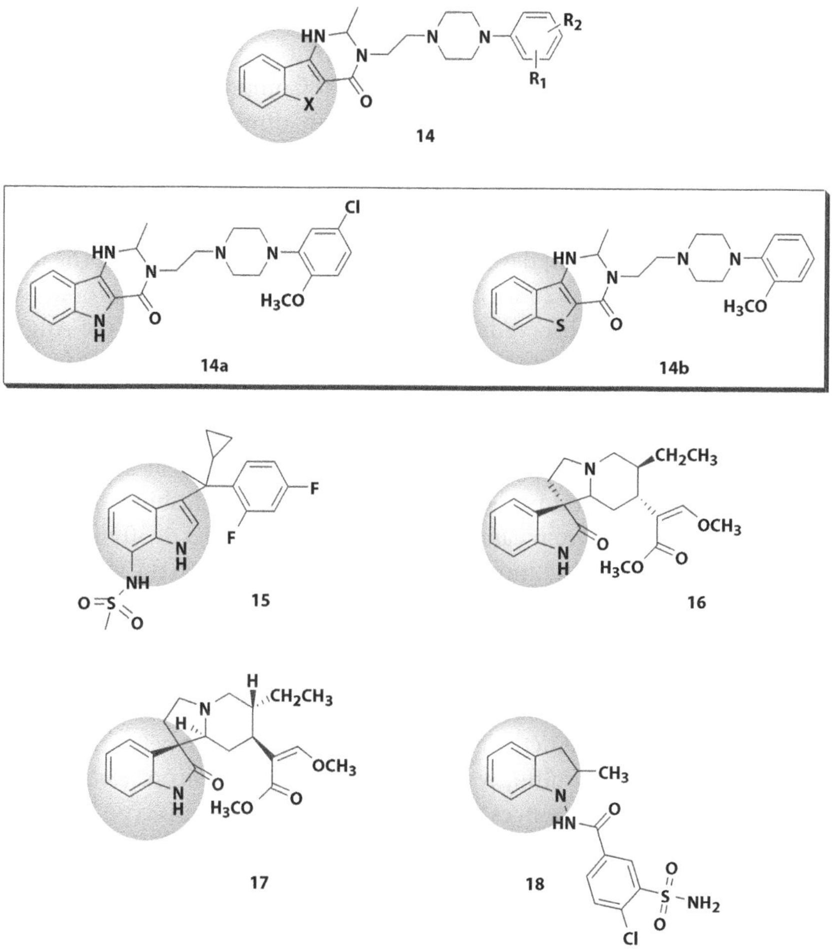

Figure 10.5 Indole derivatives (14–18) as antihypertensive agents.

and isorhynchophylline (Compound **17**) from the Chinese plant *Uncaria*. Compound **16** (Figure 10.5) exhibited antihypertensive properties, resulting in a decrease in blood pressure and ultimately a restoration to normal levels [64–66]. Administration of Compound **17** (Figure 10.5) at dosages of 10 or 20 mg/kg *via* the duodenum elicited hypotensive activity at 10 and 20 minutes postadministration, with effects persisting for a minimum of 3 hours. Compound **17**, administered intravenously at doses ranging from 10 to 20 mg/kg, reduced blood pressure in hypertensive rats and decreased heart rate in sedated dogs. Indapamide (Compound **18**) comprises an indole ring and functions as a sulfonamide diuretic. Experimental studies indicated that Compound **18** shown in Figure 10.5 exhibited a vasodilatory effect. Clinical research has demonstrated that direct vascular activities contributed to the antihypertensive effects of Compound **18**. Compound **18** is a highly lipophilic compound that can diminish transmembrane calcium influx and induce vascular dilatation [67]. Consequently, several hypotensive effects of Compound **18** may relate to calcium antagonist-like activity.

In 2014, Zhu *et al.* [68] developed a series of indole derivatives. The ultimate target Compound **20** was obtained following the hydrolysis of derivative **19** in NaOH as observed in Figure 10.6. The *in vitro* effects of Ang II and the *in vivo* antihypertensive effects were evaluated for indole derivative **20**. Radioligand binding experiments demonstrated that **20c** exhibited affinity for the Ang II Type 1 receptor (IC_{50} = 1.03 ± 0.26 nM). In SHR and renal hypertensive rats, the injection of **20c** resulted in a significant decrease in mean blood pressure in a dose-dependent fashion. The most effective responses were mean blood pressure reductions of 30 mm Hg in the 5 mg/kg group and 41 mm Hg in the 10 mg/kg group (p.o.), with significant antihypertensive activity persisting for over 24 hours. Furthermore, Compound **20c** proved to be a more efficacious antihypertensive drug than losartan. Consequently, Compound **20c** may serve as a potentially effective antihypertensive medication.

The alkaloids *N*-demethylglypetelotine (**21**) and glypetelotine [69], which contain indole sulfur, were isolated from *Glycosmis petelotii* leaf in 2014. The effect of substances to relax aortic rings on the L-type Ba^{2+} current [Iba (L)] in a single muscle cell that was separated from the tail artery was assessed. Phenylephrine-induced aortic ring shrinkage was lessened by Compounds **21** and **22** (IC_{50} = 50 and 20 lM, respectively) Figure 10.6. Furthermore, Compounds **21** and **22** had no effect on the phenylephrine-induced calcium ion delivery from the sarcoplasmic reticulum. As the membrane became less polarized, the antispasmodic action of Compound **22** diminished. Compounds **21** and **22** inhibited the shrinkage brought on by Ca^{2+} buildup in a concentration-dependent manner at K^+ concentrations of 60 mM; the inhibition was inversely proportional to the Ca^{2+} concentration. Compound **22** affected the Iba (L) kinetics and decreased the Iba (L) value [70]. Accordingly, Compound **22** was identified as a novel class of vasodilator capable of blocking L-type Ca^{2+} channels.

In 2015, Zhu *et al.* [71] created an antihypertensive counterpart of indoloquinolizidine. Target **24** was obtained by removing the Boc group from Compound **23** in an acidic environment. Compound **24** decreased the SBP and heart rate of SHRs and significantly affected the contractile response of the thoracic aorta rings in male SD rats (IC_{50} = 1.129–10^9).

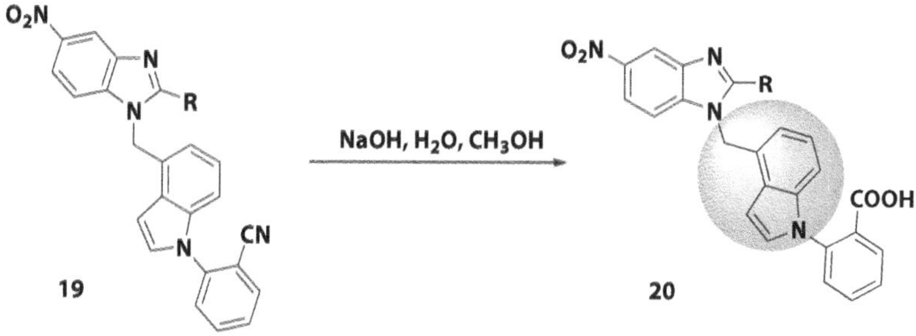

Compounds	$IC_{50} \pm$ SEM (nM)	Ki (nM)
20a	4.12 ± 1.15	3.04 ± 1.17
20b	5.43 ± 0.27	4.23 ± 0.47
20c	1.03 ± 0.26	0.97 ± 0.43
20d	15.74 ± 0.32	13.42 ± 2.78

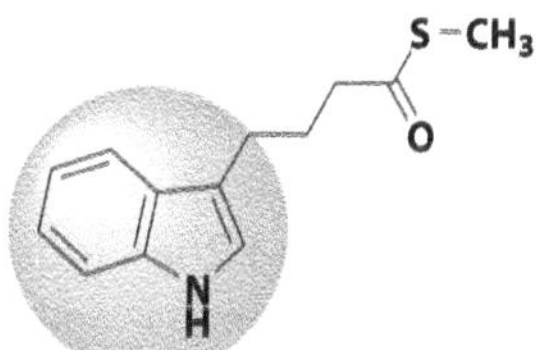

21

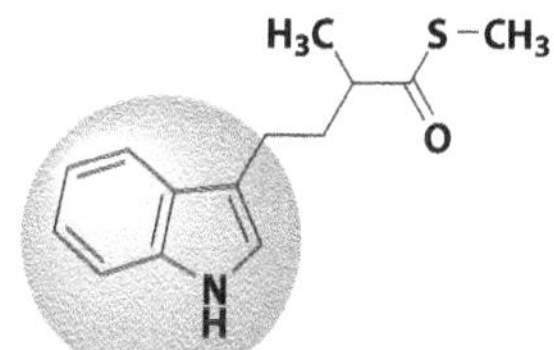

22

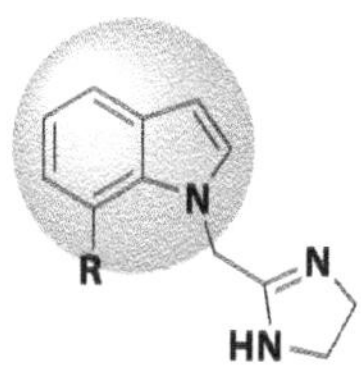

25

25a : H
25b : F
25c : Cl
25d : Br
25e : Me
25f : Et
25g : OMe
25h : OBz

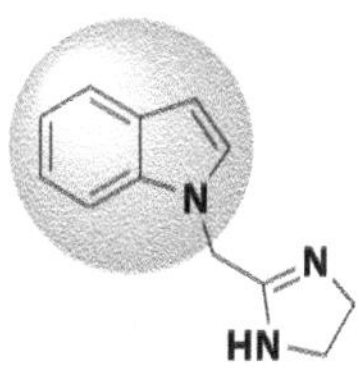

25a

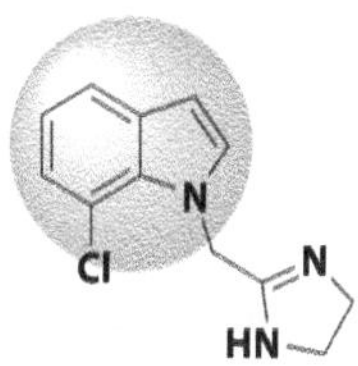

25c

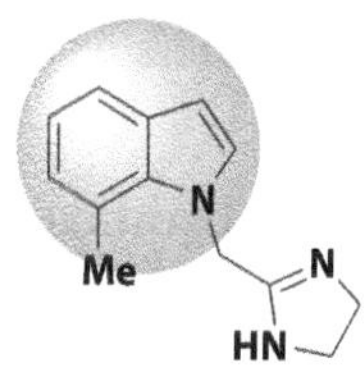

25e

Figure 10.6 Indole derivatives (19–25) as antihypertensive agents.

Compound **24** caused vasodilation in both endothelium-dependent and independent ways, according to mechanism studies. L-NAME ($1–10^6$ mol · L^{-1}) and ODQ ($1–10^6$ mol · L^{-1}) prevented the endothelium of intact aortic rings from relaxing as a result of Compound **24**. Furthermore, Compound **24** restricted Ca^{2+} inflow through L-type Ca^{2+} channels and inhibited intracellular Ca^{2+} release, but it had no effect on K^+ channels [71].

The series of derivatives Compound **25** presented in Figure 10.6 was established by Sazczewski *et al.* Next, the a1-AR and a2-AR binding of the target indole analogs was assessed through tests and through *in vitro* radioligand binding. The compounds' affinity for the a1-AR and a2-AR was influenced by the kind and location of the substituents, specifically for the a1-AR, Cl > Br > CH3 > F, and for the a2-AR, Cl > F > CH3 > Br. It has been demonstrated that clonidine-based antihypertensive medications bind to the imidazoline I_1 and I_2 receptors. Compounds **25c, 25e,** and **25a** showed promising antihypertensive properties [72].

The series of indole derivatives was synthesized by Bednarski *et al.* (26). After being administered intravenously to sedated rats with normal blood pressure, the hypotensive impact of the indole derivatives under investigation was assessed. A strong association between the hypotensive activity of spatial configuration and Compounds **26a, 26b, 26d,** and **26e** Figure 10.7 was observed. Enantiomer S is the most potent hypertensive agent, but the R-enantiomer is 2 to 32 times less potent. Adrenoceptor blockage (a1 in arteries and b1 in the heart) is most likely the cause of the hypotensive action. Radioligand binding tests only identified a difference in the enantiomers' affinity for the b1-AR, not the a1-AR. The compounds with methyl and dimethoxy groups and their enantiomers demonstrated a1-adrenolytic and b1-adrenolytic activity, respectively, and had antihypertensive pharmacological effects. Compound **26a**, in the form of a racemic mixture, decreased SBP and DBP at a dose of 0.25 to 1.0 mg/kg; hypotensive activity occurred 30 minutes after intravenous administration of 0.25 mg/kg. The S-enantiomers were the most potent hypertensive substances [73].

10.2.2 Imidazole

Smooth muscles rely on calcium influx for contraction, as other researchers have previously documented. Although their underlying mechanisms differ slightly, relaxation results from calcium-channel blockers' blocking of calcium channel influx into smooth muscles.

Kantevari *et al.* [74] synthesized 2-butyl-4-chloro-1-methylimidazoles 60 with embedded aryl- and heteroaryl-derived chalcones. These compounds were tested as inhibitors of ACE. The synthesis strategy involved

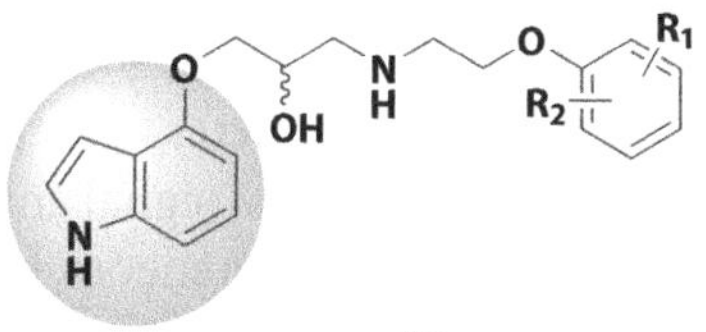

26

	R_1	R_2
26a	2-CH_3	H
26b	2-OCH_3	3-OCH_3
26c	2-OCH_3	6-OCH_3
26d	3-OCH_3	4-OCH_3
26e	3-OCH_3	5-OCH_3

Figure 10.7 Indole derivatives (26) as antihypertensive agents.

the Claisen–Schmidt condensation of several aryl or heteroaryl methyl ketones of Type **27** with imidazole-5-carbaldehyde (**28**). This procedure was catalyzed by means of 10% aqueous NaOH in methanol for 3.0 to 5.0 hours at room temperature. When screening all the new compounds with the colorimetric ACE inhibition assay, the most active inhibitors, of Type **29**, were found to have an IC_{50} of 2.24 to 3.60 μM (Figure 10.8). These values show that some derived chalcones are ~100 times more active than various chalcones and flavonoids of synthetic and natural origin [75–77].

In SHRs, the compound's effects on SBP and DBP were biologically evaluated (250–300 g), renal hypertensive rats, and rats from the Second Military Medical University in China [78]. The left renal arteries of 300 to 350 g Sprague–Dawley rats were totally clamped under sodium pentobarbital (40 mg/kg) anesthesia to prepare renal hypertensive rats. Rats with SBPs greater than 160 mm Hg were then chosen and utilized as models for renal hypertension. Rats with renal hypertension and spontaneous hypertension were then randomly assigned to 1 of 10 experimental groups (negative control group, positive control group, compound low-dose group, or high-dose group). Each substance was given orally at doses of 5 and 10 mg/kg, respectively, after being suspended in a 0.5% sodium carboxymethyl cellulose solution. SanXin Zhujiang Chemical Engineering Company's losartan (10 mg/kg) was used as a positive control. The identical amount of sodium carboxymethyl cellulose solution was given to the negative control group. The animals were sedated with sodium pentobarbital (40 mg kg, i.p.) to

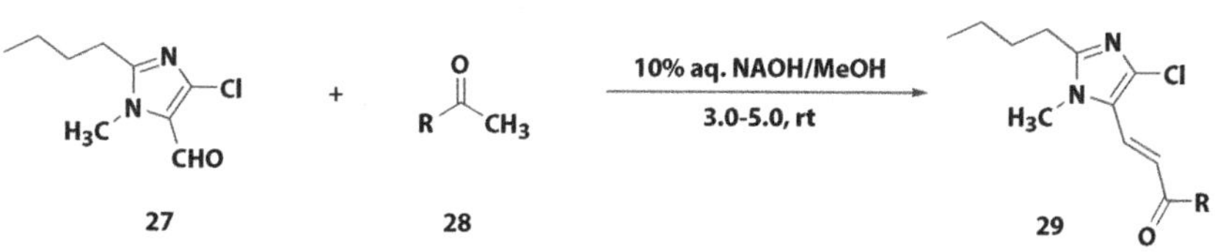

Figure 10.8 Imidazole derivatives (29) as antihypertensive agents.

take their blood pressure. For the purposes of measuring arterial pressure and administering medication, the right carotid artery and jugular vein were cannulated, respectively.

A pressure transducer was attached to the artery catheter, which was then shown on a computer and examined using a biological signal analysis system (MPA-2000; Alcott Biotech, China). Rats with spontaneous hypertension and rats with renal hypertension had their blood pressure and heart rate measured 0 to 10 and 24 hours after the medication was administered. The following formula was used to get MAP: MAP = DBP + (SBP – DBP)/3. The study's findings were presented as mean ± SD. Less than 0.05 as the probability level was deemed significant. The Institutional Animal Care and Use Committee accepted the finished animal research, which complied with the Principles of Laboratory Animal Care [78].

10.2.3 Furan

After searching, Malbranchea filamentous for bioactive substances that prevent Ca-induced vasoconstriction in rat aortic rings pretreated with high K or norepinephrine (**30**) was found. The diuretic potential of many pyrimidines (**31**) has been investigated and best exhibited a noticeable diuretic effect in contrast to furosemide, shown in Figure 10.9 [79].

Shen *et al.* [80] investigate that prazosin **32** is a sympatholytic drug used to treat high blood pressure (hypertension). It belongs to the class

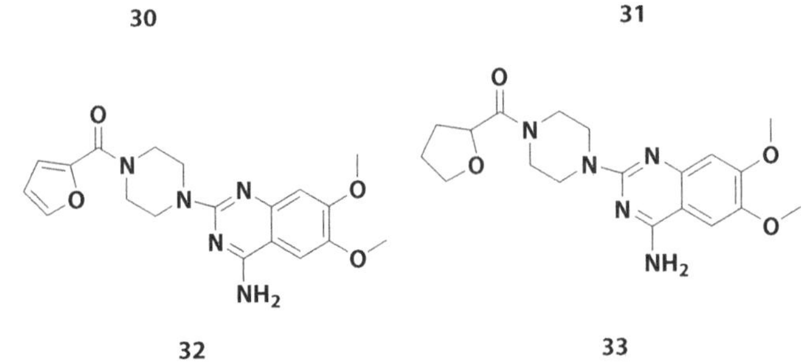

Figure 10.9 Furan derivatives (30–33) as antihypertensive agents.

of α-adrenergic blockers [81]. Tripathy *et al.* [82] reported that the drug terazosin **33** is a selective α1-antagonist used for treatment of symptoms of an enlarged prostate (BPH). It also acts to lower the blood pressure [83].

10.2.4 Tetrazole

The potential of tetrazole derivatives as putative antihypertensive medications has been demonstrated by the clinical usage of several tetrazole-containing medications, including losartan, valsartan, irbesartan, and candesartan, for the treatment of hypertension [84]. The first Ang II receptor antagonist, losartan, can prevent blood vessels from constricting, lowering blood pressure and enhancing blood flow. Additionally, it may be used to treat heart failure and hypertension, especially in individuals taking ACE inhibitors and experiencing coughing. ARBs, which include valsartan, irbesartan, and candesartan, can relax blood vessels and lessen cardiac strain by preventing the activity of several naturally occurring chemicals that constrict blood vessels. Valsartan is an oral drug used to treat diabetic renal disease, heart failure, and hypertension in adults and children over the age of 6 years. Whereas candesartan is used to treat hypertension and congestive heart failure, irbesartan is used to treat hypertension and shield the kidneys from damage caused by diabetes [81].

Chemical and spectral data were used to describe the produced compounds by Sharma *et al.* The primary goal of the current study was to identify the moieties that cause the suppression of Ang-II. It has been discovered that compounds with the same functional group at position 2 exhibit Ang II antagonism, which is dependent on the substitution at position 5. Compared to the substituted one, the presence of an amino group significantly boosted the activity. Amino-group Compounds **34, 35, 36, 37, 38, 39, 40, 41,** and **42** (Figure 10.10) have shown the strongest antihypertensive effects. This implies that the functional groups at position 2 may interact with specific places in the receptor pocket. A substituted benzimidazole nucleus connected to a tetrazole biphenyl group has been created, produced, and tested for Ang II antagonistic effects [83].

According to published research, the primary functions of the ACE are to break down BK into its inactive metabolites and to convert Ang I to Ang II. ACE and BK have a significant impact on how the human body's blood pressure varies. The literature also reports that tetrazoles and 1,5-benzothiazepines are medicinally powerful ACE inhibitors, in addition to members of the "pril" family such as captopril, lisinopril, enalapril, etc. In order to compare the *in vitro* and *in silico* ACE inhibitory effects, some tetrazole containing 1,5-benzothiazepines (**43a–e**) as presented in Figure 10.11 were

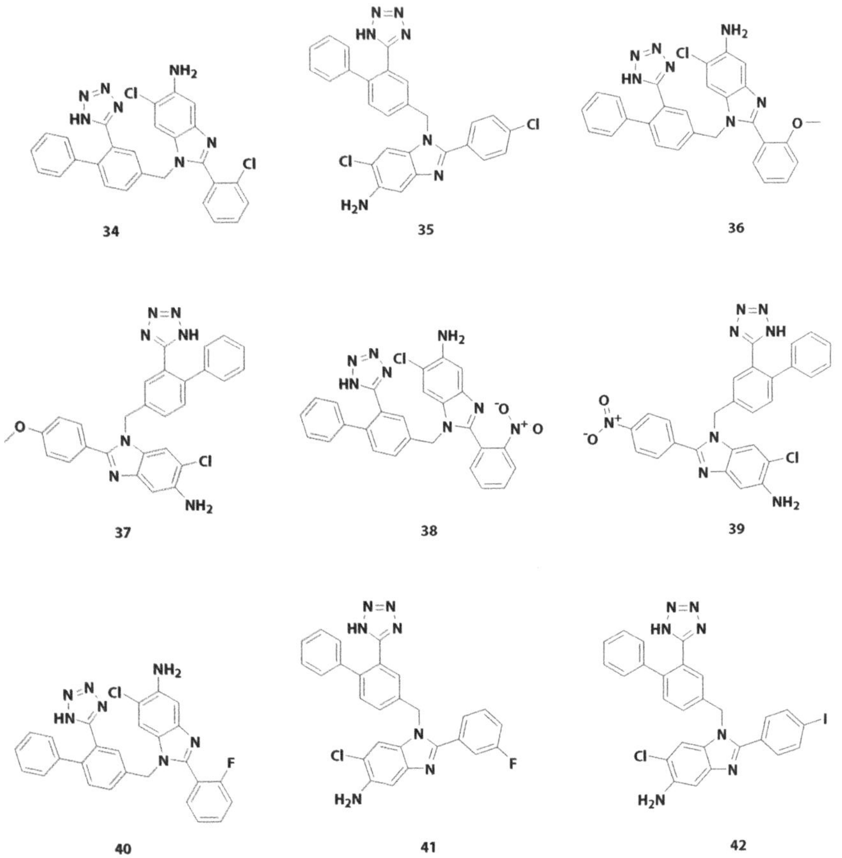

Figure 10.10 Tetrazole derivatives (34–42) as antihypertensive agents.

created by condensation of α, β-unsaturated carbonyl compounds with o-aminothiphenol, followed by the reduction of nitro group into amino group, and then reactions with sodium azide and triethyl formate (TEOF) to produce corresponding tetrazoles. The ACE inhibitory activity of each of the 1,5-benzothiazepines containing tetrazole was assessed both *in vitro* and *in silico*. Using Abz-Gly-Phe (NO_2)-Pro as a substrate, the docking scores and impact of the 1,5-benzothiazepines including tetrazole on 50% ACE inhibition (IC_{50}) were compared with that of the common medication captopril. When compared to the common medication captopril, it was found that all of the tetrazole-containing 1,5-benzothiazepines had extremely potent ACE inhibitory action. The highest ACE inhibitory activity is found in tetrazole-containing 1,5-benzothiazepines with methoxy and chloro groups at the para position. The design, synthesis, and assessment

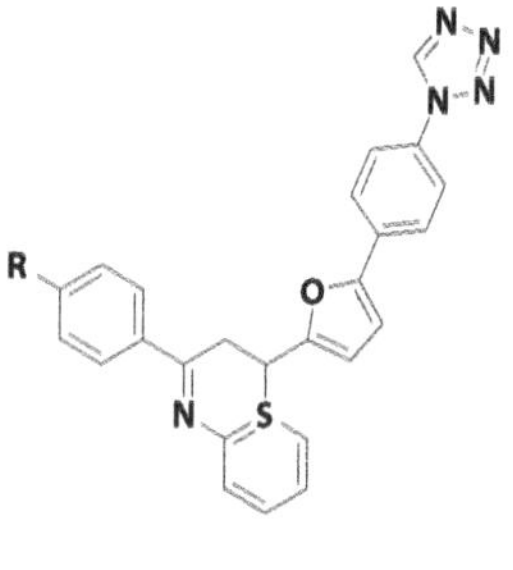

43

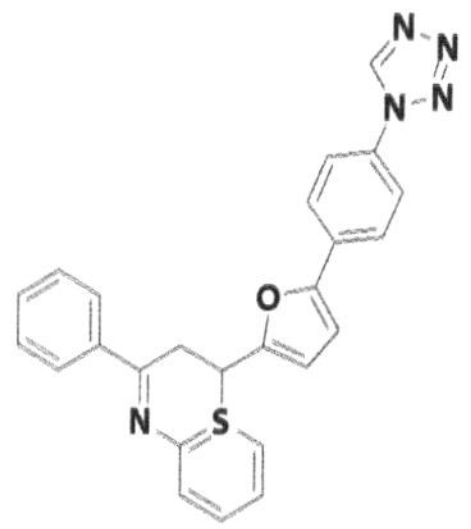

43a

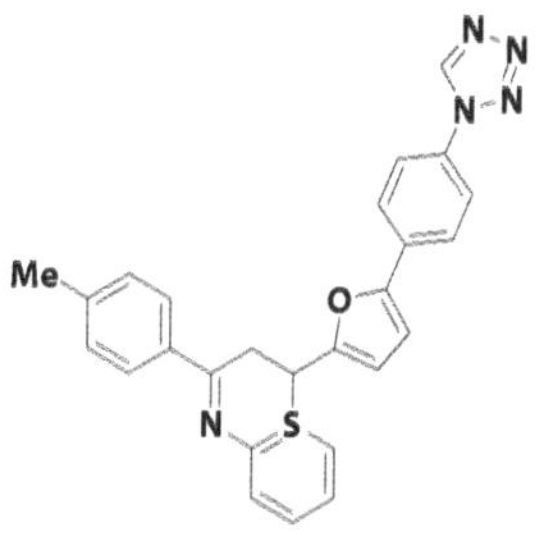

43b

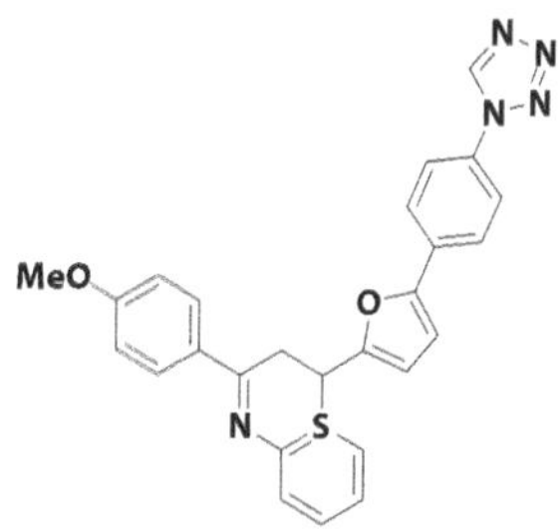

43c

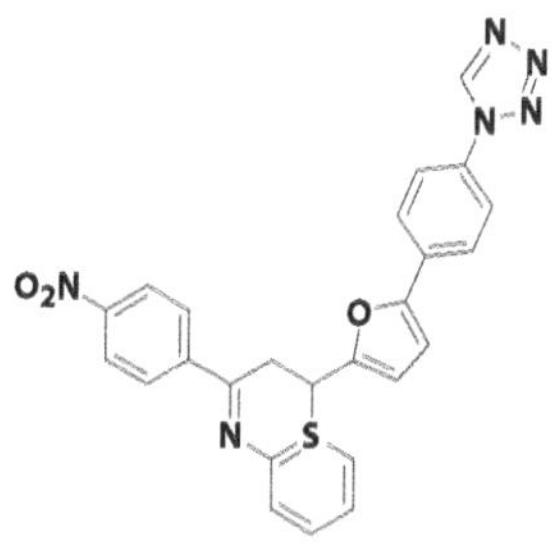

43d

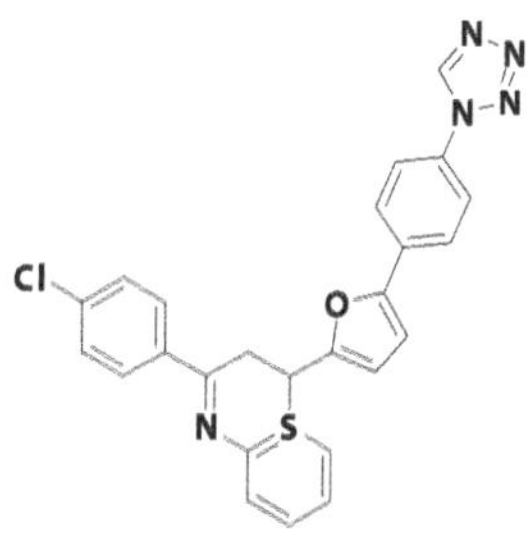

43e

Figure 10.11 Tetrazole derivatives (43) as antihypertensive agents.

of novel ACE inhibitors based on 1,5-benzothiazepines will greatly benefit from this study to provide new therapeutic agents for the treatment of hypertension and cardiovascular disorders [85].

10.2.5 Thiadiazole

Several 2-substituted phenyl derivatives (**44a–c**) illustrated in Figure 10.12 were identified by Turner *et al.* in their structure–activity studies of these thiadiazole hydrazines. These derivatives caused extended drops in blood pressure in DOCA hypertensive rats with only mild, non–dose-related tachycardia. According to research conducted on pithed rat preparations, these substances, such as hydralazine and minoxidil, worked by directly relaxing vascular smooth muscle to produce their hypotensive effects [86, 87].

2-Amino-5-aryl/alkyl-1,3,4-thiadiazoles were created by Amarish *et al.* [88] using thiosemicarbazide with aliphatic and aromatic acids. To create some naphthalenyloxy-propanol amine derivatives **45a–j**, (Figure 10.12) 2-(naphthalenyl oxymethyl) oxirane was used to condense these 2-amino-5-aryl/alkyl-1,3,4-thiadiazoles. These substances were created with the intention of acting as antihypertensive drugs [88].

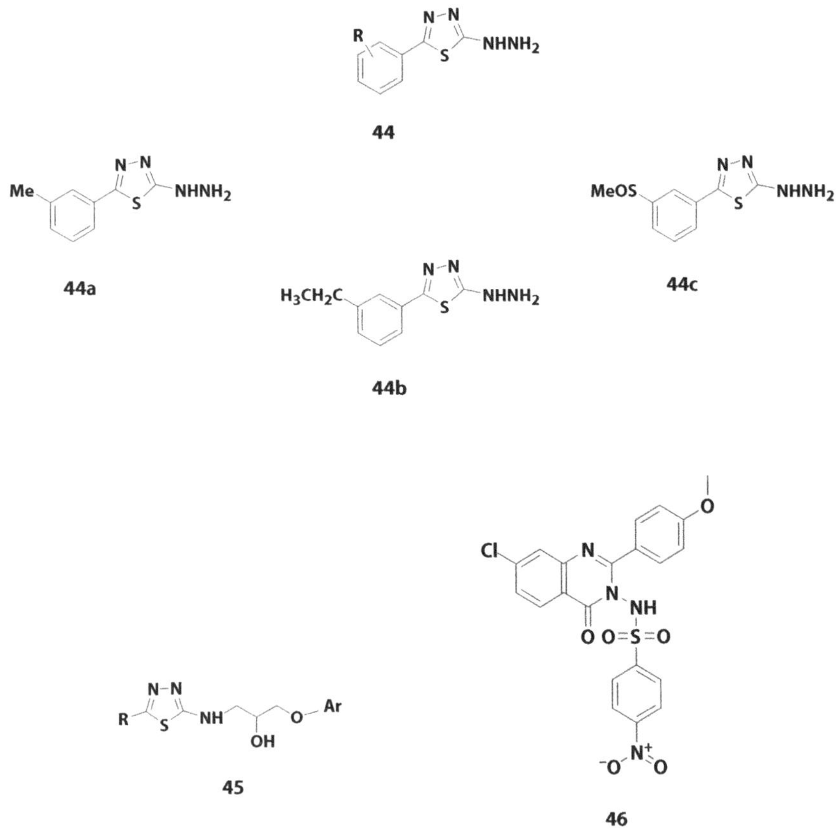

Figure 10.12 Thiadiazole derivatives (44–46) as antihypertensive agents.

10.2.6 Quinazoline

A series of *N*-substituted-(4-oxo-2-substituted-phenyl quinazolin-3-(4H)-yl) substituted benzene sulfonamide derivatives were created from substituted anthranilic acid-derived amino quinazolines and substituted benzene sulfonamides to investigate the pharmacological relationship and SAR. All of the synthesized compounds were tested for their antihypertensive activity using the tail-cuff method for noninvasive blood pressure, diuretic properties (by Lipschitz *et al.* [89] method), and antidiabetic potential in rats. Metolazone, prazosin, and diazoxide were used as benchmarks to compare six compounds exhibiting noticeably excellent activity. *N*-[7–chloro–2–(4–methoxyphenyl)–4–oxoquinoline–3(4H)–yl] is the compound of the series; 4-nitrobenzenesulfonamide (**46**) in Figure 10.12 showed the strongest effects [90].

Honkanen and associates produced and evaluated a number of substituted 2-piperidino-4-amino-6,7-dimethoxyquinazolines as possible antihypertensive drugs. After being administered intravenously to urethane-anesthetized normotensive rats, the hypotensive impact of each of the novel compounds was investigated. The prazosin molecule's furoyl piperazine moiety could be swapped out by an alternative piperidine group that is more stable without sacrificing its ability to decrease blood pressure [91].

As conformationally restricted analogs of SGB-1534 and ketanserin, a set of compounds (substituted phenyl piperazin-1-yl)methyl- and 2-[(substituted phenylpiperidinl-yl) methyl-2,3-dihydroimidazo[1,2-c] quinazolin-5(6H)-ones or -5(6H)-thiones, and 3-[(substituted phenyl piperazin-l-yl)methyl l-2,3-dihydro imidazole derivatives was synthesized for assessment as α-antagonists and antihypertensive agents. With no discernible activity at the sites, the majority of compounds with a (substituted phenyl piperazin 1) methyl side chain showed a high binding affinity for the al-AR. Compared to compounds with the identical substituent at the 2-position, compounds with a (substituted phenyl piperazin 1) methyl at the 3-position of 2,3-dihydroimidazo[1,2-c]quinazolin-5(6H)-one ring system exhibited superior activity. Compounds with substitution at the ortho position on the benzene ring of the phenylpiperazine side chain moiety are more potent than those without substitution and/or substitutions at the 3- and 4-positions, according to SAR for al-AR affinity that are shown. Little structural relationship between the quinazolinone and 2,3-dihydro imidazo [l,2-c quinazoline nucleus was shown by computer-assisted superimposition of SGB-1534 and 48b, and it is unlikely that these molecular fragments will interact specifically with the receptor protein. Intravenous

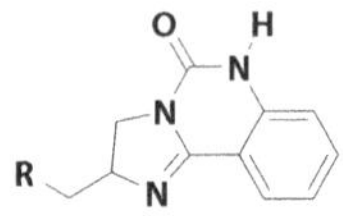

47

Compound	R
47a	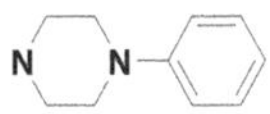
47b	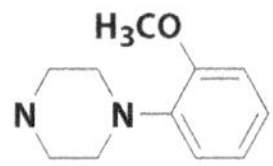

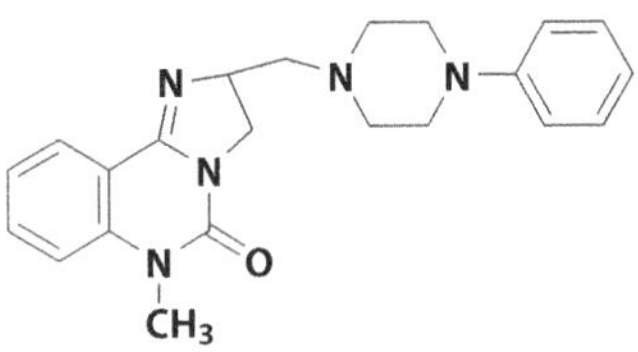

48a

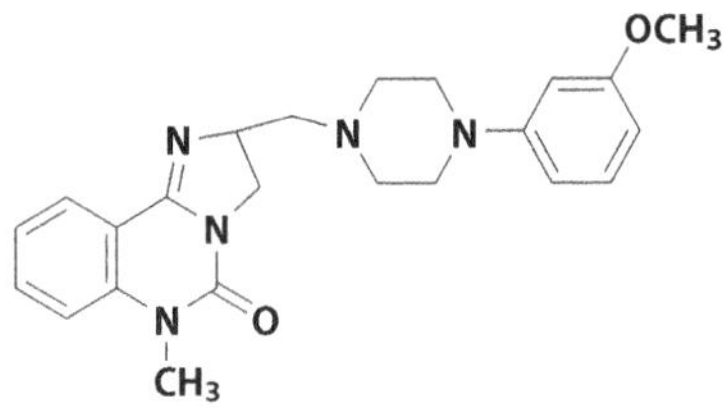

48b

Figure 10.13 Quinazoline derivatives (47,48) as antihypertensive agents.

antihypertensive activity was assessed. When administered to rats with spontaneous hypertension, Compounds **47a, 47b, 48a,** and **48b** as represented in Figure 10.13 showed comparable effectiveness to SGB-1534 after 6 hours. In the rat aorta, the pA2 value for **47a** against phenylephrine was much greater than that for prazosin. Compounds **48a** and **48b** merit additional investigation based on their al-AR affinity/selectivity *in vitro* and duration of antihypertensive effect *in vivo* [92].

10.2.7 Thiazolidine

In order to assess the antihypertensive effects of a series of tricyclic compounds with a guanidine moiety, Jen *et al.* synthesized 1,2,3,5-tetrahydro imidazo, 2, 1-b quinazoline. This series contained compounds of pyrimido[2,1–], imidazo[2,1–b]benzo-1,3-diazepine, and imidazo[1,2–a]quinazoline,

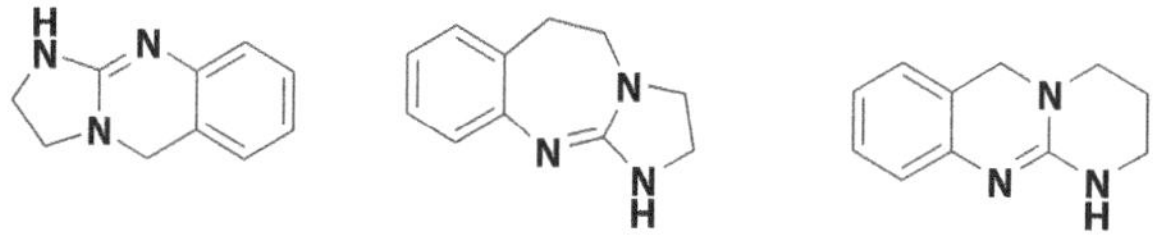

49 50 51

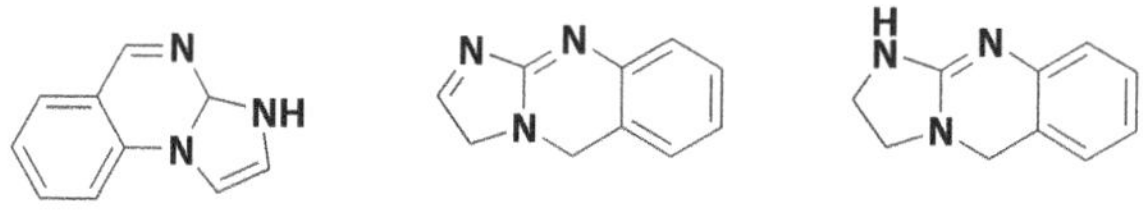

52 53 54

Figure 10.14 Thiazolidine derivatives (49–54) as antihypertensive agents.

imidazo[2,l-b quinazoline (S), imidazo[1,2-a]-benzimidazole, and b quinazoline. The assignment of tautomeric structures based on NMR spectrum data and the synthetic pathways to the novel compounds are examined. In unanesthetized neurogenic hypertensive medications, Compounds **49** to **54** (Figure 10.14) demonstrated antihypertensive action at oral dosages of 2 and 10 mg/kg, respectively [93].

Atrial flutter and atrial fibrillation are the most prevalent cardiac arrhythmias and are linked to a higher risk of heart failure, stroke, and death. One potentially atrial-selective therapy option for atrial fibrillation and atrial flutter is blocking the Kv1.5 ion channel. The production and biological assessment of thiazolidinone-based Kv1.5 blockers were described by Jackson *et al.* [94].

The strongest compounds in this series were determined to be the 3,4-dimethyl derivatives (IC_{50} = 0.069 lM) and (IC_{50} = 0.270 lM). When 2-(2-(4-(3-((5-substituted methylene)-4-oxo-2-(phenylimino)thiazolidin-3-yl)-2-hydroxypropyl amine) benzoyl) hydrazinyl)-2-oxoethyl nitrate derivatives were discovered in 2009, they were assessed for their electrocardiographic, antiarrhythmic, vasorelaxation, and antihypertensive activity as well as their capacity to release nitric oxide (NO) *in vitro*. It was discovered that some of these heterocycles were extremely powerful. The most effective compound in this series was determined to be 2-(2-(4-(3-(5-benzylidene-4-oxo-2-(phenylimino)thiazolidin-3-yl)-2-hydroxy propylamino) benzoyl) hydrazinyl)-2-oxoethyl nitrate [95].

10.2.8 Triazole

Anees and associates [96], utilizing a series of processes that began with the corresponding aryl hydrocarbons, several 6-(substituted phenyl)-2-(4-substituted phenyl-5-thioxo-4,5-dihydro-1H-1,2,4-triazol-3-yl)-4,5-dihydropyridine-3(2H)-one derivatives were created and produced. The resulting compounds' antihypertensive properties were assessed using the tail cuff technique, a noninvasive technique. Compounds **55e, 55i,** and **55k** as can be seen in Figure 10.15 demonstrated significant antihypertensive efficacy that was on par with that of propranolol and regular hydralazine [96].

Perekhoda and associates investigated compounds with 1,2,4-triazole, and morpholine or piperidine moieties have been created and synthesized as part of our ongoing investigation into possible antihypertensive agents among morpholine and piperidine derivatives. Docking studies have been carried out to identify physiologically active compounds with antihypertensive activity. According to the OECD project's classification of chemicals, all produced compounds are prospective ACE inhibitors and fall into toxicity Classes 4 and 5, according to the *in silico* tests [97].

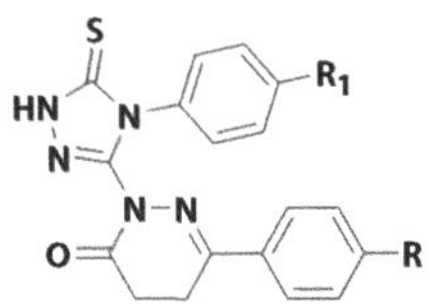

55

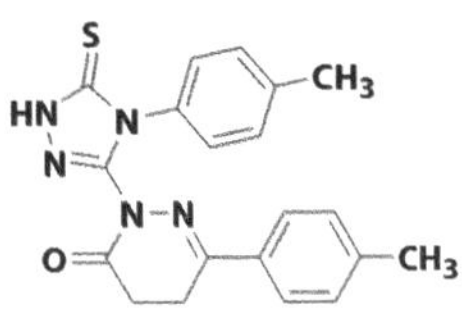

55e

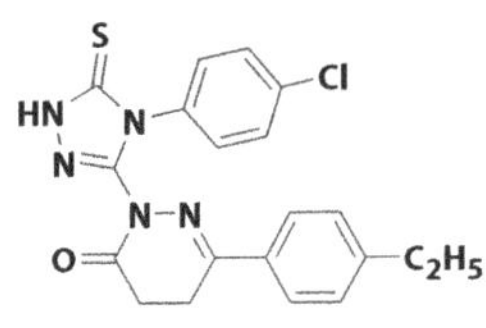

55k

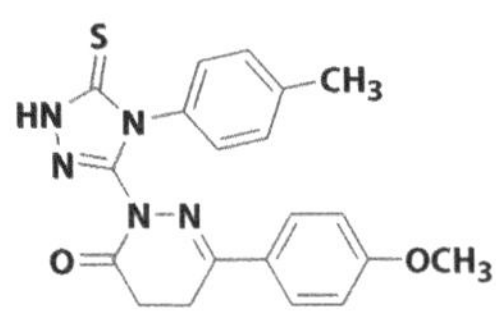

55i

Figure 10.15 Triazole derivative (55) as antihypertensive agents.

10.2.9 Pyridine

By combining various sulfonamide diazonium salts with a piperidine carbonitrile derivative, the fifth dihydropyridine derivatives (DHPD) 56a–e (Figure 10.16) substituted with a sulfonamide moiety were created. 5-((5-Cyano-1-ethyl-2-hydroxy-4-methyl-6-oxo-1,6-dihydropyridin-3-yl) arylazo) benzene-sulfonamides were discovered. Infrared, ^{1}H-NMR, ^{13}C-NMR, and elemental studies were used to clarify the chemical structures of these derivatives. The antihypertensive molecular docking test was applied to the optimized derivatives in an attempt to identify their interaction with the 6JP5 protein-coupled receptor. The derivatives 56c, 56d, and 56e were thought to have the greatest binding scores, 7.0397, 7.1905, and 7.1622 kcal/mol, respectively, based on the binding scores that were obtained for these analogs [98].

In conscious SHRs, the antihypertensive properties of 18 oxazolo[3,2-a]-pyridine, thiazolo[3,2]-pyridine, and pyrido[2,1]-bloxazine derivatives have been assessed and contrasted with the reference drug nifedipine. Eleven substances significantly decreased mean arterial blood pressure at a dose of 50 mg/kg (i.p.); 4 of the 11 were especially effective, causing severe hypotension more than 6 hours after administration and an impact that persisted for 24 hours. Six to 8 hours after being administered, the nifedipine-induced hypotension progressively subsided and vanished.

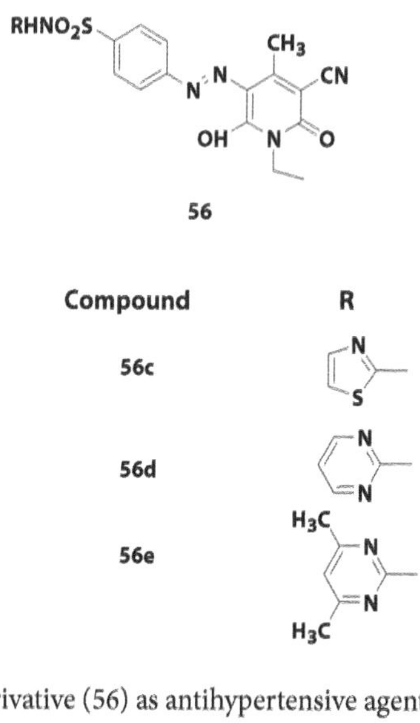

Figure 10.16 Pyridine derivative (56) as antihypertensive agents.

Generally speaking, reflex tachycardia does not accompany the prolonged activity exhibited by these substances. Two oxazolo[3,2-a]pyridine derivatives and two pyrido[2,l-b]oxazine derivatives administered intraperitoneally produced strong and enduring antihypertensive effects in SHRs [99].

A number of 2,3,5-trisubstituted 4,5-dihydro-4-oxo-3H-imidazo[4,5-c] pyridine derivatives were subjected to QSAR (quantitative SAR) investigations as Ang II AT1 receptor antagonists to determine the structural characteristics necessary for the antihypertensive action. The V-life Molecular Design Suite software was used to conduct the QSAR analysis, and the best QSAR model that was obtained using the multiple linear regression and partial least square (PLS) principal component regression methods revealed biological activity variation. The model with the highest correlation coefficient (r^2 = 0.9425) and statistical significance was chosen for more research. The model's cross-validated correlation coefficient (q^2 = 0.7786 and pred_r^2 = 0.8562) and other validation metrics demonstrate the model's strong predictive capabilities. The model demonstrated a strong correlation between the Ang II AT1 receptor antagonist activity of 2,3,5-trisubstituted 4,5-dihydro-4-oxo-3H-imidazo[4,5-c]pyridine derivatives and the parameters SdssCcount, SssNHcount, SaaaCcount, and H_ Donor count. 3D-QSAR models were created using the PLS methodology in conjunction with different feature selection techniques, such as stepwise, simulated annealing, and genetic algorithms. These models were then further verified for statistical significance and predictive power through internal and external validation. The best 3D-QSAR model-7 was built utilizing molecular field analysis and the k-nearest neighbor approach, demonstrating strong predictive and correlative skills in terms of pred_r^2 = 0.8152 and q^2 = 0.8316. These derivatives' 2D and 3D QSAR studies offer insights for future lead optimization and the development of powerful antihypertensive drugs [100].

10.3 Future Prospective

In view of their structural diversity and capacity to react with a wide range of biological targets, heterocyclic compounds still capture the interest of researchers looking to discover new antihypertensive medicines. Additional study in this area is expected to concentrate on developing heterocyclic frameworks with improved effectiveness, selectivity, and minimum side effects. Enhanced computational methods, which include docking molecules, QSAR modeling, and AI-powered drug development platforms, will be important for finding and optimizing lead compounds.

Furthermore, combination therapy incorporating heterocyclic compounds and other antihypertensive medications may provide synergistic benefits while dealing with the multifactorial character of hypertension. Investigating possibilities as dual or multitarget inhibitors, notably for RAAS components, endothelin receptors, and calcium channels, might lead to novel treatment techniques.

At last, the increased emphasis on precision medicine and personalized therapy creates a great opportunity to customize heterocyclic-based antihypertensive therapies to individual genetic profiles, resulting in improved patient results. These findings have the potential to significantly alter the function of heterocyclic compounds in hypertension management in the future decades.

10.4 Conclusion

The development of new antihypertensive drugs heavily relies on heterocyclic compounds because of their adaptable chemical structures and capacity to interact with a wide range of biological targets. The structural variety of heterocyclic frameworks offers substantial promise for influencing major blood pressure regulation mechanisms, such as the renin–angiotensin system, calcium-channel modulation, and NO signaling.

Recent advances in medicinal chemistry have resulted in novel methodologies for designing and synthesizing heterocyclic derivatives with increased potency, selectivity, and fewer adverse effects. Among the most researched heterocycles, pyridines, pyrazoles, imidazoles, and thiazoles showed outstanding activity patterns, underscoring their relevance in contemporary drug development efforts. Future research should focus on improving the pharmacokinetic and pharmacodynamic aspects of these drugs, as well as investigating new heterocyclic scaffolds using computational and high-throughput screening methods. Collaboration among chemists and pharmaceutical researchers will be required to convert these intriguing compounds into clinically viable hypertension medicines.

Thus, heterocyclic compounds constitute a viable frontier in the battle to combat hypertension, a worldwide health issue that needs novel treatment approaches.

AI disclosure statement: The author acknowledges the use of AI-based tools for minor language editing or grammatical correction. However, all intellectual contributions, analyses, and conclusions presented in this chapter are entirely the author's own.

References

1. Dong, C., Bu, X., Liu, J., Wei, L., Ma, A., Wang, T., Cardiovascular disease burden attributable to dietary risk factors from 1990 to 2019: A systematic analysis of the Global Burden of Disease study. *Nutr. Metab. Cardiovasc. Dis.*, 32, 4, 897–907, 2022.
2. James, O.O., Efosa, J.D., Romokeme, A.M., Zuobemi, A., Sotonye, D.-M., Dominance of Hypertensive Heart Disease in a Tertiary Hospital in Southern Nigeria. *Ethn. Dis.*, 22, 2, 136–139, 2012.
3. Singh, G.K., Siahpush, M., Azuine, R.E., Williams, S.D., Increasing area deprivation and socioeconomic inequalities in heart disease, stroke, and cardiovascular disease mortality among working age populations, United States, 1969-2011. *Int. J. MCH AIDS*, 3, 2, 119, 2015.
4. Bigna, J.J., *et al.*, Global burden of hypertension among people living with HIV in the era of increased life expectancy: a systematic review and meta-analysis. *J. Hypertens.*, 38, 9, 1659–1668, 2020.
5. Cao, Y., DiGiacomo, M., Du, H.Y., Ollerton, E., Davidson, P., Cardiovascular disease in Chinese women: an emerging high-risk population and implications for nursing practice. *J. Cardiovasc. Nurs.*, 23, 5, 386–394, 2008.
6. Bekele, A., *et al.*, Pattern and trend of medical admissions of patients of chronic non-communicable diseases in selected Hospitals in Addis Ababa, Ethiopia. *Am. Sci. Res. J. Eng. Technol. Sci.*, 13, 1, 34–48, 2015.
7. Wang, J., History of Development, Classification, Mechanism of Action and Rational Application of Major Anti-Hypertensive Drugs. *Int. J. Public Health Med. Res.*, 1, 2, 44–52, 2024.
8. Al Disi, S.S., Anwar, M.A., Eid, A.H., Anti-hypertensive herbs and their mechanisms of action: part I. *Front. Pharmacol.*, 6, 323, 2016.
9. da Silva, M.G., *et al.*, Bioactive Natural Products against Systemic Arterial Hypertension: A Past 20-Year Systematic and Prospective Review. *Evid.-Based Complement. Alternat. Med.*, 2022, 1, 8499625, 2022.
10. Xie, M., *et al.*, Anti-hypertensive and cardioprotective activities of traditional Chinese medicine-derived polysaccharides: A review. *Int. J. Biol. Macromol.*, 185, 917–934, 2021.
11. Rosenthal, T., Contemplating the history of drug therapy for hypertension. *Blood Press.*, 13, 5, 262–271, 2004.
12. Anthony, H., Valinsky, L., Inbar, Z., Gabriel, C., Varda, S., Perceptions of hypertension treatment among patients with and without diabetes. *BMC Fam. Pract.*, 13, 1, 24, 2012.
13. Lenfant, C., Chobanian, A.V., Jones, D.W., Roccella, E.J., Seventh report of the Joint National Committee on the Prevention, Detection, Evaluation, and Treatment of High Blood Pressure (JNC 7) resetting the hypertension sails, Lippincott Williams & Wilkins, 2003.

14. Parati, G., Ochoa, J.E., Salvi, P., Lombardi, C., Bilo, G., Prognostic value of blood pressure variability and average blood pressure levels in patients with hypertension and diabetes. *Diabetes Care*, 36, Suppl 2, S312, 2013.
15. Kjeldsen, S.E., *et al.*, Predictors of blood pressure response to intensified and fixed combination treatment of hypertension: the ACCOMPLISH study. *Blood Press.*, 17, 1, 7–17, 2008.
16. Miller, R.E., Shapiro, A.P., King, H.E., Ginchereau, E.H., Hosutt, J.A., Effect of antihypertensive treatment on the behavioral consequences of elevated blood pressure. *Hypertension*, 6, 2_pt_1, 202–208, 1984.
17. Rosendorff, C., *et al.*, REPRINT Treatment of hypertension in the prevention and management of ischemic heart disease: A scientific statement from the American Heart Association Council for high blood pressure research and the Councils on clinical cardiology and epidemiology and prevention. *Hypertension*, 50, 2, e28–e55, 2007.
18. Moser, M. and Roccella, E.J., The treatment of hypertension: a remarkable success story. *J. Clin. Hypertens.*, 15, 2, 88–91, 2013.
19. Wang, J. and Xiong, X., Control strategy on hypertension in Chinese medicine. *Evid.-Based Complement. Alternat. Med.*, 2012, 1, 284847, 2012.
20. Staessen, J.A., Wang, J., Bianchi, G., Birkenhäger, W.H., Essential hypertension. *Lancet*, 361, 9369, 1629–1641, 2003.
21. Böhm, M., La Rosée, K., Schwinger, R.H.G., Erdmann, E., Evidence for reduction of norepinephrine uptake sites in the failing human heart. *J. Am. Coll. Cardiol.*, 25, 1, 146–153, 1995.
22. Mills, K.T., Stefanescu, A., He, J., The global epidemiology of hypertension. *Nat. Rev. Nephrol.*, 16, 4, 223–237, 2020.
23. Liu N., Tang J., Xue Y., Mok V., Zhang M., Ren X., Wang Y., Fu J., EP3 receptor deficiency improves vascular remodeling and cognitive impairment in cerebral small vessel disease. *Aging Dis.*, 13, 1, 313–328, 2022.
24. Van Vuuren, M.A., An investigation into the anti-hypertensive effects of Green Rooibos Tea Extract (GRT), Stellenbosch: Stellenbosch University, 2019.
25. Altaf, R., Evaluation of Anti-hypertensive and Vasorelaxant Effects of Phaleria macrocarpa Extracts, Universiti Sains Malaysia, 2015.
26. Matsumura, E., *et al.*, Microbial production of novel sulphated alkaloids for drug discovery. *Sci. Rep.*, 8, 1, 7980, 2018.
27. Ramesh, S.S., Christopher, R., Indira Devi, B., Bhat, D.I., The vascular protective role of oestradiol: a focus on postmenopausal oestradiol deficiency and aneurysmal subarachnoid haemorrhage. *Biol. Rev.*, 94, 6, 1897–1917, 2019.
28. Mowry, F.E. and Biancardi, V.C., Neuroinflammation in hypertension: the renin-angiotensin system versus pro-resolution pathways. *Pharmacol. Res.*, 144, 279–291, 2019.
29. Banerjee, R., Kumar, H.K.S., Banerjee, M., Medicinal significance of furan derivatives: a review. *Int. J. Rev. Life Sci.*, 2, 1, 7–16, 2012.

30. Sharma, N., Rana, A.C., Bafna, P., Effect of aqueous extract of Cynodon dactylon on reserpine induced catalepsy. *Int. J. Pharm. Pharm. Sci.*, 3, 4, 424–426, 2011.
31. Witkin, J.M., Shafique, H., Smith, J.L., others, Is there a biochemical basis for purinergic P2X3 and P2X4 receptor antagonists to be considered as anti-seizure medications? *Biochem. Pharmacol.*, 222, 116046, 2024.
32. Burnstock, G., Purinergic P2 receptors as targets for novel analgesics. *Pharmacol. Ther.*, 110, 3, 433–454, 2006.
33. Martins, P., *et al.*, Heterocyclic anticancer compounds: recent advances and the paradigm shift towards the use of nanomedicine's tool box. *Molecules*, 20, 9, 16852–16891, 2015.
34. Desir, G.V., Regulation of blood pressure and cardiovascular function by renalase. *Kidney Int.*, 76, 4, 366–370, 2009.
35. Miyamoto, J., Kasubuchi, M., Nakajima, A., Irie, J., Itoh, H., Kimura, I., The role of short-chain fatty acid on blood pressure regulation. *Curr. Opin. Nephrol. Hypertens.*, 25, 5, 379–383, 2016.
36. Mao, Y., Tokudome, T., Kishimoto, I., Ghrelin and blood pressure regulation. *Curr. Hypertens. Rep.*, 18, 2, 15, 2016.
37. Heymans, C. and den Heuvel-Heymans, G., New aspects of blood pressure regulation. *Circulation*, 4, 4, 581–586, 1951.
38. Kumar, N. and Goel, N., Heterocyclic compounds: importance in anticancer drug discovery. *Anti-Cancer Agents Med. Chem.*, 22, 19, 3196–3207, 2022.
39. Alghamdi, S., Kabrah, A., Khidir, E.B., Al-Moraya, I.S., Asif, M., Diverse Heterocyclic Molecules Targeting Oxidative Stress as Therapeutic Effects Against Various Neurological Diseases. *Mini-Rev. Org. Chem.*, 22, 2, 136–150, 2025.
40. Bisi, A., Gobbi, S., Belluti, F., Rampa, A., Design of multifunctional compounds for cardiovascular disease: from natural scaffolds to 'classical' multi-target approach. *Curr. Med. Chem.*, 20, 13, 1759–1782, 2013.
41. Srour, A.M., Dawood, D.H., Saleh, D.O., Synthesis, 3D-pharmacophore modelling and 2D-QSAR study of new pyridine-3-carbonitriles as vasorelaxant active agents. *New J. Chem.*, 45, 17, 7731–7740, 2021.
42. Fotopoulou, T., *et al.*, Design and synthesis of nitrate esters of aromatic heterocyclic compounds as pharmacological preconditioning agents. *Bioorg. Med. Chem.*, 16, 8, 4523–4531, 2008.
43. Bhandari, S.V., *et al.*, Design, synthesis and pharmacological screening of novel antihypertensive agents using hybrid approach. *Bioorg. Med. Chem.*, 17, 1, 390–400, 2009.
44. Vyas, V.K. and Ghate, M., Substituted benzimidazole derivatives as angiotensin II-AT1 receptor antagonist: a review. *Mini-Rev. Med. Chem.*, 10, 14, 1366–1384, 2010.
45. Thakur, S., Arora, S., Katiyar, M.K., Joshi, G., Kumar, R., Fused Thienopyrimidines as Versatile Pharmacophores for the Development of Cyclooxygenase-2 Inhibitors. *ChemistrySelect*, 9, 22, e202400514, 2024.

46. Detke, S.J., Karwande, A.D., Ponugoti, S.S., Joshi, S.V., Kharkar, P.S., Advancing quinoxaline chemistry: Sustainable and green synthesis using L-arabinose as a catalyst. *Synth. Commun.*, 54, 23, 2024–2037, 2024.
47. Haider, M.B., *et al.*, Exploring Acyl Thiotriazinoindole Based Pharmacophores: Design, Synthesis, and SAR Studies with Molecular Docking and Biological Activity Profiling against Urease, $α$-amylase, $α$-glucosidase, Antimicrobial, and Antioxidant Targets. *Protein J.*, 43, 5, 1009–1024, 2024.
48. Hassanin, M.A., Mustafa, M., Hassan, H.A., Beshr, E.A., Elshennawy, A., Aly, O.M., Hydantoin derivatives: A review on their anticancer activities. *Octahedron Drug Res.*, 4, 1, 61–74, 2024.
49. Singla, D., Sharma, P., Luxami, V., Paul, K., *In Vitro* Cytotoxicity and Mechanistic Investigation of Quinazolin-4 (1 H)-One Linked Coumarin as a Potent Anticancer Agent. *Chem. Biol. Drug Des.*, 104, 5, e70011, 2024.
50. Quan, P., *et al.*, A novel surface modified nitrendipine nanocrystals with enhancement of bioavailability and stability. *Int. J. Pharm.*, 430, 1–2, 366–371, 2012.
51. Thapliyal, S., *et al.*, Solid lipid nanoparticles for pulmonary arterial hypertension: enhancing solubility and bioavailability of lipophilic phytoconstituents. *Biochem. Cell. Arch.*, 24, 2, 2024.
52. Kreighbaum, W.E., *et al.*, Antihypertensive indole derivatives of phenoxypropanolamines with. beta.-adrenergic receptor antagonist and vasodilating activity. *J. Med. Chem.*, 23, 3, 285–289, 1980.
53. Vega, A.M., Aldana, I., Parrado, P., Font, M., Alvarez, E.F., Antihypertensive agents: Pyridazino (4, 5-b) indole derivatives. *J. Pharm. Sci.*, 71, 12, 1406–1408, 1982.
54. Kim, S.-H., *et al.*, Discovery of a new HIV-1 inhibitor scaffold and synthesis of potential prodrugs of indazoles. *Bioorg. Med. Chem. Lett.*, 23, 10, 2888–2892, 2013.
55. Monge, A., Parrado, P., Font, M., Fernandez-Alvarez, E., Selective thromboxane synthetase inhibitors and antihypertensive agents. New derivatives of 4-hydrazino-5H-pyridazino [4, 5-b] indole, 4-hydrazinotriazino [4, 5-a] indole, and related compounds. *J. Med. Chem.*, 30, 6, 1029–1035, 1987.
56. Tan, C., Yang, S.-J., Zhao, D.-H., Li, J., Yin, L.-Q., Antihypertensive activity of indole and indazole analogues: A review. *Arab. J. Chem.*, 15, 5, 103756, 2022.
57. Squadrito, F., *et al.*, Antihypertensive activity of indolepyruvic acid: a keto analogue of tryptophan. *J. Cardiovasc. Pharmacol.*, 15, 1, 102–108, 1990.
58. Todd, P.A. and Fitton, A., Perindopril: a review of its pharmacological properties and therapeutic use in cardiovascular disorders. *Drugs*, 42, 1, 90–114, 1991.
59. El-Gendy, A.A. and El-Banna, H.A., Synthesis and antihypertensive activity of certain Mannich Bases of 2-ethoxycarbonylindoles and 5 H-pyridazino [4, 5-b] indoles. *Arch. Pharm. Res.*, 24, 1, 21–26, 2001.

60. Romeo, G., *et al.*, High affinity ligands and potent antagonists for the $α$1D-adrenergic receptor. Novel 3, 8-disubstituted [1] benzothieno [3, 2-d] pyrimidine derivatives. *Eur. J. Med. Chem.*, 83, 419–432, 2014.
61. Romeo, G., *et al.*, @article{romeo2006new, title={New pyrimido [5, 4-b] indoles and [1] benzothieno [3, 2-d] pyrimidines: High affinity ligands for the α1-adrenoceptor subtypes}, author={Romeo, Giuseppe and Materia, Luisa and Marucci, Gabriella and Modica, Maria and P. *Bioorg. Med. Chem. Lett.*, 16, 24, 6200–6203, 2006.
62. Bell, M.G., *et al.*, (S)-N-$\{$3-[1-Cyclopropyl-1-(2, 4-difluoro-phenyl)-ethyl]-1 H-indol-7-yl$\}$-methanesulfonamide: A Potent, Nonsteroidal, Functional Antagonist of the Mineralocorticoid Receptor. *J. Med. Chem.*, 50, 26, 6443–6445, 2007.
63. Zhou, J. and Zhou, S., Antihypertensive and neuroprotective activities of rhynchophylline: the role of rhynchophylline in neurotransmission and ion channel activity. *J. Ethnopharmacol.*, 132, 1, 15–27, 2010.
64. Zhang, M.-Z., *et al.*, Synthesis and antifungal activity of novel indole-replaced streptochlorin analogues. *Eur. J. Med. Chem.*, 126, 669–674, 2017.
65. Sun, Y., *et al.*, Discovery of novel anti-angiogenesis agents. Part 8: Diaryl thiourea bearing 1H-indazole-3-amine as multi-target RTKs inhibitors. *Eur. J. Med. Chem.*, 141, 373–385, 2017.
66. Chunqing, S., Yi, F., Weihui, H., Different hypotentive effects of various active constituents isolated from Uncaria rhynchophylla. *Chin. Tradit. Herb. Drugs*, 31, 10, 762–763, 2000.
67. Neglia, D., *et al.*, Treatment of hypertension with perindopril plus indapamide leads to reverse coronary microvascular remodelling and improved blood flow. *J. Hypertens.*, 29, 364–372, 2011.
68. Zhu, W., *et al.*, Design, synthesis and biological evaluation of new 5-nitro benzimidazole derivatives as AT1 antagonists with anti-hypertension activities. *Bioorg. Med. Chem.*, 22, 7, 2294–2302, 2014.
69. Sharma, S. and Singh, S., The biological and pharmacological potentials of indole-based heterocycles. *Lett. Org. Chem.*, 20, 8, 711–729, 2023.
70. Cuong, N.M., *et al.*, Vascular L-type Ca2+ channel blocking activity of sulfur-containing indole alkaloids from Glycosmis petelotii. *J. Nat. Prod.*, 77, 7, 1586–1593, 2014.
71. Zhu, K., Yang, S.-N., Ma, F.-F., Gu, X.-F., Zhu, Y.-C., Zhu, Y.-Z., The novel analogue of hirsutine as an anti-hypertension and vasodilatary agent both *in vitro* and *in vivo*. *PLoS One*, 10, 4, e0119477, 2015.
72. Sączewski, J., *et al.*, Transfer of SAR information from hypotensive indazole to indole derivatives acting at $α$-adrenergic receptors: *in vitro* and *in vivo* studies. *Eur. J. Med. Chem.*, 115, 406–415, 2016.
73. Bednarski, M., *et al.*, Synthesis and Pharmacological Activity of a New Series of 1-(1H-Indol-4-yloxy)-3-(2-(2-methoxyphenoxy) ethylamino) propan-2-ol Analogs. *Arch. Pharm. (Weinheim)*, 349, 3, 211–223, 2016.

74. Kantevari, S., Addla, D., Bagul, P.K., Sridhar, B., Banerjee, S.K., @article{kantevari2011synthesis, title={Synthesis and evaluation of novel 2-butyl-4-chloro-1-methylimidazole embedded chalcones and pyrazoles as angiotensin converting enzyme (ACE) inhibitors}, author={Kantevari, Srinivas and Addla, Dinesh and Bagul, Pank. *Bioorg. Med. Chem.*, 19, 16, 4772–4781, 2011.
75. Li, Y., Suino, K., Daugherty, J., Xu, H.E., Structural and biochemical mechanisms for the specificity of hormone binding and coactivator assembly by mineralocorticoid receptor. *Mol. Cell*, 19, 3, 367–380, 2005.
76. Bledsoe, R.K., *et al.*, A Ligand-mediated Hydrogen Bond Network Required for the Activation of the Mineralocorticoid Receptor*[boxs]. *J. Biol. Chem.*, 280, 35, 31283–31293, 2005.
77. Siwach, A. and Verma, P.K., Synthesis and therapeutic potential of imidazole containing compounds. *BMC Chem.*, 15, 1, 12, 2021.
78. Da, Y.-J., Yuan, W.-D., Zhu, L.-F., Chen, Z.-L., Synthesis, anti-hypertensive effect of a novel angiotensin II AT1 receptor antagonist and its anti-tumor activity in prostate cancer. *Arzneim.-Forsch.*, 62, 12, 637–643, 2012.
79. Nivrutti, G.P. and others, Furan: A Promising Scaffold for Biological Activity. *Int. J. Adv. Biol. Biomed. Res.*, 12, 167–181, 2024.
80. Wang, J., Shen, F., Wang, M., Su, D., Effects of nine antihypertensive drugs on blood pressure variability in sinoaortic-denervated rats. *Acta Pharmacol. Sin.*, 27, 8, 1013–1017, 2006.
81. Zou, Y., Liu, L., Liu, J., Liu, G., Bioisosteres in drug discovery: focus on tetrazole. *Future Med. Chem.*, 12, 2, 91–93, 2020.
82. Tripathi, K.D., *Essentials of medical pharmacology*, Jaypee Brothers Medical Publishers, New Delhi, 2018.
83. Sharma, M.C., Kohli, D.V., Sharma, S., Synthesis and biological evaluation 6-chloro-2-substitued-1-[2-(1h-tetrazol-5-yl)-biphenyl-4-ylmethyl]-1h-benzoimidazol-5-ylamine antihypertensive agents. *J. Optoelectron. Biomed. Mater.*, 2, 2, 59–72, 2010.
84. Kaushik, N., Kumar, N., Kumar, A., Singh, U.K., Tetrazoles: Synthesis and biological activity. *Immunol. Endocr. Metab. Agents-Med. Chem.*, 18, 1, 3–21, 2018.
85. Singh, B. and Suman, A.K., Tetrazole containing 1, 5-benzothiazepines: Design, synthesis, in silico and *in vitro* evaluation as medicinally potent ACE inhibitors. *J. Chem. Lett.*, 5, 4, 192–205, 2024.
86. Mishra, R., Sharma, P.K., Verma, P.K., Tomer, I., Mathur, G., Dhakad, P.K., Biological potential of thiazole derivatives of synthetic origin. *J. Heterocycl. Chem.*, 54, 4, 2103–2116, 2017.
87. Turner, S., *et al.*, Antihypertensive thiadiazoles. 1. Synthesis of some 2-aryl-5-hydrazino-1, 3, 4-thiadiazoles with vasodilator activity. *J. Med. Chem.*, 31, 5, 902–906, 1988.

88. Samel, A.B. and Pai, N.R., Synthesis of novel aryloxy propanoyl thiadiazoles as potential antihypertensive agents. *J. Chin. Chem. Soc.*, 57, 6, 1327–1330, 2010.
89. Lipschitz, W.L., Hadidian, Z., Kerpcsar, A., Bioassay of diuretics. *J. Pharmacol. Exp. Ther.*, 79, 2, 97–110, 1943.
90. Rahman, M.U., Rathore, A., Siddiqui, A.A., Parveen, G., Yar, M.S., Synthesis and characterization of quinazoline derivatives: search for hybrid molecule as diuretic and antihypertensive agents. *J. Enzyme Inhib. Med. Chem.*, 29, 5, 733–743, 2014.
91. Honkanen, E., Pippuri, A., Kairisalo, P., Nore, P., Karppanen, H., Paakkari, I., Synthesis and antihypertensive activity of some new quinazoline derivatives. *J. Med. Chem.*, 26, 10, 1433–1438, 1983.
92. Chern, J.W., *et al.*, Studies on quinazolines. 5. 2, 3-Dihydroimidazo [1, 2-c] quinazoline derivatives: a novel class of potent and selective. alpha. 1-adrenoceptor antagonists and antihypertensive agents. *J. Med. Chem.*, 36, 15, 2196–2207, 1993.
93. Jen, T., Bender, P., Van Hoeven, H., Dienel, B., Loev, B., Amidines. 4. Synthesis of tricyclic guanidines related to 1, 2, 3, 5-tetrahydroimidazo [2, 1-b] quinazoline, a new antihypertensive agent. *J. Med. Chem.*, 16, 4, 407–411, 1973.
94. Jackson, C.M., *et al.*, Evolution of thiazolidine-based blockers of human Kv1. 5 for the treatment of atrial arrhythmias. *Bioorg. Med. Chem. Lett.*, 17, 1, 282–284, 2007.
95. Jain, A.K., Vaidya, A., Ravichandran, V., Kashaw, S.K., Agrawal, R.K., Recent developments and biological activities of thiazolidinone derivatives: A review. *Bioorg. Med. Chem.*, 20, 11, 3378–3395, 2012.
96. Siddiqui, A.A., Mishra, R., Shaharyar, M., Husain, A., Rashid, M., Pal, P., Triazole incorporated pyridazinones as a new class of antihypertensive agents: Design, synthesis and *in vivo* screening. *Bioorg. Med. Chem. Lett.*, 21, 3, 1023–1026, 2011.
97. Perekhoda, L., Georgiyants, V., Yeromina, H., Drapak, I., Lubenets, V., Ieromina, Z., Sych, I., Severina, H., Demchenko, A., The synthesis and *in silico* antihypertensive activity prognosis of new Mannich bases containing the 1,2,4-triazole moiety. *Chem. Chem. Technol.*, 14, 2, 214–220, 2020.
98. Masaret, G.S., Synthesis, docking and antihypertensive activity of pyridone derivatives. *ChemistrySelect*, 5, 44, 13995–14003, 2020.
99. Moran, A., *et al.*, Antihypertensive Effect of some Oxazolo [3, 2-a] pyridines, Thiazolo [3, 2-a] pyridines and Pyrido [2, 1-b] oxazines in Conscious Spontaneously Hypertensive Rats. *J. Pharm. Pharmacol.*, 49, 4, 421–425, 1997.
100. Sharma, M.C. and Kohli, D.V., A comprehensive structure–activity analysis 2, 3, 5-trisubstituted 4, 5-dihydro-4-oxo-3 H-imidazo [4, 5-c] pyridine derivatives as angiotensin II receptor antagonists: using 2D-and 3D-QSAR approach. *Med. Chem. Res.*, 22, 2, 588–605, 2013.

11

In Silico Studies on Heterocyclics as an Active Pharmaceutical Agent

Anurag Paul[1], Debanjali Adhikary[1], Abdul Kareem Parchur[2], Gayatri Sharma[1*] and Apoorva Mishra[3†]

[1]*Centre for Medical Biotechnology, Amity Institute of Biotechnology, Amity University Uttar Pradesh, Noida, India*
[2]*Department of Radiation Oncology, Medical College of Wisconsin, Milwaukee, Wisconsin, USA*
[3]*Centre for Cellular and Molecular Biotechnology, Amity Institute of Biotechnology, Amity University Uttar Pradesh, Noida, India*

Abstract
Heterocyclic compounds are pivotal in pharmaceutical research due to their diverse biological activities, structural adaptability, and prominence in numerous therapeutic agents. This chapter presents an integrated *in silico* approach to the design and evaluation of heterocyclic compounds, emphasizing their potential as active pharmaceutical agents and their applications in radiation-based diagnostics and therapy. Computational techniques such as molecular docking, pharmacophore modeling, quantum chemical calculations, and ADMET (absorption, distribution, metabolism, excretion, and toxicity) profiling enable the efficient prediction of bioactivity, drug-likeness, and interaction with molecular targets. Special attention is given to radiolabelled heterocycles, which are increasingly used in nuclear medicine for targeted imaging (e.g., positron emission tomography and single-photon emission computed tomography) and radiotherapy, particularly in oncology. *In silico* models assist in optimizing these compounds for selective binding, minimal toxicity, and favorable biodistribution. Furthermore, the chapter explores the effects of ionizing radiation on heterocyclic stability and function, along with the computational prediction of radiotoxicity and dosimetry. Case studies demonstrate how such integrative computational methods accelerate the development of radiopharmaceuticals and dual-function agents that combine

**Corresponding author*: gsharma10@amity.edu
†*Corresponding author*: apoorvamshr@gmail.com

Divya Bajpai Tripathy, Priyanka Chhabra and Anjali Gupta (eds.) Heterocyclic Compounds as Pharmaceutical Ingredients, (389–406) © 2026 Scrivener Publishing LLC

therapeutic efficacy with diagnostic precision. Overall, this chapter highlights the synergistic role of heterocyclic chemistry, radiation science, and computational drug design in advancing modern pharmaceutical innovation.

***Keywords*:** Heterocyclic compounds, *in silico* drug design, molecular docking, ADMET, radiopharmaceuticals, PET imaging, radiotherapy

11.1 Introduction

Heterocyclic compounds are ring compounds with one element other than carbon. Heterocyclic compounds play a crucial role in drug discovery and the pharmaceutical industries as key pharmaceutical agents. The pharmaceutical industry is facing many challenges due to continuous rising cost of medicines and increasing expenses of research and development. There are many heterocyclic compounds that are used for drug discovery, serving as an antimalarial, anticancer, and antibacterial agent [1]. To support drug discovery efforts, various databases such as MCE (MedChemExpress) offer a specialized collection of 6164 heterocyclic compounds, designed to support drug discovery initiatives through high-throughput screening and high-content screening. This comprehensive library plays a vital role in facilitating the development of new therapeutics [2].

This book chapter discusses the role of heterocyclic compounds in drug discovery, active pharmaceutical agents, and role of *in silico* studies in evaluating their potential.

11.2 *In Silico* Approaches for Heterocyclic Drug Discovery

In silico methods refer to the computational techniques and simulations that are performed to predict and analyze biological and chemical processes. These methods have revolutionized the domain of drug discovery by accelerating processes such as identification, optimization, and validation while significantly decreasing the time associated with experimental research. These computational techniques also assist in predicting behavior, thus leading to the production of more effective and safer heterocyclic drugs [2, 3]. There are various *in silico* approaches, which we will discuss here in detail.

11.2.1 Molecular Docking

Molecular docking is a widely used computational technique in drug discovery that predicts how heterocyclic compounds bind to their biological

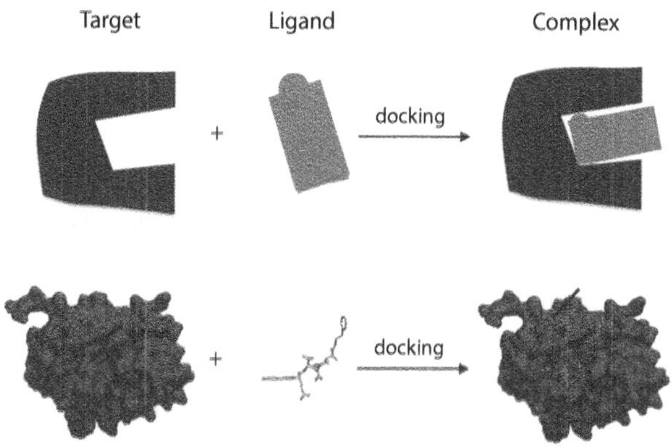

Figure 11.1 Molecular docking: Depicting the steps involved in this approach.

targets, such as enzymes or receptors. This method simulates the interaction between a ligand and active site of a protein, thereby analyzing the efficacy of the potential drug, as illustrated in Figure 11.1 [4].

Let's explore how this method works. First, the 3D structure of the biological target is obtained from Protein Data Bank, and heterocyclic ligand is prepared using molecular modeling software. Algorithms are used to generate multiple binding conformations, and the best binding pose is selected based on scoring functions that estimate the binding affinity. These docking results are further analyzed based on binding energy, hydrogen bonding, hydrophobic interactions, and molecular orientation. Various tools such as Auto Dock (for flexible docking), Glide (for high precision docking), and Gold (for genetic algorithm-based docking) are used for this analysis [5, 6].

This method has been used in investigating the bioactivity of some known heterocyclic antiviral drugs named favipiravir, amodiaquine, 2′-fluoro-2′-deoxycytidine, and ribavirin as inhibitors and nucleotide analogs for COVID-19. These nominated drugs were docked on SARS-CoV-2 main protease to predict the binding pattern and evaluate the binding affinity. Computational data of docking simulations predicted that amodiaquine (2) has the least binding energy (−7.77 kcal/mol) and might be a good inhibitor to SARS-CoV-2 as compared to approved medicines, hydroxychloroquine and remdesivir, which have binding affinity of −6.06 and −4.96 kcal/mol, respectively [7].

Similarly, in an endeavor to develop effective anticancer drugs, Ali *et al.* designed and synthesized group of 12 heterocyclic compounds embedded with oxopyrrolidine rings. They were studied using molecular docking studies carried out to determine the biological targets. The DNA binding constants of these compounds ranged from 1.5×10^4 to 4.0×10^5 M^{-1}.

The docking energies of the DNA-compound adducts varied from – 23.84 to −30.96 kJ/mol. The compounds preferred to bind minor grooves of DNA. Cytotoxicity of these drugs was also evaluated on cancer cell lines. Among these compounds, seven compounds exhibited anticancer activities [8]. Molecular docking has also been studied to predict the optimized conformation of pyrimidines to act as active ligands against the *Escherichia coli* alkaline phosphatase [9].

These studies exhibit that molecular docking plays a crucial role in drug discovery by predicting interact action potential drugs with their biological targets at the molecular targets. Thus, molecular docking serves as a powerful tool for identifying potential treatments for COVID-19, cancer, and bacterial infections by enabling precise drug–target interaction analysis and aiding in the development of more effective therapeutics.

11.2.2 Molecular Dynamic Simulations

Molecular dynamic simulations are computational methods that help to understand the dynamic behavior of heterocyclic drug molecules in biological environments. Unlike molecular docking, which gives a static snapshot, MD simulations provide an image of time-dependent molecular interactions [10, 11].

Let us understand how this approach works, first the protein–ligand complex is solvated in a simulation box, and an appropriate force field, e.g., AMBER or CHARM, is assigned. The system is stabilized by removing steric clashes, and temperature and pressure are equilibrated. The simulation runs for nanoseconds to microseconds when molecular movements and interactions are recorded. Finally, drug stability is analyzed by calculations such as root mean square deviation (RMSD) and root mean square fluctuation (RMSF) [12–14].

Pyridine is a heterocyclic compound containing nitrogen, and its derivatives have been synthesized, demonstrating significant anticancer activity against various cancer cell lines. Hence, Fawzy *et al.* designed and synthesized new pyridine derivatives and determine their anticancer activity by MTT assays on various cancer cell lines [59]. Further, *in vitro* assays were used to determine depolymerization activity of compounds toward tubulin. They further utilized molecular docking and MD simulations to explain the mechanism of action and the stability of the interaction between the new pyridine-based compounds and tubulin over time. The simulations mimicked real-life conditions by gradually increasing the temperature from 50°C to 300°C and observing the behavior of molecules inside the binding pocket. Key factors such as energy stability, structural deviations (RMSD), and flexibility (RMSF) were analyzed. The results helped researchers to identify a compound that formed a most stable interaction with tubulin and hence suggesting its potential as a strong anticancer agent [15].

Molecular dynamics simulations play a crucial role in studying drug–DNA adducts, providing insights into structural properties and binding mechanisms, which are essential for developing effective anticancer drugs [16]. MD simulations have been conducted to check the stability and flexibility of protein–ligand interaction in atomic level, to evaluate certain heterocyclic compounds as potential anti-inflammatory agents [17].

MD simulations help evaluate the stability and interaction patterns of heterocyclic compounds with biological targets, thereby aiding in the development of potent anticancer and anti-inflammatory agents.

11.2.3 Quantitative Structure–Activity Relationship

Quantitative structure–activity relationship (QSAR) models are computational techniques that establish relationships between the chemical structure of heterocyclic compounds and their biological activity by using statistical techniques and machine learning (ML) models. This model establishes mathematical relationships between the chemical structures of heterocyclic compounds and their biological activities as illustrated in Figure 11.2. QSAR can predict efficacy of new drugs by analyzing their

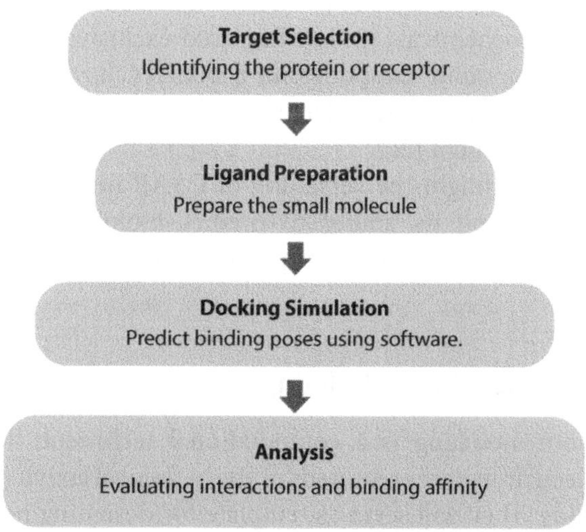

Figure 11.2 Steps involved in molecular docking simulation.

molecular descriptors. Tools such as QSAR ToolBox, CHEMPROP, and OCHEM are used here [18, 19]. These are of two types, namely:

1. **2D-QSAR**: This uses molecular descriptors such as logP, molecular weight, and topological indices.
2. **3D-QSAR**: This considers spatial molecular features using techniques such as comparative molecular field analysis (CoMFA) [20, 21].

QSAR modeling begins with the collection of experimental activity data for known heterocyclic compounds. Relevant molecular descriptors influencing the activity are then identified. ML algorithms (e.g., SVM, random forest) are used to train and test the QSAR model, which is then used to predict the activity of new heterocyclic compounds [22, 23].

In a recent study, 3D QSAR modeling has been used to analyze a set of heterocyclic diamidine compounds with potential antiparasitic activity, specifically against *Trypanosoma brucei.* Researchers applied CoMFA and Comparative Molecular Similarity Indices Analysis (CoMSIA) techniques to identify key molecular features influencing biological activity. By examining factors such as steric, electrostatic, hydrophobic, donor, and acceptor properties, they developed predictive models using partial least squares regression. The best CoMSIA model showed strong accuracy ($q^2 = 0.699$, $r^2 = 0.974$), helping to guide the design of new, more effective antiparasitic compounds [24].

QSAR has served as a crucial predictive tool in various pharmaceutical drugs and agrochemicals [25]. It has used exclusively in the study of an important compound campothecin, a pentacyclic quinoline alkaloid obtained from a native tree from Tibet, which is a lead compound in anticancer drug development [26].

These studies highlight the versatility of QSAR modeling in drug discovery, demonstrating its ability to predict biological activity across diverse therapeutic areas, from antiparasitic agents to anticancer drug development.

11.2.4 Pharmacophore Modeling

Pharmacophore modeling is a computational technique that identifies essential molecular features required for the biological activity of heterocyclic compounds. They will serve as template for designing new molecules with desired product as shown in Figure 11.3 [27, 28].

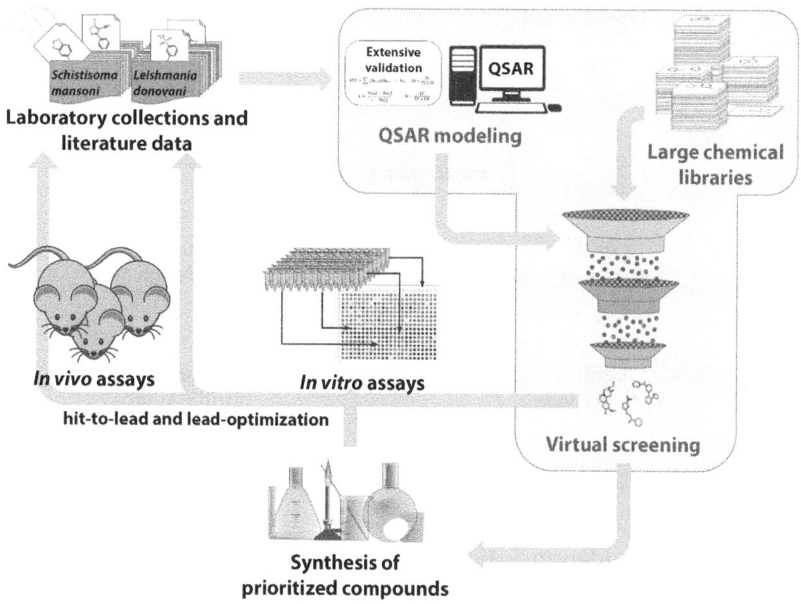

Figure 11.3 Comprehensive QSAR modeling, depicting the steps involved in QSAR approach [58].

This process is carried out by a series of steps, First, critical molecular features such as hydrogen bond donors/acceptors, hydrophobic regions, and aromatic rings are identified, and then a 3D arrangement of features is developed representing the pharmacophore. This pharmacophore model is used to virtually screen compound libraries to find compounds that match the desired features. The identified compound is experimentally tested to confirm biological activity [29].

Pharmacophore modeling is used in research to explore the structural characteristics influencing the effectiveness of heterocyclic bisphosphonates in inhibiting osteoclastic bone reference; researchers used the Catalyst program to pinpoint essential pharmacophore elements, including two negatively charged sites, one positively charged region, and an aromatic feature near the active site (Figure 11.4). The findings underscored the significance of proper protonation states in heterocyclic rings, such as thiazoles, pyrazoles, and triazoles, as incorrect protonation diminished their ability to function as reactive intermediate analogs. By integrating QSAR analysis with X-ray crystallographic data, the study enabled the prediction of bisphosphonate activity, supporting the development of more effective treatments for bone resorption, parasitic infections, and immune-related disorders [30].

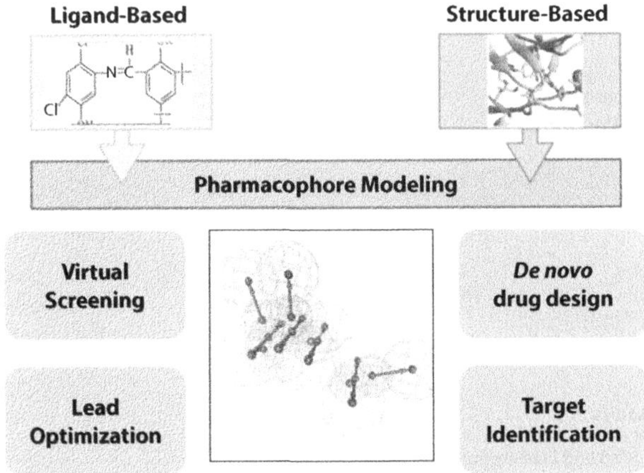

Figure 11.4 Pharmacophore modeling.

QSAR model identifies heterocyclic compounds as potential DHFR inhibitors. Virtual screening, molecular docking, and pharmacokinetic assessments helped select four promising heterocyclic candidates with strong binding affinity and drug-like properties [16].

These studies highlight the crucial role of pharmacophore modeling in drug discovery, demonstrating its effectiveness in identifying and optimizing heterocyclic compounds for diverse therapeutic applications, from bone resorption and parasitic infections to cancer treatment.

11.2.5 ADMET and Toxicity Prediction

ADMET refers to absorption, distribution, metabolism, excretion, and toxicity, which are crucial factors in assessing a drug's pharmacokinetics and safety. It plays a key role in determining a drug's effectiveness and suitability for therapeutic use. In drug discovery, ADMET modeling is used in early developmental stage to predict these properties, helping to minimize late-stage failures and refine drug candidates for improved efficacy and safety [31].

Absorption determines how a drug enters the bloodstream, influenced by factors such as solubility, lipophilicity, and membrane permeability. **Distribution** examines how the drug spreads across tissues, considering protein binding and volume of distribution. **Metabolism** focuses on how liver enzymes modify the drug, affecting its breakdown and interactions with metabolic pathways such as cytochrome P450. **Excretion** assesses the body's elimination of drugs, primarily through the kidneys or bile.

Lastly, **toxicity** prediction helps ensure drug safety by detecting harmful chemical structures and leveraging ML models to evaluate potential side effects [32, 33].

ADMET modeling has been integrated into studies to evaluate the blood–brain barrier (BBB) permeability of heterocyclic drug candidates. By combining biomimetic chromatography with IAM (immobilized artificial membrane) columns and QSAR methodologies, the researchers assess key ADMET properties such as absorption, distribution, and metabolism. Molecular descriptors, including lipophilicity, molecular weight, hydrogen bond donors/acceptors, and topological polar surface area, were analyzed to predict BBB penetration [34]. ADMET properties have also been analyzed in computational study to predict the therapeutic potential of N-heterocyclic derivatives against SARS-CoV-2 [35].

11.3 Heterocyclic Compounds as Active Pharmaceutical Agents

The heterocyclic compounds play a great role in the pharmaceuticals and medical chemistry. Heterocycles are found in enzymes, medicines, and proteins. Heterocycles played a great role in biological parameter in the modeling of new systems. The N-heterocycles are the most important in medicinal compounds [36]. The heterocycles help to decide the biological target with an organ in the cell or tissue, and they fit into special pockets which allow them to modulate enzymes and prevent the active site of the target proteins where cancer cells are restricted. Thus, the heterocycles are very important (Table 11.1) [36].

11.3.1 Antifungal Activity

The heterocyclic compounds usually show antifungal activity as they are used to treat fungal infections that are found in hair and skin. It treated athlete's foot and ringworm disease with proper medications [37].

It has helped produce various antifungal strains such as *Fusarium graminearum* and many other strains. This has caused cell death by the cell membrane to leak out the cellular components [37].

There are certain biological hydrogels that are made from indole 3 acetic acid and that takes actions against microbial and fungal infections. *Candida albicans* and fumigates were some fungi tested as antifungal agents, and they showed a sigh of improvement in the treatment [36].

Table 11.1 Heterocyclic compounds as active pharmaceutical agents.

Property	Compound name	Mechanism of action	References
Antibacterial	Quinolones	Inhibit bacterial DNA gyrase and topoisomerase IV, preventing DNA replication	[53]
	β-Lactams (penicillins)	Inhibit bacterial cell wall synthesis by targeting penicillin-binding proteins (PBPs)	[53]
Antifungal	Azoles (fluconazole)	Inhibit ergosterol synthesis by targeting cytochrome P450-dependent 14α-demethylase	[54]
	Echinocandins	Inhibit β-1,3-glucan synthase, disrupting fungal cell wall synthesis	[55]
Anticancer	Imidazoles (e.g., temozolomide)	Alkylate DNA, leading to apoptosis	[56]
	Quinazolines (e.g., gefitinib)	Inhibit EGFR tyrosine kinase signaling	[56]
Herbicidal	Triazines (e.g., atrazine)	Inhibit photosystem II, blocking electron transport in photosynthesis	[57]
	Pyrazoles (e.g., pyrazolate)	Inhibit 4-hydroxyphenylpyruvate dioxygenase (HPPD), affecting carotenoid biosynthesis	[57]
Anticonvulsant	Oxazolidinediones	Block T-type calcium channels, reducing neuronal excitability	[57]
	Barbiturates	Enhance GABAergic inhibition *via* GABA-A receptor	[57]

For example, azoles, such as fluconazole, inhibit ergosterol synthesis by targeting cytochrome P450–dependent 14α-demethylase, disrupting fungal cell membrane integrity.

11.3.2 Antibacterial Activity

These are used to treat or eradicate bacterial development. It connects the antiprotozoal effects with the antibacterial agents to prevent the bacterial infections. These are effective against the viral infections and treatment of flu also. Aromatic heterocyclic derivatives are also used as β-lactam antibiotics. For example, quinolones, such as ciprofloxacin, inhibit bacterial DNA gyrase and topoisomerase IV, preventing DNA replication [38].

11.3.3 Anticancer Activity

Cancer is the abnormal proliferation of cells that may invade to different parts of the body. Many things can be responsible for cancer that we all know well. The heterocyclic compounds containing pyrrole in its rings are fused to pyridines. Pyrazole specially possess biological anticancer property and a unique therapeutic value [39].

One example is phenanthroindolizidine [40]. It has a special property to reduce the proliferation of cells in the cell cycle. Some lipophilic prodrugs are used as membrane proteins in combination with these heterocyclic compounds are able to use certain medications [41]. A new therapy called membrane lipid therapy has been introduced where the membranes are combined with heterocycles, and this can alter the composition of the carcinogenic membranes. The vessels that are made up of lipids are deconstructed by this therapy [42]. It has proteins linked to the membranes that inhibit the downstream processes of tumor formation. Doxorubicin is a drug that is used to proliferate tumor [41]. The oxygen- and sulfur-based heterocyclic compounds act as anticancer agent; one of them is daunorubicin [43].

11.3.4 Herbicidal Activity

To eradicate unwanted plants and herbs, many medications are used that include heterocyclic compounds. N-(arylmethoxy)-2-chloronicotinamide is an example developed that is a herbicidal agent [44]. Clomazone has IC_{50} values of 125 and 2 M.

When the herbal weeds undergo treatment, they are provided with HPPD herbicides. More attention is being paid to agriculture-related heterocyclic compounds for developing crops and treatment of weeds.

These heterocyclic herbicides improve the translocation in the plants and crops [45]. For example, triazine derivatives, such as atrazine, act as herbicides by inhibiting photosystem II, thereby blocking electron transport in photosynthesis.

11.3.5 Anticonvulsant

These are those substances that fall into a number of categories as active pharmaceutical agents to treat epilepsy and other disorders. They are also called antiepileptic medicines. It enhances the mechanism of γ-aminobutyric acid (GABA), and sodium channels are blocked.

Pregabalin, progabide, vigabatrin, and gabapentin are the heterocyclic compounds that are used as anticonvulsants in drug discovery. The mechanism is a bit complex [46].

They do not bind to the receptors of GABA; they do not show any effect on the receptors; they indirectly affect the transport channels. They function as they inhibit the voltage-dependent calcium channels and have a subunit of α2δs [47].

11.4 Challenges and Future Trends

In silico studies have become integral to the development of heterocyclic drugs, offering computational methods to predict molecular interactions, optimize drug design, and streamline the discovery process; however, there are several challenges.

a) The intricate nature of biological systems means that computational models may oversimplify interactions, potentially overlooking critical factors that influence a drug's behavior in the human body [48].
b) The effectiveness of *in silico* studies heavily relies on the quality and comprehensiveness of available biological and chemical data. Incomplete or biased datasets can lead to inaccurate predictions [49].
c) Simulating the dynamic and multifaceted interactions of heterocyclic drugs requires substantial computational power [48].
d) Despite the efficiencies introduced by computational methods, the overall cost of drug development continues to rise, partly due to the complexity of validating *in silico* predictions through experimental studies [50].

Recent advancements in *in silico* methods have significantly transformed the landscape of heterocyclic drug discovery. The integration of artificial intelligence (AI) and ML has enabled more accurate predictions of biological activities and facilitated the design of novel heterocyclic compounds with optimized properties. For instance, AI-driven models can predict potential drug–target interactions, thereby streamlining the identification of promising drug candidates. Furthermore, the fusion of cheminformatics, systems biology, and bioinformatics has led to the development of AI-based methods that enhance the efficiency of drug discovery processes. Additionally, *in silico* toxicity screening using ML models allows for early prediction of potential adverse effects, reducing late-stage drug development failures. Collectively, these advancements underscore a shift toward more efficient and precise computational methods in heterocyclic drug discovery [51, 52].

11.5 Conclusion

In silico methods such as molecular docking, molecular dynamics simulations, QSAR modeling, and pharmacophore modeling have significantly enhanced the efficiency of drug discovery by enabling accurate predictions of drug–target interactions, stability, and toxicity. These approaches not only reduce the time and cost associated with experimental research but also facilitate the design of more potent and safer drugs. Advancements in AI and ML is helping *in silico* drug discovery to become even more precise and efficient. The integration of these computational techniques will continue to drive innovation in pharmaceuticals, paving the way for the development of novel heterocyclic-based drugs with enhanced therapeutic potential.

AI disclosure statement: The author acknowledges the use of AI-based tools for minor language editing or grammatical correction. However, all intellectual contributions, analyses, and conclusions presented in this chapter are entirely the author's own.

References

1. Alvarez-Builla, J. and Barluenga, J., Heterocyclic compounds: An introduction, in: *Modern Heterocyclic Chemistry*, pp. 1–9, 2011.
2. Časar, Z. (Ed.), *Synthesis of Heterocycles in Contemporary Medicinal Chemistry*, vol. 44, Springer, Berlin/Heidelberg, Germany, 2016.

3. Terstappen, G.C. and Reggiani, A., *In silico* research in drug discovery. *Trends Pharmacol. Sci.*, 22, 1, 23–26, 2001.
4. Rao, V.S. and Srinivas, K., Modern drug discovery process: An *in silico* approach. *J. Bioinform. Seq. Anal.*, 2, 5, 89–94, 2011.
5. Agarwal, S. and Mehrotra, R.J.J.C., An overview of molecular docking. *JSM Chem.*, 4, 2, 1024–1028, 2016.
6. Tripathi, A. and Misra, K., Molecular docking: A structure-based drug designing approach. *JSM Chem.*, 5, 2, 1042–1047, 2017.
7. Morris, G.M. and Lim-Wilby, M., Molecular docking, in: *Molecular Modeling of Proteins*, pp. 365–382, 2008.
8. Hagar, M., *et al.*, Investigation of some antiviral N-heterocycles as COVID-19 drug: Molecular docking and DFT calculations. *Int. J. Mol. Sci.*, 21, 11, 3922, 2020.
9. Ali, I., *et al.*, Insights into the pharmacology of new heterocycles embedded with oxopyrrolidine rings: DNA binding, molecular docking, and anticancer studies. *J. Mol. Liq.*, 234, 391–402, 2017.
10. Sanad, S.MH., Ahmed, A.AM., Mekky, A.EM., Efficient synthesis and molecular docking of novel antibacterial pyrimidines and their related fused heterocyclic derivatives. *J. Heterocycl. Chem.*, 57, 2, 590–605, 2020.
11. Karplus, M. and Petsko, G.A., Molecular dynamics simulations in biology. *Nature.*, 347, 6294, 631–639, 1990.
12. Binder, K., *et al.*, Molecular dynamics simulations. *J. Phys. Condens. Matter.*, 16, 5, S429, 2004.
13. Singh, S., Baker, Q.B., Singh, D.B., Molecular docking and molecular dynamics simulation, in: *Bioinformatics*, pp. 291–304, Academic Press, 2022.
14. Krieger, E. and Vriend, G., New ways to boost molecular dynamics simulations. *J. Comput. Chem.*, 36, 13, 996–1007, 2015.
15. Van Gunsteren, W.F. and Berendsen, H.JC., Computer simulation of molecular dynamics: Methodology, applications, and perspectives in chemistry. *Angew. Chem. Int. Ed. Engl.*, 29, 9, 992–1023, 1990.
16. Sun, X., *et al.*, Identification of some promising heterocycles useful in treatment of allergic rhinitis: Virtual screening, pharmacophore mapping, molecular docking, and molecular dynamics. *Russ. J. Bioorg. Chem.*, 48, 2, 438–456, 2022.
17. Spiegel, K. and Magistrato, A., Modeling anticancer drug–DNA interactions *via* mixed QM/MM molecular dynamics simulations. *Org. Biomol. Chem.*, 4, 13, 2507–2517, 2006.
18. Savjani, J., *et al.*, Drug design, synthesis and biological evaluation of heterocyclic molecules as anti-inflammatory agents. *Molecules*, 27, 4, 1262, 2022.
19. Veerasamy, R., *et al.*, Validation of QSAR models – strategies and importance. *Int. J. Drug Des. Discov.*, 3, 511–519, 2011.
20. Gramatica, P., Principles of QSAR models validation: Internal and external. *QSAR Comb. Sci.*, 26, 5, 694–701, 2007.

21. Bahia, M.S., *et al.*, A comparison between 2D and 3D descriptors in QSAR modeling based on bio-active conformations. *Mol. Inform.*, 42, 4, 2200186, 2023.
22. Vukovic, K., Gadaleta, D., Benfenati, E., Methodology of aiQSAR: A group-specific approach to QSAR modelling. *J. Cheminform.*, 11, 1–9, 2019.
23. Tropsha, A., Best practices for QSAR model development, validation, and exploitation. *Mol. Inform.*, 29, 6–7, 476–488, 2010.
24. Athri, P., *et al.*, 3D QSAR on a library of heterocyclic diamidine derivatives with antiparasitic activity. *Bioorg. Med. Chem.*, 14, 9, 3144–3152, 2006.
25. Patel, H.M., *et al.*, Quantitative structure–activity relationship (QSAR) studies as strategic approach in drug discovery. *Med. Chem. Res.*, 23, 4991–5007, 2014.
26. Verma, R.P. and Hansch, C., Camptothecins: A SAR/QSAR study. *Chem. Rev.*, 109, 1, 213–235, 2009.
27. Khedkar, S.A., *et al.*, Pharmacophore modeling in drug discovery and development: An overview. *Med. Chem.*, 3, 2, 187–197, 2007.
28. Caporuscio, F. and Tafi, A., Pharmacophore modelling: A forty year old approach and its modern synergies. *Curr. Med. Chem.*, 18, 17, 2543–2553, 2011.
29. Vuorinen, A. and Schuster, D., Methods for generating and applying pharmacophore models as virtual screening filters and for bioactivity profiling. *Methods*, 71, 113–134, 2015.
30. Kotsikorou, E. and Oldfield, E., A quantitative structure–activity relationship and pharmacophore modeling investigation of aryl-X and heterocyclic bisphosphonates as bone resorption agents. *J. Med. Chem.*, 46, 14, 2932–2944, 2003.
31. Ferreira, L.LG. and Andricopulo, A.D., ADMET modeling approaches in drug discovery. *Drug Discov. Today.*, 24, 5, 1157–1165, 2019.
32. Van De Waterbeemd, H. and Gifford, E., ADMET *in silico* modelling: Towards prediction paradise? *Nat. Rev. Drug Discov.*, 2, 3, 192–204, 2003.
33. Norinder, U. and Bergström, C.AS., Prediction of ADMET properties. *ChemMedChem*, 1, 9, 920–937, 2006.
34. Rashdan, H.RM., El-Naggar, M., Abdelmonsef, A.H., Synthesis, molecular docking studies and *in silico* ADMET screening of new heterocycles linked thiazole conjugates as potent anti-hepatic cancer agents. *Molecules.*, 26, 6, 1705, 2021.
35. Babalola, B.A., *et al.*, Computational study of the therapeutic potential of novel heterocyclic derivatives against SARS-CoV-2. *Covid*, 1, 4, 757–774, 2021.
36. Mishra, L., Said, M.K., Itokawa, H., Takeya, K., Antitumor and antimicrobial activities of Fe(II)/Fe(III) complexes derived from some heterocyclic compounds. *Bioorg. Med. Chem.*, 3, 9, 1241–1245, 1995.

37. Liberta, A.E. and West, D.X., Antifungal and antitumor activity of heterocyclic thiosemicarbazones and their metal complexes: Current status. *Biometals*, 5, 121–126, 1992.
38. Kabir, E. and Uzzaman, M., A review on biological and medicinal impact of heterocyclic compounds. *Results Chem.*, 4, 100606, 2022.
39. Hosseinzadeh, Z., Ramazani, A., Razzaghi-Asl, N., Anti-cancer nitrogen-containing heterocyclic compounds. *Curr. Org. Chem.*, 22, 23, 2256–2279, 2018.
40. Hassan, A.Y., Mohamed, M.A., Abdel-Aziem, A., Hussain, A.O., Synthesis and anticancer activity of some fused heterocyclic compounds containing pyrazole ring. *Polycycl. Aromat. Compd.*, 40, 4, 1280–1290, 2020.
41. Negi, B. and Kwatra, A., A review of recent progress on the anticancer activity of heterocyclic compounds. *SynOpen*, 8, 3, 185–210, 2024.
42. Kciuk, M., Gielecińska, A., Mujwar, S., Kołat, D., Kałuzińska-Kołat, Z., Celik, I., Kontek, R., Doxorubicin—an agent with multiple mechanisms of anticancer activity. *Cells*, 12, 4, 659, 2023.
43. Sachdeva, H., Khaturia, S., Saquib, M., Khatik, N., Khandelwal, A.R., Meena, R., Sharma, K., Oxygen- and sulphur-containing heterocyclic compounds as potential anticancer agents. *Appl. Biochem. Biotechnol.*, 194, 12, 6438–6467, 2022.
44. Wu, T.T., Huang, J., Arrington, N.D., Dill, G.M., Synthesis and herbicidal activity of α-heterocyclic carbinol carbamates. *J. Agric. Food. Chem.*, 35, 5, 81, 1987.
45. Grego, S., Aricò, F., Tundo, P., Highly selective phosgene-free carbamoylation of aniline by dimethyl carbonate under continuous-flow conditions. *Org. Process Res. Dev.*, 17, 4, 679–683, 2013.
46. Malawska, B., New anticonvulsant agents. *Front Med. Chem.*, 4, 805–841, 2009.
47. Al-Otaibi, F., An overview of structurally diversified anticonvulsant agents. *Acta Pharm.*, 69, 3, 321–344, 2019.
48. Sacan, A., Ekins, S., Kortagere, S., Applications and limitations of *in silico* models in drug discovery, in: *Bioinformatics and Drug Discovery*, pp. 87–124, 2012.
49. Marques, L., Costa, B., Pereira, M., Silva, A., Santos, J., Saldanha, L., Silva, I., Magalhães, P., Schmidt, S., Vale, N., Advancing precision medicine: A review of innovative *in silico* approaches for drug development, clinical pharmacology and personalized healthcare. *Pharmaceutics*, 16, 332, 2024.
50. Wouters, O.J. and Kesselheim, A.S., Quantifying research and development expenditures in the drug industry. *JAMA Netw. Open*, 7, 6, e2415407, 2024, doi: 10.1001/jamanetworkopen.2024.15407.
51. Atasever, S., *In silico* drug discovery: A machine learning-driven systematic review. *Med. Chem. Res.*, 33, 9, 1465–1490, 2024.
52. Roney, M. and Mohd Aluwi, M.F.F., The importance of *in-silico* studies in drug discovery. *Intellig. Pharm.*, 2, 4, 578–579, 2024.

53. Kuhlmann, J., Dalhoff, A., Zeiler, HJ. (Eds.), *Quinolone Antibacterials*, vol. 127, Springer Science & Business Media, 2012.
54. Saag, M.S. and Dismukes, W.E., Azole antifungal agents: Emphasis on new triazoles. *Antimicrob. Agents Chemother.*, 32, 1, 1–8, 1988.
55. Odds, F.C., Brown, A.J.P., Gow, N.A.R., Antifungal agents: Mechanisms of action. *Trends Microbiol.*, 11, 6, 272–279, 2003.
56. Shang, X.-F., *et al.*, Biology of quinoline and quinazoline alkaloids, in: *The Alkaloids: Chemistry and Biology*, vol. 88, pp. 1–47, 2022.
57. Reygaert, W.C., An overview of the antimicrobial resistance mechanisms of bacteria. *AIMS Microbiol.*, 4, 3, 482, 2018.
58. Neves, B.J., *et al.*, QSAR-based virtual screening: Advances and applications in drug discovery. *Front. Pharmacol.*, 9, 1275, 2018.
59 Abdelshaheed, M.M., El Subbagh, H.I., Tantawy, M.A., Attia, R.T., Youssef, K.M., Fawzy, I.M., Discovery of new pyridine heterocyclic hybrids: design, synthesis, dynamic simulations, and *in vitro* and *in vivo* breast cancer biological assays. *RSC Adv.*, 24, 13, 23, 15689–15703, 2023.

12

Piperazine Derivatives: Potential Antimicrobial Agents

R. Prathap[1], Rashmi Dubey[2], Anjali Gupta[1] and Divya Bajpai Tripathy[1]*

[1]*Division of Chemistry, School of Basic and Applied Science, Galgotias University, Greater Noida, India*

[2]*Department of Chemistry, School of Basic Sciences, CSJM University, Kanpur, India*

Abstract

The increasing number of antibiotic-resistant infections is a global crisis, affecting millions of people worldwide. This increase in resistance is seen in most countries and is causing alarm among medical professionals (WHO; Antimicrobial Resistance: Global Report on Surveillance). The lower susceptibility to antibiotics of common pathogens, such as *Staphylococcus aureus* (Gram-positive), *Escherichia coli* (Gram-negative), and others, is a major concern, as this makes treating infections more difficult. It is become a real challenge for scientists, pharmaceutical industries, and researchers to keep up with pathogens that rapidly evolve and adapt through various pathways. Piperazine and its derivatives possess significant antimicrobial properties against a range of bacterial, fungal, and parasitic pathogens. These compounds have the potential to serve as lead compounds for the development of novel antimicrobial agents with unique mechanisms of action, which can help to overcome the challenge of antibiotic resistance. Piperazine derivatives have been shown to exhibit a broad spectrum of antimicrobial activity against both Gram-positive and Gram-negative bacteria, as well as fungi and parasites. In addition to their antimicrobial properties, piperazine derivatives have also shown potential as anticancer agents. Some studies have shown that certain piperazine derivatives can induce apoptosis (programmed cell death) in cancer cells, making them a promising option for cancer treatment. The diverse chemical and biological properties of piperazine derivatives make them a promising class of compounds for the development of new antimicrobial and anticancer drugs. In this review, we will discuss the antimicrobial properties of the piperazine nucleus,

**Corresponding author*: divyabaj@gmail.com

Divya Bajpai Tripathy, Priyanka Chhabra and Anjali Gupta (eds.) Heterocyclic Compounds as Pharmaceutical Ingredients, (407–442) © 2026 Scrivener Publishing LLC

with a particular focus on its antibacterial activity. We will also highlight some of the recent advances in the development of piperazine-based antimicrobial agents.

***Keywords*:** Piperazine, antimicrobial, heterocycles, antifungal, antiparasitic

12.1 Introduction

The emergence of antimicrobial-resistant strains of microorganisms has led to a global health crisis. This has prompted researchers to explore alternative sources of antimicrobial agents, including natural products and other synthetic compounds. Molecules with antimicrobial properties have played a critical role in modern medicine that help to combat the spread of infectious diseases caused by bacteria, viruses, fungi, and parasites. The emergence of antibiotic resistance, however, has resulted in a significant challenge to the effectiveness of these agents. This has led to an increased interest in the search for novel antimicrobial agents with unique mechanisms of action.

Piperazine is a heterocyclic organic compound that contains a six-membered ring with two nitrogen atoms at positions 1 and 4 (Figure 12.1). It has been widely used as a pharmacophore in the design of a variety of therapeutic agents, including anthelmintics, antipsychotics, and antihistamines, due to its ability to interact with receptors and enzymes in the body. Piperazine and its derivatives have also been shown to possess significant antimicrobial properties against a range of bacterial, fungal, and parasitic pathogens. The mechanism of action of piperazine-based antimicrobial agents involves their ability to disrupt the cell membrane and inhibit essential enzymatic processes within the target organism.

Studies indicated that piperazine analogs exhibit potent antibacterial activity against a variety of Gram-positive and Gram-negative bacteria, including methicillin-resistant *Staphylococcus aureus* (MRSA) and vancomycin-resistant *Enterococcus faecium* [1]. In addition, piperazine-based compounds have been shown to be effective against multidrug-resistant (MDR) strains of *Pseudomonas aeruginosa* and *Acinetobacter baumannii* [2, 3].

Piperazine derivatives also exhibit significant antifungal activity against a range of pathogenic fungi, including *Candida albicans* and *Aspergillus fumigatus* [4, 5]. Piperazine-based antifungal agents have been shown to inhibit fungal growth by disrupting the cell wall and plasma membrane,

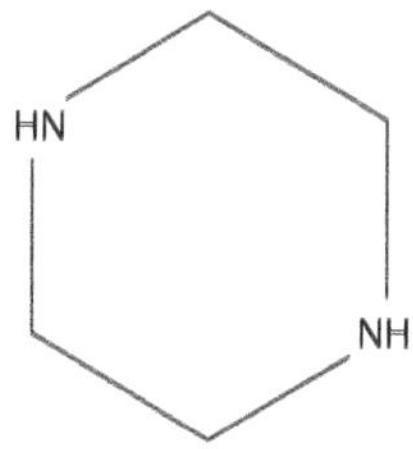

Figure 12.1 Structure of piperazine.

as well as by inhibiting essential enzymatic processes involved in fungal metabolism. In addition to its antibacterial and antifungal properties, piperazine has also been known to possess significant antiparasitic activity against a range of protozoan parasites, including *Plasmodium falciparum*, the causative agent of malaria [6]. Piperazine-based antiparasitic agents have been shown to inhibit essential enzymatic processes involved in parasite metabolism, leading to parasite death.

Furthermore, piperazine derivatives have been shown to exhibit low toxicity and good pharmacokinetic properties, making them attractive candidates for the development of new antimicrobial agents [7]. In fact, several piperazine-based compounds have already been approved for clinical use, including the anthelmintic drug praziquantel and the antipsychotic drug flupenthixol.

12.2 Synthetic Strategies for Piperazine Derivatives

Several researchers have reported different strategies for the preparation of piperazine derivatives, *via* nucleophilic substitution, ring-opening, reduction, and oxidation and cyclization reactions as well.

Vitaku *et al.* [8] and Huang *et al.* [9] reported the synthesis of piperazine derivatives by the reaction of diamines with α-halocarbonyl compounds [8, 9]. Zheng *et al.* [10] reviewed the synthesis of piperazine derivatives using a copper-catalyzed aerobic oxidation reaction. Another ring-opening reaction involves the reaction of piperazine with epoxides [10]. A review is shown in Figure 12.2 [11].

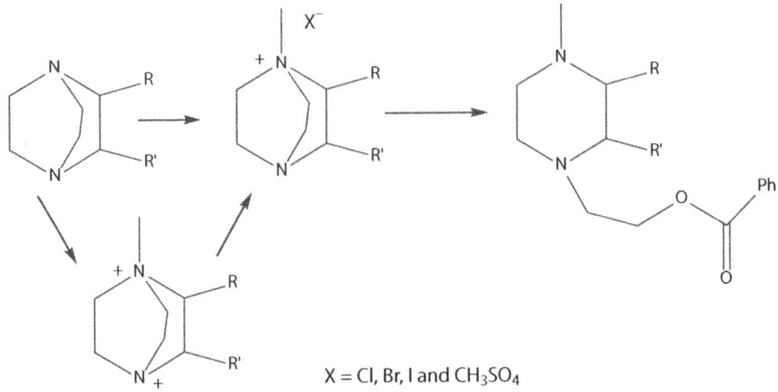

Figure 12.2 Synthesis of piperazine derivatives by the ring-opening reaction.

Imides can be reduced to piperazine using various reducing agents such as lithium aluminum hydride (LAH), sodium borohydride ($NaBH_4$), or hydrogen gas (H_2) with a catalyst. Jin *et al.* studied the use of LAH to reduce phthalimide derivatives to piperazines [12]. Aldehydes or ketones can be reacted with primary amines and a reducing agent such as sodium cyanoborohydride ($NaBH_3CN$) or sodium triacetoxyborohydride (STAB) to form piperazines [13]. Unsaturated compounds such as alkenes or alkynes can be hydrogenated to form piperazines in the presence of a catalyst such as palladium on carbon (Pd/C) [14].

Piperazine derivatives can be synthesized through various oxidation reactions, including the use of oxidizing agents such as hydrogen peroxide, ozone, and oxygen. Saleh *et al.* [15], synthesized a piperazine derivative 4-(4-methoxybenzyl)piperazine through the oxidation of 4-(4-methoxybenzyl)aniline with hydrogen peroxide in the presence of acetic acid. The reaction was carried out at room temperature, and the product was obtained in good yield [15]. Le Thanh *et al.* [16] reported the synthesis of a series of piperazine derivatives through the oxidation of benzylamines with ozone in the presence of catalysts such as CuO, NiO, and MnO_2 [16]. The reaction was carried out at low temperature, and the products were obtained in moderate to good yield. Kalsi *et al.* [17] revealed the production of a piperazine derivative called *N*-(4-(piperazin-1-yl)phenyl)-4-(trifluoromethyl)benzamide through the oxidation of *N*-(4-(piperazin-1-yl)phenyl)-4-(trifluoromethyl)benzamide with oxygen in the presence of a copper catalyst. The reaction was carried out at high temperature and pressure, and the product was obtained in good yield [17].

The most commonly used method for the synthesis of piperazine derivatives is the condensation reaction of ethylenediamine with various substituted halides or carboxylic acids. This reaction leads to the formation of *N*-substituted piperazines, which can be further functionalized to produce a wide range of derivatives [18].

One of the most common methods for the synthesis of piperazine derivatives is the reaction of piperazine with various electrophiles. For example, piperazine can react with alkyl halides, acyl halides, and sulfonyl halides to form *N*-alkyl, *N*-acyl, and *N*-sulfonyl piperazines, respectively. This method has been used to synthesize various pharmacologically active compounds, such as antipsychotics, antihistamines, and anti-inflammatory agents [19].

Another method for the synthesis of piperazine derivatives is the reduction of nitro compounds. Nitro compounds can be reduced to piperazine derivatives using a variety of reducing agents, such as H_2, Raney, nickel, or zinc dust. This method has been used to synthesize various antipsychotics, antidepressants, and antihistamines [20].

Methods for the synthesis of piperazine derivatives are the cycloaddition reaction between piperazine and various alkynes producing a wide variety of piperazine derivatives, including imidazo[1,5-a]pyridines, pyrido[1,2-a]pyrimidines, and pyrazolo[1,5-a]pyridines. These compounds have potent antifungal, antibacterial, and anticancer activities. Another common method for the synthesis of piperazine derivatives is the substitution reaction of piperazine with various electrophiles, such as alkyl halides, acyl halides, and sulfonyl chlorides. This reaction leads to the formation of *N*-alkyl, *N*-acyl, and *N*-sulfonyl piperazines, which can be further functionalized to produce a range of derivatives [21].

Cyclization reactions, such as the Gabriel synthesis and the Bischler–Napieralski reaction, can also be used for the synthesis of piperazine derivatives. These reactions involve the cyclization of diamines or amino alcohols to form piperazine rings, which can be further functionalized to produce a range of derivatives [22, 23].

12.2.1 Synthesis of 1-Benzyl-4-Phenylpiperazine

The synthesis of 1-benzyl-4-phenylpiperazine has been reported in several literature sources, including research articles and patents.

Murugesh *et al.* [24] synthesized a series of 1-benzyl-4-phenylpiperazine derivatives and evaluated their anticonvulsant activity. The synthesis of 1-benzyl-4-phenylpiperazine was carried out by the reaction of benzyl chloride with 1-(4-phenylpiperazin-1-yl)propan-2-on in the presence of a base. The yield of the product was reported to be 72%.

Jida and Ballet [25] described a method for the preparation of 1-benzyl-4-phenylpiperazine and its analogs using a one-pot synthesis. The reaction involves the condensation of benzyl chloride with 1-phenylpiperazine in the presence of a base, followed by reduction of the resulting imine using $NaBH_4$. The authors reported yields of up to 91% for the desired products.

Żesławska *et al.* [26] synthesized a series of 1-benzyl-4-phenylpiperazine derivatives and evaluated their binding affinity for the serotonin transporter. The synthesis of 1-benzyl-4-phenylpiperazine was carried out by the reaction of benzyl chloride with 1-(4-phenylpiperazin-1-yl)propan-2-on in the presence of sodium hydroxide in ethanol. The yield of the product was reported to be 85%.

12.2.2 Synthesis of 1-Phenylpiperazine

The synthesis of 1-phenylpiperazine can be achieved using various methods, including two-step, three-step, and one-pot reactions. The purity and yield of the final product can be confirmed using nuclear magnetic resonance (NMR) and infrared (IR) spectroscopy.

Bosc *et al.* [27] described the synthesis and antidepressant activity of a series of 5-(1-aryl-4-piperazino) methyl-2-amino-2-oxazolines. A series of substituted 2-amino-2-oxazoline derivatives were first prepared using one-pot synthesis of piperazine derivative by reacting the appropriate piperazine with epichlorohydrin. The synthesized compounds were evaluated for their antidepressant activity, and results showed that several of the compounds exhibited significant antidepressant-like activity, with potency comparable to that of the reference drug imipramine [27].

Andrew and Ajibade [28] describe synthesis, characterization, and anticancer studies of three metal complexes: bis(1-phenylpiperazine dithiocarbamato), Cu(II), Zn(II), and Pt(II). The crystal structures of 1-phenylpiperazine, dithiocarbamato-S,S′, zinc(II), and Pt(II) were also determined. To synthesize the complexes, 1-phenylpiperazine dithiocarbamate was first prepared by reacting 1-phenylpiperazine with carbon disulfide and sodium hydroxide. The resulting dithiocarbamate was then reacted with Cu(II), Zn(II), and Pt(II) salts to form the corresponding metal complexes. The synthesized complexes were characterized using various techniques including infrared spectroscopy, UV-visible spectroscopy, elemental analysis, and magnetic susceptibility measurements. The crystal structures of the Zn(II) and Pt(II) complexes were determined using X-ray diffraction analysis. The anticancer activities of the complexes were evaluated against three human cancer cell lines: MCF-7 (breast), A549 (lung), and HeLa (cervical). The complexes showed promising anticancer activity, with the Pt(II) complex exhibiting the highest activity against all three cell lines [28].

Jia *et al.* [29] illustrated the synthesis and evaluation of a novel class of HSP90 inhibitors containing a 1-phenylpiperazine scaffold. To synthesize the inhibitors, a series of 1-phenylpiperazine derivatives were first prepared by reacting the appropriate aniline and 1-(2-chloroethylethyl) piperazine under basic conditions. The resulting derivatives were then reacted with a variety of substituted phenylboronic acids in the presence of a palladium catalyst to form the target compounds. The synthesized compounds were evaluated for their inhibitory activity against HSP90, a protein that is known to play a critical role in cancer cell proliferation and survival. The compounds were also tested for their cytotoxicity against two cancer cell lines, A549 (lung) and MCF-7 (breast). The results showed that several of the synthesized compounds exhibited potent inhibitory activity against HSP90, with IC_{50} values in the low micromolar range. Furthermore, these compounds were found to be selectively toxic toward cancer cells, with minimal toxicity toward normal cells. This study provided evidence that the 1-phenylpiperazine scaffold can serve as a promising starting point for the development of novel HSP90 inhibitors with potential anticancer activity. The results also highlight the importance of structure–activity relationship (SAR) studies in the design and optimization of therapeutic agents.

Takahashi *et al.* [30] revealed the synthesis and SARs of phenylpiperazine derivatives exhibiting potent antagonistic activity against the neuropeptide Y, Y5 receptor. A series of phenylpiperazine compounds were first prepared and modified through various chemical transformations, including substitution at the phenyl ring and variation of the alkyl chain length and substitution at the piperazine nitrogen. The synthesized compounds were evaluated for their binding affinity and antagonistic activity against the neuropeptide Y Y5 receptor using radioligand binding assays and functional assays. The results showed that several of the compounds exhibited potent binding affinity and antagonistic activity, with subnanomolar to low nanomolar IC_{50} values. Further SAR studies revealed that substitution at the phenyl ring and variation of the alkyl chain length and substitution at the piperazine nitrogen were critical for enhancing the binding affinity and antagonistic activity of the compounds. This study provided insights into the design and optimization of potent neuropeptide Y Y5 receptor antagonists using phenylpiperazine derivatives as a starting scaffold. The results may have implications for the development of therapeutic agents targeting this receptor for the treatment of obesity, diabetes, and other metabolic disorders.

12.2.3 Synthesis of 1-[(4-Chlorophenyl)sulfonyl]piperazine

The synthesis of 1-[(4-chlorophenyl)sulfonyl]piperazine can be achieved through the reaction of piperazine with 4-chlorobenzenesulfonyl chloride in the presence of a base in various solvents such as acetonitrile or dichloromethane. The product can be obtained in good to moderate yields and purity, and it has been used as an intermediate in the synthesis of various pharmaceutical compounds.

Wang *et al.* [31] disclosed the design , synthesis, and antiallergic activities of novel (R)(−)-1-[(4-chlorophenyl)phenyl methyl]piperazine derivatives. To design the compounds, the researchers used a structure-based approach to identify the key structural features required for inhibiting the histamine H1 receptor, which is involved in the pathogenesis of allergic reactions. Based on this analysis, they designed a series of piperazine derivatives that incorporate a (R)(−)-1-[(4-chlorophenyl)phenyl methyl] group as a pharmacophore. The synthesized compounds were evaluated for their antiallergic activities using a variety of *in vitro* and *in vivo* assays, including inhibition of histamine release from mast cells and inhibition of ear swelling in a mouse model of allergic inflammation. The results showed that several of the compounds exhibited significant antiallergic activity, with potency comparable to that of the reference drug cetirizine. Further SAR studies revealed that the substitution pattern on the phenyl ring and the alkyl chain length and the substitution pattern on the piperazine ring were critical for enhancing the antiallergic activities of the compounds. Furthermore, the study provided the evidence that the (R)(−)-1-[(4-chlorophenyl)phenyl methyl]piperazine scaffold can serve as a promising starting point for the development of novel antiallergic agents. The results also highlight the importance of structure-based design and SAR studies in drug discovery.

Mallikarjuna *et al.*, [32] described the synthesis, anticancer, and antituberculosis studies for 1-(4-chlorophenyl) cyclopropyl methanone derivatives. In this research, a variety of substituted cyclopropyl ketones were first prepared by reacting the appropriate substituted benzyl chlorides with ethyl diazoacetate in the presence of a copper catalyst. The resulting cyclopropyl ketones were then reacted with a variety of substituted piperazines under basic conditions to form the target compound. The synthesized compounds were evaluated for their anticancer activities using a variety of *in vitro* assays, including cell viability assays, cell cycle analysis, and apoptosis assays. The results showed that several of the compounds exhibited significant anticancer activity against a panel of cancer cell lines, with some compounds exhibiting submicromolar IC_{50} values. In addition, the compounds were evaluated for their antituberculosis activity using the microplate alamar blue assay against *Mycobacterium tuberculosis*. The results showed that several of the compounds exhibited moderate to good antituberculosis activity , with some compounds exhibiting MIC values in the low micromolar range. Further, SAR studies revealed that the substitution pattern of the phenyl ring and the size and substitution pattern on the piperazine ring were critical for enhancing the anticancer and antituberculosis activities of the compounds.

Combretastatin-A4 (CA-4) is a natural compound that has been shown to possess potent cytotoxicity against cancer cells by inhibiting tubulin polymerization, a process necessary for cell division. However, CA-4 is unstable and difficult to synthesize, which has limited its clinical use. To overcome this limitation, the researchers used a molecular hybridization approach to design a series of CA-4 carboxamides that incorporate a sulfonyl piperazine moiety as a pharmacophore, which has been shown to enhance the stability and potency of CA-4 analogs. Jadala *et al.* [33] prepared a variety of substituted sulfonyl piperazines first by reacting the appropriate sulfonyl chlorides with a variety of substituted piperazines under basic conditions followed by their reaction with a CA-4 derivative under acidic conditions to form the target compounds. The synthesized compounds were evaluated for their *in vitro* cytotoxicity against a panel of cancer cell lines using the MTT assay. The results showed that several of the compounds exhibited significant cytotoxicity, with some compounds exhibiting IC_{50} values in the low nanomolar range. In addition, the compounds were evaluated for their ability to inhibit tubulin polymerization using the turbidity assay. The results showed that several of the compounds exhibited potent inhibition of tubulin polymerization, with some compounds exhibiting IC_{50} values in the low micromolar range.

12.2.4 Synthesis of Piperazine-Phthalimide Derivatives

Piperazine-based phthalimide derivatives have emerged as a promising class of compounds with potent antibacterial activity against MRSA. They have the potential to be developed as a new class of antibiotics to combat antibiotic-resistant bacteria. The synthesis of piperazine-based phthalimide derivatives involves the reaction of phthalic anhydride with piperazine in the presence of a catalyst. The synthesis of these compounds is relatively simple and can be carried out on a large scale, making them suitable for drug development. Several piperazine-based phthalimide derivatives have been reported to exhibit potent antibacterial activity against MRSA. The mechanism of action of these compounds involves inhibition of bacterial cell wall synthesis, which leads to bacterial cell death. The compounds have also been found to be effective against other Gram-positive and Gram-negative bacteria. The MIC (minimum inhibitory concentration) values of these compounds have been found to be in the range of 0.125 to 16 μg/mL, indicating their high potency.

Prasad *et al.* [34] synthesized piperazine-based phthalimide derivatives and studied their antimicrobial activity against MRSA, a superbug that has become resistant to most antibiotics. To synthesize the compounds, a variety of substituted phthalic anhydrides were first reacted with a variety of substituted piperazines under basic conditions to form the corresponding phthalimide derivatives. The resulting phthalimides were then evaluated for their antibacterial activities against MRSA using the agar diffusion method and the broth microdilution method. The results showed that several of the compounds exhibited significant antibacterial activity against MRSA, with some compounds exhibiting MIC values in the low micromolar range. The most active compounds were further evaluated for their cytotoxicity against human embryonic kidney cells (HEK293) using the MTT assay, and the results showed that the compounds were not toxic to the cells at the concentrations tested. Further SAR studies revealed that the substitution pattern on the phthalimide ring and the size and substitution pattern on the piperazine ring were critical for enhancing the antibacterial activities of the compounds. In addition, molecular docking studies were performed to investigate the binding modes of the compounds to the active site of the bacterial enzyme, dihydrofolate reductase, which is essential for bacterial survival.

12.2.5 Synthesis of Sulfonyl Piperazine

Sulfonyl piperazines are a class of organic compounds that have received considerable attention in medicinal chemistry due to their potential as therapeutic agents. The synthesis of sulfonyl piperazines has been studied extensively, with various methods reported in the literature. The choice of the method will depend on factors such as the desired product, yield, and reaction conditions.

One commonly used method for the synthesis of sulfonyl piperazines involves the reaction of piperazine with sulfonyl chloride in the presence of a base. For example, the reaction of piperazine with p-toluenesulfonyl chloride in the presence of triethylamine has been reported to give the corresponding sulfonyl piperazine in good yield [35]. Another method involves the reaction of piperazine with a sulfonyl isocyanate in the presence of a base. For example, the reaction of piperazine with p-toluenesulfonyl isocyanate in the presence of triethylamine has been reported to give the corresponding sulfonyl piperazine in good yield [35].

Mohan *et al.* [36] suggested that methylpyrimidine sulfonyl piperazine derivatives can serve as a promising scaffold for the development of antibacterial, anthelmintic, and anti-inflammatory agents. However, further optimization and modification of the synthesized compounds are required to improve their potency and selectivity. The researchers synthesized a series of compounds using a multistep synthetic route starting with 4-methylpyrimidine-5-carboxylic acid. The synthesized compounds were evaluated for their antibacterial activity against several bacterial strains, including *S. aureus* and *Escherichia coli*, using the disk diffusion method. The results showed that some of the compounds exhibited moderate to good antibacterial activity, with one compound showing activity comparable to that of the standard drug ampicillin. The anthelmintic activity of the synthesized compounds was evaluated against the parasitic worm *Caenorhabditis elegans*. Some of the compounds exhibited moderate anthelmintic activity, with one compound showing activity comparable to that of the standard drug levamisole. The anti-inflammatory activity of the synthesized compounds was evaluated using the carrageenan-induced paw edema model in rats. Some of the compounds exhibited significant anti-inflammatory activity, with one compound showing activity comparable to that of the standard drug indomethacin [36].

Bungard *et al.* [37] designed and synthesized piperazine sulfonamide cores leading to highly potent HIV-1 protease inhibitors. HIV-1 protease is an essential enzyme for the replication of the HIV, and therefore, it is a key target for the development of antiretroviral drugs. The researchers designed and synthesized a library of piperazine sulfonamide derivatives using a structure-based approach. The compounds were designed to interact with key residues in the active site of HIV-1 protease, specifically the catalytic aspartic acid residues Asp25 and Asp25′. The synthesized compounds were evaluated for their inhibitory activity against HIV-1 protease using a fluorometric assay. The most potent compounds were further evaluated for their inhibitory activity against a panel of HIV-1 protease variants, including those with mutations associated with drug resistance. The results showed that several of the compounds exhibited high potency against HIV-1 protease, with some compounds exhibiting nanomolar inhibitory activity. The most potent compound had an IC_{50} value of 1.1 nM against wild-type HIV-1 protease and exhibited subnanomolar activity against several drug-resistant HIV-1 protease variants. Molecular modeling studies were performed to investigate the binding modes of the compounds to the active site of HIV-1 protease. The results suggested that the compounds interacted with the catalytic aspartic acid residues through hydrogen bonding and electrostatic interactions.

12.2.6 Synthesis of 1-(2,5-Dimethoxyphenyl)piperazine

Synthesis of 1-(2,5-dimethoxyphenyl)piperazine is a well-established procedure in organic chemistry for the preparation of a useful intermediate in the synthesis of various biologically active compounds. The synthesis of 1-(2,5-dimethoxyphenyl)piperazine has been described in several research articles, and the procedure generally involves the reaction of 2,5-dimethoxyaniline with piperazine followed by *N*-methylation.

Lee *et al.* demonstrated the synthesis of 1-[(5- or 6-substituted 2-alkoxyquinoxalin-3-yl) aminocarbonyl]-4-(hetero) arylpiperazine derivatives exhibiting promising anticancer activity, [38]. The synthesis of the compounds was performed using a multistep procedure involving the reaction of 2-aminobenzonitrile with an appropriate aldehyde to yield the corresponding quinoxaline derivative. The quinoxaline derivative was then coupled with an appropriate acid chloride and heteroaryl piperazine to obtain the final product. The compounds were characterized using various spectroscopic techniques, such as IR, NMR, and mass spectrometry. The *in vitro* anticancer activity of the compounds was evaluated using a panel of human cancer cell lines, including A549, HCT-15, etc., using the sulforhodamine B assay. The outcomes revealed that some of the compounds exhibited profound anticarcinogenic activity against the tested cell lines, with IC_{50} values ranging from 11 to 21 nM. The article also reported the SAR analysis of the compounds, which indicated that the presence of electron-withdrawing groups on the heteroaryl piperazine ring and the substitution pattern on the quinoxaline ring significantly influenced the anticancer activity of the compounds.

Hafeez *et al.* [39] invented the ultrasound-assisted synthesis of some methanesulfonyl-piperazine-based dithiocarbamates and their evaluation as potential anticancer, thrombolytic, and hemolytic agents. The created molecules were characterized using numerous spectroscopic methods, such as NMR, IR, and mass spectrometry. The study evaluated the thrombolytic activity of the compounds using the *in vitro* clot lysis assay.

The results indicated that the piperazine with the 2,5-dimethoxyphenyl moiety displayed the best thrombolytic effect among the tested compounds.

In a research done by Darji *et al.*, synthesis and spectral characterization of some novel *N*-(substituted benzoyl)piperidine and related molecules and their possible antibacterial agents were described. In this work, 14 new compounds with modified substituents on the thiazole ring were synthesized. The compounds were characterized using techniques such as Fourier transform IR spectroscopy, ^{1}H-NMR, and mass spectrometry. The study explored the antibacterial activity of the compounds against three bacterial strains: *E. coli*, *S. aureus*, and *P. aeruginosa*. The results showed that some of the compounds exhibited significant antibacterial activity, with MICs [40].

12.3 Structural Relationship of Piperazine Derivatives

The structural relationship of piperazine derivatives is an important factor in determining their biological activity and pharmacokinetic properties. The structure of piperazine derivatives can be modified by varying the substituents on the nitrogen atoms and on the ring, as well as by altering the size and shape of the ring itself (Figure 12.3). Its unique structure is responsible for the diverse pharmacological properties of piperazine and its derivatives, including its significant antimicrobial activity. Piperazine-based compounds have ability to interact with a variety of biological targets, including receptors and enzymes. These interactions are largely dependent on the spatial arrangement of functional groups around the piperazine ring, which can significantly affect the compound's biological activity and selectivity. In terms of antibacterial activity, the structure of piperazine-based compounds can be modified to optimize their ability to disrupt bacterial cell membranes and inhibit essential enzymatic processes within the target organism. For example, the addition of hydrophobic substituents to the piperazine ring has been shown to increase the lipophilicity and cell membrane penetration of piperazine-based antibacterial agents [41]. This is thought to enhance their ability to disrupt bacterial cell membranes and ultimately lead to bacterial death.

The nature and position of substituents on the nitrogen atoms of the piperazine ring can significantly affect the biological activity and pharmacokinetic properties of piperazine derivatives. For example, *N*-alkyl and *N*-aryl substituents have been shown to enhance the antimicrobial activity of piperazine derivatives by increasing their lipophilicity and membrane permeability [42].

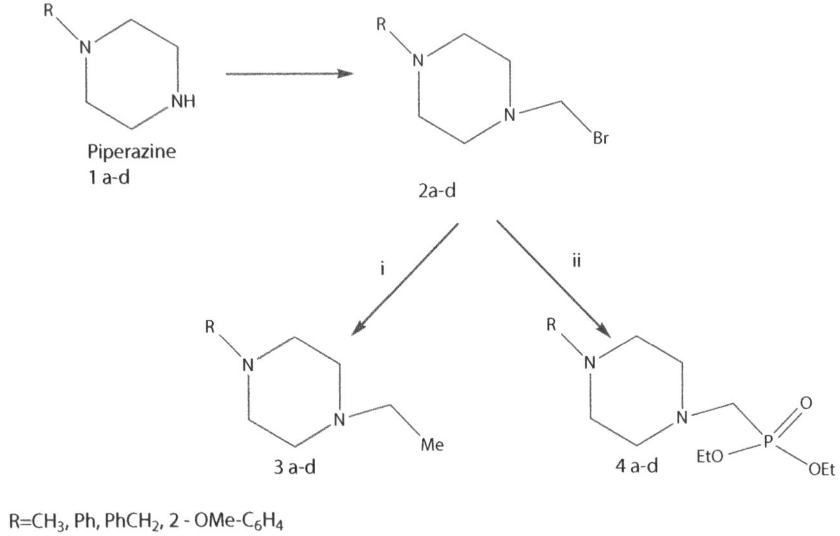

Figure 12.3 Substituted piperazines.

The nature and position of substituents on the ring of the piperazine derivative can also affect its biological activity and pharmacokinetic properties. For example, substitution at the 1- and 4-positions of the piperazine ring has been shown to significantly enhance the antimicrobial activity of piperazine derivatives by increasing their membrane permeability and enzymatic activity [43]. Additionally, substitution at the 2- and 3-positions of the piperazine ring has been shown to increase the selectivity of piperazine derivatives toward specific targets, such as enzymes or receptors [44].

The diverse pharmacological properties of piperazine and its derivatives are largely dependent on the spatial arrangement of functional groups around the piperazine ring. The ability to modify the structure of piperazine-based compounds in a targeted manner makes them attractive candidates for the development of novel antimicrobial agents with unique mechanisms of action.

Presence of specific functional groups around the piperazine ring can also significantly affect the antibacterial activity of piperazine-based compounds. For example, the presence of a carboxylic acid group at the 3-position of the piperazine ring has been shown to increase the antibacterial activity of piperazine-based compounds against Gram-negative bacteria [45]. This is considered due to the ability of the carboxylic acid group to form hydrogen bonds with the bacterial cell wall, which enhances the ability of the compound to disrupt the cell membrane.

Similarly, the presence of specific functional groups around the piperazine ring can also affect the antifungal activity of piperazine-based compounds. For example, the addition of an imidazole ring to the piperazine nucleus has been shown to increase the antifungal activity of piperazine-based compounds against *C. albicans* and *A. fumigatus* [46]. This is considered due to the ability of the imidazole ring to interact with fungal cell membranes and inhibit essential enzymatic processes within the fungal cell.

The structure of piperazine-based compounds can also be optimized to enhance their antiparasitic activity. Patel *et al.* [47] claimed the increase antimalarial activity of piperazine-based compounds against *P. falciparum* due to the presence of pyridine ring on the piperazine [48]. This is considered due to the ability of the pyridine ring to interact with specific target enzymes within the parasite, leading to parasite death.

The size and shape of the piperazine ring can also affect the biological activity and pharmacokinetic properties of piperazine derivatives. For example, piperazine derivatives with smaller ring sizes, such as azetidine and pyrrolidine, have been shown to exhibit enhanced antimicrobial activity by increasing their membrane permeability and enzymatic activity [49]. Similarly, piperazine derivatives with larger ring sizes, such as homopiperazines and diazepanes, have been shown to exhibit increased selectivity toward specific targets, such as receptors or transporters [50].

12.4 Antimicrobial Activities of Piperazine Derivatives

Piperazine and its derivatives have been found to exhibit a broad spectrum of antimicrobial activity against a range of bacterial, fungal, and parasitic pathogens. The effectiveness of piperazine-based antimicrobial agents depends on their ability to interact with specific targets within the cell, leading to inhibition of essential enzymatic processes and eventual cell death. In this section, we will explore the relationship between the antimicrobial spectrum and the chemical structure of piperazine derivatives.

12.4.1 Antibacterial Activity

Piperazine derivatives have been shown to exhibit potent antibacterial activity against a variety of Gram-positive and Gram-negative bacteria. The mechanism of action of these compounds involves their ability to disrupt the cell membrane and inhibit essential enzymatic processes within the target organism. The antibacterial spectrum of piperazine derivatives depends on their chemical structure, which can be modified to improve their activity against specific bacterial species.

Tahir *et al.* [51] synthesized a series of piperazine derivatives and evaluated their antibacterial activity against a panel of bacterial strains [51]. They found that the substitution of the piperazine ring with hydrophobic groups such as methyl, ethyl, or propyl groups led to increased activity against Gram-positive bacteria such as *S. aureus* and *Streptococcus pneumoniae.* On the other hand, substitution with hydrophilic groups such as hydroxyl, carboxyl, or amino groups led to increased activity against Gram-negative bacteria such as *E. coli* and *P. aeruginosa.*

In addition, piperazine derivatives have been shown to exhibit activity against MDR bacterial strains. Brown and Dawson [52] explored a series of piperazine-based compounds and evaluated their antibacterial activity against a panel of MDR bacterial strains. They found that the substitution of the piperazine ring with a variety of functional groups such as hydroxyl, carboxyl, and amino groups led to increased activity against MDR strains of *P. aeruginosa* and *A. baumannii* [52].

12.4.2 Antifungal Activity

Piperazine derivatives have also been found to possess significant antifungal activity against a range of pathogenic fungi. The mechanism of action of piperazine-based antifungal agents involves their ability to disrupt the cell wall and plasma membrane, as well as by inhibiting essential enzymatic processes involved in fungal metabolism.

The antifungal spectrum of piperazine derivatives is also dependent on their chemical structure. Kathiravan *et al.* [53] reviewed a series of piperazine-based compounds and evaluated their antifungal activity against *C. albicans* and *A. fumigatus*. They found that the substitution of the piperazine ring with hydrophobic groups such as benzyl or phenyl groups led to increased activity against *Candida parapsilosis*, whereas substitution with hydrophilic groups such as hydroxyl, carboxyl, or amino groups led to increased activity against plant fungi such as *Xanthomonas axonopodis*, *Xanthomonas oryzae*, *Ralstonia solanacearum*, *Erwinia aroideae*, etc. [53].

12.4.3 Antiparasitic Activity

Piperazine derivatives have shown significant antiparasitic activity against protozoan parasites. Earlier studies have shown that piperazine derivatives exhibit significant antimalarial activity by inhibiting essential enzymatic processes involved in the metabolism of *P. falciparum*, leading to parasite death. Some of these derivatives have also been found to be effective against MDR strains of the parasite, which is a major concern in the treatment of malaria. These findings suggest that piperazine derivatives have the potential to be developed into effective antimalarial agents [54, 55].

Kuettel *et al.* [56] described the synthesis and evaluation of a series of new compounds for their antiparasitic activity. The compounds are derivatives of 4-[5-(4-phenoxyphenyl)-2H-pyrazol-3-yl] morpholine, which has previously been shown to have activity against *Trypanosoma brucei rhodesiense*, the parasite that causes African sleeping sickness. The authors synthesized a series of new compounds by modifying the phenoxyphenyl moiety of the molecule and evaluated their activity against several parasitic organisms, including *T. brucei rhodesiense*, *Trypanosoma cruzi*, *Leishmania donovani*, and *P. falciparum* [56].

Silva *et al.* [57] synthesized and evaluated a series of piperazine derivatives of betulinic acid for their antimalarial activity, cytotoxicity, and mechanism of action. Betulinic acid is a natural product with known antimalarial activity, but it has low solubility and bioavailability, which limits its use as a drug. The authors synthesized a series of piperazine derivatives of betulinic acid with the aim of improving its solubility and bioavailability. The compounds were evaluated for their antimalarial activity against the chloroquine-sensitive *P. falciparum* strain 3D7 and for their cytotoxicity against human erythrocytes and two human cell lines. The mechanism of action was also investigated through *in vitro* assays. The results showed that some of the piperazine derivatives exhibited higher antimalarial activity than betulinic acid, with some compounds showing submicromolar activity. The derivatives also showed lower cytotoxicity than betulinic acid, with some compounds showing selectivity for the parasite. The mechanism of action studies suggested that the compounds may act by disrupting the parasite's membrane integrity [57].

12.4.4 Anticancerous Activities of Piperazine Derivatives

Piperazine derivatives have been extensively investigated for their potential anticancer activities. Various studies demonstrate the potential of piperazine derivatives as anticancer agents, with various mechanisms of action depending on the specific compound studied.

Sheena Mary *et al.* [58] illustrated the computational and spectroscopic study of a piperazine derivative with potential anticancer activity. Characterizations of molecules were achieved by density functional theory calculations to study the molecular structure, electronic properties, and vibrational spectra of the compound. Molecular docking studies were also performed to investigate its potential binding to the active site of the cancer target protein, and molecular dynamics simulations were used to explore the stability of the complex. The results showed that the benzoxazole derivative of piperazine has a stable molecular structure and favorable electronic properties for anticancer activity. The docking studies revealed potential binding interactions with the active site of the cancer target protein, and the molecular dynamics simulations suggested that the complex is stable over time. The SERS analysis confirmed the calculated vibrational frequencies and provided insights into the adsorption behavior of the compound on a silver surface [58].

Ahmed *et al.* [59] investigated the design, synthesis, and evaluation of novel compounds as potential anticancer agents targeting the VEGFR-2 kinase enzyme. In this work, a series of piperazine–chalcone hybrids and related pyrazoline analogs were synthesized and evaluated their ability to inhibit VEGFR-2 kinase enzyme activity. The compounds were also subjected to molecular docking studies to investigate their binding interactions with the active site of the enzyme. The results revealed that some of the synthesized compounds exhibited promising inhibitory activity against VEGFR-2 kinase, with IC_{50} values in the micromolar range. The molecular docking studies suggested that the compounds may interact with the enzyme through hydrogen bonding and hydrophobic interactions. The most active compound was further evaluated for its anticancer activity against three human cancer cell lines, and it showed significant cytotoxicity against these cell lines. The compound also exhibited low cytotoxicity against normal human fibroblast cells. Studies concluded that the piperazine–chalcone hybrids and related pyrazoline analogs have potential as anticancer agents targeting the VEGFR-2 kinase enzyme. The molecular docking studies provided valuable insights into the binding interactions between the compounds and the enzyme, which can aid in the rational design of new compounds with improved activity and selectivity [59].

Raju *et al.* [60] explored the synthesis and characterization of a series of novel 5-amino pyrazole derivatives and their evaluation for antiproliferative effects against breast cancer cell lines, and a total of 13 compounds were synthesized and evaluated their cytotoxicity against MCF-7 and MDA-MB-231 breast cancer cell lines using MTT assay. The results disclosed that some of the compounds exhibited potent antiproliferative activity against these cell lines, with IC_{50} values in the micromolar range. The results of the mechanism of action studies suggested that the compound induced cell cycle arrest at the G2/M phase and activated apoptosis through the mitochondrial pathway. The authors also performed molecular docking studies to investigate the binding interactions of the compounds with the active site of cyclin-dependent kinase 2 (CDK2), which is known to play a crucial role in cell cycle regulation. The molecular docking studies suggested that the compound may interact with CDK2 through hydrogen bonding and hydrophobic interactions and that the presence of the 5-amino pyrazole moiety may be important for the activity of the compound [60].

Mao *et al.* [61] synthesized and evaluated a series of hybrid compounds between benzofuran and *N*-aryl piperazine as potential anticancer agents. A total of 12 compounds were synthesized and evaluated for their cytotoxicity against three human cancer cell lines (A549, HepG2, and MCF-7) using MTT assay. The results showed that some of the compounds exhibited potent anticancer activity, with IC_{50} values in the micromolar range. The most active compound was further evaluated for its mechanism of action using flow cytometry and Western blot analysis. The results of the mechanism of action studies suggested that the compound induced apoptosis through the mitochondrial pathway and inhibited the proliferation of cancer cells by arresting the cell cycle at the G1 phase. The authors also performed molecular docking studies to investigate the binding interactions of the compound with the active site of the target enzyme, DNA topoisomerase II. The molecular docking studies suggested that the compound may interact with DNA topoisomerase II through hydrogen bonding and hydrophobic interactions, and that the benzofuran moiety may be important for the activity of the compounds. In this work, it has been concluded that the synthesized hybrid compounds between benzofuran and *N*-aryl piperazine have potential as anticancer agents and that further optimization and evaluation of these compounds are warranted. They also suggest that the molecular docking studies provided valuable insights into the binding interactions of the compounds with DNA topoisomerase II, which can aid in the rational design of new compounds with improved activity and selectivity [61].

Gong *et al.* [62] revealed the synthesis and crystal structure of a new pyrazolo[1,5-a]pyrimidine derivative called 7-(4-methylpiperazin-1-yl)-5-[4-(trifluoromethyl)phenyl]pyrazolo[1,5-a]pyrimidine-3-carbonitrile. This compound was characterized using various spectroscopic techniques, and its crystal structure was determined using X-ray diffraction and found to have remarkable biological activity, particularly its cytotoxicity against several cancer cell lines. The results showed that the compound exhibited significant anticancer activity against several cell lines, including A549, K562, and MKN45 [62].

Current research on the potential anticancer activity of piperazine-containing heterocyclic compounds was reviewed by Al-Ghorbani *et al.* less than a year ago. In their article, they have provided an overview of the chemistry and biology of piperazine derivatives and discussed their mechanisms of action against cancer cells. Synthesis and biological evaluation of a variety of piperazine-containing compounds, including pyrazoles, pyridazines, pyrimidines, and other heterocycles, were investigated. SARs of these compounds were also explored, and the most promising compounds in terms of their anticancer activity were highlighted. The article also concluded piperazine derivatives as potential anticancer agents [63].

12.5 Synergistic Activity of Piperazine Derivatives with Other Antimicrobial Agents, Enhancing Their Effectiveness against Certain Pathogens

Piperazine derivatives have the potential to enhance the effectiveness of existing antimicrobial agents, which could help to address the growing problem of antimicrobial resistance. Various researches [64–66] disclosed the synergistic activity of piperazine with other antimicrobial agents such as antibiotics and antifungal agents. The combination of piperazine derivatives with these agents has been found to enhance their effectiveness against certain pathogens, such as *S. aureus* and *C. albicans*. In particular, piperazine has been found to enhance the activity of antibiotics against MDR strains of *S. aureus*. These findings suggest that piperazine derivatives have the potential to be used in combination therapy with other antimicrobial agents for the treatment of certain infections.

Synergistic activity of piperazine derivatives with fluconazole against *C. albicans*: Piperazine derivatives, when combined with fluconazole, showed significant synergistic activity against *C. albicans*, a common fungal pathogen that can cause serious infections in immunocompromised patients. The authors concluded that piperazine derivatives have the potential to enhance the effectiveness of fluconazole in the treatment of candidiasis [67].

Research on the synergistic activity of piperazine derivatives with ciprofloxacin against *P. aeruginosa* disclosed that some piperazine derivatives, when combined with ciprofloxacin, showed significant synergistic activity against MDR strains of *P. aeruginosa*, a common Gram-negative bacterium that can cause serious infections in hospital settings. The authors suggested that the combination of piperazine derivatives with ciprofloxacin could be a promising strategy for the treatment of *P. aeruginosa* infections [68].

Various studies suggested [69–71] that the combination of piperazine derivatives with rifampicin could be a promising strategy for the treatment of drug-resistant tuberculosis. Furthermore, piperazine derivatives, when combined with rifampicin, showed significant synergistic activity against *M. tuberculosis*, the causative agent of tuberculosis.

12.6 Mechanism of Action

Antimicrobial activity of piperazine derivatives can be attributed to their ability to target essential cellular processes in microorganisms, leading to growth inhibition or cell death. By targeting multiple mechanisms of action, piperazine derivatives have the potential to be effective against a wide range of microorganisms, including those that are resistant to other antimicrobial agents. The exact mechanism of action of antimicrobial piperazine compounds can vary depending on the specific derivative and the type of microorganism being targeted. However, some general mechanisms of action have been proposed based on experimental evidence. Here are some of the proposed mechanisms of action [72]:

12.6.1 Inhibition of Cell Wall Synthesis

Some piperazine derivatives have been shown to inhibit the synthesis of bacterial cell walls, leading to cell lysis and death. This mechanism of action is similar to that of beta-lactam antibiotics, which also target cell wall synthesis [73].

12.6.2 Inhibition of Protein Synthesis

Few piperazine derivatives were found to have the potential to inhibit bacterial protein synthesis by binding to the 70S ribosome, which is responsible for translating messenger RNA into proteins. This mechanism of action is similar to that of macrolide antibiotics, which also target bacterial protein synthesis [74].

12.6.3 Disruption of Membrane Integrity

Some piperazine derivatives also showed the ability to disrupt the integrity of bacterial and fungal cell membranes, leading to leakage of cellular contents and cell death. This mechanism of action is similar to that of polymyxin antibiotics, which also target bacterial and fungal cell membranes [64].

12.6.4 Inhibition of Nucleic Acid Synthesis

Some piperazine derivatives showed efficacy to inhibit the synthesis of bacterial and fungal nucleic acids, leading to impaired DNA replication and RNA transcription. This mechanism of action is similar to that of quinolone antibiotics, which also target nucleic acid synthesis [75].

12.7 Applications of Piperazine Derivatives as Microbial Agents

12.7.1 In human medicine

Antimicrobial piperazine agents have a wide range of applications in human medicine. They can be used to treat common bacterial, fungal, and parasitic infections occurring in human beings. They can also be exploited for disinfection and sanitation and show versatility and importance in human medicine [76].

12.7.2 Treatment of Bacterial Infections

Piperazine derivatives have been used in the treatment of various bacterial infections, including respiratory tract infections, urinary tract infections, and skin and soft tissue infections. For example, the piperazine derivative cefpodoxime proxetil is used to treat respiratory tract infections, whereas ciprofloxacin is used to treat urinary tract infections [77].

12.7.3 Treatment of Fungal Infections

Some piperazine derivatives, such as fluconazole, have been used in the treatment of fungal infections. Fluconazole is commonly used to treat infections caused by *Candida* species, including oral thrush and vaginal yeast infections [78].

12.7.4 Treatment of Parasitic Infections

Piperazine derivatives have also been used in the treatment of parasitic infections, including helminth infections and protozoan infections such as malaria. For example, the piperazine derivative praziquantel is used to treat schistosomiasis, a parasitic infection caused by trematode worms, whereas chloroquine, another piperazine derivative, is used to treat malaria [79].

12.7.5 Disinfection and Sanitation

Piperazine derivatives have also been used in disinfection and sanitation applications, such as in the treatment of water and wastewater. For example, piperazine-based disinfectants have been shown to be effective against a wide range of microorganisms, including bacteria, viruses, and fungi [80].

12.7.6 As Anticancerous Agents

Piperazine derivatives have shown promise as potential anticancer agents due to their ability to induce apoptosis, inhibit cell proliferation, and inhibit tumor growth. They can be used to treat solid tumors in targeted and combination therapy and as a personalized medicine.

Treatment of solid tumors: Piperazine derivatives have shown activity against a variety of solid tumors, including lung, breast, colon, ovarian, and pancreatic cancer. These compounds may be used as single agents or in combination with other anticancer drugs to enhance their effectiveness [81].

Samala *et al.* [82] developed a one-pot synthesis process to create the compounds, and multiple spectroscopic methods were used to establish their structural details. Three human cancer cell lines, MCF-7 (breast cancer), A549 (lung cancer), and HepG2 (hepatocellular carcinoma), were used to test the anticancer activity of the synthesized compounds (liver cancer). With IC_{50} values in the micromolar range, the findings demonstrated that some of the compounds had strong anticancer action [82].

Akl *et al.* [83] synthesized a series of compounds and evaluated their CDK1 inhibitory activity using a biochemical assay. The most promising compounds were then tested for their *in vitro* anticancer activity against two pancreatic cancer cell lines, AsPC-1 and BxPC-3. In this study, molecular docking studies were exploited to investigate the binding mode of the compounds to the active site of CDK1. The results of the docking studies suggested that the synthesized compounds could inhibit the activity of CDK1 by binding to its active site. The results showed that the compounds exhibited potent anticancer activity with IC_{50} values in the nanomolar range [83].

Targeted therapy: Some piperazine derivatives have shown activity against specific cancer targets, such as the Src/FAK and PI3K/Akt/mTOR signaling pathways. These compounds may be used as targeted therapies to inhibit specific cancer pathways and reduce side effects associated with traditional chemotherapy. Sahin and Durdagi [84] identified new piperazine-based PARP1 inhibitors using text mining and integrated molecular modeling approaches. In the current study, a set of piperazine-based compounds as potential PARP1 inhibitors was synthesized, followed by further analyses to validate their inhibitory activity. They found that one of the selected compounds showed strong inhibitory activity against PARP1 [84].

Combination therapy: Piperazine derivatives have shown synergistic effects when used in combination with other anticancer agents, such as cisplatin and doxorubicin. These combinations may enhance the effectiveness of existing treatments and improve outcomes for cancer patients [85].

12.8 In Veterinary Medicine

Piperazine derivatives have the potential to be effective against bacterial infections in livestock and companion animals, particularly those caused by Gram-positive bacteria such as *S. aureus*. Additionally, some piperazine derivatives have shown activity against MDR bacteria, making them promising candidates for the treatment of antibiotic-resistant infections. Studies suggested [86] that piperazine derivatives have potential applications in veterinary medicine and could help to address the growing problem of antibiotic resistance in animal health. Piperazine antimicrobial agents have shown potential applications in veterinary medicine, particularly in the treatment of bacterial infections in livestock and companion animals. Piperazine derivatives have been used as antimicrobial agents in veterinary medicine due to their broad-spectrum activity against a range of microorganisms. The following are some examples of applications of piperazine derivatives in veterinary medicine:

12.8.1 Treatment of Bacterial Infections in Livestock

Piperazine derivatives such as enrofloxacin and ciprofloxacin have been used to treat bacterial infections in livestock, including respiratory infections in pigs and poultry, and urinary tract infections in cattle. These drugs are effective against a range of Gram-negative and Gram-positive bacteria and are often used as a first-line treatment in veterinary medicine [87].

12.8.2 Treatment of Parasitic Infections in Pets

Piperazine derivatives such as pyrantel, pamoate, and praziquantel have been used to treat parasitic infections in pets, including roundworms, hookworms, and tapeworms. These drugs are effective against a range of parasitic worms and are often used in combination with other drugs for more effective treatment [88].

12.8.3 Use as a Disinfectant

Piperazine derivatives such as benzalkonium chloride and dodecyl dimethyl ammonium chloride have been used as disinfectants in veterinary medicine to prevent the spread of disease-causing microorganisms in animal housing and equipment. These compounds are effective against a wide range of bacteria, fungi, and viruses and are often used in combination with other disinfectants for more effective control of microbial contamination [89].

12.9 In Agriculture

Piperazine derivatives have also shown potential as antimicrobial agents in agriculture, particularly in the control of plant pathogens and in the preservation of crops. Studies suggest that piperazine derivatives have potential applications as antimicrobial agents in agriculture, particularly in the control of plant pathogens and preservation of crops. For example, in avocado crops, rhizospheric and endophytic bacteria containing piperazine derivatives have been shown to reduce the incidence of white root rot disease. Piperazine derivatives have also been investigated as quorum sensing inhibitors in the control of plant pathogenic bacteria such as *P. aeruginosa.* Furthermore, piperazine derivatives have been studied for their potential to enhance the activity of existing antifungal agents, such as azoles, against drug-resistant strains of *C. albicans* [90].

12.9.1 Control of Plant Pathogens

A study revealed that a piperazine derivative called 1-(3-methoxypropyl)-piperazine (MPP) showed potent antifungal activity against several plant pathogenic fungi, including *Fusarium oxysporum* and *Botrytis cinerea.* The authors suggested that MPP could be a promising alternative to conventional fungicides for the control of plant pathogens [91].

12.9.2 Preservation of Crops

Noor *et al.* [92] reported a piperazine derivative called 1-(2-hydroxypropyl)-4-methylpiperazine (HMP) showed potent antimicrobial activity against several food spoilage microorganisms, including *E. coli* and *S. aureus*. The authors suggested that HMP could be used as a natural preservative in the food industry to extend the shelf life of fresh produce [92].

12.9.3 Control of Nematodes

Nicolay *et al.* [93] investigated the mode of action of the compounds and found that PF1022A and piperazine act on different targets within the nematode nervous system. PF1022A targets GABA receptors, whereas piperazine acts on nicotinic acetylcholine receptors. This study suggests that the combination of PF1022A and piperazine has potential as a novel anthelmintic therapy for nematode infections, particularly those caused by drug-resistant strains. The findings also highlight the importance of understanding the mode of action of anthelmintic compounds to optimize their effectiveness and minimize the development of drug resistance [93].

12.10 Challenges and Future Prospects

Piperazine derivatives have shown promising antimicrobial activity against a wide range of pathogens, including bacteria, fungi, and protozoa. However, there are several challenges that essentially are addressed to fully comprehend their efficacy as antimicrobial agents [48].

One major challenge is the development of resistance to piperazine derivatives. As with other antimicrobial agents, prolonged and widespread use of piperazine derivatives can lead to the emergence of resistant strains of microorganisms. Hence, it is significant to develop approaches to minimize the development of resistance, such as the use of combination therapies and the progress of novel derivatives with different mechanisms of action.

Another challenge is the toxicity of piperazine derivatives. Some piperazine derivatives have been found to have toxic effects on mammalian cells, which can limit their clinical use. Therefore, it is important to identify derivatives with low toxicity profiles and to conduct thorough safety studies before their use in humans.

Additionally, there is a requirement to grow more competent synthetic procedures for the development of piperazine derivatives. The synthesis of piperazine derivatives can be complex and time-consuming, which can limit their availability for research and development. Therefore, it is important to develop new synthetic routes that are more efficient and cost-effective.

Despite these challenges, the potential applications of piperazine derivatives in the field of antimicrobial therapy are promising. Its derivatives have shown synergistic activity with existing antimicrobial agents, which could enhance their effectiveness and potentially reduce the development of resistance. Additionally, piperazine derivatives have potential applications in agriculture for the control of plant diseases caused by bacteria and fungi [94–96].

12.11 Conclusion

Piperazine derivatives have demonstrated broad-spectrum antimicrobial activity against different microorganisms, such as bacteria, fungi, and protozoan parasites. The mechanisms of action of these derivatives are complex and involve inhibition of cell wall synthesis, disruption of cell membrane integrity, inhibition of nucleic acid synthesis, and interference with essential metabolic processes. Additionally, piperazine derivatives have shown potential as synergistic agents, enhancing the effectiveness of existing antimicrobial agents against certain pathogens. Moreover, there is a growing interest in the use of piperazine derivatives in agriculture for the control of plant diseases caused by bacteria and fungi, and they have potential as quorum sensing inhibitors and as enhancers of the activity of existing antifungal agents. Piperazine derivatives have great potential as antimicrobial agents in various fields, including medicine, agriculture, and industry. However, further research is needed to fully understand the mechanisms underlying their activity, optimize their efficacy and safety, and identify the most promising compounds for further development.

AI disclosure statement: The author acknowledges the use of AI-based tools for minor language editing or grammatical correction. However, all intellectual contributions, analyses, and conclusions presented in this chapter are entirely the author's own.

References

1. Kumar, A., Khan, I.A., Koul, S., Piperazine: A valuable scaffold for the development of diverse therapeutic agents. *Eur. J. Med. Chem.*, 159, 56–88, 2018 Nov 5. doi: 10.1016/j.ejmech.2018.08.038. Epub 2018 Aug 21. PMID: 30196044.
2. Bax, B.D., Chan, P.F., Eggleston, D.S., Fosberry, A., Gentry, D.R., Gorrec, F., Giordano, I., Hann, M.M., Hennessy, A., Hibbs, M., Huang, J., Jones, E., Jones, J., Brown, K.K., Lewis, C.J., May, E.W., Saunders, M.R., Type IIA topoisomerase inhibition by a new class of antibacterial agents. *Nature*, 466, 7309, 935–940, 2010. doi:10.1038/nature09197.
3. Singh, O., Spitzfaden, C.E., Shen, C., Shillings, A.J., Theobald, A.J., Wohlkonig, A., Pearson, N.D., Gwynn, M.N., Type IIA topoisomerase inhibition by a new class of antibacterial agents. *Nature*, 466, 7309, 935–40, 2010 Mar 25. doi: 10.1038/nature09197. PMID: 20725040.
4. Dorababu, A., Recent update on antibacterial and antifungal activity of quinoline scaffolds. *Arch. Pharm.*, *354*, 3, 2000232, 2021.
5. Shinde, R.R., Gaikwad, D., Farooqui, M., Synthesis and antimicrobial activity of 2-(4-(benzo [d] thiazol-5-ylsulfonyl) piperazine-1-yl)-N-substituted acetamide derivatives. *J. Heterocycl. Chem.*, *57*, 11, 3907–3917, 2020.
6. Kahlon, G., Lira, R., Masvlov, N., Pompa, E., Brar, N., Eagon, S., Mallari, J.P., Structure guided development of potent piperazine-derived hydroxamic acid inhibitors targeting falcilysin. *Bioorg. Med. Chem. Lett.*, *32*, 127683, 2021.
7. Prasad, H.N., Ananda, A.P., Lohith, T.N., Prabhuprasad, P., Jayanth, H.S., Krishnamurthy, N.B., Mallu, P., Design, synthesis, molecular docking and DFT computational insight on the structure of Piperazine sulfynol derivatives as a new antibacterial contender against superbugs MRSA. *J. Mol. Struct.*, *1247*, 131333, 2022.
8. Vitaku, E., Smith, D.T., Njardarson, J.T., Analysis of the structural diversity, substitution patterns, and frequency of nitrogen heterocycles among US FDA approved pharmaceuticals: miniperspective. *J. Med. Chem.*, *57*, 24, 10257–10274, 2014.
9. Huang, X., Liu, L., Liu, Y., Wang, H., Synthesis of piperazine derivatives *via* α-halocarbonyl compounds. *Synth. Commun.*, 49, 7, 839–848, 2019.
10. Zheng, Y.N., Zheng, H., Li, T., Wei, W.T., Recent Advances in Copper-Catalyzed C– N Bond Formation Involving N-Centered Radicals. *ChemSusChem*, *14*, 24, 5340–5358, 2021.
11. Wu, H.Q., Yang, K., Chen, X.Y., Arulkumar, M., Wang, N., Chen, S.H., Wang, Z.Y., A 3, 4-dihalo-2 (5 H)-furanone initiated ring-opening reaction of DABCO in the absence of a metal catalyst and additive and its application in a one-pot two-step reaction. *Green Chem.*, *21*, 14, 3782–3788, 2019.
12. Jin, L., Wang, Y., Gao, H., Yang, H., Recent advances in heterocyclic compounds as antimicrobial agents. *Eur. J. Med. Chem.*, 138, 761–777, 2017.

13. Alinezhad, H., Yavari, H., Salehian, F., Recent advances in reductive amination catalysis and its applications. *Curr. Org. Chem.*, *19*, 11, 1021–1049, 2015.
14. Timsina, Y.N., Gupton, B.F., Ellis, K.C., Palladium-catalyzed C–H amination of C (sp2) and C (sp3)–H bonds: Mechanism and scope for N-based molecule synthesis. *ACS Catal.*, *8*, 7, 5732–5776, 2018.
15. Saleh, N.A., *et al.*, Synthesis of some novel benzylpiperazine derivatives with expected anticancer activity. *J. Heterocycl. Chem.*, 56, 2, 536–541, 2019.
16. Le Thanh, L., *et al.*, Synthesis of new piperazine derivatives *via* ozonolysis of benzylamines. *J. Chem.*, 2018, 1–8, 2018.
17. Kalsi, S., *et al.*, Copper-catalyzed oxidative synthesis of piperazine derivatives under ambient conditions. *J. Org. Chem.*, 82, 21, 11640–11649, 2017.
18. Sajadikhah, S.S. and Nassiri, M., Recent developments in the synthesis of piperazines (microreview). *Chem. Heterocycl. Compd.*, *57*, 905–907, 2021.
19. Boas, U., Brask, J., Jensen, K.J., Backbone amide linker in solid-phase synthesis. *Chem. Rev.*, *109*, 5, 2092–2118, 2009.
20. Wu, H., Li, H., Fang, Z., Hydrothermal amination of biomass to nitrogenous chemicals. *Green Chem.*, *23*, 18, 6675–6697, 2021.
21. Arias-Gómez, A., Godoy, A., Portilla, J., Functional pyrazolo [1, 5-a] pyrimidines: Current approaches in synthetic transformations and uses as an antitumor scaffold. *Molecules*, *26*, 9, 2708, 2021.
22. Franzén, R.G., Recent advances in the preparation of heterocycles on solid support: a review of the literature. *J. Comb. Chem.*, *2*, 3, 195–214, 2000.
23. Dandia, A., Khan, S., Parewa, V., Sharma, A., Kumawat, B., Modern Synthesis of Bioactive Heterocycles *via* IMCR Modification, in: *Multicomponent Reactions*, pp. 229–280, CRC Press, Boca Raton, FL, USA, 2017.
24. Murugesh, V., Bruneau, C., Achard, M., Sahoo, A.R., Sharma, G.V., Suresh, S., Ruthenium catalyzed β-C (sp 3)–H functionalization on the 'privileged' piperazine nucleus. *Chem. Commun.*, *53*, 75, 10448–10451, 2017.
25. Jida, M. and Ballet, S., Efficient one-pot synthesis of enantiomerically pure N-protected-α-substituted piperazines from readily available α-amino acids. *New J. Chem.*, *42*, 3, 1595–1599, 2018.
26. Żesławska, E., Kincses, A., Spengler, G., Nitek, W., Wyrzuc, K., Kieć-Kononowicz, K., Handzlik, J., The 5-aromatic hydantoin-3-acetate derivatives as inhibitors of the tumour multidrug resistance efflux pump P-glycoprotein (ABCB1): Synthesis, crystallographic and biological studies. *Bioorg. Med. Chem.*, 24, 12, 2815–2822, 2016.
27. Bosc, J.J., Jarry, C., Carpy, A., Panconi, E., Descas, P., Synthesis and antidepressant activity of 5-(1-aryl-4-piperazino) methyl-2-amino-2-oxazolines. *Eur. J. Med. Chem.*, *27*, 5, 437–442, 1992.
28. Andrew, F.P. and Ajibade, P.A., Synthesis, characterization and anticancer studies of bis (1-phenylpiperazine dithiocarbamato) Cu (II), Zn (II) and Pt (II) complexes: Crystal structures of 1-phenylpiperazine dithiocarbamato-S, S′ zinc (II) and Pt (II). *J. Mol. Struct.*, *1170*, 24–29, 2018.

29. Jia, J.M., Liu, F., Xu, X.L., Guo, X.K., Jiang, F., Cherfaoui, B., You, Q.D., Synthesis and evaluation of a novel class Hsp90 inhibitors containing 1-phenylpiperazine scaffold. *Bioorg. Med. Chem. Lett.*, *24*, 6, 1557–1561, 2014.
30. Takahashi, T., Sakuraba, A., Hirohashi, T., Shibata, T., Hirose, M., Haga, Y., Sato, N., Novel potent neuropeptide Y Y5 receptor antagonists: Synthesis and structure–activity relationships of phenylpiperazine derivatives. *Bioorg. Med. Chem.*, *14*, 22, 7501–7511, 2006.
31. Wang, L., Wang, T., Yang, B., Chen, Z., Yang, H., Design, synthesis, and anti-allergic activities of novel (R)(–)-1-[(4-chlorophenyl) phenyl methyl] piperazine derivatives. *Med. Chem. Res.*, 21, 124–132, 2012.
32. Mallikarjuna, S.M., Padmashali, M.S.M.B., Chandrashekharappa, S., Sandeep, C., Synthesis, anticancer and antituberculosis studies for [1-(4-chlorophenyl) Cyclopropyl](Piperazine-yl) Methanone derivates. *Int. J. Pharm. Pharm. Sci.*, 6, 7, 423–427, 2014.
33. Jadala, R.S., Singh, N., Sharma, V., Kumar, S., Synthesis and antimicrobial evaluation of novel piperazine derivatives. *J. Pharm. Anal. Res.*, 8, 4, 205–214, 2019.
34. Majer, J., Paljevac, M., Žagar, E., Kovačič, S., Krajnc, P., Functionalization of 2-hydroxyethyl methacrylate-based polyHIPEs: effect of the leaving group. *React. Funct. Polym.*, *109*, 99–103, 2016.
35. Boido, C.C. and Sparatore, F., Synthesis and preliminary pharmacological evaluation of some cytisine derivatives. *Il Farmaco*, *54*, 7, 438–451, 1999.
36. Mohan, N.R., Sreenivasa, S., Manojkumar, K.E., Rao, T., Thippeswamy, B.S., Suchetan, P.A., Synthesis, antibacterial, anthelmintic and anti-inflammatory studies of novel methylpyrimidine sulfonyl piperazine derivatives. *J. Braz. Chem. Soc.*, *25*, 1012–1020, 2014.
37. Bungard, C.J., Williams, P.D., Schulz, J., Wiscount, C.M., Holloway, M.K., Loughran, H.M., McCauley, J.A., Design and synthesis of piperazine sulfonamide cores leading to highly potent HIV-1 protease inhibitors. *ACS Med. Chem. Lett.*, 8, 12, 1292–1297, 2017.
38. Lee, Y.B., Gong, Y.D., Yoon, H., Ahn, C.H., Jeon, M.K., Kong, J.Y., Synthesis and anticancer activity of new 1-[(5 or 6-substituted 2-alkoxyquinoxalin-3-yl) aminocarbonyl]-4-(hetero) arylpiperazine derivatives. *Bioorg. Med. Chem.*, 18, 22, 7966–7974, 2010.
39. Hafeez, F., Zahoor, A.F., Rasul, A., Mansha, A., Noreen, R., Raza, Z., El-Hiti, G.A., Ultrasound-Assisted Synthesis and *In Silico* Modeling of Methanesulfonyl-Piperazine-Based Dithiocarbamates as Potential Anticancer, Thrombolytic, and Hemolytic Structural Motifs. *Molecules*, *27*, 15, 4776, 2022.
40. Mishra, R., Sharma, P.K., Verma, P.K., Tomer, I., Mathur, G., Dhakad, P.K., Biological potential of thiazole derivatives of synthetic origin. *J. Heterocycl. Chem.*, 54, 4, 2103–2116, 2017.

41. Zhang, R.H., Guo, H.Y., Deng, H., Li, J., Quan, Z.S., Piperazine skeleton in the structural modification of natural products: A review. *J. Enzyme Inhib. Med. Chem.*, *36*, 1, 1165–1197, 2021.
42. Hu, Y.Q., Gao, C., Zhang, S., Xu, L., Xu, Z., Feng, L.S., Zhao, F., Quinoline hybrids and their antiplasmodial and antimalarial activities. *Eur. J. Med. Chem.*, *139*, 22–47, 2017.
43. Ostrowska, K., Coumarin-piperazine derivatives as biologically active compounds. *Saudi Pharm. J.*, *28*, 2, 220–232, 2020.
44. Singh, N., Shreshtha, A.K., Thakur, M.S., Patra, S., Xanthine scaffold: Scope and potential in drug development. *Heliyon*, *4*, 10, e00829, 2018.
45. Liu, L., Zou, D., Zhang, Y., Zhang, Q., Feng, Y., Guo, Y., Liu, M., Pharmaceutical salts/cocrystals of enoxacin with dicarboxylic acids: Enhancing *in vitro* antibacterial activity of enoxacin by improving the solubility and permeability. *Eur. J. Pharm. Biopharm.*, *154*, 62–73, 2020.
46. Gan, L.L., Fang, B., Zhou, C.H., Synthesis of azole-containing piperazine derivatives and evaluation of their antibacterial, antifungal and cytotoxic activities. *Bull. Korean Chem. Soc.*, *31*, 12, 3684–3692, 2010.
47. Patel, V., Keum, Y.S., Won Park, S., Medicinal chemistry discoveries among 1, 3, 5-triazines: Recent advances (2000-2013) as antimicrobial, anti-TB, anti-HIV and antimalarials. *Mini-Rev. Med. Chem.*, *14*, 9, 768–789, 2014.
48. Jalageri, M.D., Nagaraja, A., Puttaiahgowda, Y.M., Piperazine based antimicrobial polymers: A review. *RSC Adv.*, *11*, 25, 15213–15230, 2021.
49. Zhang, Y., Zhan, Y.Z., Ma, Y., Hua, X.W., Wei, W., Zhang, X., Wang, B.L., Synthesis, crystal structure and 3D-QSAR studies of antifungal (bis-) 1, 2, 4-triazole Mannich bases containing furyl and substituted piperazine moieties. *Chin. Chem. Lett.*, *29*, 3, 441–446, 2018.
50. Naidu, K.M., Srinivasarao, S., Agnieszka, N., Ewa, A.K., Kumar, M.M.K., Sekhar, K.V.G.C., Seeking potent anti-tubercular agents: Design, synthesis, anti-tubercular activity and docking study of various ((triazoles/indole)-piperazin-1-yl/1, 4-diazepan-1-yl) benzo [d] isoxazole derivatives. *Bioorg. Med. Chem. Lett.*, *26*, 9, 2245–2250, 2016.
51. Tahir, S., Mahmood, T., Dastgir, F., Haq, I.U., Waseem, A., Rashid, U., Design, synthesis and anti-bacterial studies of piperazine derivatives against drug resistant bacteria. *Eur. J. Med. Chem.*, *166*, 224–231, 2019.
52. Brown, P. and Dawson, M.J., Development of new polymyxin derivatives for multi-drug resistant Gram-negative infections. *J. Antibiot.*, *70*, 4, 386–394, 2017.
53. Kathiravan, M.K., Salake, A.B., Chothe, A.S., Dudhe, P.B., Watode, R.P., Mukta, M.S., Gadhwe, S., The biology and chemistry of antifungal agents: a review. *Bioorg. Med. Chem.*, *20*, 19, 5678–5698, 2012.
54. Yamamoto, Y., Shimizu, E., Ban, K., Wada, Y., Mizusaki, T., Yoshimura, M., Sajiki, H., Facile hydrogenative deprotection of N-benzyl groups using a mixed catalyst of palladium and niobic acid-on-carbon. *ACS Omega*, *5*, 6, 2699–2709, 2020.

55. Ibezim, E., Duchowicz, P.R., Ortiz, E.V., Castro, E.A., QSAR on arylpiperazine derivatives with activity on malaria. *Chemom. Intell. Lab. Syst.*, *110*, 1, 81–88, 2012.
56. Kuettel, S., Zambon, A., Kaiser, M., Brun, R., Scapozza, L., Perozzo, R., Synthesis and evaluation of antiparasitic activities of new 4-[5-(4-phenoxyphenyl)-2 H-pyrazol-3-yl] morpholine derivatives. *J. Med. Chem.*, *50*, 23, 5833–5839, 2007.
57. Silva, G.N., Schuck, D.C., Cruz, L.N., Moraes, M.S., Nakabashi, M., Gosmann, G., Gnoatto, S.C., Investigation of antimalarial activity, cytotoxicity and action mechanism of piperazine derivatives of betulinic acid. *Trop. Med. Int. Health*, *20*, 1, 29–39, 2015.
58. Sheena Mary, Y., Shyma Mary, Y., Temiz-Arpaci, O., Yadav, R., Celik, I., DFT, docking, MD simulation, and vibrational spectra with SERS analysis of a benzoxazole derivative: An anti-cancerous drug. *Chem. Pap.*, *75*, 8, 4269–4284, 2021.
59. Ahmed, M.F., Santali, E.Y., El-Haggar, R., Novel piperazine–chalcone hybrids and related pyrazoline analogues targeting VEGFR-2 kinase; design, synthesis, molecular docking studies, and anticancer evaluation. *J. Enzyme Inhib. Med. Chem.*, *36*, 1, 308–319, 2021.
60. Raju, H., Chandrappa, S., S Prasanna, D., Ananda, H., S Nagamani, T., M Byregowda, S., S Rangappa, K., Synthesis, characterization and *in vitro* antiproliferative effects of novel 5-amino pyrazole derivatives against breast cancer cell lines. *Recent Pat. Anti-Cancer Drug Discovery*, *6*, 2, 186–195, 2011.
61. Mao, Z.W., Zheng, X., Lin, Y.P., Hu, C.Y., Wang, X.L., Wan, C.P., Rao, G.X., Design, synthesis and anticancer activity of novel hybrid compounds between benzofuran and N-aryl piperazine. *Bioorg. Med. Chem. Lett.*, *26*, 15, 3421–3424, 2016.
62. Gong, Y.L., Dai, P., Chen, Y., Ding, S., Liu, J., Synthesis, Crystal Structure and Biological Activity of 7-(4-Methylpiperazin-1-Yl)-5-[4-(Trifluoromethyl) Phenyl] pyrazolo [1, 5-a] Pyrimidine-3-Carbonitrile. *Crystallogr. Rep.*, *65*, 1111–1116, 2020.
63. Al-Ghorbani, M., Gouda, M.A., Baashen, M., Alharbi, O., Almalki, F.A., Ranganatha, L.V., Piperazine Heterocycles as Potential Anticancer Agents: A Review. *Pharm. Chem. J.*, *56*, 1, 29–37, 2022.
64. Kumar, S.D., Park, J.H., Kim, H.S., Seo, C.D., Ajish, C., Kim, E.Y., Shin, S.Y., Cationic, amphipathic small molecules based on a triazine-piperazine-triazine scaffold as a new class of antimicrobial agents. *Eur. J. Med. Chem.*, *243*, 114747, 2022.
65. Li, L., Li, Z., Guo, N., Jin, J., Du, R., Liang, J., Yu, L., Synergistic activity of 1-(1-naphthylmethyl)-piperazine with ciprofloxacin against clinically resistant Staphylococcus aureus, as determined by different methods. *Lett. Appl. Microbiol.*, *52*, 4, 372–378, 2011.
66. Chen, Y.Y., Gopala, L., Bheemanaboina, R.R.Y., Liu, H.B., Cheng, Y., Geng, R.X., Zhou, C.H., Novel naphthalimide aminothiazoles as potential

multitargeting antimicrobial agents. *ACS Med. Chem. Lett.*, *8*, 12, 1331–1335, 2017.

67. Sui, Y.F., Ansari, M.F., Fang, B., Zhang, S.L., Zhou, C.H., Discovery of novel purinylthiazolylethanone derivatives as anti-Candida albicans agents through possible multifaceted mechanisms. *Eur. J. Med. Chem.*, *221*, 113557, 2021.
68. Siriyong, T., Srimanote, P., Chusri, S., Yingyongnarongkul, B.E., Suaisom, C., Tipmanee, V., Voravuthikunchai, S.P., Conessine as a novel inhibitor of multidrug efflux pump systems in Pseudomonas aeruginosa. *BMC Complementary Altern. Med.*, *17*, 1–7, 2017.
69. Zhou, X., Zhao, L., Liu, X., Li, X., Jia, F., Zhang, Y., Wang, Y., Antimycobacterial and Synergistic Effects of 18β-Glycyrrhetinic Acid or Glycyrrhetinic acid-30-piperazine in Combination with Isoniazid, Rifampicin or Streptomycin against Mycobacterium bovis. *Phytother. Res.*, *26*, 2, 253–258, 2012.
70. Makarov, V., Lechartier, B., Zhang, M., Neres, J., van der Sar, A.M., Raadsen, S.A., Cole, S.T., Towards a new combination therapy for tuberculosis with next generation benzothiazinones. *EMBO Mol. Med.*, *6*, 3, 372–383, 2014.
71. Kuang, W., Zhang, H., Wang, X., Yang, P., Overcoming Mycobacterium tuberculosis through small molecule inhibitors to break down cell wall synthesis. *Acta Pharm. Sin. B.*, 2022.
72. Bolla, J.M., Alibert-Franco, S., Handzlik, J., Chevalier, J., Mahamoud, A., Boyer, G., Pagès, J.M., Strategies for bypassing the membrane barrier in multidrug resistant Gram-negative bacteria. *FEBS Lett.*, *585*, 11, 1682–1690, 2011.
73. Piton, J., Vocat, A., Lupien, A., Foo, C.S., Riabova, O., Makarov, V., Cole, S.T., Structure-based drug design and characterization of sulfonyl-piperazine benzothiazinone inhibitors of DprE1 from Mycobacterium tuberculosis. *Antimicrob. Agents Chemother.*, *62*, 10, e00681–18, 2018.
74. Hong, M.N., Nam, K.Y., Kim, K.K., Kim, S.Y., Kim, I., The small molecule '1-(4-biphenylylcarbonyl)-4-(5-bromo-2-methoxybenzyl) piperazine oxalate'and its derivatives regulate global protein synthesis by inactivating eukaryotic translation initiation factor 2-alpha. *Cell Stress Chaperones*, *21*, 485–497, 2016.
75. Bassetto, M., Leyssen, P., Neyts, J., Yerukhimovich, M.M., Frick, D.N., Courtney-Smith, M., Brancale, A., *In silico* identification, design and synthesis of novel piperazine-based antiviral agents targeting the hepatitis C virus helicase. *Eur. J. Med. Chem.*, *125*, 1115–1131, 2017.
76. Rathi, A.K., Syed, R., Shin, H.S., Patel, R.V., Piperazine derivatives for therapeutic use: a patent review (2010-present). *Expert Opin. Ther. Pat.*, *26*, 7, 777–797, 2016.
77. Sanad, S.M. and Mekky, A.E., Synthesis, *in-vitro* antibacterial and anticancer screening of novel nicotinonitrile-coumarin hybrids utilizing piperazine citrate. *Synth. Commun.*, *50*, 10, 1468–1485, 2020.

78. Kharb, R., Bansal, K., Sharma, A.K., A valuable insight into recent advances on antimicrobial activity of piperazine derivatives. *Der Pharma Chem.*, *4*, 6, 2470–2488, 2012.
79. Elmeligy, E., Abdelbaset, A., Elsayed, H.K., Bayomi, S.A., Hafez, A., Abu-Seida, A.M., Khalphallah, A., Oxidative stress in Strongylus spp. infected donkeys treated with piperazine citrate versus doramectin. *Open Vet. J.*, *11*, 2, 238–250, 2021.
80. Kagira, J.M., Kanyari, P.W.N., Munyua, W.K., Waruiru, R.M., Relationship between the prevalence of gastrointestinal nematode infections and management practises in pig herds in Thika District, Kenya, 2017.
81. Eilon, G.F., Gu, J., Slater, L.M., Hara, K., Jacobs, J.W., Tumor apoptosis induced by epoxide-containing piperazines, a new class of anti-cancer agents. *Cancer Chemother. Pharmacol.*, *45*, 3, 2000.
82. Samala, R., Nukala, S.K., Thirukovela, N.S., Dasari, G., Bandari, S., One-Pot Synthesis of Some New Phthalazine-Piperazine-1, 2, 4-Oxadiazole Hybrids: Anticancer Evaluation, Molecular Docking and ADMET Studies. *Polycyclic Aromat. Compd.*, 1–18, 2022.
83. Akl, L., Abd El-Hafeez, A.A., Ibrahim, T.M., Salem, R., Marzouk, H.M.M., El-Domany, R.A., Abou-Seri, S.M., Identification of novel piperazine-tethered phthalazines as selective CDK1 inhibitors endowed with *in vitro* anticancer activity toward the pancreatic cancer. *Eur. J. Med. Chem.*, *243*, 114704, 2022.
84. Sahin, K. and Durdagi, S., Identifying new piperazine-based PARP1 inhibitors using text mining and integrated molecular modeling approaches. *J. Biomol. Struct. Dyn.*, *39*, 2, 681–690, 2021.
85. Kano, Y., Narita, T., Suzuki, K., Akutsu, M., Suda, K., Sakamoto, S., Miura, Y., The effects of ICRF-154 in combination with other anticancer agents *in vitro*. *Br. J. Cancer*, *66*, 2, 281–286, 1992.
86. Bártíková, H., Podlipná, R., Skálová, L., Veterinary drugs in the environment and their toxicity to plants. *Chemosphere*, *144*, 2290–2301, 2016.
87. Magotra, A., Sharma, A., Singh, S., Ojha, P.K., Kumar, S., Bokolia, N., Nandi, U., Physicochemical, pharmacokinetic, efficacy and toxicity profiling of a potential nitrofuranyl methyl piperazine derivative IIIM-MCD-211 for oral tuberculosis therapy *via in-silico–in-vitro–in-vivo* approach. *Pulm. Pharmacol. Ther.*, *48*, 151–160, 2018.
88. Kang, S., Jang, H., Jeong, W., Prevalence of intestinal parasites from dogs in Korea. *Korean J. Vet. Public Health*, *24*, 3, 195–202, 2000.
89. Sayid, A.S.A. and Mohammed, B.A., Prevalence and treatment of Toxoascaris leonina in naturally infected carnivores in old Khartoum Zoo. *Sudan J. Vet. Res.*, *15*, 31–38, 1997.
90. Serban, G., Stanasel, O., Serban, E., Bota, S., 2-Amino-1, 3, 4-thiadiazole as a potential scaffold for promising antimicrobial agents. *Drug Des. Dev. Ther.*, 1545–1566, 2018.

91. Xiang, M., Song, Y.L., Ji, J., Zhou, X., Liu, L.W., Wang, P.Y., Yang, S., Synthesis of novel 18 β-glycyrrhetinic piperazine amides displaying significant *in vitro* and *in vivo* antibacterial activities against intractable plant bacterial diseases. *Pest. Manage. Sci.*, *76*, 9, 2959–2971, 2020.
92. Noor, H., David, I.G., Jinga, M.L., Popa, D.E., Buleandra, M., Iorgulescu, E.E., Ciobanu, A.M., State of the Art on Developments of (Bio) Sensors and Analytical Methods for Rifamycin Antibiotics Determination. *Sensors*, *23*, 2, 976, 2023.
93. Nicolay, F., Harder, A., von Samson-Himmelstjerna, G., Mehlhorn, H., Synergistic action of a cyclic depsipeptide and piperazine on nematodes. *Parasitol. Res.*, *86*, 982–992, 2000.
94. World Health Organization, *Antimicrobial Resistance: Global Report on Surveillance 2014*. WHO Press, Geneva, Switzerland, 2014.
95. Richard, A., Jayakumar, R., Suresh, G., Devi, K., Piperazine derivatives as promising antimicrobial agents: A comprehensive review. *J. Heterocycl. Chem.*, 56, 10, 2850–2865, 2019.
96. Prasad, H., Ananda, A., Mukarambi, A., Gaonkar, N., Sumathi, S., Spoorthy, H., Mallu, P., Design, synthesis, and anti-bacterial activities of piperazine based phthalimide derivatives against superbug-Methicillin-Resistant Staphylococcus aureus. *Curr. Chem.*, 12, 1, 65–78, 2023.

13

Synthetic Drug Discovery Strategy of Benzimidazole Derivatives and Their Applications

Prasad Vijay Korake, Diwakar Chauhan and Subhalaxmi Pradhan*

Department of Chemistry and Sustainability, School of Basic Sciences, Galgotias University, Greater Noida, U.P., India

Abstract

In the extensive world of hetero compound, benzimidazole analogs are discovered as fashionable structures, which are used for drug development in the field of pharmaceutical chemistry. The benzimidazole is a preferred compound in medicine development due to its distinctive structural characteristics and a wide spectrum of biological activity such as antibacterial, antifungal, and anticancer activities of its derivatives. Due to that biological activity, benzimidazole analogs have a significant impact on medicine. This medicinally valuable medications' ability to effectively treat various infections and other conditions sparked the creation of several more potent and noteworthy molecules from benzimidazole. In a recent study, benzimidazole derivatives show receptor-binding properties against COVID-19 variant. Parasitic drugs such as mebendazole and flubendazole are now accessible therapy options for SARS-CoV-2. Mebendazole and flubendazole show better receptor-binding properties than standard medication remdesivir and hydroxychloroquine. The current chapter provides broad information on the synthesis and chemistry of various derivatives of benzimidazoles as well as their biological properties in context of drug discovery approach.

Keywords: Heterocyclic, benzimidazole, antibacterial, antifungal, anticancer

**Corresponding author*: subhalaxmipradhan@gmail.com

Divya Bajpai Tripathy, Priyanka Chhabra and Anjali Gupta (eds.) Heterocyclic Compounds as Pharmaceutical Ingredients, (443–474) © 2026 Scrivener Publishing LLC

Abbreviations

ACN	acetonitrile
AcOH	acetic acid
CNBr	cyanogen bromide
DCM	dichloromethane
DDQ	dichloro dicyano benzoquinone
PPA	polyphosphoric acid
DMF	dimethyl formamide
TBAB	tetra butyl ammonium bromide
THF	tetrahydrofuran
$Zn(OTf)_2$	zinc trifluoromethanesulfonate
AD	Alzheimer disease
TEAB	tetraethylammonium bromide
IC_{50}	inhibitory concentration half-maximal
EtOH	ethanol
DIPEA	diisopropylethylamine
DCM	dichloromethane
EAC	Ehrlich ascites carcinoma
TP	trypan blue
ATP	adenosine triphosphate
CAM	chorioallantoic membrane
SRB	sulforhodamine-B

13.1 Introduction

Heterocyclic compounds play a promising role in the metabolism of all living cells. Nitrogenous heterocyclic compounds are very essential for human beings. Benzimidazole has a significant impact on industrial and biological processes, particularly on the whole class of heterocyclic compounds based on nitrogen. It is quite usual to use the benzimidazole ring system. Due of their widespread occurrence in biologically active chemicals, these substructures are frequently referred to as "privileged." Hoebrecker was the first to synthesize benzimidazole. The fourth and fifth positions (Figure 13.1) of the imidazole ring fused with benzene to form the benzimidazole molecule with chemical formula $C_7H_6N_2$. It is soluble in polar solvent. They possess characteristics of both bases and acids; NH group in this molecule is weakly basic and extremely acidic, which include the capacity to create salts [1]. Goodman and Nancy published their initial

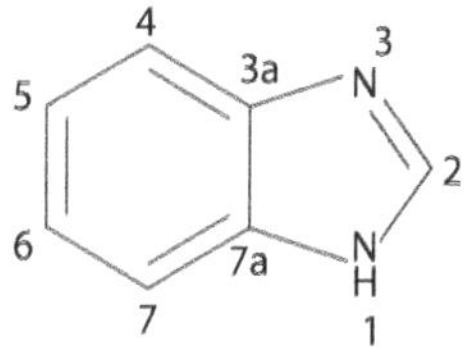

Figure 13.1 Structure of benzimidazole.

investigation into the biological properties of benzimidazole. It is helpful in the pharmaceutical industry for the creation of innovative therapeutic molecules. Additionally, benzimidazole is a crucial pharmacophore, a favored substructure in medicinal chemistry that serves as a crucial component for a variety of natural functions [2].

It is also reserved special position in medicinal chemistry field for a variety of natural products, important constituents of which include benzimidazole derivatives, such as purine, histamine, histidine, and nucleic acid. As a result, incorporating the benzimidazole nucleus into the preparation or synthesis of novel benzimidazole derivatives has always attracted the interest of many medicinal chemists and has shown to be an essential analog. Benzimidazole derivatives are used against the inflammation, bacteria, microbes, fungus, virus infection, antihelminthic, anticonvulsant, anticancer, antihypertensive, and antiphrastic agents in synthetic drug research [3]. Mosquitoes are insects found all over the world. *Aedes* mosquitoes are responsible for spreading diseases, including bronze john, dengue, chikungunya, and other dangerous arbo infections. Bandopadhyay *et al.*, in 2014, synthesized the derivatives of benzimidazole and benzothiazole-linked phosphoramidates, phosphoramidothioates, phenylphosphoramidates, and phenylphosphoramidothioates (Figure 13.2) and studied their mosquitoes-killing ability [4]. The synthesized molecules were initially tested at concentrations ranging from 40-50 mg/L for their ability to kill mosquito larvae of *Aedes albopictus* and *Culex quinquefasciatus*. A 100% mortality rate for three of the investigated substances points to potential larvicidal effects. Compound (1) has been found very effective to kill *A. albopictus* and *C. quinquefasciatus*. It is found that Compound (1) showed LC_{50} of 6.42 and 5.25 mg/L after 24 and 48 hours for *A. albopictus*; however, it was found to be 7.01 and 3.88 mg/L in *C. quinquefasciatus* [4].

Alpan *et al.* revealed the study that acetylcholinesterase (AChE) is a crucial molecule involved in the termination of nerve signals by hydrolyzing acetylcholine. Combating neuromuscular disorders such as glaucoma, myasthenia gravis, and Alzheimer disease (AD) is a goal of drug

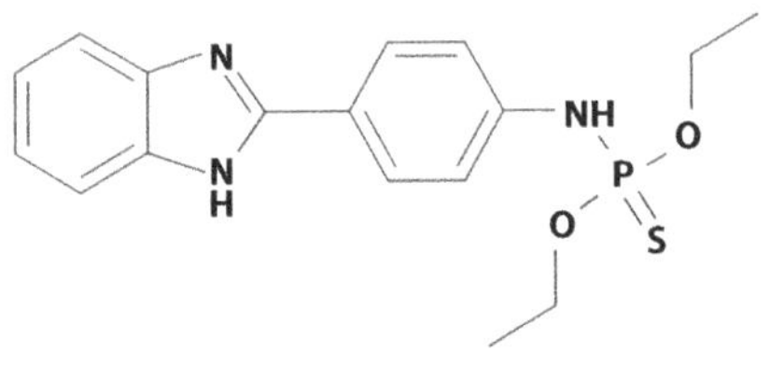

1

Figure 13.2 Benzimidazole-phenyl phosphomidothioate.

development [5]. In order to treat AD, a progressive neurological disease is characterized by neurofibrillary tangles, amyloid plaques, and a loss of localized cholinergic function, in which AChE has been a primary focus. Reminiscence deficits in AD patients are thought to be caused, in part, by a deficiency in cholinergic neurotransmission. Use of AChE inhibitors is one of the compelling methods for enhancing cholinergic transmission. Molecules (2) and (3) in Figure 13.3 were most effective against *Electrophorus electricus* AChE and human AChE, respectively. Additionally, molecular docking experiments and simulations of molecular dynamics were performed by above researchers [5].

Variants of the coronavirus disease, which poses a rare threat to world health, were discovered, and there are changes in the receptor-binding area of spike glycoprotein that resulted in the two versions, known as SARS-CoV-2 B.1.1.7 (alpha) and P.1 (gamma) (SGP) [6]. It is well known that these two variations have a strong affinity for the enzyme, which converts angiotensin. The fast proliferation of these mutant strains makes it difficult and labor-intensive to conduct research and produce new compounds. Vidyasrilekha *et al.* [6] have chosen imidazole and benzimidazole scaffolds

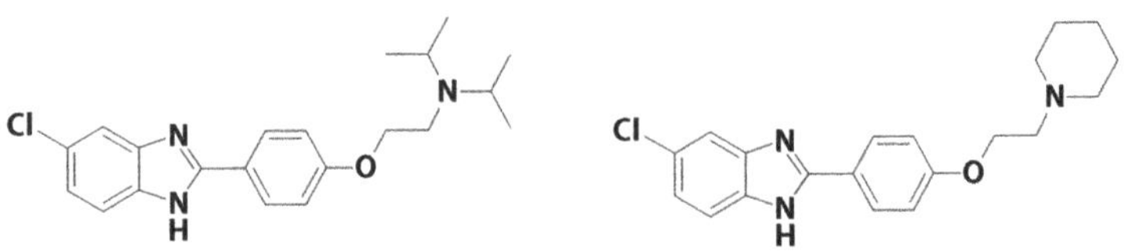

2 3

Figure 13.3 Structure of acetylcholinesterase inhibitors.

for this inquiry because of their improved efficacy against a number of therapeutic targets and specific structural properties and electron-rich potential [6]. Imidazole and benzimidazole-based antiparasitic drugs are repurposed against SARS-CoV-2 alpha and gamma mutant spike glycoproteins in the new work using computational approaches. According to their gliding score and binding free energy, 2 of the 15 compounds that were evaluated—flubendazole and mebendazole—showed efficient bonding qualities to the two receptors (PDB ID: 7NEH and 7NXC). The results are compared to those of hydroxychloroquine and remdesivir, two commonly used drugs. For SARS-CoV-2 mutant lineages, metronidazole and flubendazole have emerged as viable treatment choices. The stability of the flubendazole throughout a 100-ns MD simulation using GROMACS software further confirmed the drug's cutting edge. The current review work will focus on different novel analogs of benzimidazole, synthesis methods, and their versatile application as antifungal, antimicrobial, and anticancer agent in the present scenario.

13.2 Benzimidazole as an Antifungal Agent

Khabnadideh *et al.* discovered N-alkylated series of benzimidazole [7]. The benzimidazole rings refluxed with appropriate bromo nonane in acetonitrile in presence of sodium hydroxide, potassium carbonate, and TEAB for 48 hours to produce nonyl benzimidazole (Scheme 13.1).

The pharmacological activities of the synthesized derivative were assessed for a total of 11 standard strains of fungi, including *Candida albicans* (ATCC 10261, CBS 1912), *Candida krusei* (ATCC 6258), *Candida glabrata* (ATCC 90030), *Candida parapsilosis* (ATCC 4344), *Candida dubliniensis* (CBS 8500), *Candida tropicalis* (ATCC 750), *Cryptococcus neoformans* (ATCC 11). Additionally, tests using morphological and physiological data were used to identify the dermatophytes *Microsporum canis*

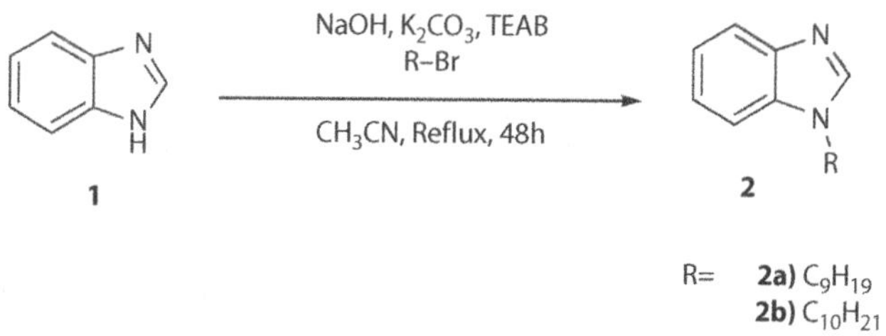

Scheme 13.1 Synthesis of N-alkylated series of benzimidazole derivatives.

and *Epidermophyton floccosum*, and the inhibitory actions of the aforementioned substances against these organisms were also investigated [7]. When tested against the clinical species of *Candida*, Compounds **2a** and **2b**, at 0.5 to 256 and 2 to 256 g/mL (Scheme 13.1) demonstrated the strongest antifungal activity. Aside from fluconazole-resistant species, these two compounds were also efficient. It also demonstrated the strongest antifungal activity against *Aspergillus* species with a dose of 16 to 256 μg/mL.

Nishad *et al.* have synthesized 18 analogs of benzimidazole to investigate the connection between the antifungal action of benzimidazoles and the number of 1,3-diazacyclopenta-2,4-diene rings and also the nature of the substituted alkyl group of an analog [8]. They introduced a novel approach using benzimidazole surrounded by linkers with various structural moieties. The bis-benzimidazole derivatives were prepared by condensation of Compounds A and B (Scheme 13.2). Compound A (Path 1) is synthesized by using following step, dinitrobenzoic acid on reaction with sulfurous oxychloride to make dinitrobenzene acid chloride after N,O-dimethyl hydroxylamine hydrochloric acid was added to prepare Weinreb amide. The 3,4-diamino-N-methoxy-N-methylbenzamide was prepared by the reduction of Weinreb amide using Pd/C. The cyclization has been carried out by condensation of 3,4-diamino-N-methoxy-N-methylbenzamide and 4-O-alkylated benzaldehyde to get 2-(4-O-alkylated phenyl)-N-methoxy-N-methyl-^{1}H-benzo[d]imidazole-6-carboxamide. The 2-(4-O-alkylated phenyl)-N-methoxy-N-methyl-^{1}H-benzo[d]imidazole-6-carboxamide was reacted with lithium aluminum hydride to get A (Path 1). Compound B was synthesized by using number of reactions such as alkylation, deprotection, cyclization, and reduction (Path 2). Then, **18** analogs were synthesized by reaction of A with B. R_1 and R_2 groups in different analogs are given below in Scheme 13.2. The fungicidal activity of produced analogs is tested and MIC values obtained ranging from 15.6 to 0.975 μg/mL

2: $R_1 = H$, $R_2 = CH_3$, **3:** $R_1 = H$, $R_2 = C_3H_7$ **4:** $R_1 = H$, $R_2 = C1_2H_{25}$ **5:** $R_1 = CH_3$, $R_2 = CH_3$ **6:** $R_1 = CH_3$, $R_2 = C_2H_5$ **7:** $R_1 = CH_3$, $R_2 = C_3H_7$ **8:** $R_1 = CH_3$, $R_2 = C_5H_{11}$ **9:** $R_1 = C_2H_5$, $R_2 = CH_3$ **10:** $R_1 = C_2H_5$, $R_2 = C_2H_5$ **11:** $R_1 = C_2H_5$, $R_2 = C_3H_7$ **12:** $R_1 = C_2H_5$, $R_2 = C_5H_{11}$ **13:** $R_1 = C_{12}H_{25}$, $R_2 = CH_3$ **14:** $R_1 = CH_2C\,CH$, $R_2 = CH_3$ **15:** $R_1 = CH_2C\,CH$, $R_2 = C_3H_7$ **16:** $R_1 = (CH_2)_2OH$, $R_2 = CH_3$ **17:** $R_1 = (CH_2CH_2O)_2H$, $R_2 = CH_3$ **18:** $R_1 = (C_2HCH_2O)_2H$, $R_2 = CH_3$

The synthesized analogs show MIC values ranging from 0.975 to 15.6 μg/mL. Compound **3** in which $R_1 = H$, $R_2 = C_3H_7$ (Scheme 13.2) shows improved efficacy (7.8 μg/mL) against *C. glabrata* ATCC 2001 (strain H), *C. parapsilosis* ATCC 22019 (strain J), and *Aspergillus nidulans* ATCC 38163 (strain L), and Compounds **2** to **4**, having small chain length in piperazine ring, showed decreased antifungal activity. Compound **5**, having

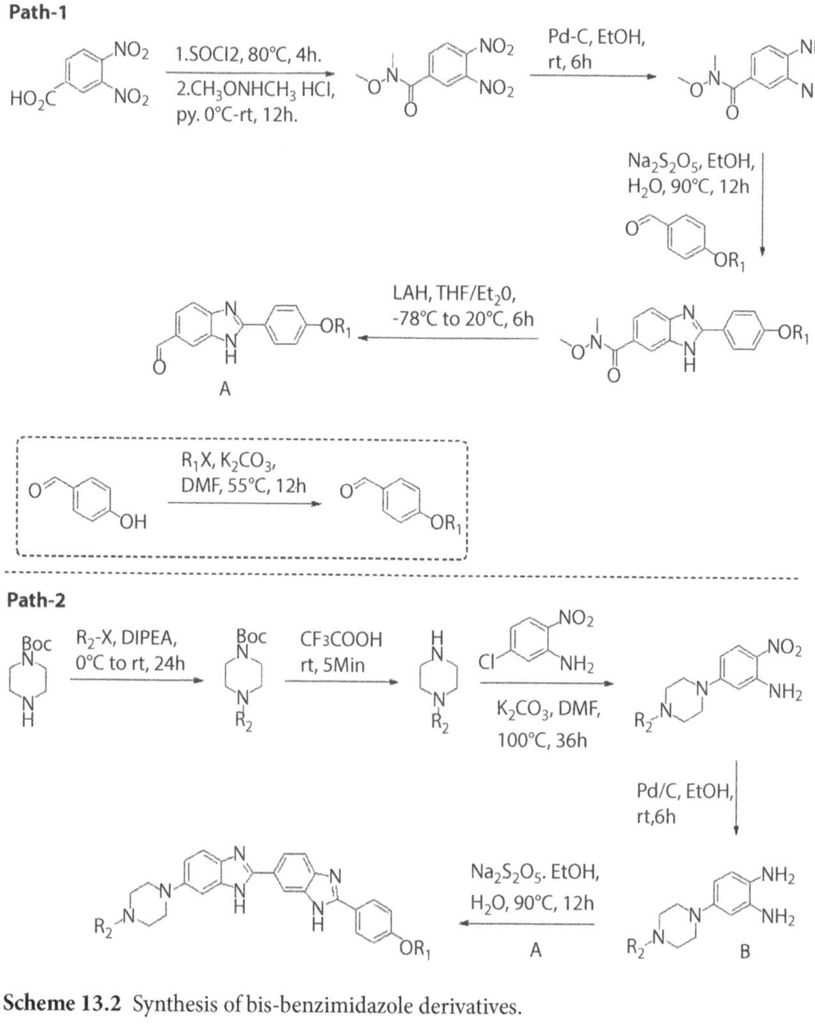

Scheme 13.2 Synthesis of bis-benzimidazole derivatives.

methyl group at piperazine ring, exhibited high efficacy (3.9–7.8 μg/mL) against four of the tested strains (*C. albicans* ATCC MYA-2310(S) (strain E), *C. glabrata* ATCC 2001 (strain H), and *C. parapsilosis* ATCC 22019 (strain J)) [8]. But from Compounds 5 to 8, keeping methoxy group at R_1 in phenyl ring and increasing the carbon chain length of piperazine ring from C1 to C5 exhibited decrease in antifungal activity. (15.6 μg/mL) against *Aspergillus terreus* ATCC MYA-3633 (strain M). In this study they observed that introducing short alkyl chains in place of R_2 and keeping –OH group at phenyl ring exhibited improved activity against different fungal strains. They also observed that increasing alkyl.

Nafiz *et al.* [9] synthesized new ergosterol inhibitors, a new series of 1,3-Benzodiazole-triazole analog taking 4-methyl-1,2-phenylenediamine and methyl-4-formyl benzoate as starting material (Scheme 13.3).

The condensation of 4-methylbenzene-1,2-diamine and methyl 4-formylbenzoate in DMF was carried out in microwave irradiation to get Compound **3**. Compound **3** was heated in microwave with hydrazine hydride and thioisocynate to get Compound **6**. The cyclization is carried out by heating Compound **6** with sodium hydroxide to obtain Compound **7**. Compound **9** achieved by reacting different types of derivatives of Compound **8** in potassium carbonate and dimethyl formamide as solvent. Spectroscopic techniques were used to characterize the chemical structures of the target substances. *In vitro* antifungal test was carried out on synthesized compound against C. glabrata (ATCC 90030), *C. krusei* (ATCC 6258), C. parapsilosis (ATCC 22019), and *C. albicans* (ATCC 24433). Some analogs showed the antifungal activity with MIC values *C. glabrata*: 1.56 μg/mL, *C. krusei*: 1.56 μg/mL, *C. parapsilosis*: 0.78 μg/mL, and *C. albicans*: 0.78 μg/mL. Compound **9** in which $R_1 = -CH_3$, $R_2 = H$, $R_3 = -OH$, $R_4 = OH$ and $R_1 = -C_2H_5$, $R_2 = H$, $R_3 = -OH$, $R_4 = -OH$ significantly inhibited the growth of *Candida* strains (Scheme 13.3). According to cytotoxicity data, some moiety has IC_{50} values against NIH/3T3 that are much greater than their MIC_{50} values [9].

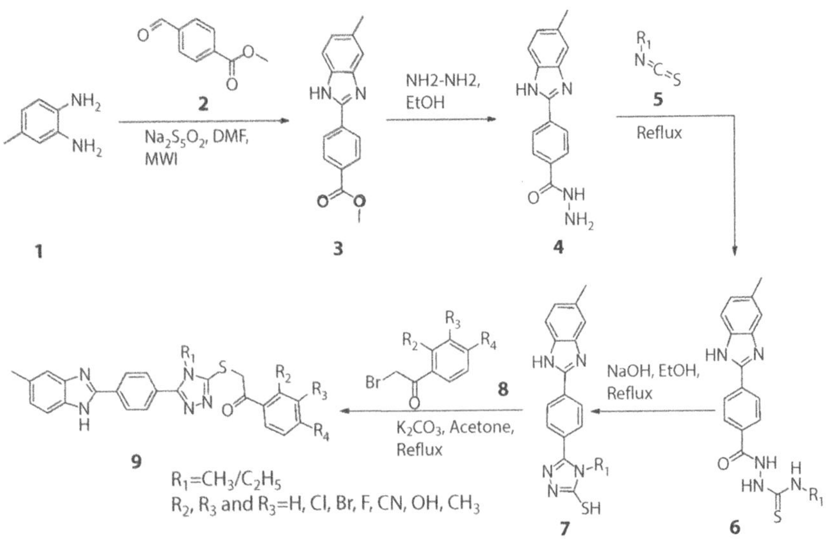

Scheme 13.3 Synthesis of series of 1,3-benzodiazole-triazole analog.

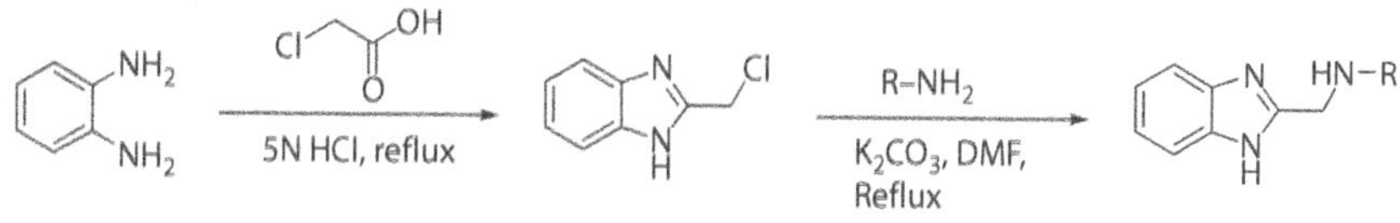

R=$-C_{14}H_{12}N_3Cl$, $-C_{11}H_9N_5O_2$, $-C_{10}H_{10}N_6$, $-C_{15}H_{13}N_3O_2$, $-C_{15}H_{15}N_3$, $-C_{15}H_{12}ClN_3O_2$, $-C_{15}H_{15}N_3$, $-C_{14}H_{12}N_3Cl$,

Scheme 13.4 Synthesis of 2-chloromethyl-1H-benzimidazole analog.

Vijey and Anbhule reported the synthesis of series of 2-chloromethyl-^{1}H-benzimidazole [10]. The chloroacetic acid and o-phenylenediamine refluxed in presence of 5N HCl to get 2-chloromethyl-^{1}H-benzimidazole. After that varies heterocyclic amines were synthesized by reacting with variety of heterocyclic amines in presence of K_2CO_3 and DMF (Scheme 13.4).

The synthesized compounds were tested taking different fungal strains such as *Aspergillus niger*, *C. albicans*, *Aspergillus fumigatus*, and *Penicillium notatum*. The compound R = $-C_{15}H_{13}N_3O_2$, $-C_{15}H_{12}ClN_3O_2$, $-C_{15}H_{15}N_3$ showed good biological activity and R = $-C_1{}^1H_9N_5O_2$, $-C_{10}H_{10}N_6$ showed excellent activity (Scheme 13.4). The researchers concluded that benzimidazole nucleus serves as the structural foundation for a unique class of antifungal drugs in the synthesized analogs [10].

13.3 Benzimidazole as an Antimicrobial Agent

Desai *et al.* [11] reported synthesis pathway of brand-new series of benzimidazole-pyrazol-pyridin methanones derivatives, shown in Scheme 13.5. 2-Acetylbenzimidazole (Compound **4**) was used in the production of benzimidazolyl-chalcones (Compound **6**). Compound 4 was produced in two steps of reaction. In first step 2-hydroxyethylbenzimidazole (3) was produced by condensation of Compound 1 with Compound 2 using the Phillips technique. In second step 2-acetylbenzimidazole was produced by oxidation with potassium dichromate in presence of acetic acid, then latter neutralizing it with ammonia. The Claisen–Schmidt reaction was used to condense this ketone with derivatives of aryl or heteroaryl aldehydes (5). Compound (**6**) was produced by neutralizing the reaction mixture with diluted acetic acid and recrystallization. Finally, the mixture of variously substituted benzimidazolyl chalcones (6) and isonicotinohydrazide (7) was refluxed in acetic acid and then mixture crystalized to get pure form of Compound **8** (Scheme 13.5).

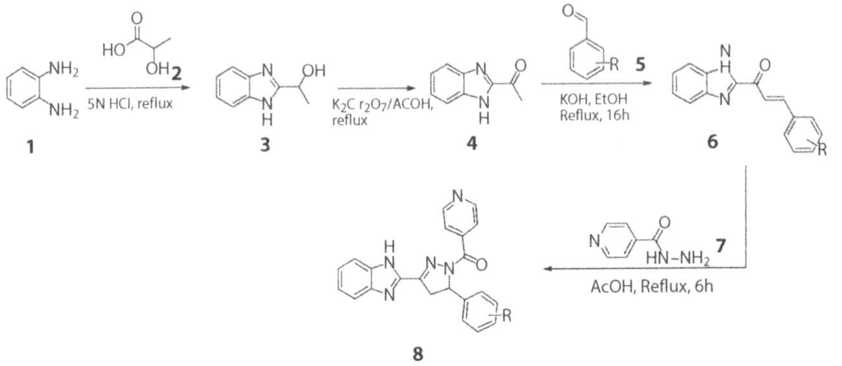

Scheme 13.5 Synthesis pathways of brand-new series of benzimidazole-pyrazol-pyridine methanones derivatives.

All synthesized derivatives were tested *in vitro* against fungal and bacterial strain [11]. The antimicrobial activity was tested against bacterial strains *Escherichia coli*, *Pseudomonas aeruginosa*, *Staphylococcus aureus*, *Streptococcus pyogenes*, and fungi *C. albicans*, *A. niger* and *Aspergillus clavatus*. Compounds in which –OH at C3, $-OCH_3$ at C3 and –OH at C4, $-CH_3$ at C4 and –OH at C4 were found to be most active with MIC of 25 μg/mL against selected bacterial strains. Similarly compounds having –OH at C2, –OH at C4, and $-N(CH_3)_2$ at C4 were found to be most active with MIC of 100 μg/mL against selected fungal strains [11].

Suram *et al.* created benzimidazole–pyrazole derivatives from chlororacetylpyrazole–benzimidazole and benzimidazoles to get analogs 9a-9d (Scheme 13.6). The synthesis of S-([1]H-benzo[d]imidazol-2-yl) (E)-3-phenylprop-2-enethioate (3) was carried out by the reaction of [1]H-benzo[d]imidazole-2-thiol and cinnamoyl chloride in presence of toluene at reflux. The cyclization was done with reaction of S-([1]H-benzo[d] imidazol-2-yl) (E)-3-phenylprop-2-enethioate (3) and hydrazine hydride at reflux in presence of ethanol and then heated chloranil in xylene at reflux to obtain 2-((5-phenyl-[1]H-pyrazol-3-yl)thio)-[1]H-benzo[d]imidazole (5). Compound 5 reacted with 2-chloroacetyl chloride to obtain 1-(3-(([1]H-benzo[d]imidazol-2-yl)thio)-5-phenyl-[1]H-pyrazol-1-yl)-2-chloroethan-1-one (7). Compound 7 reacted with different types of benzimidazole and benzothiazole to obtain bis(benzimidazolyl)pyrazole derivatives(9a–9d).

According to *in vitro* study the Gram-negative bacteria were shown to be more vulnerable to the investigated chemicals than Gram-positive bacteria [12]. Compounds **9a, 9b, 9c** and **9d** showed greater activity. When compared to the reference medicine ketoconazole, Compound 9b (X = S)

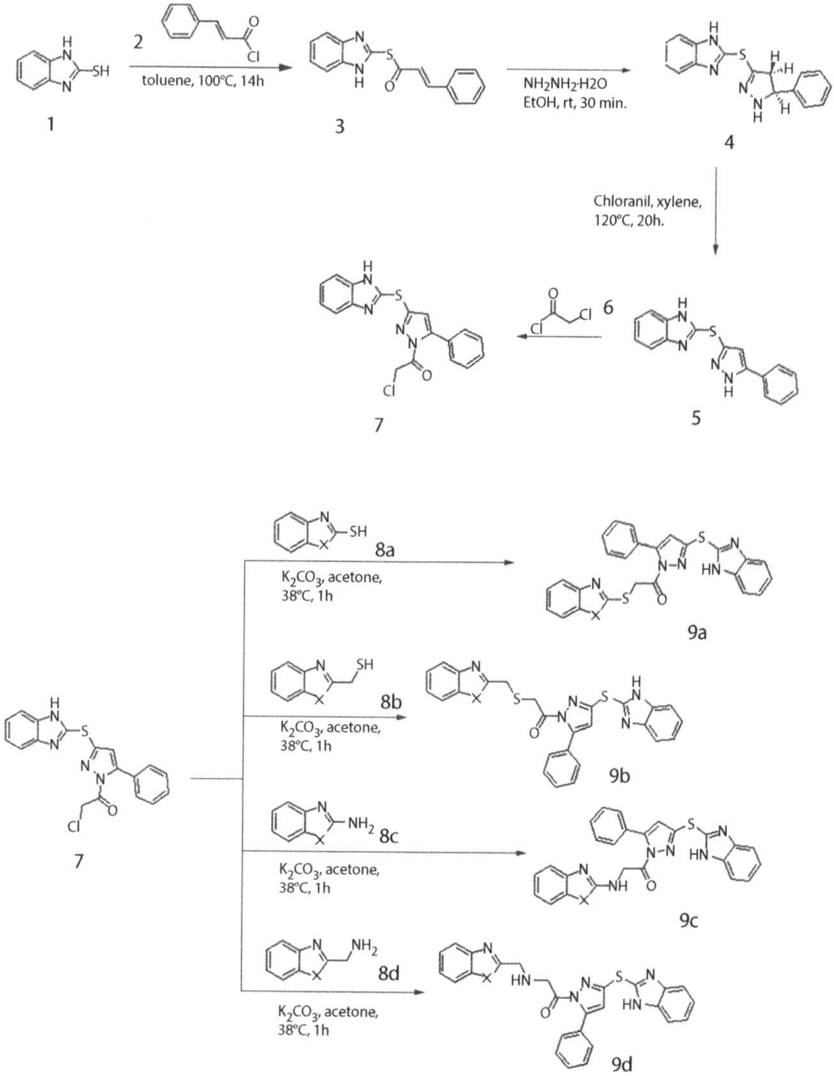

Scheme 13.6 Synthesis of bis(benzimidazolyl)pyrazole derivatives.

and Compound 9d (X = NH) showed better inhibitory action with MIC 100 μg/mL against *Penicillium chrysogenum* (Scheme 13.6).

N-substituted derivatives of Benzimidazole were synthesized by Afaf and his research group taking 2-aminoaniline [13]. The orthophenylene diamine (1) reacted with acetic acid in presence of butylacrylate and pyridine to obtain butyl 3-(2-methyl-^{1}H-benzo[d]imidazol-1-yl propanoate (2).

3-(2-methyl-[1]H-benzo[d]imidazol-1-yl)propanehydrazide (3) was obtained by reaction of butyl 3-(2-methyl-[1]H-benzo[d]imidazol-1-yl)propanoate with hydrazine hydride in ethanol.

Compound **3** was reacted with carbon disulfide and potassium hydroxide to get Compound 4. Then different types of derivatizations were carried out shown in Scheme 13.7.

All compounds underwent *in vitro* testing for their antibacterial and antifungal properties [13]. The data show that Compound **4** was highly effective against *Bacillus cereus* whereas Compound 7 (R = –2Cl phenyl) shows moderate activity against *Bacillus cereus* and **Compounds 4** and **9** (*R* = 2thiophene) moderately effective against *E. coli* (Scheme 13.7). It is also cleared that all synthesizes compound shown inactive against *Saccharomyces cerevisae* and *Aspergillus niger.*

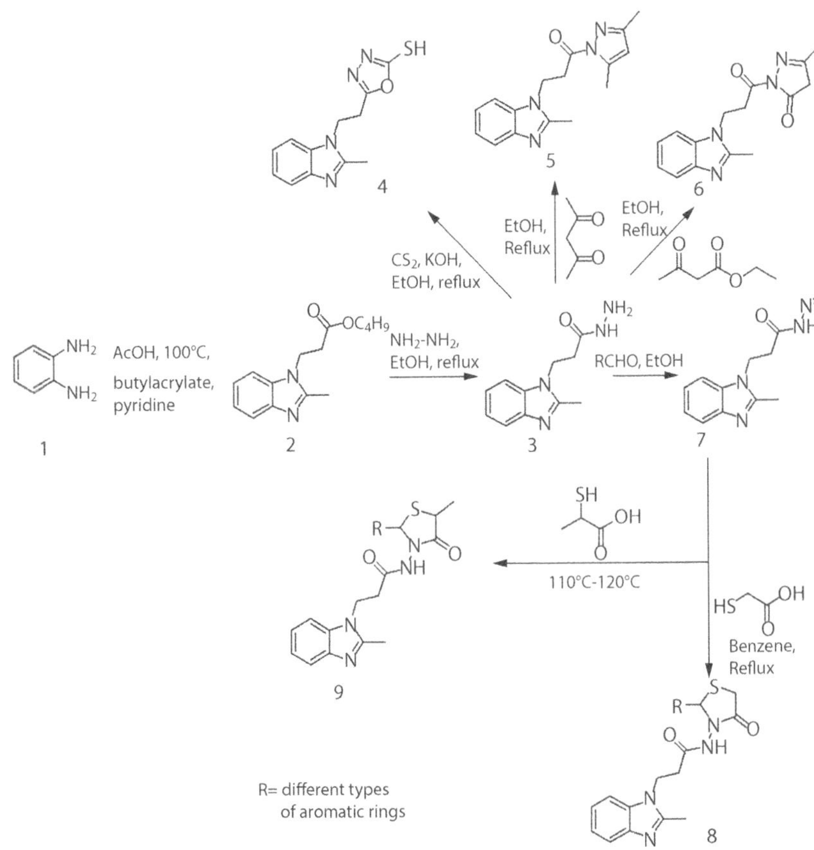

Scheme 13.7 Synthesis of benzimidazole derivatives.

Malleshappa *et al.* synthesized a sequence of 1-methyl-N-[(substituted-phenylmethylidene)-[1]H-benzimidazol-2-amines [14]. In order to create different analogs of 1-methyl-N-[(substituted-phenylmethylidene)-[1]H-benzimidazol-2-amines (5a–5f) first 1-methyl-[1]H-benzimidazol-2-amine (3) was obtained by cycloaddition reaction of o-phenylenediamine (1) with cyanogen bromide in the presence of aqueous base, followed by N-methylation reaction was done using methyl iodide and triethylamine to get Compound 3. Compound 3 was reacted with different types of aldehydes to get the series of (5a–5f). Finally the series of Compound (5a–5f) reacted with chloroacetyl chloride to produce 3-chloro-1-(1-methyl-[1]H-benzimidazol-2-yl)-(substituted)-phenylazetidin-2-one (6a–6f), which introduced the four-membered b-lactam ring is show in Scheme 13.8.

The synthesized compounds were characterized by IR, ^{1}H NMR, ^{13}C NMR and Mass spectral technique and additionally underwent tests to determine their cytotoxic and antibacterial activity. By measuring the zone of inhibition at a concentration of 100 μg/mL, the synthesized Compounds 5a–5f and 6a–6f were evaluated for their *in vitro* antimicrobial activity against Gram-positive bacteria (*S. aureus, Bacillus pumillus*) and Gram-negative bacteria (*P. aeruginosa, E. coli*). In comparison to the standard Ampicillin (13.8, 11.8, 13.2 and 12.4 mm), Compound **4(a)** (12.4, 10.2, 11.9 and 10.8 mm), 13 (12.9, 11.3, 11.8 and 10.8 mm), **5(a)** (13.2, 11.2, 13.6, 11.4 mm), 15 (13.6, 12.4, 13.4 and 11.7 mm), and **5(f)** (13.2, 11.5, 12.8, 11.3 mm) each demonstrated good antimicrobial activities against *S. aureus, Bacillus pumillus, E. coli* and *Pseudomonas aeruginosa*, respectively [14].

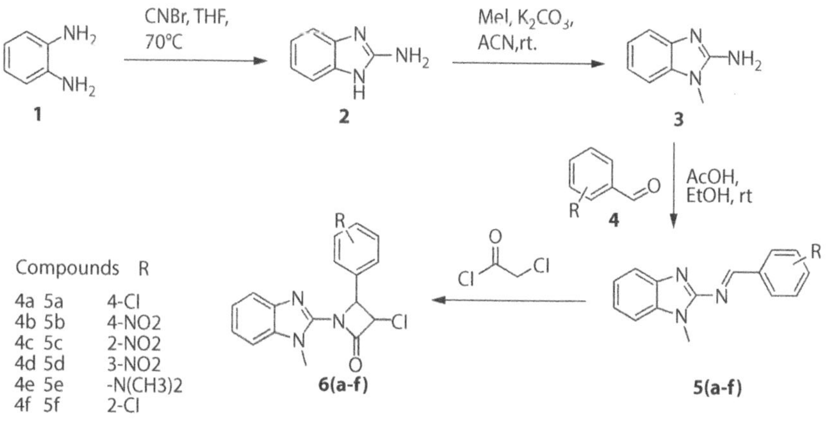

Scheme 13.8 Synthesis of azetidine-2-one analogs of [1]H-benzimidazole derivatives.

Ansari *et al.* synthesized series of azetidine 2 one based compounds [15]. The 2 methyl benzimidazole reacted with ethylchloroacetate in presence of anhydrous potassium carbonate and acetone to obtain ethyl (2-methyl-^{1}H-benzimidazol-1-yl) acetate (3). Intermediate 3 was refluxed with thiosemicarbazide in presence of 1, 4-dioxane and water to get 2-[(2-methyl-^{1}H-benzimidazol-1-yl) acetyl] hydrazine carbothioamide (5). The cyclization is carried out by reacting intermediate 5 and concentrated sulfuric acid to obtain 5-[(2-methyl-^{1}H-benzimidazol-1-yl) methyl]-1,3,4-thiadiazol-2-amine. Then imine formation done by using different types of aromatic aldehyde to obtain different derivatives of Compounds 7a–7i. Finally Compound 7 treated with chloroacetyl chloride in presence of triethylamine to give different derivatives of Compounds 8a–8i (Scheme 13.9).

The antimicrobial activity of the produced chemicals 7a–7d and 8a–8d were evaluate *in vitro*. **8d, 8a**, and **8** for concentration 10 mg/mL against

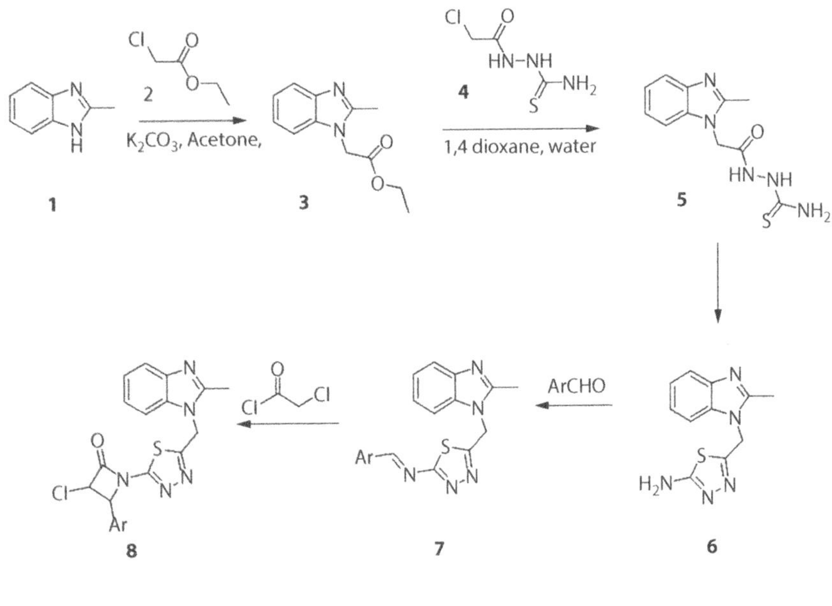

Compound	R
7a 8a	-4Br
7b 8b	-4Me
7c 8c	-4OMe
7d 8d	-2Cl

Scheme 13.9 Synthesis of derivatives of ^{1}H-benzimidazol-1,3,4-thiadiazol-imine and ^{1}H-benzimidazole-1,3,4-thiadiazol-azetidin-2-one derivatives.

Bacillus subtilis shows good activity and **8c, 7d**, and **7a** for concentration 100 mg/mL against *E. coli* shows good activity [15].

13.4 Benzimidazole as an Anticancer Agent

Zhe *et al.* synthesized a series of chrysin benzimidazole compounds [16]. The synthesis of chrysin benzimidazole has been carried out in single step. First step, compound react with different types of dibromo alkane in presence of base potassium carbonate and acetone to obtain 5,7-bromo alkylated-dihydroxy-2-phenyl-4H-chromen-4-one. In second step derivatives of benzimidazole were reacted with 5,7-bromo alkylated-dihydroxy-2-phenyl-4H-chromen-4-one in presence of base potassium carbonate and acetone to obtain different analogs of N-substituted benzimidazole shown in Scheme 13.10.

In vitro anticancer activity of synthesized compound was evaluated. Compound **3** (R_1-Me, R_2-H, R_3-H and n = 4) shown strongest anti-proliferative effects on MFC cells, with IC_{50} values of 25.72 ± 3.95 μM in dose dependent manner. Additionally, the preliminary anticancer efficacy *in vivo* was investigated in mice that had tumors, and Compound 1 had a strong inhibitory effect on tumor growth. These findings revealed that Compound **3** was good antitumor active and, with additional optimization and evaluation, would be a promising anticancer drug [16].

Morais *et al.* discovered the series of N-unsubstituted and methyl substituted aryl benzimidazole [17]. The two series were synthesized one by reacting N-methyl-N-(2′-hydroxyethyl)-4-aminobenzaldehyde and sodium metabisulfite in dimethylfarmamide to get Compounds 1, 2, and 3. And second by reacting N-methyl-N-(2′-O-tosyloxyethyl)-4-aminobenzaldehyde and sodium metabisulfite in dimethylfarmamide to get 6 and 7. In the second series 2-((4-([1]H-benzo[d]imidazol-2-yl)phenyl)

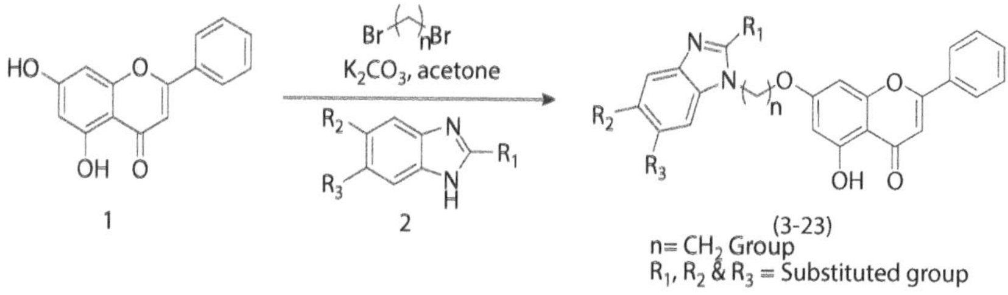

Scheme 13.10 Synthesis of chrysin benzimidazole derivatives.

(methyl)amino)ethyl tert-butyl carbonate was reacted with methyl iodide and sodium hydroxide in acetone to get Compound **8**. The hydroxy Compound (9) was synthesized by deprotection of OBoc Compound (8) with trifluoroacetic acid in dichloromethane. Compound **9** was reacted with tosyl chloride in triethylamine in dichloromethane to get Compound **10**. Then it was heated in Tetra-n-butylammonium fluoride and tetrahydrofuran to get Compound **11**, which is shown in Scheme 13.11.

All synthesized compound were tested *in vitro* for human tumor cell including MCF7, A375, HeLa, and U87 from breast, melanoma, cervix carcinoma, and glioblastoma, respectively. Compounds **4** (R = H) and **5** (R = Br) were more favorable against all cancer cells, notably against glioblastoma in dose dependent manner [17]. The MTT assay (IC_{50} value) for Compound 4 was MCF7: 71.1 ± 18.0, A375: 106 ± 44.0, HeLa: 62.7 ± 15.0,

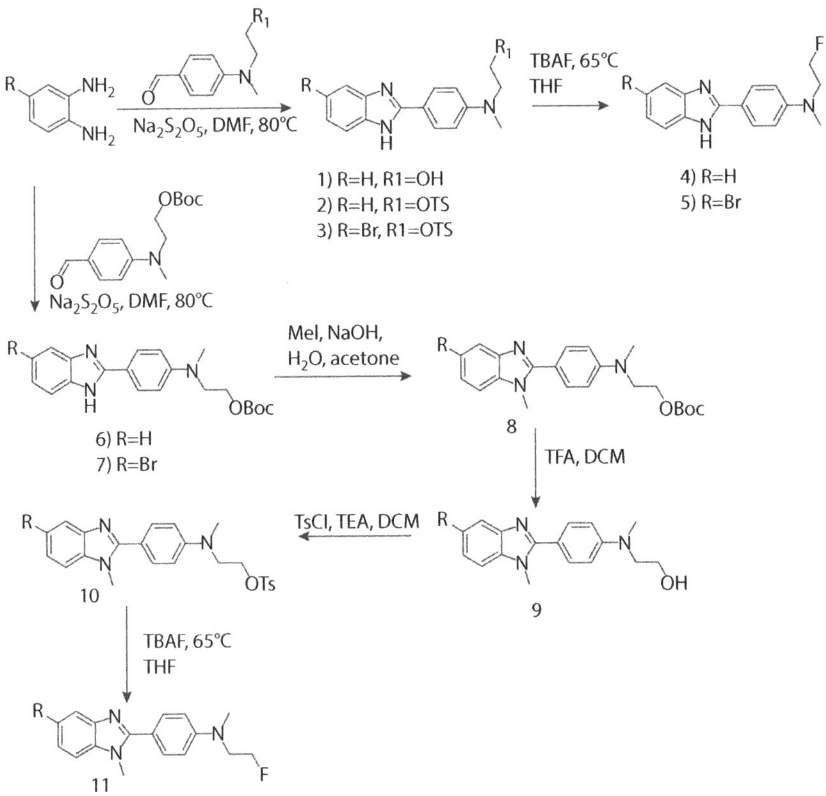

Scheme 13.11 Synthesis of N-unsubstituted and N-methylated aryl benzimidazole derivatives.

and U87: 45.2 ± 13.0 and for Compound 5 was MCF7: 90.2 ± 15.0, A375: 91.6 ± 31.0, HeLa: 17.8 ± 3.00, and U87: 62.8 ± 18.0.

Li *et al.* synthesized two series and evaluated their anticancer activity [18]. As shown in Scheme 13.12 para nitro ortho phenylenediamine subjected to cyclization with different types of benzaldehyde 2 and 1-methyl-^{1}H-pyrrole-2-carbaldehyde **2a** gives **3** and **3a**, respectively. The nitro group of 3 and 3a was reduced under tin chloride and ethanol to give **4** and **4a**. Then different types of chloroacetyl chloride were reacted to give **6** and **6a**.

The created derivatives were assessed *in vitro* cell cytotoxicity, kinase inhibition and apoptosis test. The synthesized derivatives were initially tested by MTT assay against HepG-2 cells as a positive control for their *in vitro* cell cytotoxicity for IC_{50} value. Compounds **6** (n = 1) and **6a** (n = 1) showed best activity 2.0 and 1.8 μM, respectively. The *in vitro* kinase assay was used to assess the kinase inhibitory activity of compounds at a dose of

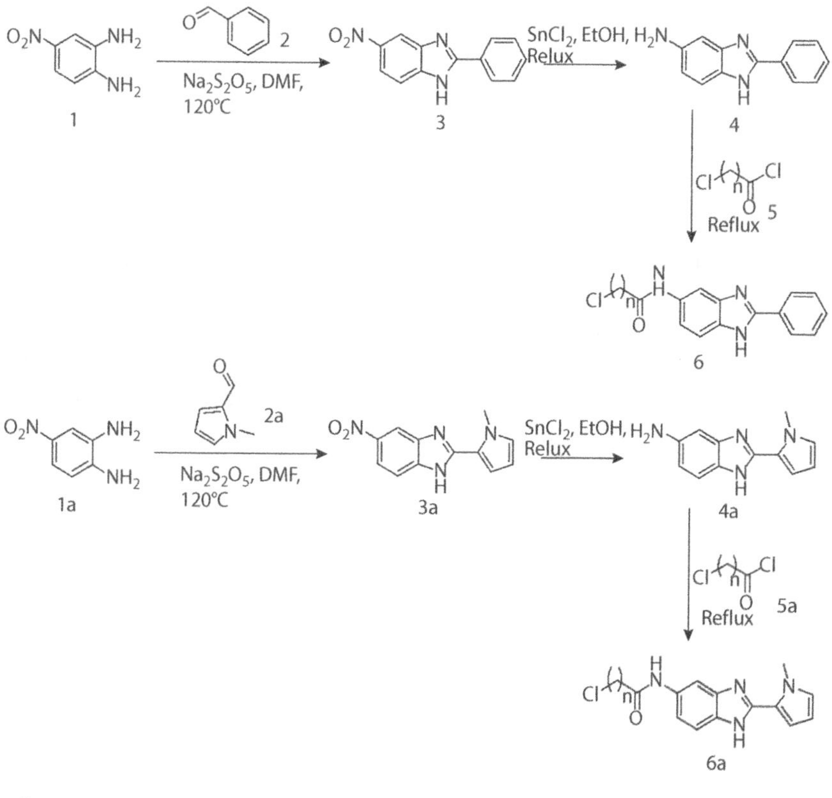

Scheme 13.12 Synthesis of 2 aryl benzimidazole derivatives.

50 μM. In which Compounds 6 (n = 1) and **6a** (n = 1) had good cytotoxicity and also inhibited EGFR activity by 88.03% and 86.90%, respectively. In order to further investigate apoptosis utilizing the annexin V/propidium iodide (PI) binding assay, Compounds **5a** and **5e** were chosen due to their high cytotoxicity and potent kinase inhibitory activity. Phosphatidylserine residues and DNA are both stained by annexin V and PI, respectively. These findings unambiguously showed that Compounds 4 and 5 induced apoptosis in HepG-2 cells at low μM concentrations [18].

Karpinska *et al.* created series of novel benzimidazole benzene diols derivatives [19]. Synthesis has been carried out by heating aryl-modified sulfinyl bis[(2,4-dihydroxyphenyl)-methanethione] and 3-methylbenzene-1,2-diamine in MeOH to get different types of diol benzimidazole derivative (Scheme 13.13).

The created analogs were tested against human tumor cell lines such as SW707 (rectal), HCV29T (bladder), A549 (lung), and T47D (breast). Compound 2 (R_1 = Me, R_2 = Me and R_3 = H) antiproliferative ability was found to be superior to that of the widely prescribed medication cisplatin. Anticancer activity Compound **6** was measured by their IC_{50} (μg/mL) value and which was HCV29T-0.53 ± 0.17, A549-3.51 ± 0.37, T47D-0.65 ± 0.17, and SW707-1.72 ± 0.67, whereas anticancer activity of standard drug cisplatin for IC_{50} value was HCV29T: 2.66 ± 1.15, A549: 5.29 ± 1.85, T47D: 2.03 ± 1.61, and SW707: 3.13 ± 1.49 [19].

Romero-Castro *et al.* synthesized Molecules **3** and **3a** [20]. The synthesis of Molecules **3** and **3a** (Scheme 13.14) was achieved by heating to reflux of 4 nitro ortho phenylenediamine with 4-chloro-3-nitro benzoic acid (**2**) in phosphorous oxychloride and 4 nitro ortho phenylenediamine with 3,5-dinitro benzoic acid (**2a**) in phosphorous oxychloride, respectively.

Lab tests of Compounds 3 and 3a (Scheme 13.14) through the ATP assay evaluated their inhibitory effects on human neoplastic cell growth in lung carcinoma (A549), bone marrow, chronic myelogenous leukemia

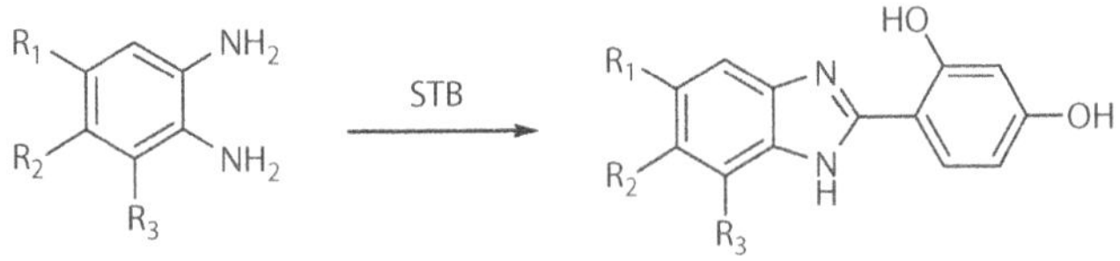

1 **2**

R_1, R_2 & R_3= Substituted group

Scheme 13.13 Synthesis route of 4-(^{1}H-benzimidazol-2-yl) benzene-1,3-diols derivatives.

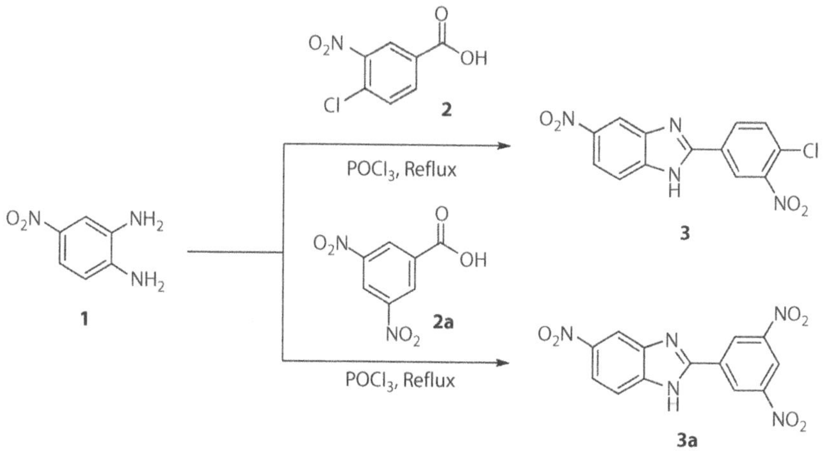

Scheme 13.14 Synthesis route of nitro benzimidazole derivatives.

(K562), HeLa contaminant (KB), peripheral blood, acute promyelocytic leukemia (HL60), breast, adenocarcinoma (MDA231), breast, adenocarcinoma (MCF-7) and colon adenocarcinoma (HT29). The antiproliferative potential of Compound 3 matched the results of the well-known therapeutic drug carboplatin. Anticancer activity of Compound 3 was measured by their IC_{50} (μg/mL) value and which was K562: 2.4 ± 0.2, HL60: 1.1 ± 0.5, MCF7: 0.6 ± 0.2, MDA231: 0.4 ± 0.1, A549: 0.8 ± 0.10, HT29: 0.4 ± 0.3, KB: 2.67 ± 0.2, HACAT: 1.2 ± 0.8 whereas anticancer activity of standard drug carboplatin for IC_{50} value was K562: 1.2 ± 0.2, HL60: 0.3 ± 0.1, MCF7: 8.9 ± 0.1, MDA231: 2.6 ± 0.4, A549: 0.05 ± 0.0, HT29: 0.9 ± 0.1, KB: 4.1 ± 0.7, HACAT: 0.6 ± 0.1. However, Compound 3 also shown cytotoxicity against the noncancer cell line HACAT, indicating that it is not only hazardous to cancer cell lines. The activity of Compound **3a** had comparable to carboplatin and was the most potent antitumor derivative for the A549 cell line while causing the low cytotoxic activity for HACAT [20].

Ranganatha *et al.* synthesized a benzophenone benzimidazole derivatives [21]. 4-Hydroxybenzophenones were synthesized by benzoylating substituted phenols with substituted benzoyl chloride, and the resulting ester derivative was subsequently obtained by esterification with ethylchloroacetate to produce acid (7) analog (Scheme 13.15).

The anticancer activity of synthesized compounds was tested against EAC by using three different test techniques (TP, ATP, and LDH release). The active compounds were also tested for their reducing the growth of

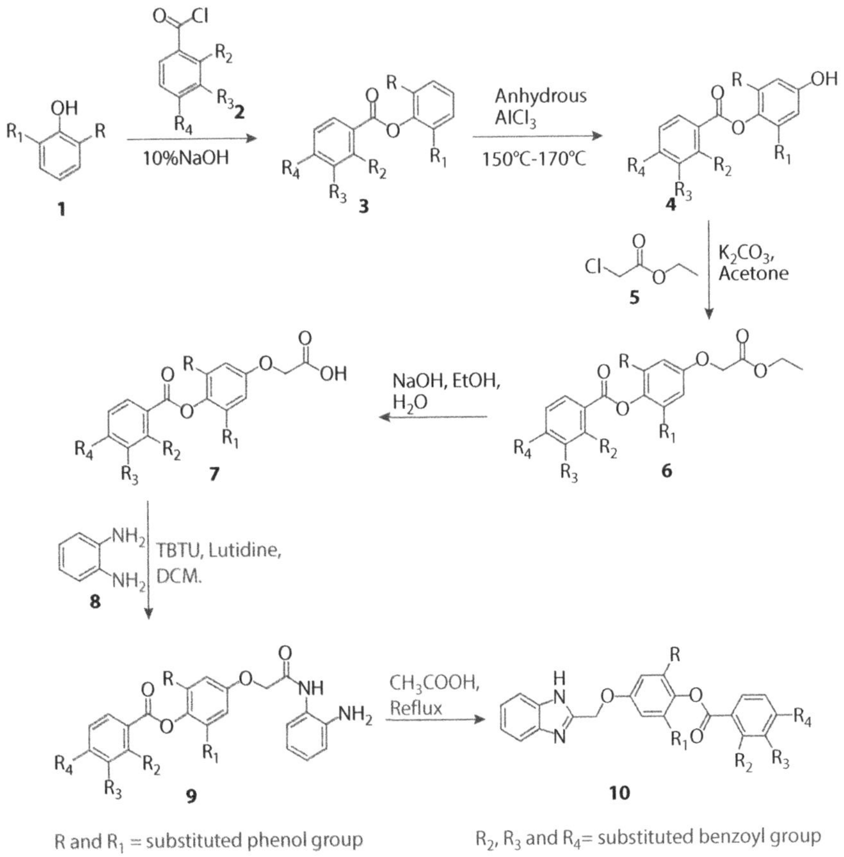

Scheme 13.15 Synthesis route of benzophenone benzimidazole derivatives.

new blood vessels properties in a CAM model and a tumor-induced vasculature *in vivo* model. The reducing growth of new blood vessels effect on tumor growth was confirmed in a mouse model. According to results from the trypan blue, MTT, and LDH release tests, Compound 10 (R = Me, R_1 = Me, R_2 = H, R_3 = H and R_4 = OMe) exhibited the highest value of cytotoxicity, with IC_{50} values of 12, 10, and 10 M, respectively [21].

Sontakke *et al.* were created a chain of 2-alkylated benzodiazole analogs by reacting 1,2-diaminobenzene and aromatic formyl group in catalyst H_5IO_6–SiO_2, which is given in Scheme 13.16 [22].

The synthesized chemicals were studied for their ability to prevent cell line proliferation of MCF7 and HL60 cells through ATP assessment experiments. The series possesses better anticancer capacity than the conventional

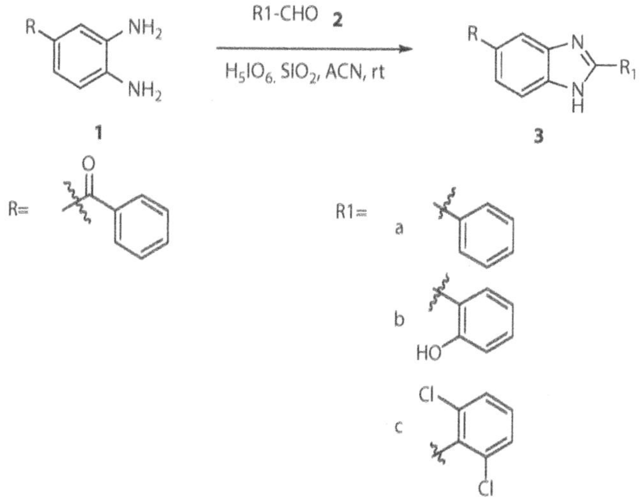

Scheme 13.16 Synthesis route of 2 aryl benzimidazole derivatives by using hydrogen periodate.

medicine cisplatin. Among the series Compounds 3a and 3b demonstrated maximal activity against MCF-7 cell lines with IC_{50} values of 17.32 μM per compound whereas Compound 3c showed best activity against HL60 cell lines with IC_{50} of 18.13 μM even though cisplatin exhibited IC_{50} of 40.45 and 41.08 μM toward both MCF-7 and HL60 cell lines [22].

Harkala *et al.* created new benzimidazole linked triazole series [23] *via* key intermediate 2-(4-azidophenyl)-^{1}H-benzo[d]imidazole (**4**) and which was achieved by cyclization of 1,2-phenylene diamine with para-amino benzoic acid in polyphosphoric acid followed by sodium nitrate and sodium azide (Scheme 13.17). The click reaction was performed by using different types of alkyne and zinc triflate to obtain benzimidazole-triazole Compound (**5**).

The molecules were evaluated against capacity to inhibit cell proliferation in some tumor cell lines, including human melanoma cell line, Mouse melanoma, human colon adenocarcinoma cell line, Human breast cancer, and hypotriploid alveolar basal epithelial cells by MTT assay method. The most of the compounds had strong antiproliferative activity with IC_{50} values from 0.1 to 43 μM. while standard drug Adriamycin showed activity for value IC_{50} in between 0.03 to 3.5 μM in the cell lines. Adriamycin was found to be less effective than Compound **5a** and Compound **5b** against the colon-205 and MCF-7 cell lines. Anticancer activity of Compound 5a and **5b** was measured by their IC_{50} (μg/mL) value and which was Colon-205 = 0.12 μM, MCF-7 = 1.9 μM; and Colon-205 = 1.3 μM, MCF-7 = 0.3

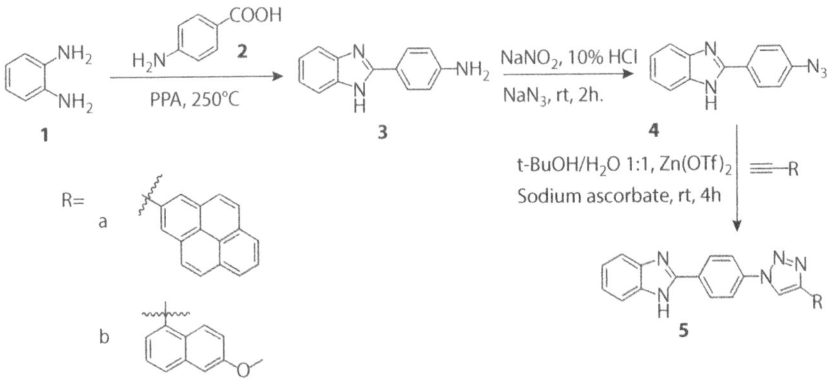

Scheme 13.17 Synthesis of benzimidazole 1,2,3-triazole derivatives.

μM, respectively, whereas anticancer activity of standard drug adriamycin for IC_{50} value was Colon-205 = 3.5 μM, MCF-7 = 3.5 μM [23].

Risley *et al.* created novel series of benzimidazole blended with 1,3-Diazabenzene derivatives [24] by condensing 2-aminobenzodiazole, cyanoacetonitrile, and different aromatic formyl group and followed by oxidization with DDQ in acetonitrile to get Compound **5** (Scheme 13.18).

The *in vitro* anticancer study evaluated the human pancreatic (PANC-1) and breast cancer MCF-7 and Hs578T cell lines using the synthesized compound. Compound 5a displayed the strongest anticancer effect on PANC-1 cells where it achieved an IC_{50} value of 5.5 ± 1.4 μM. While Molecule **5c** was shown high antileukemic against Hs578T cell line (IC_{50} = 3.2 ± 1.6 μM), Compound **5b** had the giant active compound against MCF-7 cell line with IC_{50} = 11.85 ± 2.6 μM [24].

Abdel-Aziz *et al.* synthesized 3-methylthiazolo-benzimidazole derivatives [25]. The key intermediate 1 (Scheme 13.19) was synthesized by the

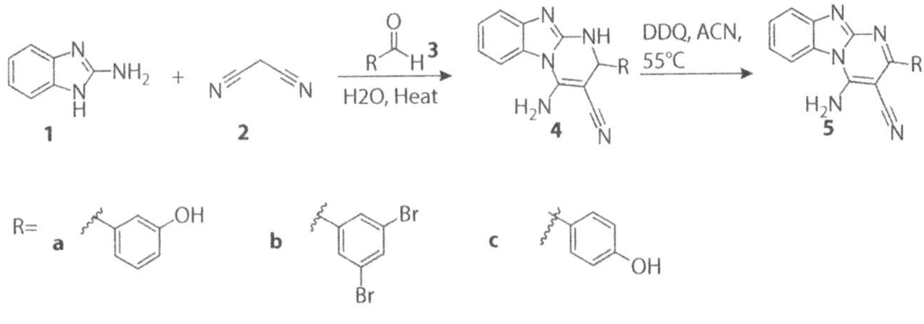

Scheme 13.18 Synthesis of benzimidazole–pyrimidine derivatives.

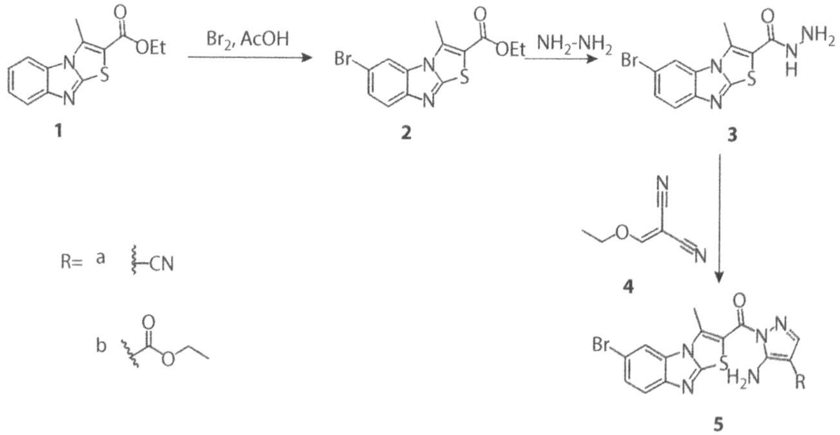

Scheme 13.19 Synthesis of 3-methylthiazolo-benzodiazole molecule.

D'Amico *et al.* method [26]. The bromo ester Molecule 2 was obtain by bromination of intermediate (**1**) which was then reacted with hydrazine hydrate to get the carbohydrazide Molecule **3** followed by cyclization with 2-(ethoxymethylene) malononitrile (**4**) to get final derivatives (**5**) [25].

Synthesized molecules were tested against anticancer efficacy against the colon tumor and hepatocellular carcinoma cells by using the ATP assay method. Results were compared with the standard anticancer drug paclitaxel. Anticancer activity of Compound **5a** was measured by their IC_{50} (µg/mL) value and which was highly cytotoxic but nonspecific against Hep-G2 and HCT-116 cells (IC_{50} = 18.4 and 12.5 g/mL, respectively); however, it was less toxic than paclitaxel (IC_{50} = 710 µg/mL for Hep-G2 and 825 µg/mL for HCT-116 cells). Ethylcarboxy substituted Compound **5b** exhibited modest but distinct cytotoxic action towards HCT-116 cells.

Hranjec *et al.* synthesized a series of acrylonitrile and quinolines fused benzimidazole derivatives [27]. Synthesis was achieved by reaction aromatic aldehyde with Compound **1** (Scheme 13.20) in piperidine and ethanol followed by cyclization with iodine and oxygen in presence of ethanol to afford final derivatives (**3**).

Synthesized compounds were tested for the antiproliferative potential against tumor cell lines such as cervical tumor (HeLa), pancreatic tumor (MiaPaCa-2), colon tumor (SW 620), breast tumor (MCF-7) and lung tumor (H460). Except few all compounds exhibited strong antiproliferative activity. Compounds **2a** and **3a** for value IC_{50} = 0.8 and 0.7 µM were less active than standard drug doxorubicin (IC_{50} = 0.040.009 µM) but

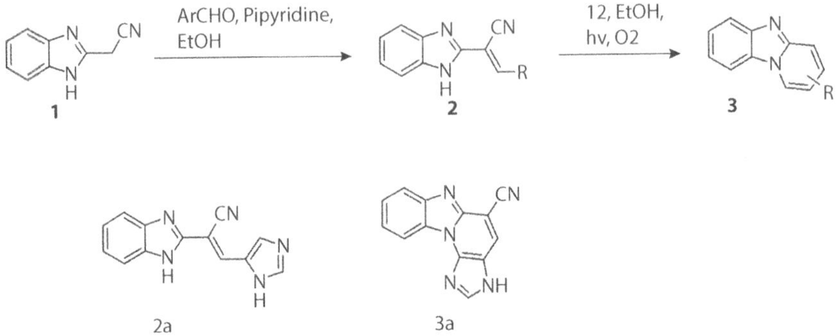

Scheme 13.20 Synthesis of acrylonitrile and quinolines fused benzimidazole derivatives.

approximately four times as active against HeLa cells when compared to standard drug cisplatin (IC_{50} = 30.6 μM) [27].

Hranjec *et al.* synthesized different types of benzimidazole Schiff base [28]. The synthesis achieved by cyclization of substituted ortho phenylenediamine with cyanogen bromide and followed by reaction with different types of aromatic aldehyde in ethanol to afford substituted benzimidazole Schiff bases derivatives (Schemes 13.21 and 13.22).

The IC_{50} value was determined through testing the antiproliferative behavior of Schiff bases on various tumor cell lines *in vitro* experiments. The Compounds 4a and 4b demonstrated maximum nonselective antiproliferative activity across all examined cell lines including HeLa (cervical tumor) and SW620 (colorectal adenocarcinoma, metastatic) and MiaPaCa-2 (pancreatic tumor) and MCF-7 (breast epithelial adenocarcinoma, metastatic), besides normal human fibroblast (WI38). The activity (μM) of Compounds 22 and 23 were HeLa: 4.73, MCF-7: 9.23, SW620:

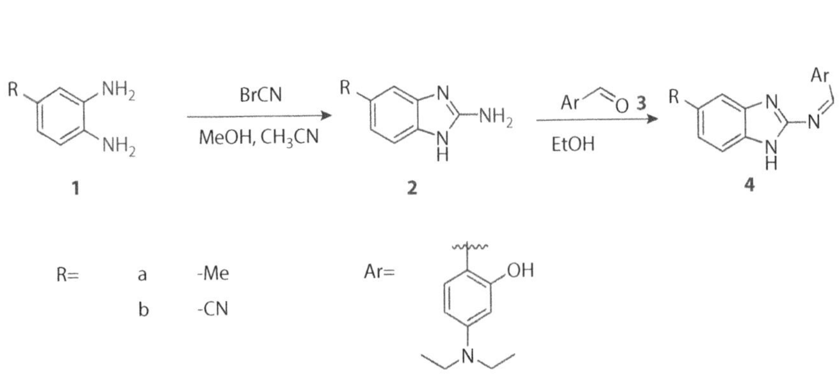

Scheme 13.21 Synthesis of benzimidazole Schiff bases derivatives.

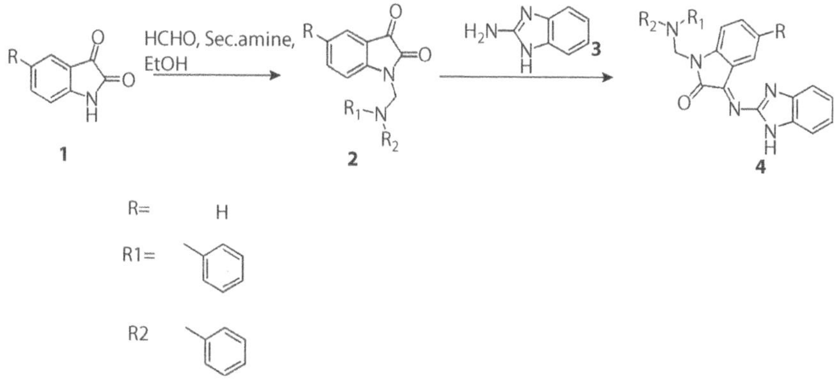

Scheme 13.22 Synthesis of isatin–benzimidazole derivative.

49.15, MiaPaCa-2: 27.92, W138: 0.96; and HeLa: 3.24, MCF: 7 = 15.27, SW620: 52.04, MiaPaCa-2 = 22.24, W138: 1.67, respectively [28].

Taher *et al.* created isatin–benzimidazole derivative [29]. Synthesis of Compound 2 was achieved by formation of imine indoline-2,3-dione with formaldehyde followed by alkylation with secondary amine and finally 2 amino benzimidazole reacted with Compound 2 to afford isatin–benzimidazole derivatives.

The molecule has been tested for anticancer activity by *in vitro* against MCF-7 (human breast cancer) cell line by using SRB assay method. In that finding Molecule **4** (IC_{50} = 22.59 μM) had substantial cell inhibition growth potent against human breast cancer cells, this was almost four times less than the dosage of the common cancer treatment doxorubicin (IC_{50} = 5.46 μM) [29].

Singla *et al.* synthesized triazine–benzimidazole analog [30]. The key intermediate (**4**) was synthesized by cyclization of o-phenalene diamine with acetic acid followed by nitration using sulfuric acid and nitric acid and alkylation with benzyl chloride then reduction with Tin(II) chloride. Then reaction of benzyl substituted benzodiazole with triazine intermediate in presence of sodium bicarbonate to afford intermediate 6 followed by displacement reaction with tetrahydro-1,4-oxazine 1-Oxa-4-azacyclohexane to get triazine–benzimidazole derivative **8** (Scheme 13.23).

The synthesized compound was tested for anticancer activity against human tissue cancer cell lines at different dose concentration. The Molecule **8** showed most effective mammalian dihydrofolate reductase (DHFR) inhibitor (IC_{50} = 1.05 μM), and a high GI_{50} value of 9.79 suggested that it exhibited promise for wide-ranging antitumor potential. According to

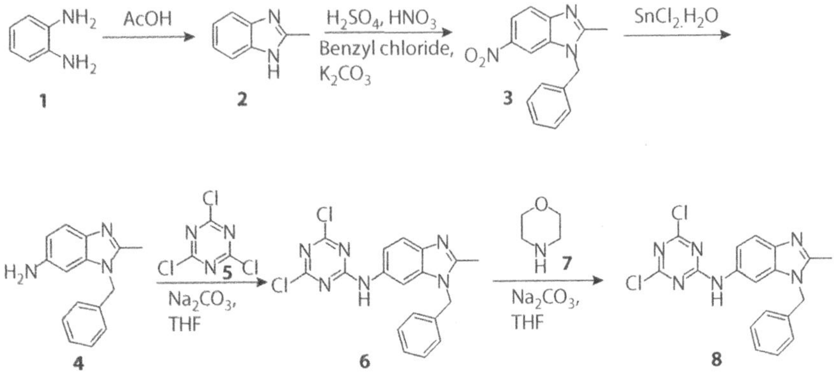

Scheme 13.23 Synthesis of triazine-benzimidazole.

molecular modelling studies and DNA binding experiments, Compound **8** to the active site of the DHFR enzyme is stabilized and strongly bonded, respectively [30].

Shi *et al.* created a of quinazoline benzodiazole derivative series [31]. Synthesis is achieved by cyclization (Scheme 13.24) of 4-nitrobenzene-1,2-diamine with substituted benzoic acid in polyphosphoric acid followed by reaction of nitro group by H_2 Pd/C and then displacement with 4 quinazoline to afford final derivatives of quinazoline–benzimidazole (**6**).

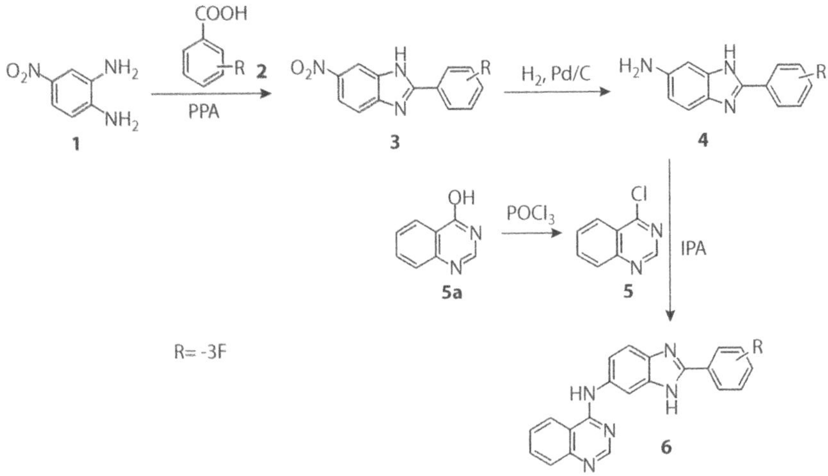

Scheme 13.24 Synthesis of quinazoline benzimidazole derivatives as an anticancer agent.

The synthesized molecule was tested for cytotoxic capability against the human breast adenocarcinoma and hepatocellular cell lines using the ATP method and the enzymatic activity against vascular endothelial growth factor receptor 2 and c-mesenchymal epithelial transition factor were also evaluated. Nearly all of the compounds showed superior cytotoxic activity towards MCF-7 compared to Hep-G2. Compound **6** was shown to have IC_{50} values of 1.5×10^{-6} M and 8.7×10^{-6} M for MCF-7 and HepG-2, respectively, which was the most cytotoxic molecule than the standard anticancer drug, golvatinib (IC_{50} values of 65.5 and 49.6 μM for HepG-2 and MCF-7, respectively). Compound 6 was equally effective at inhibiting the enzymes VEGFR-2 and c-Met (IC_{50} values of 0.02 and 0.05 μM, respectively) as golvatinib (IC_{50} values of 0.02 μM against c-Met and 0.04 μM, respectively) [31].

Azam *et al.* synthesized the 2 aromatized benzo diazole analog by cyclization of 1,2-benzodiamine with substituted aromatic aldehyde (Scheme 13.25) to afford final Compound (**3**) [32].

In vitro cytotoxic effects of all the synthesized derivatives were evaluated against human tissue tumor cell line such as lung cancer, MCF-7, THP-1, and PC-3. In comparison with other synthesized compound, 2-(4-bromophenyl)-^{1}Hbenzo[d]imidazole **3** (-4Br), shown a marginally greater affinity against cancer cell types. Compound **3** (-4Br) showed activity for value IC_{50} (μM) A-549 = 43 ± 2, MCF-7 = 35 ± 2, THP-1 = 48 ± 2, and PC-3 = 46 ± 1 [32].

Malleshappa *et al.* synthesized a series of Azetidine 2-one derivatives (Scheme 13.8). The best antibacterial compounds were chosen, and the anticancer efficacy against MCF-7 cell lines was tested *in vitro* using the MTT assay. In comparison with standard drug cyclophosphamide (IC_{50} = 8.7 M), Compound **6** (2,5-dimethoxy) showed promising cytotoxic activity (IC_{50} = 6.0 μM), with percentage viability of 51.3% and inhibition of 48.7% at 5.0 μM solution [14].

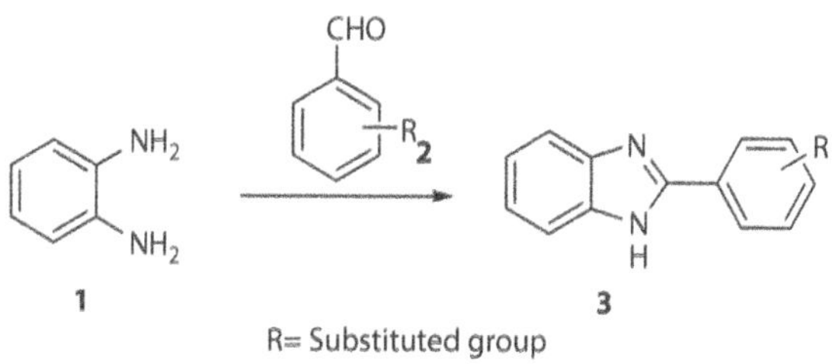

Scheme 13.25 Synthesis 2 substituted benzimidazole derivatives.

Kishore Babu *et al.* created dibenzodiazole series [33]. Synthesis has been achieved by cyclization of ortho phenalene diamine with 2-hydroxy butanoic acid followed by oxidation with chromic acid obtained ketone Compound (**4**) which were further alkylation using TBAB and potassium carbonate followed by reaction with phenyl hydrazine hydrate in acetic acid to afford Compound (**7**). Then final compound was obtained by reaction of intermediate 7 with polyphosphoric acid, potassium carbonate and TBAB reagent (Scheme 13.26).

The created molecules were examined *in vitro* for antitumor activity by ATP assay method against four different carcinoma cells line. These cell lines were breast cancer cell, prostate tumor cell, liver tumor and colon cancer cells. cytotoxicity of test compounds on human cell lines measured for IC_{50} values (μM)) and compared with standard drug Doxorubicin. Compound **8b** shown selective action against the HepG2 cell line (HepG2 = 51.8 ± 0.09), while Compound **8a** demonstrated potential cytotoxicity against all of the examined cell lines (MDA-MB-231 = 10.3 ± 0.13, DU145 = 12.3 ± 0.12, HepG2 = 14.4 ± 0.22, COLO 205 = 16.1 ± 0.41), whereas activity of doxorubicin (MDA-MB-231 = 0.6 ± 0.18, DU145 = 0.6 ± 0.1, HepG2 = 0.5 ± 0.12, COLO 205 = 0.6 ± 0.21) [33].

Yoon *et al.* synthesized ethyl piperazine benzimidazole derivatives [34]. Ester intermediate **2** was produced by esterifying substituted benzoic acid, which was followed by a reaction with amine intermediate **3** and DIPEA to afford Compound **4**. The nitro group was reduced by hydrogen and Pd/C followed by cyclization with substituted aromatic aldehyde to obtain ethyl piperazine benzimidazole derivatives (**6**), which is shown in Scheme 13.27.

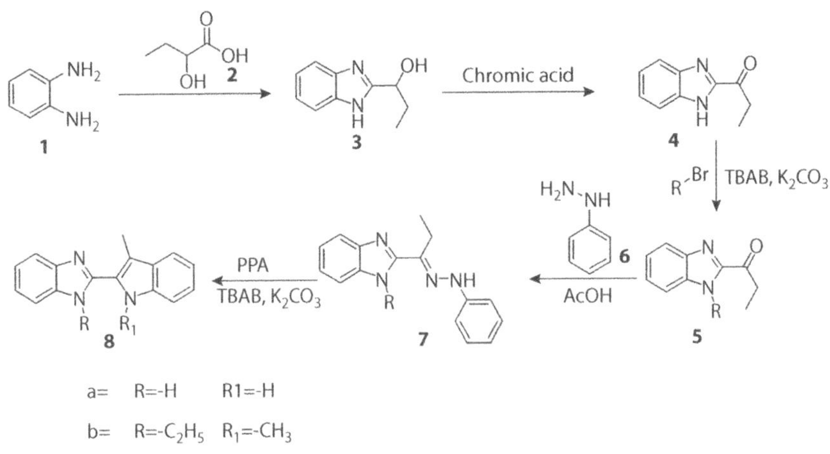

Scheme 13.26 Synthesis of dibenzimidazole derivatives.

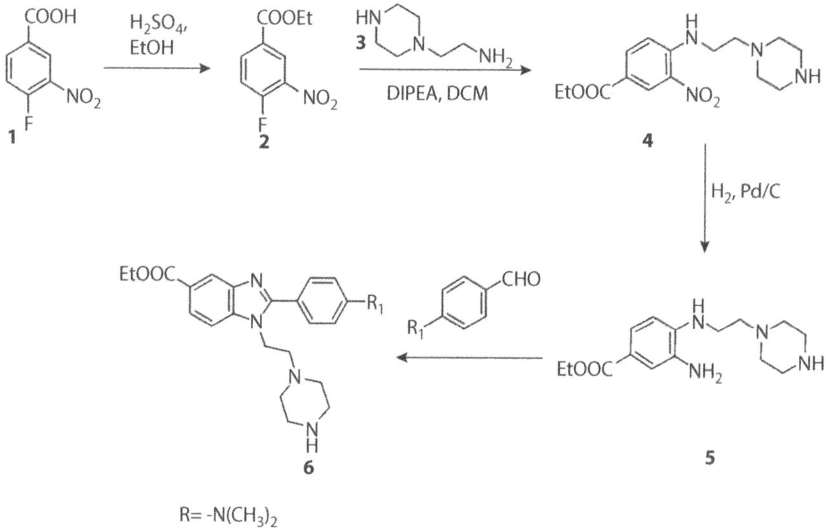

Scheme 13.27 Synthesis of ethyl piperazine benzimidazole derivatives.

The ability of the synthesized derivatives to inhibit sirtuins was compared with standard drug sirtinol. It was found that the substituent having negative inductive effect showed better sirtuin inhibitory activities. Compound **6** was most effective sirtuin inhibitor [34].

13.5 Conclusion

The current research work shows benzimidazole as a promising heterocyclic compound for application in the field of medicinal chemistry and drug discovery. According to the available literature, benzimidazole derivatives can be used as an alternative to conventional therapy to treat issues such as medication resistance. Many studies have shown how benzimidazole derivatives can be used to treat pain and inflammation by acting on many clinically recognized therapeutic targets. So, there is a lot of potential to find novel, improved, secure, and more effective cancer drugs using benzimidazole.

AI disclosure statement: The author confirms that no artificial intelligence (AI)-generated content or AI-assisted writing tools were used in the preparation of this chapter (including summaries and the abstract). All text, analysis, and conclusions are the result of the author's original work.

References

1. Hoebrecker, F., Benzimidazole. *Ber*, 5, 920–926, 1872.
2. Goodman, L.A. and Nancy, H., Preliminary investigations on the pharmacology of benzimidazole. *Fed. Proc.*, 2, 1, 1943.
3. Woolley, D.W., Some biological effects produced by benzimidazole and their reversal by purines. *J. Biol. Chem.*, 152–225, 1944.
4. Bandyopadhyay, P., Sathe, M., Sachin, N., Tikar, B., Yadav, R., Sharma, P., Ashok, K., Kaushik, M., Synthesis of some novel phosphorylated and thiophosphorylated benzimidazoles and benzothiazoles and their evaluation for larvicidal potential to Aedes albopictus and Culex quinquefasciatus. *Bioorg. Med. Chem. Lett.*, 24, 2934–2939, 2014.
5. Alpan, A.S., Parlar, S., Carlino, L., Tarikogullari, A.H., Vildan, A., Hasan, S.G., Synthesis, biological activity and molecular modelling studies on 1H-benzimidazole derivatives as acetylcholinesterase inhibitors. *Bioorg. Med. Chem. Lett.*, 21, 4928–4937, 2013.
6. Vidyasrilekha, Y., Reddy, B.K., Mohammaed, A.A., Imidazoles and benzimidazoles as putative inhibitors of SARS-CoV-2 B.1.1.7 (Alpha) and P.1 (Gamma) variant spike glycoproteins A computational approach. *Chem Zvesti.*, 76, 2, 1107–1117, 2022.
7. Khabnadideh, S., Rezaei, S., Pakshir, K., Zomorodian, K., Ghafari, N., Synthesis and antifungal activity of benzimidazole, benzotriazole and aminothiazole derivatives. *Res. Pharm. Sci.*, 7, 2, 65–72, 2012.
8. Nishad, C.T., Shrestha, S.K., Ngo, H.X., Tsodikova, S.G., Synthesis and investigation of novel benzimidazole derivatives as antifungal agents. *Bioorg. Med. Chem.*, 24, 3680–3686, 2016.
9. Nafiz, O.C., Ulviye, A.C., Begum, N.S., Serkan, L., Bugra, K., Yusuf, O., Zafer, A., Ali, S.K., Synthesis, Molecular Docking Studies, and Antifungal Activity Evaluation of New Benzimidazole-Triazoles as Potential Lanosterol 14α-Demethylase Inhibitors. *J. Chem.*, 1–15, 2017.
10. Vijey, A.M. and Anbhule, S.J., Synthesis, Characterization and evaluation of antifungal properties of substituted benzimidazole analogue. *Asian J. Pharma. Clin. Res.*, 11, 4, 66–69, 2018.
11. Desai, N.C., Pandya, D.D., Joshi, V.V., Rajpara, K.M., Vaghani, H., V; Satodiya, H. M.; Synthesis, characterization and antimicrobial screening of hybrid molecules containing benzimidazole-pyrazole and pyridine nucleus. *Med. Chem. Res.*, 21, 4463–4472, 2012.
12. Suram, D., Thatha, S., Venkatapuram, P., Adivireddy, P., Synthesis and Antimicrobial Activity of a New Class of Benzazolyl Pyrazoles. *J. Heterocycl. Chem.*, 54, 6, 3152–3162, 2017.
13. Afaf, H.E., Fahmy, H.H., Ali Abdelwahed, S.H., Synthesis and Antimicrobial Activity of Some New Benzimidazole Derivatives. *Molecules*, 5, 1429–1438, 2000.

14. Malleshappa, N., Agrawal, S., Patel, H., Badiger, A., Gaba, M., Zambre, A., Synthesis, antimicrobial and cytotoxic activity of novel azetidine-2-one derivatives of 1H-benzimidazole. *Arab. J. Chem.*, 7, 2, 219–226, 2014.
15. Ansari, K.F. and Lal, C., Synthesis and evaluation of some new benzimidazole derivatives as potential antimicrobial agents. *Eur. J. Med. Chem.*, 44, 2294–2299, 2009.
16. Zhe, W., Xiangping, D., Shujuan, X., Runde, X., Juan, L., Liu, Z., Xiaoyong, L., Xua, C., Zhizhong, X., Yanming, C., Yunmei, L., Xing, Z., Guotao, T., Design, synthesis and biological evaluation of chrysin benzimidazole derivatives as potential anticancer agents. *Nat. Prod. Res.*, 32, 1–10, 2017.
17. Morais, G.R., Elisa, P., Fernanda, M., Lurdes, G., Maria, C.O., Antero, A., Hugo, V.M., Tiago, F.O., Isabel, S., Antonio, P., Synthesis and Biological Evaluation of Novel 2-Aryl Benzimidazoles as Chemotherapeutic Agents. *J. Heterocycl. Chem.*, 54, 1, 255–267, 2015.
18. Li, Y., Tan, C., Gao, C., Zhang, C., Luan, X., Chen, X., Liu, H., Chen, Y., Jiang, Y., Discovery of benzimidazole derivatives as novel multi-target EGFR, VEGFR-2 and PDGFR kinase inhibitors. *Bioorg. Med. Chem.*, 19, 15, 4529–4535, 2011.
19. Karpinska, M.M., Matysiak, J., Niewiadomy, A., Synthesis of novel 4-(1H-benzimidazol-2-yl)benzene-1,3-diols and their cytotoxic activity against human cancer cell lines. *Arch. Pharm. Res.*, 34, 10, 1639–1647, 2011.
20. Romero-Castro, A., Leon-Rivera, I., Avila-Rojas, L.C., Navarrete-Vazquez, G., Nieto-Rodriguez, A., Synthesis and preliminary evaluation of selected 2-aryl-5(6)-nitro-^{1}H-benzimidazolederivatives as potential anticancer agents. *Arch. Pharm. Res.*, 34, 2, 181–189, 2011.
21. Ranganatha, V.L., Vijay, B.R.V., Thirusangu, P., Prashanth, T., Prabhakar, B.T., Khanum, S.A., Synthesis, angiopreventive activity, and *in vivo* tumor inhibition of novel benzophenone-benzimidazole analogs. *Life Sci.*, 93, 23, 904–911, 2013.
22. Sontakke, V.A., Ghosh, S., Lawande, P.P., Chopade, B.A., Shinde, V.S.A., simple, efficient synthesis of 2-aryl benzimidazoles using silica supported periodic acid catalyst and evaluation of anticancer activity. *ISRN Org. Chem.*, 2013, 1–7, 2013.
23. Harkala, K., Eppakayala, L., Maringanti, T.C., Synthesis and biological evaluation of benzimidazole-linked 1,2,3-triazole congeners as agents. *Org. Med. Chem. Lett.*, 4, 14–18, 2014.
24. Risley, V.A., Henry, S., Kosyrikhina, M.V., Manzanares, M.R., Payan, I., Downer, C.D., Hellmann, C.C., Slambrouck, S.V., Frolova, L.V., 4-Amino-2-aryl-3-cyano-1,2-dihydropyrimido- [1,2–a]benzimidazoles and their pyrimidine analogs as new anticancer agents. *Chem. Heterocycl. Compd.*, 50, 2, 185–194, 2014.
25. Abdel-Aziz, H.A., Gamal-Eldeen, A.M., Hamdy, N.A., Fakhr, I.M.I., Immunomodulatory and anticancer activities of some novel

2-substituted-6-bromo-3-methylthiazolo[3,2-a]benzimidazole derivatives. *Arch. Pharm. Chem. Life Sci.*, 342, 230–237, 2009.
26. D'Amico, J.J., Campbell, R.H. Guinn, E.C., Derivatives of 3-Methylthiazolo [3, 2-a] benzimidazole1, 2. *J. Org. Chem.*, 29, 4, 865–869, 1964.
27. Hranjec, M., Pavlovic, G., Marjanovic, M., Kralj, M., Karminski- Zamola, G., Benzimidazole derivatives Related to 2,3-acrylonitriles, benzimidazo[1,2-a] quinolines and fluorenes: Synthesis, antitumor evaluation *in vitro* and crystal structure determination. *Eur. J. Med. Chem.*, 45, 6, 2405–2417, 2010.
28. Hranjec, M., Starcevic, G., Pavelic, S.K., Lucin, P., Pavelic, K., Zamola, G.K., Synthesis, spectroscopic characterization and antiproliferative evaluation *in vitro* of novel Schiff bases related tobenzimidazoles. *Eur. J. Med. Chem.*, 46, 6, 2274–2279, 2011.
29. Taher, A.T., Khalil, N.A., Ahmed, E.M., Synthesis of novel isatinthiazoline and isatin-benzimidazole conjugates as anti-breast cancer agents. *Arch. Pharm. Res.*, 34, 10, 1615–1621, 2011.
30. Singla, P., Luxami, V., Paul, K., Triazine-benzimidazole hybrids: Anticancer activity, DNA interaction and dihydrofolate reductase inhibitors. *Bioorg. Med. Chem.*, 23, 8, 1691–700, 2015.
31. Shi, L., Wu, T., Wang, Z., Xue, J., Xu, Y., Discovery of quinazolin- 4-amines bearing benzimidazole fragments as dual inhibitors of c-Met and VEGFR-2. *Bioorg. Med. Chem.*, 22, 17, 4735–4744, 2014.
32. Azam, M., Khan, A.A., Al-Resayes, S.I., Islam, M.S., Saxena, A.K., Dwivedi, S., Musarrat, J., Trzesowska-Kruszynska, A., Kruszynski, R., Synthesis and characterization of 2-substituted benzimidazoles and their evaluation as anticancer agent. *Spectro Chem. Acta Part A*, 142, 286–291, 2015.
33. Kishore Babu, P.N., Ramadevi, B., Poornachandra, Y., Kumar, C.G., Synthesis antimicrobial and anticancer evaluation of novel 2- (3-methylindolyl) benzimidazole derivatives. *Med. Chem. Res.*, 23, 9, 3970–3978, 2014.
34. Yoon, Y.K., Ali, M.S., Wei, A.C., Choon, T.S., Facile synthesis and evaluation of sirtuin inhibitory activity of novel benzimidazole dervatives. *Pharmacologia*, 617–622, 2013.

14

Synthesis and Biological Evaluation of Thiophene- and Piperazine-Based Active Compounds

Ayyagari Subramanya Sharma[1], Subhalaxmi Pradhan[1]* and Prashant Kumar[2]

[1]*Department of Chemistry and Sustainability, School of Basic Sciences, Galgotias University, Greater Noida, UP, India*
[2]*Department of Chemical Engineering, RMIT University, Melbourne, Australia*

Abstract

Heterocyclic compounds serve numerous applications in medicinal chemistry; therefore, scientists extensively examine them for drug development purposes because of their active biological mechanisms. Thiophene, a sulfur-containing five-membered heterocyclic compound, has the ability to exhibit antibacterial action in combination with anticancer properties and anti-inflammatory effects attributed due to its. Thiophene derivatives hold a privileged position in drug designs because their electron-rich properties enable stronger biological target contacts. Piperazine represents a common six-membered heterocycle containing nitrogen, which appears widely in pharmaceutical development mainly as anti-cancer and antihistamine, as well as central nervous system agents. Medicinal chemistry relies on piperazine as an important structural component because of its dual capability to adjust its shape and form hydrogen bonds. The present chapter evaluates various piperazine derivatives featuring significant biological effects focused on enzyme blocking abilities as well as neurotransmitter managing capabilities. The current study investigates the crucial role of thiophene and piperazine derivatives in medicinal chemistry regarding their various structures and pharmaceutical potentials. The heterocyclic systems demonstrate useful potential in therapeutic development through their dual effect that strengthens drug properties while improving receptor interactions. The research delivers structure–activity relationship information while demonstrating functionalization and synthesis approaches for thiophene and piperazine derivatives with an emphasis on modern

**Corresponding author*: subhalaxmipradhan@gmail.com

Divya Bajpai Tripathy, Priyanka Chhabra and Anjali Gupta (eds.) Heterocyclic Compounds as Pharmaceutical Ingredients, (475–508) © 2026 Scrivener Publishing LLC

developments. The chapter shows applications of heterocycles for handling significant healthcare issues while revealing new possibilities for treating multiple diseases.

***Keywords*:** Heterocyclic compounds, thiophene, piperazine, anticancer, antidiabetic, neurotransmitter

14.1 Introduction

Heterocyclic compounds are significant both in medicinal chemistry and organic chemistry because of their substantial activity. Heterocyclic compounds having hetero atom such as sulfur, nitrogen, and oxygen fetch the attention of medicinal chemists and researchers due to their various pharmacological profile [1–3]. Due to its presence in a number of pharmacologically active compounds, it has been hypothesized to be a crucial scaffold. The Greek words *thi-on*, which means "sulfur," and *ph-en*, which means "shining," are the origins of the name "thiophene" [4]. Thiophene has been dubbed the wonder heterocycle because of its various biological properties, including anticancer [5–9], antimicrobial [10–12], anti-inflammatory [13, 14], antidepressant [15], analgesic [16], and anticonvulsant [17]. The anticancer properties of compounds (Figure 14.1) based on thiophene have been extensively studied among all diseases. One of the most horrible illnesses is cancer, and new anticancer drugs are constantly needed [18]. Of the 14 million new cases of cancer worldwide each year, more than 1.1 million are reported in India. It is predicted that, by 2030, there will be approximately 26 million new cases of cancer and 17 million annual deaths from the disease [19]. Many researchers are making sincere efforts to use important heterocyclic moieties to meet the urgency for the detection and progress of powerful chemotherapeutic agents as well as to lessen the issues with presently existing anticancer drugs, such as toxicity and drug resistance [20, 21]. Researchers are always trying to create new anticancer drugs taking thiophene as main skeleton. Derivatives of mono-, di-, or tri-substituted thiophene have been recognized as anticancer agents for many years. As shown in Figure 14.1, it was found that there are numerous commercially available anticancer medications with thiophene nuclei, showing their effect *via* multiple pathways [22–32]. Chemically speaking, piperazine is 1,4-hexahydropyrazine, a versatile nitrogen-containing six-membered heterocyclic ring that can be chemically synthesized or found in a variety of naturally occurring bioactive products. In this sense, it is among the most crucial components of the range of clinically beneficial

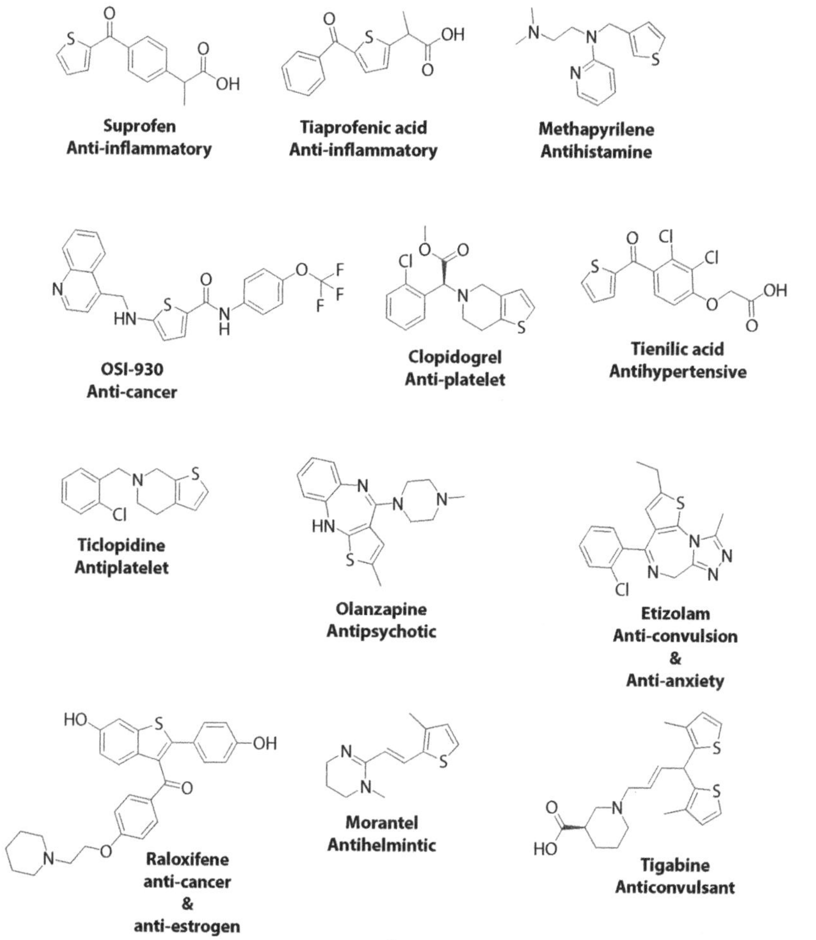

Figure 14.1 Anticancer drugs containing thiophene as main skeleton [22–32].

drugs used to treat erectile dysfunction, including anthelminthic, antiallergic, antihelminthic, anxiolytic, antipsychotic, and antidepressant medications. Among the most crucial are analgesic and anti-inflammatory drugs, which are always in high demand because they are used to treat a variety of conditions, including cancer, heart disease, gout, osteoarthritis, Alzheimer disease, and more. Nevertheless, they are linked to inevitable adverse effects such as renal problems, gastrointestinal bleeding, ulcers, etc. Therefore, there is an urgent need to develop new compounds that are more effective and less toxic.

14.2 Chemistry of Thiophene and Synthesis of Its Derivative

Thiophene is a five-membered heteroaromatic compound having a molar mass of 84.14 and a molecular formula of C_4H_4S. This compound was discovered as contaminant in benzene; it is practically soluble in almost all the organic solvents and insoluble in water. Electron pairs on sulfur are significantly delocalized in the π electron system and behave extremely reactive like benzene compounds. Thiophene forms an azeotrope with ethanol and benzene. Thiophene can be easily subjected to various types of organic reactions such as sulfonation, nitration, halogenation, and acylation. However, these compounds cannot be alkylated or oxidized. Well-known methods for synthesis of thiophene are Pall–Knorr synthesis, Gewald reaction, and Volhard–Erdmann reaction. Additionally, there are some traditional methods, which involve cyclizing β-keto sulfide to form 2-aryl benzothiophenes, another method involved Knoevenagel condensation, and some other methods are as depicted in the below general schematic representation [33]. Scientists are continuously searching for more efficient synthetic methods (Figure 14.2) to prepare thiophene-based analog compounds.

14.2.1 Metal-Catalyzed Synthetic Approaches

Metal-catalyzed synthetic approaches have, in general, transformed the synthetic chemical conversions, and same has been used quite extensively in synthesis of thiophene. In this regard, it has been found that copper-based catalysts such as copper acetate, copper iodide, indium, and rhodium-based catalysts were found to be giving good conversion as well as regioselectivity and allow introducing a wide variety of functional groups. Some of the synthetic methods reported [33] in literature are depicted in Figure 14.3.

14.2.2 Metal-Free Synthetic Approaches

Numerous green chemistry–based synthetic methods were used for preparation of thiophene; some of the examples as reported [33] are depicted in Figure 14.4.

14.2.3 Multicomponent Synthetic Approaches

Multicomponent reactions are generally one-pot synthetic methods involving multiple reactants with different substituents (Figure 14.5).

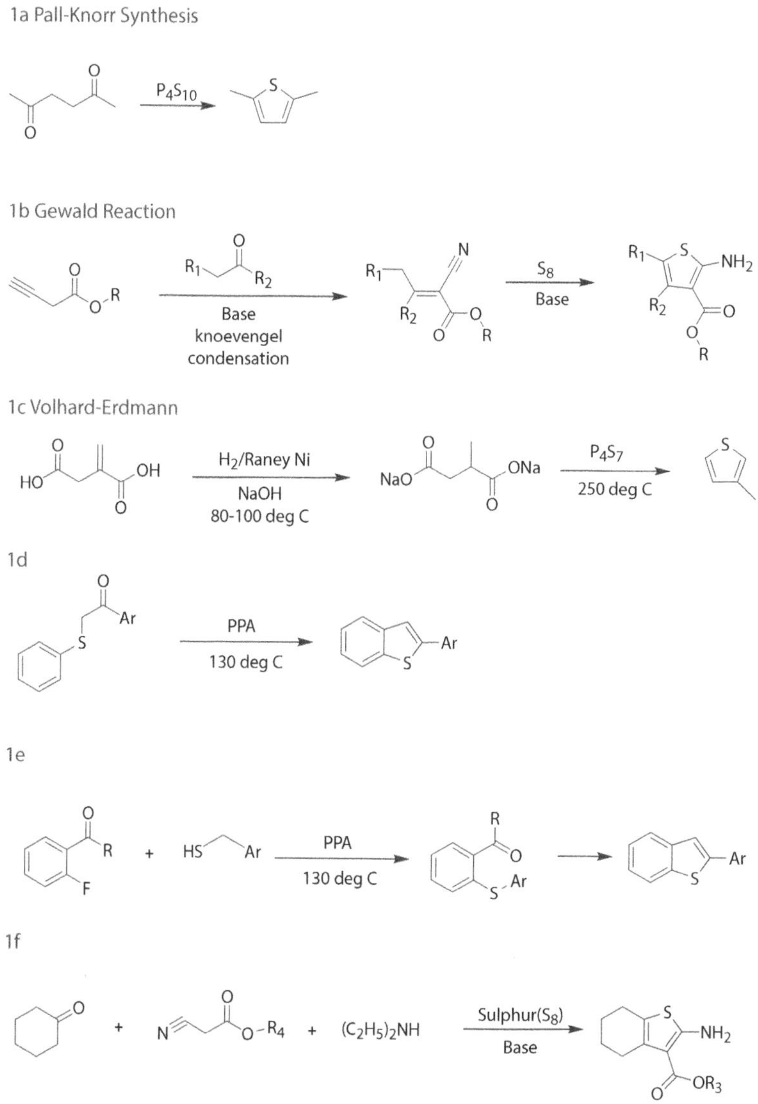

Figure 14.2 General synthesis methods of thiophene [33].

This type of reactions reduces waste and requires minimum operations enhancing overall efficiency; such processes are of great industrial importance because of its lower number of operations. Some of the synthetic methods reported [33] in literature are depicted as below.

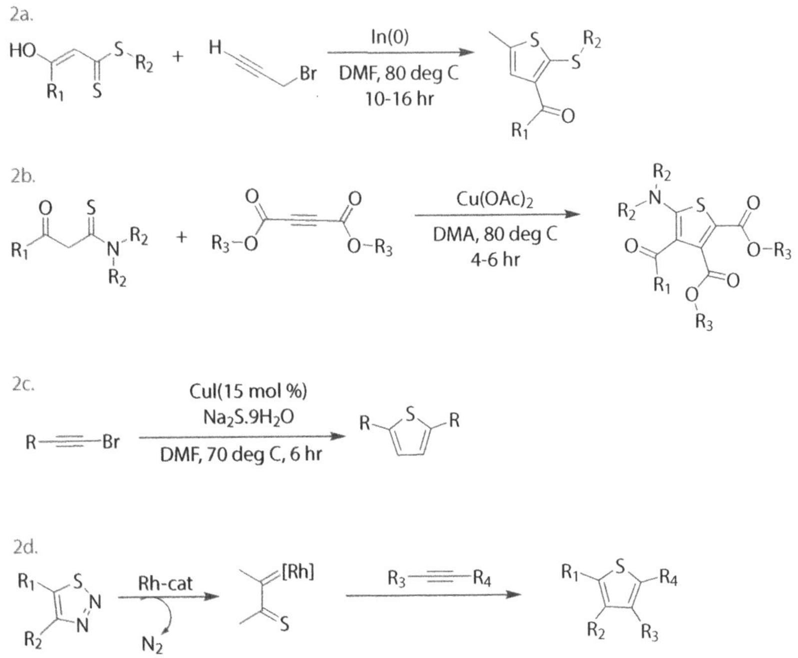

Figure 14.3 Metal-mediated synthesis of thiophene [33].

14.3 Biological Application of Thiophene and Its Derivatives

Thiophenes are known for their electron-rich properties like benzene, which enable them to interact with multiple biological targets, and this property makes them a favored molecule with receptor site responsible for molecule's biological activity. The US Food and Drug Administration approved 26 drugs having thiophene backbone in their structure; five of them were meant for treating inflammation, and four were in the cardiovascular category. Besides these 2 drugs, each was approved for anticancer activity, antimicrobial activity, and in chronic obstructive pulmonary disease category, one each for Parkinson disease, antipsychotic, antiepileptic, and antianxiety activity. Thiophene-based drugs are also favored in neurological drug category because their lipophilicity enables them to penetrate the blood–brain barrier.

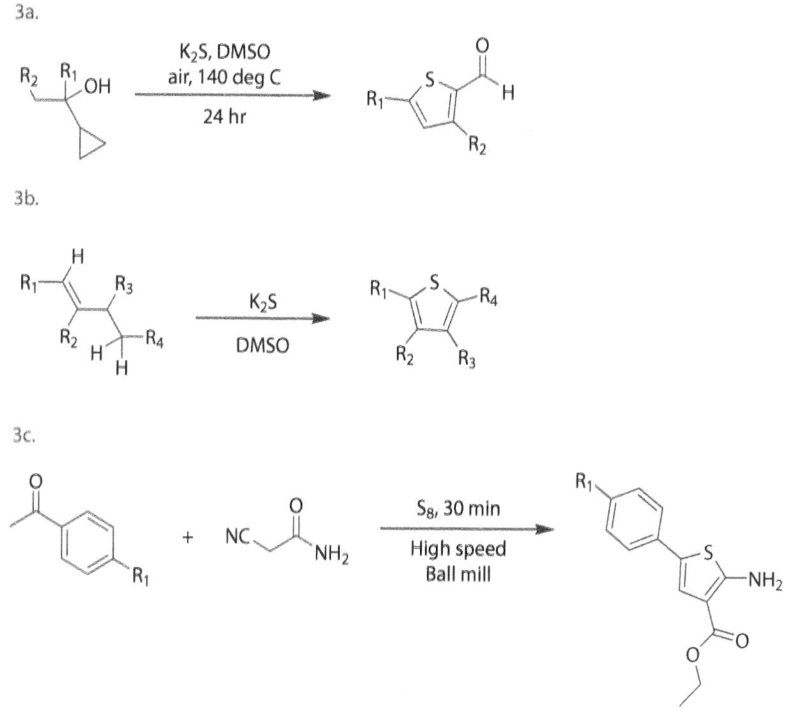

Figure 14.4 Metal-free synthesis of thiophene [33].

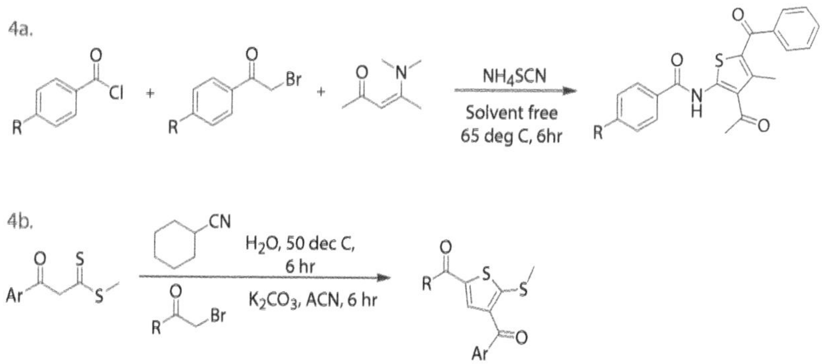

Figure 14.5 Synthesis of thiophene having multicomponents [33].

14.3.1 Mono-Substituted Thiophene Derivatives as Anticancer Agents

It has been suggested that a number of mono-substituted thiophene derivatives, primarily those at Positions 2 and 3, have anticancer properties. Saad *et al.* [34] synthesized thiophene derivatives with 1,2,4-triazino [3,4–b][1,3,4] thiadiazinones, 1,3,4-thiadiazole[2,3-c]triazinone and pyrazolo[5,1-c] [1,2,4] in 2011. As seen in Figure 14.5, the initial reaction between 3 and acid chlorides produced 4; similarly, treatment with various reagents and conditions, such as CS_2 in ethanolic KOH, aromatic aldehydes, and DCAA in ethanolic KOH, produced various products shown in Figure 14.6, such as 6, 7, and 8, respectively. Among all derivatives, Compound 8 showed the highest cytotoxic activity against the human cancer cell line HCT-116. Generally speaking, the compounds' antiproliferative properties were linked to their free thiol or thioether groups [34].

A collection of thiophene derivatives with substituents at the second position was also compiled by Ghorab *et al.* in 2014 [35]. Sulfonamides-substituted thiophene derivatives (11 and 12) were produced by the synthetic reaction, which was carried out by refluxing 3-(dimethylamine)-1-(thiophen-2-yl)prop 2-en-1-one (9) and substituted sulfonamides (10) in

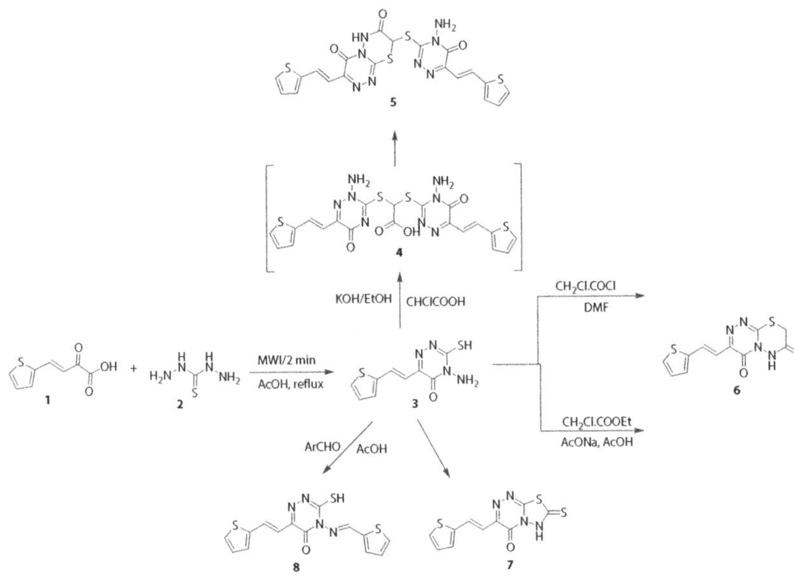

Figure 14.6 Various thiophene derivatives reported cytotoxic against HCT-116 cancer cell line [34].

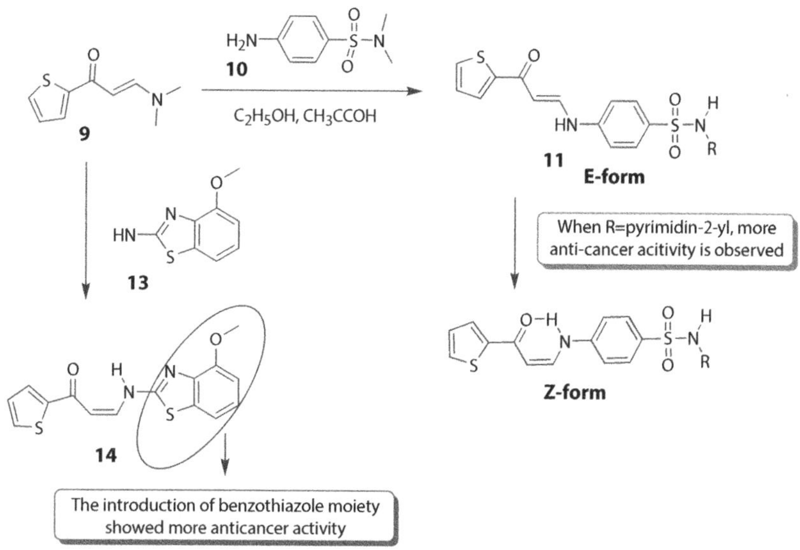

Figure 14.7 Sulfonamides and benzothiazole-substituted thiophene derivatives as anticancer agents [35].

ethanol and acetic acid (Figure 14.7). Likewise, under similar circumstances, Compound 9 was refluxed with 2-amino 4-methoxy benzothiazole 13 to create 14, a benzothiazole-substituted thiophene derivative. Using doxorubicin as a standard, these synthetic compounds were further examined for their anticancer activity against the human breast cancer cell line MCF7. When compared to the standard drug, which has an IC_{50} of 32.00 μmol L^{-1}, Compounds 12 and 14 demonstrated four times the anticancer activity, with respective IC_{50} values of 9.55 and 9.39 μmol L^{-1}.

It was discovered recently using the structure–activity relationship (SAR) approach that adding a benzothiazole fragment enhanced the anticancer effect and that modifying sulfonamides pyrimidine-based also increased the activity [35].

14.3.2 Di-Substituted Thiophene Derivatives as Anticancer Agents

Bichalcophene monoamides were synthesized in 2012 by Youssef *et al.*, [36] as possible effective antiproliferative agents. Figure 14.8 [36] shows how LiN $(TMS)_2$ (lithium bis (trimethylsilyl)amide) was used to prepare the desired compounds from corresponding mononitrile bithiophene derivatives.

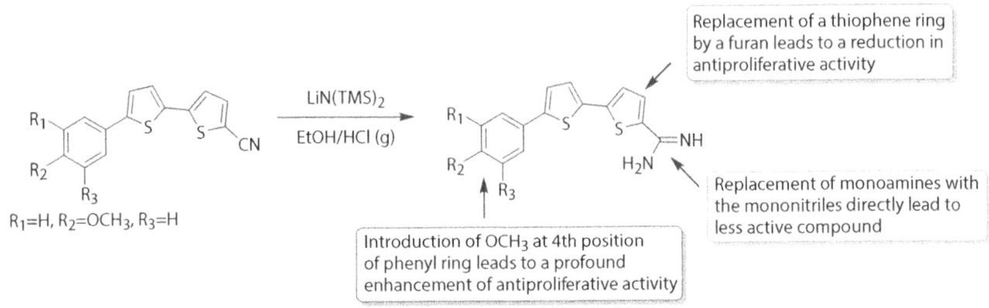

Figure 14.8 SAR profile of thiophene-based bichalcophene monoamides as antiproliferative agents [36].

Compounds were tested on 60 distinct cancer cell lines using a broad-based anticancer assay. The majority of the synthesized compounds demonstrated good activity against the A549 cell line of non–small cell lung cancer. The most active of all the compounds tested was the one created in Scheme 7, with a GI_{50} value of 0.3 μM, indicating submicromolar activity. DNA nuclease-like and cleavage efficacy was found to be the key components of the activity. Interestingly, replacing the thiophene ring with furan resulted in a decrease in the antiproliferative effect; in non–small cell lung and breast cancer cell lines, the effect decreased even more when furans were used to replace both thiophene rings.

14.3.3 Tri-Substituted Thiophene Derivatives as Anticancer Agents

Thomas and colleagues (2017) synthesized methyl 2-alkyl-3-amino-3-carboxylate substituted thiophene, which has the capacity to eradicate various myeloid and other cancer cell lines, including T-lymphoma CEM and Molt-4, prostate PC-3, kidney Caki-1, and hepatoma Huh-7 tumor cells [37]. The targeted derivatives were synthesized using the Gewald reaction, which involved reacting the aliphatic carbonyl compound with activated nitrile, elemental sulfur, and triethylamine (NEt3) to produce a new compound (Figure 14.9). This compound, with IC_{50} values of 0.18 ± 0.08 μM and 0.057 ± 0.026 μM, respectively, possesses promising cytotoxicity against CEM and Molt/4.

According to the SAR data analyzed here, substituting higher alkyl derivatives such as pentyl, hexyl, heptyl, octyl, and nonyl for propyl and butyl is toxic to all sensitive tumor cell lines at submicromolar concentrations. Propyl and butyl substitutions were observed to exhibit low micromolar inhibition, whereas methyl and ethyl groups at the C-5 position did

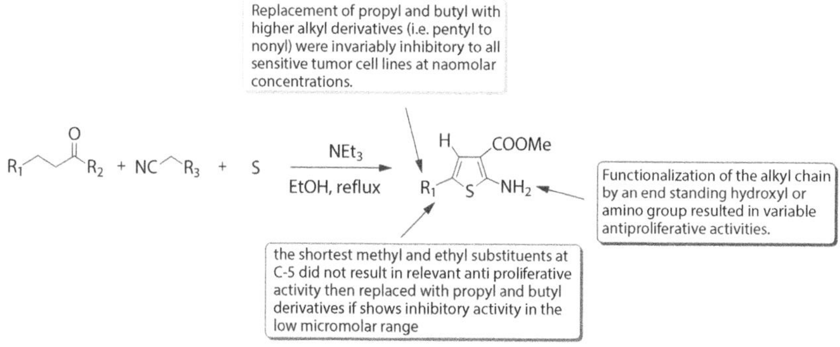

Figure 14.9 SAR profile of 5-alkyl-substituted methyl-2-amino-3-carboxylate thiophenes as antiproliferative agents [37].

not exhibit antiproliferative effects. Variable and moderate antiproliferative effects were shown by the final series of functionalization of the alkyl chain by an end-standing hydroxyl or amino group.

There are many derivatives of thiophene, which showed anticancer activity, which is reported in Figure 14.10.

14.3.4 Biological Application of Thiophene Derivatives as Antimalarial Agent

Novel compounds with strong antimalarial activity are being investigated by researchers due to the pressing need for new therapeutic agents. Because of their distinct chemical characteristics and biological activities, thiophene derivatives have demonstrated the most promise among them [38]. Numerous investigations have demonstrated the effectiveness of thiophene derivatives against *Plasmodium* species, with different modifications improving their specificity and bioactivity [39]. The structural flexibility of thiophene derivatives provide a wide range of synthetic modifications, leading to the creation of extremely effective antimalarial compounds. Benzophenone derivatives have been successfully synthesized by researchers and have shown notable activity against strains of *Plasmodium* that are resistant to chloroquine [40]. These compounds' effectiveness has been further enhanced by the addition of carboxamide groups and other substituents that withdraw electrons. Thiophene derivatives work against malaria in a number of ways. Research shows that these substances disrupt vital biological processes of the *Plasmodium* parasite, including ATPase inhibition and heme detoxification [41]. Additionally, molecular docking

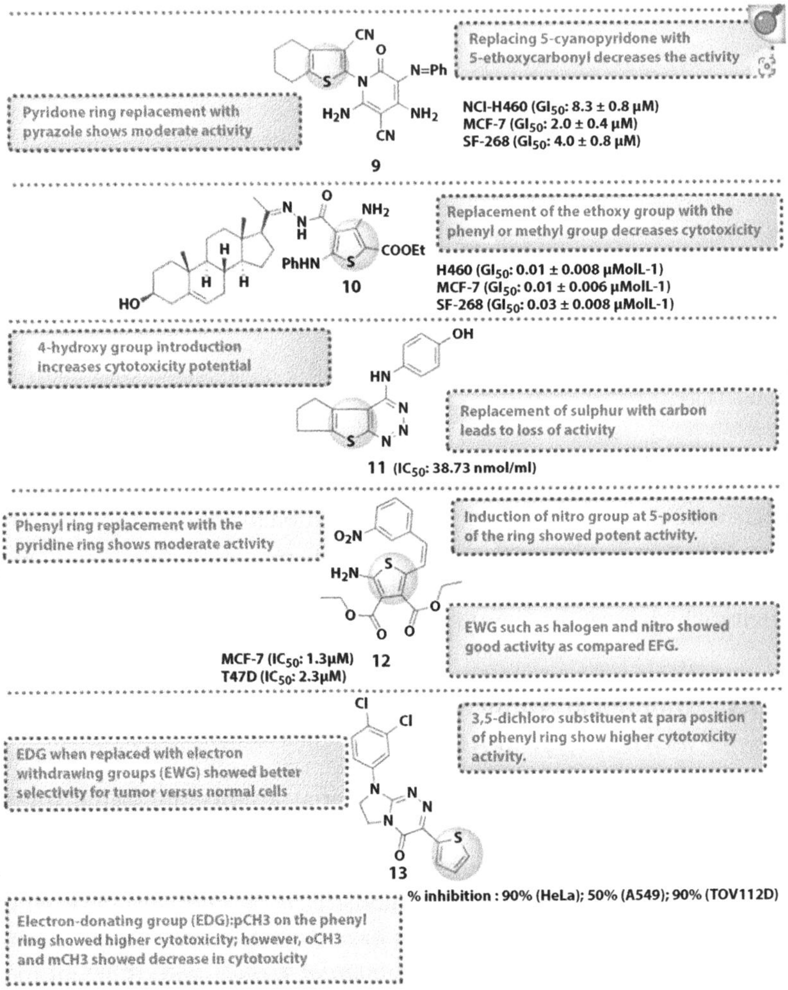

Figure 14.10 Derivatives of thiophene having anticancer activity [37].

studies have demonstrated that thiophene-based compounds and essential *Plasmodium* enzymes have strong binding interactions [42]. A chemically and efficiently straightforward approach is described for the synthesis of tetra-substituted thiophene compounds (Figure 14.11). The synthesized compounds were characterized by IR (Infrared), 1H NMR (Nuclear magnetic resonance), 13 C NMR, spectra and Mass Spectra (MS) and found to exhibit antimalarial activity.

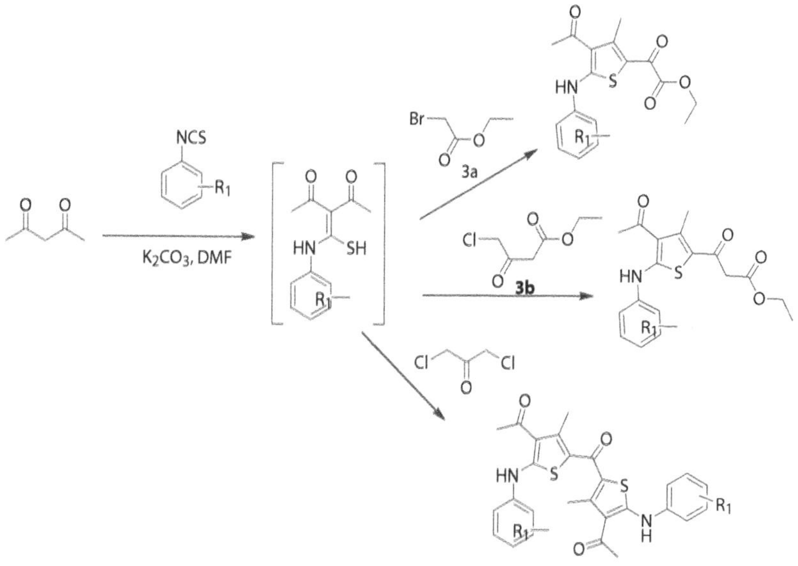

Figure 14.11 Synthesis of tetra-substituted thiophene derivatives as antimalarial agent [10].

14.3.5 Biological Application of Thiophene-Based Compounds as Anti-Inflammatory Agent

Thiophenes are important heterocyclic compounds very relevant in the synthesis of many agrochemical and pharmaceutical agents as observed from the examples in Figure 14.12, including the nonsteroidal anti-inflammatory drug, lornoxicam, the thiophene derivative of piroxicam. Originally, the strategy of researchers was to prepare new series of substituted cinnoline derivatives condensed with thiophene fragments. All the synthesized compounds were tested for their efficacy to inhibit inflammation.

14.3.6 Biological Applications Thiophene-Based Compounds as Antidepressant

Compounds based on thiophene have attracted a lot of attention in medicinal chemistry because of their various biological activities, which include the possibility that they could act as antidepressants. To modulate neurotransmitter activity, which is essential for treating depression, thiophene moieties have shown promise in drug design. With reference to current research, this paper investigates the synthesis methods, biological significance, and antidepressant potential of thiophene derivatives.

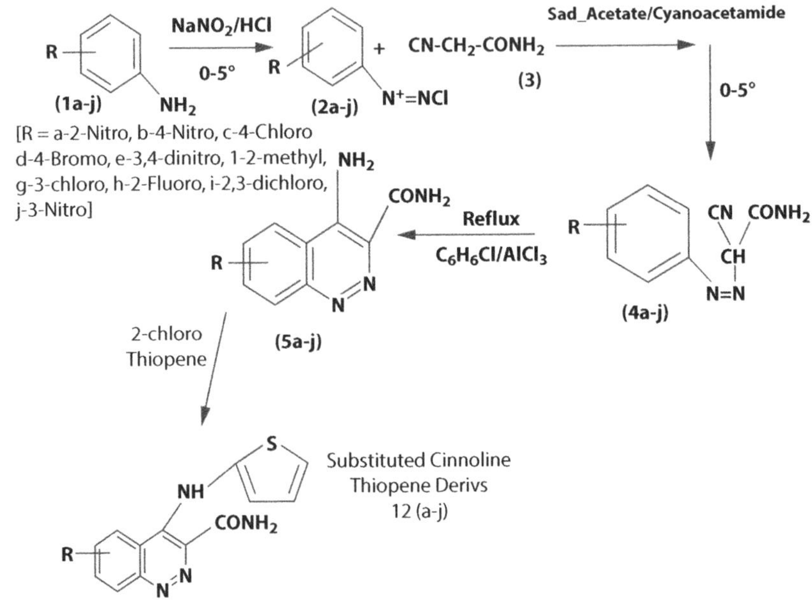

Figure 14.12 Synthesis of substituted cinnoline thiophene derivatives [13].

Thiophene derivatives are thought to have antidepressant properties because they can alter important neurotransmitters such as dopamine, serotonin, and norepinephrine. 3-(Dimethylamino)-1-(thiophen-2-yl) propan-1-ol (DMATP) interacts with neurotransmitter receptors and inhibits reuptake mechanisms to produce notable antidepressant activity [43]. The binding affinity of DMATP to common antidepressant medication targets is further supported by *in silico* investigations. Using both *in vitro* and *in vivo* models, several studies have evaluated the antidepressant effectiveness of compounds based on thiophene. In their thorough review of sulfur-containing heterocycles, including thiophene derivatives, Kapoor *et al.* [44] show how promising these compounds are as lead compounds in the development of antidepressant medications [44]. Changes made to the thiophene ring at particular locations improve its pharmacodynamic characteristics, according to the SAR analysis. A promising approach to the development of antidepressant medications is the use of thiophene-based compounds. They are positioned as viable candidates for future clinical applications due to their capacity to interact with neurotransmitter pathways and advancements in synthetic methodologies. To improve their pharmacokinetic profiles and guarantee safety in long-term therapeutic use, more investigation is required.

14.3.7 Biological Applications of Thiophene Analogs as Anti-Inflammatory and Analgesic

Thiophene-based compounds have shown significant anti-inflammatory effects, mainly by inhibiting important inflammatory mediators such as lipoxygenase (LOX), cyclooxygenase (COX), and tumor necrosis factor α (TNF-α). According to research, thiophene derivatives can successfully control inflammatory pathways by lowering the generation of reactive oxygen species (ROS) and proinflammatory cytokines [45]. A number of derivatives based on thiophene-2-ylmethylene were created, and their anti-inflammatory properties assessed. According to the results, these substances significantly reduced the production of TNF-α, a key mediator in chronic inflammatory diseases [45]. Furthermore, it has been noted that alterations to the thiophene ring structure, such as halogen substitutions, increase the anti-inflammatory potency [46]. Additionally, the molecular basis of thiophene-based anti-inflammatory agents was reviewed recently, with a focus on their potential to function as dual COX-2 and 5-LOX inhibitors, which are important targets for inflammation control [47]. The molecular docking studies further validated the therapeutic relevance of thiophene analogs by confirming their strong binding affinities with inflammatory enzymes.

Researchers synthesized a new series of thieno[2′,3′:4,5] pyrimido[1,2-b][1,2,4]triazines and thieno[2,3-d] [1,2,4]triazolo[1,5-a]pyrimidines [15, 16]. Diclofenac Na, a common medication, was used to test the newly synthesized compounds' effectiveness and potency against pain and inflammation. In acute and subacute models, the Compounds 9, 13, and 14a from the thienotriazolopyrimidine series in Figure 14.13 demonstrated

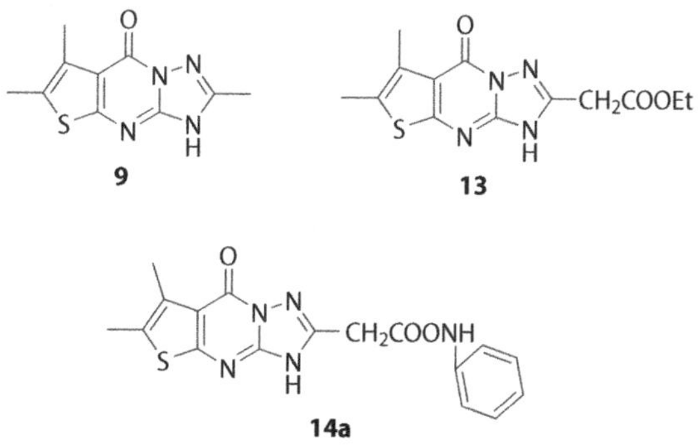

Figure 14.13 Thieno triazolo pyrimidine derivatives with promising anti-inflammatory and analgesic activities [15].

a distinctive anti-inflammatory effect, in addition to a satisfactory analgesic activity, albeit with a 16-hour onset time. Thiophene analogs' capacity to target various inflammatory pathways and pain receptors makes them a promising class of analgesic and anti-inflammatory drugs.

14.3.8 Biological Applications of Thiophene Analogs as Anticonvulsant

The sulfur-containing heterocycle thiophene has attracted a lot of interest in medicinal chemistry because of its many biological activities, most notably its anticonvulsant qualities. Thiophene moieties have been incorporated into a variety of chemical scaffolds to create compounds that are more effective at managing seizures [38]. The type and location of substitutions on the thiophene ring affect the anticonvulsant potential of thiophene derivatives. In preclinical models, for example, some thiophene carboxamide analogs have shown notable anticonvulsant activity [48]. Furthermore, because of their distinct structural characteristics, benzothiophene derivatives have been investigated for their therapeutic uses, such as seizure control [49]. Derivatives of thiophene have been shown to have a variety of biological properties, including anticonvulsant properties. The synthesis of novel anticonvulsant agents has been reported by Kulandasamy and colleagues [17]. They synthesized and evaluated the anticonvulsant efficacy of 3,4-di(substitute)oxy-N2,N5-bis(substituted)thiophene-2,5-dicarbohydrazides with active groups (Figure 14.14).

Furthermore, thiophene's fusion with other heterocyclic systems has been a calculated strategy in drug development. After being synthesized and tested, compounds such as chalcone-based 1,5-benzodiazepines with thiophene rings have shown promise as possible anticonvulsant agents [50]. These results highlight how thiophene can be used to improve the pharmacological profiles of anticonvulsant medications.

Figure 14.14 Thiophene derivative as anticonvulsant agents [17].

14.3.9 Biological Applications of Thiophene-Based Compound as Antidiabetic Agent

Diabetes is a chronic metabolic disorder illustrated by insufficient insulin production. Diabetes increases the risk of various diseases such as cardiovascular diseases, kidney diseases, retinopathy, cancer, and peripheral neuropathy. The primary strategy used for targeting diabetes includes the modulation of insulin receptors and signaling. Considering thiophene in antidiabetic activity summary of antidiabetic drugs bearing the thiophene ring is presented in Figure 14.15 [18]. Joshi *et al.* [51] synthesized and characterized a number of new benzo[b]thiophene-2-carboxylic acid derivatives (3a-p) [51]. These substances were assessed for their ability to prevent diabetes, with a focus on α-amylase inhibition. In contrast to the standard medication acarbose, which had an IC_{50} of 18.0 ± 0.2 μM, Compound 3b demonstrated the strongest α-amylase inhibitory activity among the synthesized derivatives, with an IC_{50} value of 12.5 ± 0.5 μM. These results were corroborated by molecular docking studies, which showed that Compound 3b can effectively interact with the α-amylase enzyme's active site, indicating that it could be a lead compound for the development of antidiabetic drugs.

In their thorough review of thiophene and its derivatives, Thakur *et al.* [52] highlighted a variety of biological activities, including antidiabetic effects. In addition to discussing SAR studies that investigate the chemical groups responsible for therapeutic activities, the review highlighted the importance of structural bioinformatics in drug discovery [52]. Furthermore, Ullah *et al.* [53] looked into the potential of novel thiazole derivatives based on thiophene carbaldehyde in the treatment of diabetes [53]. To comprehend the compounds' binding interactions with the enzyme, *in silico* studies were carried out, and their *in vitro* α-amylase inhibitory activity was assessed. The results indicated that these derivatives based on thiophene might be good options for creating novel antidiabetic medications. The potential of thiophene-based compounds in the creation of potent antidiabetic treatments is highlighted by these studies taken together.

14.4 Biological Application of Piperazine-Based Compounds

Piperazine nucleus has attracted considerable attention of medicinal chemist due to its broad spectrum of biological activity in the field of

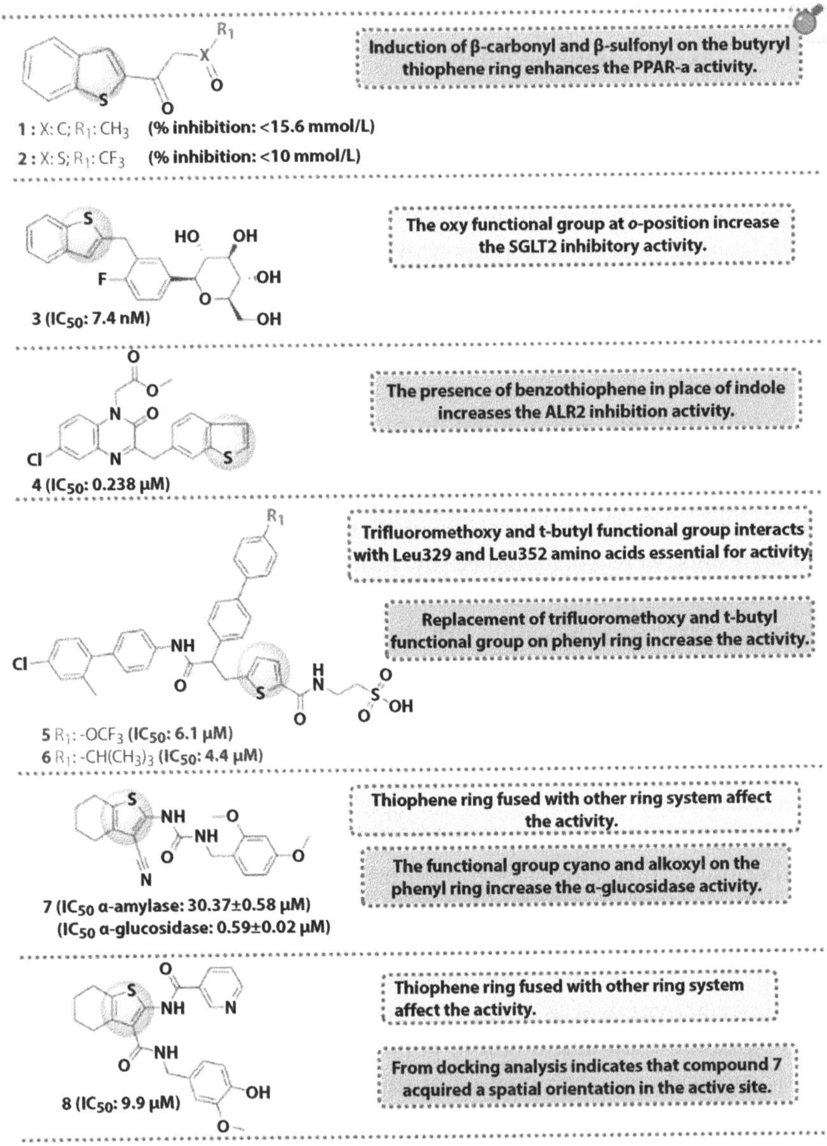

Figure 14.15 Thiophene derivative as antidiabetic agents [18].

pharmaceutical science. Piperazine scaffolds are a six-membered heterocyclic compound with two nitrogen atoms in 1,4 positions. In 1954 piperazine was introduced as anthelmintic drug. Piperazine is a white solid, which is obtained as leaflets when crystallized from alcohols. Piperazine ranks as the third most common nitrogen heterocycle in drug discovery, and it is the key component of many blockbuster drugs such as imatinib, dasatinib, levofloxacin, quetiapine, and sitagliptin and across the drug categories. The incorporation of piperazine is an important synthetic strategy in drug discovery due to its versatility, water solubility, alkaline nature, capacity for formation of hydrogen bonding, and change physicochemical properties.

14.4.1 Piperazine as an Antimicrobial Agent

Microbial infections are life-threatening concerns in several areas, which include healthcare products, biomedical sector, water purification systems, and food packaging industry. In 2019, Jalageri *et al.* [54] designed novel piperazine methacrylate monomer and its homopolymers. Synthesized piperazine-based polymer is used in wound dressing, textile field, water purification systems, and biomedical field (Figure 14.16). These polymers were tested for antimicrobial activity against bacteria such as *Escherichia coli* and *Staphylococcus aureus*, using drugs such as ciprofloxacin and fluconazole.

14.4.2 Anticancer Activity of Piperazine

The compounds based on piperazine have drawn a lot of interest in medicinal chemistry because of their potential as anticancer agents. These heterocyclic compounds have a variety of pharmacological characteristics, especially when it comes to preventing the growth of cancer cells. Numerous investigations have examined the effectiveness of piperazine derivatives in targeting different cancer cell lines, showing encouraging cytotoxic

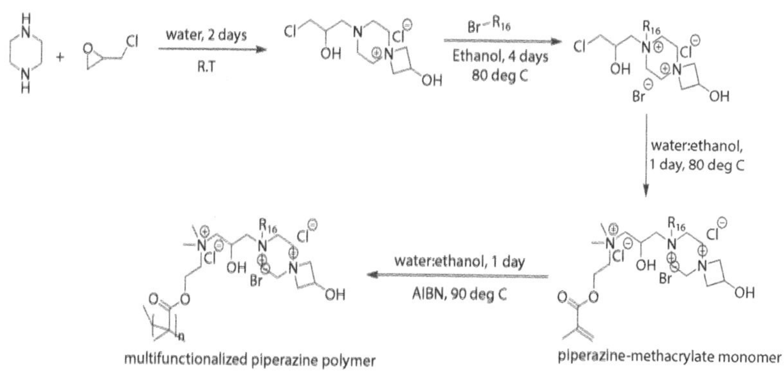

Figure 14.16 Piperazine derivative as antimicrobial agents [54].

and antiproliferative effects. Piperazine derivatives' capacity to disrupt important cellular functions is thought to be the source of their anticancer properties. By triggering apoptosis, interfering with the progression of the cell cycle, and blocking particular enzymatic pathways, a variety of substances demonstrate cytotoxicity. Interestingly, it has been discovered that piperazine-linked compounds selectively inhibit cyclin-dependent kinases (CDKs), which stops the growth of tumor cells [55]. Further clarifying their antiproliferative potential, docking studies have also revealed strong interactions with androgen receptors [56]. The anticancer potency of piperazine has been considerably increased by the synthesis of piperazine bound to different heterocyclic moieties. Piperazine–amide hybrids have been successfully developed, according to studies, and they show exceptional cytotoxicity against a variety of human cancer cell lines [57]. The selectivity and potency of these compounds have been further enhanced by structural changes, such as the addition of triazole or phthalazine groups [58].

The efficacy of piperazine-based compounds has been confirmed by numerous preclinical analyses. Novel vindoline–piperazine conjugates, for example, showed potent inhibitory effects on 60 human cancer cell lines, highlighting their broad-spectrum activity [59]. Comparably, substituted piperazines have demonstrated strong cytotoxicity against models of breast cancer, with IC_{50} values falling within the therapeutic range [60]. Additionally, it has been determined that piperazine-linked 1,8-naphthalimide derivatives are promising multitarget antitumor agents with low toxicity profiles. A derivative of piperazine amide was synthesized, characterized, and tested for anticancer activity by Sagar *et al.* [57]. By coupling 1-piperonyl piperazine derivative with benzoic acid, a novel 4-(benzo[1,3] dioxol-5-ylmethyl)piperazine amide derivative was created. Compound 3 (Figure 14.17) showed promise for anticancer activity after being synthesized; characterized by IR, NMR, and MS; and tested for *in vitro* cytotoxic studies.

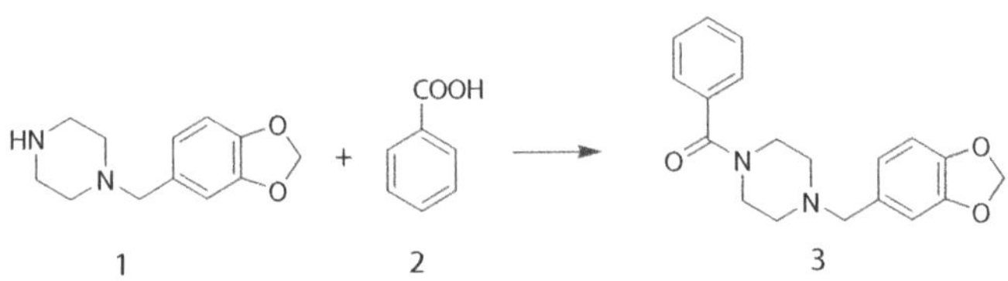

Figure 14.17 Piperazine derivative as anticancer agents [54].

14.4.3 Antioxidant Activity of Piperazine

The synthesis and assessment of compounds containing piperazine for possible antioxidant effects have been the subject of recent research. The antioxidant activity of α-aminophosphonate derivatives containing the piperazine-1,2,3-triazole moiety was also assessed by Kandula *et al.* [61] using the DPPH and H_2O_2 methods [61]. According to the study, these substances successfully donated hydrogen atoms, which enhanced their antioxidant capacity. Another study by da Silva Moreira *et al.* [62] investigated the piperazine derivative's antioxidant properties. It also demonstrated neuropharmacological advantages by reducing inflammatory and behavioral changes brought on by oxidative stress [62]. The capacity of compounds based on piperazine to donate electrons or hydrogen atoms stabilizes ROS and inhibits oxidative damage, which is the basis for their antioxidant activity. The protective function of these compounds in biological systems was further supported by Kilic-Kurt and colleagues' [63] report that piperazine-linked heterocycles with a 1,3,4-oxadiazole linker demonstrated significant inhibition of oxidative stress–related enzymes [63]. Al-Masoudi *et al.* [64] synthesized and assessed 1-(4-chlorobenzhydryl) piperazine derivatives, demonstrating through computational and *in vitro* investigations their capacity to reduce oxidative damage [64].

The antimicrobial and antioxidant properties of 1-[1,4-benzodioxane-2-carbonyl] piperazine derivatives were synthesized and reported by Mallesha and Mohana in 2011 [65]. Catechol and ethyl 2,3-dibromopropionate were condensed in dry acetone to create 1,4-benzodioxan-2-carboxylic acid (1), which is exhibited in Figure 14.18. In methylene dichloride (MDC), (1) was reacted with thionyl chloride to produce 1,4-benzodioxan-2-carbonyl chloride (2). Using piperazine in dry DMF at 80°C, 1-(1,4-benzodioxane-2-carbonyl) piperazine (3) was created from (2). In MDC, intermediate (3) is further reacted with different sulfonyl and acid chlorides to produce 1,4-benzodioxan-2-carboxylic acid derivatives, which yield high yields of 4 and 5.

Future research should focus on optimizing the SAR of piperazine derivatives to enhance their antioxidant efficacy. *In silico* studies, molecular docking, and *in vivo* evaluations will further validate their therapeutic potential.

14.4.4 Antitumor Activity of Piperazine

The ability of piperazine-containing heterocycles to stop the growth of cancer cells and trigger apoptosis in various cancer cell lines has been demonstrated in a number of studies. In 2020, Wen Li and colleagues [66] used

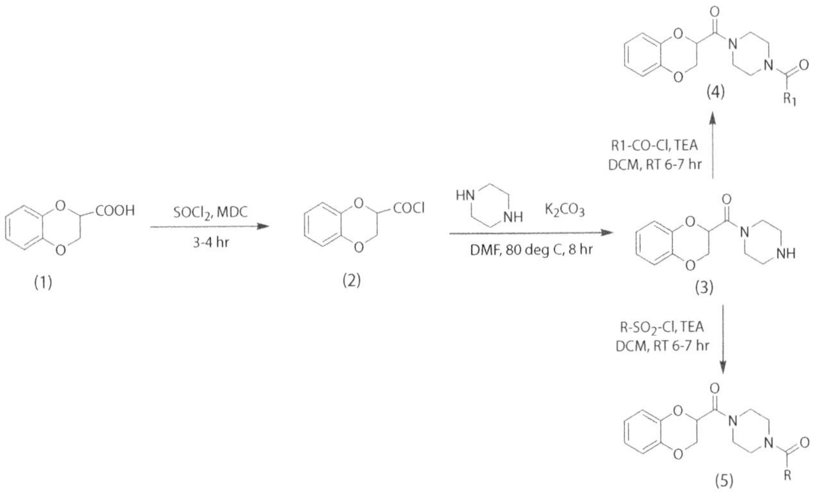

Figure 14.18 Synthesis of piperazine derivative as an antioxidant agent [65].

a substitution reaction with 6,7 di-substituted-4-chloroquinazoline and benzyl piperazine to create a novel quinazoline derivative that contained piperazine analogs. Because of their capacity to interact with important biological targets, compounds based on piperazine display a broad range of biological activities, including anticancer properties. Benzothiazole–piperazine derivatives demonstrated strong antitumor activity against MIAPACA pancreatic cancer cells, according to Al-Ghorbani *et al.* [67] [67]. To improve their potency and selectivity, these derivatives underwent N-terminal modifications. Similarly, Sharma *et al.* [56] discussed the use of green chemistry methods to produce piperazine-containing anticancer medications. They emphasized that strong compounds with low environmental impact and high efficacy could be produced using environmentally friendly synthetic techniques [56]. New piperazine-tethered phthalazines were discovered by Akl *et al.* [55] to be selective CDK1 inhibitors. Their research also showed that p-methoxy benzyl substitution increased anticancer activity, especially against pancreatic cancer [55]. When Wang *et al.* [68] synthesized dehydroabietyl piperazine dithiocarbamate ruthenium (II) polypyridyl complexes, they found that cancer cells experienced cell cycle arrest and severe DNA damage [68]. These results imply that piperazine complexes based on metals may be useful as chemotherapeutic agents. Additionally, Rasheed *et al.* [69] investigated the dual function of piperazine derivatives as anticancer and antihistamine agents, showcasing their adaptability in pharmacological applications [69].

The structural modification of natural products with a piperazine skeleton was also reviewed by Zhang *et al.* [70], who highlighted chalcone-piperazine hybrids that have strong anti–cervical and anti–lung cancer cell activity [70].

14.4.5 Piperazine: Insecticidal Activity

Zhang *et al.* found that piperazine and its derivatives have strong insecticidal properties, which makes them useful for agricultural protection and pesticide discovery. According to studies, the ability of compounds containing piperazine to disrupt insects' nervous systems—often by targeting acetylcholinesterase and other vital enzymes—is linked to their insecticidal efficacy [71]. Li *et al.* [72] investigated new heterocyclic piperazine-containing mono-, di-, and tri-amide derivatives and discovered that certain structural alterations improved insecticidal efficacy [72]. It was discovered that the presence of functional groups such as amides, pyridine rings, and halogen substitutions greatly increased bioactivity [71]. New developments show that several derivatives of piperazine have potent insecticidal effects. He *et al.* [73] confirmed the potential of piperazine in the development of biopesticides by discovering piperazine-based dimeric indole alkaloids that were highly effective against *Galleria mellonella* [73]. N-substituted piperazine-containing phenylpyridines were synthesized by Xu *et al.* [74] and showed encouraging insecticidal activity against important agricultural pests [74]. By using N-alkylation and addition reactions, Li and colleagues [75] created piperazine derivatives in 2008 that contained pyridinemethyl/thiazolemethyl derivatives. Certain compounds exhibited good insecticidal activity against *Culex pipiens* but weak herbicidal and insecticidal activity against *Aphis fabae.*

14.4.6 Piperazine: Antidiabetic Activity

The potential therapeutic uses of piperazine and its derivatives, especially in the management of diabetes mellitus, have drawn a lot of interest in medicinal chemistry. Several changes to the piperazine core have been investigated in recent research to improve its antidiabetic effectiveness. The mechanism of action and potential as dipeptidyl peptidase-4 (DPP-4) inhibitors are highlighted in this review of the antidiabetic activity of piperazine derivatives. The primary mechanism by which piperazine derivatives reduce blood sugar is by inhibiting DPP-4, an enzyme that breaks down incretin hormones such as glucose-dependent insulinotropic polypeptide and glucagon-like peptide-1. Piperazine-based compounds prolong the activity of these hormones by inhibiting DPP-4, which lowers blood glucose levels and increases insulin secretion [76]. Novel piperazine derivatives with encouraging antidiabetic activity have been designed and synthesized, according to several studies. A number of piperazine sulfonamide derivatives were synthesized and assessed by Kalli and Velmurugan [76] as DPP-4 inhibitors. *In vivo*, these substances showed notable glucose-lowering effects. To improve antidiabetic efficacy, Özil *et al.* [77] investigated the combination of piperazine and benzimidazole moieties. According to their research, these hybrid compounds had strong glucose-lowering and antioxidant properties [77]. Paliwal *et al.* [78] identified N-arylsulfonyl-N-2-pyridinyl-piperazines as putative DPP-4 inhibitors using molecular modeling techniques [78].

Chalcone–sulfonyl piperazine derivatives were created by Nasab *et al.* in 2025, and they showed potent α-amylase and α-glucosidase inhibitory properties [79]. Piperazine derivatives have better antidiabetic qualities as a result of structural changes, which make them promising candidates for additional clinical research. They may be used as substitutes for current antidiabetic medications such as sitagliptin because of their capacity to block important enzymes involved in glucose metabolism.

Devine *et al.* produced 41 compounds derived from 3a (RTC1) to complete a detailed SAR analysis in 2020 [80]. The study discovered several compounds, which showed both inhibitory action against NADH:ubiquinone oxidoreductase and stimulating properties for glucose uptake *in vitro* testing. Defining factor among the examined compounds was Compound 3a (RTC1), which elevated glucose uptake by 2.57 times compared to vehicle control and reduced NADH:ubiquinone oxidoreductase by micromolar concentrations.

14.4.7 Piperazine Derivatives as Anti-Inflammatory and Analgesic Agents

The potential of piperazine as a scaffold for the creation of innovative therapeutic agents that target inflammation and pain has been investigated in a number of studies. By inhibiting COX enzymes, modifying the production of nitric oxide, and interacting with opioid receptors, piperazine derivatives have anti-inflammatory and analgesic effects. The piperazine derivative LQFM202 showed promise as a pain reliever by lowering the nociceptive response in animal models, according to Martins *et al.* [81]. Similarly, Buran *et al.* [82] observed that piperazine-substituted coumarins showed inhibition of prostaglandin E_2, a crucial pathway in the regulation

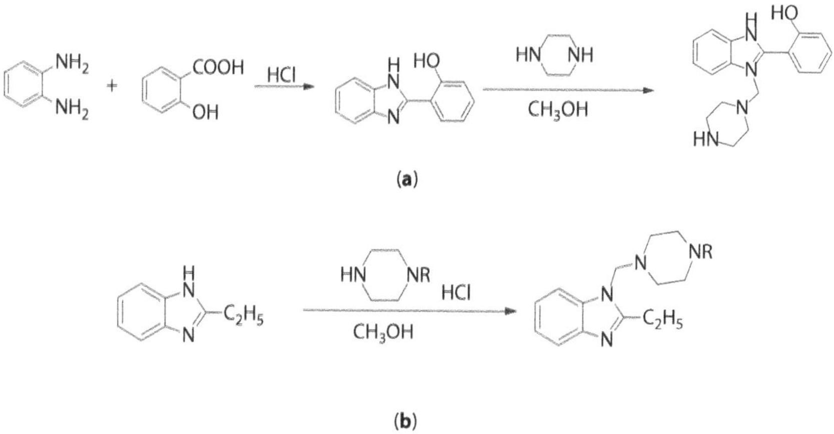

Figure 14.19 Synthesis of piperazine derivatives of (a) 2-hydroxyphenyl benzimidazole and (b) 2-ethyl substituted benzimidazole [83].

of inflammation [82]. Several *in vitro* and *in vivo* models have been used in studies to evaluate the analgesic and anti-inflammatory properties of piperazine derivatives. The efficiency of several piperazine-based compounds in lowering inflammatory markers was examined by Jain *et al.* in 2021 [83]. They prepared and evaluated a number of piperazine derivatives to determine their suitability for managing inflammation. As all the compounds synthesized were reported to be highly potent than standard drugs such as aspirin and indomethacin, the compounds of series of 5-substituted-^{1}H-benzimidazol-2-yl-thioacetyl piperazine and thioacetylbenzhydryl piperazine derivatives have been confirmed to have maximum activity through the Western blot assays [83]. The synthesis protocol is given in Figure 14.19. There are different thiazole and tetrazole derivatives which shows potential activity as anti-inflamatory and analgesic agents [84, 85].

14.5 Conclusion

Thiophene and piperazine derivatives are rather important reagents exhibiting diverse biological activity, especially in medicinal chemistry. They also showed moderate to fair to good activity against several biological targets. Variations on the substituents at both the thiophene and piperazine nucleus can lead to changes to the biological profile from bacterial and fungal disorders to viral and other cancers. These derivatives bind to enzymes and proteins to solve multiple biological challenges.

This review focuses on the multifarious biological potentials of thiophene- and piperazine-containing derivatives useful for the treatment of infections such as cancer, tumor growth, inflammation, etc.

The world is facing a critical challenge of developing resistance to drugs, hence the need to look at other new structural classes such as thiophene- and piperazine-based drugs. These compounds act as multidisease chemicals, which can help fight different diseases.

AI disclosure statement: The author acknowledges the use of AI-based tools for minor language editing or grammatical correction. However, all intellectual contributions, analyses, and conclusions presented in this chapter are entirely the author's own.

References

1. García-Valverde, M. and Torroba, T., Sulfur-nitrogen heterocycles. *Molecules.*, 10, 318–320, 2005, https://doi.org/10.3390/10020318.
2. Singh, P.K. and Silakari, O., The current status of *O*-heterocycles: A synthetic and medicinal overview. *ChemMedChem.*, 13, 1071–1087, 2018, https://doi.org/10.1002/cmdc.201800119.
3. Pathania, S., Narang, R.K., Rawal, R.K., Role of sulphur-heterocycles in medicinal chemistry: An update. *Eur. J. Med. Chem.*, 180, 486–508, 2019, https://doi.org/10.1016/j.ejmech.2019.07.043.
4. Mishra, R., Jha, K., Kumar, S., Tomer, I., Synthesis, properties and biological activity of thiophene: A review. *Der Pharma Chem*, 3, 38–54, 2011, http://derpharmachemica.com/archive.html.
5. de Vasconcelos, A., Campos, V.F., Nedel, F., Seixas, F.K., Dellagostin, O.A., Smith, K.R., de Pereira, C.M.P., Stefanello, F.M., Collares, T., Barschak, A.G., Cytotoxic and apoptotic effects of chalcone derivatives of 2-acetyl thiophene on human colon adenocarcinoma cells. *Cell Biochem. Funct.*, 31, 289–297, 2013, https://doi.org/10.1002/cbf.2897.
6. Schmitt, Kail, D., Mariano, M., Empting, M., Weber, N., Paul, T., Hartmann, R.W., Engel, M., Design and synthesis of a library of lead-like 2, 4-bisheterocyclic substituted thiophenes asselective Dyrk/Clk inhibitors. *PLoS One*, 9, e87851, 2014, 10.1371/journal.pone.0087851.
7. Al-Said, M.S., Bashandy, M.S., Al-Qasoumi, S.I., Ghorab, M.M., Anti-breast cancer activity of some novel 1, 2-dihydropyridine, thiophene and thiazole derivatives. *Eur. J. Med. Chem.*, 46, 137–141, 2011, https://doi.org/10.1016/j.ejmech.2010.10.024.
8. Dallemagne, P., Khanh, L.P., Alsaïdi, A., Renault, O., Varlet, I., Collot, V., Bureau, R., Rault, S., Synthesis and biological evaluation of cyclopenta [*c*]

thiophene related compounds as new antitumor agents. *Bioorg. Med. Chem.*, 10, 2185–2191, 2002, https://doi.org/10.1016/S0968- 0896(02)00070-6.

9. Venkataramireddy, V., Shankaraiah, V., Rao, A.T., Kalyani, C., Narasu, M.L., Varala, R., Jayashree, A., Synthesis and anti-cancer activity of novel 3-aryl thiophene-2-carbaldehydes and their aryl/heteroaryl chalcone derivatives. *Rasayan J. Chem.*, 9, 31–39, 2016, http://www.rasayanjournal.co.in.
10. Sable, P.N. and Ganguly, S., P.D. Chaudhari, an efficient one-pot three-component synthesis and antimicrobial evaluation of tetra substituted thiophene derivatives. *Chin. Chem. Lett.*, 25, 1099–1103, 2014, https://doi.org/10.1016/j.cclet.2014.03.044.
11. Harit, T., Bellaouchi, R., Asehraou, A., Rahal, M., Bouabdallah, I., Malek, F., Synthesis, characterization, antimicrobial activity and theoretical studies of new thiophene-based tripodal ligands. *J. Mol. Struct.*, 1133, 74–79, 2017, https://doi.org/10.1016/j.molstruc.2016.11.051.
12. Sulthana, S.S., Antony, S.A., Balachandran, C., Shafi, S.S., Thiophene and benzodioxole appended thiazolyl-pyrazoline compounds: Microwave assisted synthesis, antimicrobial and molecular docking studies. *Bioorg. Med. Chem. Lett.*, 25, 2753–2757, 2015, https://doi.org/10.1016/j.bmcl.2015.05.033.
13. Mishra, P., Middha, A., Saxena, V., Saxena, A., Synthesis and evaluation of anti-inflammatory activity of some cinnoline derivatives-4 (-2-amino-thiophene) cinnoline-3-carboxamide. *J. Pharm. Biosci*, 4, 64–68, 2016, http://dx.doi.org/10.20510/ukjpb/4/i3/108388.
14. Murakami, N., Takase, H., Saito, T., Iwata, K., Miura, H., Naruse, T., Effects of a novel nonsteroidal anti-inflammatory drug (M-5011) on bone metabolism in rats with collagen-induced arthritis. *Eur. J. Pharmacol.*, 352, 81–90, 1998, https://doi.org/10.1016/S0014-2999(98)00342-2.
15. Mathew, B., Suresh, J., Anbazhagan, S., Synthesis, *in silico* preclinical evaluation, antidepressant potential of 5-substituted phenyl-3-(thiophen-2-yl)-4, 5-dihydro-1h-pyrazole-1-carboxamides. *Biomed. Aging Pathol*, 4, 327–333, 2014, https://doi.org/10.1016/j.biomag.2014.08.002.
16. Ashour, H.M., Shaaban, O.G., Rizk, O.H., El-Ashmawy, I.M., Synthesis and biological evaluation of thieno [2′, 3′: 4, 5] pyrimido [1, 2-*b*][1, 2, 4] triazines and thieno [2, 3-*d*][1, 2, 4]triazolo [1, 5-a] pyrimidines as anti-inflammatory and analgesic agents. *Eur. J. Med. Chem.*, 62, 341–351, 2013, https://doi.org/10.1016/j.ejmech.2012.12.003.
17. Kulandasamy, R., Adhikari, A.V., Stables, J.P., A new class of anticonvulsants possessing 6 Hz activity: 3, 4-dialkyloxy thiophene bishydrazones. *Eur. J. Med. Chem.*, 44, 4376–4384, 2009, https://doi.org/10.1016/j.ejmech.2009.05.026.
18. Singh, P.K., Histone methyl transferases: A class of epigenetic opportunities to counter uncontrolled cell proliferation. *Eur. J. Med. Chem.*, 166, 351–368, 2019, https://doi.org/10.1016/j.ejmech.2019.01.069.
19. Roy, P. and Saikia, B., Cancer and cure: a critical analysis. *Indian J. Cancer*, 53, 441–442, 2016. 10.4103/0019-509X.200658.

20. Kumar Singh, P. and Silakari, O., *In silico* guided development of imine-based inhibitors for resistance-deriving kinases. *J. Biomol. Struct. Dyn.*, 37, 2593–2599, 2019, https://doi.org/10.1080/07391102.2018.1491893.
21. Singh, P.K., Chaudhari, D., Jain, S., Silakari, O., Structure based designing of triazolopyrimidone-based reversible inhibitors for kinases involved in NSCLC. *Bioorg. Med. Chem. Lett.*, 29, 1565–1571, 2019, https://doi.org/10.1016/j.bmcl.2019.05.004.
22. Gul, H.I., Yamali, C., Sakagami, H., Angeli, A., Leitans, J., Kazaks, A., Tars, K., Ozgun, D.O., Supuran, C.T., New anticancer drug candidates' sulfonamides as selective hCA IX or hCA XII inhibitors. *Bioorg. Chem.*, 77, 411–419, 2018, doi: 10.1016/j.bioorg.2018.01.021.
23. Zhang, H.Z., Kasibhatla, S., Kuemmerle, J., Kemnitzer, W., Ollis-Mason, K., Qiu, L., Crogan-Grundy, C., Tseng, B., Drewe, J., Cai, S.X., Discovery and structure-activity relationship of 3-aryl-5-aryl-1, 2, 4-oxadiazoles as a new series of apoptosis inducers and potential anticancer agents. *J. Med. Chem.*, 48, 5215–5223, 2005, https://doi.org/10.1021/jm050292k.
24. Agarwal, M., Singh, V., Sharma, S.K., Sharma, P., Ansari, M.Y., Jadav, S.S., Yasmin, S., Sreenivasulu, R., Hassan, M.Z., Saini, V., Design and synthesis of new 2, 5-disubstituted-1, 3, 4-oxadiazole analogues as anticancer agents. *Med. Chem. Res.*, 25, 2289–2303, 2016, https://doi.org/10.1007/s00044-016-1672-1.
25. Elmetwally, S.A., Saied, K.F., Eissa, I.H., Elkaeed, E.B., Design, synthesis and anticancer evaluation of thieno [2, 3-*d*] pyrimidine derivatives as dual EGFR/HER2 inhibitors and apoptosis inducers. *Bioorg. Chem.*, 88, 102944, 2019, https://doi.org/10.1016/j.bioorg.2019.102944.
26. Milik, S.N., Abdel-Aziz, A.K., Lasheen, D.S., Serya, R.A., Minucci, S., Abouzid, K.A., Surmounting the resistance against EGFR inhibitors through the development of thieno [2, 3-*d*]pyrimidine-based dual EGFR/HER2 inhibitors. *Eur. J. Med. Chem.*, 155, 316–336, 2018, https://doi.org/10.1016/j.ejmech.2018.06.011.
27. Hirsch, F.R. and Witta, S., Biomarkers for prediction of sensitivity to EGFR inhibitors in nonsmall cell lung cancer. *Curr. Opin. Oncol.*, 17, 118–122, 2005, 10.1097/01.cco.0000155059.39733.9d.
28. Giri, R.S., Thaker, H.M., Giordano, T., Williams, J., Rogers, D., Vasu, K.K., Sudarsanam, V., Design, synthesis and evaluation of novel 2-thiophen-5-yl-3H-quinazolin-4-one analogues as inhibitors of transcription factors NF-κB and AP-1 mediated transcriptional activation: Their possible utilization as anti-inflammatory and anti-cancer agents. *Bioorg. Med. Chem.*, 18, 2796–2808, 2010, https://doi.org/10.1016/j.bmc.2010.01.007.
29. Gulipalli, K.C., Bodige, S., Ravula, P., Endoori, S., Vanaja, G., Babu, G.S., Chandra, J.N.S., Seelam, N., Design, synthesis, *in silico* and *in vitro* evaluation of thiophene derivatives: A potent tyrosine phosphatase 1B inhibitor and anticancer activity. *Bioorg. Med. Chem. Lett.*, 27, 3558–3564, 2017, https://doi.org/10.1016/j.bmcl.2017.05.047.

30. AbdElhameid, M.K., Labib, M.B., Negmeldin, A.T., Al-Shorbagy, M., Mohammed, M.R., Design, synthesis, and screening of ortho-amino thiophene carboxamide derivatives on hepatocellular carcinomaas VEGFR-2Inhibitors. *J. Enzyme Inhib. Med. Chem.*, 33, 1472–1493, 2018, https://doi.org/10.1080/14756366.2018.1503654.
31. Ashok, A., Thanukrishnan, K., Bhojya Naik, H.S., Shaik, A.G., 6, 7-Dimethoxy-quinazolin-4-yl-amino-thiophene-2-carboxamides as Potent Inhibitors of VEGFReceptors 1 and 2. *J. Heterocycl. Chem.*, 54, 1065–1070, 2017, https://doi.org/10.1002/jhet.2675.
32. Ghith, A., Youssef, K.M., Ismail, N.S., Abouzid, K.A., Design, synthesis and molecular modeling study of certain VEGFR-2 inhibitors based on thienopyrimidne scaffold as cancer targeting agents. *Bioorg. Chem.*, 83, 111–128, 2019, https://doi.org/10.1016/j.bioorg.2018.10.008.
33. Thakur, S., Kumar, D., Jaiswal, S., Goel, K.K., Rawat, P., Srivastava, V., Dhiman, S., Jadhav, H.R., Dwivedi, A.R., Medicinal chemistry-based perspectives on thiophene and its derivatives: exploring structural insights to discover plausible druggable leads. *RSC Med. Chem.*, 16, 481–510, 2025.
34. Saad, H.A., Youssef, M.M., Mosselhi, M.A., Microwave assisted synthesis of some new fused 1, 2, 4-triazines bearing thiophene moieties with expected pharmacological activity. *Molecules*, 16, 4937–4957, 2011, https://doi.org/10.3390/molecules16064937.
35. Ghorab, M.M., Bashandy, M.S., Alsaid, M.S., Novel thiophene derivatives with sulfonamide, isoxazole, benzothiazole, quinoline and anthracene moieties as potential anticancer agents. *Acta Pharm.*, 64, 419–431, 2014, https://doi.org/10.2478/acph-2014-0035.
36. Youssef, M.M., Al-Omair, M.A., Ismail, M.A., Synthesis, D.N.A., affinity, and antimicrobial activity of 4-substituted phenyl-2, 2′-bichalcophenes and aza-analogues. *Med. Chem. Res.*, 21, 4074–4082, 2012, https://doi.org/10.1007/s00044-011-9964-y.
37. Thomas, J., Jecic, A., Vanstreels, E., van Berckelaer, L., Romagnoli, R., Dehaen, W., Liekens, S., Balzarini, J., Pronounced anti-proliferative activity and tumor cell selectivity of 5-alkyl-2-amino- 3-methylcarboxylate thiophenes. *Eur. J. Med. Chem.*, 132, 219–235, 2017, https://doi.org/10.1016/j.ejmech.2017.03.044.
38. Mishra, I., Sharma, V., Kumar, N., Krishna, G., Sethi, V.A., Mittal, R., Mishra, R., Exploring Thiophene Derivatives: Synthesis Strategies and Biological Significance. *Med. Chem.*, *21*, 1, 11–31, 2025.
39. Dhanya, T.M., Anjali Krishna, G., Savitha, D.P., Shanty, A.A., Divya, K.M., Priya, S.K., Mohanan, P.V., A review on the synthesis and biological relevance of benzo [b] thiophene derivatives. *Phosphorus Sulfur Silicon Relat. Ele.*, *198*, 4, 283–299, 2023.
40. Pieroni, M., Azzali, E., Basilico, N., Parapini, S., Zolkiewski, M., Beato, C., Costantino, G., Accepting the invitation to open innovation in malaria drug discovery: synthesis, biological evaluation, and investigation on the

structure–activity relationships of benzo [b] thiophene-2-carboxamides as antimalarial agents. *J. Med. Chem.*, *60*, 5, 1959–1970, 2017.
41. Shah, R. and Verma, P.K., Therapeutic importance of synthetic thiophene. *Chem. Cent. J.*, *12*, 1–22, 2018.
42. Mohamed, M.A., Roy, N., Benjamin, I., Joo, S.W., Musthafa, Y.M., Ghfar, A.A., Louis, H., Investigating the antimicrobial and anti-inflammatory activities of novel thiophene derivatives by *in-vitro* studies, molecular docking, spectral analysis, and quantum chemical calculations. *J. Mol. Struct.*, *1306*, 137810, 2024.
43. Poojith, N., Kigga, M., Rose, J.J., Potla, K.M., Vankayalapati, S., Chinnam, S., Pillai, R.R., Structural, spectroscopic, and *in silico* studies of 3-(dimethylamino)-1-(thiophen-2-yl) propan-1-ol: A potential antidepressant agent. *J. Mol. Struct.*, *1250*, 131859, 2022.
44. Kapoor, K., Kaur, N., Sohal, H.S., Kaur, M., Singh, K., Kumar, A., Drugs and Their Mode of Action: A Review on Sulfur-Containing Heterocyclic Compounds. *Polycyclic Aromat. Compd.*, *45*, 1, 136–175, 2025.
45. Dawood, D.H., Nossier, E.S., Abdelhameed, M.F., Asaad, G.F., Abd El-Rahman, S.S., Design, synthesis, anti-inflammatory evaluation and molecular docking of novel thiophen-2-ylmethylene-based derivatives as potential TNF-α production inhibitors. *Bioorg. Chem.*, *122*, 105726, 2022.
46. Mahnashi, M.H., Rashid, U., Almasoudi, H.H., Nahari, M.H., Ahmad, I., Binshaya, A.S., Sadiq, A., Modification of 4-(4-chlorothiophen-2-yl) thiazol-2-amine derivatives for the treatment of analgesia and inflammation: synthesis and *in vitro*, *in vivo*, and *in silico* studies. *Front. Pharmacol.*, *15*, 1366695, 2024.
47. Khadri, M.N., Ramu, R., Simha, N.A., Khanum, S.A., Synthesis, molecular docking, analgesic, anti-inflammatory, and ulcerogenic evaluation of thiophene-pyrazole candidates as COX, 5-LOX, and TNF-α inhibitors. *Inflammopharmacology*, *32*, 1, 693–713, 2024.
48. Pal, R., Singh, K., Paul, J., Khan, S.A., Naim, M.J., Akhtar, M.J., Overview of chemistry and therapeutic potential of non-nitrogen heterocyclics as anticonvulsant agents. *Curr. Neuropharmacol.*, *20*, 8, 1519–1553, 2022.
49. Pathak, S., Singh, A.P., Sharma, R., Pandey, R., An overview of the pharmacological activities and synthesis of benzothiophene derivatives. *Med. Chem.*, *20*, 9, 839–854, 2024.
50. Rajkumar, T., Kumar, S.V.S., Srinivasan, N., Ultrasonic Assisted Synthesis, Biological Evaluation, and Molecular Docking of Chalcone-based 1, 5-benzodiazepine as Potential Anticonvulsant Agents. *Curr. Enzyme Inhib.*, *18*, 1, 32–39, 2022.
51. Joshi, R.J., Dholariya, M.P., Chothani, S.R., Chamakiya, C.A., Varu, H.L., Karmur, M.B., Kapuriya, N.P., Synthesis, antidiabetic activity and *in silico* studies of benzo [b] thiophene based small molecule α-amylase inhibitors. *J. Mol. Struct.*, *1312*, 138570, 2024.

52. Thakur, S., Kumar, D., Goel, K.K., Rawat, P., Srivastava, V., Dhiman, S., Dwivedi, A.R., Medicinal chemistry-based perspective on thiophene and its derivatives: Exploring the structural insights to discover plausible druggable leads. *RSC Med. Chem.*, 2024.
53. Ullah, N., Alam, A., Elhenawy, A.A., Naz, S., Islam, M.S., Ahmad, M., Investigating novel thiophene carbaldehyde based thiazole derivatives as potential hits for diabetic management: Synthesis, *in vitro* and *in silico* approach. *Chem. Select*, *9*, 8, e202304601, 2024.
54. Jalageri, M.D., Malgar Puttaiahgowda, Y., Parambil, A.M., Kulal, A., *J. Appl. Polym. Sci.*, 4752, 2019.
55. Akl, L., Abd El-Hafeez, A.A., Ibrahim, T.M., Salem, R., Marzouk, H.M.M., El-Domany, R.A., Abou-Seri, S.M., Identification of novel piperazine-tethered phthalazines as selective CDK1 inhibitors endowed with *in vitro* anticancer activity toward the pancreatic cancer. *Eur. J. Med. Chem.*, *243*, 114704, 2022.
56. Sharma, V., Das, R., Sharma, D., Mujwar, S., Mehta, D.K., Green chemistry approach towards Piperazine: Anticancer agents. *J. Mol. Struct.*, *1292*, 136089, 2023.
57. Sagar, B.K., Kumar, M.D., Bhat, M., Sekhar, E.V., Supreeth, M., Synthesis, Characterization, and Anticancer Activity of Piperazine Amide Derivatives. *Int. J. Pharm. Sci. Res.*, *12*, 10, 5424–5430, 2021.
58. Roy, L.D., Remady, S., Kumar, J., Piperazine tethered heterocyclic hybrids: Synthesis, characterization, and *in vitro* studies from potent proteinase inhibitors to antiproliferative activity. *Eur. J. Med. Chem. Rep.*, *10*, 100132, 2024.
59. Zsoldos, B., Nagy, N., Donkó-Tóth, V., Keglevich, P., Weber, M., Dékány, M., Hazai, L., Novel Piperazine Derivatives of Vindoline as Anticancer Agents. *Int. J. Mol. Sci.*, *25*, 14, 7929, 2024.
60. Singh, S., Kumar, R., Tripathi, S., Mazumder, A., Singh, N., Fused and Substituted Piperazines as Anticancer Agents: A Review. *Chem. Biol. Drug Des.*, *105*, 3, e70077, 2025.
61. Kandula, M.K.R., Gundluru, M., Nemallapudi, B.R., Gundala, S., Kotha, P., Zyryanov, G.V., Cirandur, S.R., Synthesis, antioxidant activity, and α-glucosidase enzyme inhibition of α-aminophosphonate derivatives bearing piperazine-1, 2, 3-triazole moiety. *J. Heterocycl. Chem.*, *58*, 1, 172–181, 2021.
62. da Silva Moreira, L.K., Turones, L.C., Campos, H.M., Nazareth, A.M., Thomaz, D.V., de Souza Gil, E., Costa, E.A., LQFM212, a piperazine derivative, exhibits potential antioxidant effect as well as ameliorates LPS-induced behavioral, inflammatory and oxidative changes. *Life Sci.*, *312*, 121199, 2023.
63. Kilic-Kurt, Z., Konyar, D., Okur, H., Kaplan, A., Boga, M., Some heterocycles connected to substituted piperazine by 1, 3, 4-oxadiazole linker: Design, synthesis, anticholinesterase and antioxidant activity. *J. Mol. Struct.*, *1321*, 139854, 2025.
64. Al-Masoudi, N.A., Jihad, R.S., Abdul-Rida, N.A., Al-Shamari, A.M., Saeed, B.A., Al-Masoudi, W.A., Al-Soud, Y.A., Synthesis, antiproliferative,

antioxidant activities, *in silico* studies, DFT analysis and molecular dynamics simulation of novel 1-(4-chlorobenzhydryl) piperazine derivatives. *J. Mol. Struct., 1298*, 137028, 2024.
65. Mallesha, L. and Mohana, K.N., Synthesis, antimicrobial and antioxidant activities of 1-(1, 4-benzodioxane-2-carbonyl) piperazine derivatives. *Eur. J. Chem., 2*, 2, 193–199, 2011.
66. Li, W., Chen, S.Y., Hu, W.N., Zhu, M., Liu, J.M., Fu, Y.H., ... & OuYang, G.P., Design, synthesis, and biological evaluation of quinazoline derivatives containing piperazine moieties as antitumor agents. *J. Chem. Res.*, 44, 9–10, 536–542, 2020.
67. Al-Ghorbani, M., Gouda, M.A., Baashen, M., Alharbi, O., Almalki, F.A., Ranganatha, L.V., Piperazine heterocycles as potential anticancer agents: A review. *Pharm. Chem. J., 56*, 1, 29–37, 2022.
68. Wang, H., Wei, J., Jiang, H., Zhang, Y., Jiang, C., Ma, X., Design, synthesis and pharmacological evaluation of three novel dehydroabietyl piperazine dithiocarbamate ruthenium (II) polypyridyl complexes as potential antitumor agents: DNA damage, cell cycle arrest and apoptosis induction. *Molecules, 26*, 5, 1453, 2021.
69. Rasheed, A.M., Shetty, K.K., Roy, L.D., Kumar, J., Novel Piperazine Derivatives as Potent Antihistamine, Anti-Inflammatory, and Anticancer Agents, their Synthesis and Characterization. *Anti-Cancer Agents Med. Chemistry-Anti-Cancer Agents, 24*, 14, 1063–1073, 2024.
70. Zhang, R.H., Guo, H.Y., Deng, H., Li, J., Quan, Z.S., Piperazine skeleton in the structural modification of natural products: A review. *J. Enzyme Inhib. Med. Chem., 36*, 1, 1165–1197, 2021.
71. Zhang, W., Guo, S., Yu, L., Wang, Y., Chi, Y.R., Wu, J., Piperazine: Its role in the discovery of pesticides. *Chin. Chem. Lett., 34*, 8, 108123, 2023.
72. Li, H., Liu, H., Zhang, Y., Yang, N., Xiong, L., Li, Z., Synthesis, insecticidal activities, and SAR studies of novel piperazine-containing heterocyclic mono-/di-/tri-amide derivatives. *Chin. Chem. Lett., 32*, 8, 1020–1035, 2021.
73. He, Q., Zhang, H.R., Zou, Y., A cytochrome P450 catalyzes oxidative coupling formation of insecticidal dimeric indole piperazine alkaloids. *Angew. Chem. Int. Ed., 63*, 5, 11234–11248, 2024.
74. Xu, Y., Wang, H., Du, X., Design, synthesis, and fungicidal activity of novel N-substituted piperazine-containing phenylpyridines against cucumber downy mildew. *Pest. Manage. Sci., 78*, 9, 1342–1354, 2022.
75. Li, G.-Y., Yan, S.-G., Jiang, S., Qian, X.-H., Huang, Q.-C., Zhang, R., *Basics Appl. Sci.*, 2001–2006, 2008.
76. Kalli, S.B. and Velmurugan, V., Design, synthesis and anti-diabetic activity of piperazine sulphonamide derivatives as dipeptidyl peptidase-4 inhibitors. *Pharmacia, 69*, 987–993, 2022.
77. Özil, M., Khan, K.M., Baltaş, N., Wadood, A., Samad, A., Kahveci, B., Synthesis of benzimidazoles containing piperazine ring as potential

therapeutic agents against diabetes mellitus and antioxidant activities. *J. Mol. Struct.*, *1304*, 137714, 2024.
78. Paliwal, A., Jain, S., Paliwal, S., Molecular modeling, docking, and QSAR studies on a series of N-arylsulfonyl-N-2-pyridinyl-piperazines analogs acting as anti-diabetic agents. *Trends Sci.*, *20*, 7, 5472–5472, 2023.
79. Hosseini Nasab, N., Raza, H., Eom, Y.S., Shah, F.H., Kwak, J.H., Kim, S.J., Exploring chalcone-sulfonyl piperazine hybrids as anti-diabetes candidates: Design, synthesis, biological evaluation, and molecular docking study. *Mol. Diversity*, *29*, 1, 43–59, 2025.
80. Devine, R., Kelada, M., Leonard, S., Martin, D.S.D., Walsh, J.M.D., Breen, C.J., Driver, R.B., Kinsella, G.K., Findlay, J.B.C., Stephens, J.C., *Eur. J. Med. Chem.*, 2020. ISSN 0223-5234.
81. Martins, A.N., de Souza Almeida, D., Florentino, I.F., da Silva Moreira, L.K., Turones, L.C., Batista, D.C., da Silva, D.P., Pharmacological evaluation of antinociceptive and anti-inflammatory activities of LQFM202: a new piperazine derivative. *Inflammopharmacology*, *31*, 1, 411–422, 2023.
82. Buran, K., Reis, R., Sipahi, H., Önen Bayram, F.E., Piperazine and piperidine-substituted 7-hydroxy coumarins for the development of anti-inflammatory agents. *Arch. Pharm.*, *354*, 7, 2000354, 2021.
83. Jain, A., Chaudhary, J., Khaira, H., Chopra, B., Dhingra, A., Piperazine: A promising scaffold with analgesic and anti-inflammatory potential. *Drug Res.*, *71*, 02, 62–72, 2021.
84. Mishra, R., Sharma, P.K., Verma, P.K., Tomer, I., Mathur, G., Dhakad, P.K., Biological potential of thiazole derivatives of synthetic origin. *Journal of Heterocyclic Chemistry*, 54, 4, 2103–2116, 2017.
85. Kaushik, N., Kumar, N., Kumar, A., Singh, U.K., Tetrazoles: Synthesis and biological activity. *Immunology, Endocrine & Metabolic Agents-Medicinal Chemistry Current Medicinal Chemistry-Immunology, Endocrine & Metabolic Agents*, 18, 1, 3–21, 2018.

15

Heterocyclic Compounds as Antileprotic Agents

Sanjita Das[1]*, Anjali Singh[1], Shalini Gautam[1] and Gyas Khan[2]

[1]Department of Pharmacy, SMAS, Galgotias University, Greater Noida, Uttar Pradesh, India
[2]Department of Pharmacology and Toxicology, College of Pharmacy, Jazan University, Jazan, KSA

Abstract

Heterocyclic compounds are a highly unique and valuable class of compounds that are present in abundance in our surroundings. Recently different heterocyclic compounds and their derivative are found to have various pharmacological properties attracting a strong interest from researchers around the world. Different heterocyclic analogs have been assessed for a variety of biological activities such as antiviral, anti-inflammatory, antihypertensive, antileprotic agents, etc. Leprosy, also referred to as Hansen disease, is brought on by the bacteria *Mycobacterium leprae*, which is treated using antileprotic drugs. These agents are effectively used for controlling, preventing, and reducing the risk of complications in leprosy. Leprosy-associated physical deformities and disabilities can be averted using essential early diagnosis and consistent treatment. Recent advancement in understanding the disease helps in improving drug regimens that can continue to enhance the management of leprosy. Following the recent developments in antileprotic drug discovery through newer targets against leprosy, this chapter discusses the pathophysiology, drug classification, synthesis, mechanism of action, metabolism, and adverse drug reactions of antileprotic drugs.

Keywords: *Mycobacterium laprae*, heterocyclic, clofazimine, ethionamide, fluoroquinolones, dapsone, antileprotic

**Corresponding author*: sanjita.das@galgotiasuniversity.edu.in

Divya Bajpai Tripathy, Priyanka Chhabra and Anjali Gupta (eds.) Heterocyclic Compounds as Pharmaceutical Ingredients, (509–530) © 2026 Scrivener Publishing LLC

15.1 Introduction

Leprosy, often known as Hansen disease, is a chronic infectious disease caused by the *Mycobacterium leprae* bacteria. The skin, mucosal surfaces, and peripheral nerves are the main areas affected by this bacterium. If treatment is not received, it can cause serious deformities and nerve damage. Leprosy has historically caused a great deal of social stigma, with those who have the disease frequently being shunned by society. These substances are essential to contemporary antileprotic treatment and have been thoroughly investigated for their antibacterial qualities.

Heterocyclic substances that target important metabolic pathways, such as rifampicin, dapsone, and clofazimine, have strong antibacterial activity against *M. leprae*. They are essential parts of multidrug therapy (MDT), the accepted treatment for leprosy, due to their great efficacy and structural variety [1].

An extensive examination of the function of heterocyclic compounds as antileprotic drugs is given in this chapter. It discusses their synthesis, pharmacokinetics, modes of action, resistance mechanisms, historical development, chemistry, clinical applications, and possibilities for drug discovery in the future. This chapter delves into the significance of heterocyclic compounds in the development of antileprotic drugs, examining their mechanisms of action, recent progress, and future outlooks [2].

15.2 Historical Perspectives

Records of leprosy date back to ancient civilizations such as Egypt, China, and India. Herbal cures and formulations based on minerals were among the early treatments, which were mostly empirical. The 1940s saw the development of dapsone, the first potent antibiotic medication, significantly transforming the way leprosy was treated. However, dapsone monotherapy caused dapsone-resistant strains of *M. leprae* to appear, which made combination therapy necessary [3].

Rifampicin and clofazimine were added to MDT in the 1980s, which greatly enhanced treatment results and aided in the prevention of the emergence of resistance [4]. The World Health Organization (WHO) continues to support MDT as the mainstay of leprosy therapy. It is impossible to overestimate the importance of heterocyclic chemicals in these therapies. They have proven invaluable in the fight against leprosy because of their capacity to interact with bacterial DNA, impede enzyme activity, and disrupt vital metabolic pathways [5].

15.3 Pathophysiology

Gram-positive, acid-fast bacillus *M. leprae* is an obligate intracellular parasite that mostly infects phagocytic cells, including skin macrophages and Schwann cells in nerves. This bacillus differs from other gram-positive and gram-negative bacteria due to its lipid-rich cell wall, which contains phenolic glycolipid-1 and mycolic acid [6]. By using the CR1, CR3, and CR4 receptors, PGL-1 facilitates the bacillus' entrance into macrophages *via* the C3 component of the complement, resulting in phagocytosis. It is essential to the pathophysiology of the disease and contributes to the lysosomal escape mechanisms. It works as an immunosuppressive drug that allows *M. leprae* to survive inside the host cell, particularly the macrophage [7]. Reversal reactions, sometimes called Type 1 responses (T1Rs), and erythema nodosum leprosum (ENL), also called Type 2 reactions (T2Rs), are the two main forms of leprosy reactions [8].

Type 1 reaction: In leprosy border spectrum, T1R causes T-cell hypersensitivity to *M. leprae.* When *M. leprae* is injected into the nasal mucosa, antigen-presenting cells (APCs) phagocytose the antigen, triggering an immunological reaction. Following their absorption of peripheral tissue antigens, APCs go into the lymph nodes. APCs are MHC II molecules that send two signals to naive $CD4^+$ helper T cells and trigger the adaptive immune response. APC emits interleukin 12 (IL-12), which stimulates the immune system by promoting the development of effector helper T (T_H) cells from $CD4^+$ T cells. Effector T cells release IL-2, which promotes the growth of T_H cells and $CD8^+$ T cells [9]. $CD8^+$ T cells release granzymes and perforin upon activation of T cells returning to peripheral tissues, destroying Schwann cells and the myelin sheath. Transforming growth factor β and IL-6, which are released by the APCs, encourage naive T cells to develop into T_H17 cells. Neutrophils are drawn in by the IL-17 released by T_H17 cells that have been activated. APC generates interferon γ, which activates macrophages and triggers the release of proinflammatory cytokines such as tumor necrosis factor α, IL-1β, and IL-6. Secretion of IL-16 by activated macrophages maintains this inflammatory response (Figure 15.1) [10].

Type 2 reaction: Type III hypersensitivity brought on by insufficient antigen–antibody complex clearance causes inflammation and leukocyte

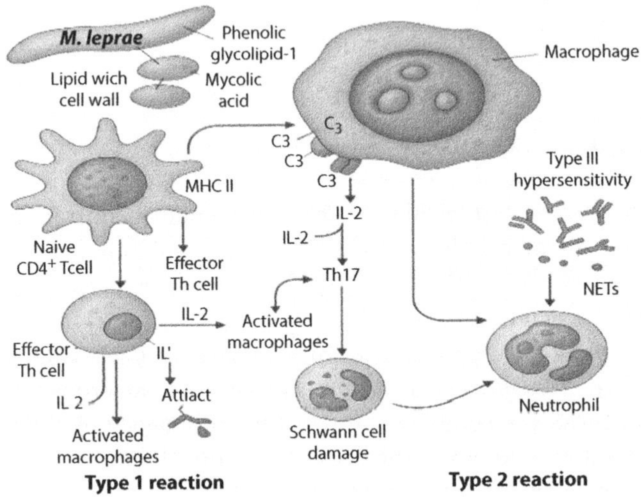

Figure 15.1 Pathophysiology of leprosy: Type 1 and Type 2 immune reaction.

chemotaxis in the pathophysiology of T2R (Figure 15.2). Neutrophils play a central role in ENL pathogenesis. Neutrophil extracellular traps, significant in severe autoimmune disorders, are abundant in ENL patients [11].

15.4 Classification of Heterocyclic Antileprotic Drugs

15.4.1 Characteristics of Structure

At least one heteroatom can be found in the ring structures of heterocyclic compounds. Heteroatoms such as sulfur, oxygen, or nitrogen provide them special electrical characteristics that affect their biological activity. Important heterocyclic compounds found in antileprotic medications include the following:

- Dapsone-like sulfone rings
- Ring of phenazine (such as clofazimine)
- Rings that are macrocyclic (such as rifampicin) [12] Phenazine derivatives: clofazimine
- Antitubercular drugs: rifampin, ethionamide
- Other antimicrobials: ofloxacin, moxifloxacin, minocycline, clarithromycin
- Chaulmoogric acid [13]

15.4.2 Heterocyclic Ring's Role in Mechanism of Action

Heterocyclic compounds are known for their strong antimycobacterial properties and have become promising candidates for treating infections caused by *M. leprae*. These compounds hinder the growth of *M. leprae* through various mechanisms [14]. A key method involves disrupting folate biosynthesis. Specifically, heterocyclic compounds target enzymes such as dihydropteroate synthase (DHPS) and dihydrofolate reductase, which are crucial for producing tetrahydrofolate, a cofactor necessary for DNA and amino acid synthesis (Figure 15.2). By inhibiting these enzymes, heterocyclic compounds deprive the bacteria of essential metabolites required for their survival and reproduction. Disrupting the bacterial DNA replication is another method of action. This is mainly accomplished by targeting bacterial DNA gyrase and topoisomerases, which are vital for maintaining DNA supercoiling and enabling the unwinding of DNA strands during replication (Figure 15.2). Inhibiting these enzymes disrupts the replication process, resulting in DNA damage and ultimately bacterial cell death. Heterocyclic compounds can also compromise a critical structure for *M. leprae*'s survival, i.e., cell wall by interfering with the bacterial cell wall structure. A large number of these substances block the enzymes that produce mycolic acids, which are long-chain fatty acids that are essential for the formation of mycobacterial cell wall (Figure 15.2) [15, 16]. Without proper cell wall formation, *M. leprae* becomes structurally weakened and susceptible to host immune defenses. Finally, some heterocyclic agents trigger the generation of ROS (reactive oxygen species) within bacterial cells. Oxidative stress brought on by elevated ROS levels damages lipids, proteins, and DNA in cells. This oxidative damage overwhelms the bacterial defense systems and leads to cell death. Together, these diverse

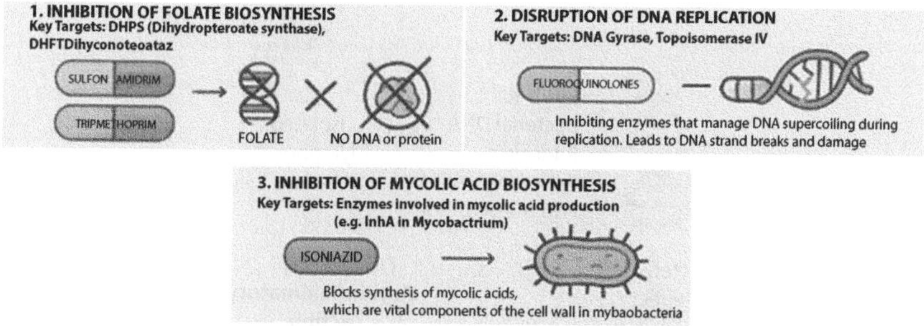

Figure 15.2 Suggested mode of action of various heterocyclic compounds in their antileprotic effect.

mechanisms make heterocyclic compounds highly effective in controlling and potentially eliminating *M. leprae* infections [17].

15.5 Important Heterocyclic Antileprotic Drugs

15.5.1 Clofazimine

Clofazimine binds to bacterial DNA and interferes with its replication and transcription resulting in a bactericidal effect against *M. leprae*. By decreasing mitogen and antigen-driven proliferative responses of isolated T lymphocytes and limiting lymphocyte transformation, it also functions as anti-inflammatory drug, lowering the immunological response linked to leprosy (Figure 15.3) [18].

Pharmacokinetics: Absorption in the range of 45% to 62% is observed after oral medication in leprosy patients based on whether drug is administered with or without food. When coadministration of drug with food is conducted with 200 mg clofazimine, C_{max} is found to be 0.41 mg/L with T_{max} at 8 hours; when fasting, C_{max} is reduced by 30% with T_{max} at 12 hours. Clofazimine is very lipophilic in nature, so predominantly deposits in fat tissue as well as reticuloendothelial cells. Clofazimine in serum is bound

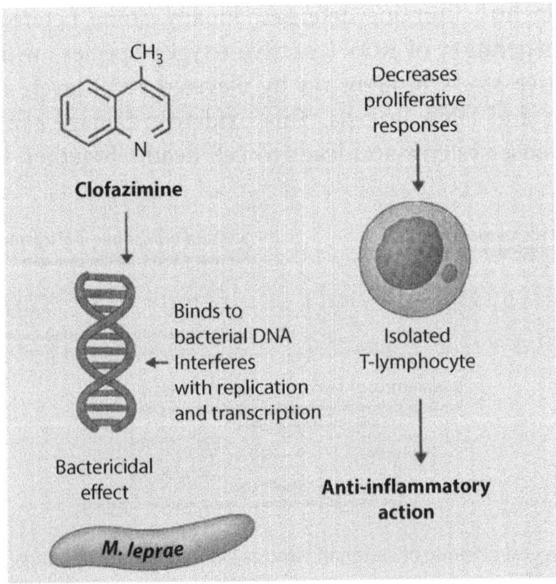

Figure 15.3 Mechanism of action of clofazimine.

mainly to β-lipoproteins (~10 μg/mL saturation) as well as to lesser extent to α-lipoproteins; binding to γ-globulin and albumin is minimal. In leprosy patients, daily oral administration of 100, 300, or 400 mg clofazimine resulted in average plasma levels of 0.7, 1.0, and 1.41 mg/L, respectively. Daily intake of 600 mg results in peak serum levels of 4 mg/L. In leprosy patients, the concentration of clofazimine in the fat has been reported to be as high as 5.3 mg/g, whereas concentration of 0.1 mg/g was found in bile juice, kidney, pancreas, skin, liver, spleen, lymph nodes, eyes, gallbladder, and lung. Clofazimine's metabolism is found to be hepatic [19, 20].

Adverse effect: Within a few weeks, 75% to 100% of individuals who take clofazimine develop pink to brownish black pigmentation. This is a significant negative cosmetic consequence. In addition to rash and itching (1%–5%), ichthyosis and dryness (8%–28%) can be caused by clofazimine. After therapy concludes, skin discoloration usually goes away in a few months. Most side effects are mild and do not necessitate stopping treatment. However, skin discoloration might cause despair in certain people. Although it does not impair vision, clofazimine, which is released into bodily fluids, can color the cornea and conjunctiva. Similar to skin pigmentation, ocular staining usually goes away following treatment [21].

Forty percent to 50% of people may experience mild gastrointestinal side effects, such as diarrhea, nausea, vomiting, or gastrointestinal intolerance. It is uncommon to administer a high dosage (>100 mg/d) for several months or longer. Severe gastrointestinal side effects, including bleeding, bowel obstruction, and splenic infarction, can happen and occasionally be lethal [22].

15.5.2 Rifampin

Rifampin has bactericidal antimicrobial effects because it inhibits DNA-dependent RNA polymerase (RNAP). Either reducing the RNAP's affinity for short RNA transcripts or inhibiting the elongating RNA at its 5′ end causes inhibition (Figure 15.4). Rifampin stops continuous RNA creation by selectively targeting microbial RNAP. Because rifampin does not interfere with the mammalian RNAP enzyme, there are less possible adverse effects in people [23]. Furthermore, rifampin has bactericidal effects against both intracellular and extracellular *Mycobacterium tuberculosis*.

Pharmacokinetics: Rifampin is quickly absorbed in the gastrointestinal tract. However, when provided with food, the drug's absorption is reduced by 30%. Throughout the body, rifampin is widely distributed, including in cerebrospinal fluid, which caused sputum, urine, sweat, saliva, and tears to turn orange–red. At 1.6 L/kg in adults and 1.1 L/kg in

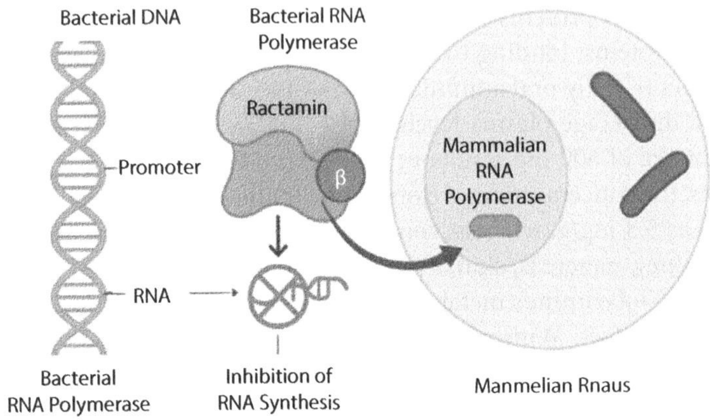

Figure 15.4 Mechanism of action of rifampin.

neonates, the drug has a large volume of distribution. Rifampin binds to plasma proteins around 80% of the time. At constant plasma concentration, rifampin has a half-life of 2 to 3 hours. Urine eliminates about 30% of the amount taken [24].

Adverse effect: Due to its excretion in bodily fluids, rifampin can cause dose-dependent side effects such as orange discoloration of tears, which can discolor contact lenses, sweat, saliva, urine, and feces. Additionally linked are gastrointestinal symptoms such as diarrhea, anorexia, and nausea [25].

15.5.3 Ethionamide

M. leprae is susceptible to the bacteriostatic effect of ethionamide. Inside the bacterium, the mycobacterial enzyme EthaA helps to activate the prodrug ethionamide to 2-ethyl-4-aminopyridine [26]. The identical enoylacyl carrier protein reductase of fatty acid synthase II (encoded by the inhA gene) is blocked by the activated drug, which stops the manufacture of mycolic acid and eventually hinders the development of cell walls (Figure 15.5). Ethionamide resistance may be caused by mutations in the ethA, etaR (an EthA transcriptional repressor gene), and inhA. Efflux pump also plays a role [27].

Pharmacokinetic: Bioavailability of ethionamide is 100% [28].

Adverse effect: Ethionamide frequently causes anorexia, nausea, vomiting, and stomach discomfort. Additionally, it can cause a number of neurological symptoms, which can be lessened by taking 100 mg of pyridoxine daily. Hepatitis is another possible side effect (5%) [29].

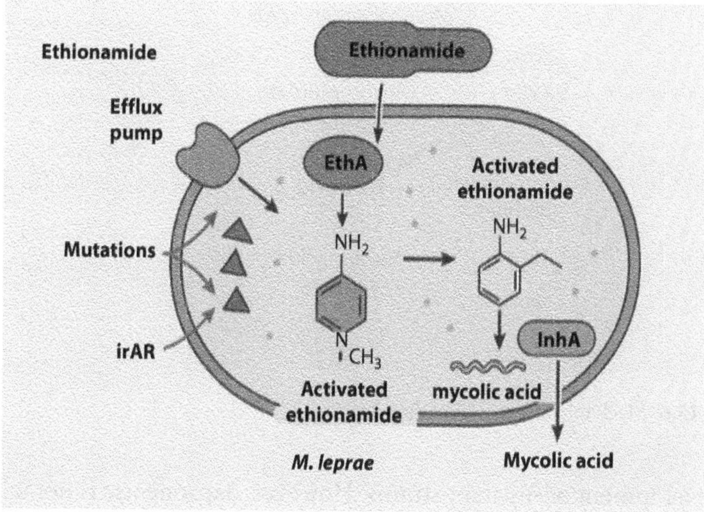

Figure 15.5 Mechanism of action of ethionamide.

15.5.4 Dapsone or Di-Amino-Diphenyl-Sulfone

Dapsone, an essential element of MDT for leprosy, is a synthetic compound from the sulfone class with a heterocyclic structure. Its chemical composition features a sulfone ($-SO_2-$) group linked to two aromatic rings, which enhances its pharmacological characteristics and therapeutic effectiveness [30]. This sulfone group gives dapsone a distinct aromatic smell and improves its interaction with bacterial enzymes. Dapsone's primary mechanism of action involves disrupting folic acid metabolism in *M. leprae*. It serves as a competitive inhibitor of the enzyme DHPS, which is essential in the bacterial folate synthesis pathway. By imitating para-aminobenzoic acid (PABA), dapsone prevents the incorporation of PABA into dihydropteroate, a precursor of folic acid (Figure 15.6). This inhibition ultimately hinders nucleic acid synthesis, stopping bacterial replication and growth [31].

Pharmacokinetics: Clinically, dapsone is taken orally and is well absorbed in the gastrointestinal tract, reaching peak plasma levels within 2 to 8 hours postingestion. Because *M. leprae* primarily affects the skin, liver, and peripheral nerves, its lipophilic nature guarantees that its broadly dispersed in these tissues. Although its relatively long half-life allows for once-daily dosing, dapsone is never used alone due to the risk of resistance [30, 31]. Instead, it is combined with rifampicin and clofazimine in MDT regimens to ensure synergistic effectiveness and minimize

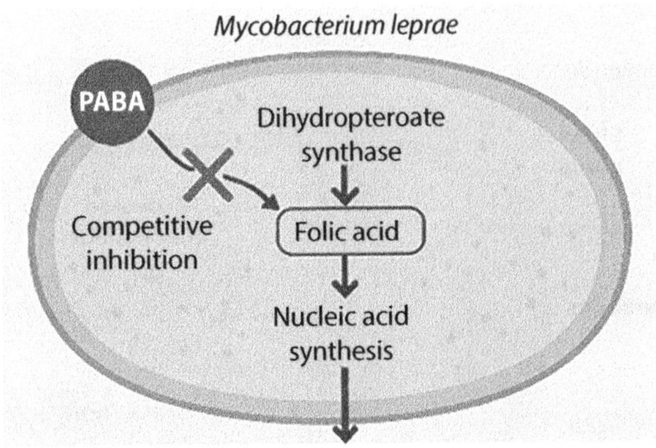

Figure 15.6 Mechanism of action of dapsone.

the development of resistant strains. However, dapsone use is not without challenges [30].

Adverse effect: Hemolysis in people with glucose-6-phosphate dehydrogenase insufficiency, a hereditary condition that impairs red blood cells' resistance to oxidative damage, is a significant concern. Dapsone-induced oxidative stress can cause severe hemolysis in these patients. Other side effects include hypersensitivity reactions such as skin rashes, methemoglobinemia (an abnormal increase in methemoglobin levels), and, in rare instances, dapsone hypersensitivity syndrome, a potentially life-threatening condition marked by fever, hepatitis, and lymphadenopathy. When resistance to dapsone arises or adverse reactions occur, alternative or additional antimicrobial therapies are necessary to ensure effective disease management [32].

15.5.5 Fluoroquinolones

Ofloxacin and moxifloxacin are manmade antimicrobial drugs characterized by a quinolone ring with a fluorine atom substitution, which greatly boosts their antibacterial effectiveness and ability to penetrate cells. Ofloxacin and moxifloxacin have proven especially effective in managing rifampicin-resistant leprosy and are frequently included in multidrug regimens for resistant or relapsed cases. These drugs are regarded as second-line treatments for leprosy, especially beneficial when first-line medications such as rifampicin are unsuitable or ineffective due to resistance [33, 34]. The addition of fluorine to the quinolone structure enhances the drugs' lipophilicity, facilitating better passage through bacterial membranes and greater accumulation in infected tissues. Furthermore, fluoroquinolones possess broad-spectrum activity, effectively targeting a variety of gram-positive and gram-negative bacteria, including *M. leprae*. The main action mechanism of fluoroquinolones is the inhibition of two essential bacterial enzymes: DNA gyrase (also known as topoisomerase II) and topoisomerase IV (Figure 15.7) [34]. These enzymes are vital for bacterial DNA replication, transcription, and repair. DNA gyrase introduces negative supercoils into DNA to relieve torsional stress during replication, whereas topoisomerase IV is responsible for separating daughter DNA strands after replication. Fluoroquinolones stabilize the DNA–enzyme complex following cleavage, preventing the re-ligation step and causing DNA breaks that ultimately result in bacterial cell death [35].

Pharmacokinetics: Fluoroquinolones are rapidly and almost completely absorbed orally, with excellent bioavailability and extensive tissue distribution, including penetration into the skin, lungs, and nerves—common sites affected in leprosy [35, 36].

Adverse effect: Despite their effectiveness, fluoroquinolones are linked to certain side effects and limitations. One of the most significant concerns

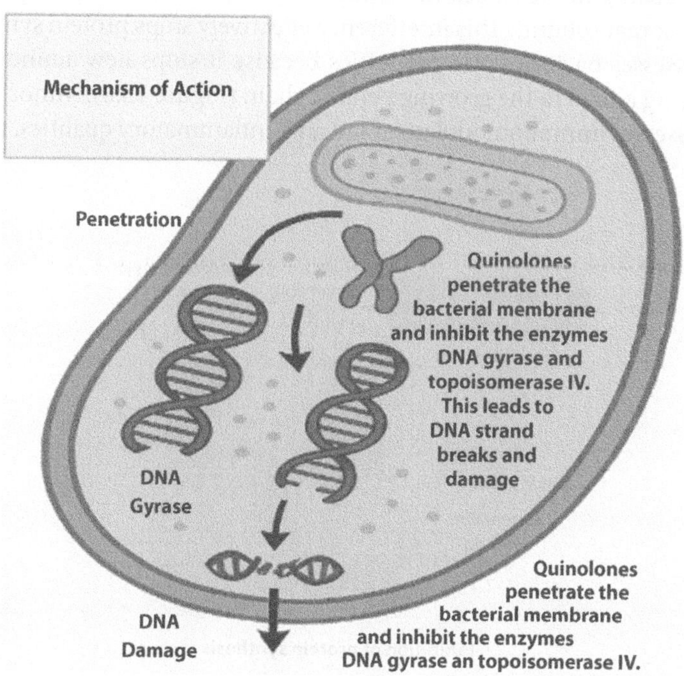

Figure 15.7 Mechanism of action of fluroquinolones.

is the risk of tendinopathy and tendon rupture, particularly in old patients or those taking corticosteroids. Peripheral neuropathy, another serious adverse effect, may manifest as numbness, tingling, or weakness. Cardiac toxicity is also a concern, as fluoroquinolones can prolong the QT interval on an electrocardiogram, potentially leading to arrhythmias in susceptible individuals. Additionally, the growing bacterial resistance to fluoroquinolones poses a major challenge, often resulting from mutations in the target enzymes or efflux pump overexpression. Therefore, their use should be carefully monitored and restricted to cases where they are clearly indicated [36].

15.5.6 Minocycline

Semisynthetic tetracycline antibiotic has demonstrated notable effectiveness, especially in treating infections caused by *M. leprae*. In MDT regimens, it is frequently used as second-line medication, particularly when patients show resistance or intolerance to first-line medications such as rifampicin or dapsone [35]. The suppression of bacterial protein synthesis is the mechanism of action of minocycline. It prevents aminoacyl-tRNA from attaching to the ribosome's acceptor site by reversibly binding to the 30S ribosomal subunit. This interference effectively stops protein synthesis and possesses bacteriostatic properties because it stops new amino acids from being added to the growing peptide chain (Figure 15.8). Minocycline also possesses immunomodulatory and anti-inflammatory qualities, which

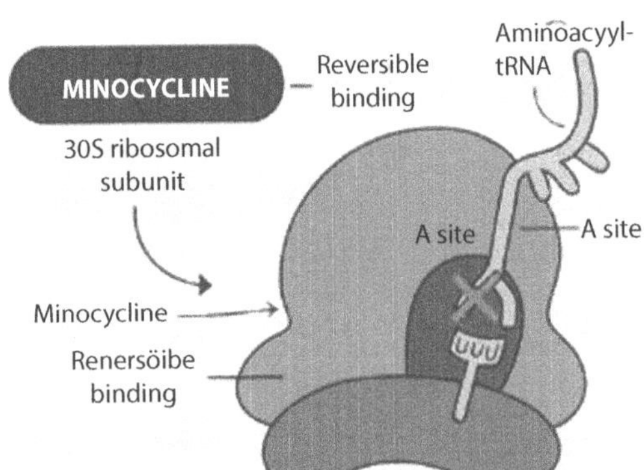

Figure 15.8 Mechanism of action of minocycline.

may help explain its therapeutic advantages in treating leprosy-related neuritis and responses [37, 38].

Pharmacokinetics: Pharmacokinetically, minocycline has a bioavailability of roughly 90% to 100% and is well absorbed after oral dosing. Food and dairy items have no effect on its absorption, in contrast to other tetracyclines. The medication can enter deep into tissues, such as skin, neurons, and macrophages—the main sites of *M. leprae* infection—because it is highly lipophilic. The simple once- or twice-daily dosing of minocycline is made possible by its comparatively lengthy half-life of around 11 to 22 hours. It is mostly broken down in the liver and eliminated through the biliary and renal systems [37].

Adverse effect: Minocycline has a number of adverse effects despite its efficacy. Gastrointestinal problems including nausea, vomiting, and diarrhea are frequent side responses. Hyperpigmentation of the skin, nails, or mucous membranes is a common side effect of minocycline medication that can occur with prolonged use. Hepatotoxicity, hypersensitivity reactions, vestibular toxicity (which manifests as vertigo or dizziness), and, in rare cases, drug-induced lupus and autoimmune hepatitis are more severe adverse effects. Its usage is contraindicated in pregnant women and children younger than 8 years due to the possibility of permanent tooth discoloration. However, minocycline is still a useful tool in the pharmaceutical toolbox for treating leprosy when taken sparingly [38].

15.5.7 Clarithromycin

The macrolide antibiotic clarithromycin has been known as a helpful second-line treatment for leprosy, particularly when resistance to main medications such as dapsone or rifampicin has developed. Its capacity to prevent bacterial protein production is the main reason for its efficacy against *M. leprae* [39]. Clarithromycin selectively disrupts the translocation phase during protein elongation by attaching itself to the 50S subunit of the bacterial ribosome. This stops the peptidyl-tRNA from moving from the A site to the P site, which stops the production of vital proteins needed for the growth and survival of bacteria (Figure 15.9). Although it may have bactericidal activity against *M. leprae* at high concentrations or after extended exposure, the end effect is bacteriostatic [38, 39].

Pharmacokinetics: Clarithromycin has a 50% bioavailability when taken orally and is well absorbed when taken with food. In the liver, it is first-pass metabolized to produce 14-hydroxyclarithromycin, an active metabolite with antibacterial properties. The medication is especially powerful against intracellular pathogens such as *M. leprae* because it is widely

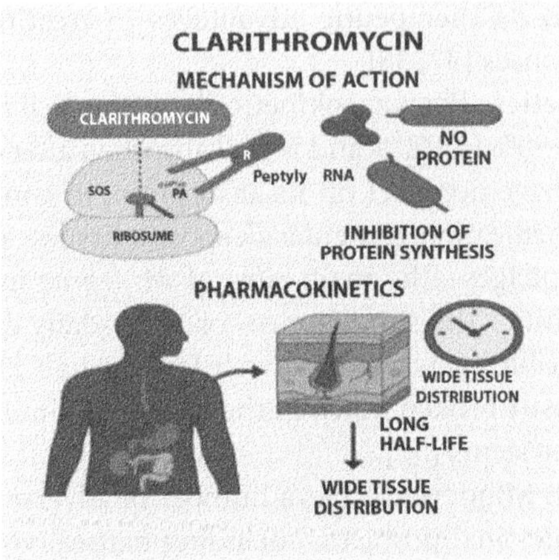

Figure 15.9 Mechanism of action of Clarithromycin.

distributed in bodily tissues and fluids, such as the skin and respiratory system, and it reaches higher intracellular concentrations than serum levels. The twice-daily dosage of clarithromycin is supported by its comparatively long elimination half-life of roughly 3 to 7 hours. It is mostly eliminated by the kidneys and liver, and dosage modifications are required in situations of renal or hepatic impairment [39].

Adverse effect: Clarithromycin has a number of adverse effects despite its therapeutic advantages. Typical adverse effects of digestive disorders include nausea, vomiting, stomach pain, and diarrhea [38]. Along with headaches and, less commonly, skin rashes, it can also produce taste abnormalities. Hepatotoxicity, QT-interval prolongation, and cardiac arrhythmias are more severe but less frequent hazards, especially for people who already have heart problems or are taking other drugs that lengthen the QT. Clarithromycin inhibits cytochrome P450 enzymes, which can raise the plasma levels of medications taken together, including theophylline, warfarin, and several antiepileptics. This raises concerns about drug interactions. However, clarithromycin is still a useful therapy choice for leprosy, particularly when first-line medicines are contraindicated or in instances that are resistant to several drugs [40].

Table 15.1 Shows the list of important anti-leprotic agents with their structure.

Sequence	Compounds	Structure
1	Clofazimine	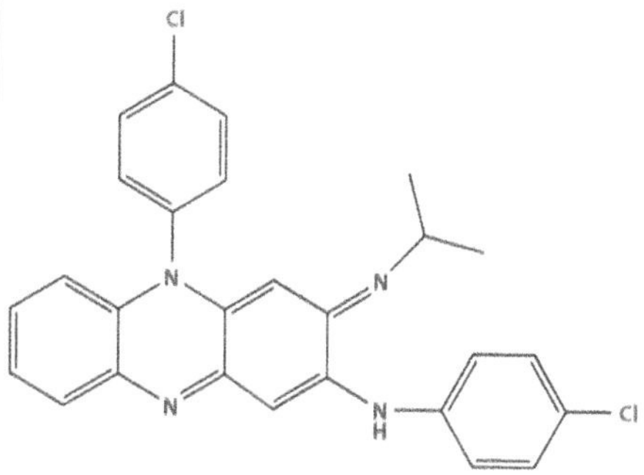
2	Rifampin	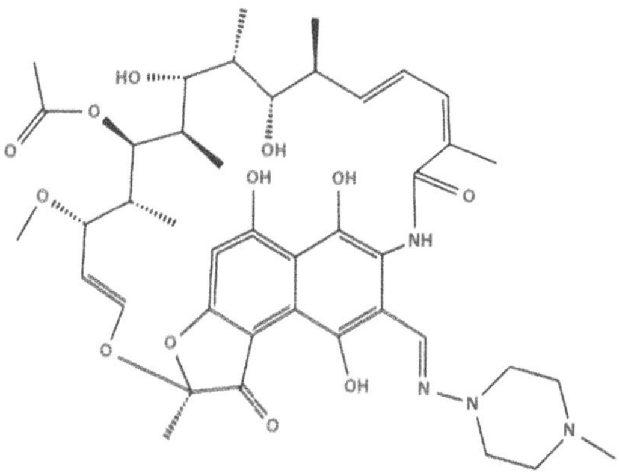
3	Ethionamide	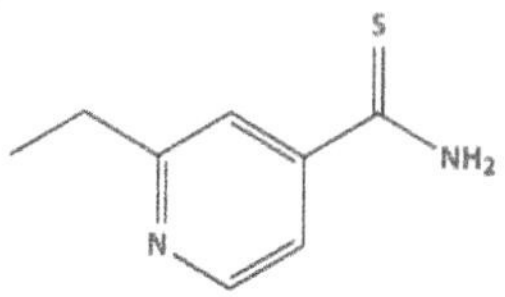
4	Ofloxacin	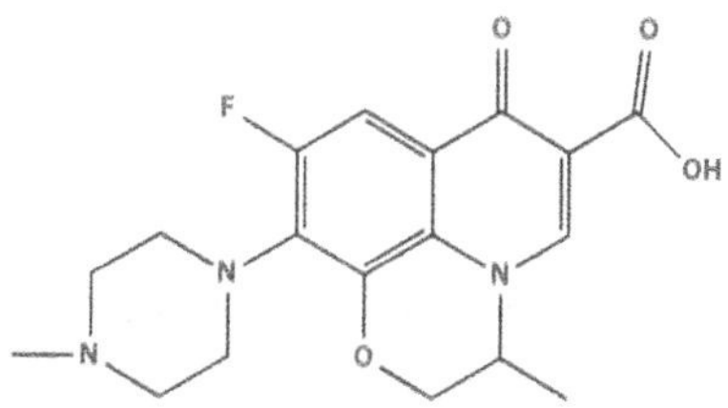

(*Continued*)

Table 15.1 Shows the list of important anti-leprotic agents with their structure. (*Continued*)

Sequence	Compounds	Structure
5	Moxifloxacin	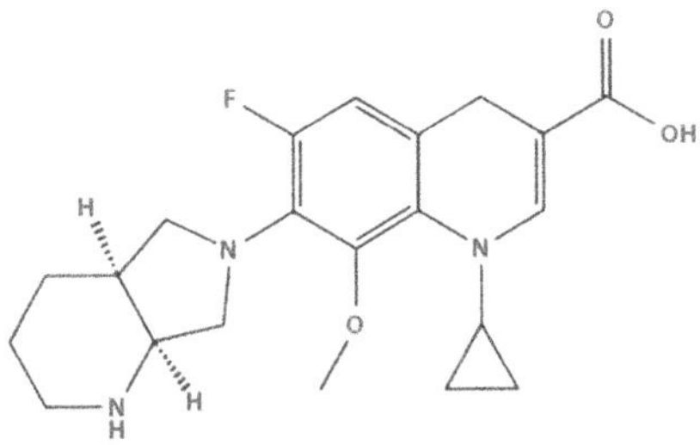
6	Minocycline	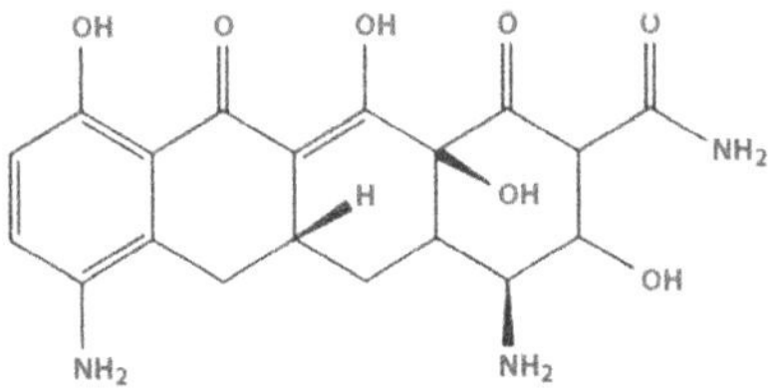
7	Clarithromycin	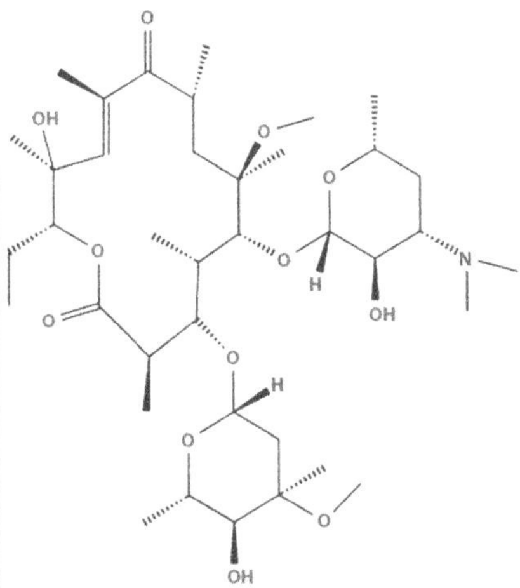

15.6 Advances in Antileprotic Agents

Heterocyclic drugs are fundamental to MDT regimens recommended by the WHO and are necessary for the therapeutic treatment of leprosy. Combining dapsone and rifampicin is used to treat paucibacillary leprosy, which is characterized by a low bacterial burden. The regimen consists of dapsone, rifampicin, and clofazimine for multibacillary leprosy, where the bacterial burden is much larger. In order to lower the bacterial load, stop resistance, and produce long-term remission, these heterocyclic compounds cooperate [41]. Research into novel drug delivery systems has been prompted by their shortcomings, which include low solubility, tissue penetration problems, and systemic side effects, despite their demonstrated effectiveness.

The distribution of heterocyclic medications in the treatment of leprosy has been transformed by recent developments in nanotechnology, which seek to get beyond these restrictions and improve therapeutic results. For example, clofazimine nanoformulations have been created to increase its bioavailability and lessen pigmentation adverse effects by allowing for more regulated medication release. Table 15.1 shows the list of important antileprotic agents with their structure. Heterocyclic compounds can be delivered precisely using liposomal encapsulation, which minimizes systemic toxicity by enabling the medicine to accumulate at the site of infection. Drug carriers based on polymers offer prolonged drug release, which minimizes the need for frequent dosing and guarantees steady plasma concentrations throughout time. Micelle-based delivery methods also improve the pharmacokinetic characteristics of poorly water-soluble heterocyclic medicines by increasing their solubility and stability. With its localized, controlled medication release, enhanced patient compliance, and fewer side effects, hydrogel-based carriers are another attractive platform [42, 43]. These novel drug delivery strategies hold great promise for improving leprosy patients' overall quality of care and optimizing the effectiveness of heterocyclic antileprotic drugs.

15.7 Obstacles and Prospects for the Development of Heterocyclic Antileprotic Drugs

The growing prevalence of drug-resistant *M. leprae*, especially to first-line medications such as dapsone, is one of the main obstacles facing the fight against leprosy. The efficiency of conventional MDT is jeopardized by resistance, which calls for the investigation of alternate therapeutic approaches [44]. The creation of novel heterocyclic compounds with enhanced potency and resistance characteristics has become the focus of research in response. In an effort to restore therapeutic efficacy and stop the spread of resistant strains, combination therapies involving these new drugs are being investigated.

Future medication research is increasingly utilizing cutting-edge technology such as artificial intelligence and computational drug design to address these issues. Using molecular docking, pharmacophore modeling, and machine learning techniques, structure-based drug design is making it possible to identify promising heterocyclic inhibitors with high specificity for crucial bacterial targets [45]. In addition to speeding up the drug discovery process, these tools help cut down on the time and expense of conventional screening techniques. Preclinical and clinical trials are being conducted on a number of these computationally designed drugs to evaluate their pharmacokinetic, safety, and efficacy in practical situations [46, 47].

Moreover, interdisciplinary cooperation is essential for the future of antileprotic treatment. By combining developments in pharmacology, molecular biology, medicinal chemistry, and nanotechnology, more efficient, patient-friendly, and tailored treatments can be created. To improve their bioavailability, lower their toxicity, and get beyond the present restrictions on tissue penetration and drug administration, researchers are also concentrating on altering heterocyclic scaffolds. All things considered, a concentrated international effort combining creative drug design methodologies with cooperative scientific research is opening the door for a new generation of heterocyclic medications that can treat leprosy more precisely and effectively over the long term [48].

AI disclosure statement: The author acknowledges the use of AI-based tools for minor language editing or grammatical correction. However, all intellectual contributions, analyses, and conclusions presented in this chapter are entirely the author's own.

References

1. Britton, W.J., Lockwood DN. Leprosy. *Lancet*, 363, 1209—19, 2004.
2. Gelber, R.H., Activity of minocycline in Mycobacterium leprae-infected mice. *J7 Infect Dis.*, 156, 236–9, 1987.
3. Fitness, J., Tosh, K., Hill, A., Genetics of susceptibility to leprosy. *Genes Immune*, 3, 441–53, 2002.
4. Rees, R.J.W., Pearson, J.M.H., Waters, M.F.R., Experimental and clinical studies on rifampicin in treatment of leprosy. *BMJ*, i, 89–92, 1970.
5. WHO, Global leprosy (Hansen disease) update, 2019: time to step-up prevention initiatives. *Wkly Epidemiol. Rec.*, 95, 417–40, 2020.
6. Mennuru, N.B., Birudala, R., Birudala, G., Pathogenesis, clinical manifestations, and treatment of leprosy, in: *Elsevier eBooks*, pp. 609–615, 2023, https://doi.org/10.1016/b978-0-323-85730-7.00042-4.

7. Bhat, R.M. and Prakash, C., Leprosy: An Overview of Pathophysiology. *Interdiscip. Perspect. Infect. Dis.*, 2012, 1–6, 2012, https://doi.org/10.1155/2012/181089.
8. Froes Junior, L.A., Jr., Sotto, M.N., Trindade, M.A.B. Leprosy: Clinical and immunopathological characteristics, in: *Sociedade Brasileira de Dermatologia*, vol. 97–97, Issue 3, pp. 338–347, Anais Brasileiros De Dermatologia, Sociedade Brasileira de Dermatologia, Brazil, 2022. https://doi.org/10.1016/j.abd.2021.08.006.
9. Mehta H., Jain, S., Narang, T., Chhabra, S., Dogra, S., Leprosy reactions: New knowledge on pathophysiology, diagnosis, treatment and prevention. *Indian J. Dermatol. Venereol. Leprol.*, 2024. doi: 10.25259/IJDVL_915_2024.
10. Froes, L.A.R., Trindade, M.A.B., Sotto, M.N., Immunology of leprosy. *Int. Rev. Immunol.*, 1–21, 2020. doi:10.1080/08830185.2020.18513.
11. Rosa, T.L.S.A., Mendes, M.A., Linhares, N.R.C., Rodrigues, T.F., Dias, A.A., Leal-Calvo, T., Gandini, M., Ferreira, H., Da Mota Ramalho Costa, F., Sales, A.M., Amadeu, T.P., Schmitz, V., Pinheiro, R.O., Rodrigues, L.S., Moraes, M.O., Pessolani, M.C.V., The type I interferon pathway is upregulated in the cutaneous lesions and blood of multibacillary leprosy patients with erythema nodosum leprosum. *Front. Med.*, 9, 2022, https://doi.org/10.3389/fmed.2022.899998.
12. Kyatham, R., Bathula, R., Palle, S., Vantikommu, J., Hemalatha, S., Pasha, M.A., Ethnopharmacological significance of nitrogen containing heterocyclic compounds. In Chapter, https://doi.org/10.22271/int.book.149, 2024.
13. A Review Article on: Natural Anti-Leprotic Drugs. 2020. [Review Article]. World Journal of Pharmaceutical Research, 9, 3, 1708–1715, https://doi.org/10.20959/wjpr20203-17078.
14. Slman, D.K., Satar, H.A., Ketan, Z.A., Jawad, A.A., Heterocyclic Compounds: A Study of its Biological Activity. *Al-Nahrain J. Sci.*, 27, 5, 19–24, 2024, https://doi.org/10.22401/anjs.27.5.03.
15. Kabir, E. and Uzzaman, M., A review on biological and medicinal impact of heterocyclic compounds. *Results Chem.*, 4, 100606, 2022, doi: 10.1016/j.rechem.2022.100606.
16. Qadir, T., Amin, A., Sharma, P.K., Jeelani, I., Abe, H., A review on medicinally important heterocyclic compounds. *Open Med. Chem. J.*, 1, 2022, [Journal-article], https://doi.org/10.2174/18741045-v16-e2202280.
17. Jampilek, J., Heterocycles in medicinal chemistry. *Molecules*, 24, 21, 3839, 2019, https://doi.org/10.3390/molecules24213839.
18. Raj, G.M., Antimycobacterial drugs, in: *Introduction to Basics of Pharmacology and Toxicology*, Paul, A. (Ed.), p. 869, 2021, https://doi.org/10.1007/978-981-33-6009-9_56.
19. Butlin, C.R. and Maghanoy, A., Benefits of high dose clofazimine. *Leprosy Rev.*, 92, 4, 427–432, 2021, https://doi.org/10.47276/lr.92.4.427.

20. Garg, S., Rajiv Gandhi Institute of Pharmacy, Faculty of Pharmaceutical Science & Technology, AKS University Satna, Antileprotic agents. In Chapter-10 (pp. 93–96) [Book-chapter], n.d.
21. Nugraha, R.V., Yunivita, V., Santoso, P., Aarnoutse, R.E., Ruslami, R., Clofazimine as a Treatment for Multidrug-Resistant Tuberculosis: A review, in: *Scientia Pharmaceutica*, vol. 89, p. 19, 2021, [Journal-article], https://doi.org/10.3390/scipharm89020019.
22. Mendes, A.F.M., Gomes, C.M., Kurizky, P.S., Ianhez, M., Case Report: A case series of immunobiological therapy (Anti-TNF-A) for patients with Erythema nodosum leprosum. *Front. Med.*, 9, 2022, https://doi.org/10.3389/fmed.2022.879527.
23. Gerstein, W., Kucers' The use of antibiotics. Chapter-Rifampicin. *Clin. Infect. Dis.*, 52, 11, 1397, 2011, https://doi.org/10.1093/cid/cir204.
24. Lyons, M.A., Reisfeld, B., Yang, R.S.H., Lenaerts, A.J., A physiologically based pharmacokinetic model of rifampin in mice. *Antimicrob. Agents Chemother.*, 57, 4, 1763–1771, 2013, https://doi.org/10.1128/aac.01567-12.
25. Richardus, J.H., Tiwari, A., Barth-Jaeggi, T., Arif, M.A., Banstola, N.L., Baskota, R., Blaney, D., Blok, D.J., Bonenberger, M., Budiawan, T., Cavaliero, A., Gani, Z., Greter, H., Ignotti, E., Kamara, D.V., Kasang, C., Manglani, P.R., Mieras, L., Njako, B.F., Novartis Foundation. (n.d.). Leprosy post-exposure prophylaxis with single-dose rifampicin (LPEP): an international feasibility programme, in: *The Lancet Global Health*, vol. 9, p. e81, https://doi.org/10.1016/S2214-109X(20)30396-X.
26. Singh, P. and Sharma, M., Repurposing Drugs to Combat Drug Resistance in Leprosy: A Review of opportunities. *Comb. Chem. High Throughput Screening*, 25, 10, 1578–1586, 2021, https://doi.org/10.2174/1386207325666211007110638.
27. Wang, F., Langley, R., Gulten, G., Dover, L.G., Besra, G.S., Jacobs, W.R., Sacchettini, J.C., Mechanism of thioamide drug action against tuberculosis and leprosy. *J. Exp. Med.*, 204, 1, 73–78, 2007, https://doi.org/10.1084/jem.20062100.
28. Kumar, G., Sharma, S., Shafiq, N., Pandhi, P., Khuller, G.K., Malhotra, S., Pharmacokinetics and tissue distribution studies of orally administered nanoparticles encapsulated ethionamide used as potential drug delivery system in management of multi-drug resistant tuberculosis. *Drug Delivery*, 18, 1, 65–73, 2010, https://doi.org/10.3109/10717544.2010.509367.
29. Scardigli, A., Caminero, J.A., Sotgiu, G., Centis, R., D'Ambrosio, L., Migliori, G.B., Efficacy and tolerability of ethionamideversusprothionamide: A systematic review. *Eur. Respir. J.*, 48, 3, 946–952, 2016, https://doi.org/10.1183/13993003.00438-2016.
30. Eichelmann, K., *et al.*, Lepra: Puestaaldía. Definición, patogénesis, clasificación, diagnóstico y tratamiento. *Actas Dermosifiliogr*, 104, 554–63, 2013.

31. D'Cunha, R., Kartha, V., Gurnani, S., Raman and i.r. studies of the antileprotic drug Dapsone. *Spectrochim. Acta Part Mol. Spectrosc.*, *39*, 4, 331–336, 1983b, https://doi.org/10.1016/0584-8539(83)80007-5.
32. Guragain, S., Upadhayay, N., Bhattarai, B.M., Adverse reactions in leprosy patients who underwent dapsone multidrug therapy: A retrospective study. *Clin. Pharmacol.: Adv. Appl.*, 9, 73–78, 2017, https://doi.org/10.2147/CPAA.S135846.
33. Li, X., Li, G., Yang, J., Jin, G., Shao, Y., Li, Y., Wei, P., Zhang, L., DrugResistance (Dapsone, Rifampicin, Ofloxacin) and Resistance-Related Gene Mutation Features in Leprosy Patients: A Systematic Review and Meta-Analysis. *Int. J. Mol. Sci.*, 23, 12443, 2022, https://doi.org/10.3390/ ijms232012443.
34. Chaves, L.L., Patriota, Y., Soares-Sobrinho, J.L., Vieira, A.C.C., Lima, S.A.C., Reis, S., Drug Delivery Systems on Leprosy Therapy: Moving Towards Eradication? *Pharmaceutics*, 12, 12, 1202, 2020 Dec 11. doi: 10.3390/pharmaceutics12121202. PMID: 33322356; PMCID: PMC7763250.
35. *2023 Antibacterial agents in clinical and preclinical development: An overview and analysis*, World Health Organization, Geneva, 2024, Licence: CC BY-NC-SA 3.0 IGO.
36. Fish, D.N., Fluoroquinolone adverse effects and drug interactions. *Pharmacother. J. Hum. Pharmacol. Drug Ther.*, 21, 10P2, 2001, https://doi.org/10.1592/phco.21.16.253s.33993.
37. Lenz, S.M., Collins, J.H., Ray, N.A., Hagge, D.A., Lahiri, R., Adams, L.B., Post-exposure prophylaxis (PEP) efficacy of rifampin, rifapentine, moxifloxacin, minocycline, and clarithromycin in a susceptible-subclinical model of leprosy. *PLoS Negl.Trop. Dis.*, 14, 9, e0008583, 2020, https://doi.org/10.1371/journal.pntd.0008583.
38. Maymone, Mayra, B.C. *et al.*, Leprosy: Treatment and management of complications, *Journal of the American Academy of Dermatology*, 83, 1, 17–30, 2020.
39. Kurosawa, K., Rossenu, S., Biewenga, J., Ouwerkerk-Mahadevan, S., Willems, W., Ernault, E., Kambili, C., Population pharmacokinetic analysis of Bedaquiline-Clarithromycin for dose selection against pulmonary nontuberculous mycobacteria based on a phase 1, randomized, pharmacokinetic study. *J. Clin. Pharmacol.*, 61, 10, 1344–1355, 2021, https://doi.org/10.1002/jcph.1887.
40. Chaves, L.L., Patriota, Y., Soares-Sobrinho, J.L., Vieira, A.C.C., Lima, S.A.C., Reis, S., Drug Delivery Systems on Leprosy Therapy: Moving Towards Eradication? *Pharmaceutics*, 12, 1202, 2020. https://doi.org/10.3390/pharmaceutics12121202.
41. Andrade, E.S.N., Brandão, J.G., Da Silva, J.S., Kurizky, P.S., Rosa, P.S., De Araújo, W.N., Gomes, C.M., A systematic review and meta-analysis of studies on the diagnostic accuracy and screening of tests to detect antimicrobial resistance in leprosy. *Diagn. Microbiol. Infect. Dis.*, 100, 1, 115325, 2021, https://doi.org/10.1016/j.diagmicrobio.2021.115325.

42. Da Rocha, N.P., Barbosa, E.J., De Araujo, G.L.B., Bou-Chacra, N.A., Innovative drug delivery systems for leprosy treatment. *Indian J. Dermatol. Venereology Leprology*, 88, 437–442, 2022, https://doi.org/10.25259/ijdvl_1119_20.
43. Chaves, L. L., Silveri, A., Vieira, A. C., Ferreira, D., Cristiano, M. C., Paolino, D., Di Marzio, L., Lima, S. C., Reis, S., Sarmento, B., & Celia, C. (2019). pH-responsive chitosan based hydrogels affect the release of dapsone: Design, set-up, and physicochemical characterization. *International Journal of Biological Macromolecules*, 133, 1268–1279,The American Society for Experimental NeuroTherapeutics, Inc., Haddonfield, NJ, USA. https://doi.org/10.1016/j.ijbiomac.2019.04.178.
44. Ebenezer, G.J. and Scollard, D.M., Treatment and evaluation advances in leprosy neuropathy, in: *Neurotherapeutics*, vol. 18–18, pp. 2337–2350, The American Society for Experimental NeuroTherapeutics, Inc, 2021, [Journal-article], https://doi.org/10.1007/s13311-021-01153-z.
45. Artificial intelligence and machine learning technology driven modern drug discovery and development. *Int. J. Mol. Sci.*, 24, 2026, 2023, https://doi.org/10.3390/ijms24032026.
46. Gautam, S., Sharma, D., Singh, S., Deo, N., Goel, A., Gupta, V.K., Bisht, D., Increscent journey of anti-leprosy drug development. *Curr. Sci.*, 253–267, 2023.
47. Duthie, M.S., Frevol, A., Day, T., Coler, R.N., Vergara, J., Rolf, T., Sagawa, Z.K., Beckmann, A.M., Casper, C., Reed, S.G., A phase 1 antigen dose escalation trial to evaluate safety, tolerability and immunogenicity of the leprosy vaccine candidate LepVax (LEP-F1 + GLA–SE) in healthy adults. *Vaccine*, 38, 7, 1700–1707, 2019, https://doi.org/10.1016/j.vaccine.2019.12.050.
48. Kaushik, N., Kumar, N., Kumar, A., Singh, U.K., Tetrazoles: Synthesis and biological activity. *Immunology, Endocrine & Metabolic Agents-Medicinal Chemistry Current Medicinal Chemistry-Immunology, Endocrine & Metabolic Agents*, 1, 18, 1, 3–21, 2018 May.

16

Bromodomains: Biomedical Applications of Active Ingredients

Sweta Raj[1], Anjan Kumar Nayak[1,2], Harsh Vardhan[1] and Divya Bajpai Tripathy[1*]

[1]Division of Chemistry, School of Basic Sciences, Galgotias University, Greater Noida, India
[2]Integral Bioscience Pvt. Ltd., Noida, India

Abstract
Bromodomains are evolutionarily conserved protein motifs that recognize and bind acetylated lysine residues on histone tails, playing a pivotal role in regulating gene expression through epigenetic mechanisms. Dysregulation of bromodomain-containing proteins has been implicated in various diseases, including cancer, inflammation, and cardiovascular disorders. Recent advances in medicinal chemistry have led to the development of small-molecule inhibitors and other active ingredients targeting bromodomains, particularly the BET (bromodomain and extra-terminal) family, showing promising therapeutic potential. These inhibitors modulate transcriptional programs associated with disease progression, offering a novel strategy for precision medicine. This review highlights the structure and function of bromodomains, their pathological significance, and the biomedical applications of active compounds designed to modulate their activity, emphasizing current clinical advances and future perspectives in drug development.

Keywords: Bromodomains, epigenetics, histone acetylation, BET inhibitors, small-molecule inhibitors, cancer therapy, gene regulation, drug discovery

**Corresponding author*: divyabaj@gmail.com

Divya Bajpai Tripathy, Priyanka Chhabra and Anjali Gupta (eds.) Heterocyclic Compounds as Pharmaceutical Ingredients, (531–546) © 2026 Scrivener Publishing LLC

16.1 Introduction

Bromodomain (BRD) is a 110-amino-acid protein module that recognizes lysines acetylated by histones and other proteins [1]. These proteins affect how tight or loose DNA is, which is close or loose with chromatin, which controls access to genetic information. Acetylation of lysis to nucleosome histones has long been recognized as a responsibility for weakening DNA histone interactions, thereby affecting the activation of gene transcription [2]. The Bromodomain and Extra-Terminal (BET) protein family includes the key members BRD2, BRD3, BRD4, and the testis-specific BRDT. These proteins and two tandem BRDs have additional terminal (AT) domains. Because acetylated lysine residues on histone proteins are crucial in the control of gene expression, these proteins are known as epigenetic readers because they can identify and bind to them [3]. BRD shows its presence when researchers examined HATs and HDACs as transcriptional activators and inhibitors. In 1992, common motifs of *Drosophila* BRAHMA (BRM) and homeostatic (FSH) proteins from female sterilization were found, referring to "bromodomain" [4].

16.1.1 Significance in Medical Researches

We thoroughly inspected BET families. It consists of BRD2, BRD3, BRD4, and BRDT, all of which are ubiquitous except for BRDT, which is expressed only in the testis. Their main features are two BRDs at the N-terminal (BD1 and BD2) and AT domain at the C-terminal [5]. The ability of BRD2, BRD3, and BRD4 to bind to acetylated histones and other proteins involved in gene regulation is a crucial aspect of their functions. Dysregulation of BRD2, BRD3, and BRD4 in cancer has been linked to encouraging tumor development and survival, which makes them desirable targets for cancer treatment, [6] and BRDs also play a role in inflammatory gene expression. By inhibiting BRD activity, it may be possible to reduce the expression of proinflammatory cytokines and treat conditions such as rheumatoid arthritis or inflammatory bowel disease. Some research has shown that the proteins that contain BRDs are involved in the expression of genes linked to neurodegenerative diseases such as Alzheimer disease. Modulating BRD activity might offer new therapeutic avenues for managing these conditions.

16.2 History and Discovery of BRDs

The BRD was discovered in 1992 by John M. Bustin and colleagues. They discovered this domain as part of a family of proteins, which are capable of binding to acetylated lysine residues on histones. The name "bromodomain" arises from the *Drosophila* Brahma (BRM), where the domain was first found. Brahma is a key component of chromatin-remodeling complexes that regulate gene transcription. The BET family of proteins, which includes BRD2, BRD3, BRD4, and BRDT, became an interest of research. These proteins were found to regulate critical genes involved in cell growth and inflammation.

BRD4, in particular, was recognized for its role in promoting the transcription of oncogenes, such as MYC, which are essential for cancer cell proliferation.

16.2.1 Evolution of BRD Research

BRD research has developed rapidly since its discovery, driven by advances in molecular biology, epigenetics, and drug development. The discovery of acetylated lysine as a remaining protein domain in histones in the early 1990s quickly became central to chromatin biology and epigenetics testing. Their structure studies demonstrated the unique ability to regulate transcription by reading histone modifications and marking them as key regulators of gene expression. Identification of the BET family of proteins containing BRD4, particularly the BRD, halts the role of researchers in maintaining cancer transcription, leading to the development of BET inhibitors such as JQ1. These inhibitors were of great interest due to their ability to inhibit cancer growth by disrupting tumor regulation. Over time, research on BRD has developed beyond cancer to include cancer, cardiovascular disease, and neurodegenerative diseases. This field will further develop, creating new therapeutic techniques such as selective inhibitors and proteins to improve resistance mechanisms and expand the scope of BRD-targeted treatments. This evolution highlights the importance of BRD as epigenetic regulators and their potential in precision medicine.

16.2.2 Structure of BRDs

The initial nuclear structure of BRD from the HAT-CO activator PCAF (P300/CBP-related factor) was determined by the Ming-Ming-Zhou laboratory in both unbound forms, in both Ming-Ming-Zhou complexes with acetyl histamine, acetyl–lysine analogs [7]. The BRD structure consists of four alpha helices, forming a left-handed bundle with two loops that create deep columns. The four helices form the hydrophobic core of the domain. The two loops form a deep column that binds to the acetylated lysine of the histone tail sequence. The long Z-A loop connects the helices alpha-Z and alpha-A with a short BC loop that connects the helix alpha-B and alpha-C. Hydrophobic bags (binding bags) are formed by ZA and BC loops. BRD structures are conserved in all BRDs, but the different regions are ZA and BC loops.

The human proteome includes 46 proteins that contain BRD motifs, encompassing a total of 61 unique BRDs. These BRDs are classified into eight major families based on their sequence and structural similarities. High-resolution structures, either from X-ray crystallography or nuclear magnetic resonance, have been determined for 47 of these BRDs, many through the efforts of the Structural Genomics Consortium (SGC). Among these, 23 BRDs have structural data available in complexes with peptide ligands or small molecule inhibitors. To date, full structural characterization has been achieved for BRDs in families I, VI, and VIII, whereas families II and IV are almost entirely covered. However, structural information remains incomplete for several members of families III, V, and VII. The most extensively researched subgroup of BRDs belongs to family II, specifically the BET domain proteins. The BET BRDs, both unbound and ligand-bound, represent 39% of all human BRD structures available in the Protein Data Bank (PDB), with 108 PDB entries for BRD4 and 32 for BRD2 [1].

16 2.3 Interaction with Acetylated Lysines

BRD is an epigenetic reader that marks the artifacts of acetylated lysine for transcription. They are highly conserved protein modules developed to specifically recognize and bind acetylated lysine residues. These types of acetylated lysine are in histone or other protein domains [8]. Proteins containing BRD carry a variety of functions, including histone acetyltransferase activity, chromatin remodeling, and transcriptional activation. As a "reader" of lysine acetylation, it plays an important role in interpreting signals carried by acetylated lysine residues and is converted into a variety of physiological or pathological outcomes [9].

Disruption of acetylation regulation has been linked to multiple sclerosis, an inflammatory demyelinating disorder of the central nervous system, which leads to substantial neurological disability, particularly in young adults [10].

16.3 BRDs as Drug Targets

BRDs have shown potential in treating cancer, inflammatory disorders, and parasitic diseases. Researchers have created powerful inhibitors and degraders targeting BRD-containing proteins, with some of these compounds already entering clinical trials. However, the development of these drugs remains challenging, primarily due to issues with dose-limiting toxicity. BET proteins are especially promising targets for drug development due to their involvement in cancers and inflammatory diseases. BET proteins contain two BRDs, whereas non-BET proteins typically have only one. JQ1 and OTX015 are BET inhibitors that have shown capability in treating diseases driven by JAK/STAT signaling. RVX-208, a selective inhibitor of BET BD2, has undergone clinical trials for conditions such as diabetes, coronary artery disease, and other related disorders.

16.3.1 Drug Design Strategies Targeting BRDs

One key strategy in BRD drug design is to create molecules that mimic the acetyl–lysine binding motif. Because BRDs recognize and bind to acetylated lysines, inhibitors are designed to either block this interaction or disrupt the structural integrity of the BRD. Small molecules, such as JQ1 and OTX015, have identified the ability to bind particularly to the acetyl–lysine recognition site in BET BRDs, thereby inhibiting the recruitment of transcriptional machinery and modulating gene expression. The specificity of these inhibitors has made them potential therapeutic candidates for diseases where BRD-driven pathways are dysregulated, such as in cancers driven by MYC overexpression or inflammatory conditions such as rheumatoid arthritis.

Another strategy in drug design is the use of "selective targeting" of specific BRDs within different protein families. BET proteins have two BRDs (BD1 and BD2); many non-BET proteins contain only a single BRD. Researchers have focused on designing inhibitors that particularly target one BRD over the other, with the goal of improving specificity and reducing off-target effects. For instance, RVX-208, a selective BD2 inhibitor, has been tested for its efficacy in treating diseases such as coronary artery disease and diabetes.

However, the development of drugs targeting BRDs faces significant challenges, one of the most pressing being "dose-limiting toxicity." As many BRDs are involved in important cellular processes such as gene regulation, chromatin remodeling, and protein stability, systemic inhibition can have unintended consequences. For example, broad inhibition of BET BRDs can lead to nondesirable effects on gene expression that may cause toxicity or disrupt normal cellular functions. Therefore, achieving a balance between potency and specificity is essential. Advances in structure-based drug design, coupled with high-throughput screening and computational modeling, are helping researchers better understand the precise interactions between BRDs and small molecules. This knowledge is guiding the design of more selective and less toxic compounds that can target specific BRDs or cellular pathways with minimal side effects.

16.4 BRD Inhibitors

The first known inhibitors of the BET BRD family were the (+)-JQ1, developed by the Dana-Farber Cancer Institute, reported by the SGC and GlaxoSmithKline (GSK), and by IBET762. (+)-JQ1, thieno-triazolo-1,4-diazepine connections were initially tested in nut midline cancer (NMC), a rare and aggressive form of squamous cell carcinoma. In NMC, Brd4 protein forms a fusion with the core protein of the testis round (groove) protein, leading to Brd4-NUT oncoprotein, which plays a crucial role in tumor growth and differentiation [11]. JQ1 functions through competitive binding to the acetyl–lysine detection center of BRD4 BRD, which suppresses BRD4-NUT fusion proteins from chromatin. This disorder induced squamous epithelialization, resulting in specific antiproliferative effects in both heterologous induction models and NMC-derived cell lines. This breakthrough determined the potential of BET BRD inhibitors as therapeutically active ingredients in the treatment of cancer types driven by abnormal BET protein activity [12]. According to the first report, many studies have been published identifying the efficacy of JQ1 in most cancers. Due to hematologic malignant tumors, JQ1 showed promising results in conditions such as leukemia and lymphoma. Several solid tumors were also tested, including glioblastoma, medulloblastoma, and hepatocellular carcinoma. JQ1 was also featured in antitumor activity in colon, pancreatic, prostate, lung, and breast cancer [13].

Current studies on colon cancer showed an important role for BRD4 in tumor growth, with JQ1 showing excellent antigrowth effects, particularly in tumors that exhibited the methylation factor phenotype (CIMP) of CPG islands. This effect is related to the presence of CIMP-related superamps that regulate MYC transcription. This superamp is transcribed by the long nonencoded RNA CCAT1 (colon cancer–associated transcript 1), whose expression significantly improves the susceptibility of these tumors compared to JQ1 treatment. Furthermore, in another study, the combination of JQ1 and arsenic sulfide in gastrointestinal and intestinal cancer cells resulted in inhibition of BRD4 and C-MYC, thereby activating the synergistic effect of tumor suppressor protein P53 [14].

16.5 Next-Generation BRDs Inhibitors

16.5.1 Two Classes of BET Inhibitors

Irreversible and subdomain selective
Irreversible

1. Extended pharmacological effects compared to systemic exposure, even after "nonbinding inhibitors" leave the circulation.
2. Pharmacology is independent of PK.
3. Covalent bond formation precludes competition from natural ligands, regardless of concentration.
4. Not affected by changes in affinity for binding to mutated enzymes—meaning of cancer resistance.
5. Time-dependent inactivation gives the possibility that irreversible inhibitors overcome resistance.
6. Improved *in vivo* selectivity, because only those targets that form the covalent bond will retain the inhibitor's subdomain selective.
7. Pan-BET inhibitors are reported to have a major impact but have a close therapeutic index.
8. Preclinical studies have shown that selective BRD2 inhibitors act primarily on acute myeloid leukemia (AML) and prostate cancer cell lines, with effective doses of mouse xenotransplants in the MTD group [15, 16].

16.6 BET Inhibitors in Clinical Trials

A summary of cancer-related clinical trials is shown in Table 16.1 (Monography) and Table 16.2 (Combination studies).

An opportunity for irreversible inhibition can be seen in the molecular docking results (Figure 16.1 and Figure 16.2).

Table 16.1 Cancer-related clinical trials (Monotherapy).

CPI-0610	Malignant peripheral nerve sheath tumors, hematologic malignancies, previously treated multiple myeloma, progressive lymphoma phase 2
MK-8628	Hematologic malignancies, progressive solid tumors
GSK2820151	Progressive and recurrent solid tumor
GSK525762	Relapsed hematologic malignancies, solid tumors
ZEN003694	Metastatic castration-resistant prostate cancer (mCRPC)
FT-1101	Hematologic malignancies, phase 1/1b
BMS-986158	Advanced solid tumors, phase 1/2
ABBV-075	Advanced cancer
ABBV-744	mCRPC, AML
GS-5829	mCRPC, progressive solid tumors and lymphomas
PLX51107	Advanced solid tumors, AML, myelodysplastic syndrome
NUV-868	Advanced solid tumors

16.6.1 Key Insights on Selectivity between BRD4 (BRD1) and BRD3 (BRD2)

- N-methyl in amphipathic water pocket
- Ethyl amide buries itself in His 437 (BD II) and not Asp 144 (BD I)
- 2-6-Dimethyl group clashes with Ile162 (BD I) and not with Val 435 (BD II)

16.6.2 BD2 Selectivity (Why Target Selective Compounds?)

Despite the similarity of the sequence, BD1 and BD2 have different functions due to interactions with various lysine-acetylated histone or transfer proteins [18]. BD1 is associated with H4K5CK8AC, which was moved to fix the relevant protein in the enhancer sites of chromatin. The BD2 does not interact with one acetylated H3K4AC, but shows a strong interaction with the rationalized H3K4CK9AC. Initial studies show that selective internal suppression can provide a variety of pharmacology compared to

Table 16.2 Cancer-related clinical trials (Combination studies).

GSK525762	
Fulvestrant	ER+ breast cancer, phase 2
Abiraterone or enzalutamide	Castration-resistant prostate cancer
ZEN003694	
Eenzalutamide	mCRPC
ABBV-075	
Venetoclax	Advanced cancer
GS-5829	
Fulvestrant or exemestane	Advanced ER$^+$ HER2$^-$ breast cancer
Enzalutamide	mCRPC
NUV-868	
Olaparib	Ovarian, TNBC, pancreatic, mCRPC, and other solid tumors
Enzalutamide	mCRPC

Data courtesy of ClinicalTrials.gov.

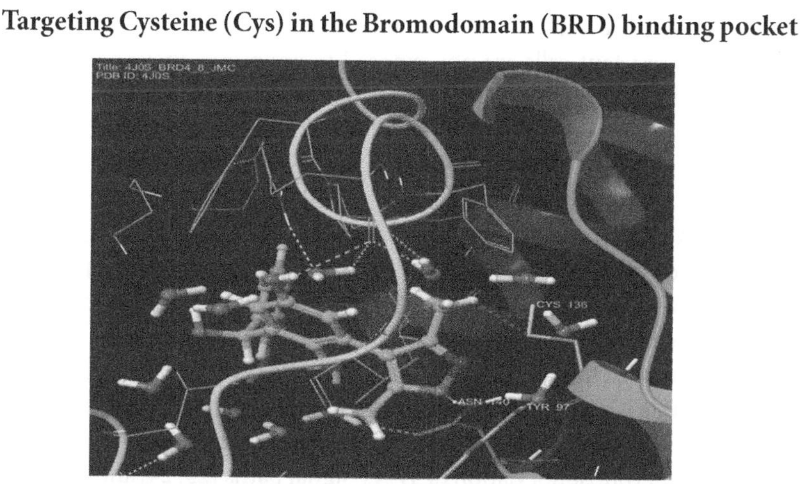

Figure 16.1 Molecular docking results of interaction of drug with target [17].

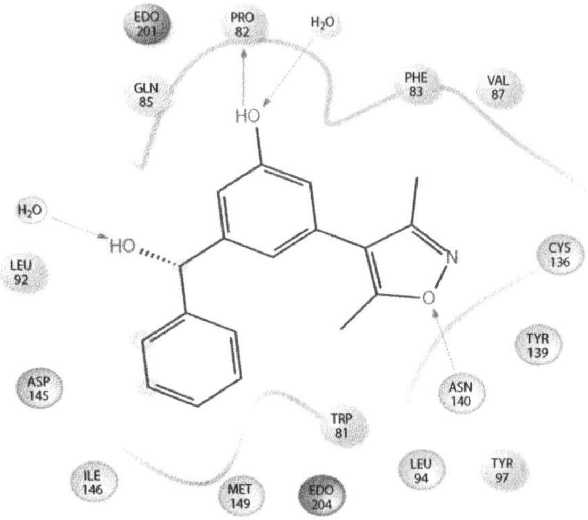

Figure 16.2 Cys136 in BRD4 is within 4 Å from the ligand in the x-ray crystal structure (4J0S.pdb) and could be easily approached by an electrophile at the methyl position of the molecule.

pan-BET inhibitors [11]. BD2 selectivity helps to correct toxicity associated with betting inhibitors in relation to inhibitors [15].

16.7 Resistance toward BET Inhibitors

So far, the only mechanisms of resistance to BET inhibitors that have been demonstrated are the BRD2/3/4 mutations, not genetic targets. Although AML-JQ1–resistant cells are not exposed to apoptosis, they show high regulation of beclin1, increased LC3 lipidation, and the formation of autophagosomes. Activation of AMPK (P-THR172)ULK1 (P-SER555) PAD was associated with JQ1-induced autophagy in resistant cells, regardless of MTOR signaling. To support pharmacological AMPK inhibition or degradation of PRKAA/AMPKA, autophagy was inhibited and promoted JQ1-induced apoptosis of AML stem cells [19, 20]. Similarly, bortezomib induced JQ1 treatment and induced resistant cells of JQ1-bortezomibe, and colon cancer cells induced G2/M starvation. Inhibition of NF-κB by bortezomib or NF-κB inhibitors using crazy cells of JQ1 [21] also affected the phosphorylation of Bcl-2 and mitotic products of bortezomib, leading to G2/M arrest and apoptosis [22]. Another resistance mechanism is compensatory regulation of MYC by hyperphosphorylation of WNT signal paths [23, 24] and Brd4 due to increased expression of PP2A phosphatase [25] and BCL2L1/BCL-XL [25].

Limitations involved in resistance to small shift inhibitors aim to eliminate the actual inhibition and mechanism of actual inhibition (i.e., Hemerocallis germination–degradation of methyl acyl pyrrole) [26]. Other clinical tests are required to fully determine the role of BET inhibitors in tumors.

16.8 Toxicity of BET Inhibitors Used Clinically

Due to its effects on cellular proteins, including other BRD-containing sensors, BET inhibition is associated with dose-limiting toxicity. Finally, a systematic review of the security, efficacy, and pharmacodynamics of 12 BET inhibitors in clinical research was conducted [27]. All BET inhibitors showed similar toxicity profiles with gastrointestinal toxicity (diarrhea, nausea, vomiting), thrombocytopenia, anemia, fatigue, and hyperbilirubinemia [28]. Undesired toxicity considers BRD to be a target and an extraordinary inhibition or dismantling.

16.9 BRD Inhibition and Human Diseases

Over the past 5 years, BRDi has attracted considerable attention as a promising therapeutic strategy to treat many diseases. The majority of translational studies focused on BET mainly due to the presence of highly potent BET inhibitors such as JQ1 and I-BET762, as well as the presence of numerous cellular pathways affected by the loss of BET function. Nevertheless, CBP and P300 were tested individually as potential drug targets for cardiovascular disease [29].

16.9.1 Cancer

JQ1 was first tested in a mouse model of carcinoma [1]. This study showed that JQ1 BRD4-NUT, a fusion tumor protein formed by the combination of BRD4 and NUT (the core protein of testicle), is effectively inhibited in both laboratory environments and organisms. BRD4 mothers are attributed to chromosomal translocation in NMCs. This is a particularly aggressive form of human squamous cell carcinoma. By creating fluorescence, the researchers showed that JQ1 can bind to the BRD4 nut and remove chromatin. Treatment of NMC cells with JQ1 also covered differentiation and halted growth. This indicates successful inhibition of the BRD4 nut. This treatment also led to improved tumor contraction and survival in a mouse implant model from NMC.

16.9.2 Cardiovascular Disease

BRDI has examined many studies as a potential treatment for cardiovascular diseases such as cerebrovascular disease, heart failure, atherosclerosis, and hypertension. In the case of ischemic heart disease, a key event is myocardial myocyte apoptosis as the disease progresses, driving p53 in response to DNA damage *via* myocardial chemicals. The p53 response to genotoxic tone is regulated by CBP. As mentioned previously, CBP P53 has REST K382, and FRG binds to acetylated P53. This interaction promotes further acetylation of p53 in some lysine by CBP. Acetylation of p53 promotes nuclear translocation and activates genes involved in cellular stress responses that cause apoptosis [30].

16.10 Conclusion and Future Perspectives

BET inhibitors are recognized as a promising class of novel cancer therapies. Recent clinical studies have confirmed the potential for antitumor, but their effectiveness appears to be limited as individual active ingredients. All available BET inhibitors work by preventing the interaction of BRD with histone acetylation band. However, targeting of BRD proteins outside the target leads to dose-limiting toxicity that affects its clinical success. Furthermore, primary and acquired resistance to these drugs, driven by a variety of mechanisms, has already been observed. Currently, BET inhibitors are not approved for clinical use.

Could improve selectivity be the key to overcoming these challenges? Understanding BET protein biology, the specificity of BET inhibitors, and the associated drug pharmacokinetics and pharmacodynamic biomarkers more is crucial for enhancing the therapeutic efficacy of anticancer treatments. Although more potential and more selective BRD inhibitors have been identified, the diverse and well-understood roles of proteins containing BRD are challenged in view of cellular effects and selectivity, taking into account the structural similarity of these inhibitors. To minimize off-tart effect, improved selectivity toward specific BRD subdomains is required. Furthermore, irreversible inhibition of cysteine relics may provide a promising option worthy of further research.

In addition to BRDs, the BET protein also contains the C-terminal ET domain. It interacts with several cellular proteins, such as NSD331 and JMJD63 [30], both of which are involved in the development of cancers such as AML and lung/intestinal cancer. At this time, no therapeutic strategies have been proposed to target the ET domain of the BET protein. This is a potential new direction in cancer treatment. Of the four BET family members, BRD4 was most frequently tested, but the role of BRD2 and BRD3 was unknown to the survival of cancer cells. Finally, there is still much to learn, and the search for answers continues.

AI disclosure statement: The author acknowledges the use of AI-based tools for minor language editing or grammatical correction. However, all intellectual contributions, analyses, and conclusions presented in this chapter are entirely the author's own.

References

1. Ferri, E., Petosa, C., McKenna, C.E., Bromodomains: Structure, function and pharmacology of inhibition. *Biochem. Pharmacol.*, 106, 1–18, 2016.
2. Gil, J., Ramírez-Torres, A., Encarnación-Guevara, S., Lysine acetylation and cancer: A proteomics perspective. *J. Proteomics, 150*, 297–309, 2017.
3. Meslamani, J., Smith, S.G., Sanchez, R., Zhou, M.M., Structural features and inhibitors of bromodomains. *Drug Discovery Today: Technol.*, *19*, 3–15, 2016.
4. Tamkun, J.W., Deuring, R., Scott, M.P., Kissinger, M., Pattatucci, A.M., Kaufman, T.C., Kennison, J.A., Brahma: A Regulator of Drosophila Homeotic Genes Structurally Related to the Yeast Transcriptional Activator Snf2/Swi2. *Cell*, 68, 561–572, 1992.
5. Iudin, M.S., Khodarovich, Y.M., Varizhuk, A.M., Tsvetkov, V.B., Severov, V.V., A Minireview on BET Inhibitors: Beyond Bromodomain Targeting. *Biomedicines*, *13*, 3, 594, 2025.
6. Wang, Z.Q., Zhang, Z.C., Wu, Y.Y., Pi, Y.N., Lou, S.H., Liu, T.B., Lou, G., Yang, C., Bromodomain and extra terminal (BET) proteins: Biological functions, diseases, and targeted therapy. *Signal Transduct Target Ther*, 8, 1, 420, 2023 Nov 6. doi: 10.1038/s41392-023-01647-6. PMID: 37926722, PMCID: PMC10625992.
7. Dhalluin, C., Carlson, J.E., Zeng, L., He, C., Aggarwal, A.K., Zhou, M.M., Structure and ligand of a histone acetyltransferase bromodomain. *Nature*, 399, 491–496, 1999.
8. Ntranos, A. and Casaccia, P., Bromodomains: Translating the words of lysine acetylation into myelin injury and repair. *Neurosci. Lett.*, 625, 4–10, 2016.
9. Gold, S. and Shilatifard, A., Therapeutic targeting of BET bromodomain and other epigenetic acetylrecognition domain–containing factors. *Curr. Opin. Genet. Dev.*, 86, 102181, 2024.
10. Pedre, X., Mastronardi, F., Bruck, W., Lopez-Rodas, G., Kuhlmann, T., Casaccia, P., Changed histone acetylation patterns in normal-appearing

white matter and early multiple sclerosis lesions. *J. Neurosci.*, 31, 3435–3445, 2011.
11. Cochran, A.G., Conery, A.R., Sims III, R.J., Bromodomains: A new target class for drug development. *Nat. Rev. Drug Discovery*, 18, 8, 609–628, 2019.
12. Leigh, R.S., Kaynak, B.L., Ruskoaho, H., Välimäki, M.J., Development and comparison of single FLT3-inhibitors to dual FLT3/TAF1-inhibitors as an anti-leukemic approach. *PLoS One*, 20, 3, e0320443, 2025.
13. Pérez-Salvia, M. and Esteller, M., Bromodomain inhibitors and cancer therapy: From structures to applications. *Epigenetics*, 12, 5, 323–339, 2017.
14. Boyson, S.P., Gao, C., Quinn, K., Boyd, J., Paculova, H., Frietze, S., Glass, K.C., Functional roles of bromodomain proteins in cancer. *Cancers*, 13, 14, 3606, 2021.
15. Kati, W., *et al.*, BET AACR Presentation ABBV-744, 2018.
16. Ramadoss, M. and Mahadevan, V., Targeting the cancer epigenome: Synergistic therapy with bromodomain inhibitors. *Drug Discovery Today*, 23, 1, 76–89, 2018.
17. Hewings, D.S., Fedorov, O., Filippakopoulos, P., Martin, S., Picaud, S., Tumber, A., Wells, C., Olcina, M.M., Freeman, K., Gill, A., Ritchie, A.J., Sheppard, D.W., Russell, A.J., Hammond, E.M., Knapp, S., Brennan, P.E., Conway, S.J., Optimization of 3,5-Dimethylisoxazole Derivatives as Potent Bromodomain Ligands. *J. Med. Chem.*, 56, 8, 3217–3227, 2013.
18. Liu, Z., Wang, P., Chen, H., A Wold, E., Tian, B., R Brasier, A., Zhou, J., Liu, Z., Drug Discovery Targeting Bromodomain-Containing Protein 4. *J. Med. Chem.*, 60, 11, 4533–4558, 2017.
19. Jang, J.E., Eom, J.I., Jeung, H.K., *et al.*, Targeting AMPK-ULK1-mediated autophagy for combating BET inhibitor resistance in acute myeloid leukemia stem cells. *Autophagy*, 13, 4, 761–762, 2017.
20. Jang, J.E., Eom, J.I., Jeung, H.K., *et al.*, AMPK-ULK1-mediated autophagy confers resistance to BET inhibitor JQ1 in acute myeloid leukemia stem cells. *Clin. Cancer Res.*, 23, 11, 2781–2794, 2017.
21. Wu, T., Wang, G., Chen, W., *et al.*, Co-inhibition of BET proteins and NF-kappaB as a potential therapy for colorectal cancer through synergistic inhibiting MYC and FOXM1 expressions. *Cell Death Dis.*, 9, 3, 315, 2018.
22. Ling, Y.H., Liebes, L., Ng, B., *et al.*, PS-341, a novel proteasome inhibitor, induces Bcl-2 phosphorylation and cleavage in association with G2-M phase arrest and apoptosis. *Mol. Cancer Ther.*, 1, 10, 841–849, 2002.
23. Rathert, P., Roth, M., Neumann, T., *et al.*, Transcriptional plasticity promotes primary and acquired resistance to BET inhibition. *Nat.*, 525, 7570, 543–547, 2015.
24. Kumar, K., Raza, S.S., Knab, L.M., *et al.*, GLI2-dependent c-MYC upregulation mediates resistance of pancreatic cancer cells to the BET bromodomain inhibitor JQ1. *Sci. Rep.*, 5, 9489, 2015.

25. Shu, S., Lin, C.Y., He, H.H., *et al.*, Response and resistance to BET bromodomain inhibitors in triple-negative breast cancer. *Nature*, 529, 7586, 413–417, 2016.
26. Lucas, X., Wohlwend, D., Hugle, M., *et al.*, 4-Acyl pyrroles: mimicking acetylated lysines in histone code reading. *Angew. Chem. Int. Ed. Engl.*, 52, 52, 1405514059, 2013.
27. Sun, Y., Han, J., Wang, Z., Li, X., Sun, Y., Hu, Z., Safety and efficacy of bromodomain and extra-terminal inhibitors for the treatment of hematological malignancies and solid tumors: A systematic study of clinical trials. *Front. Pharmacol.*, 11, 621093, 2021.
28. Sarnik, J., Poplawski, T., Tokarz, P., BET proteins as attractive targets for cancer therapeutics. *Int. J. Mol. Sci.*, 22, 11102, 2021.
29. Borah, J.C., Mujtaba, S., Karakikes, I., Zeng, L., Muller, M., Patel, J., *et al.*, A small molecule binding to the coactivator CREB-binding protein blocks apoptosis in cardiomyocytes. *Chem. Biol.*, 18, 531–541, 2011.
30. Liu, W., Ma, Q., Wong, K., Li, W., Ohgi, K., Zhang, J., Aggarwal, A.K., Rosenfeld, M.G., Brd4 and JMJD6-associated anti-pause enhancers in regulation of transcriptional pause release. *Cell*, 155, 1581–1595, 2013.

Index

α-amylase inhibition, 491
α-glucosidase inhibitors, 47, 491

β-adrenergic receptor, 361
β-lactams, 62, 63

Absorption, 396
ACE inhibitors, 351, 354, 361, 363, 371, 373, 378
Acetylated lysine, 534
Acetylation (histone), 532, 534
Acetylcholine sterase, 445, 446
Acetylcholinesterase (AChE), 35, 36
Acrylonitrile, 466
Active pharmaceutical ingredients (APIs), 1, 2, 249, 250
Acute myeloid leukemia (AML), 537, 540
Acylation reactions, 478
Adenocarcinoma, 461
ADMET, 121, 369–397
Aerobic glycolysis, 137
Aflatoxins, 323
Agrochemical, 214, 227
Aldosterone, 353, 358
Alkaloids, 27, 28
Allergic rhinitis, 272, 274, 278, 280, 292
Allergy, 279, 280
Amino intermediates, 456
Aminomethyl indole derivatives, 35
Aminopyralid, 324
Amphoteric, 71
Ampicillin, 66, 67, 69, 70
AMPK pathway, 540
AMPK/mTOR signaling pathway, 138
Analgesic activity, 489, 490
Angiotensin enzyme, 446, 447
Angiotensin II, 353–355
Anthocyanins, 318
Anthracyclines, 131–135
Antibacterial activity, 10, 11, 258, 259, 398, 399, 408–441, 451, 452, 454, 475
Anticancer activity, 399, 407, 408, 437–441, 457, 458, 461, 475, 482–486
Anticancer agents, 9, 10, 257, 258
Anticonvulsant activity, 400, 490, 491
Antidepressant activity, 488
Antidiabetic activity, 491, 447, 448, 450, 500
Antifungal activity, 397, 398, 408, 439
Antifungal agents, 10, 258
Antihelmintic drugs, 477
Antihypertensive activity, 359, 366, 376
Anti-inflammatory activity, 475, 487–490
Anti-inflammatory agents, 179, 181, 183, 185, 187, 189, 191–195, 197, 199–201
Antimalarial activity, 486, 487
Antimalarial drugs, 10, 258
Antimicrobial activity, 451, 452, 455, 480
Antimicrobial agents, 62, 63
Antimicrobial resistance, 407, 408

Antioxidant activity (piperazine), 495, 496
Antiparasitic activity, 408
Antiparasitic agents, 447
Antiplatelet drugs, 477
Anti-VEGFR, 91
Antiviral agents, 10, 258
Anxiolytic agents, 477
Apoptosis, 460, 540, 542
Apoptosis induction, 140, 146, 407, 494
Arachidonic acid (AA), 181
Arsenic sulfide, 537
Aryl substitution, 458, 466
ATPase inhibition, 486
Auxin, 317, 336
Azeotrope formation, 478
Azetidine, 30
Azetidinones, 455, 456
Azole–piperazine derivatives, 439

Ball-mill synthesis, 481
BCL-2 phosphorylation, 540
BCL-XL, 540
BD1 selectivity, 538
BD2 selectivity, 538–540
Beclin1, 540
Benzimidazole, 443, 444, 445
Benzimidazole derivatives, 6, 43–45, 254, 447, 448, 450, 456
Benzimidazole hybrids, 452, 453, 464
Benzo[b]furan synthesis, 5, 253
Benzodiazoles, 450
Benzofuran, 224
Benzophenone analogs, 462
Benzopyrans, 223
Benzothiazepines, 371, 372, 373
Benzothiazole substitution, 483
Benzothiazole–piperazine hybrids, 435
Benzothiazoles, 179, 200
Benzothiophene derivatives, 491
BET inhibitors, 535–542
BET proteins, 532, 535
Binding affinity, 391
Bioactive heterocycles, 435
Bioavailability, 357, 358
Biofilm, 74, 89
Biological activities, 34–48
Biological activity, 443, 444
Biological targets interaction, 480
Biphenyl, 227
Bisbenzimidazoles, 448, 449
Blocks H1 receptor, 273–276
Blood pressure regulation, 357, 381
Blood–brain barrier (BBB), 397
Blood–brain barrier penetration, 480
BRD inhibitors, 536–540
BRD resistance, 540, 541
BRD toxicity, 541
BRD2, 532, 534, 543
BRD3, 532, 538, 543
BRD4, 533, 536–542
BRDT, 532
Breast cancer, 458, 463
Broad-spectrum activity, 407, 408
Bromination reactions, 478
Bromodomain (BRD), 531–543
Bromodomain discovery, 533
Bromodomain structure, 533, 534
Butyrylcholinesterase (BuChE), 35, 36

Calcium-channel blockers, 350, 368
cAMP, 278, 280–286
Cancer, 399
Cancer cell lines, 407, 437
Cancer epidemiology statistics, 476
Cancer therapy, 535–542
Candesartan, 354, 371
Candida albicans, 447, 450
Candida glabrata, 447, 448
Candida krusei, 447, 450
Candida parapsilosis, 447, 450
Candida tropicalis, 447
Canpothecin, 119, 128
Captopril, 354, 359, 361, 371, 372
Carbazole derivatives, 35, 38
Carbothioamides, 456
Carboxamide substitution, 486

Cardiovascular disease, 542
Cardiovascular sickness, 225
Catalysis, 4–6, 252–254
Catechol, 91
CCAT1, 537
CDK inhibition, 494
Celecoxib, 181, 183, 188
Cell cycle arrest, 494
Cell proliferation, 463, 464
Cephalosporins, 62, 73
C–H activation, 4, 252
Chalcone–piperazine hybrids, 497
Chelation, 211, 223
Chloroacetyl chloride, 455, 457
Chloroquine-resistant malaria, 486
Cholinergic transmission, 446
Chromatin remodeling, 532, 534
Ciprofloxacin, 63
Ciprofloxacin synergy, 439
Cleavable complex, 132
Click chemistry, 464
Clinical trials (BET inhibitors), 537–539
Clonidine, 350, 351, 368
Colon cancer, 537
Compatibility, 325
Competition assay, 105
Computational drug design, 389
Computational drug design (CADD), 524
Condensation reactions, 448, 450, 455
Copper-catalyzed C–N bond formation, 435
Copper-catalyzed synthesis, 480
Coronavirus binding, 446, 447
Coumarin–piperazine derivatives, 437
COVID-19 drugs (docking studies), 391
COX inhibition, 489
CREB-binding protein (CBP), 541, 542
Crystalline, 223, 225
Cyclization, 448, 456, 466
Cyclization reactions, 478
Cyclooxygenase (COX), 179–181
Cyclooxygenase-2 (COX-2) inhibitors, 44, 45
CYP51, 84, 81
Cys136 (BRD4), 540
Cytotoxic activity, 437
Cytotoxicity, 89, 95, 450, 459, 462
Cytotoxicity assays, 482–486

DABCO ring opening, 435
Dapsone and rifampicin combination treatment, 523
Delocalization of sulfur electrons, 478
Diabetes mellitus, 46
Dimeric indole alkaloids, 497
Distribution, 396
Diuretics, 350, 358
DNA adduct formation, 133
DNA binding, 468
DNA cleavage activity, 484
DNA gyrase, 63, 75
DNA intercalation, 131, 134, 135
DNA replication, 64, 63
Docking studies, 446, 447, 486, 491
Dopamine modulation, 488
Dose-limiting toxicity, 535, 541
Doxazosin, 356
Drug design, 2, 32, 250
Drug design strategies, 535, 536
Drug discovery, 2, 3, 250, 251, 390, 391, 443, 444
Drug resistance, 407, 408
Drug resistance challenges, 500
Drug-resistant antibiotic resistance, 524
Drug–target interaction, 391

Efflux pump inhibition, 439
EGFR inhibition, 460
EGFR-TK inhibitors, 141, 142
Electron-rich heterocycles, 475, 480
Embryonic kidney cells, 89
Endothelin-1 (ET-1), 353

Enkephalinase inhibitor, 352
Environmental safety, 326, 337
Enzyme inhibition, 480
Epigenetic readers, 532
Epigenetics, 531
Ergosterol biosynthesis, 90, 91, 450
ET domain, 542
Ethanol azeotrope, 478
Ethanol reflux, 447, 451
Excretion, 396
Exploration, 237

Favonoids, 209
FDA-approved nitrogen heterocycles, 435
Fibroblast cells, 74
Fluoroquinolones, 62, 63
Fluoroquinolones anti-bacterial, 510
Fulvestrant, 539
Functional group modification, 486
Fundamental, 233
Fungal inhibition, 448, 450
Fungicidal activity, 448
Furanone, 194, 195

Gene expression, 531, 532
Genetic engineering, 337
Gewald reaction, 478, 485
Glioblastoma, 458
Glucuronic acid, 232
Glutamate synthase, 317
Glyphosate, 313
Gram-negative bacteria, 407, 452, 453
Gram-positive bacteria, 407, 455
Gram-positive causative agent, 510
Green synthesis, 4, 252
Green synthesis approaches, 481
Groove binding, 135
GSK525762, 538, 539

H1 blocker, 280
H2 receptor, 281
H3 receptor, 278, 284
H3K4 acetylation, 538
H4 receptor, 286
H4K5K8 acetylation, 538
Halogen substitution effects, 489
Hematologic malignancies, 538
Heme detoxification inhibition, 486
Hepatocellular carcinoma, 465, 469
Herbicidal activity, 400
Heterocycles, 308, 310
Heterocyclic compounds, 1–3, 28, 29, 31, 179–201, 249–251, 389, 390, 407, 408, 444
Heterocyclic drug scaffolds, 476
Heterocyclic leprosy, 509, 510, 512, 513, 523, 525
Heterocyclic scaffolds, 8–10, 256–258
Histamine, 269
Histone acetylation, 531–534
HIV-1 protease inhibitors, 437
Hybrid scaffold design, 437–439
Hydralazine, 361, 374, 378
Hydrazine hydrate, 450, 456
Hydrogen bonding, 446
Hydrogen bonding capability, 475, 492
Hydroxyl group, 211, 219, 232, 235

I-BET762, 536, 541
IL-12, 272, 282
Imidazole, 28, 31, 192, 193
Imidazole derivatives, 8, 256
Imidazole scaffold, 447
Imidazolinones, 321, 326
Imine formation, 457
In silico methods, 390–392
In vitro testing, 451, 454
In vivo studies, 457
Indium-catalyzed synthesis, 480
Indole derivatives, 7, 34, 35, 51, 255
Indoles, 179, 190, 191
Indomethacin, 183, 191, 192, 199
Inflammation, 269, 272, 277, 274, 282, 532

Inflammatory cytokine reduction, 489
Insecticidal activity (piperazine), 497
Introduction, 390
Irreversible inhibitors, 537
Ischemic heart disease, 542
Isoquinoline, 28, 31
Isoxazole, 196

JAK/STAT signaling, 535
JQ1, 533, 535–542

Kinase inhibition, 460

Larvicidal activity, 445
LC3 lipidation, 540
Leprosy heterocyclic agent, 509, 510, 513, 524–527
Leukotriene, 179, 181, 183
Lipoxygenase (LOX), 179, 181
Lipoxygenase inhibition, 489
Lithium aluminium, 448
Lithium bis(trimethylsilyl)amide, 484
Lonazolac, 31
Lung carcinoma, 461
Lycorine, 93
Lysine acetylation, 531, 534

Malaria parasite enzyme targeting, 486
Mast cells, 270, 278
Mebendazole, 447
Medicinal chemistry, 2, 3, 250, 251
Medicinal chemistry applications, 475, 476
Melanoma cells, 464
Metabolic syndrome, 208
Metabolism, 396
Metal-catalyzed synthesis, 478–480
Metal-free synthesis, 481
Metastasis modulation, 119
Methicillin-resistant *Staphylococcus aureus* (MRSA), 441
Metronidazole, 63
Microbial strains, 451, 452
Microorganisms, 226
Microwave synthesis, 450
Microwave-assisted synthesis, 5, 6, 253, 254
Molecular docking, 69, 78, 84, 98, 391, 392, 435, 446, 447, 539
Molecular docking studies, 138
Molecular dynamics, 447
Molecular dynamics simulation, 392, 393
Monoamine oxidase (MAO-B), 42
Mosquito larvae, 445
Multicomponent reactions, 479
Multidrug resistance, 436
Multidrug therapy treatment, 510, 516, 523, 524, 526
Multiple sclerosis, 534
MYC oncogene, 533, 537, 540
Mycobacterium leprae, 510
Mycophenolic acid (MPA), 94

N-alkylation, 447
Nanocatalysts, 7, 255
Nanoformulations drug delivery, 523
Natriuretic peptide, 351, 352
Natural product modification, 437
Neurofibrillary tangles, 446
Neurotransmitter modulation, 488
NF-κB, 179, 199
NF-κB pathway, 540
Nitration reactions, 478
Nitro derivatives, 461
Nitrogen heterocycles, 492
Nonnucleoside reverse transcriptase inhibitors, 93, 100, 103
Non-small cell lung cancer activity, 484
Nonsteroidal anti-inflammatory drugs (NSAIDs), 179, 180
Norepinephrine, 350, 351, 370
NUT midline carcinoma (NMC), 536, 541
Nutrients, 312

Orthopoxvirus, 103
Oteseconazole, 81
OTX015, 535
Oxazoles, 179, 200
Oxidation, 228
Oxidative cyclization, 466
Oxidative stress, 44
Oxidative stress reduction, 495

P450, 87
P53 tumor suppressor, 537, 542
Pall–Knorr synthesis, 478
Pan-BET inhibitors, 537
Pancreatic cancer inhibition, 496
Parkinson disease therapeutics, 480
Peniclin, 62, 63, 76
Peptic ulcer, 278
Peptidoglycan, 63
Peroxidation, 208
Pharmaceutical agents, 397
Pharmaceutical applications, 9–11, 257–259
Pharmacodynamics, 541, 542
Pharmacokinetics, 80, 542
Pharmacological activity, 444, 445
Pharmacophore, 357, 358, 363
Pharmacophore modeling, 394–396
Pharmacophores, 84
Phenyl carbamates, 316
Photosynthetic, 330, 336
Piperazine anticancer hybrids, 494
Piperazine antimicrobial polymers, 493
Piperazine antioxidant derivatives, 495
Piperazine derivatives, 407–441
Piperazine nucleus, 408
Piperazine polymers, 439
Piperazine scaffold properties, 492
Piperazine sulfonamide derivatives, 498
Piperazine-amide conjugates, 494
Piperazine-based insecticides, 497
Piperazine-linked heterocycles, 495
Piperidine, 28
Piroxicam, 198, 199
Plasmodium enzyme inhibition, 486
Polyfunctionalized polymers, 493
Polyphenolic chemicals, 228, 229, 236
Precision medicine, 531
Privileged scaffold, 444
Proinflammatory cytokines, 489
Prostaglandins, 179, 182
Protein–ligand interaction, 392
Protoporphyrin, 318, 320
Purine, 29, 31
Purine analogs, 445
Pyrazole, 28, 31, 44, 179, 186, 187, 194, 200
Pyrazole substitution, 486
Pyrazolidine, 186
Pyridine, 31, 37, 40, 189, 198, 314
Pyridine derivatives, 9, 257
Pyrimidine, 28, 31, 37, 181, 189, 190
Pyrimidine derivatives, 9, 10, 257, 258
Pyrimidine glucoside derivatives, 49, 50
Pyrrole, 191, 193

Quantitative structure-activity relationship (QSAR), 393, 394
Quinazoline hybrids, 468, 469
Quinazoline–piperazine derivatives, 496
Quinoline, 28, 31, 41, 46, 200
Quinoline hybrids, 437
Quinone-hydroquinone redox cycling, 132

RAAS, 358, 359, 381
Radiopharmaceuticals, 389
Reactive oxygen species, 44, 54
Reactive oxygen species reduction, 489, 495
Receptor binding, 447
Receptor binding affinity, 480
Reduction reactions, 448, 466
Reductive amination, 435
Regioselectivity, 480

Renal hypertension, 369, 370
R-enantiomers, 86
Renin inhibitors, 350
Resistance mechanisms, 540, 541
Rheumatoid arthritis, 188, 194, 196
Rhodium-catalyzed reactions, 480
RMSD/RMSF, 392
Rofecoxib, 188
RVX-208, 535

S-adenosyl-L-methionine binding assay, 93
SAR studies, 483–486, 491
Schiff base adduct, 132
Schiff bases, 467
Selective inhibitors, 535, 538
Serotonin, 361
Serotonin reuptake inhibition, 488
Small-molecule inhibitors, 531, 535
Solubility properties, 478
Solvent-free synthesis, 6, 254
Structure–activity, 448, 449
Structure–activity relationship (SAR), 47, 437, 483–486
Substituted cinnoline derivatives, 487
Substitution, 226, 223
Sulfonamide derivatives, 483
Sulfonyl piperazine derivatives, 435
Sulfur derivatives, 454
Sulfur heterocycles, 476
Sulphuric acid, 309
Superbug targeting, 441
Super-enhancers, 537
Synthesis strategies, 435–439
Synthetic methods, 443, 444
Syphilis, 62

Tail-cuff method, 375
Tetracyclins, 62
Tetra-substituted thiophenes, 486
Tetrazole derivatives, 5, 253
Thiabendazole, 104
Thiadiazoles, 456, 457
Thiazole derivatives, 437
Thiazoles, 179, 200
Thieno-triazolopyrimidine derivatives, 489
Thiocarbamate, 319
Thiophene, 314
Thiophene anticancer derivatives, 482–486
Thiophene anticonvulsants, 490
Thiophene antidiabetic agents, 491
Thiophene anti-inflammatory drugs, 487–490
Thiophene antimalarial compounds, 486, 487
Thiophene aromaticity, 478
Thiophene discovery in benzene, 478
Thiophene heteroaromatic, 478
Thiophene synthesis, 478–481
Thiophene-based antidepressants, 488
Thiophene-based neurological drugs, 480
Thiophene-containing pharmaceuticals, 477
Thiophene–piperazine hybrids, 492
Thiosemicarbazides, 456
Thrombocytopenia, 541
Topoisomerase 4, 75
Toxicity prediction, 396, 397
Toxicity profile, 541
Transcription regulation, 531–533
Treponema pallidum, 62
Triazine analogs, 468
Triazine–piperazine scaffold, 439
Triazole, 31, 37
Trypan blue, 462
Tuberculosis therapy, 439
Tumor angiogenesis, 462
Tumor necrosis factor, 489
Tumor suppression, 461, 463

Ubiquitin-proteasome pathway, 141
Ulcer, 294
Ultrasound-assisted synthesis, 6, 254
Urticaria, 273, 275, 280, 292

Valsartan, 355, 371
Vasodilation, 351, 358, 368
VEGFR inhibition, 469
VEGFR inhibitors, 126, 142
Viral targeting, 447
Volhard–Erdmann reaction, 478

Warburg effect, 137
Water purification, 493
West Nile virus, 93
WHO antimicrobial report, 407

Yellow fever virus, 97

Also of Interest

Check out these other related titles from Scrivener Publishing

Psychopathology Early Prediction and Classification: A Deep Generative AI Approach, Edited by Abhishek Kumar, Pramod Singh Rathore, Sachin Ahuja, and Pankaj Rahi, ISBN: 9781394336739. Unlock the full potential of artificial intelligence in mental health with this definitive guide to combining multiple machine learning models for reproducible biomarker discovery and predictive treatment modeling.

Genomics and Precision Healthcare: Innovations, Applications and Challenges, Edited by Swetha Kumari Koduru, Sarad Pawar Naik Bukke, Hussaini Bello, Muhamad Mustafa Mundu, and Ayush Madan, ISBN: 9781394367528. Unlock the future of individualized patient care with this guide to mastering the interdisciplinary synergy of next-generation sequencing, clinical diagnostics, and ethical frameworks that define the new era of precision medicine.

Generative Artificial Intelligence in Ophthalmology, Edited by Binod Kumar Mishra, Abhishek Kumar, K. Mariyappan, Vibha Tiwari, Pramod Singh Rathore, and Gaur Hari Das, ISBN: 9781394358175. Empower your clinical practice to lead the future of vision science with this guide to mastering generative AI strategies that transform complex imaging data into high-precision, ethical diagnostic solutions.

Physiotherapy Using Artificial Intelligence: Enhancing Biomechanics for Optimal Rehabilitation, Edited by Abhishek Kumar, T. Ananth Kumar, Sachin Ahuja, J.P. Ananth, and S. Oswalt Manoj, ISBN: 9781394391530. Empower your practice with this definitive resource that bridges the gap between artificial intelligence and biomechanics, providing the essential tools and knowledge to optimize assessments, personalize treatment plans, and predict recovery outcomes in the rapidly evolving landscape of modern physiotherapy.

Environmental Toxicology and Human Health: Current Challenges and Future Perspectives, Edited by Ashish Kumar, Veer Singh, and Maulin P. Shah, ISBN: 9781394399826. Equip yourself to navigate environmental health with this essential resource that integrates fundamentals with the cutting-edge to provide a definitive guide on how chronic exposure to modern pollutants shapes human biological outcomes and regulatory policy.

Advances in Integrative Medicine, Edited by Mohd. Shahnawaz, Sajjid H. Batt, Damodaran Sruthi, Mubashir Javed Mintoo, M.K. Sangale, and Khalid Rehman Hakeem, ISBN: 9781394336777. *Advances in Integrative Medicine* introduces integrative medicine as a holistic approach to health that can effectively address the limitations and side effects of traditional allopathic treatments, providing valuable insights for practitioners and patients alike.

Artificial Intelligence in Healthcare for the Elderly, Edited by Sandeep Sharma, Preeti Nagrath, Bhawna Singla, and Joel J.P.C. Rodrigues, ISBN: 9781394275366. *Artificial Intelligence in Healthcare for the Elderly* provides valuable insights into how artificial intelligence can transform healthcare through personalized monitoring, ethical considerations, and real-world applications.

Design and Forecasting Models for Disease Management, Edited by Pijush Dutta, Sudip Mandal, Korhan Cengiz, Arindam Sadhu, and Gour Gopal Jana, ISBN: 9781394234042. The book provides an essential overview of AI techniques in disease management and how these computational methods can lead to further innovations in healthcare.

Smart Hospitals: 5G, 6G, and Moving Beyond Connectivity, Edited by Arun Kumar, Manoj Gupta, Sanjeev Sharma, Er. Himanshu Sharma, and Khursheed Aurangzeb, ISBN: 9781394275441. Discover the potential of 5G, 6G, and smart hospitals beyond connectivity in Smart Hospitals: 5G, 6G, and Moving Beyond Connectivity and learn how these advancements are revolutionizing healthcare and the digital world.

Sustainable Nanomaterials for Treatment and Diagnosis of Infectious Diseases, Edited by Rajasekhar Chokkareddy, Suvardhan Kanchi and Gan G. Redhi, ISBN: 9781394200016. The book focuses on the design and novel synthetic routes of sustainable nanomaterials in diagnosing and treating infectious diseases offering potential benefits in terms of efficiency, biocompatibility, and environmental impact.

Federated Intelligent System for Healthcare: A Practical Guide, Edited by S. Rakesh Kumar, N. Gayathri and Seifedine Kadry, ISBN: 9781394271351. This practical guide gives valuable insights for integrating advanced technologies in healthcare, empowering researchers to effectively navigate and implement federated systems to enhance patient care.

Internet of Medicine for Smart Healthcare, Edited by Abhishek Kumar, Narayan Vyas, Pramod Singh Rathore, Abhineet Anand, and Pooja Dixit, ISBN: 9781394272235. This book provides in-depth explanations and discussions of the latest applications of AI, machine learning, and the Internet of Medicine, offering readers the cutting edge on this rapidly growing technology that has the potential to transform healthcare and improve patient outcomes.

Viral Vectors for Vaccine Delivery, Edited by Vivek Chavda and Vasso Apostolopoulos, ISBN: 9781394271535. The book is essential for anyone interested in vaccine development, as it highlights the unique advantages of viral vector vaccines in triggering robust, long-lasting immunity and provides an in-depth exploration of the technology and advancements shaping the future of healthcare.

Artificial Intelligence and Cybersecurity in Healthcare, Edited by Rashmi Agrawal, Pramod Singh Rathore, Ganesh Gopal Devarajan, and Rajiva Ranjan Divivedi, ISBN: 9781394229796. *Artificial Intelligence and Cybersecurity in Healthcare* offers a vital exploration of the intersection of artificial intelligence and cybersecurity within healthcare Cyber Physical Systems, equipping readers with knowledge on how to navigate the transformative yet complex technological landscape shaping modern patient care and data protection.

Explainable Artificial Intelligence in the Healthcare Industry, Edited by Abhishek Kumar, T. Ananth Kumar, Prasenjit Das, Chetan Sharma, and Ashutosh Kumar Dubey, ISBN: 9781394249268. Discover the essential insights and practical applications of explainable AI in healthcare that will empower professionals and enhance patient trust with Explainable AI in Healthcare, a must-have resource for anyone looking to navigate the future of medical practices with confidence.

Role of Flavonoids in Chronic Metabolic Diseases: From Bench to Clinic, Edited by Neeraj Mishra, Sumel Ashique, B.H. Jaswanth Gowda, Arshad Farid, Ashish Garg, ISBN: 9781394238040. This comprehensive volume covers the entire field of flavonoids by explaining their complex functions in reducing chronic metabolic illnesses, from the early stages of laboratory research to the development of therapeutic uses.

Printed and bound by CPI Group (UK) Ltd, Croydon, CR0 4YY

17/09/2026

14964262-0008